3D Math Primer for Graphics and Game Development

Second Edition

3D Math Primer for Graphics and Game Development

Second Edition

Fletcher Dunn
Ian Parberry

CRC Press
Taylor & Francis Group
Boca Raton London New York

CRC Press is an imprint of the
Taylor & Francis Group, an **informa** business

AN A K PETERS BOOK

A K Peters/CRC Press
Taylor & Francis Group
6000 Broken Sound Parkway NW, Suite 300
Boca Raton, FL 33487-2742

Printed in India by Replika Press Pvt. Ltd.
10 9 8 7 6 5 4 3 2

International Standard Book Number: 978-1-56881-723-1 (Hardback)

Library of Congress Cataloging-in-Publication Data

Dunn, Fletcher.
3D math primer for graphics and game development / Fletcher Dunn, Ian Parberry. -- 2nd ed.
p. cm.
Summary: "This book presents the essential math needed to describe, simulate, and render a 3D world. It provides an introduction to mathematics for game designers, including fundamentals of coordinate spaces, vectors, and matrices, orientation in three dimensions introduction to calculus and dynamics, graphics, and parametric curves"-- Provided by publisher.
Includes bibliographical references and index.
ISBN 978-1-56881-723-1 (hardback)
1. Computer graphics. 2. Computer games--Programming. 3. Computer science--Mathematics. I. Parberry, Ian. II. Title.

T385.D875 2011
006.6--dc22
2010053535

Visit the Taylor & Francis Web site at
http://www.taylorandfrancis.com

and the A K Peters Web site at
http://www.akpeters.com

To A'me

—F.D.

To Maggie

in the hope that she continues
her interest in math

—I. P.

Contents

So much time, and so little to do!
Strike that, reverse it.

— Willy Wonka

Acknowledgments

Fletcher would like to thank his wife, A'me, who endured the absolute *eternity* that it took to produce this book, and his general tendency to generate lots of interesting ideas for large-scale projects that are initiated and then dropped a quarter of the way through. (No more gigantic projects for at least two or three weeks, I promise!)

Ian would like to thank his wife and children for not whining too loudly, and Fletcher for putting up with his procrastination. He would also like to thank Douglas Adams for the herring sandwich scoop, the bowl of petunias, and countless other references to the *Hitchhiker's Guide to the Galaxy* trilogy that you will find in this book.

Mike Pratcher gets a very huge thanks for his detailed and knowledgable critique, and for writing a very large portion of the exercises. Matt Carter made the robot and kitchen and agreed to numerous requests to pose the robot one way or another. Thanks to Glenn Gamble for the dead sheep. Eric Huang created the cover illustration and all other 2D artwork that required any artistic talent whatsoever. (The authors made the rest.) Pavel Krajcevski provided helpful criticism.

Gratitude is merely the secret hope of further favors.
— Francois de La Rochefoucauld (1613–1680)

Always look and smell your best.
— Riley Dunn (1945–)

Introduction

Who Should Read This Book

This book is about 3D math, the geometry and algebra of 3D space. It is designed to teach you how to describe objects and their positions, orientations, and trajectories in 3D using mathematics. This is not a book about computer graphics, simulation, or even computational geometry, although if you plan on studying those subjects, you will definitely need the information here.

This is not just a book for video game programmers. We do assume that a majority of our readers are learning for the purpose of programming video games, but we expect a wider audience and we have designed the book with a diverse audience in mind. If you're a programmer or interested in learning how to make video games, welcome aboard! If you meet neither of these criteria, there's still plenty for you here. We have made every effort to make the book useful to designers and technical artists. Although there are several code snippets in the book, they are (hopefully) easy to read even for nonprogrammers. Most important, even though it is always necessary to understand the surrounding concepts to make sense of the code, the reverse is never true. We use code samples to illustrate how ideas can be implemented on a computer, not to explain the ideas themselves.

The title of this book says it is for "game development," but a great deal of the material that we cover is applicable outside of video games. Practically anyone who wants to simulate, render, or understand a three-dimensional world will find this book useful. While we do try to provide motivating examples from the world of video game development, since that is our area of expertise and also our primary target audience, you won't be left out if the last game you completed was *Space Quest*.[1] If your interests

---

[1] Well, you may be left out of a few jokes, like that one. Sorry.

lie in more "grown up" things than video games, rest assured that this book is *not* filled with specific examples from video games about head-shots or severed limbs or how to get the blood spurt to look just right.

Why You Should Read This Book

This book has many unique features, including its topic, approach, authors, and writing style.

Unique topic. This book fills a gap that has been left by other books on graphics, linear algebra, simulation, and programming. It's an introductory book, meaning we have focused our efforts on providing thorough coverage on fundamental 3D concepts—topics that are normally glossed over in a few quick pages or relegated to an appendix in other books (because, after all, you already know all this stuff). We have found that these very topics are often the sticking points for beginners! In a way, this book is the mirror image of gluing together books on graphics, physics, and curves. Whereas that mythical conglomeration would begin with a brief overview of the mathematical fundamentals, followed by in-depth coverage of the application area, we start with a thorough coverage of the math fundamentals, and then give compact, high-level overviews of the application areas.

This book does try to provide a graceful on-ramp for beginners, but that doesn't mean we'll be stuck in the slow lane forever. There is plenty of material here that is traditionally considered "advanced" and taught in upper-level or graduate classes. In reality, these topics are specialized more than they are difficult, and they have recently become important prerequisites that need to be taught earlier, which is part of what has driven the demand for a book like this.

Unique approach. All authors think that they strike the perfect balance between being pedantic and being chatty in order to best reach their audience, and we are no exception. We recognize, however, that the people who disagree with this glowing self-assessment will mostly find this book too informal (see the index entry for "stickler alert"). We have focused on perspicuous explanations and intuition, and sometimes we have done this at the expense of rigor. Our aim is to simplify, but not to oversimplify. We lead readers to the goal through a path that avoids the trolls and dragons, so why begin the journey by pointing them all out before we've even said what our destination is or why we're going there? However, since we know readers will be crossing the field on their own eventually, after we reach our goal we will turn around to point out where the dangers lie. But we may sometimes need to leave certain troll-slaying to another source, especially if we expect that your usual path won't take you near the danger. Those who intend to be on that land frequently should consult with a local for more intimate knowledge. This is not to say that we think rigor is unimportant; we just think it's easier to get rigor after intuition about the big picture has been established, rather than front-loading every discussion with definitions and axioms needed to handle the edge cases. Frankly, nowadays a reader can pursue concise and formal presentations *free* on wikipedia.org or Wolfram MathWorld (mathworld.wolfram.com), so we don't think any book offers much worth paying for by dwelling excessively on definitions, axioms, proofs, and edge cases, especially for introductory material targeted primarily to engineers.

Unique authors. Our combined experience brings together academic authority with in-the-trenches practical advice. Fletcher Dunn has 15 years of professional game programming experience, with around a dozen titles under his belt on a variety of gaming platforms. He worked at Terminal Reality in Dallas, where as principal programmer he was one of the architects of the Infernal engine and lead programmer on *BloodRayne*. He was a technical director for The Walt Disney Company at Wideload Games in Chicago and the lead programmer for *Disney Guilty Party*, IGN's E3 2010 Family Game of the Year. He now works for Valve Software in Bellevue, Washington. But his biggest claim to fame by far is as the namesake of Corporal Dunn from *Call of Duty: Modern Warfare 2*.

Dr. Ian Parberry has more than a quarter century of experience in research and teaching in academia. This is his sixth book, his third on game programming. He is currently a tenured full professor in the Department of Computer Science & Engineering at the University of North Texas. He is nationally known as one of the pioneers of game programming in higher education, and has been teaching game programming classes at the University of North Texas continuously since 1993.

Unique writing style. We hope you will enjoy reading this math book (say *what?*) for two reasons. Most important, we want you to *learn* from this book, and learning something you are interested in is fun. Secondarily, we want you to enjoy *reading* this book in the same way that you enjoy reading a work of literature. We have no delusions that we're in the same class as Mark Twain, or that this book is destined to become a classic like, say, *The Hitchhikers Guide to the Galaxy*. But one can always have aspirations. Honestly, we are just silly people. At the same time, no writing style should stand in the way of the first priority: clear communication of mathematical knowledge about video games.[2]

---

[2]Which is why we've put most of the jokes and useless trivia in footnotes like this. Somehow, we felt like we could get away with more that way.

What You Should Know before Reading This Book

We have tried to make the book accessible to as wide an audience as possible; no book, however, can go back all the way to first principles. We expect from the reader the following basic mathematical skills:

- Manipulating algebraic expressions, fractions, and basic algebraic laws such as the associative and distributive laws and the quadratic equation.

- Understanding what variables are, what a function is, how to graph a function, and so on.

- Some very basic 2D Euclidian geometry, such as what a point is, what a line is, what it means for lines to be parallel and perpendicular, and so forth. Some basic formulas for area and circumference are used in a few places. It's OK if you have temporarily forgotten those—you will hopefully recognize them when you see them.

- Some prior exposure to trigonometry is best. We give a brief review of trigonometry in the front of this book, but it is not presented with the same level of paced explanation found most elsewhere in this book.

- Readers with some prior exposure to calculus will have an advantage, but we have restricted our use of calculus in this book to very basic principles, which we will (attempt to) teach in Chapter 11 for those without this training. Only the most high-level concepts and fundamental laws are needed.

Some programming knowledge is helpful, but *not* required. In several places, we give brief code snippets to show how the ideas being discussed get translated into code. (Also certain procedures are just easier to explain in code.) These snippets are extremely basic, well commented, and require only the most rudimentary understanding of C language syntax (which has been copied to several other languages). Most technical artists or level designers should be able to interpret these snippets with ease.

Overview

- *Chapter 1* gets warmed up with some groundwork that it is needed in the rest of the book and which you probably already know. It reviews the Cartesian coordinate system in 2D and 3D and discusses how to use the Cartesian coordinate system to locate points in space. Also included is a very quick refresher on trigonometry and summation notation.

- *Chapter 2* introduces vectors from a mathematical and geometric perspective and investigates the important relationship between points and vectors. It also discusses a number of vector operations, how to do them, what it means geometrically to do them, and situations for which you might find them useful.

- *Chapter 3* discusses examples of coordinate spaces and how they are nested in a hierarchy. It also introduces the central concepts of basis vectors and coordinate-space transformations.

- *Chapter 4* introduces matrices from a mathematical and geometric perspective and shows how matrices are a compact notation for the math behind linear transformations.

- *Chapter 5* surveys different types of linear transformations and their corresponding matrices in detail. It also discusses various ways to classify transformations.

- *Chapter 6* covers a few more interesting and useful properties of matrices, such as affine transforms and perspective projection, and explains the purpose and workings of four-dimensional vectors and matrices within a three-dimensional world.

- *Chapter 7* discusses how to use polar coordinates in 2D and 3D, why it is useful to do so, and how to convert between polar and Cartesian representations.

- *Chapter 8* discusses different techniques for representing orientation and angular displacement in 3D: Euler angles, rotation matrices, exponential maps, and quaternions. For each method, it explains how the method works and presents the advantages and disadvantages of the method and when its use is recommended. It also shows how to convert between different representations.

- *Chapter 9* surveys a number of commonly used geometric primitives and discusses how to represent and manipulate them mathematically.

- *Chapter 10* is a whirlwind lesson on graphics, touching on a few selected theoretical as well as modern practical issues. First, it presents a high-level overview of "how graphics works," leading up to the rendering equation. The chapter then walks through a few theoretical topics of a mathematical nature. Next it discusses two contemporary topics that are often sources of mathematical difficulty and should be of particular interest to the reader: skeletal animation and bump mapping. Finally, the chapter presents an overview of the real-time graphics pipeline, demonstrating how the theories from the first half of the chapter are implemented in the context of current rendering hardware.

- *Chapter 11* crams two rather large topics into one chapter. It interleaves the highest-level topics from first-semester calculus with a discussion of rigid body kinematics—how to describe and analyze the motion of a rigid body without necessarily understanding its cause or being concerned with orientation or rotation.

- *Chapter 12* continues the discussion of rigid body mechanics. It starts with a condensed explanation of classical mechanics, including Newton's laws of motion and basic concepts such as inertia, mass, force, and momentum. It reviews a few basic force laws, such as gravity, springs, and friction. The chapter also considers the rotational analogs of all of the linear ideas discussed up to this point. Due attention is paid to the important topic of collisions. The chapter ends with a discussion of issues that arise when using a computer to simulate rigid bodies.

- *Chapter 13* explains parametric curves in 3D. The first half of the chapter explains how a relatively short curve is represented in some common, important forms: monomial, Bézier, and Hermite. The second half is concerned with fitting together these shorter pieces into a longer curve, called a *spline*. In understanding each system, the chapter considers what controls the system presents to a designer of curves, how to take a description of a curve made by a designer and recreate the curve, and how these controls can be used to construct a curve with specific properties.

- *Chapter 14* inspires the reader to pursue greatness in video games.

- *Appendix A* is an assortment of useful tests that can be performed on geometric primitives. We intend it to be a helpful reference, but it can also make for interesting browsing.

- *Appendix B* has all the answers.[3]

Find a Bug in This Book?

We calculated the odds that we could write an 800+ page math book free of mistakes. The result was a negative number, which we know can't be right, but is probably pretty close. If you find a bug in this book, please

---

[3]To the exercises, that is.

visit gamemath.com. Most likely, the error is already listed in the errata, in which case you have our profound apologies. Otherwise, send us an email, and you will have (in addition to our profound thanks) everlasting fame via credit in the errata for being the first to find the mistake.

Careful. We don't want to learn from this.
— Bill Watterson (1958–) from *Calvin and Hobbes*

Chapter 1

Cartesian Coordinate Systems

Before turning to those moral and mental aspects of the matter
which present the greatest difficulties, let the inquirer begin by
mastering more elementary problems.

— Sherlock Holmes from *A Study in Scarlett* (1887)

3D math is all about measuring locations, distances, and angles precisely
and mathematically in 3D space. The most frequently used framework to
perform such calculations using a computer is called the *Cartesian coordinate system*. Cartesian mathematics was invented by (and is named after)
a brilliant French philosopher, physicist, physiologist, and mathematician
named René Descartes, who lived from 1596 to 1650. René Descartes is
famous not just for inventing Cartesian mathematics, which at the time
was a stunning unification of algebra and geometry. He is also well-known
for making a pretty good stab of answering the question "How do I know
something is true?"—a question that has kept generations of philosophers
happily employed and does not necessarily involve dead sheep (which will
perhaps disturbingly be a central feature of the next section), unless you
really want it to. Descartes rejected the answers proposed by the Ancient
Greeks, which are *ethos* (roughly, "because I told you so"), *pathos* ("because it would be nice"), and *logos* ("because it makes sense"), and set
about figuring it out for himself with a pencil and paper.

This chapter is divided into four main sections.

- Section 1.1 reviews some basic principles of number systems and the
  first law of computer graphics.

- Section 1.2 introduces 2D Cartesian mathematics, the mathematics
  of flat surfaces. It shows how to describe a 2D cartesian coordinate
  space and how to locate points using that space.

- Section 1.3 extends these ideas into three dimensions. It explains left-
  and right-handed coordinate spaces and establishes some conventions
  used in this book.

- Section 1.4 concludes the chapter by quickly reviewing assorted pre-requisites.

1.1   1D Mathematics

You're reading this book because you want to know about 3D mathematics, so you're probably wondering why we're bothering to talk about 1D math. Well, there are a couple of issues about number systems and counting that we would like to clear up before we get to 3D.

Figure 1.1
One dead sheep

The *natural numbers*, often called the *counting numbers*, were invented millennia ago, probably to keep track of dead sheep. The concept of "one sheep" came easily (see Figure 1.1), then "two sheep," "three sheep," but people very quickly became convinced that this was too much work, and gave up counting at some point that they invariably called "many sheep." Different cultures gave up at different points, depending on their threshold of boredom. Eventually, civilization expanded to the point where we could afford to have people sitting around thinking about numbers instead of doing more survival-oriented tasks such as killing sheep and eating them. These savvy thinkers immortalized the concept of zero (no sheep), and although they didn't get around to naming all of the natural numbers, they figured out various systems whereby they could name them if they really wanted to using digits such as 1, 2, etc. (or if you were Roman, M, X, I, etc.). Thus, mathematics was born.

The habit of lining sheep up in a row so that they can be easily counted leads to the concept of a number line, that is, a line with the numbers marked off at regular intervals, as in Figure 1.2. This line can in principle go on for as long as we wish, but to avoid boredom we have stopped at five sheep and used an arrowhead to let you know that the line can continue. Clearer thinkers can visualize it going off to infinity, but historical purveyors of dead sheep probably gave this concept little thought outside of their dreams and fevered imaginings.

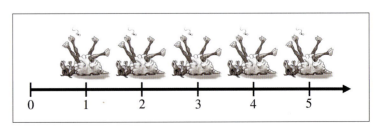

Figure 1.2. A number line for the natural numbers

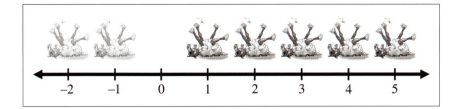

Figure 1.3. A number line for integers. (Note the ghost sheep for negative numbers.)

At some point in history, it was probably realized that sometimes, particularly fast talkers could sell sheep that they didn't actually own, thus simultaneously inventing the important concepts of debt and negative numbers. Having sold this putative sheep, the fast talker would in fact own "negative one" sheep, leading to the discovery of the *integers*, which consist of the natural numbers and their negative counterparts. The corresponding number line for integers is shown in Figure 1.3.

The concept of poverty probably predated that of debt, leading to a growing number of people who could afford to purchase only half a dead sheep, or perhaps only a quarter. This led to a burgeoning use of fractional numbers consisting of one integer divided by another, such as 2/3 or 111/27. Mathematicians called these *rational numbers*, and they fit in the number line in the obvious places between the integers. At some point, people became lazy and invented decimal notation, writing "3.1415" instead of the longer and more tedious 31415/10000, for example.

After a while it was noticed that some numbers that appear to turn up in everyday life were not expressible as rational numbers. The classic example is the ratio of the circumference of a circle to its diameter, usually denoted π (the Greek letter pi, pronounced "pie"). These are the so-called *real numbers*, which include the rational numbers and numbers such as π that would, if expressed in decimal notation, require an infinite number of decimal places. The mathematics of real numbers is regarded by many to be the most important area of mathematics—indeed, it is the basis of most forms of engineering, so it can be credited with creating much of modern civilization. The cool thing about real numbers is that although rational numbers are countable (that is, can be placed into one-to-one correspondence with the natural numbers), the real numbers are uncountable. The study of natural numbers and integers is called *discrete mathematics*, and the study of real numbers is called *continuous mathematics*.

The truth is, however, that real numbers are nothing more than a polite fiction. They are a relatively harmless delusion, as any reputable physicist will tell you. The universe seems to be not only discrete, but also finite.

If there are a finite amount of discrete things in the universe, as currently appears to be the case, then it follows that we can only count to a certain fixed number, and thereafter we run out of things to count on—not only do we run out of dead sheep, but toasters, mechanics, and telephone sanitizers, too. It follows that we can describe the universe using only discrete mathematics, and only requiring the use of a finite subset of the natural numbers at that (large, yes, but finite). Somewhere, someplace there may be an alien civilization with a level of technology exceeding ours who have never heard of continuous mathematics, the fundamental theorem of calculus, or even the concept of infinity; even if we persist, they will firmly but politely insist on having no truck with π, being perfectly happy to build toasters, bridges, skyscrapers, mass transit, and starships using 3.14159 (or perhaps 3.14159265358979323846264338332795 if they are fastidious) instead.

So why do we use continuous mathematics? Because it is a useful tool that lets us do engineering. But the real world is, despite the cognitive dissonance involved in using the term "real," discrete. How does that affect you, the designer of a 3D computer-generated virtual reality? The computer is, by its very nature, discrete and finite, and you are more likely to run into the consequences of the discreteness and finiteness during its creation than you are likely to in the real world. C++ gives you a variety of different forms of number that you can use for counting or measuring in your virtual world. These are the `short`, the `int`, the `float` and the `double`, which can be described as follows (assuming current PC technology). The short is a 16-bit integer that can store 65,536 different values, which means that "many sheep" for a 16-bit computer is 65,537. This sounds like a lot of sheep, but it isn't adequate for measuring distances inside any reasonable kind of virtual reality that take people more than a few minutes to explore. The `int` is a 32-bit integer that can store up to 4,294,967,296 different values, which is probably enough for your purposes. The `float` is a 32-bit value that can store a subset of the rationals (slightly fewer than 4,294,967,296 of them, the details not being important here). The `double` is similar, using 64 bits instead of 32.

The bottom line in choosing to count and measure in your virtual world using `ints`, `floats`, or `doubles` is not, as some misguided people would have it, a matter of choosing between discrete `shorts` and `ints` versus continuous `floats` and `doubles`; it is more a matter of precision. They are all discrete in the end. Older books on computer graphics will advise you to use integers because floating-point hardware is slower than integer hardware, but this is no longer the case. In fact, the introduction of dedicated floating point vector processors has made floating-point arithmetic faster than integer in many common cases. So which should you choose? At this point, it is probably best to introduce you to the first law of computer graphics and leave you to think about it.

The First Law of Computer Graphics

If it *looks* right, it *is* right.

We will be doing a lot of trigonometry in this book. Trigonometry involves real numbers such as π, and real-valued functions such as sine and cosine (which we'll get to later). Real numbers are a convenient fiction, so we will continue to use them. How do you know this is true? Because, Descartes notwithstanding, we told you so, because it would be nice, and because it makes sense.

1.2   2D Cartesian Space

You probably have used 2D Cartesian coordinate systems even if you have never heard the term "Cartesian" before. "Cartesian" is mostly just a fancy word for "rectangular." If you have ever looked at the floor plans of a house, used a street map, seen a football[1] game, or played chess, you have some exposure to 2D Cartesian coordinate spaces.

This section introduces 2D Cartesian mathematics, the mathematics of flat surfaces. It is divided into three main subsections.

- Section 1.2.1 provides a gentle introduction to the concept of 2D Cartesian space by imagining a fictional city called Cartesia.

- Section 1.2.2 generalizes this concept to arbitrary or abstract 2D Cartesian spaces. The main concepts introduced are

  - the origin

  - the x- and y-axes

  - orienting the axes in 2D

- Section 1.2.3 describes how to specify the location of a point in the 2D plane using Cartesian (x, y) coordinates.

---

[1]This sentence works no matter which sport you think we are referring to with the word "football." Well, OK, it works a little better with American football because of the clearly marked yard lines.

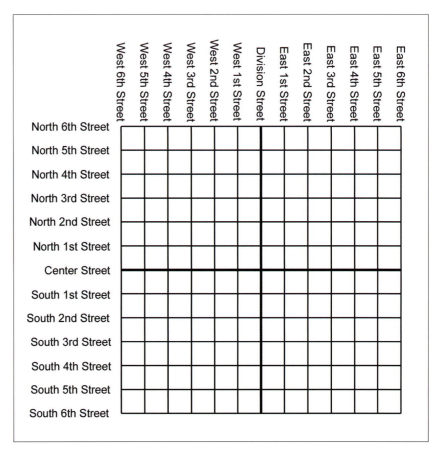

Figure 1.4. Map of the hypothetical city of Cartesia

1.2.1   An Example: The Hypothetical City of Cartesia

Let's imagine a fictional city named Cartesia. When the Cartesia city planners were laying out the streets, they were very particular, as illustrated in the map of Cartesia in Figure 1.4.

As you can see from the map, Center Street runs east-west through the middle of town. All other east-west streets (parallel to Center Street) are named based on whether they are north or south of Center Street, and how far they are from Center Street. Examples of streets that run east-west are North 3rd Street and South 15th Street.

The other streets in Cartesia run north-south. Division Street runs north-south through the middle of town. All other north-south streets (parallel to Division Street) are named based on whether they are east or west of Division Street, and how far they are from Division Street. So we have streets such as East 5th Street and West 22nd Street.

The naming convention used by the city planners of Cartesia may not be creative, but it certainly is practical. Even without looking at the map, it is easy to find the donut shop at North 4th and West 2nd. It's also easy to determine how far you will have to drive when traveling from one place to another. For example, to go from that donut shop at North 4th and West 2nd, to the police station at South 3rd and Division, you would travel seven blocks south and two blocks east.

1.2.2 Arbitrary 2D Coordinate Spaces

Before Cartesia was built, there was nothing but a large flat area of land. The city planners arbitrarily decided where the center of town would be, which direction to make the roads run, how far apart to space the roads, and so forth. Much like the Cartesia city planners laid down the city streets, we can establish a 2D Cartesian coordinate system anywhere we want—on a piece of paper, a chessboard, a chalkboard, a slab of concrete, or a football field.

Figure 1.5 shows a diagram of a 2D Cartesian coordinate system.

As illustrated in Figure 1.5, a 2D Cartesian coordinate space is defined by two pieces of information:

- Every 2D Cartesian coordinate space has a special location, called the origin, which is the "center" of the coordinate system. The origin is analogous to the center of the city in Cartesia.

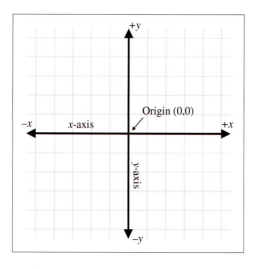

Figure 1.5
A 2D Cartesian coordinate space

- Every 2D Cartesian coordinate space has two straight lines that pass through the origin. Each line is known as an axis and extends infinitely in two opposite directions. The two axes are perpendicular to each other. (Actually, they don't have to be, but most of the coordinate systems we will look at will have perpendicular axes.) The two axes are analogous to Center and Division streets in Cartesia. The grid lines in the diagram are analogous to the other streets in Cartesia.

At this point it is important to highlight a few significant differences between Cartesia and an abstract mathematical 2D space:

- The city of Cartesia has official city limits. Land outside of the city limits is not considered part of Cartesia. A 2D coordinate space, however, extends infinitely. Even though we usually concern ourselves with only a small area within the plane defined by the coordinate space, in theory this plane is boundless. Also, the roads in Cartesia go only a certain distance (perhaps to the city limits) and then they stop. In contrast, our axes and grid lines extend potentially infinitely in two directions.

- In Cartesia, the roads have thickness. In contrast, lines in an abstract coordinate space have location and (possibly infinite) length, but no real thickness.

- In Cartesia, you can drive only on the roads. In an abstract coordinate space, every point in the plane of the coordinate space is part of the coordinate space, not just the "roads." The grid lines are drawn only for reference.

In Figure 1.5, the horizontal axis is called the x-axis, with positive x pointing to the right, and the vertical axis is the y-axis, with positive y pointing up. This is the customary orientation for the axes in a diagram. Note that "horizontal" and "vertical" are terms that are inappropriate for many 2D spaces that arise in practice. For example, imagine the coordinate space on top of a desk. Both axes are "horizontal," and neither axis is really "vertical."

The city planners of Cartesia could have made Center Street run north-south instead of east-west. Or they could have oriented it at a completely arbitrary angle. For example, Long Island, New York, is reminiscent of Cartesia, where for convenience the "streets" (1st Street, 2nd Street etc.) run across the island, and the "avenues" (1st Avenue, 2nd Avenue, etc.) run along its long axis. The geographic orientation of the long axis of the island is an arbitrary result of nature. In the same way, we are free to orient our axes in any way that is convenient to us. We must also decide for each axis which direction we consider to be positive. For example, when working with images on a computer screen, it is customary to use the coordinate system shown in Figure 1.6. Notice that the origin is in the upper left-hand corner, $+x$ points to the right, and $+y$ points down rather than up.

Unfortunately, when Cartesia was being laid out, the only mapmakers were in the neighboring town of Dyslexia. The minor-level functionary who sent the contract out to bid neglected take into account that the dyslectic mapmaker was equally likely to draw his maps with north pointing up, down, left, or right. Although he always drew the east-west line at right angles to the north-south line, he often got east and west backwards. When his boss realized that the job had gone to the lowest bidder, who happened to live in Dyslexia, many hours were spent in committee meetings trying to figure out what to do. The paperwork had been done, the purchase order had been is-

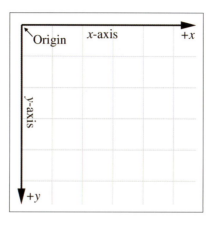

Figure 1.6
Screen coordinate space

sued, and bureaucracies being what they are, it would be too expensive and time-consuming to cancel the order. Still, nobody had any idea what the mapmaker would deliver. A committee was hastily formed.

The committee fairly quickly decided that there were only eight possible orientations that the mapmaker could deliver, shown in Figure 1.7. In

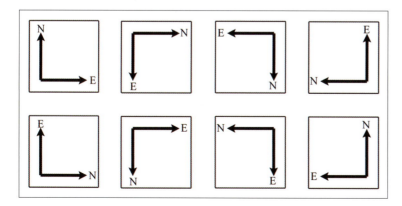

Figure 1.7. Possible map axis orientations in 2D

the best of all possible worlds, he would deliver a map oriented as shown in the top-left rectangle, with north pointing to the top of the page and east to the right, which is what people usually expect. A subcommittee formed for the task decided to name this the normal orientation.

After the meeting had lasted a few hours and tempers were beginning to fray, it was decided that the other three variants shown in the top row of Figure 1.7 were probably acceptable too, because they could be transformed to the normal orientation by placing a pin in the center of the page and rotating the map around the pin. (You can do this, too, by placing this book flat on a table and turning it.) Many hours were wasted by tired functionaries putting pins into various places in the maps shown in the second row of Figure 1.7, but no matter how fast they twirled them, they couldn't seem to transform them to the normal orientation. It wasn't until everybody important had given up and gone home that a tired intern, assigned to clean up the used coffee cups, noticed that the maps in the second row can be transformed into the normal orientation by holding them up against a light and viewing them from the back. (You can do this, too, by holding Figure 1.7 up to the light and viewing it from the back—you'll have to turn it, too, of course.) The writing was backwards too, but it was decided that if Leonardo da Vinci (1452–1519) could handle backwards writing in 15th century, then the citizens of Cartesia, though by no means his intellectual equivalent (probably due to daytime TV), could probably handle it in the 21st century.

In summary, no matter what orientation we choose for the x- and y-axes, we can always rotate the coordinate space around so that $+x$ points to our right and $+y$ points up. For our example of screen-space coordinates, imagine turning the coordinate system upside down and looking at the screen from behind the monitor. In any case, these rotations do not distort the original shape of the coordinate system (even though we may be looking at it upside down or reversed). So in one particular sense, all 2D coordinate systems are "equal." In Section 1.3.3, we discover the surprising fact that this is not the case in 3D.

1.2.3   Specifying Locations in 2D Using Cartesian Coordinates

A coordinate space is a framework for specifying location precisely. A gentleman of Cartesia could, if he wished to tell his lady love where to meet him for dinner, for example, consult the map in Figure 1.4 and say, "Meet you at the corner of East 2nd Street and North 4th Street." Notice that he specifies two coordinates, one in the horizontal dimension (East 2nd Street, listed along the top of the map in Figure 1.4) and one in the vertical dimension (North 4th Street, listed along the left of the map). If he wished to be concise he could abbreviate the "East 2nd Street" to "2" and the "North 4th Street" to "4" and say to his lady love, somewhat cryptically, "Meet you at $(2, 4)$."

The ordered pair $(2, 4)$ is an example of what are called *Cartesian coordinates*. In 2D, two numbers are used to specify a location. (The fact that we use two numbers to describe the location of a point is the reason it's called *two*-dimensional space. In 3D, we will use three numbers.) The first coordinate (the 2 in our example $(2, 4)$) is called the x-coordinate, and the second coordinate (the 4 in our example $(2, 4)$) is called the y-coordinate.

Analogous to the street names in Cartesia, each of the two coordinates specifies which side of the origin the point is on and how far away the point is from the origin in that direction. More precisely, each coordinate is the *signed distance* (that is, positive in one direction and negative in the other) to one of the axes, measured along a line parallel to the other axis. Essentially, we use positive coordinates for east and north streets and negative coordinates for south and west streets. As shown in Figure 1.8, the x-coordinate designates the signed distance from the point to the y-axis, measured along a line parallel to the x-axis. Likewise, the y-coordinate designates the signed distance from the point to the x-axis, measured along a line parallel to the y-axis.

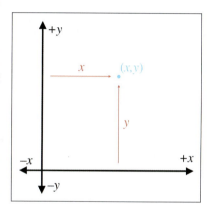

Figure 1.8
How to locate a point using 2D Cartesian coordinates

Figure 1.9 shows several points and their Cartesian coordinates. Notice that the points to the left of the y-axis have negative x values, and those to the right of the y-axis have positive x values. Likewise, points with positive y are located above the x-axis, and points with negative y are below the x-axis. Also notice that *any* point can be specified, not just the points at grid line intersections. You should study this figure until you are sure that you understand the pattern.

Let's take a closer look at the grid lines usually shown in a diagram. Notice that a vertical grid line is composed of points that all have the same x-coordinate. In other words, a vertical grid line (actually *any* vertical line) marks a line of constant x. Likewise, a horizontal grid line marks a line of constant y; all the points on that line have the same y coordinate. We'll come back to this idea in a bit when we discuss polar coordinate spaces.

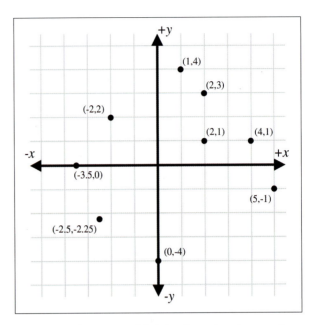

Figure 1.9. Example points labeled with 2D Cartesian coordinates

1.3   3D Cartesian Space

The previous sections have explained how the Cartesian coordinate system works in 2D. Now it's time to leave the flat 2D world and think about 3D space.

It might seem at first that 3D space is only "50% more complicated" than 2D. After all, it's just *one* more dimension, and we already had *two*. Unfortunately, this is not the case. For a variety of reasons, 3D space is *more* than incrementally more difficult than 2D space for humans to visualize and describe. (One possible reason for this difficulty could be that our physical world is 3D, whereas illustrations in books and on computer screens are 2D.) It is frequently the case that a problem that is "easy" to solve in 2D is much more difficult or even undefined in 3D. Still, many concepts in 2D do extend directly into 3D, and we frequently use 2D to establish an understanding of a problem and develop a solution, and then extend that solution into 3D.

This section extends 2D Cartesian math into 3D. It is divided into four major subsections.

- Section 1.3.1 begins the extension of 2D into 3D by adding a third axis. The main concepts introduced are

  - o the z-axis

  - o the xy, xz, and yz planes

- Section 1.3.2 describes how to specify the location of a point in the 3D plane using Cartesian (x, y, z) coordinates.

- Section 1.3.3 introduces the concepts of left-handed and right-handed 3D coordinate spaces. The main concepts introduced are

  - o the hand rule, an informal definition for left-handed and right-handed coordinate spaces

  - o differences in rotation in left-handed and right-handed coordinate spaces

  - o how to convert between the two

  - o neither is better than the other, only different

- Section 1.3.4 describes some conventions used in this book.

1.3.1   Extra Dimension, Extra Axis

In 3D, we require three axes to establish a coordinate system. The first two axes are called the x-axis and y-axis, just as in 2D. (However, it is not accurate to say that these are the *same* as the 2D axes; more on this later.) We call the third axis (predictably) the z-axis. Usually, we set things up so that all axes are mutually perpendicular, that is, each one is perpendicular to the others. Figure 1.10 shows an example of a 3D coordinate space.

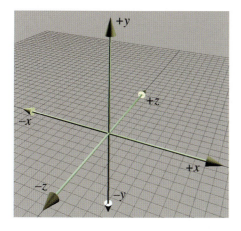

Figure 1.10
A 3D Cartesian coordinate space

As discussed in Section 1.2.2, it is customary in 2D for $+x$ to point to the right and $+y$ to point up. (Or sometimes $+y$ may point down, but in either case, the x-axis is horizontal and the y-axis is vertical.) These conventions in 2D are fairly standardized. In 3D, however, the conventions for arrangement of the axes in diagrams and the assignment of the axes onto physical dimensions (left, right, up, down, forward, back) are not very standardized. Different authors and fields of study have different conventions. Section 1.3.4 discusses the conventions used in this book.

As mentioned earlier, it is not entirely appropriate to say that the x-axis and y-axis in 3D are the "same" as the x-axis and y-axis in 2D. In 3D, any pair of axes defines a plane that contains the two axes and is perpendicular to the third axis. For example, the plane containing the x- and y-axes is the xy plane, which is perpendicular to the z-axis. Likewise, the xz plane is perpendicular to the y-axis, and the yz plane is perpendicular to the x-axis. We can consider any of these planes a 2D Cartesian coordinate space in its own right. For example, if we assign $+x$, $+y$, and $+z$ to point right, up, and forward, respectively, then the 2D coordinate space of the "ground" is the xz plane, as shown in Figure 1.10.

1.3.2   Specifying Locations in 3D

In 3D, points are specified using three numbers, x, y, and z, which give the signed distance to the yz, xz, and xy planes, respectively. This distance is measured along a line parallel to the axis. For example, the x-value is the signed distance to the yz plane, measured along a line parallel to the x-axis. Don't let this precise definition of how points in 3D are located confuse you. It is a straightforward extension of the process for 2D, as shown in Figure 1.11.

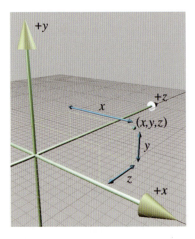

Figure 1.11
Locating points in 3D

1.3.3   Left-handed versus Right-handed Coordinate Spaces

As we discussed in Section 1.2.2, all 2D coordinate systems are "equal" in the sense that for any two 2D coordinate spaces A and B, we can rotate coordinate space A such that $+x$ and $+y$ point in the same direction as they do in coordinate space B. (We are assuming perpendicular axes.) Let's examine this idea in more detail.

Figure 1.5 shows the "standard" 2D coordinate space. Notice that the difference between this coordinate space and "screen" coordinate space shown Figure 1.6 is that the y-axis points in opposite directions. However, imagine rotating Figure 1.6 clockwise 180 degrees so that $+y$ points up and $+x$ points to the left. Now rotate it by "turning the page" and viewing the diagram from behind. Notice that now the axes are oriented in the "standard" directions like in Figure 1.5. No matter how many times we flip an axis, we can always find a way to rotate things back into the standard orientation.

Let's see how this idea extends into 3D. Examine Figure 1.10 once more. We stated earlier that $+z$ points into the page. Does it have to be this way? What if we made $+z$ point out of the page? This is certainly allowed, so let's flip the z-axis.

Now, can we rotate the coordinate system around such that things line up with the original coordinate system? As it turns out, we cannot. We can rotate things to line up two axes at a time, but the third axes always points in the wrong direction! (If you have trouble visualizing this, don't worry. In just a moment we will illustrate this principle in more concrete terms.)

All 3D coordinate spaces are not equal, in the sense that some pairs of coordinate systems cannot be rotated to line up with each other. There are exactly two distinct types of 3D coordinate spaces: *left-handed* coordinate spaces and *right-handed* coordinate spaces. If two coordinate spaces have the same handedness, then they can be rotated such that the axes are aligned. If they are of opposite handedness, then this is not possible.

What exactly do "left-handed" and "right-handed" mean? The most intuitive way to identify the handedness of a particular coordinate system is to use, well, your hands! With your left hand, make an 'L' with your thumb and index finger.[2] Your thumb should be pointing to your right, and your index finger should be pointing up. Now extend your third finger[3] so it points directly forward. You have just formed a left-handed coordinate system. Your thumb, index finger, and third finger point in the $+x$, $+y$, and $+z$ directions, respectively. This is shown in Figure 1.12.

---

[2] You may have to put the book down.

[3] This may require some dexterity. The authors advise that you *not* do this in public without first practicing privately, to avoid offending innocent bystanders.

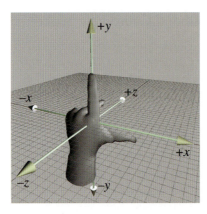

Figure 1.12
Left-handed coordinate space

Now perform the same experiment with your right hand. Notice that your index finger still points up, and your third finger points forward. However, with your right hand, your thumb will point to the left. This is a right-handed coordinate system. Again, your thumb, index finger, and third finger point in the $+x$, $+y$, and $+z$ directions, respectively. A right-handed coordinate system is shown in Figure 1.13.

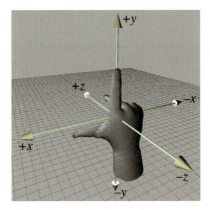

Figure 1.13
Right-handed coordinate space

Try as you might, you cannot rotate your hands into a position such that all three fingers simultaneously point the same direction on both hands. (Bending your fingers is not allowed.)

Left-handed and right-handed coordinate systems also differ in the definition of "positive rotation." Let's say we a have line in space and we need to rotate about this line by a specified angle. We call this line an *axis of rotation*, but don't think that the word *axis* implies that we're talking only about one of the cardinal axes (the x-, y-, or z-axis). An axis of rotation can be arbitrarily oriented. Now, if you tell me to "rotate 30° about the

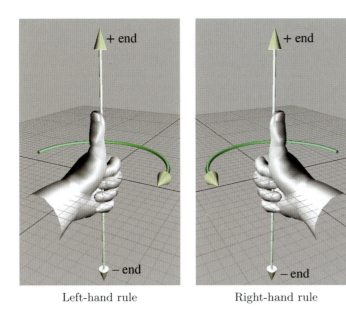

Left-hand rule   Right-hand rule

Figure 1.14. The left-hand rule and right-hand rule define which direction is considered "positive" rotation.

axis," how do I know which way to rotate? We need to agree between us that one direction of rotation is the positive direction, and the other direction is the negative direction. The standard way to tell which is which in a left-handed coordinate system is called the *left-hand rule*. First, we must define which way our axis "points." Of course, the axis of rotation is theoretically infinite in length, but we still consider it having a positive and negative end, just like the normal cardinal axes that define our coordinate space. The left-hand rule works like this: put your left hand in the "thumbs up" position, with your thumb pointing towards the positive end of the axis of rotation. Positive rotation about the axis of rotation is in the direction that your fingers are curled. There's a corresponding rule for right-handed coordinate spaces; both of these rules are illustrated in Figure 1.14.

As you can see, in a left-handed coordinate system, positive rotation rotates *clockwise* when viewed from the positive end of the axis, and in a right-handed coordinate system, positive rotation is *counterclockwise*. Table 1.1 shows what happens when we apply this general rule to the specific case of the cardinal axes.

Any left-handed coordinate system can be transformed into a right-handed coordinate system, or vice versa. The simplest way to do this is by swapping the positive and negative ends of one axis. Notice that if we

When looking towards the origin from...	Positive rotation Left-handed: Clockwise Right-handed: Counterclockwise	Negative rotation Left-handed: Counterclockwise Right-handed: Clockwise
$+x$	$+y \rightarrow +z \rightarrow -y \rightarrow -z \rightarrow +y$	$+y \rightarrow -z \rightarrow -y \rightarrow +z \rightarrow +y$
$+y$	$+z \rightarrow +x \rightarrow -z \rightarrow -x \rightarrow +z$	$+z \rightarrow -x \rightarrow -z \rightarrow +x \rightarrow +z$
$+z$	$+x \rightarrow +y \rightarrow -x \rightarrow -y \rightarrow +x$	$+x \rightarrow -y \rightarrow -x \rightarrow +y \rightarrow +x$

Table 1.1. Rotation about the cardinal axes in left- and right-handed coordinate systems

flip *two* axes, it is the same as rotating the coordinate space 180° about the third axis, which does not change the handedness of the coordinate space. Another way to toggle the handedness of a coordinate system is to exchange two axes.

Both left-handed and right-handed coordinate systems are perfectly valid, and despite what you might read in other books, neither is "better" than the other. People in various fields of study certainly have preferences for one or the other, depending on their backgrounds. For example, some newer computer graphics literature uses left-handed coordinate systems, whereas traditional graphics texts and more math-oriented linear algebra people tend to prefer right-handed coordinate systems. Of course, these are gross generalizations, so always check to see what coordinate system is being used. The bottom line, however, is that in many cases it's just a matter of a negative sign in the z-coordinate. So, appealing to the first law of computer graphics in Section 1.1, if you apply a tool, technique, or resource from another book, web page, or article and it doesn't look right, try flipping the sign on the z-axis.

1.3.4   Some Important Conventions Used in This Book

When designing a 3D virtual world, several design decisions have to be made beforehand, such as left-handed or right-handed coordinate system, which direction is $+y$, and so forth. The map makers from Dyslexia had to choose from among eight different ways to assign the axes in 2D (see Figure 1.7). In 3D, we have a total of 48 different combinations to choose from; 24 of these combinations are left-handed, and 24 are right-handed. (Exercise 3 asks you to list all of them.)

Different situations can call for different conventions, in the sense that certain tasks can be easier if you adopt the right conventions. Usually, however, it is not a major deal as long as you establish the conventions early in your design process and stick to them. (In fact, the choice is most likely thrust upon you by the engine or framework you are using, because very few people start from scratch these days.) All of the basic principles discussed in this book are applicable regardless of the conventions used. For the most part, all of the equations and techniques given are applicable regardless of convention, as well.[4] However, in some cases there are some slight, but critical, differences in application dealing with left-handed versus right-handed coordinate spaces. When those differences arise, we will point them out.

We use a left-handed coordinate system in this book. The $+x$, $+y$, and $+z$ directions point right, up, and forward, respectively, as shown in Figure 1.15. In situations where "right" and "forward" are not appropriate terms (for example, when we discuss the world coordinate space), we assign $+x$ to "east" and $+z$ to "north."

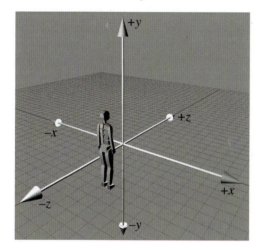

Figure 1.15
The left-handed coordinate system conventions used in this book

1.4   Odds and Ends

In this book, we spend a lot of time focusing on some crucial material that is often relegated to a terse presentation tucked away in an appendix in the books that consider this material a prerequisite. We, too, must assume a nonzero level of mathematical knowledge from the reader, or else every book would get no further than a review of first principles, and so we also have our terse presentation of some prerequisites. In this section we present a few bits of mathematical knowledge with which most readers are probably familiar, but might need a quick refresher.

---

[4]This is due to a fascinating and surprising symmetry in nature. You might say that nature doesn't know if we are using left- or right-handed coordinates. There's a really interesting discussion in *The Feynman Lectures on Physics* about how it is impossible without very advanced physics to describe the concepts of "left" or "right" to someone without referencing some object you both have seen.

1.4.1   Summation and Product Notation

Summation notation is a shorthand way to write the sum of a list of things. It's sort of like a mathematical `for` loop. Let's look at an example:

$$\sum_{i=1}^{6} a_i = a_1 + a_2 + a_3 + a_4 + a_5 + a_6.$$

The variable i is known as the *index variable*. The expressions above and below the summation symbol tell us how many times to execute our "loop" and what values to use for i during each iteration. In this case, i will count from 1 to 6. To "execute" our loop, we iterate the index through all the values specified by the control conditions. For each iteration, we evaluate the expression on the right-hand side of the summation notation (substituting the appropriate value for the index variable), and add this to our sum.

Summation notation is also known as *sigma notation* because that cool-looking symbol that looks like an E is the capital version of the Greek letter sigma.

A similar notation is used when we are taking the product of a series of values, only we use the symbol Π, which is the capital version of the letter π:

$$\prod_{i=1}^{n} a_i = a_1 \times a_2 \times \cdots \times a_{n-1} \times a_n.$$

1.4.2   Interval Notation

Several times in this book, we refer to a subset of the real number line using *interval notation*. The notation $[a, b]$ means, "the portion of the number line from a to b." Or, more formally, we could read $[a, b]$ as "all numbers x such that $a \leq x \leq b$." Notice that this is a *closed* interval, meaning that the endpoints a and b are included in the interval. An *open* interval is one in which the endpoints are excluded. It is denoted using parentheses instead of square brackets: (a, b). This interval contains all x such that $a < x < b$. Sometimes a closed interval is called *inclusive* and an open interval called *exclusive*.

Occasionally, we encounter *half-open* intervals, which include one endpoint but exclude the other. These are denoted with a lopsided[5] notation such as $[a, b)$ or $(a, b]$, with the square bracket being placed next to the endpoint that is included. By convention, if an endpoint is infinite, we consider that end to be open. For example, the set of all nonnegative numbers is $[0, \infty)$.

---

[5] And confusing to the delimiter matching feature of your text editor.

Notice that the notation (x, y) could refer to an open interval or a 2D point. Likewise, $[x, y]$ could be a closed interval or a 2D vector (discussed in the next chapter). The context will always make clear which is the case.

1.4.3   Angles, Degrees, and Radians

An angle measures an amount of rotation in the plane. Variables representing angles are often assigned the Greek letter θ.[6] The most important units of measure used to specify angles are degrees (°) and radians (rad).

Humans usually measure angles using *degrees*. One degree measures 1/360th of a revolution, so 360° represents a complete revolution.[7] Mathematicians, however, prefer to measure angles in *radians*, which is a unit of measure based on the properties of a circle. When we specify the angle between two rays in radians, we are actually measuring the length of the intercepted arc of a unit circle (a circle centered at the origin with radius 1), as shown in Figure 1.16.

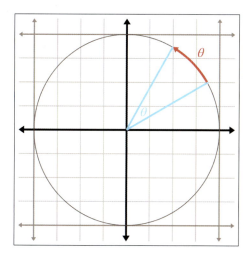

Figure 1.16
A radian measures arc length on a unit circle

---

[6] One prerequisite that we do *not* assume in this book is familiarity with the Greek alphabet. The symbol θ is the lowercase theta, pronounced "THAY-tuh."

[7] The number 360 is a relatively arbitrary choice, which may have had its origin in primitive calendars, such as the Persian calendar, which divided the year into 360 days. This error was never corrected to 365 because the number 360 is so darn convenient. The number 360 has a whopping 22 divisors (not counting itself and 1): 2, 3, 4, 5, 6, 8, 9, 10, 12, 15, 18, 20, 24, 30, 36, 40, 45, 60, 72, 90, 120, and 180. This means 360 can be divided evenly in a large number of cases without needing fractions, which was apparently a good thing to early civilizations. As early as 1750 BCE the Babylonians had devised a sexagesimal (base 60) number system. The number 360 is also large enough so that precision to the nearest whole degree is sufficient in many circumstances.

The circumference of a unit circle is 2π, with π approximately equal to 3.14159265359. Therefore, 2π radians represents a complete revolution.

Since $360° = 2\pi$ rad, $180° = \pi$ rad. To convert an angle from radians to degrees, we multiply by $180/\pi \approx 57.29578$, and to convert an angle from degrees to radians, we multiply by $\pi/180 \approx 0.01745329$. Thus,

**Converting between
radians and degrees**

$$1 \text{ rad} = (180/\pi)° \approx 57.29578°,$$
$$1° = (\pi/180) \text{ rad} \approx 0.01745329 \text{ rad}.$$

In the next section, Table 1.2 will list several angles in both degree and radian format.

1.4.4   Trig Functions

There are many ways to define the elementary trig functions. In this section, we define them using the unit circle. In two dimensions, if we begin with a unit ray pointing towards $+x$, and then rotate this ray counterclockwise by an angle θ, we have drawn the angle in the *standard position*. (If the angle is negative, rotate the ray in the other direction.) This is illustrated in Figure 1.17.

The (x, y) coordinates of the endpoint of a ray thus rotated have special properties and are so significant mathematically that they have been assigned special functions, known as the *cosine* and *sine* of the angle:

**Defining sine and cosine
using the unit circle**

$$\cos\theta = x, \qquad\qquad \sin\theta = y.$$

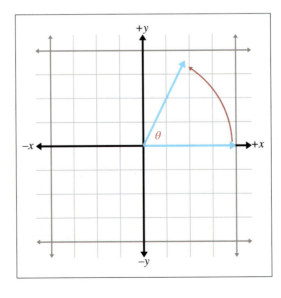

Figure 1.17
An angle in standard
position

You can easily remember which is which because they are in alphabetical order: x comes before y, and cos comes before sin.

The *secant, cosecant, tangent,* and *cotangent* are also useful trig functions. They can be defined in terms of the the sine and cosine:

$$\sec\theta = \frac{1}{\cos\theta}, \qquad\qquad \tan\theta = \frac{\sin\theta}{\cos\theta},$$

$$\csc\theta = \frac{1}{\sin\theta}, \qquad\qquad \cot\theta = \frac{1}{\tan\theta} = \frac{\cos\theta}{\sin\theta}.$$

If we form a right triangle using the rotated ray as the hypotenuse (the side opposite the right angle), we see that x and y give the lengths of the legs (those sides that form the right angle). The length of the adjacent leg is x, and the length of the opposite leg is y, with the terms "adjacent" and "opposite" interpreted relative to the angle θ. Again, alphabetical order is a useful memory aid: "adjacent" and "opposite" are in the same order as the corresponding "cosine" and "sine." Let the abbreviations *hyp, adj,* and *opp* refer to the lengths of the hypotenuse, adjacent leg, and opposite leg, respectively, as shown in Figure 1.18.

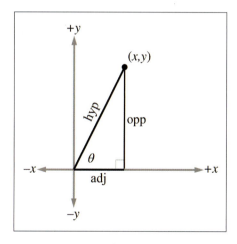

Figure 1.18
The hypotenuse and the adjacent and opposite legs

The primary trig functions are defined by the following ratios:

$$\cos\theta = \frac{adj}{hyp}, \qquad \sin\theta = \frac{opp}{hyp}, \qquad \tan\theta = \frac{opp}{adj},$$

$$\sec\theta = \frac{hyp}{adj}, \qquad \csc\theta = \frac{hyp}{opp}, \qquad \cot\theta = \frac{adj}{opp}.$$

Because of the properties of similar triangles, the above equations apply even when the hypotenuse is not of unit length. However, they do not

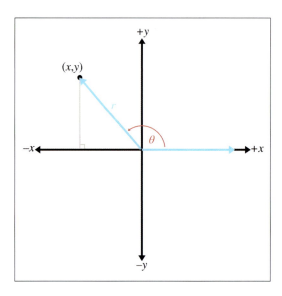

Figure 1.19
A more general
interpretation using (x, y)
coordinates rather than side
lengths

apply when θ is obtuse, since we cannot form a right triangle with an
obtuse interior angle. But by showing the angle in standard position and
allowing the rotated ray to be of any length r (Figure 1.19), we can express
the ratios using x, y, and r:

$$\cos\theta = x/r, \qquad\qquad \sin\theta = y/r, \qquad\qquad \tan\theta = y/x,$$

$$\sec\theta = r/x, \qquad\qquad \csc\theta = r/y, \qquad\qquad \cot\theta = x/y.$$

Table 1.2 shows several different angles, expressed in degrees and radians, and the values of their principal trig functions.

1.4.5   Trig Identities

In this section we present a number of basic relationships between the trig
functions. Because we assume in this book that the reader has some prior
exposure to trigonometry, we do not develop or prove these theorems. The
proofs can be found online or in any trigonometry textbook.

A number of identities can be derived based on the symmetry of the
unit circle:

**Basic identities related
to symmetry**

$$\sin(-\theta) = -\sin\theta, \qquad \cos(-\theta) = \cos\theta, \qquad \tan(-\theta) = -\tan\theta,$$

$$\sin\left(\frac{\pi}{2} - \theta\right) = \cos\theta, \qquad \cos\left(\frac{\pi}{2} - \theta\right) = \sin\theta, \qquad \tan\left(\frac{\pi}{2} - \theta\right) = \cot\theta.$$

$\theta°$	θ rad	$\cos\theta$	$\sin\theta$	$\tan\theta$	$\sec\theta$	$\csc\theta$	$\cot\theta$
0	0	1	0	0	1	undef	undef
30	$\frac{\pi}{6} \approx 0.5236$	$\frac{\sqrt{3}}{2}$	$\frac{1}{2}$	$\frac{\sqrt{3}}{3}$	$\frac{2\sqrt{3}}{3}$	2	$\sqrt{3}$
45	$\frac{\pi}{4} \approx 0.7854$	$\frac{\sqrt{2}}{2}$	$\frac{\sqrt{2}}{2}$	1	$\sqrt{2}$	$\sqrt{2}$	1
60	$\frac{\pi}{3} \approx 1.0472$	$\frac{1}{2}$	$\frac{\sqrt{3}}{2}$	$\sqrt{3}$	2	$\frac{2\sqrt{3}}{3}$	$\frac{\sqrt{3}}{3}$
90	$\frac{\pi}{2} \approx 1.5708$	0	1	undef	undef	1	0
120	$\frac{2\pi}{3} \approx 2.0944$	$-\frac{1}{2}$	$\frac{\sqrt{3}}{2}$	$-\sqrt{3}$	-2	$\frac{2\sqrt{3}}{3}$	$-\frac{\sqrt{3}}{3}$
135	$\frac{3\pi}{4} \approx 2.3562$	$-\frac{\sqrt{2}}{2}$	$\frac{\sqrt{2}}{2}$	-1	$-\sqrt{2}$	$\sqrt{2}$	-1
150	$\frac{5\pi}{6} \approx 2.6180$	$-\frac{\sqrt{3}}{2}$	$\frac{1}{2}$	$-\frac{\sqrt{3}}{3}$	$-\frac{2\sqrt{3}}{3}$	2	$-\sqrt{3}$
180	$\pi \approx 3.1416$	-1	0	0	-1	undef	undef
210	$\frac{7\pi}{6} \approx 3.6652$	$-\frac{\sqrt{3}}{2}$	$-\frac{1}{2}$	$\frac{\sqrt{3}}{3}$	$-\frac{2\sqrt{3}}{3}$	-2	$-\sqrt{3}$
225	$\frac{5\pi}{4} \approx 3.9270$	$-\frac{\sqrt{2}}{2}$	$-\frac{\sqrt{2}}{2}$	1	$-\sqrt{2}$	$-\sqrt{2}$	-1
240	$\frac{4\pi}{3} \approx 4.1888$	$-\frac{1}{2}$	$-\frac{\sqrt{3}}{2}$	$\sqrt{3}$	-2	$-\frac{2\sqrt{3}}{3}$	$-\frac{\sqrt{3}}{3}$
270	$\frac{3\pi}{2} \approx 4.7124$	0	-1	undef	undef	-1	0
300	$\frac{5\pi}{3} \approx 5.2360$	$\frac{1}{2}$	$-\frac{\sqrt{3}}{2}$	$-\sqrt{3}$	2	$-\frac{2\sqrt{3}}{3}$	$-\frac{\sqrt{3}}{3}$
315	$\frac{7\pi}{4} \approx 5.4978$	$\frac{\sqrt{2}}{2}$	$-\frac{\sqrt{2}}{2}$	-1	$\sqrt{2}$	$-\sqrt{2}$	-1
330	$\frac{11\pi}{6} \approx 5.7596$	$\frac{\sqrt{3}}{2}$	$-\frac{1}{2}$	$-\frac{\sqrt{3}}{3}$	$\frac{2\sqrt{3}}{3}$	-2	$-\sqrt{3}$
360	$2\pi \approx 6.2832$	1	0	0	1	undef	undef

Table 1.2. Common angles in degrees and radians, and the values of the principal trig functions

Perhaps the most famous and basic identity concerning the right triangle, one that most readers learned in their primary education, is the *Pythagorean theorem*. It says that the sum of the squares of the two legs of a right triangle is equal to the square of the hypotenuse. Or, more famously, as shown in Figure 1.20,

$$a^2 + b^2 = c^2.$$

Pythagorean theorem

By applying the Pythagorean theorem to the unit circle, one can deduce the identities

$$\sin^2\theta + \cos^2\theta = 1, \qquad 1 + \tan^2\theta = \sec^2\theta, \qquad 1 + \cot^2\theta = \csc^2\theta.$$

Pythagorean identities

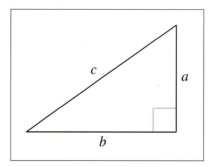

Figure 1.20
The Pythagorean theorem

The following identities involve taking a trig function on the sum or difference of two angles:

Sum and difference identities

$$\sin(a + b) = \sin a \cos b + \cos a \sin b,$$
$$\sin(a - b) = \sin a \cos b - \cos a \sin b,$$
$$\cos(a + b) = \cos a \cos b - \sin a \sin b, \qquad (1.1)$$
$$\cos(a - b) = \cos a \cos b + \sin a \sin b,$$
$$\tan(a + b) = \frac{\tan a + \tan b}{1 - \tan a \tan b},$$
$$\tan(a - b) = \frac{\tan a - \tan b}{1 + \tan a \tan b}.$$

If we apply the sum identities to the special case where a and b are the same, we get the following double angle identities:

Double angle identities

$$\sin 2\theta = 2 \sin \theta \cos \theta,$$
$$\cos 2\theta = \cos^2 \theta - \sin^2 \theta = 2 \cos^2 \theta - 1 = 1 - 2 \sin^2 \theta,$$
$$\tan 2\theta = \frac{2 \tan \theta}{1 - \tan^2 \theta}.$$

We often need to solve for an unknown side length or angle in a triangle, in terms of the known side lengths or angles. For these types of problems the *law of sines* and *law of cosines* are helpful. The formula to use will depend on which values are known and which value is unknown. Figure 1.21 illustrates the notation and shows that these identities hold for any triangle, not just right triangles:

Law of sines

$$\frac{\sin A}{a} = \frac{\sin B}{b} = \frac{\sin C}{c},$$

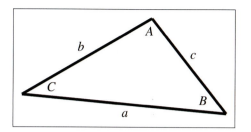

Figure 1.21
Notation used for the law of sines
and law of cosines

$$a^2 = b^2 + c^2 - 2bc \cos A,$$
$$b^2 = a^2 + c^2 - 2ac \cos B,$$
$$c^2 = a^2 + b^2 - 2ab \cos C.$$

Law of cosines

1.5   Exercises

(Answers on page 745.)

1. Give the coordinates of the following points. Assume the standard 2D
   conventions. The darker grid lines represent one unit.

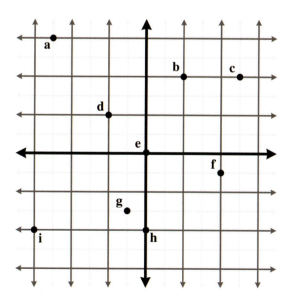

2. Give the coordinates of the following points:

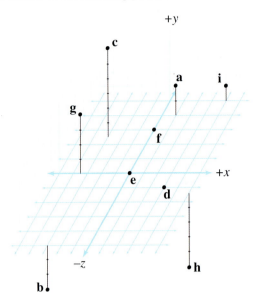

3. List the 48 different possible ways that the 3D axes may be assigned to the directions "north," "east," and "up." Identify which of these combinations are left-handed, and which are right-handed.

4. In the popular modeling program *3DS Max*, the default orientation of the axes is for $+x$ to point right/east, $+y$ to point forward/north, and $+z$ to point up.

    (a) Is this a left- or right-handed coordinate space?

    (b) How would we convert 3D coordinates from the coordinate system used by *3DS Max* into points we could use with our coordinate conventions discussed in Section 1.3.4?

    (c) What about converting from our conventions to the *3DS Max* conventions?

5. A common convention in aerospace is that $+x$ points forward/north, $+y$ points right/east, and z points down.

    (a) Is this a left- or right-handed coordinate space?

    (b) How would we convert 3D coordinates from these aerospace conventions into our conventions?

    (c) What about converting from our conventions to the aerospace conventions?

6. In a left-handed coordinate system:

(a) when looking from the positive end of an axis of rotation, is positive rotation clockwise (CW) or counterclockwise (CCW)?

(b) when looking from the negative end of an axis of rotation, is positive rotation CW or CCW?

In a right-handed coordinate system:

(c) when looking from the positive end of an axis of rotation, is positive rotation CW or CCW?

(d) when looking from the negative end of an axis of rotation, is positive rotation CW or CCW?

7. Compute the following:

$$\text{(a) } \sum_{i=1}^{5} i \quad \text{(b) } \sum_{i=1}^{5} 2i \quad \text{(c) } \prod_{i=1}^{5} 2i \quad \text{(d) } \prod_{i=0}^{4} 7(i+1) \quad \text{(e) } 8 \sum_{i=1}^{100} i$$

8. Convert from degrees to radians:

(a) $30°$ (b) $-45°$ (c) $60°$ (d) $90°$ (e) $-180°$
(f) $225°$ (g) $-270°$ (h) $167.5°$ (i) $527°$ (j) $-1080°$

9. Convert from radians to degrees:

(a) $-\pi/6$ (b) $2\pi/3$ (c) $3\pi/2$ (d) $-4\pi/3$ (e) 2π
(f) $\pi/180$ (g) $\pi/18$ (h) -5π (i) 10π (j) $\pi/5$

10. In *The Wizard of Oz*, the scarecrow receives his degree from the wizard and blurts out this mangled version of the Pythagorean theorem:

> The sum of the square roots of any two sides of an isosceles triangle is equal to the square root of the remaining side.

Apparently the scarecrow's degree wasn't worth very much, since this "proof that he had a brain" is actually wrong in at least two ways.[9] What should the scarecrow have said?

---

[8]There is a well-known story about the mathematician Karl Friedrich Gauss solving this problem in only a few seconds while a student in primary school. As the story goes, his teacher wanted to occupy the students by having them add the numbers 1 to 100 and turn in their answers at the end of class. However, mere seconds after being given this assignment, Gauss handed the correct answer to his teacher as the teacher and the rest of the class gazed in astonishment at the young Gauss.

[9]Homer Simpson repeated the same jibberish after putting on a pair of glasses found in a toilet. A man in a nearby stall corrected him on one of his errors. So if you saw that episode of *The Simpsons*, then you have a headstart on this question, but not the whole answer.

11. Confirm the following:

(a) $(\sin(\alpha)/\csc(\alpha)) + (\cos(\alpha)/\sec(\alpha)) = 1$

(b) $(\sec^2(\theta) - 1)/\sec^2(\theta) = \sin^2(\theta)$

(c) $1 + \cot^2(t) = \csc^2(t)$

(d) $\cos(\phi)(\tan(\phi) + \cot(\phi)) = \csc(\phi)$

People, places, science, odds and ends, and things you should have learned in school had you been paying attention.

— Categories from Michael Feldman's
weekend radio show *Whaddya know?*

Chapter 2

Vectors

Yellow fever vaccine should be administered 10 to 12 days before exposure to the vector.

— *The United States Dispensatory* (1978)

Vectors are the formal mathematical entities we use to do 2D and 3D math. The word *vector* has two distinct but related meanings. Mathematics books, especially those on linear algebra, tend to focus on a rather abstract definition, caring about the numbers in a vector but not necessarily about the context or actual meaning of those numbers. Physics books, on the other hand, tend towards an interpretation that treats a vector as a geometric entity to the extent that they avoid any mention of the coordinates used to measure the vector, when possible. It's no wonder that you can sometimes find people from these two disciplines correcting one another on the finer points of "how vectors *really* work." Of course the reality is that they are both right,[1] and to be proficient with 3D math, we need to understand both interpretations of vectors and how the two interpretations are related.

This chapter introduces the concept of vectors. It is divided into the following sections.

- Section 2.1 covers some of the basic mathematical properties of vectors.

- Section 2.2 gives a high-level introduction to the geometric properties of vectors.

- Section 2.3 connects the mathematical definition with the geometric one and discusses how vectors work within the framework of Cartesian coordinates.

---

[1]But the perspective taken by physics textbooks is probably the one that's more appropriate for video game programming, at least in the beginning.

31

- Section 2.4 discusses the often confusing relationship between points and vectors and considers the rather philosophical question of why it is so difficult to make absolute measurements.

- Sections 2.5–2.12 discuss the fundamental calculations we can perform with vectors, considering both the algebra and geometric interpretations of each operation.

- Section 2.13 presents a list of helpful vector algebra laws.

2.1　Mathematical Definition of Vector, and Other Boring Stuff

To mathematicians, a vector is a list of numbers. Programmers will recognize the synonymous term *array*. Notice that the STL template array class in C++ is named `vector`, and the basic Java array container class is `java.util.Vector`. So mathematically, a vector is nothing more than an array of numbers.

Yawn... If this abstract definition of a vector doesn't inspire you, don't worry. Like many mathematical subjects, we must first introduce some terminology and notation before we can get to the "fun stuff."

Mathematicians distinguish between vector and *scalar* (pronounced "SKAY-lur") quantities. You're already an expert on scalars—scalar is the technical term for an ordinary number. We use this term specifically when we wish to emphasize that a particular quantity is not a vector quantity. For example, as we will discuss shortly, "velocity" and "displacement" are vector quantities, whereas "speed" and "distance" are scalar quantities.

The *dimension* of a vector tells how many numbers the vector contains. Vectors may be of any positive dimension, including one. In fact, a scalar can be considered a 1D vector. In this book, we primarily are interested in 2D, 3D, and (later) 4D vectors.

When writing a vector, mathematicians list the numbers surrounded by square brackets, for example, $[1, 2, 3]$. When we write a vector inline in a paragraph, we usually put commas between the numbers. When we write it out in an equation, the commas are often omitted. In either case, a vector written horizontally is called a *row vector*. Vectors are also frequently written vertically:

$$\begin{bmatrix} 1 \\ 2 \\ 3 \end{bmatrix}.$$

A 3D column vector

A vector written vertically is known as a *column vector*. This book uses both notations. For now, the distinction between row and column vectors won't matter. However, in Section 4.1.7 we discuss why in certain circumstances the distinction is critical.

When we wish to refer to the individual components in a vector, we use subscript notation. In math literature, integer indices are used to access the elements. For example v_1 refers to the first element in $\mathbf{v}$. However, we are specifically interested in 2D, 3D, and 4D vectors rather than vectors of arbitrary dimension n, so we rarely use this notation. Instead, we use x and y to refer to the elements in a 2D vector; x, y, and z to refer to the elements in a 3D vector; and x, y, z, and w to refer to the elements in a 4D vector. This notation is shown in Equation (2.1).

$$
\mathbf{a} = \begin{bmatrix} 1 \\ 2 \end{bmatrix} \qquad \begin{aligned} a_1 &= a_x = 1 \\ a_2 &= a_y = 2 \end{aligned}
$$

$$
\mathbf{b} = \begin{bmatrix} 3 \\ 4 \\ 5 \end{bmatrix} \qquad \begin{aligned} b_1 &= b_x = 3 \\ b_2 &= b_y = 4 \\ b_3 &= b_z = 5 \end{aligned} \tag{2.1}
$$

$$
\mathbf{c} = \begin{bmatrix} 6 \\ 7 \\ 8 \\ 9 \end{bmatrix} \qquad \begin{aligned} c_1 &= c_x = 6 \\ c_2 &= c_y = 7 \\ c_3 &= c_z = 8 \\ c_4 &= c_w = 9 \end{aligned}
$$

Vector subscript notation

Notice that the components of a 4D vector are not in alphabetical order. The fourth value is w. (Hey, they ran out of letters in the alphabet!)

Now let's talk about some important typeface conventions that are used in this book. As you know, *variables* are placeholder symbols used to stand for unknown quantities. In 3D math, we work with scalar, vector, and (later) matrix quantities. In the same way that it's important in a C++ or Java program to specify what type of data is stored by a variable, it is important when working with vectors to be clear what type of data is represented by a particular variable. In this book, we use different fonts for variables of different types:

- *Scalar* variables are represented by lowercase Roman or Greek letters in italics: a, b, x, y, z, θ, α, ω, γ.

- *Vector* variables of any dimension are represented by lowercase letters in boldface: $\mathbf{a}$, $\mathbf{b}$, $\mathbf{u}$, $\mathbf{v}$, $\mathbf{q}$, $\mathbf{r}$.

- *Matrix* variables are represented using uppercase letters in boldface: $\mathbf{A}$, $\mathbf{B}$, $\mathbf{M}$, $\mathbf{R}$.

Note that other authors may use different conventions. One common convention used frequently when writing vectors by hand, is to draw an arrow over the vector, for example, $\vec{a}$.

Before we go any further, a bit of context is in order concerning the perspective that we are adopting about vectors. The branch of mathematics that deals primarily with vectors and matrices is called *linear algebra*, a subject that assumes the abstract definition given previously: a vector is an array of numbers. This highly generalized approach allows for the exploration of a large set of mathematical problems. In linear algebra, vectors and matrices of dimension n are used to solve a system of n linear equations for n unknowns, without knowing or caring what physical significance, if any, is attached to any of the numbers. This is certainly a fascinating and highly practical study, but it is not of primary interest to our investigation of 3D math. For 3D math, we are mostly concerned with the geometric interpretations of vectors and vector operations.

Our focus is geometric, so we omit many details and concepts of linear algebra that do not further our understanding of 2D or 3D geometry. Even though we occasionally discuss properties or operations for vectors of an arbitrary dimension n, we will usually focus on 2D, 3D, and (later) 4D vectors and matrices. Even when the numbers in a vector *do not* have any physical significance, the ability to visualize the linear algebra operations is of some utility, so learning how to interpret the operations geometrically is useful even in nonphysical applications. Some more context about how the topics in this book fit into the bigger picture of linear algebra can be found in Section 4.3.

2.2   Geometric Definition of Vector

Now that we have discussed what a vector is mathematically, let's look at a more geometric interpretation of vectors. Geometrically speaking, a vector is a directed line segment that has *magnitude* and *direction*.

- The *magnitude* of a vector is the length of the vector. A vector may have any nonnegative length.

- The *direction* of a vector describes which way the vector is pointing in space. Note that "direction" is not exactly the same as "orientation," a distinction we will reexamine in Section 8.1.

Let's look at a vector. Figure 2.1 shows an illustration of a vector in 2D. It looks like an arrow, right? This is the standard way to represent a vector graphically, since the two defining characteristics of a vector are captured: its magnitude and direction.

Figure 2.1
A 2D vector

We sometimes refer to the *head* and *tail* of a vector. As shown in Figure 2.2, the head is the end of the vector with the arrowhead on it (where the vector "ends"), and the tail is the other end (where the vector "starts").

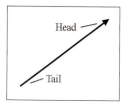

Figure 2.2
A vector has a head and a tail

Where is this vector? Actually, that is not an appropriate question. Vectors do not have position, only magnitude and direction. This may sound impossible, but many quantities we deal with on a daily basis have magnitude and direction, but no position. Consider how the two statements below could make sense, regardless of the location where they are applied.

- *Displacement.* "Take three steps forward." This sentence seems to be all about positions, but the actual quantity used in the sentence is a relative displacement and does not have an absolute position. This relative displacement consists of a magnitude (3 steps) and a direction (forward), so it could be represented by a vector.

- *Velocity.* "I am traveling northeast at 50 mph." This sentence describes a quantity that has magnitude (50 mph) and direction (northeast), but no position. The concept of "northeast at 50 mph" can be represented by a vector.

Notice that *displacement* and *velocity* are technically different from the terms *distance* and *speed*. Displacement and velocity are vector quantities and therefore entail a direction, whereas distance and speed are scalar quantities that do not specify a direction. More specifically, the scalar quantity distance is the magnitude of the vector quantity displacement, and the scalar quantity speed is the magnitude of the vector quantity velocity.

Because vectors are used to express displacements and relative differences between things, they can describe relative positions. ("My house is 3 blocks east of here.") However, you should not think of a vector as having an absolute position itself, instead, remember that it is describing the displacement from one position to another, in this case from "here" to "my house." (More on relative versus absolute position in Section 2.4.1.) To help enforce this, when you imagine a vector, picture an arrow. Remember that the length and direction of this arrow are significant, but not the position.

Since vectors do not have a position, we can represent them on a diagram anywhere we choose, provided that the length and direction of the vector are represented correctly. We often use this fact to our advantage by sliding the vector around into a meaningful location on a diagram.

Now that we have the big picture about vectors from a mathematical and geometric perspective, let's learn how to work with vectors in the Cartesian coordinate system.

2.3   Specifying Vectors with Cartesian Coordinates

When we use Cartesian coordinates to describe vectors, each coordinate measures a *signed displacement* in the corresponding dimension. For example, in 2D, we list the displacement parallel to the x-axis, and the displacement parallel to the y-axis, as illustrated in Figure 2.3.

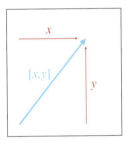

Figure 2.3
Vectors are specified by giving the signed displacement in each dimension.

Figure 2.4 shows several 2D vectors and their values. Notice that the position of each vector on the diagram is irrelevant. (The axes are conspicuously absent to emphasize this fact, although we do assume the standard convention of $+x$ pointing to the right and $+y$ pointing up.) For example, two of the vectors in Figure 2.4 have the value $[1.5, 1]$, but they are not in the same place on the diagram.

3D vectors are a simple extension of 2D vectors. A 3D vector contains three numbers, which measure the signed displacements in the x, y, and z directions, just as you'd expect.

We are focusing on Cartesian coordinates for now, but they are not the only way to describe vectors mathematically. *Polar coordinates* are also common, especially in physics textbooks. Polar coordinates are the subject of Chapter 7.

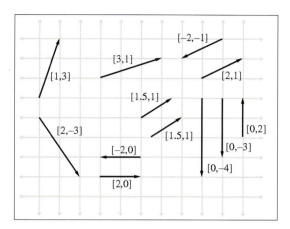

Figure 2.4. Examples of 2D vectors and their values

2.3.1 Vector as a Sequence of Displacements

One helpful way to think about the displacement described by a vector is to break out the vector into its axially aligned components. When these axially aligned displacements are combined, they cumulatively define the displacement defined by the vector as a whole.

For example, the 3D vector $[1, -3, 4]$ represents a single displacement, but we can visualize this displacement as moving 1 unit to the right, 3 units down, and then 4 units forward. (Assume our convention that $+x$, $+y$, and $+z$ point right, up, and forward, respectively. Also note that we do

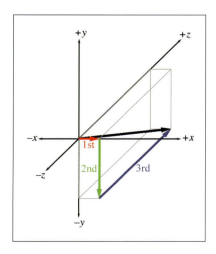

Figure 2.5
Interpreting a vector as a sequence of displacements

not "turn" between steps, so "forward" is always parallel to $+z$.) This displacement is illustrated in Figure 2.5.

The order in which we perform the steps is not important; we could move 4 units forward, 3 units down, and then 1 unit to the right, and we would have displaced by the same total amount. The different orderings correspond to different routes along the axially aligned bounding box containing the vector. Section 2.7.2 mathematically verifies this geometric intuition.

2.3.2 The Zero Vector

For any given vector dimension, there is a special vector, known as the *zero vector*, that has zeroes in every position. For example, the 3D zero vector is $[0, 0, 0]$. We denote a zero vector of any dimension using a boldface zero: $\mathbf{0}$. In other words,

The zero vector

$$\mathbf{0} = \begin{bmatrix} 0 \\ 0 \\ \vdots \\ 0 \end{bmatrix}.$$

The zero vector is special because it is the only vector with a magnitude of zero. All other vectors have a positive magnitude. The zero vector is also unique because it is the only vector that does not have a direction.

Since the zero vector doesn't have a direction or length, we don't draw it as an arrow like we do for other vectors. Instead, we depict the zero vector as a dot. But don't let this make you think of the zero vector as a "point" because a vector does not define a location. Instead, think of the zero vector as a way to express the concept of "no displacement," much as the scalar zero stands for the concept of "no quantity."

Like the scalar zero you know, the zero vector of a given dimension is the *additive identity* for the set of vectors of that dimension. Try to take yourself back to your algebra class, and retrieve from the depths of your memory the concept of the additive identity: for any set of elements, the additive identity of the set is the element x such that for all y in the set, $y + x = y$.[2] In other words, when we add the zero vector to any other vector, we get that vector: $\mathbf{0} + \mathbf{a} = \mathbf{a}$. Section 2.7 deals with vector addition.

---

[2] The typeface used here is not intended to limit the discussion to the set of scalars. We are talking about elements in any set. Also, we request leniency from the abstract algebra sticklers for our use of the word "set," when we should use "group." But the latter term is not as widely understood, and we could only afford this footnote to dwell on the distinction.

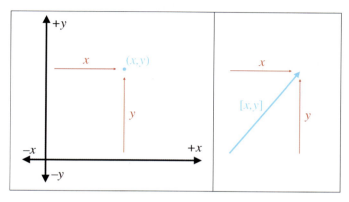

Figure 2.6. Locating points versus specifying vectors

2.4 Vectors versus Points

Recall that a "point" has a location but no real size or thickness. In this chapter, we have learned how a "vector" has magnitude and direction, but no position. So "points" and "vectors" have different purposes, conceptually: a "point" specifies a position, and a "vector" specifies a displacement.

But now examine Figure 2.6, which compares an illustration from Chapter 1 (Figure 1.8), showing how 2D points are located, with a figure from earlier in this chapter (Figure 2.3), showing how 2D vectors are specified. It seems that there is a strong relationship between points and vectors. This section examines this important relationship.

2.4.1 Relative Position

Section 2.2 discussed the fact that because vectors can describe displacements, they can describe relative positions. The idea of a relative position is fairly straightforward: the position of something is specified by describing where it is in relation to some other, known location.

This begs the questions: Where are these "known" locations? What is an "absolute" position? It is surprising to realize that there is no such thing! Every attempt to describe a position requires that we describe it relative to something else. Any description of a position is meaningful only in the context of some (typically "larger") reference frame. Theoretically, we could establish a reference frame encompassing everything in existence and select a point to be the "origin" of this space, thus defining the "absolute" coordinate space. However, even if such an absolute coordinate space were possible, it would not be practical. Luckily for us, absolute positions in the universe aren't important. Do you know *your* precise position in the universe right now? We don't know ours, either.[3]

2.4.2   The Relationship between Points and Vectors

Vectors are used to describe displacements, and therefore they can describe relative positions. Points are used to specify positions. But we have just established in Section 2.4.1 that *any* method of specifying a position must be relative. Therefore, we must conclude that points are relative as well—they are relative to the origin of the coordinate system used to specify their coordinates. This leads us to the relationship between points and vectors.

Figure 2.7 illustrates how the point (x, y) is related to the vector $[x, y]$, given arbitrary values for x and y.

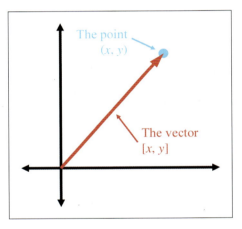

As you can see, if we start at the origin and move by the amount specified by the vector $[x, y]$, we will end up at the location described by the point (x, y). Another way of saying this is that the vector $[x, y]$ gives the displacement from the origin to the point (x, y).

This may seem obvious, but it is important to understand that points and vectors are conceptually distinct, but mathematically equivalent. This confusion between "points" and "vectors" can be a stumbling block for beginners, but it needn't be a problem for you.

Figure 2.7
The relationship between points and vectors

When you think of a location, think of a point and visualize a dot. When you think of a displacement, think of a vector and visualize an arrow.

In many cases, displacements are from the origin, and so the distinction between points and vectors will be a fine one. However, we often deal with quantities that are not relative to the origin, or any other point for that matter. In these cases, it is important to visualize these quantities as an arrow rather than a point.

The math we develop in the following sections operates on "vectors" rather than "points." Keep in mind that any point can be represented as a vector from the origin.

---

[3]But we do know our position relative to the nearest Taco Bell.

Actually, now would be a good time to warn you that a lot of people take a much firmer stance on this issue and would not approve of our cavalier attitude in treating vectors and points as mathematical equals.[4] Such hardliners will tell you, for example, that while you can add two vectors (yielding a third vector), and you can add a vector and a point (yielding a point), you cannot add two points together. We admit that there is some value in understanding these distinctions in certain circumstances. However, we have found that, especially when writing code that operates on points and vectors, adherence to these ethics results in programs that are almost always longer and never faster.[5] Whether it makes the code cleaner or easier to understand is a highly subjective matter. Although this book does not use different notations for points and vectors, in general it will be clear whether a quantity is a point or a vector. We have tried to avoid presenting results with vectors and points mixed inappropriately, but for all the intermediate steps, we might not have been quite as scrupulous.

2.4.3   It's All Relative

Before we move on to the vector operations, let's take a brief philosophical intermission. Spatial position is not the only aspect of our world for which we have difficulty establishing an "absolute" reference, and so we use relative measurements. There are also temperature, loudness, and velocity.

Temperature. One of the first attempts to make a standard temperature scale occurred about AD 170, when Galen proposed a standard "neutral" temperature made up of equal quantities of boiling water and ice. On either side of this temperature were four degrees of "hotter" and four degrees of "colder." Sounds fairly primitive, right? In 1724, Gabriel Fahrenheit suggested a bit more precise system. He suggested that mercury be used as the liquid in a thermometer, and calibrated his scale using two reference points: the freezing point of water, and the temperature of a healthy human being. He called his scale the *Fahrenheit* scale, and measurements were in °F. In 1745, Carolus Linnaeus of Uppsala, Sweden, suggested that things would be simpler if we made the scale range from 0 (at the freezing point of water) to 100 (water's boiling point), and called this scale the *centigrade* scale. (This scale was later abandoned in favor of the *Celsius* scale, which is technically different from centigrade in subtle ways that are not important here.) Notice that all of these scales are *relative*—they are based on the freezing point of water, which is an arbitrary (but highly practical) reference point. A temperature reading of x°C basically means "x degrees hotter than the temperature at which water freezes." It wasn't until 1848, with

---

[4]If you *are* one of those people, then this is a warning of a slightly different sort!
[5]Indeed, sometimes slower, depending on your compiler.

the invention of the Kelvin scale by Lord Kelvin, that mankind finally had an absolute temperature scale. 0 K is the coldest possible temperature, corresponding to -273°C.

Loudness. Loudness is usually measured in *decibels* (abbreviated dB). To be more precise, decibels are used to measure the *ratio* of two power levels. If we have two power levels P_1 and P_2, then the *difference in decibels* between the two power levels is

$$10 \log_{10}(P_2/P_1) \text{ dB.}$$

So, if P_2 is about twice the level of P_1, then the difference is about 3 dB. Notice that this is a relative system, providing a precise way to measure the *relative strength* of two power levels, but not a way to assign a number to one power level. In other words, we haven't established any sort of absolute reference point. (It's also a logarithmic scale, but that isn't important here.) You may have used a mixer board, volume control knob, or digital audio program that measures volume in dB. Normally, there's an arbitrary point marked 0 dB, and then most of the readings have negative values. In other words, 0 dB is the loudest volume, and all other volume settings are softer.

None of these values are absolute—but how could they be? How could your digital audio program know the absolute loudness you will experience, which depends not only on the audio data, but also the volume setting on your computer, the volume knob on your amplifier, the power supplied by the amplifier to your speakers, the distance you are from the speakers, and so on.

Sometimes people describe how loud something is in terms of an absolute dB number. Following in the footsteps of Gabriel Fahrenheit, this scale uses a reference point based on the human body. "Absolute" dB numbers are actually relative to the threshold of hearing for a normal human.[6] Because of this, it's actually possible to have an "absolute" dB reading that is negative. This simply means that the intensity is below the threshold where most people are able to hear it.

At this point, we should probably mention that there *is* a way to devise an absolute scale for loudness, by measuring a physical quantity such as pressure, energy, or power, all of which have an absolute minimum value

---

[6]About 20 micropascals. However, this number varies with frequency. It also increases with age. One author remembers that when he was young, his father would never turn the radio in the car completely off, but rather would turn the volume down below the (father's) threshold of hearing. The son's threshold of hearing was just low enough for this to be irritating. Today the son owns his own car and car radio, and has realized, with some degree of embarrassment, that he *also* often turns the radio volume down without turning it off. However, he offers in his defense that he turns it *all the way* down, below *everyone's* threshold of hearing. (The other author wishes to suggest that clearly even the term "normal human" is relative.)

of zero. The point is that these absolute systems aren't used in many cases—the relative system is the one that's the most useful.

Velocity. How fast are you moving right now? Perhaps you're sitting in a comfy chair, so you'd say that your speed was zero. Maybe you're in a car and so you might say something like 65 mph. (Hopefully someone *else* is driving!) Actually, you are hurtling through space at almost 30 km *per second*! That's about the speed that Earth travels in order to make the 939-million-km trek around the sun each year. Of course, even this velocity is relative to the sun. Our solar system is moving around within the Milky Way galaxy. So then how fast are we actually moving, in absolute terms? Galileo told us back in the 17th century that this question doesn't have an answer—all velocity is relative.

Our difficulty in establishing absolute velocity is similar to the difficulty in establishing position. After all, velocity is displacement (difference between positions) over time. To establish an absolute velocity, we'd need to have some reference location that would "stay still" so that we could measure our displacement from that location. Unfortunately, everything in our universe seems to be orbiting something else.

2.5   Negating a Vector

The previous sections have presented a high-level overview of vectors. The remainder of this chapter looks at specific mathematical operations we perform on vectors. For each operation, we first define the mathematical rules for performing the operation and then describe the geometric interpretations of the operation and give some practical uses for the operation.

The first operation we'd like to consider is that of vector negation. When discussing the zero vector, we asked you to recall from group theory the idea of the *additive identity*. Please go back to wherever it was in your brain that you found the additive identity, perhaps between the metaphorical couch cushions, or at the bottom of a box full of decade-old tax forms. Nearby, you will probably find a similarly discarded obvious-to-the-point-of-useless concept: the *additive inverse*. Let's dust it off. For any group, the additive inverse of x, denoted by $-x$, is the element that yields the additive identity (zero) when added to x. Put simply, $x + (-x) = 0$. Another way of saying this is that elements in the group can be negated.

The negation operation can be applied to vectors. Every vector $\mathbf{v}$ has an additive inverse $-\mathbf{v}$ of the same dimension as $\mathbf{v}$ such that $\mathbf{v} + (-\mathbf{v}) = \mathbf{0}$. (We will learn how to add vectors in Section 2.7.)

2.5.1   Official Linear Algebra Rules

To negate a vector of any dimension, we simply negate each component of the vector. Stated formally,

Negating a vector

$$-\begin{bmatrix} a_1 \\ a_2 \\ \vdots \\ a_{n-1} \\ a_n \end{bmatrix} = \begin{bmatrix} -a_1 \\ -a_2 \\ \vdots \\ -a_{n-1} \\ -a_n \end{bmatrix}.$$

Applying this to the specific cases of 2D, 3D, and 4D vectors, we have

Negating 2D, 3D, and 4D vectors

$$-\begin{bmatrix} x & y \end{bmatrix} = \begin{bmatrix} -x & -y \end{bmatrix},$$
$$-\begin{bmatrix} x & y & z \end{bmatrix} = \begin{bmatrix} -x & -y & -z \end{bmatrix},$$
$$-\begin{bmatrix} x & y & z & w \end{bmatrix} = \begin{bmatrix} -x & -y & -z & -w \end{bmatrix}.$$

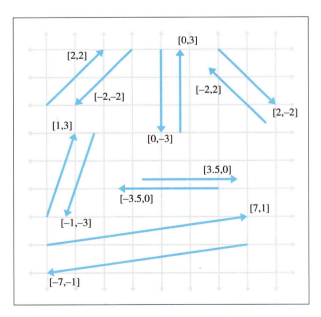

Figure 2.8. Examples of vectors and their negatives. Notice that a vector and its negative are parallel and have the same magnitude, but point in opposite directions.

A few examples are

$$-\begin{bmatrix} 4 & -5 \end{bmatrix} = \begin{bmatrix} -4 & 5 \end{bmatrix},$$
$$-\begin{bmatrix} -1 & 0 & \sqrt{3} \end{bmatrix} = \begin{bmatrix} 1 & 0 & -\sqrt{3} \end{bmatrix},$$
$$-\begin{bmatrix} 1.34 & -3/4 & -5 & \pi \end{bmatrix} = \begin{bmatrix} -1.34 & 3/4 & 5 & -\pi \end{bmatrix}.$$

2.5.2 Geometric Interpretation

Negating a vector results in a vector of the same magnitude but opposite direction, as shown in Figure 2.8.

Remember, the position of a vector on a diagram is irrelevant—only the magnitude and direction are important.

2.6 Vector Multiplication by a Scalar

Although we cannot add a vector and a scalar, we can multiply a vector by a scalar. The result is a vector that is parallel to the original vector, with a different length and possibly opposite direction.

2.6.1 Official Linear Algebra Rules

Vector-times-scalar multiplication is straightforward; we simply multiply each component of the vector by the scalar. Stated formally,

$$k \begin{bmatrix} a_1 \\ a_2 \\ \vdots \\ a_{n-1} \\ a_n \end{bmatrix} = \begin{bmatrix} a_1 \\ a_2 \\ \vdots \\ a_{n-1} \\ a_n \end{bmatrix} k = \begin{bmatrix} ka_1 \\ ka_2 \\ \vdots \\ ka_{n-1} \\ ka_n \end{bmatrix}.$$

Multiplying a vector by a scalar

Applying this rule to 3D vectors, as an example, we get

$$k \begin{bmatrix} x \\ y \\ z \end{bmatrix} = \begin{bmatrix} x \\ y \\ z \end{bmatrix} k = \begin{bmatrix} kx \\ ky \\ kz \end{bmatrix}.$$

Multiplying a 3D vector by a scalar

Although the scalar and vector may be written in either order, most people choose to put the scalar on the left, preferring $k\mathbf{v}$ to $\mathbf{v}k$.

A vector may also be divided by a nonzero scalar. This is equivalent to multiplying by the reciprocal of the scalar:

$$\frac{\mathbf{v}}{k} = \left(\frac{1}{k}\right)\mathbf{v} = \begin{bmatrix} v_x/k \\ v_y/k \\ v_z/k \end{bmatrix} \quad \text{for 3D vector } \mathbf{v} \text{ and nonzero scalar } k.$$

Dividing a 3D vector by a scalar

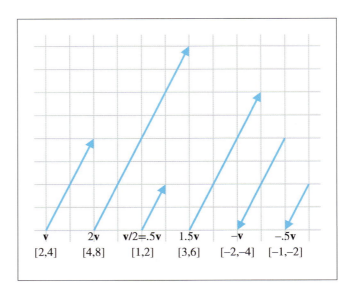

Figure 2.9
A 2D vector
multiplied by
various scalars

Some examples are

$$2\begin{bmatrix} 1 & 2 & 3 \end{bmatrix} = \begin{bmatrix} 2 & 4 & 6 \end{bmatrix},$$
$$-3\begin{bmatrix} -5.4 & 0 \end{bmatrix} = \begin{bmatrix} 16.2 & 0 \end{bmatrix},$$
$$\begin{bmatrix} 4.7 & -6 & 8 \end{bmatrix}/2 = \begin{bmatrix} 2.35 & -3 & 4 \end{bmatrix}.$$

Here are a few things to notice about multiplication of a vector by a scalar:

- When we multiply a vector and a scalar, we do not use any multiplication symbol. The multiplication is signified by placing the two quantities side-by-side (usually with the vector on the right).

- Scalar-times-vector multiplication and division both occur before any addition and subtraction. For example $3\mathbf{a} + \mathbf{b}$ is the same as $(3\mathbf{a}) + \mathbf{b}$, not $3(\mathbf{a} + \mathbf{b})$.

- A scalar may not be divided by a vector, and a vector may not be divided by another vector.

- Vector negation can be viewed as the special case of multiplying a vector by the scalar -1.

2.6.2   Geometric Interpretation

Geometrically, multiplying a vector by a scalar k has the effect of scaling the length by a factor of $|k|$. For example, to double the length of a vector we would multiply the vector by 2. If $k < 0$, then the direction of the vector is flipped. Figure 2.9 illustrates a vector multiplied by several different scalars.

2.7 Vector Addition and Subtraction

We can add and subtract two vectors, provided they are of the same dimension. The result is a vector quantity of the same dimension as the vector operands. We use the same notation for vector addition and subtraction as is used for addition and subtraction of scalars.

2.7.1 Official Linear Algebra Rules

The linear algebra rules for vector addition are simple: to add two vectors, we add the corresponding components:

$$
\begin{bmatrix} a_1 \\ a_2 \\ \vdots \\ a_{n-1} \\ a_n \end{bmatrix} + \begin{bmatrix} b_1 \\ b_2 \\ \vdots \\ b_{n-1} \\ b_n \end{bmatrix} = \begin{bmatrix} a_1 + b_1 \\ a_2 + b_2 \\ \vdots \\ a_{n-1} + b_{n-1} \\ a_n + b_n \end{bmatrix}.
$$

Adding two vectors

Subtraction can be interpreted as adding the negative, so $\mathbf{a} - \mathbf{b} = \mathbf{a} + (-\mathbf{b})$:

$$
\begin{bmatrix} a_1 \\ a_2 \\ \vdots \\ a_{n-1} \\ a_n \end{bmatrix} - \begin{bmatrix} b_1 \\ b_2 \\ \vdots \\ b_{n-1} \\ b_n \end{bmatrix} = \begin{bmatrix} a_1 \\ a_2 \\ \vdots \\ a_{n-1} \\ a_n \end{bmatrix} + \left(- \begin{bmatrix} b_1 \\ b_2 \\ \vdots \\ b_{n-1} \\ b_n \end{bmatrix} \right) = \begin{bmatrix} a_1 - b_1 \\ a_2 - b_2 \\ \vdots \\ a_{n-1} - b_{n-1} \\ a_n - b_n \end{bmatrix}.
$$

Subtracting two vectors

For example, given

$$
\mathbf{a} = \begin{bmatrix} 1 \\ 2 \\ 3 \end{bmatrix}, \qquad \mathbf{b} = \begin{bmatrix} 4 \\ 5 \\ 6 \end{bmatrix}, \qquad \mathbf{c} = \begin{bmatrix} 7 \\ -3 \\ 0 \end{bmatrix},
$$

then

$$\mathbf{a} + \mathbf{b} = \begin{bmatrix} 1 \\ 2 \\ 3 \end{bmatrix} + \begin{bmatrix} 4 \\ 5 \\ 6 \end{bmatrix} = \begin{bmatrix} 1+4 \\ 2+5 \\ 3+6 \end{bmatrix} = \begin{bmatrix} 5 \\ 7 \\ 9 \end{bmatrix},$$

$$\mathbf{a} - \mathbf{b} = \begin{bmatrix} 1 \\ 2 \\ 3 \end{bmatrix} - \begin{bmatrix} 4 \\ 5 \\ 6 \end{bmatrix} = \begin{bmatrix} 1-4 \\ 2-5 \\ 3-6 \end{bmatrix} = \begin{bmatrix} -3 \\ -3 \\ -3 \end{bmatrix},$$

$$\mathbf{b} + \mathbf{c} - \mathbf{a} = \begin{bmatrix} 4 \\ 5 \\ 6 \end{bmatrix} + \begin{bmatrix} 7 \\ -3 \\ 0 \end{bmatrix} - \begin{bmatrix} 1 \\ 2 \\ 3 \end{bmatrix} = \begin{bmatrix} 4+7-1 \\ 5+(-3)-2 \\ 6+0-3 \end{bmatrix} = \begin{bmatrix} 10 \\ 0 \\ 3 \end{bmatrix}.$$

A vector cannot be added or subtracted with a scalar, or with a vector of a different dimension. Also, just like addition and subtraction of scalars, vector addition is commutative,

$$\mathbf{a} + \mathbf{b} = \mathbf{b} + \mathbf{a},$$

whereas vector subtraction is anticommutative,

$$\mathbf{a} - \mathbf{b} = -(\mathbf{b} - \mathbf{a}).$$

2.7.2 Geometric Interpretation

We can add vectors $\mathbf{a}$ and $\mathbf{b}$ geometrically by positioning the vectors so that the head of $\mathbf{a}$ touches the tail of $\mathbf{b}$ and then drawing a vector from

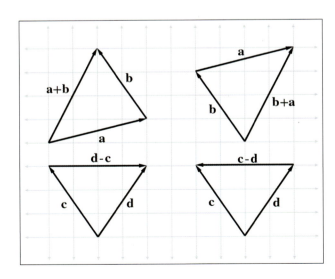

Figure 2.10
2D vector addition and subtraction using the triangle rule.

the tail of **a** to the head of **b**. In other words, if we start at a point and apply the displacements specified by **a** and then **b**, it's the same as if we had applied the single displacement **a** + **b**. This is known as the *triangle rule* of vector addition. It also works for vector subtraction, as shown in Figure 2.10.

Figure 2.10 provides geometric evidence that vector addition is commutative but vector subtraction is not. Notice that the vector labeled **a** + **b** is identical to the vector labeled **b** + **a**, but the vectors **d** − **c** and **c** − **d** point in opposite directions because **d** − **c** = −(**c** − **d**).

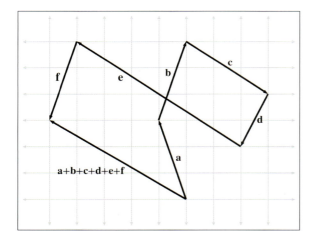

Figure 2.11
Extending the triangle rule to more than two vectors

The triangle rule can be extended to more than two vectors. Figure 2.11 shows how the triangle rule verifies something we stated in Section 2.3.1: a vector can be interpreted as a sequence of axially aligned displacements.

Figure 2.12 is a reproduction of Figure 2.5, which shows how the vector $[1, -3, 4]$ may be interpreted as a displacement of 1 unit to the right, 3 units down, and then 4 units forward, and can be verified mathematically by using vector addition:

$$\begin{bmatrix} 1 \\ -3 \\ 4 \end{bmatrix} = \begin{bmatrix} 1 \\ 0 \\ 0 \end{bmatrix} + \begin{bmatrix} 0 \\ -3 \\ 0 \end{bmatrix} + \begin{bmatrix} 0 \\ 0 \\ 4 \end{bmatrix}.$$

This seems obvious, but this is a very powerful concept. We will use a similar technique in Section 4.2 to transform vectors from one coordinate space to another.

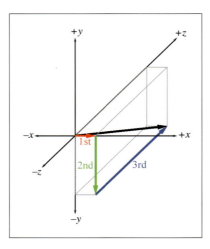

Figure 2.12
Interpreting a vector as a sequence of displacements

2.7.3   Displacement Vector from One Point to Another

It is very common that we will need to compute the displacement from one point to another. In this case, we can use the triangle rule and vector subtraction. Figure 2.13 shows how the displacement vector from **a** to **b** can be computed by subtracting **a** from **b**.

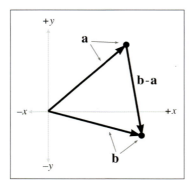

Figure 2.13
Using 2D vector subtraction to compute the vector from point **a** to point **b**

As Figure 2.13 shows, to compute the vector from **a** to **b**, we interpret the points **a** and **b** as vectors from the origin, and then use the triangle rule. In fact, this is how vectors are defined in some texts: the subtraction of two points.

Notice that the vector subtraction **b** − **a** yields a vector *from* **a** *to* **b**. It doesn't make any sense to simply find the vector "between two points," since the language in this sentence does not specify a direction. We must always form a vector that goes *from* one point *to* another point.

2.8   Vector Magnitude (Length)

As we have discussed, vectors have magnitude and direction. However, you might have noticed that neither the magnitude nor the direction is expressed explicitly in the vector (at least not when we use Cartesian coordinates). For example, the magnitude of the 2D vector $[3, 4]$ is neither 3 nor 4; it's 5. Since the magnitude of the vector is not expressed explicitly, we must compute it. The magnitude of a vector is also known as the *length* or *norm* of the vector.

2.8.1   Official Linear Algebra Rules

In linear algebra, the magnitude of a vector is denoted by using double vertical bars surrounding the vector. This is similar to the single vertical bar notation used for the absolute value operation for scalars. This notation and the equation for computing the magnitude of a vector of arbitrary dimension n are shown in Equation (2.2):

$$\|\mathbf{v}\| = \sqrt{\sum_{i=1}^{n} v_i{}^2} = \sqrt{v_1{}^2 + v_2{}^2 + \cdots + v_{n-1}{}^2 + v_n{}^2}. \qquad (2.2)$$

Magnitude of a vector of arbitrary dimension

Thus, the magnitude of a vector is the square root of the sum of the squares of the components of the vector. This sounds complicated, but the magnitude equations for 2D and 3D vectors are actually very simple:

$$\|\mathbf{v}\| = \sqrt{v_x{}^2 + v_y{}^2} \qquad \text{(for a 2D vector } \mathbf{v}\text{),} \qquad (2.3)$$

$$\|\mathbf{v}\| = \sqrt{v_x{}^2 + v_y{}^2 + v_z{}^2} \qquad \text{(for a 3D vector } \mathbf{v}\text{).}$$

Vector magnitude for 2D and 3D vectors

The magnitude of a vector is a nonnegative scalar quantity. An example of how to compute the magnitude of a 3D vector is

$$\left\| \begin{bmatrix} 5 & -4 & 7 \end{bmatrix} \right\| = \sqrt{5^2 + (-4)^2 + 7^2} = \sqrt{25 + 16 + 49} = \sqrt{90}$$
$$\approx 9.4868.$$

Some books use a single bar notation to indicate vector magnitude: $|\mathbf{v}|$

One quick note to satisfy all you sticklers who already know about vector norms and at this moment are pointing your web browser to gamemath.com, looking for the email address for errata. The term *norm* actually has a very general definition, and basically any equation that meets a certain set of criteria can call itself a norm. So to describe Equation (2.2) as *the* equation for the vector norm is slightly misleading. To be more accurate, we should say that Equation (2.2) is the equation for the *2-norm*, which is one specific way to calculate a norm. The 2-norm belongs to a class of norms known as the *p-norms*, and the *p*-norm is not the only way to define a norm. Still, omitting this level of generality isn't too harmful of a delusion; because the 2-norm measures Euclidian distance, it is by far the most commonly used norm in geometric applications. It is also widely used in situations even where a geometric interpretation is not directly applicable. Readers interested in such exotica should check out Exercise 15.

2.8.2   Geometric Interpretation

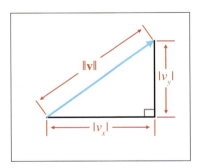

Let's try to get a better understanding of why Equation (2.3) works. For any vector $\mathbf{v}$ in 2D, we can form a right triangle with $\mathbf{v}$ as the hypotenuse, as shown in Figure 2.14.

Notice that to be precise we had to put absolute value signs around the components v_x and v_y. The components of the vector may be negative, since they are *signed displacements*, but *length* is always positive.

The Pythagorean theorem states that for any right triangle, the square of the length of the hypotenuse is equal to the sum of the squares of the lengths of the other two sides. Applying this theorem to Figure 2.14, we have

Figure 2.14
Geometric interpretation of the magnitude equation

$$\|\mathbf{v}\|^2 = |v_x|^2 + |v_y|^2.$$

Since $|x|^2 = x^2$, we can omit the absolute value symbols:

$$\|\mathbf{v}\|^2 = v_x{}^2 + v_y{}^2.$$

Then, by taking the square root of both sides and simplifying, we get

$$\sqrt{\|\mathbf{v}\|^2} = \sqrt{v_x{}^2 + v_y{}^2},$$

$$\|\mathbf{v}\| = \sqrt{v_x{}^2 + v_y{}^2},$$

which is the same as Equation (2.3). The proof of the magnitude equation in 3D is only slightly more complicated.

For any positive magnitude m, there are an infinite number of vectors of magnitude m. Since these vectors all have the same length but different directions, they form a circle when the tails are placed at the origin, as shown in Figure 2.15.

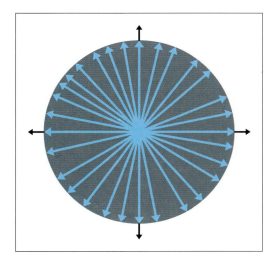

Figure 2.15
For any positive magnitude, there are an infinite number of vectors with that magnitude

2.9   Unit Vectors

For many vector quantities, we are concerned only with direction and not magnitude: "Which way am I facing?" "Which way is the surface oriented?" In these cases, it is often convenient to use unit vectors. A *unit vector* is a vector that has a magnitude of one. Unit vectors are also known as *normalized vectors*.

Unit vectors are also sometimes simply called *normals*; however, a warning is in order concerning terminology. The word "normal" carries with it the connotation of "perpendicular." When most people speak of a "normal" vector, they are usually referring to a vector that is perpendicular to something. For example, a *surface normal* at a given point on an object is a vector that is perpendicular to the surface at that location. However, since the concept of perpendicular is related only to the direction of a vector and not its magnitude, in most cases you will find that unit vectors are used for normals instead of a vector of arbitrary length. When this book refers to a vector as a "normal," it means "a unit vector perpendicular to something else." This is common usage, but be warned that the word "normal" primarily means "perpendicular" and not "unit length." Since it is so common for normals to be unit vectors, we will take care to call out any situation where a "normal" vector does not have unit length.

In summary, a "normalized" vector always has unit length, but a "normal" vector is a vector that is perpendicular to something and by convention usually has unit length.

2.9.1 Official Linear Algebra Rules

For any nonzero vector $\mathbf{v}$, we can compute a unit vector that points in the same direction as $\mathbf{v}$. This process is known as *normalizing* the vector. In this book we use a common notation of putting a hat symbol over unit vectors; for example, $\hat{\mathbf{v}}$ (pronounced "v hat"). To normalize a vector, we divide the vector by its magnitude:

Normalizing a vector

$$\hat{\mathbf{v}} = \frac{\mathbf{v}}{\|\mathbf{v}\|} \qquad \text{for any nonzero vector } \mathbf{v}.$$

For example, to normalize the 2D vector $[12, -5]$,

$$\frac{\begin{bmatrix} 12 & -5 \end{bmatrix}}{\left\|\begin{bmatrix} 12 & -5 \end{bmatrix}\right\|} = \frac{\begin{bmatrix} 12 & -5 \end{bmatrix}}{\sqrt{12^2 + 5^2}} = \frac{\begin{bmatrix} 12 & -5 \end{bmatrix}}{\sqrt{169}} = \frac{\begin{bmatrix} 12 & -5 \end{bmatrix}}{13} = \begin{bmatrix} \frac{12}{13} & \frac{-5}{13} \end{bmatrix}$$
$$\approx \begin{bmatrix} 0.923 & -0.385 \end{bmatrix}.$$

The zero vector cannot be normalized. Mathematically, this is not allowed because it would result in division by zero. Geometrically, it makes sense because the zero vector does not define a direction—if we normalized the zero vector, in what direction should the resulting vector point?

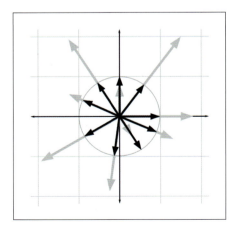

Figure 2.16
Normalizing vectors in 2D

2.9.2   Geometric Interpretation

In 2D, if we draw a unit vector with the tail at the origin, the head of the vector will touch a unit circle centered at the origin. (A unit circle has a radius of 1.) In 3D, unit vectors touch the surface of a unit sphere. Figure 2.16 shows several 2D vectors of arbitrary length in gray, beneath their normalized counterparts in black.

Notice that normalizing a vector makes some vectors shorter (if their length was greater than 1) and some vectors longer (if their length was less than 1).

2.10   The Distance Formula

We are now prepared to derive one of the oldest and most fundamental formulas in computational geometry: the distance formula. This formula is used to compute the distance between two points.

First, let's define distance as the length of the line segment between the two points. Since a vector is a directed line segment, geometrically it makes sense that the distance between the two points would be equal to the length of a vector from one point to the other. Let's derive the distance formula in 3D. First, we will compute the vector $\mathbf{d}$ from $\mathbf{a}$ to $\mathbf{b}$. We learned how to do this in 2D in Section 2.7.3. In 3D, we use

$$\mathbf{d} = \mathbf{b} - \mathbf{a} = \begin{bmatrix} b_x - a_x \\ b_y - a_y \\ b_z - a_z \end{bmatrix}.$$

The distance between $\mathbf{a}$ and $\mathbf{b}$ is equal to the length of the vector $\mathbf{d}$, which we computed in Section 2.8:

$$\text{distance}\,(\mathbf{a}, \mathbf{b}) = \|\mathbf{d}\| = \sqrt{d_x{}^2 + d_y{}^2 + d_z{}^2}.$$

Substituting for $\mathbf{d}$, we get

$$\text{distance}\,(\mathbf{a}, \mathbf{b}) = \|\mathbf{b} - \mathbf{a}\| = \sqrt{(b_x - a_x)^2 + (b_y - a_y)^2 + (b_z - a_z)^2}.$$

The 3D distance formula

Thus, we have derived the distance formula in 3D. The 2D equation is even simpler:

$$\text{distance}\,(\mathbf{a}, \mathbf{b}) = \|\mathbf{b} - \mathbf{a}\| = \sqrt{(b_x - a_x)^2 + (b_y - a_y)^2}.$$

The 2D distance formula

Let's look at an example in 2D:

$$\begin{aligned} \text{distance}\,\left(\begin{bmatrix} 5 & 0 \end{bmatrix}, \begin{bmatrix} -1 & 8 \end{bmatrix}\right) &= \sqrt{(-1 - 5)^2 + (8 - 0)^2} \\ &= \sqrt{(-6)^2 + 8^2} = \sqrt{100} = 10. \end{aligned}$$

Notice that it doesn't matter which point we call **a** and which point we call **b**. If we define **d** to be the vector from **b** to **a** instead of from **a** to **b**, we will derive a slightly different, but mathematically equivalent, equation.

2.11 Vector Dot Product

Section 2.6 showed how to multiply a vector by a scalar. We can also multiply two vectors together. There are two types of vector products. The first vector product is the *dot product* (also known as the *inner product*), the subject of this section. We talk about the other vector product, the *cross product*, in Section 2.12.

The dot product is ubiquitous in video game programming, useful in everything from graphics, to simulation, to AI. Following the pattern we used for the operations, we first discuss the algebraic rules for computing dot products in Section 2.11.1, followed by some geometric interpretations in Section 2.11.2.

The dot product formula is one of the few formulas in this book worth memorizing. First of all, it's really easy to memorize. Also, if you understand what the dot product does, the formula makes sense. Furthermore, the dot product has important relationships to many other operations, such as matrix multiplication, convolution of signals, statistical correlations, and Fourier transforms. Understanding the formula will make these relationships more apparent.

Even more important than memorizing a formula is to get an intuitive grasp for what the dot product *does*. If there is only enough space in your brain for either the formula or the geometric definition, then we recommend internalizing the geometry, and getting the formula tattooed on your hand. You need to understand the geometric definition in order to *use* the dot product. When programming in computer languages such as C++, HLSL, or even Matlab and Maple, you won't need to know the formula anyway, since you will usually tell the computer to do a dot product calculation not by typing in the formula, but by invoking a high-level function or overloaded operator. Furthermore, the geometric definition of the dot product does not assume any particular coordinate frame or even the use of Cartesian coordinates.

2.11.1 Official Linear Algebra Rules

The name "dot product" comes from the dot symbol used in the notation: **a** · **b**. Just like scalar-times-vector multiplication, the vector dot product is performed before addition and subtraction, unless parentheses are used to override this default order of operations. Note that although we usually omit the multiplication symbol when multiplying two scalars or a scalar and a vector, we must not omit the dot symbol when performing a vector dot product. If you ever see two vectors placed side-by-side with no symbol in between, interpret this according to the rules of *matrix multiplication*, which we discuss in Chapter 4.[7]

The dot product of two vectors is the sum of the products of corresponding components, resulting in a *scalar*:

$$\begin{bmatrix} a_1 \\ a_2 \\ \vdots \\ a_{n-1} \\ a_n \end{bmatrix} \cdot \begin{bmatrix} b_1 \\ b_2 \\ \vdots \\ b_{n-1} \\ b_n \end{bmatrix} = a_1 b_1 + a_2 b_2 + \cdots + a_{n-1} b_{n-1} + a_n b_n.$$

Vector dot product

This can be expressed succinctly by using the summation notation

$$\mathbf{a} \cdot \mathbf{b} = \sum_{i=1}^{n} a_i b_i.$$

Dot product using summation notation

Applying these rules to the 2D and 3D cases yields

$$\mathbf{a} \cdot \mathbf{b} = a_x b_x + a_y b_y \qquad \text{(}\mathbf{a} \text{ and } \mathbf{b} \text{ are 2D vectors)},$$
$$\mathbf{a} \cdot \mathbf{b} = a_x b_x + a_y b_y + a_z b_z \qquad \text{(}\mathbf{a} \text{ and } \mathbf{b} \text{ are 3D vectors)}.$$

2D and 3D dot products

Examples of the dot product in 2D and 3D are

$$\begin{bmatrix} 4 & 6 \end{bmatrix} \cdot \begin{bmatrix} -3 & 7 \end{bmatrix} = (4)(-3) + (6)(7) = 30,$$

$$\begin{bmatrix} 3 \\ -2 \\ 7 \end{bmatrix} \cdot \begin{bmatrix} 0 \\ 4 \\ -1 \end{bmatrix} = (3)(0) + (-2)(4) + (7)(-1) = -15.$$

It is obvious from inspection of the equations that vector dot product is commutative: $\mathbf{a} \cdot \mathbf{b} = \mathbf{b} \cdot \mathbf{a}$. More vector algebra laws concerning the dot product are given in Section 2.13.

2.11.2 Geometric Interpretation

Now let's discuss the more important aspect of the dot product: what it means geometrically. It would be difficult to make too big of a deal

---

[7]One notation you will probably bump up against is treating the dot product as an ordinary matrix multiplication, denoted by $\mathbf{a}^{\mathrm{T}}\mathbf{b}$ if $\mathbf{a}$ and $\mathbf{b}$ are interpreted as column vectors, or $\mathbf{a}\mathbf{b}^{\mathrm{T}}$ for row vectors. If none of this makes sense, don't worry, we will repeat it after we learn about matrix multiplication and row and column vectors in Chapter 4.

out of the dot product, as it is fundamental to almost every aspect of 3D math. Because of its supreme importance, we're going to dwell on it a bit. We'll discuss two slightly different ways of thinking about this operation geometrically; since they are really equivalent, you may or may not think one interpretation or the other is "more fundamental," or perhaps you may think we are being redundant and wasting your time. You might especially think this if you already have some exposure to the dot product, but please indulge us.

The first geometric definition to present is perhaps the less common of the two, but in agreement with the advice of Dray and Manogue [15], we believe it's actually the more useful. The interpretation we first consider is that of the dot product performing a *projection*.

Assume for the moment that $\hat{\mathbf{a}}$ is a unit vector, and $\mathbf{b}$ is a vector of any length. Now take $\mathbf{b}$ and *project* it onto a line parallel to $\hat{\mathbf{a}}$, as in Figure 2.17.

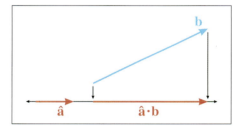

Figure 2.17
The dot product as a projection

(Remember that vectors are displacements and do not have a fixed position, so we are free to move them around on a diagram anywhere we wish.) We can define the dot product $\hat{\mathbf{a}} \cdot \mathbf{b}$ as the signed length of the projection of $\mathbf{b}$ onto this line. The term "projection" has a few different technical meanings (see Section 5.3) and we won't bother attempting a formal definition here.[8] You can think of the projection of $\mathbf{b}$ onto $\hat{\mathbf{a}}$ as the "shadow" that $\mathbf{b}$ casts on $\hat{\mathbf{a}}$ when the rays of light are perpendicular to $\hat{\mathbf{a}}$.

We have drawn the projections as arrows, but remember that the result of a dot product is a scalar, not a vector. Still, when you first learned about negative numbers, your teacher probably depicted numbers as arrows on a number line, to emphasize their sign, just as we have. After all, a scalar is a perfectly valid one-dimensional vector.

What does it mean for the dot product to measure a *signed* length? It means the value will be negative when the projection of $\mathbf{b}$ points in the opposite direction from $\hat{\mathbf{a}}$, and the projection has zero length (it is a single point) when $\hat{\mathbf{a}}$ and $\mathbf{b}$ are perpendicular. These cases are illustrated in Figure 2.18.

---

[8]Thus shirking our traditional duties as mathematics authors to make intuitive concepts sound much more complicated than they are.

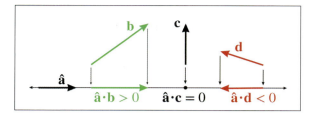

Figure 2.18. Sign of the dot product

In other words, the sign of the dot product can give us a rough classification of the relative directions of the two vectors. Imagine a line (in 2D) or plane (in 3D) perpendicular to the vector $\hat{\mathbf{a}}$. The sign of the dot product $\hat{\mathbf{a}} \cdot \mathbf{b}$ tells us which half-space $\mathbf{b}$ lies in. This is illustrated in Figure 2.19.

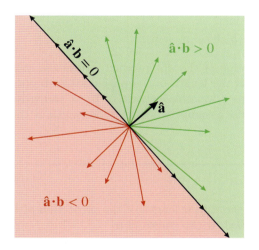

Figure 2.19
The sign of the dot product gives a rough classification of the relative orientation of two vectors.

Next, consider what happens when we scale $\mathbf{b}$ by some factor k. As shown in Figure 2.20, the length of the projection (and thus the value of the dot product) increases by the same factor. The two triangles have equal interior angles and thus are similar. Since the hypotenuse on the right is longer than the hypotenuse on the left by a factor of k, by the properties of similar triangles, the base on the right is also longer by a factor of k.

Let's state this fact algebraically and prove it by using the formula:

$$\hat{\mathbf{a}} \cdot (k\mathbf{b}) = a_x(kb_x) + a_y(kb_y) + a_z(kb_z)$$
$$= k(a_xb_x + a_yb_y + a_zb_z)$$
$$= k(\hat{\mathbf{a}} \cdot \mathbf{b}).$$

Dot product is associative with multiplication by a scalar

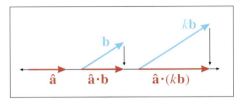

Figure 2.20
Scaling one operand of the dot product

The expanded scalar math in the middle uses three dimensions as our example, but the vector notation at either end of the equation applies for vectors of any dimension.

We've seen what happens when we scale **b**: the length of its projection onto **â** increases along with the value of the dot product. What if we scale **a**? The algebraic argument we just made can be used to show that the value of the dot product scales with the length of **a**, just like it does when we scale **b**. In other words,

$$(k\mathbf{a}) \cdot \mathbf{b} = k(\mathbf{a} \cdot \mathbf{b}) = \mathbf{a} \cdot (k\mathbf{b}).$$

Dot product is associative with multiplication by a scalar for either vector

So scaling **a** scales the numeric value of the dot product. However, this scale has no affect geometrically on the length of the projection of **b** onto **a**. Now that we know what happens if we scale either **a** or **b**, we can write our geometric definition without any assumptions about the length of the vectors.

Dot Product as Projection

The dot product $\mathbf{a} \cdot \mathbf{b}$ is equal to the signed length of the projection of **b** onto any line parallel to **a**, multiplied by the length of **a**.

As we continue to examine the properties of the dot product, some will be easiest to illustrate geometrically when either **a**, or both **a** and **b**, are unit vectors. Because we have shown that scaling either **a** or **b** directly scales the value of the dot product, it will be easy to generalize our results after we have obtained them. Furthermore, in the algebraic arguments that accompany each geometric argument, unit vector assumptions won't be necessary. Remember that we put hats on top of vectors that are assumed to have unit length.

You may well wonder why the dot product measures the projection of the second operand onto the first, and not the other way around. When the two vectors **â** and **b̂** are unit vectors, we can easily make a geometric argument that the projection of $\hat{\mathbf{a}}$ onto $\hat{\mathbf{b}}$ has the same length as the projection of $\hat{\mathbf{b}}$ onto $\hat{\mathbf{a}}$. Consider Figure 2.21. The two triangles have equal interior angles and thus are similar. Since $\hat{\mathbf{a}}$ and $\hat{\mathbf{b}}$ are corresponding sides and have the same length, the two triangles are reflections of each other.

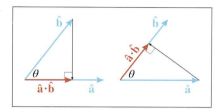

Figure 2.21
Dot product is commutative

We've already shown how scaling either vector will scale the dot product proportionally, so this result applies for $\mathbf{a}$ and $\mathbf{b}$ with arbitrary length. Furthermore, this geometric fact is also trivially verified by using the formula, which does not depend on the assumption that the vectors have equal length. Using two dimensions as our example this time,

$$\mathbf{a} \cdot \mathbf{b} = a_x b_x + a_y b_y = b_x a_x + b_y a_y = \mathbf{b} \cdot \mathbf{a}.$$

Dot product is commutative

The next important property of the dot product is that it distributes over addition and subtraction, just like scalar multiplication. This time let's do the algebra before the geometry. When we say that the dot product "distributes," that means that if one of the operands to the dot product is a sum, then we can take the dot product of the pieces individually, and then take their sum. Switching back to three dimensions for our example,

$$
\begin{aligned}
\mathbf{a} \cdot (\mathbf{b} + \mathbf{c}) &= \begin{bmatrix} a_x \\ a_y \\ a_z \end{bmatrix} \cdot \begin{bmatrix} b_x + c_x \\ b_y + c_y \\ b_z + c_z \end{bmatrix} \\
&= a_x(b_x + c_x) + a_y(b_y + c_y) + a_z(b_z + c_z) \\
&= a_x b_x + a_x c_x + a_y b_y + a_y c_y + a_z b_z + a_z c_z \\
&= (a_x b_x + a_y b_y + a_z b_z) + (a_x c_x + a_y c_y + a_z c_z) \\
&= \mathbf{a} \cdot \mathbf{b} + \mathbf{a} \cdot \mathbf{c}.
\end{aligned}
$$

Dot product distributes over addition and subtraction

By replacing $\mathbf{c}$ with $-\mathbf{c}$, it's clear that the dot product distributes over vector subtraction just as it does for vector addition. Figure 2.22 shows how the dot product distributes over addition.

Now let's look at a special situation in which one of the vectors is the unit vector pointing in the $+x$ direction, which we'll denote as $\hat{\mathbf{x}}$. As shown in Figure 2.23, the signed length of the projection is simply the x-coordinate

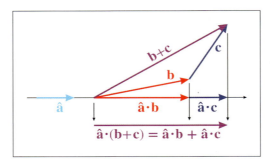

Figure 2.22
The dot product distributes over addition.

of the original vector. In other words, taking the dot product of a vector with a cardinal axis "sifts" out the coordinate for that axis.

If we combine this "sifting" property of the dot product with the fact that it distributes over addition, which we have been able to show in purely geometric terms, we can see why the formula has to be what it is.

Because the dot product measures the length of a projection, it has an interesting relationship to the vector magnitude calculation. Remember that the vector magnitude is a scalar measuring the amount of displacement (the length) of the vector. The dot product also measures the amount of displacement, but only the displacement *in a particular direction* is counted; perpendicular displacement is discarded by the projecting process. But what if we measure the displacement in the same direction that the vector is pointing? In this case, *all* of the vector's displacement is in the di-

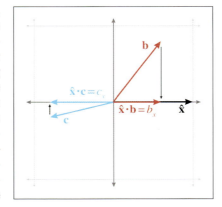

Figure 2.23
Taking the dot product with a cardinal axis sifts out the corresponding coordinate.

rection being measured, so if we project a vector onto itself, the length of that projection is simply the magnitude of the vector. But remember that $\mathbf{a} \cdot \mathbf{b}$ is equal to the length of the projection of $\mathbf{b}$ onto $\mathbf{a}$, scaled by $\|\mathbf{a}\|$. If we dot a vector with itself, such as $\mathbf{v} \cdot \mathbf{v}$, we get the length of the projection, which is $\|\mathbf{v}\|$, times the length of the vector we are projecting onto, which is also $\|\mathbf{v}\|$. In other words,

Relationship between vector magnitude and the dot product

$$\mathbf{v} \cdot \mathbf{v} = \|\mathbf{v}\|^2, \qquad\qquad \|\mathbf{v}\| = \sqrt{\mathbf{v} \cdot \mathbf{v}}.$$

Before we switch to the second interpretation of the dot product, let's check out one more very common use of the dot product as a projection. Assume once more that $\hat{\mathbf{a}}$ is a unit vector and $\mathbf{b}$ has arbitrary length. Using the dot product, it's possible to separate $\mathbf{b}$ into two values, $\mathbf{b}_{\parallel}$ and $\mathbf{b}_{\perp}$ (read "$\mathbf{b}$ parallel" and "$\mathbf{b}$ perp"), which are parallel and perpendicular to $\hat{\mathbf{a}}$, respectively, such that $\mathbf{b} = \mathbf{b}_{\parallel} + \mathbf{b}_{\perp}$. Figure 2.24 illustrates the geometry involved.

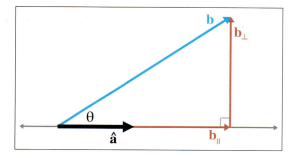

Figure 2.24
Projecting one vector onto another

We've already established that the length of $\mathbf{b}_{\parallel}$ will be equal to $\hat{\mathbf{a}} \cdot \mathbf{b}$. But the dot product yields a scalar, and $\mathbf{b}_{\parallel}$ is a vector, so we'll take the direction specified by the unit vector $\hat{\mathbf{a}}$ and scale it up:

$$\mathbf{b}_{\parallel} = (\hat{\mathbf{a}} \cdot \mathbf{b})\hat{\mathbf{a}}.$$

Once we know $\mathbf{b}_{\parallel}$, we can easily solve for $\mathbf{b}_{\perp}$:

$$\mathbf{b}_{\perp} + \mathbf{b}_{\parallel} = \mathbf{b},$$
$$\mathbf{b}_{\perp} = \mathbf{b} - \mathbf{b}_{\parallel},$$
$$\mathbf{b}_{\perp} = \mathbf{b} - (\hat{\mathbf{a}} \cdot \mathbf{b})\hat{\mathbf{a}}.$$

It's not too difficult to generalize these results to the case where $\mathbf{a}$ is not a unit vector.

In the rest of this book, we make use of these equations several times to separate a vector into components that are parallel and perpendicular to another vector.

Now let's examine the dot product through the lens of trigonometry. This is the more common geometric interpretation of the dot product, which places a bit more emphasis on the angle between the vectors. We've been thinking in terms of projections, so we haven't had much need for this angle. Less experienced and conscientious authors [16] might give you just one of the two important viewpoints, which is probably sufficient to interpret an equation that contains the dot product. However, a more valuable skill is to recognize situations for which the dot product is the correct tool for the job; sometimes it helps to have other interpretations pointed out, even if they are "obviously" equivalent to each other.

Consider the right triangle on the right-hand side of Figure 2.25. As the figure shows, the length of the hypotenuse is 1 (since $\hat{\mathbf{b}}$ is a unit vector) and the length of the base is equal to the dot product $\hat{\mathbf{a}} \cdot \hat{\mathbf{b}}$. From elementary trig (which was reviewed in Section 1.4.4), remember that the cosine of an angle is the ratio of the length of the adjacent leg divided by the length of the hypotenuse. Plugging in the values from Figure 2.25, we have

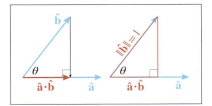

Figure 2.25
Interpreting the dot product by using the trigonometry of the right triangle

$$\cos \theta = \frac{\text{adjacent}}{\text{hypotenuse}} = \frac{\hat{\mathbf{a}} \cdot \hat{\mathbf{b}}}{1} = \hat{\mathbf{a}} \cdot \hat{\mathbf{b}}.$$

In other words, the dot product of two unit vectors is equal to the cosine of the angle between them. This statement is true even if the right triangle in Figure 2.25 cannot be formed, when $\hat{\mathbf{a}} \cdot \hat{\mathbf{b}} \le 0$ and $\theta > 90°$. Remember that the dot product of any vector with the vector $\hat{\mathbf{x}} = [1, 0, 0]$ will simply extract the x-coordinate of the vector. In fact, the x-coordinate of a unit vector that has been rotated by an angle of θ from standard position is one way to *define* the value of $\cos \theta$. Review Section 1.4.4 if this isn't fresh in your memory.

By combining these ideas with the previous observation that scaling either vector scales the dot product by the same factor, we arrive at the general relationship between the dot product and the cosine.

Dot Product Relation to Intercepted Angle

The dot product of two vectors **a** and **b** is equal to the cosine of the angle θ between the vectors, multiplied by the lengths of the vectors (see Figure 2.26). Stated formally,

$$\mathbf{a} \cdot \mathbf{b} = \|\mathbf{a}\| \|\mathbf{b}\| \cos \theta. \tag{2.4}$$

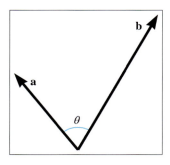

Figure 2.26
The dot product is related to the angle between two vectors.

What does it mean to measure the angle between two vectors in 3D? Any two vectors will always lie in a common plane (place them tail to tail to see this), and so we measure the angle in the plane that contains both vectors. If the vectors are parallel, the plane is not unique, but the angle is either $0°$ or $\pm 180°$, and it doesn't matter which plane we choose.

The dot product provides a way for us to compute the angle between two vectors. Solving Equation (2.4) for θ,

$$\theta = \arccos \left(\frac{\mathbf{a} \cdot \mathbf{b}}{\|\mathbf{a}\|\|\mathbf{b}\|} \right). \qquad (2.5)$$

Using the dot product to compute the angle between two vectors

We can avoid the division in Equation (2.5) if we know that $\mathbf{a}$ and $\mathbf{b}$ are unit vectors. In this very common case, the denominator of Equation (2.5) is trivially 1, and we are left with

$$\theta = \arccos \left(\hat{\mathbf{a}} \cdot \hat{\mathbf{b}} \right) \qquad \text{(assume } \hat{\mathbf{a}} \text{ and } \hat{\mathbf{b}} \text{ are unit vectors).}$$

Computing the angle between two unit vectors

If we do not need the exact value of θ, and need only a classification of the relative orientation of $\mathbf{a}$ and $\mathbf{b}$, then we need only the *sign* of the dot product. This is the same idea illustrated in Figure 2.18, only now we can relate it to the angle θ, as shown in Table 2.1.

$\mathbf{a} \cdot \mathbf{b}$	θ	Angle is	a and b are
> 0	$0° \leq \theta < 90°$	acute	pointing mostly in the same direction
0	$\theta = 90°$	right	perpendicular
< 0	$90° < \theta \leq 180°$	obtuse	pointing mostly in the opposite direction

Table 2.1. The sign of the dot product can be used as a rough classification of the angle between two vectors.

Since the magnitude of the vectors does not affect the sign of the dot product, Table 2.1 applies regardless of the lengths of $\mathbf{a}$ and $\mathbf{b}$. However, notice that if either $\mathbf{a}$ or $\mathbf{b}$ is the zero vector, then $\mathbf{a} \cdot \mathbf{b} = 0$. Thus, when we use the dot product to classify the relationship between two vectors, the dot product acts as if the zero vector is perpendicular to any other vector. As it turns out, the cross product behaves differently.

Let's summarize the dot product's geometric properties.

- The dot product $\mathbf{a} \cdot \mathbf{b}$ measures the length of the projection of $\mathbf{b}$ onto $\mathbf{a}$, multiplied by the length of $\mathbf{a}$.

- The dot product can be used to measure displacement in a particular direction.

- The projection operation is closely related to the cosine function. The dot product $\mathbf{a} \cdot \mathbf{b}$ also is equal to $\|\mathbf{a}\|\|\mathbf{b}\| \cos \theta$, where θ is the angle between the vectors.

We review the commutative and distributive properties of the dot product at the end of this chapter along with other algebraic properties of vector operations.

2.12   Vector Cross Product

The other vector product, known as the *cross product*, can be applied only in 3D. Unlike the dot product, which yields a scalar and is commutative, the vector cross product yields a 3D vector and is not commutative.

2.12.1   Official Linear Algebra Rules

Similar to the dot product, the term "cross" product comes from the symbol used in the notation $\mathbf{a} \times \mathbf{b}$. We always write the cross symbol, rather than omitting it as we do with scalar multiplication. The equation for the cross product is

Cross product

$$\begin{bmatrix} x_1 \\ y_1 \\ z_1 \end{bmatrix} \times \begin{bmatrix} x_2 \\ y_2 \\ z_2 \end{bmatrix} = \begin{bmatrix} y_1 z_2 - z_1 y_2 \\ z_1 x_2 - x_1 z_2 \\ x_1 y_2 - y_1 x_2 \end{bmatrix}.$$

For example,

$$\begin{bmatrix} 1 \\ 3 \\ 4 \end{bmatrix} \times \begin{bmatrix} 2 \\ -5 \\ 8 \end{bmatrix} = \begin{bmatrix} (3)(8) - (4)(-5) \\ (4)(2) - (1)(8) \\ (1)(-5) - (3)(2) \end{bmatrix} = \begin{bmatrix} 24 - (-20) \\ 8 - 8 \\ -5 - 6 \end{bmatrix} = \begin{bmatrix} 44 \\ 0 \\ -11 \end{bmatrix}.$$

The cross product enjoys the same level of operator precedence as the dot product: multiplication occurs before addition and subtraction. When dot product and cross product are used together, the cross product takes precedence: $\mathbf{a} \cdot \mathbf{b} \times \mathbf{c} = \mathbf{a} \cdot (\mathbf{b} \times \mathbf{c})$. Luckily, there's an easy way to remember this: it's the only way it could work. The dot product returns a scalar, and so $(\mathbf{a} \cdot \mathbf{b}) \times \mathbf{c}$ is undefined, since you cannot take the cross product of a

scalar and a vector. The operation $\mathbf{a} \cdot (\mathbf{b} \times \mathbf{c})$ is known as the *triple product*. We present some special properties of this computation in Section 6.1.

As mentioned earlier, the vector cross product is not commutative. In fact, it is *anticommutative*: $\mathbf{a} \times \mathbf{b} = -(\mathbf{b} \times \mathbf{a})$. The cross product is not associative, either. In general, $(\mathbf{a} \times \mathbf{b}) \times \mathbf{c} \neq \mathbf{a} \times (\mathbf{b} \times \mathbf{c})$. More vector algebra laws concerning the cross product are given in Section 2.13.

2.12.2 Geometric Interpretation

The cross product yields a vector that is perpendicular to the original two vectors, as illustrated in Figure 2.27.

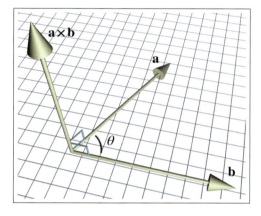

Figure 2.27
Vector cross product

The length of $\mathbf{a} \times \mathbf{b}$ is equal to the product of the magnitudes of $\mathbf{a}$ and $\mathbf{b}$ and the sine of the angle between $\mathbf{a}$ and $\mathbf{b}$:

$$\|\mathbf{a} \times \mathbf{b}\| = \|\mathbf{a}\|\|\mathbf{b}\| \sin \theta.$$

The magnitude of the cross product is related to the sine of the angle between the vectors

As it turns out, this is also equal to the area of the parallelogram formed with two sides $\mathbf{a}$ and $\mathbf{b}$. Let's see if we can verify why this is true by using Figure 2.28.

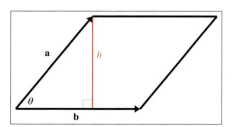

Figure 2.28
A parallelogram with sides $\mathbf{a}$ and $\mathbf{b}$

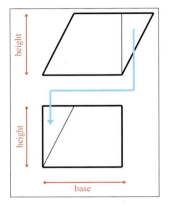

First, from planar geometry, we know that the area of the parallelogram is bh, the product of the base and the height. (In Figure 2.28, the base is $b = \|\mathbf{b}\|$.) We can verify this rule by "clipping" off a triangle from one end and moving it to the other end, forming a rectangle, as shown in Figure 2.29.

The area of a rectangle is given by its length and width. In this case, this area is the product bh. Since the area of the rectangle is equal to the area of the parallelogram, the area of the parallelogram must also be bh.

Returning to Figure 2.28, let a and b be the lengths of $\mathbf{a}$ and $\mathbf{b}$, respectively, and

Figure 2.29
Area of a parallelogram

note that $\sin\theta = h/a$. Then

$$
\begin{aligned}
A &= bh \\
&= b(a\sin\theta) \\
&= \|\mathbf{a}\|\|\mathbf{b}\|\sin\theta \\
&= \|\mathbf{a} \times \mathbf{b}\|.
\end{aligned}
$$

If $\mathbf{a}$ and $\mathbf{b}$ are parallel, or if $\mathbf{a}$ or $\mathbf{b}$ is the zero vector, then $\mathbf{a} \times \mathbf{b} = \mathbf{0}$. So the cross product interprets the zero vector as being parallel to every other vector. Notice that this is different from the dot product, which interprets the zero vector as being *perpendicular* to every other vector. (Of course, it is ill-defined to describe the zero vector as being perpendicular or parallel to any vector, since the zero vector has no direction.)

We have stated that $\mathbf{a} \times \mathbf{b}$ is perpendicular to $\mathbf{a}$ and $\mathbf{b}$. But there are two directions that are perpendicular to $\mathbf{a}$ and $\mathbf{b}$—which of these two directions does $\mathbf{a} \times \mathbf{b}$ point? We can determine the direction of $\mathbf{a} \times \mathbf{b}$ by placing the tail of $\mathbf{b}$ at the head of $\mathbf{a}$, and examining whether we make a clockwise or counterclockwise turn from $\mathbf{a}$ to $\mathbf{b}$. In a left-handed coordinate system, $\mathbf{a} \times \mathbf{b}$ points towards you if the vectors $\mathbf{a}$ and $\mathbf{b}$ make a clockwise turn from your viewpoint, and away from you if $\mathbf{a}$ and $\mathbf{b}$ make a counterclockwise turn. In a right-handed coordinate system, the exact opposite occurs: if $\mathbf{a}$ and $\mathbf{b}$ make a counterclockwise turn, $\mathbf{a} \times \mathbf{b}$ points towards you, and if $\mathbf{a}$ and $\mathbf{b}$ make a clockwise turn, $\mathbf{a} \times \mathbf{b}$ points away from you.

Figure 2.30 shows clockwise and counterclockwise turns. Notice that to make the clockwise or counterclockwise determination, we must align the

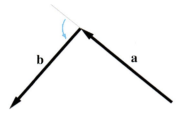

Clockwise turn	Counterclockwise turn
In a left-handed coordinate system, $\mathbf{a} \times \mathbf{b}$ (not shown) points towards you. In a right-handed coordinate system, $\mathbf{a} \times \mathbf{b}$ points away from you.	In a left-handed coordinate system, $\mathbf{a} \times \mathbf{b}$ (not shown) points away from you. In a right-handed coordinate system, $\mathbf{a} \times \mathbf{b}$ points towards you.

Figure 2.30. Determining clockwise versus counterclockwise turns

head of **a** with the tail of **b**. Compare this to Figure 2.26, where the tails are touching. The tail-to-tail alignment shown in Figure 2.26 is the correct way to position the vectors to measure the angle between them, but to judge whether the turn is clockwise or counterclockwise, the vectors should be aligned head-to-tail, as shown in Figure 2.30.

Let's apply this general rule to the specific case of the cardinal axes. Let $\hat{\mathbf{x}}$, $\hat{\mathbf{y}}$, and $\hat{\mathbf{z}}$ be unit vectors that point in the $+x$, $+y$, and $+z$ directions, respectively. The results of taking the cross product of each pair of axes are

$$\hat{\mathbf{x}} \times \hat{\mathbf{y}} = \hat{\mathbf{z}}, \qquad \hat{\mathbf{y}} \times \hat{\mathbf{x}} = -\hat{\mathbf{z}},$$
$$\hat{\mathbf{y}} \times \hat{\mathbf{z}} = \hat{\mathbf{x}}, \qquad \hat{\mathbf{z}} \times \hat{\mathbf{y}} = -\hat{\mathbf{x}},$$
$$\hat{\mathbf{z}} \times \hat{\mathbf{x}} = \hat{\mathbf{y}}, \qquad \hat{\mathbf{x}} \times \hat{\mathbf{z}} = -\hat{\mathbf{y}}.$$

Cross product of the cardinal axes

You can also remember which way the cross product points by using your hand, similar to the way we distinguished between left-handed and right-handed coordinate spaces in Section 1.3.3. Since we're using a left-handed coordinate space in this book, we'll show how it's done using your left hand. Let's say you have two vectors, **a** and **b**, and you want to figure out which direction $\mathbf{a} \times \mathbf{b}$ points. Point your thumb in the direction of **a**, and your index finger (approximately) in the direction of **b**. If **a** and **b** are pointing in nearly the opposite direction, this may be difficult. Just make sure that if your thumb points exactly in the direction of **a**; then your index finger is on the same side of **a** as the vector **b** is. With your fingers in this position, extend your third finger to be perpendicular to your thumb and index finger, similar to what we did in Section 1.3.3. Your third finger now points in the direction of $\mathbf{a} \times \mathbf{b}$.

Of course, a similar trick works with your right hand for right-handed coordinate spaces.

One of the most important uses of the cross product is to create a vector that is perpendicular to a plane (see Section 9.5), triangle (Section 9.6), or polygon (Section 9.7).

2.13   Linear Algebra Identities

The Greek philosopher Arcesilaus reportedly said, "Where you find the laws most numerous, there you will find also the greatest injustice." Well, nobody said vector algebra was fair. Table 2.2 lists some vector algebra laws that are occasionally useful but should not be memorized. Several identities are obvious and are listed for the sake of completeness; all of them can be derived from the definitions given in earlier sections.

Identity	Comments
$\mathbf{a} + \mathbf{b} = \mathbf{b} + \mathbf{a}$	Commutative property of vector addition
$\mathbf{a} - \mathbf{b} = \mathbf{a} + (-\mathbf{b})$	Definition of vector subtraction
$(\mathbf{a} + \mathbf{b}) + \mathbf{c} = \mathbf{a} + (\mathbf{b} + \mathbf{c})$	Associative property of vector addition
$s(t\mathbf{a}) = (st)\mathbf{a}$	Associative property of scalar multiplication
$k(\mathbf{a} + \mathbf{b}) = k\mathbf{a} + k\mathbf{b}$	Scalar multiplication distributes over vector addition
$\|k\mathbf{a}\| = \|k\|\|\mathbf{a}\|$	Multiplying a vector by a scalar scales the magnitude by a factor equal to the absolute value of the scalar
$\|\mathbf{a}\| \geq 0$	The magnitude of a vector is nonnegative
$\|\mathbf{a}\|^2 + \|\mathbf{b}\|^2 = \|\mathbf{a} + \mathbf{b}\|^2$	The Pythagorean theorem applied to vector addition.
$\|\mathbf{a}\| + \|\mathbf{b}\| \geq \|\mathbf{a} + \mathbf{b}\|$	Triangle rule of vector addition. (No side can be longer than the sum of the other two sides.)
$\mathbf{a} \cdot \mathbf{b} = \mathbf{b} \cdot \mathbf{a}$	Commutative property of dot product
$\|\mathbf{a}\| = \sqrt{\mathbf{a} \cdot \mathbf{a}}$	Vector magnitude defined using dot product
$k(\mathbf{a} \cdot \mathbf{b}) = (k\mathbf{a}) \cdot \mathbf{b} = \mathbf{a} \cdot (k\mathbf{b})$	Associative property of scalar multiplication with dot product
$\mathbf{a} \cdot (\mathbf{b} + \mathbf{c}) = \mathbf{a} \cdot \mathbf{b} + \mathbf{a} \cdot \mathbf{c}$	Dot product distributes over vector addition and subtraction
$\mathbf{a} \times \mathbf{a} = \mathbf{0}$	The cross product of any vector with itself is the zero vector. (Because any vector is parallel with itself.)
$\mathbf{a} \times \mathbf{b} = -(\mathbf{b} \times \mathbf{a})$	Cross product is anticommutative.
$\mathbf{a} \times \mathbf{b} = (-\mathbf{a}) \times (-\mathbf{b})$	Negating both operands to the cross product results in the same vector.
$k(\mathbf{a} \times \mathbf{b}) = (k\mathbf{a}) \times \mathbf{b} = \mathbf{a} \times (k\mathbf{b})$	Associative property of scalar multiplication with cross product.
$\mathbf{a} \times (\mathbf{b} + \mathbf{c}) = \mathbf{a} \times \mathbf{b} + \mathbf{a} \times \mathbf{c}$	Cross product distributes over vector addition and subtraction.

Table 2.2
Table of vector algebra identities

2.14   Exercises

(Answers on page 746.)

1. Let

$$\mathbf{a} = \begin{bmatrix} -3 & 8 \end{bmatrix}, \quad \mathbf{b} = \begin{bmatrix} 4 \\ 0 \\ 5 \end{bmatrix}, \quad \mathbf{c} = \begin{bmatrix} 16 \\ -1 \\ 4 \\ 6 \end{bmatrix}.$$

   (a) Identify $\mathbf{a}$, $\mathbf{b}$, and $\mathbf{c}$, as row or column vectors, and give the dimension of each vector.

   (b) Compute $b_y + c_w + a_x + b_z$.

2. Identify the quantities in each of the following sentences as scalar or vector. For vector quantities, give the magnitude and direction. (Note: some directions may be implicit.)

   (a) How much do you weigh?

   (b) Do you have any idea how fast you were going?

   (c) It's two blocks north of here.

   (d) We're cruising from Los Angeles to New York at 600 mph, at an altitude of 33,000 ft.

3. Give the values of the following vectors. The darker grid lines represent one unit.

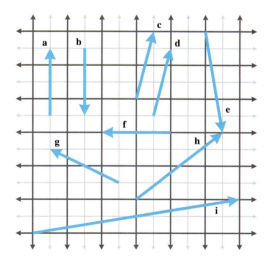

4. Identify the following statements as true or false. If the statement is false, explain why.

   (a) The size of a vector in a diagram doesn't matter; we just need to draw it in the right place.

   (b) The displacement expressed by a vector can be visualized as a sequence of axially aligned displacements.

   (c) These axially aligned displacements from the previous question must occur in order.

   (d) The vector $[x, y]$ gives the displacement from the point (x, y) to the origin.

5. Evaluate the following vector expressions:

   (a) $-\begin{bmatrix} 3 & 7 \end{bmatrix}$

   (b) $\left\| \begin{bmatrix} -12 & 5 \end{bmatrix} \right\|$

   (c) $\left\| \begin{bmatrix} 8 & -3 & 1/2 \end{bmatrix} \right\|$

   (d) $3 \begin{bmatrix} 4 & -7 & 0 \end{bmatrix}$

   (e) $\begin{bmatrix} 4 & 5 \end{bmatrix} / 2$

6. Normalize the following vectors:

   (a) $\begin{bmatrix} 12 & 5 \end{bmatrix}$

   (b) $\begin{bmatrix} 0 & 743.632 \end{bmatrix}$

   (c) $\begin{bmatrix} 8 & -3 & 1/2 \end{bmatrix}$

   (d) $\begin{bmatrix} -12 & 3 & -4 \end{bmatrix}$

   (e) $\begin{bmatrix} 1 & 1 & 1 & 1 \end{bmatrix}$

7. Evaluate the following vector expressions:

   (a) $\begin{bmatrix} 7 & -2 & -3 \end{bmatrix} + \begin{bmatrix} 6 & 6 & -4 \end{bmatrix}$

   (b) $\begin{bmatrix} 2 & 9 & -1 \end{bmatrix} + \begin{bmatrix} -2 & -9 & 1 \end{bmatrix}$

   (c) $\begin{bmatrix} 3 \\ 10 \\ 7 \end{bmatrix} - \begin{bmatrix} 8 \\ -7 \\ 4 \end{bmatrix}$

   (d) $\begin{bmatrix} 4 \\ 5 \\ -11 \end{bmatrix} - \begin{bmatrix} -4 \\ -5 \\ 11 \end{bmatrix}$

   (e) $3 \begin{bmatrix} a \\ b \\ c \end{bmatrix} - 4 \begin{bmatrix} 2 \\ 10 \\ -6 \end{bmatrix}$

8. Compute the distance between the following pairs of points:

   (a) $\begin{bmatrix} 10 \\ 6 \end{bmatrix}, \begin{bmatrix} -14 \\ 30 \end{bmatrix}$

   (b) $\begin{bmatrix} 0 \\ 0 \end{bmatrix}, \begin{bmatrix} -12 \\ 5 \end{bmatrix}$

   (c) $\begin{bmatrix} 3 \\ 10 \\ 7 \end{bmatrix}, \begin{bmatrix} 8 \\ -7 \\ 4 \end{bmatrix}$

   (d) $\begin{bmatrix} -2 \\ -4 \\ 9 \end{bmatrix}, \begin{bmatrix} 6 \\ -7 \\ 9.5 \end{bmatrix}$

   (e) $\begin{bmatrix} 4 \\ -4 \\ -4 \\ 4 \end{bmatrix}, \begin{bmatrix} -6 \\ 6 \\ 6 \\ -6 \end{bmatrix}$

9. Evaluate the following vector expressions:

   (a) $\begin{bmatrix} 2 \\ 6 \end{bmatrix} \cdot \begin{bmatrix} -3 \\ 8 \end{bmatrix}$

   (b) $-7\begin{bmatrix} 1 & 2 \end{bmatrix} \cdot \begin{bmatrix} 11 & -4 \end{bmatrix}$

   (c) $10 + \begin{bmatrix} -5 \\ 1 \\ 3 \end{bmatrix} \cdot \begin{bmatrix} 4 \\ -13 \\ 9 \end{bmatrix}$

   (d) $3\begin{bmatrix} -2 \\ 0 \\ 4 \end{bmatrix} \cdot \left(\begin{bmatrix} 8 \\ -2 \\ 3/2 \end{bmatrix} + \begin{bmatrix} 0 \\ 9 \\ 7 \end{bmatrix} \right)$

10. Given the two vectors

$$\mathbf{v} = \begin{bmatrix} 4 \\ 3 \\ -1 \end{bmatrix}, \quad \hat{\mathbf{n}} = \begin{bmatrix} \sqrt{2}/2 \\ \sqrt{2}/2 \\ 0 \end{bmatrix},$$

   separate $\mathbf{v}$ into components that are perpendicular and parallel to $\hat{\mathbf{n}}$. (As the notation implies, $\hat{\mathbf{n}}$ is a unit vector.)

11. Use the geometric definition of the dot product

$$\mathbf{a} \cdot \mathbf{b} = \|\mathbf{a}\|\|\mathbf{b}\| \cos \theta$$

   to prove the law of cosines.

12. Use trigonometric identities and the algebraic definition of the dot product in 2D

$$\mathbf{a} \cdot \mathbf{b} = a_x b_x + a_y b_y$$

   to prove the geometric interpretation of the dot product in 2D. (Hint: draw a diagram of the vectors and all angles involved.)

13. Calculate $\mathbf{a} \times \mathbf{b}$ and $\mathbf{b} \times \mathbf{a}$ for the following vectors:

    (a) $\mathbf{a} = \begin{bmatrix} 0 & -1 & 0 \end{bmatrix}, \mathbf{b} = \begin{bmatrix} 0 & 0 & 1 \end{bmatrix}$

    (b) $\mathbf{a} = \begin{bmatrix} -2 & 4 & 1 \end{bmatrix}, \mathbf{b} = \begin{bmatrix} 1 & -2 & -1 \end{bmatrix}$

    (c) $\mathbf{a} = \begin{bmatrix} 3 & 10 & 7 \end{bmatrix}, \mathbf{b} = \begin{bmatrix} 8 & -7 & 4 \end{bmatrix}$

14. Prove the equation for the magnitude of the cross product

$$\|\mathbf{a} \times \mathbf{b}\| = \|\mathbf{a}\| \|\mathbf{b}\| \sin \theta.$$

(Hint: make use of the geometric interpretation of the dot product and try to show how the left and right sides of the equation are equivalent, rather than trying to derive one side from the other.)

15. Section 2.8 introduced the norm of a vector, namely, a scalar value associated with a given vector. However, the definition of the norm given in that section is not the only definition of a norm for a vector. In general, the *p-norm* of an *n*-dimensional vector is defined as

$$\|\mathbf{x}\|_p \equiv \left(\sum_{i=1}^{n} |x_i|^p \right)^{1/p}.$$

Some of the more common *p*-norms include:

- The L^1 norm, a.k.a. Taxicab norm ($p = 1$):

$$\|\mathbf{x}\|_1 \equiv \sum_{i=1}^{n} |x_i|.$$

- The L^2 norm, a.k.a. Euclidean norm ($p = 2$). This is the most common and familiar norm, since it measures geometric length:

$$\|\mathbf{x}\|_2 \equiv \sqrt{\sum_{i=1}^{n} x_i^2}.$$

- The infinity norm, a.k.a. Chebyshev norm ($p = \infty$):

$$\|\mathbf{x}\|_\infty \equiv \max \left(|x_1|, \ldots, |x_n| \right).$$

Each of these norms can be thought of as a way to assigning a length or size to a vector. The Euclidean norm was discussed in Section 2.8. The Taxicab norm gets its name from how a taxicab would measure distance driving the streets of a city laid out in a grid (e.g., Cartesia from Section 1.2.1). For example, a taxicab that drives 1 block east and 1 block north drives a total distance of 2 blocks, whereas a bird flying "as the crow flies" can fly in a straight line from start to finish and travel only $\sqrt{2}$ blocks (Euclidean norm). The Chebyshev norm is simply the absolute value of the vector component with the largest absolute value. An example of how this norm can be used is to consider the number of moves required to move a king in a game of chess from one square to another. The immediately surrounding squares require 1 move, the squares surrounding those require 2 moves, and so on.

(a) For each of the following find $\|\mathbf{x}\|_1$, $\|\mathbf{x}\|_2$, $\|\mathbf{x}\|_3$, and $\|\mathbf{x}\|_\infty$:

    (1) $\begin{bmatrix} 3 & 4 \end{bmatrix}$

    (2) $\begin{bmatrix} 5 & -12 \end{bmatrix}$

    (3) $\begin{bmatrix} -2 & 10 & -7 \end{bmatrix}$

    (4) $\begin{bmatrix} 6 & 1 & -9 \end{bmatrix}$

    (5) $\begin{bmatrix} -2 & -2 & -2 & -2 \end{bmatrix}$

*(b) Draw the unit circle (i.e., the set of all vectors with $\|\mathbf{x}\|_p = 1$) centered at the origin for the L^1 norm, L^2 norm, and infinity norm.

16. A man is boarding a plane. The airline has a rule that no carry-on item may be more than two feet long, two feet wide, or two feet tall. He has a very valuable sword that is three feet long, yet he is able to carry the sword on board with him.[9] How is he able to do this? What is the longest possible item that he could carry on?

17. Verify Figure 2.11 numerically.

18. Is the coordinate system used in Figure 2.27 a left-handed or right-handed coordinate system?

19. One common way of defining a bounding box for a 2D object is to specify a center point $\mathbf{c}$ and a *radius vector* $\mathbf{r}$, where each component of $\mathbf{r}$ is half the length of the side of the bounding box along the corresponding axis.

(a) Describe the four corners $\mathbf{p}_{\text{UpperLeft}}$, $\mathbf{p}_{\text{UpperRight}}$, $\mathbf{p}_{\text{LowerLeft}}$, and $\mathbf{p}_{\text{LowerRight}}$.

(b) Describe the eight corners of a bounding cube, extending this idea into 3D.

---

[9]Please ignore the fact that nowadays this could never happen for security reasons. You can think of this exercise as taking place in a Quentin Tarantino movie.

20. A nonplayer character (NPC) is standing at location $\mathbf{p}$ with a forward direction of $\mathbf{v}$.

    (a) How can the dot product be used to determine whether the point $\mathbf{x}$ is in front of or behind the NPC?

    (b) Let $\mathbf{p} = \begin{bmatrix} -3 & 4 \end{bmatrix}$ and $\mathbf{v} = \begin{bmatrix} 5 & -2 \end{bmatrix}$. For each of the following points $\mathbf{x}$ determine whether $\mathbf{x}$ is in front of or behind the NPC:

        (1) $\mathbf{x} = \begin{bmatrix} 0 & 0 \end{bmatrix}$
        (2) $\mathbf{x} = \begin{bmatrix} 1 & 6 \end{bmatrix}$
        (3) $\mathbf{x} = \begin{bmatrix} -6 & 0 \end{bmatrix}$
        (4) $\mathbf{x} = \begin{bmatrix} -4 & 7 \end{bmatrix}$
        (5) $\mathbf{x} = \begin{bmatrix} 5 & 5 \end{bmatrix}$
        (6) $\mathbf{x} = \begin{bmatrix} -3 & 0 \end{bmatrix}$
        (7) $\mathbf{x} = \begin{bmatrix} -6 & -3.5 \end{bmatrix}$

21. Extending the concept from Exercise 20, consider the case where the NPC has a limited field of view (FOV). If the total FOV angle is ϕ, then the NPC can see to the left or right of its forward direction by a maximum angle of $\phi/2$.

    (a) How can the dot product be used to determine whether the point $\mathbf{x}$ is visible to the NPC?

    (b) For each of the points $\mathbf{x}$ in Exercise 20 determine whether $\mathbf{x}$ is visible to the NPC if its FOV is $90°$.

    (c) Suppose that the NPC's viewing distance is also limited to a maximum distance of 7 units. Which points are visible to the NPC then?

22. Consider three points labeled $\mathbf{a}$, $\mathbf{b}$, and $\mathbf{c}$ in the xz plane of our left-handed coordinate system, which represent waypoints on an NPC's path.

    (a) How can the cross product be used to determine whether, when moving from $\mathbf{a}$ to $\mathbf{b}$ to $\mathbf{c}$, the NPC makes a clockwise or counterclockwise turn at $\mathbf{b}$, when viewing the path from above?

    (b) For each of the following sets of three points, determine whether the NPC is turning clockwise or counterclockwise when moving from $\mathbf{a}$ to $\mathbf{b}$ to $\mathbf{c}$:

        (1) $\mathbf{a} = \begin{bmatrix} 2 & 0 & 3 \end{bmatrix}$, $\mathbf{b} = \begin{bmatrix} -1 & 0 & 5 \end{bmatrix}$, $\mathbf{c} = \begin{bmatrix} -4 & 0 & 1 \end{bmatrix}$
        (2) $\mathbf{a} = \begin{bmatrix} -3 & 0 & -5 \end{bmatrix}$, $\mathbf{b} = \begin{bmatrix} 4 & 0 & 0 \end{bmatrix}$, $\mathbf{c} = \begin{bmatrix} 3 & 0 & 3 \end{bmatrix}$
        (3) $\mathbf{a} = \begin{bmatrix} 1 & 0 & 4 \end{bmatrix}$, $\mathbf{b} = \begin{bmatrix} 7 & 0 & -1 \end{bmatrix}$, $\mathbf{c} = \begin{bmatrix} -5 & 0 & -6 \end{bmatrix}$
        (4) $\mathbf{a} = \begin{bmatrix} -2 & 0 & 1 \end{bmatrix}$, $\mathbf{b} = \begin{bmatrix} 1 & 0 & 2 \end{bmatrix}$, $\mathbf{c} = \begin{bmatrix} 4 & 0 & 4 \end{bmatrix}$

23. In the derivation of a matrix to scale along an arbitrary axis, we reach a step where we have the vector expression

$$\mathbf{p}' = \mathbf{p} + (k-1)\,(\mathbf{p} \cdot \mathbf{n})\,\mathbf{n},$$

where $\mathbf{n}$ is an arbitrary vector $[n_x, n_y, n_z]$ and k is an arbitrary scalar, but $\mathbf{p}$ is one of the cardinal axes. Plug in the value $\mathbf{p} = [1, 0, 0]$ and simplify the resulting expression for $\mathbf{p}'$. The answer is not a vector expression, but a single vector, where the scalar expressions for each coordinate have been simplified.

24. A similar problem arises with the derivation of a matrix to rotate about an arbitrary axis. Given an arbitrary scalar θ and a vector $\mathbf{n}$, substitute $\mathbf{p} = [1, 0, 0]$ and simplify the value of $\mathbf{p}'$ in the expression

$$\mathbf{p}' = \cos\theta\,(\mathbf{p} - (\mathbf{p} \cdot \mathbf{n})\,\mathbf{n}) + \sin\theta\,(\mathbf{n} \times \mathbf{p}) + (\mathbf{p} \cdot \mathbf{n})\,\mathbf{n}.$$

What's our vector, Victor?

— Captain Oveur in *Airplane!* (1980)

Chapter 3

Multiple Coordinate Spaces

The boundary lines have fallen for me in pleasant places;
surely I have a delightful inheritance.

— Psalm 16:6 (New International Version)

Chapter 1 discussed how we can establish a coordinate space anywhere we want simply by picking a point to be the origin and deciding how we want the axes to be oriented. We usually don't make these decisions arbitrarily; we form coordinate spaces for specific reasons (one might say "different spaces for different cases"). This chapter gives some examples of common coordinate spaces that are used for graphics and games. We will then discuss how coordinate spaces are nested within other coordinate spaces.

This chapter introduces the idea of multiple coordinate systems. It is divided into five main sections.

- Section 3.1 justifies the need for multiple coordinate systems.

- Section 3.2 introduces some common coordinate systems. The main concepts introduced are

  - world space
  - object space
  - camera space
  - upright space

- Section 3.3 describes coordinate-space transformations.

  - Section 3.3.1 exposes a duality between two ways of thinking about coordinate-space transformations.

  - Section 3.3.2 describes how to specify one coordinate system in terms of another.

  - Section 3.3.3 discusses the very important concept of *basis vectors*.

- Section 3.4 discusses nested coordinate spaces, commonly used for animating hierarchically segmented objects in 3D space.

- Section 3.5 is a political campaign for more human readable code.

3.1   Why Bother with Multiple Coordinate Spaces?

Why do we need more than one coordinate space? After all, any *one* 3D coordinate system extends infinitely and thus contains all points in space. So we could just pick a coordinate space, declare it to be the "world" coordinate space, and all points could be located using this coordinate space. Wouldn't that be easier? In practice, the answer to this is "no." Most people find it more convenient to use different coordinate spaces in different situations.

The reason multiple coordinate spaces are used is that certain pieces of information are known only in the context of a particular reference frame. It might be true that theoretically all points could be expressed using a single "world" coordinate system. However, for a certain point **a**, we may not know the coordinates of **a** in the "world" coordinate system. But we may be able to express **a** relative to some *other* coordinate system.

For example, the residents of Cartesia (see Section 1.2.1) use a map of their city with the origin centered quite sensibly at the center of town and the axes directed along the cardinal points of the compass. The residents of Dyslexia use a map of their city with the coordinates centered at an arbitrary point and the axes running in some arbitrary directions that probably seemed a good idea at the time. The citizens of both cities are quite happy with their respective maps, but the State Transportation Engineer assigned the task of running up a budget for the first highway between Cartesia and Dyslexia needs a map showing the details of both cities, which therefore introduces a third coordinate system that is superior to him, though not necessarily to anybody else. Each major point on both maps needs to be converted from the local coordinates of the respective city to the new coordinate system to make the new map.

The concept of multiple coordinate systems has historical precedent. While Aristotle (384–322 BCE), in his books *On the Heavens* and *Physics*, proposed a geocentric universe with Earth at the origin, Aristarchus (ca. 310–230 BCE) proposed a heliocentric universe with the sun at the origin. So we can see that more than two millennia ago the choice of coordinate system was already a hot topic for discussion. The issue wasn't settled for another couple of millennia until Nicolaus Copernicus (1473–1543) observed in his book *De Revolutionibus Orbium Coelestium* (*On the Revolutions of the Celestial Orbs*) that the orbits of the planets can be explained more simply in a heliocentric universe without all the mucking about with wheels within wheels in a geocentric universe.

In *Sand-Reckoner*, Archimedes (d. 212 BCE), perhaps motivated by some of the concepts introduced in Section 1.1, developed a notation for writing down very large numbers—numbers much larger than anybody had ever counted at that time. Instead of choosing to count dead sheep, as in Section 1.1, he chose to count the number of grains of sand it would take to fill the universe. (He estimated that it would take 8×10^{63} grains of sand, but he did not, however, address the question of where he would get the sand.) In order to make the numbers larger, he chose not the geocentric universe generally accepted at the time, but Aristarchus' revolutionary new heliocentric universe. In a heliocentric universe, Earth orbits the sun, in which case the fact that the stars show no parallax means that they must be much farther away than Aristotle could ever have imagined. To make his life more difficult, Archimedes deliberately chose the coordinate system that would produce larger numbers. We will use the direct opposite of his approach. In creating our virtual universe inside the computer we will choose coordinate systems that make our lives *easier*, not *harder*.

In today's enlightened times, we are accustomed to hearing in the media about cultural relativism, which promotes the idea that it is incorrect to consider one culture or belief system or national agenda to be superior to another. It's not too great a leap of the imagination to extend this to what we might call "transformational relativism"—the contention that no place or orientation or coordinate system can be considered superior to others. In a certain sense that's true, but to paraphrase George Orwell in *Animal Farm*: "All coordinate systems are considered equal, but some are more equal than others." Now let's look at some examples of common coordinate systems that you will meet in 3D graphics.

3.2   Some Useful Coordinate Spaces

Different coordinate spaces are needed because some information is meaningful or available only in a particular context. In this section, we give some examples of common coordinate spaces.

3.2.1   World Space

The authors wrote this book from Chicago, Illinois, and Denton, Texas. More precisely, their locations are as shown in Table 3.1.

These latitude and longitude values express our "absolute" position in the world. You don't need to know where Denton, Chicago, Texas, Illinois,

Author	City	Latitude		Longitude	
Fletcher	Chicago	41°57′	North	87°39′	West
Ian	Denton	33°11′	North	97°	West

Table 3.1. Locations of authors, including a random offset introduced to protect us from our many obsessive stalker fans.

or even the United States is to use this information because the position is absolute. The origin, or $(0,0)$ point, in the world was decided for historical reasons to be located on the equator at the same longitude as the Royal Observatory in the town of Greenwich, England.

(The astute reader will note that these coordinates are not Cartesian coordinates, but rather they are spherical coordinates—see Section 7.3.2. That is not significant for this discussion. We live in a flat 2D world wrapped around a sphere, a concept that supposedly eluded most people until Christopher Columbus verified it experimentally.)

The *world coordinate system* is a special coordinate system that establishes the "global" reference frame for all other coordinate systems to be specified. In other words, we can express the position of other coordinate spaces in terms of the world coordinate space, but we cannot express the world coordinate space in terms of any larger, outer coordinate space.

In a nontechnical sense, the world coordinate system establishes the "biggest" coordinate system that we care about, which in most cases is not actually the entire world. For example, if we wanted to render a view of Cartesia, then for all practical purposes Cartesia would be "the world," since we wouldn't care where Cartesia is located (or even if it exists at all). To find the optimal way to pack automobile parts into a box, we might write a physics simulation that "jiggles" a box full of parts around until they settle. In this case we confine our "world" to the inside of a box. So in different situations the world coordinate space will define a different "world."

We've said that world coordinate space is used to describe absolute positions. We hope you pricked up your ears when you heard this, and you knew we weren't being entirely truthful. We already discussed in Section 2.4.1 that there's really no such thing as "absolute position." In this book, we use the term "absolute" to mean "absolute with respect to the largest coordinate space we care about." In other words, "absolute" to us actually means "expressed in the world coordinate space."

The world coordinate space is also known as the *global* or *universal* coordinate space.

3.2.2   Object Space

Object space is the coordinate space associated with a particular object. Every object has its own independent object space. When an object moves or changes orientation, the object coordinate space associated with that object is carried along with it, so it moves or changes orientation too. For example, we all carry our own personal coordinate system around with us. If we were to ask you to "take one step forward," we are giving you an instruction in your object space. (Please forgive us for referring to you as an object—you know what we mean.) We have no idea which way you will move in absolute terms. Some of you will move north, some south, and those wearing magnet boots on the side of a building might move upward! Concepts such as "forward," "back," "left," and "right" are meaningful in object coordinate space. When someone gives you driving directions, sometimes you will be told to "turn left" and other times you will be told to "go east." "Turn left" is a concept that is expressed in object space, and "go east" is expressed in world space.

Locations as well as directions can be specified in object space. For example, if I asked you where the muffler on your car was, you wouldn't tell me "Cambridge,[1] MA," even if you were Tom or Ray Magliozzi and your car was actually *in* Cambridge. In this case, an answer expressed with a global perspective like this is totally useless;[2] I want you to express the location of your muffler *in the object space of your car.*

In the context of graphics, object space is also known as *model* space, since the coordinates for the vertices of a model are expressed in model space. Object space is also known as *body* space, especially in physics contexts. It's also common to use a phrase like "with respect to the body axes," which means the same thing as "expressed using body space coordinates."

3.2.3   Camera Space

One especially important example of an object space is *camera space*, which is the object space associated with the viewpoint used for rendering. In camera space, the camera is at the origin, with $+x$ pointing to the right, $+z$ pointing forward (into the screen, the direction the camera is facing), and $+y$ pointing "up." (Not "up" with respect to the world, "up" with respect to the top of the camera.) Figure 3.1 shows a diagram of camera space.

These are the traditional left-handed conventions; others are common. In particular, the OpenGL tradition is right-handed, with $-z$ pointing into the screen and $+z$ coming out of the screen towards the viewer.

---

[1]Our fair city.
[2]Come to think of it, this is *exactly* what Tom or Ray Magliozzi would say.

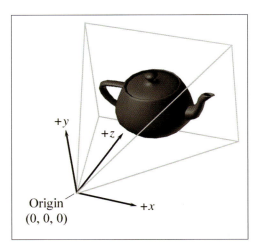

Origin
(0, 0, 0)

Figure 3.1
Camera space using left-handed conventions

Note carefully the differences between camera space, which is a 3D space, and screen space, which is a 2D space. The mapping of camera-space coordinates to screen-space coordinates involves an operation known as *projection*. We'll discuss camera space in more detail, and this conversion process in particular, when we talk about coordinate spaces used in rendering in Section 10.3.

3.2.4   Upright Space

Sometimes the right terminology is the key to unlocking a better understanding of a subject. Don Knuth coined the phrase "name and conquer" to refer to the common and important practice in mathematics and computer science of giving a name to a concept that is used frequently. The goal is to avoid repeating the details of this idea each time it is invoked, resulting in a reduction of clutter and an easier time focusing on the larger issue, for which the thing being named is only one part. It has been our experience that to conquer coordinate space transformations, when communicating either to human beings via words or to computers via code, it is helpful to associate with each object a new coordinate space, which we call the *upright* coordinate space of the object. An object's upright space is, in a certain sense, "halfway" between world space and its object space. The axes of upright space are parallel with the axes of world space, but the origin of upright space is coincident with the origin of object space. Figure 3.2 illustrates this principle in 2D. (Notice that we have made an arbitrary choice to place the origin between the robot's feet, rather than at her center of mass.)

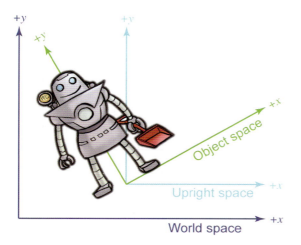

+y

+y

+y

+x

Object space

+x

Upright space

+x

World space

Figure 3.2
Object, upright, and world space.

Why is upright space interesting? To transform a point between object space and upright space requires only rotation, and to transform a point between upright space and world space requires only a change of location, which is usually called a translation. Thinking about these two things independently is easier than trying to cope with them both at once. This is

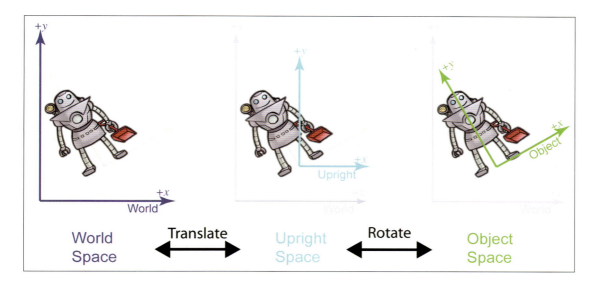

Figure 3.3
Conversion between world and upright space via translation; upright and object space are related by rotation

shown in Figure 3.3. World space (on the left) is transformed into upright space (in the center) by translating the origin. To transform upright space into object space, we rotate the axes until they align with the object-space axes. In this example, the robot thinks that her y-axis points from her feet to her head and that her x-axis points to her left.[3] We will return to this concept in Section 3.3.

The term "upright" is of our own invention and is not (yet!) a standard you are likely to find elsewhere. But it's a powerful concept in search of a good name. In physics, the term "center of mass coordinates" is sometimes used to describe coordinates expressed in the space that we are calling upright space. In the first edition of this book, we used the term "inertial space" to refer to this space, but we have changed it to avoid confusion with inertial reference frames in physics, which have some similar connotations but are different. We'll have a bit more philosophical pleas regarding upright space at the end of this chapter.

3.3   Basis Vectors and Coordinate Space Transformations

We said that a major justification for the existence of more than one coordinate space is because certain positions or directions are known only in a particular coordinate space. Likewise, sometimes certain *questions* can be answered only in particular coordinate spaces. When the question is best asked in one space, and the information we need in order to answer that question is known in a different space, we have a problem to solve.

For example, suppose that our robot is attempting to pick up a herring sandwich in our virtual world. We initially know the position of the sandwich and the position of the robot in *world coordinates*. World coordinates can be used to answer questions like "Is the sandwich north or south of me?" A different set of questions could be answered if we knew the position of the sandwich in the *object space* of the robot—for example, "Is the sandwich in front of me or behind me?" "Which way should I turn to face the sandwich?" "Which way do I move my herring sandwich scoop to get in position to pick up the sandwich?" Notice that to decide how to manipulate the gears and circuits, the object-space coordinates are the rel-

---

[3]Please forgive us for turning the robot around to face you, which caused us to break from our usual conventions where $+x$ is "right" in object space. In our defense, this is a 2D diagram, and we're not really sure if people living in a flat world would have any concept of "front" and "back" (though they would probably be able to tell between "regular" and "reflected" states—just as in 3D we have left- and right-handed coordinate systems). So who's to say if a 2D robot is really facing away from you or towards you, or which direction she thinks is her left or right?

evant ones. Furthermore, any data provided by sensors would be expressed in object space. Of course, our own bodies work under similar principles. All of us are capable of seeing a tasty morsel in front of us and putting it into our mouths without knowing which direction is "north." (And thank goodness, too, or many of us would starve to death.)

Further, suppose that we wish to render an image of the robot picking up the sandwich, and that the scene is illuminated by the light mounted on her shoulder. We know the position of the light within the robot's object space, but to properly light the scene, we must know the position of the light in world space.

These problems are two sides of the same coin: we know how to express a point in one coordinate space, and we need to express that point in some other coordinate space. The technical term for this computation is a *coordinate space transformation*. We need to *transform* the position from world space to object space (in the example of the sandwich) or from object space to world space (in the example of the light). Notice that in this example, neither the sandwich nor the light really move, we are just expressing their locations in a different coordinate space.

The remainder of this section describes how to perform coordinate space transformations. Because this topic has such fundamental importance, and it can be so darn confusing, please allow us to present a very gradual transition from the warm fluffy high level to the cold hard math. Section 3.3.1 considers transformations in the very context they are often encountered for beginning video game programmers: graphics. Using the most ridiculous example we could think of, we show the basic need of transformations, and also demonstrate the duality between two useful ways of visualizing transformations. Section 3.3.2 makes sure we are clear about what it means to specify a coordinate space in terms of another space. Finally, Section 3.3.3 presents the key idea of *basis vectors*.

3.3.1 Dual Perspectives

In our robot example, the discussion was phrased in a way such that the process of transforming a point didn't really "move" the point, we just changed our frame of reference and were able to describe the point using a different coordinate space. In fact, you might say that we really didn't transform the point, we transformed the coordinate space! But there's another way of looking at coordinate space transformations. Some people find it easier in certain situations to imagine the coordinate space staying still while the point moves from one place to another. When we develop the math for actually *calculating* these transformations, this is the paradigm that is more natural. Coordinate space transforms are such an important tool, and the confusion that can arise because of an incomplete awareness of these two perspectives is so common that we will take a little bit of extra space to work through some examples.

Now for that ridiculous example. Let's say that we are working for a advertising agency that has just landed a big account with a food manufacturer. We are assigned to the project to make a slick computer-generated ad promoting one of their most popular items, *Herring Packets*, which are microwaveable herring food products for robots.

Figure 3.4
One serving contains 100% of a robot's recommended daily allowance of essential oils.

Of course, the client has a tendency to want changes made at the last minute, so we might need models of the product and robot in all possible positions and orientations. Our first attempt to accomplish this is to request, from the art department, the robot model and the product model in every possible configuration of positions and orientations. Unfortunately, they estimate that since this is an infinite amount, it will take all of eternity to produce this many assets, even after factoring in Moore's law and the fact that the product model is just a box. The director suggests increasing the art staff in order to achieve her vision, but unfortunately, after crunching the numbers, the producer discovers that this doesn't decrease the time required to finish the project.[4] In fact, the company can afford resources to produce only *one* robot model and *one* box of microwaveable herring food product.

Although you may regret spending the past 60 seconds of your life reading the preceding paragraph, this example does illustrate the fundamental necessity of coordinate space transformations. It's also a relatively accurate depiction of the creative process. Time estimates are always padded, project managers will throw more people at a project in desperation, projects must be done by a certain date to meet a quarter, and, most pertinent to this book, artists will deliver only *one* model, leaving it up to *us* to move it around in the world.

The 3D model we get from the artist is a mathematical representation of a robot. This description likely includes control points, called *vertices*, and some sort of surface description, which tells how to connect the vertices together to form the surface of the object. Depending on what tools were used by the artist to create the model, the surface description might be a polygon mesh or a subdivision surface. We're not too concerned about the

---

[4]Although this is an extreme example, it illustrates a well-known principle that, in most creative projects, total project time is not simply the amount of work divided by the number of workers. As the saying goes, "Nine women can't make a baby in a month."

surface description here; what's important is that we can move the model around by moving its vertices around. Let's ignore for now the fact that the robot is an articulated creature, and assume we can only move it around in the world like a chess piece, but not animate it.

The artist who built our robot model decided (quite reasonably) to create her at the origin of the world space. This is depicted in Figure 3.5.

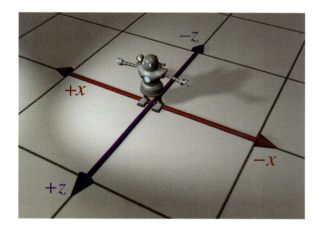

Figure 3.5
The robot model was created with the world origin at her feet.

To simplify the rest of this example, we're going to look at things from above. Although this is basically a 2D example, we're going to use our 3D conventions, ignoring the y-axis for now. In this book the convention is for $+z$ to point "forward" in object space and "north" in upright space, whereas $+x$ points "right" in object space and "east" in upright space.

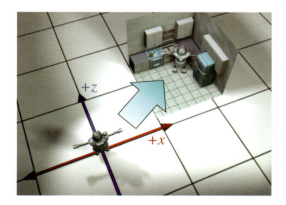

Figure 3.6
Moving the model into position

For now, because the model is in its home position, object space and world space (and upright space) are all the same *by definition*. For all practical purposes, in the scene that the artist built containing only the model of the robot, world space *is* object space.

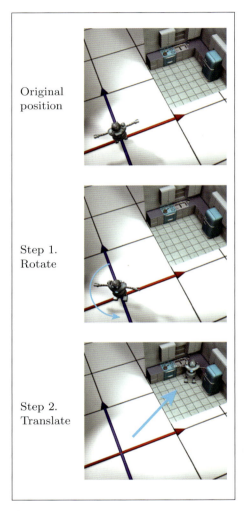

Original position

Step 1. Rotate

Step 2. Translate

Back to advertising. Our goal is to transform the vertices of the model from their "home" location to some new location (in our case, into a make-believe kitchen), according to the desired position and orientation of the robot based on the executive whims at that moment, as shown in Figure 3.6.

Let's talk a bit about how to accomplish this. We won't get too far into the mathematical details—that's what the rest of this chapter is for. Conceptually, to move the robot into position we first rotate her clockwise 120° (or, as we'll learn in Section 8.3, by "heading left 120°"). Then we translate 18 ft east and 10 ft north, which according, to our conventions, is a 3D displacement of $[18, 0, 10]$. This is shown in Figure 3.7.

At this time, please allow us a brief digression to answer a question that some readers may be asking: "Do we have to rotate first, and then translate?" The answer to this question is basically "yes." Although it may seem more natural to translate before rotating, it's usually easier to rotate first. Here's why. When we rotate the object first, the center of rotation is the origin. Rotation about the origin and translation are two primitive tools we have at our disposal, and each is easy. (Recall our motivation for introducing upright space in Section 3.2.4.) If we rotate second, then that rotation will occur about a point that is *not*

Figure 3.7
Transforming the robot from object space to world space by rotating, then translating

the origin. Rotation about the origin is a linear transform, but rotation about any other point is an *affine* transform. As we show in Section 6.4.3, to perform an affine transformation, we compose a sequence of primitive operations. For rotation about an arbitrary point, we translate the center of rotation to the origin, rotate about the origin, and then translate back. In other words, if we want to move the robot into place by translating first and rotating second, we likely go through the following process:

1. Translate.

2. Rotate. Because we're rotating about a point that's not the origin, this is a three step process:

    a. Translate the center of rotation to the origin. (This undoes step 1.)

    b. Perform the rotation about the origin.

    c. Translate to put the center of rotation in place.

Notice that steps 1 and 2a cancel each other out, and we're left with the two steps: rotate first, then translate.

So we've managed to get the robot model into the right place in the world. But to render it, we need to transform the vertices of the model into *camera space*. In other words, we need to express the coordinates of the vertices relative to the camera. For example, if a vertex is 9 ft in front of the camera and 3 ft to the right, then the z- and x-coordinates of that vertex in camera space would be 9 and 3, respectively. Figure 3.8 shows a

Translating first and then rotating

Overhead view   Camera's view

Figure 3.8
The layout of the camera and robot in the scene

particular shot we might want to capture. On the left, we see the layout of the shot from an external perspective, and on the right is what the camera sees.

It was easy to visualize transforming the model into world space. We literally "moved" it into place.[5] But how do we transform from world space to camera space? The objects are both already "in place," so where do we "move" them? For situations like this, it's helpful to think about transforming the coordinate space rather than transforming the objects, a technique we'll discuss in the next section. However, let's see if we can keep the coordinate space stationary, and still achieve the desired result by only "moving objects."

When we transformed from object space to world space, we were able to do so because we imagined the robot starting out at the origin in world space. Of course, the robot never really was at the location in world space, but we imagined it. Since we transformed from object space to world space by moving the object, perhaps we can transform from world space to camera space by moving the world! Imagine taking the entire world, including the robot, the camera, and the kitchen, and moving everything around. Clearly, such operations wouldn't affect what the camera would "see," because they don't change the relative relationship between the camera and the objects in the world. If we moved the world and camera together, such that the camera moved to the origin, then world-space coordinates and camera-space coordinates would be the same. Figure 3.9 shows the two-step process that we would use to achieve this.

Notice that, in this case, it's easier to translate before rotating. That's because we want to rotate about the origin. Also, we use the *opposite* translation and rotation amounts, compared to the camera's position and orientation. For example, in Figure 3.9 the coordinates of the camera are approximately $(13.5, 4, 2)$. (The grid lines represent 10 units.) So to move the camera to the origin, we translate everything by $[-13.5, -4, -2]$. The camera is facing roughly northeast and thus has a clockwise heading compared to north; a counterclockwise rotation is required to align camera-space axes with the world-space axes.

After picking up and moving around an entire robot in the first step, and then the *entire world*[6] in the second step, we finally have the coordinates of the vertices in camera space, and can proceed with rendering. If all this imaginary heavy lifting has worn you out, don't worry; in just a moment we will discuss an alternate way to think about this process.

Before we move on, a few important notes about this example. First, the world-to-camera transform is usually done in a vertex shader; you can

[5]OK, since this is all taking place in our imagination, the word *literally* might be a bit out of place.

[6]Yes, including the kitchen sink.

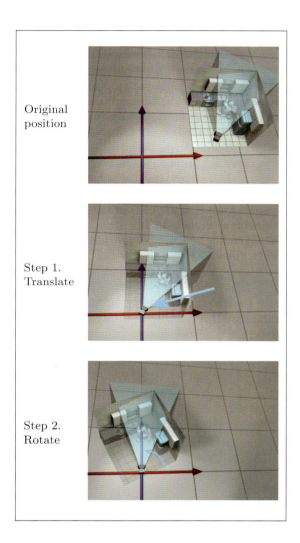

Original
position

Step 1.
Translate

Step 2.
Rotate

Figure 3.9
Transforming *everything*
from world space to camera
space by translating, then
rotating

leave this to the graphics API if you are working at a high level and not writing your own shaders. Second, camera space isn't the "finish line" as far as the graphics pipeline is concerned. From camera space, vertices are transformed into *clip space* and finally projected to *screen space*. These details are covered in Section 10.2.3.

So we've seen how we can compute world-space coordinates from object-space coordinates by imagining moving the model from the origin to its position in the world. Then we could compute camera-space coordinates from world-space coordinates by shifting the entire world to put the camera at the origin. The point to emphasize is that the coordinate space used to describe the points remained constant (even if we called it different names at different times), while we imaged the points moving in space. A transformation thus interpreted is sometimes called an *active transformation*.

Alternatively, we can think of the same process as a *passive transformation*. In the passive paradigm, we imagine the points being stationary while we move the coordinate space used to describe those points. *In either case, the coordinates of the points are the same at each step.* It's all in how we choose to view the situation. Earlier, our perspective was fixed with the coordinate space, because we were thinking of the transformation in active terms. Now we show the dual perspective, which is fixed relative to the object.

Absolute perspective   **Local perspective**

Robot object space

Robot upright space

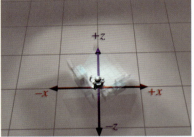

World space

Absolute perspective   **Local perspective**

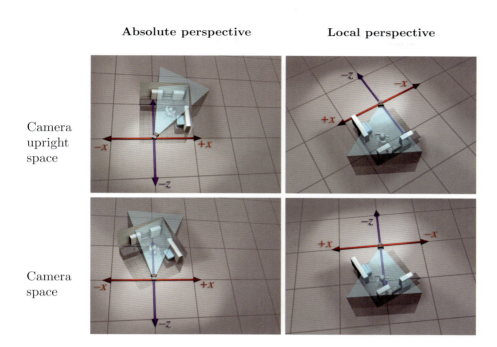

Camera
upright
space

Camera
space

Figure 3.10
The same sequence of coordinate space transformations is viewed from two perspectives. On
the left, it appears as if the objects are moving, and the coordinate axes are stationary. On the
right, the objects appear to be stationary, and the coordinate space axes are transformed.

Figure 3.10 reviews the four-step sequence from the robot's object space
to the camera's object space from both perspectives. On the left, we repeat
the presentation just given, where the coordinate space is stationary and the
robot is moving around. On the right, we show the *same process* as a passive
transformation, from a perspective that remains fixed relative to the robot.
Notice how the coordinate space appears to move around. Also, notice that
when we perform a certain transformation to the vertices, it's equivalent
to performing the *opposite* transformation to the coordinate space. The
duality between active and passive transformations is a frequent a source
of confusion. Always make sure when you are turning some transformation
into math (or code), to be clear in your mind whether the object or the
coordinate space is being transformed. We consider a classic example of this
confusion from graphics in Section 8.7.1, when we discuss how to convert
Euler angles to the corresponding rotation matrix.

Note that for clarity, the first two rows in Figure 3.10 have the kitchen
and camera mostly transparent. In reality, each individual object—the

pots, the refrigerator, and so on—could have been made by an artist at the center of some scene, and conceptually each undergoes its own unique transform from object space to world space.

We discussed two useful ways of imagining coordinate space transformations. One way is to fix our perspective with the coordinate space. This is the active transformation paradigm: the vectors and objects move around as their coordinates change. In the passive transformation paradigm, we keep our perspective fixed relative to the thing being transformed, making it appear as if we are transforming the coordinate space used to measure the coordinates. Transforming an object has the same effect on the coordinates as performing the opposite transformation to the coordinate space. Both the active and passive paradigms are quite useful, and an inadequate appreciation of the difference between them is a common cause of mistakes.

3.3.2 Specifying Coordinate Spaces

We are almost ready to talk about transformations. But there's actually one more basic question we should answer first: exactly how do we specify a coordinate space relative to another coordinate space?[7] Recall from Section 1.2.2 that a coordinate system is defined by its origin and axes. The origin defines the *position* of the coordinate space, and the axes describe its *orientation*. (Actually, the axes can describe other information, such as scale and skew. For the moment, we assume that the axes are perpendicular and the units used by the axes are the same as the units used by the parent coordinate space.) So if we can find a way to describe the origin and the axes, then we have fully documented the coordinate space.

Specifying the position of the coordinate space is straightforward. All we have to do is describe the location of the origin. We do this just like we do for any other point. Of course, we must express this point relative to the *parent coordinate space*, not the local child space. The origin of the child space, by definition, is always $(0,0,0)$ when expressed in child coordinate space. For example, consider the position of the 2D robot in Figure 3.2. To establish a scale for the diagram, let's say the robot is around 5 1/2 feet tall. Then the world-space coordinates of her origin are close to $(4.5, 1.5)$.

---

[7]We imagine that if this chapter were an episode of *Elmo's World*, this very obvious and important question would be the one that Elmo's goldfish, Dorothy, would have asked right off the bat.

Specifying the orientation of a coordinate space in 3D is only slightly more complicated. The axes are vectors (directions), and can be specified like any other direction vector. Going back to our robot example, we could describe her orientation by telling what directions the green vectors labeled $+x$ and $+y$ were pointing—these are the axes of the robot's object space. (Actually, we would use vectors with unit length. The axes in the diagrams were drawn as large as possible, but, as we see in just a moment, unit vectors are usually used to describe the axes.) Just as with position, we do not use the object space itself to describe the object-space axis directions, since those coordinates are $[1,0]$ and $[0,1]$ by definition. Instead, the coordinates are specified in upright space. In this example, unit vectors in the $+x$ and $+y$ object-space directions have upright-space coordinates of $[0.87, 0.50]$ and $[-0.50, 0.87]$, respectively.

What we have just described is one way to specify the orientation of a coordinate space, but there are others. For example, in 2D, rather than listing two 2D vectors, we could give a single angle. (The robot's object axes are rotated clockwise 30° relative to the upright axes.) In 3D, describing orientation is considerably more complicated, and in fact we have devoted all of Chapter 8 to the subject.

We specify a coordinate space by describing its origin and axes. The origin is a point that defines the position of the space and can be described just like any other point. The axes are vectors and describe the orientation of the space (and possibly other information such as scale), and the usual tools for describing vectors can be used. The coordinates we use to measure the origin and axes must be relative to some *other* coordinate space.

3.3.3   Basis Vectors

Now we are ready to actually compute some coordinate space transforms. We start with a concrete 2D example. Let's say that we need to know the world-space coordinates of the light that is attached to the robot's right shoulder. We start with the object-space coordinates, which are $(-1, 5)$. How do we get the world-space coordinates? To do this, we must go back to the beginning and poke deeper into some ideas that are so fundamental as to be taken for granted. How do we locate a point indicated by a given set of Cartesian coordinates? Let's say we needed to give step-by-step directions for how to locate the light to somebody who didn't know how Cartesian coordinates worked. We would say:

1. Start at the origin.

2. Move to the right 1 foot.

3. Move up 5 feet.

We assume this person has a tape measure, and understands that when we say "right" and "up," we mean the *robot's* "right" and "up," the directions that we enlightened people know as being parallel to the object-space axes.

Now here's the key point: we already know how to describe the origin, the direction called "the robot's right," and the direction called "the robot's up" in world coordinates! They are part of the specification of the coordinate space, which we just gave in the previous section. So all we have to do is follow our own instructions, and at each step, keep track of the world-space coordinates. Examine Figure 3.2 again.

1. *Start at the origin.* No problem, we previously determined that her origin was at

$$(4.5, 1.5).$$

2. *Move to the right 1 foot.* We know that the vector "the robot's left" is $[0.87, 0.50]$, and so we scale this direction by the distance of -1 unit, and add on the displacement to our position, to get

$$(4.5, 1.5) + (-1) \times [0.87, 0.50] = (3.63, 1).$$

3. *Move up 5 feet.* Once again, we know that "the robot's up" direction is $[-0.50, 0.87]$, so we just scale this by 5 units and add it to the result, yielding

$$(4.5, 1.5) + (-1) \times [0.87, 0.50] + 5 \times [-0.50, 0.87] = (1.13, 5.35).$$

If you look again at Figure 3.2, you'll see that, indeed, the world-space coordinates of the light are approximately $(1.13, 5.35)$.

Now let's remove the numbers specific to this example, and make some more abstract statements. Let $\mathbf{b}$ be some arbitrary point whose body-space coordinates $\mathbf{b} = (b_x, b_y)$ are known. Let $\mathbf{w} = (w_x, w_y)$ represent the world-space coordinates of that same point. We know the world-space coordinates for the origin $\mathbf{o}$ and the left and up directions, which we denote as $\mathbf{p}$ and $\mathbf{q}$, respectively. Now $\mathbf{w}$ can be computed by

$$\mathbf{w} = \mathbf{o} + b_x \mathbf{p} + b_y \mathbf{q}. \tag{3.1}$$

Now let's be even more general. In order to do so, it will help greatly to remove translation from our consideration. One way to do this is to discard "points" and think exclusively of vectors, which, as geometric entities, do not have a position (only magnitude and direction); thus translation does not really have a meaning for them. Alternatively, we can simply restrict the object space origin to be the same as the world-space origin.

Remember that in Section 2.3.1 we discussed how any vector may be decomposed geometrically into a sequence of axially-aligned displacements. Thus, an arbitrary vector $\mathbf{v}$ can be written in "expanded" form as

$$\mathbf{v} = x\mathbf{p} + y\mathbf{q} + z\mathbf{r}. \tag{3.2}$$

Expressing a 3D vector as a linear combination of basis vectors

Here, $\mathbf{p}$, $\mathbf{q}$, and $\mathbf{r}$ are *basis vectors* for 3D space. The vector $\mathbf{v}$ could have any possible magnitude and direction, and we could uniquely determine the coordinates x, y, z (unless $\mathbf{p}$, $\mathbf{q}$, and $\mathbf{r}$ are chosen poorly; we discuss this key point in just a moment). Equation (3.2) expresses $\mathbf{v}$ as a *linear combination* of the basis vectors.

Here is a common, but a bit incomplete, way to think about basis vectors: most of the time, $\mathbf{p} = [1, 0, 0]$, $\mathbf{q} = [0, 1, 0]$, and $\mathbf{r} = [0, 0, 1]$; in other unusual circumstances, $\mathbf{p}$, $\mathbf{q}$, and $\mathbf{r}$ have different coordinates. This is not quite right. When thinking about $\mathbf{p}$, $\mathbf{q}$, and $\mathbf{r}$, we must distinguish between the vectors as geometric entities (earlier, $\mathbf{p}$ and $\mathbf{q}$ were the physical directions of "left" and "up") and the particular coordinates used to describe those vectors. The former is inherently immutable; the latter depends on the choice of basis. Plenty of books emphasize this by defining all vectors in terms of the "world basis vectors," which are often denoted $\mathbf{i}$, $\mathbf{j}$, and $\mathbf{k}$ and are interpreted as elemental geometric entities that cannot be further decomposed. They do not have "coordinates," although certain axioms are taken to be true, such as $\mathbf{i} \times \mathbf{j} = \mathbf{k}$. In this framework, a coordinate triple $[x, y, z]$ is a mathematical entity, which does not have a geometric meaning until we take the linear combination $x\mathbf{i} + y\mathbf{j} + z\mathbf{k}$. Now, in response to the assertion $\mathbf{i} = [1, 0, 0]$, we might argue that since $\mathbf{i}$ is a geometric entity, it cannot be compared against a mathematical object, in the same way that the equation "kilometer = 3.2" is nonsense. Because the letters $\mathbf{i}$, $\mathbf{j}$, and $\mathbf{k}$ carry this weighty elemental connotation, we instead use the less presumptuous symbols $\mathbf{p}$, $\mathbf{q}$, and $\mathbf{r}$, and whenever we use these symbols to name our basis vectors, the message is: "we're using these as our basis vectors for now, but we might know how to express $\mathbf{p}$, $\mathbf{q}$, $\mathbf{r}$ relative to some other basis, so they aren't necessarily the 'root' basis."

The coordinates of $\mathbf{p}$, $\mathbf{q}$, and $\mathbf{r}$ are always equal to $[1, 0, 0]$, $[0, 1, 0]$, and $[0, 0, 1]$, respectively, when expressed using the coordinate space for which they are the basis, but relative to some other basis they will have arbitrary coordinates. When we say that we are using the standard basis, this is equivalent to saying that we are concerning ourselves with only a single coordinate space. What we call that coordinate space makes no difference, since we have no way to reference any other coordinate space without introducing basis vectors. When we do consider any alternative basis, we have implicitly introduced another coordinate space: the space used to measure the coordinates of the basis vectors!

The coordinates of basis vectors are measured in terms of a reference frame that is different from the one for which they are a basis. Thus basis vectors are intimately linked with coordinate space transformations.

We said earlier that $\mathbf{p}$, $\mathbf{q}$, and $\mathbf{r}$ could be chosen "poorly." This begs the question: what makes a good basis? We are accustomed to having basis vectors that are mutually perpendicular. We are also used to them having the same length: we expect the displacements $5\mathbf{p}$ and $5\mathbf{q}$ to be in different directions, but we ordinarily would assume that they have the same length. Finally, when multiple coordinate spaces are involved, we are also used to them all having the same scale. That is, the vector $\mathbf{v}$ has the same numeric magnitude, no matter what coordinate system we use to measure it. But as we're about to see, that isn't necessarily the case. These properties are certainly desirable; in fact, we might say that this is the "best basis" in many cases. But they may not always be immediately available, they are often not necessary, and there are some situations for which we purposefully choose basis vectors without these properties.

We briefly mention here two examples, both from the world of graphics. Imagine we want to animate a squishing or stretching motion of our robot model. To do so, we would modify the coordinate space used to interpret the coordinates of our vertices. We would animate the basis vectors of the robot's object space, probably in ways that caused them to have different lengths from one another or to cease to be perpendicular. As we squish or stretch the object-space vectors, the object-space coordinates of the vertices remain constant, but the resulting camera-space coordinates change, producing the desired animation.

Another example arises with basis vectors for texture mapping. (We're getting a bit ahead of ourselves, since we won't talk about texture mapping until Section 10.5; however we are aware that our readers are not a *tabula rasa*, and we suspect you have at least heard of these concepts. We are also aware that many readers' first introduction to the term *basis vector* is in the context of bump mapping; hopefully this example will help place that particular usage of the term in the proper context.) It's often helpful to establish a local coordinate space on the surface of an object where one axis (we'll use $+z$) is parallel to the surface normal, and the other axes point in the direction of increasing u and v in the texture. These latter two basis vectors are sometimes called the *tangent* and *binormal.* Motion in 3D space in the direction of the tangent basis vector corresponds to horizontal motion in the 2D image space of the texture, while a displacement in 3D space in the direction of the binormal would correspond to vertical image-space displacement. The key fact is that the flat 2D texture often must be warped to wrap it around an irregular surface, and the basis vectors are not guaranteed to be perpendicular.[8]

Figure 3.11 shows a situation in which the basis vectors **p** and **q** have the same length, but are not perpendicular. Although we've shown only two example vectors, **a** and **b**, the set of vectors that can be described as a linear combination $x\mathbf{p} + y\mathbf{q}$ fill an infinite plane, and for any vector in this plane, the coordinates $[x, y]$ are uniquely determined.

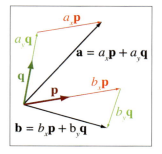

The set of vectors that can be expressed as a linear combination of the basis vectors is called the *span* of the basis. In the example in Figure 3.11, the span is an infinite 2D plane. This might seem at first like it's the only possible scenario, but let's examine some more interesting situations. First of all,

Figure 3.11
Basis vectors don't have to be perpendicular.

note that we said that the vectors fill "an" infinite plane, not "the" plane. Just because we have two coordinates and basis vectors does *not* mean that **p** and **q** must be 2D vectors! They could be 3D vectors, in which case their span will be some arbitrary plane within 3D space, as depicted in Figure 3.12.

Figure 3.12 illustrates several key points. Note that we have chosen **a** and **b** to have the same coordinates from Figure 3.11, at least relative to the basis vectors **p** and **q**. Second, when working within the space of **p** and **q**, our example vectors **a** and **b** are 2D vectors; they have only two coordinates, x and y. We might also be interested in their 3D "world" coordinates; these are obtained simply by expanding the linear combination $x\mathbf{p} + y\mathbf{q}$; the result of this expression is a 3D vector.

---

[8]Note that it is a common optimization to ignore this possibility and assume that they are perpendicular, even when they aren't. This assumption introduces some error in some cases, but it permits a reduction in storage and bandwidth, and the error is usually not noticeable in practice. We'll discuss this in greater detail in Section 10.9.

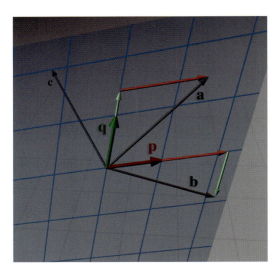

Figure 3.12
The two basis vectors **p** and **q** span a 2D subset of the 3D space.

Consider the vector **c**, which lies behind the plane in Figure 3.12. This vector is *not* in the span of **p** and **q**, which means we cannot express it as a linear combination of the basis. In other words, there are no coordinates $[c_x, c_y]$ such that $\mathbf{c} = c_x\mathbf{p} + c_y\mathbf{q}$.

The term used to describe the number of dimensions in the space spanned by the basis is the *rank* of the basis. In both of the examples so far, we have two basis vectors that span a two-dimensional space. Clearly, if we have n basis vectors, the best we can hope for is *full rank*, meaning the span is an n-dimensional space. But is it possible for the rank to be less than n? For example, if we have three basis vectors, is it possible that the span of those basis vectors is only 2D or 1D? The answer is "yes," and this situation corresponds to what we meant earlier by a "poor choice" of basis vectors.

For example, let's say we add a third basis vector **r** to our set **p** and **q**. If **r** lies in the span of **p** and **q** (for example, let's say we chose **r** = **a** or **r** = **b** as our third basis vector), then the basis vectors are *linearly dependent*, and do not have full rank. Adding in this last vector did not allow us to describe any vectors that could not already be described with just **p** and **q**. Furthermore, now the coordinates $[x, y, z]$ for a given vector in the span of the basis are not uniquely determined. The basis vectors span a space with only two degrees of freedom, but we have three coordinates. The blame doesn't fall on **r** in particular, he just happened to be the new guy. We could have chosen *any pair* of vectors from **p**, **q**, **a**, and **b**, as a valid basis for this same space. The problem of linear dependence is a problem with the set as a whole, not just one particular vector. In contrast, if our third basis vector was chosen to be *any other vector* that didn't lie in the plane spanned by **p** and **q** (for example, the vector **c**), then the basis would be linearly independent and have full rank. If a set of basis vectors are linearly independent, then it is not possible to express any one basis vector as a linear combination of the others.

So a set of linearly dependent vectors is certainly a poor choice for a basis. But there are other more stringent properties we might desire of a basis. To see this, let's return to coordinate space transformations. Assume, as before, that we have an object whose basis vectors are **p**, **q**, and **r**, and we know the coordinates of these vectors in world space. Let **b** = $[b_x, b_y, b_z]$ be the coordinates of some arbitrary vector in body space, and **u** = $[u_x, u_y, u_z]$ be the coordinates of that same vector, in upright space. From our robot example, we already know the relationship between **u** and **b**:

$$\mathbf{u} = b_x\mathbf{p} + b_y\mathbf{q} + b_z\mathbf{r}, \quad \text{or equivalently,} \quad \begin{aligned} u_x &= b_xp_x + b_yq_x + b_zr_x, \\ u_y &= b_xp_y + b_yq_y + b_zr_y, \\ u_z &= b_xp_z + b_yq_z + b_zr_z. \end{aligned}$$

Make sure you understand the relationship between these equations and Equation (3.1) before moving on.

Now here's the key problem: what if **u** is known and **b** is the vector we're trying to determine? To illustrate the profound difference between these two questions, let's write the two systems side-by-side, replacing the unknown vector with "?":

$$\begin{aligned} ?_x &= b_xp_x + b_yq_x + b_zr_x, \\ ?_y &= b_xp_y + b_yq_y + b_zr_y, \\ ?_z &= b_xp_z + b_yq_z + b_zr_z, \end{aligned} \qquad \begin{aligned} u_x &= ?_xp_x + ?_yq_x + ?_zr_x, \\ u_y &= ?_xp_y + ?_yq_y + ?_zr_y, \\ u_z &= ?_xp_z + ?_yq_z + ?_zr_z. \end{aligned}$$

The system of equations on the left is not really much of a "system" at all, it's just a list; each equation is independent, and each unknown quantity can be immediately computed from a single equation. On the right, however, we have three *interrelated* equations, and none of the unknown quantities can be determined without all three equations. In fact, if the basis vectors are linearly dependent, then the system on the right may have zero solutions (**u** is not in the span), or it might have an infinite number of solutions (**u** is in the span and the coordinates are not uniquely determined). We hasten to add that the critical distinction is not between upright or body space; we are just using those spaces to have a specific example. The important fact is whether the known coordinates of the vector being transformed are expressed relative to the basis (the easy situation on the left), or if the coordinates of the vector and the basis vectors are all expressed in the same coordinate space (the hard situation on the right).

Linear algebra provides a number of general-purpose tools for solving systems of linear equations like this, but we don't need to delve into these topics, because the solution to this system is not our primary aim. For now, we're interested in understanding one special situation for which the solution is easy. (In Section 6.2, we show how to use the *matrix inverse* to solve the general case.)

The dot product is the key. Remember from Section 2.11.2 that the dot product can be used to measure distance in a particular direction. As we observed in that same section, when using the standard basis $\mathbf{p} = [1, 0, 0]$, $\mathbf{q} = [0, 1, 0]$, and $\mathbf{r} = [0, 0, 1]$, corresponding to the object axes being parallel with the world axes in our robot example, we can dot the vector with a basis vector to "sift out" the corresponding coordinate.

$$b_x = \mathbf{u} \cdot \mathbf{p} = \mathbf{u} \cdot \begin{bmatrix} 1 & 0 & 0 \end{bmatrix} = u_x$$
$$b_y = \mathbf{u} \cdot \mathbf{q} = \mathbf{u} \cdot \begin{bmatrix} 0 & 1 & 0 \end{bmatrix} = u_y$$
$$b_z = \mathbf{u} \cdot \mathbf{r} = \mathbf{u} \cdot \begin{bmatrix} 0 & 0 & 1 \end{bmatrix} = u_z$$

Algebraically, this is rather obvious. But does this "sifting" action work for any arbitrary basis? Sometimes, but not always. In fact, we can see that it doesn't work for the example we have been using. Figure 3.13 compares the correct coordinates a_x and a_y with the dot products $\mathbf{a} \cdot \mathbf{p}$ and $\mathbf{a} \cdot \mathbf{q}$. (The illustration is completely correct only if $\mathbf{p}$ and $\mathbf{q}$ are unit vectors.)

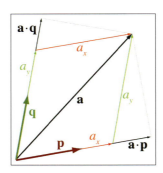

Figure 3.13
The dot product doesn't "sift out" the coordinate in this case.

Notice that, in each case, the result produced by the dot product is larger than the correct coordinate value. To understand what's wrong, we need to go back and correct a little lie that we told in Chapter 1. We said that a coordinate measures the displacement from the origin in a given direction; this displacement is exactly what the dot product is measuring. While that is the simplest way to explain coordinates, it works only under special circumstances. (Our lie isn't that harmful because these circumstances are very common!) Now that we understand basis vectors, we're ready for the more complete description.

The numeric coordinates of a vector with respect to a given basis are the coefficients in the expansion of that vector as a linear combination of the basis vectors. For example, $\mathbf{a} = a_x\mathbf{p} + a_y\mathbf{q}$.

The reason the dot product doesn't "sift out" the coordinates in Figure 3.13 is because we are ignoring the fact that $y\mathbf{q}$ will cause some displacement parallel to $\mathbf{p}$. To visualize this, imagine that we increased a_x while holding a_y constant. As $\mathbf{a}$ moves to the right and slightly upwards, its projection onto $\mathbf{q}$, which is measured by the dot product, increases.

The problem is that *the basis vectors are not perpendicular*. A set of basis vectors that are mutually perpendicular is called an *orthogonal* basis.

When the basis vectors are orthogonal, the coordinates are uncoupled. Any given coordinate of a vector $\mathbf{v}$ can be determined solely from $\mathbf{v}$ and the corresponding basis vector. For example, we can compute v_x knowing only $\mathbf{p}$, provided that the other basis vectors are perpendicular to $\mathbf{p}$.

Although we won't investigate it further in this book, the idea of orthonormal basis is a broadly powerful one with applications outside our immediate concerns. For example, it is the idea behind Fourier analysis.

If it's good when basis vectors are orthogonal, then it's best when they all have unit length. Such a set of vectors are known as an *orthonormal* basis. Why is the unit length helpful? Remember the geometric definition of the dot product: $\mathbf{a} \cdot \mathbf{p}$ is equal to the signed length of $\mathbf{a}$ projected onto $\mathbf{p}$, *times the length of* $\mathbf{p}$. If the basis vector doesn't have unit length, but it is perpendicular to all the others, we can still determine the corresponding coordinate with the dot product; we just need to divide by the square of the length of the basis vector.

In an orthonormal basis, each coordinate of a vector $\mathbf{v}$ is the signed displacement $\mathbf{v}$ measured in the direction of the corresponding basis vector. This can be computed directly by taking the dot product of $\mathbf{v}$ with that basis vector.

Thus, in the special circumstance of an orthonormal basis, we have a simple way to determine the body space coordinates, knowing only the world coordinates of the body axes. Thus, assuming $\mathbf{p}$, $\mathbf{q}$, and $\mathbf{r}$ form an orthonormal basis,

$$b_x = \mathbf{u} \cdot \mathbf{p}, \qquad\qquad b_y = \mathbf{u} \cdot \mathbf{q}, \qquad\qquad b_z = \mathbf{u} \cdot \mathbf{r}.$$

Although our example uses body space and upright space for concreteness, these are general ideas that apply to any coordinate space transformation.

Orthonormal bases are the special circumstances under which our lie from Chapter 1 is harmless; fortunately they are extremely common. At the beginning of this section, we mentioned that most of the coordinate spaces we are "accustomed to" have certain properties. All of these "customary" coordinate spaces have an orthonormal basis, and in fact they meet an even further restriction: the coordinate space is not mirrored. That is, $\mathbf{p} \times \mathbf{q} = \mathbf{r}$, and the axes obey the prevailing handedness conventions (in this book, we use the left-hand conventions). A mirrored basis where $\mathbf{p} \times \mathbf{q} = -\mathbf{r}$ can still be an orthonormal basis.

3.4   Nested Coordinate Spaces

Each object in a 3D virtual universe has its own coordinate space—its own origin and its axes. Its origin could be located at its center of mass, for example. Its axes specify which directions it considers to be "up," "right," and "forward" relative to its origin. A 3D model created by an artist for a virtual world will have its origin and axes decided by the artist, and the points that make up the polygon mesh will be relative to the object space defined by this origin and axes. For example, the center of a sheep could be placed at $(0,0,0)$, the tip of its snout at $(0,0,1.5)$, the tip if its tail at $(0,0,-1.2)$, and the tip of its right ear at $(0.5,0.2,1.2)$. These are the locations of these parts in sheep space.

The position and orientation of an object at any point in time needs be specified in world coordinates so that we can compute the interactions between nearby objects. To be precise, we must specify the location and orientation of the object's axes in world coordinates. To specify the city of Cartesia's position (see Section 1.2.1) in world space, we could state that the origin is at longitude $p°$ and latitude $q°$ and that the positive x-axis points east and the positive y-axis points north. To locate the sheep in a virtual world, it is sufficient to specify the location of its origin and the orientation of its axes in world space. The world location of the tip of its snout, for example, can be worked out from the relative position of its snout to the world coordinates of its origin. But if the sheep is not actually being drawn, we can save effort by keeping track of only the location and orientation of its object space in world space. It becomes necessary to compute the world coordinates of its snout, tail, and right ear at only certain times—for example, when it moves into view of the camera.

Since the object space moves around in world space, it is convenient to view the world space as a "parent" space, and the object space as a "child" space. It is also convenient to break objects into subobjects and to animate them independently. A model decomposed into a hierarchy like this is sometimes called an *articulated* model. For example, as the sheep walks, its head swings back and forth and its ears flap up and down. In the coordinate space of the sheep's head, the ears appear to be flapping up and down—the motion is in the y-axis only and so is relatively easy to understand and animate. In the sheep's coordinate space its head is swinging from side to side along the sheep's x-axis, which is again relatively easy to understand. Now, suppose that the sheep is moving along the world's z-axis. Each of the three actions—ears flapping, head swinging, and sheep moving forwards—involves a single axis and is easy to understand in isolation from the others. The motion of the tip of the sheep's right ear, however, traces a complicated path through the world coordinate space, truly a nightmare for a programmer to compute from scratch. By breaking the sheep into a hierarchically organized sequence of objects with nested coordinate spaces, however, the motion can be computed in separate components and combined relatively easily with linear algebra tools such as matrices and vectors, as we see in later chapters.

For example, let's say we need to know the world coordinates of the tip of the ear of the sheep. To compute these coordinates, we might first use what we know about the relationship of the sheep's ear relative to its head to compute the coordinates of that point in "head space." Next, we use the position and orientation of the head relative to the body to compute the coordinates in "body space." Finally, since we know the position and orientation of the sheep's body relative to the world's origin and axes, we can compute the coordinates in world space. The next few chapters go deeper into the details of how to do this.

It's convenient to think of the sheep's coordinate space moving relative to world space, the sheep's head coordinate space moving relative to the sheep's space, and the sheep's ear space moving relative to the sheep's head space. Thus we view the head space as a child of the sheep space, and the ear space as a child of the head space. Object space can be divided into many different subspaces at many different levels, depending on the complexity of the object being animated. We can say that the child coordinate space is *nested* in the parent coordinate space. This parent-child relationship between coordinate spaces defines a hierarchy, or tree, of coordinate spaces. The world coordinate space is the root of this tree. The nested coordinate space tree can change dynamically during the lifetime of a virtual world; for example, the sheep's fleece can be sheared and taken away from the sheep, and thus the fleece coordinate space goes from being a child of the sheep body's coordinate space to being a child of the world space. The hierarchy of nested coordinate spaces is dynamic and can be arranged in a manner that is most convenient for the information that is important to us.

3.5 In Defense of Upright Space

Finally, please allow us a few paragraphs to try to persuade you that the concept of upright space is highly useful, even though the term may not be standard. Lots[9] of people don't bother distinguishing between world space and upright space. They would just talk about rotating a vector from object space to "world space." But consider the common situation in code when the same data type, say `float3`, is used to store both "points" and "vectors." (See Section 2.4 if you don't remember why those terms were just put in quotes.) Let's say we have one `float3` that represents the position of a vertex in object space, and we wish to know the position of that vertex in world space. The transformation from object to world space must involve translation by the object's position. Now compare this to a different `float3` that describes a direction, such as a surface normal or the direction a gun is aiming. The conversion of coordinates of the direction vector from object space to "world space" (what we would call "upright space") should *not* contain this translation.

When you are communicating your intentions to a person, sometimes the other person is able to understand what you mean and fill in the blanks as to whether the translation happens or not when you say "world space." This is because they can visualize what you are talking about and implicitly know whether the quantity being transformed is a "point" or a "vector." But a computer does not have this intuition, so we must find a way to be explicit. One strategy for communicating this explicitly to a computer is to use two different data types, for example, one class named `Point3` and another named `Vector3`. The computer would know that vectors should never be translated but points should be because you would write two different transformation routines. This is a strategy adopted in some sources of academic origin, but in production code in the game industry, it is not common. (It also doesn't really work well in HLSL/Cg, which greatly encourages the use of the generic `float3` type.) Thus, we must find some other way to communicate to the computer whether, when transforming

---

[9]Here, the word "lots" means "almost everybody."

some given `float3` from object space to "world space," the translation should occur.

What seems to be the norm in a lot of game code is to simply have vector math details strewn about everywhere that multiplies by a rotation matrix (or its inverse) and explicitly has (or does not have) a vector addition or subtraction, as appropriate. We advocate giving a name to this intermediate space to differentiate it from "world space," in order to facilitate code that uses human-readable words such as "object," "upright," and "world," rather than explicit math tokens such as "add," "subtract," and "inverse." It is our experience that this sort of code is easier to read and write. We also hope this terminology will make this book easier to read as well! Decide for yourself if our terminology is of use to you, but please make sure you read the rest of our small crusade for more human-readable code in Section 8.2.1.

3.6   Exercises

(Answers on page 758.)

1. What coordinate space (object, upright, camera, or world) is the most appropriate in which to ask the following questions?

    (a) Is my computer in front of me or behind me?

    (b) Is the book east or west of me?

    (c) How do I get from one room to another?

    (d) Can I see my computer?

2. Suppose the world axes are transformed to our object axes by rotating them counterclockwise around the y-axis by $42°$, then translating 6 units along the z-axis and 12 units along the x-axis. Describe this transformation from the perspective of a point on the object.

3. For the following sets of basis vectors, determine if they are linearly independent. If not, describe why not.

    (a) $\begin{bmatrix} 1 \\ 0 \\ 0 \end{bmatrix}, \begin{bmatrix} 0 \\ 0 \\ 0 \end{bmatrix}, \begin{bmatrix} 0 \\ 2 \\ 0 \end{bmatrix}$

    (b) $\begin{bmatrix} 1 \\ 0 \\ 2 \end{bmatrix}, \begin{bmatrix} -1 \\ 1 \\ 2 \end{bmatrix}, \begin{bmatrix} 0 \\ 1 \\ 2 \end{bmatrix}$

    (c) $\begin{bmatrix} 1 \\ 2 \\ 3 \end{bmatrix}, \begin{bmatrix} -1 \\ 2 \\ 3 \end{bmatrix}, \begin{bmatrix} 1 \\ -2 \\ 3 \end{bmatrix}, \begin{bmatrix} 1 \\ 2 \\ -3 \end{bmatrix}$

(d) $\begin{bmatrix} 1 \\ 2 \\ 3 \end{bmatrix}, \begin{bmatrix} 0 \\ 1 \\ 5 \end{bmatrix}, \begin{bmatrix} -2 \\ -4 \\ -6 \end{bmatrix}$

(e) $\begin{bmatrix} 1 \\ 1 \\ 5 \end{bmatrix}, \begin{bmatrix} 0 \\ -5 \\ 4 \end{bmatrix}, \begin{bmatrix} 1 \\ -4 \\ 9 \end{bmatrix}$

(f) $\begin{bmatrix} 1 \\ 2 \\ 3 \end{bmatrix}, \begin{bmatrix} -1 \\ 2 \\ 3 \end{bmatrix}, \begin{bmatrix} 1 \\ -2 \\ 3 \end{bmatrix}$

4. For the following sets of basis vectors, determine if they are orthogonal. If not, tell why not.

(a) $\begin{bmatrix} 1 \\ 0 \\ 0 \end{bmatrix}, \begin{bmatrix} 0 \\ 0 \\ 4 \end{bmatrix}, \begin{bmatrix} 0 \\ 2 \\ 0 \end{bmatrix}$

(b) $\begin{bmatrix} 1 \\ 2 \\ 3 \end{bmatrix}, \begin{bmatrix} -1 \\ 2 \\ 3 \end{bmatrix}, \begin{bmatrix} 1 \\ -2 \\ 3 \end{bmatrix}$

(c) $\begin{bmatrix} 0 \\ 4 \\ 1 \end{bmatrix}, \begin{bmatrix} 0 \\ -1 \\ 4 \end{bmatrix}, \begin{bmatrix} 8 \\ 0 \\ 0 \end{bmatrix}$

(d) $\begin{bmatrix} 4 \\ -6 \\ 2 \end{bmatrix}, \begin{bmatrix} -4 \\ -2 \\ 2 \end{bmatrix}, \begin{bmatrix} -3 \\ -6 \\ -12 \end{bmatrix}$

(e) $\begin{bmatrix} 7 \\ -1 \\ 5 \end{bmatrix}, \begin{bmatrix} 5 \\ 5 \\ -6 \end{bmatrix}, \begin{bmatrix} -2 \\ 0 \\ 1 \end{bmatrix}$

5. Are these basis vectors orthonormal? If not, tell why not.

(a) $\begin{bmatrix} 1 \\ 0 \\ 0 \end{bmatrix}, \begin{bmatrix} 0 \\ 0 \\ 4 \end{bmatrix}, \begin{bmatrix} 0 \\ 2 \\ 0 \end{bmatrix}$

(b) $\begin{bmatrix} 1 \\ 2 \\ 3 \end{bmatrix}, \begin{bmatrix} -1 \\ 2 \\ 3 \end{bmatrix}, \begin{bmatrix} 1 \\ -2 \\ 3 \end{bmatrix}$

(c) $\begin{bmatrix} 1 \\ 0 \\ 0 \end{bmatrix}, \begin{bmatrix} 0 \\ 0 \\ -1 \end{bmatrix}, \begin{bmatrix} 0 \\ 1 \\ 0 \end{bmatrix}$

(d) $\begin{bmatrix} 0 \\ 1 \\ 0 \end{bmatrix}, \begin{bmatrix} 0 \\ .707 \\ .707 \end{bmatrix}, \begin{bmatrix} 1 \\ 0 \\ 0 \end{bmatrix}$

(e) $\begin{bmatrix} 0 \\ .707 \\ -.707 \end{bmatrix}, \begin{bmatrix} 0 \\ .707 \\ .707 \end{bmatrix}, \begin{bmatrix} 1 \\ 0 \\ 0 \end{bmatrix}$

(f) $\begin{bmatrix} .921 \\ .294 \\ -.254 \end{bmatrix}$, $\begin{bmatrix} -.254 \\ .951 \\ .178 \end{bmatrix}$, $\begin{bmatrix} .294 \\ -.100 \\ .951 \end{bmatrix}$

(g) $\begin{bmatrix} .995 \\ 0 \\ -.100 \end{bmatrix}$, $\begin{bmatrix} .840 \\ .810 \\ .837 \end{bmatrix}$, $\begin{bmatrix} .054 \\ -1.262 \\ .537 \end{bmatrix}$

6. Assume that the robot is at the position $(1, 10, 3)$, and her right, up, and forward vectors expressed in upright space are $[0.866, 0, -0.500]$, $[0, 1, 0]$, and $[0.500, 0, 0.866]$, respectively. (Note that these vectors form an orthonormal basis.) The following points are expressed in object space. Calculate the coordinates for these points in upright and world space.

   (a) $(-1, 2, 0)$

   (b) $(1, 2, 0)$

   (c) $(0, 0, 0)$

   (d) $(1, 5, 0.5)$

   (e) $(0, 5, 10)$

   The coordinates below are in world space. Transform these coordinates from world space to upright space and object space.

   (f) $(1, 10, 3)$

   (g) $(0, 0, 0)$

   (h) $(2.732, 10, 2.000)$

   (i) $(2, 11, 4)$

   (j) $(1, 20, 3)$

7. Name five examples of nested coordinate-space hierarchies.

Many a small thing has been made large
by the right kind of advertising.

— Mark Twain (1835–1910),
A Connecticut Yankee in King Arthur's Court

Chapter 4

Introduction to Matrices

Unfortunately, no one can be told what the matrix is.
You have to see it for yourself.

— Morpheus in *The Matrix* (1999)

Matrices are of fundamental importance in 3D math, where they are primarily used to describe the relationship between two coordinate spaces. They do this by defining a computation to transform vectors from one coordinate space to another.

This chapter introduces the theory and application of matrices. Our discussion will follow the pattern set in Chapter 2 when we introduced vectors: mathematical definitions followed by geometric interpretations.

- Section 4.1 discusses some of the basic properties and operations of matrices strictly from a mathematical perspective. (More matrix operations are discussed in Chapter 6.)

- Section 4.2 explains how to interpret these properties and operations geometrically.

- Section 4.3 puts the use of matrices in this book in context within the larger field of linear algebra.

4.1   Mathematical Definition of Matrix

In linear algebra, a matrix is a rectangular grid of numbers arranged into *rows* and *columns*. Recalling our earlier definition of vector as a one-dimensional array of numbers, a matrix may likewise be defined as a *two-dimensional array* of numbers. (The "two" in "two-dimensional array" comes from the fact that there are rows and columns, and should not be confused with 2D vectors or matrices.) So a vector is an array of scalars, and a matrix is an array of vectors.

113

This section presents matrices from a purely mathematical perspective. It is divided into eight subsections.

- Section 4.1.1 introduces the concept of matrix dimension and describes some matrix notation.

- Section 4.1.2 describes square matrices.

- Section 4.1.3 interprets vectors as matrices.

- Section 4.1.4 describes matrix transposition.

- Section 4.1.5 explains how to multiply a matrix by a scalar.

- Section 4.1.6 explains how to multiply a matrix by another matrix.

- Section 4.1.7 explains how to multiply a vector by a matrix.

- Section 4.1.8 compares and contrasts matrices for row and column vectors.

4.1.1 Matrix Dimensions and Notation

Just as we defined the dimension of a vector by counting how many numbers it contained, we will define the size of a matrix by counting how many rows and columns it contains. We say that a matrix with r rows and c columns is an $r \times c$ (read "r by c") matrix. For example, a 4×3 matrix has 4 rows and 3 columns:

A 4 × 3 matrix

$$\begin{bmatrix} 4 & 0 & 12 \\ -5 & \sqrt{4} & 3 \\ 12 & -4/3 & -1 \\ 1/2 & 18 & 0 \end{bmatrix}.$$

This 4×3 matrix illustrates the standard notation for writing matrices: the numbers are arranged in a grid enclosed by square brackets. Note that some authors may enclose the grid of numbers with parentheses rather than brackets, and other authors use straight vertical lines. We reserve this last notation for an entirely separate concept related to matrices, the *determinant* of a matrix. (We discuss determinants in Section 6.1.)

As we mentioned in Section 2.1, in this book we represent a matrix variable with uppercase letters in boldface, for example, **M**, **A**, **R**. When we wish to refer to the individual elements within a matrix, we use subscript notation, usually with the corresponding lowercase letter in italics. This is shown below for a 3×3 matrix:

$$\begin{bmatrix} m_{11} & m_{12} & m_{13} \\ m_{21} & m_{22} & m_{23} \\ m_{31} & m_{32} & m_{33} \end{bmatrix}$$

Subscript notation for matrix elements

The notation m_{ij} denotes the element in $\mathbf{M}$ at row i and column j. Matrices use 1-based indices, so the first row and column are numbered 1. For example, m_{12} (read "m one two," not "m twelve") is the element in the first row, second column. Notice that this is different from programming languages such as C++ and Java, which use 0-based array indices. A matrix does not have a column 0 or row 0. This difference in indexing can cause some confusion if matrices are stored using an actual array data type. For this reason, it's common for classes that store small, fixed size matrices of the type used for geometric purposes to give each element its own named member variable, such as `float m11`, instead of using the language's native array support with something like `float elem[3][3]`.

4.1.2 Square Matrices

Matrices with the same number of rows and columns are called *square* matrices and are of particular importance. In this book, we are interested in 2×2, 3×3, and 4×4 matrices.

The *diagonal elements* of a square matrix are those elements for which the row and column indices are the same. For example, the diagonal elements of the 3×3 matrix $\mathbf{M}$ are m_{11}, m_{22}, and m_{33}. The other elements are *non-diagonal* elements. The diagonal elements form the *diagonal* of the matrix:

$$\begin{bmatrix} m_{11} & m_{12} & m_{13} \\ m_{21} & m_{22} & m_{23} \\ m_{31} & m_{32} & m_{33} \end{bmatrix}$$

The diagonal of a 3×3 matrix

If all nondiagonal elements in a matrix are zero, then the matrix is a *diagonal matrix*. The following 4×4 matrix is diagonal:

$$\begin{bmatrix} 3 & 0 & 0 & 0 \\ 0 & 1 & 0 & 0 \\ 0 & 0 & -5 & 0 \\ 0 & 0 & 0 & 2 \end{bmatrix}.$$

A diagonal 4×4 matrix

A special diagonal matrix is the *identity matrix*. The identity matrix of
dimension n, denoted $\mathbf{I}_n$, is the $n \times n$ matrix with 1s on the diagonal and
0s elsewhere. For example, the 3×3 identity matrix is

The 3D identity matrix

$$\mathbf{I}_3 = \begin{bmatrix} 1 & 0 & 0 \\ 0 & 1 & 0 \\ 0 & 0 & 1 \end{bmatrix}.$$

Often, the context will make clear the dimension of the identity matrix
used in a particular situation. In these cases we omit the subscript and
refer to the identity matrix simply as $\mathbf{I}$.

The identity matrix is special because it is the multiplicative identity
element for matrices. (We discuss matrix multiplication in Section 4.1.6.)
The basic idea is that if you multiply a matrix by the identity matrix, you
get the original matrix. In this respect, the identity matrix is for matrices
what the number 1 is for scalars.

4.1.3   Vectors as Matrices

Matrices may have any positive number of rows and columns, including
one. We have already encountered matrices with one row or one column:
vectors! A vector of dimension n can be viewed either as a $1 \times n$ matrix, or
as an $n \times 1$ matrix. A $1 \times n$ matrix is known as a *row vector*, and an $n \times 1$
matrix is known as a *column vector*. Row vectors are written horizontally,
and column vectors are written vertically:

Row and column vectors

$$\begin{bmatrix} 1 & 2 & 3 \end{bmatrix} \qquad \begin{bmatrix} 4 \\ 5 \\ 6 \end{bmatrix}$$

Until now, we have used row and column notations interchangeably.
Indeed, geometrically they are identical, and in many cases the distinction
is not important. However, for reasons that will soon become apparent,
when we use vectors with matrices, we must be very clear if our vector is a
row or column vector.

4.1.4   Matrix Transposition

Given an $r \times c$ matrix $\mathbf{M}$, the *transpose* of $\mathbf{M}$, denoted $\mathbf{M}^T$, is the $c \times r$
matrix where the columns are formed from the rows of $\mathbf{M}$. In other words,
$\mathbf{M}^T{}_{ij} = \mathbf{M}_{ji}$. This "flips" the matrix diagonally. Equations (4.1) and (4.2)
show two examples of transposing matrices:

$$\begin{bmatrix} 1 & 2 & 3 \\ 4 & 5 & 6 \\ 7 & 8 & 9 \\ 10 & 11 & 12 \end{bmatrix}^{T} = \begin{bmatrix} 1 & 4 & 7 & 10 \\ 2 & 5 & 8 & 11 \\ 3 & 6 & 9 & 12 \end{bmatrix}, \tag{4.1}$$

Transposing matrices

$$\begin{bmatrix} a & b & c \\ d & e & f \\ g & h & i \end{bmatrix}^{T} = \begin{bmatrix} a & d & g \\ b & e & h \\ c & f & i \end{bmatrix}. \tag{4.2}$$

For vectors, transposition turns row vectors into column vectors and vice versa:

$$\begin{bmatrix} x & y & z \end{bmatrix}^{T} = \begin{bmatrix} x \\ y \\ z \end{bmatrix} \qquad \begin{bmatrix} x \\ y \\ z \end{bmatrix}^{T} = \begin{bmatrix} x & y & z \end{bmatrix}$$

Transposing converts between row and column vectors

Transposition notation is often used to write column vectors inline in a paragraph, such as $[1, 2, 3]^{T}$.

Let's make two fairly obvious, but significant, observations regarding matrix transposition.

- $(\mathbf{M}^{T})^{T} = \mathbf{M}$, for a matrix $\mathbf{M}$ of any dimension. In other words, if we transpose a matrix, and then transpose it again, we get the original matrix. This rule also applies to vectors.

- Any diagonal matrix $\mathbf{D}$ is equal to its transpose: $\mathbf{D}^{T} = \mathbf{D}$. This includes the identity matrix $\mathbf{I}$.

4.1.5 Multiplying a Matrix with a Scalar

A matrix $\mathbf{M}$ may be multiplied with a scalar k, resulting in a matrix of the same dimension as $\mathbf{M}$. We denote matrix multiplication with a scalar by placing the scalar and the matrix side-by-side, usually with the scalar on the left. No multiplication symbol is necessary. The multiplication takes place in a straightforward fashion: each element in the resulting matrix $k\mathbf{M}$ is the product of k and the corresponding element in $\mathbf{M}$. For example,

$$k\mathbf{M} = k \begin{bmatrix} m_{11} & m_{12} & m_{13} \\ m_{21} & m_{22} & m_{23} \\ m_{31} & m_{32} & m_{33} \\ m_{41} & m_{42} & m_{43} \end{bmatrix} = \begin{bmatrix} km_{11} & km_{12} & km_{13} \\ km_{21} & km_{22} & km_{23} \\ km_{31} & km_{32} & km_{33} \\ km_{41} & km_{42} & km_{43} \end{bmatrix}.$$

Multiplying a 4 × 3 matrix by a scalar

4.1.6   Multiplying Two Matrices

In certain situations, we can take the product of two matrices. The rules that govern when matrix multiplication is allowed and how the result is computed may at first seem bizarre.

An $r \times n$ matrix $\mathbf{A}$ may be multiplied by an $n \times c$ matrix $\mathbf{B}$. The result, denoted $\mathbf{AB}$, is an $r \times c$ matrix. For example, assume that $\mathbf{A}$ is a 4×2 matrix, and $\mathbf{B}$ is a 2×5 matrix. Then $\mathbf{AB}$ is a 4×5 matrix:

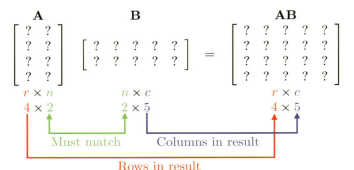

If the number of columns in $\mathbf{A}$ does not match the number of rows in $\mathbf{B}$, then the multiplication $\mathbf{AB}$ is not defined (although $\mathbf{BA}$ may be possible).

Matrix multiplication is computed as follows: let the matrix $\mathbf{C}$ be the $r \times c$ product $\mathbf{AB}$ of the $r \times n$ matrix $\mathbf{A}$ with the $n \times c$ matrix $\mathbf{B}$. Then each element c_{ij} is equal to the vector dot product of row i of $\mathbf{A}$ with column j of $\mathbf{B}$:

$$c_{ij} = \sum_{k=1}^{n} a_{ik} b_{kj}.$$

This looks complicated, but there is a simple pattern. For each element c_{ij} in the result, locate row i in $\mathbf{A}$ and column j in $\mathbf{B}$. Multiply the corresponding elements of the row and column, and sum the products. c_{ij} is equal to this sum, which is equivalent to the dot product of row i in $\mathbf{A}$ with column j in $\mathbf{B}$.

Let's look at an example of how to compute c_{24}. The element in row 2 and column 4 of $\mathbf{C}$ is equal to the dot product of row 2 of $\mathbf{A}$ with the column 4 of $\mathbf{B}$:

$$
\begin{bmatrix}
c_{11} & c_{12} & c_{13} & c_{14} & c_{15} \\
c_{21} & c_{22} & c_{23} & c_{24} & c_{25} \\
c_{31} & c_{32} & c_{33} & c_{34} & c_{35} \\
c_{41} & c_{42} & c_{43} & c_{44} & c_{45}
\end{bmatrix}
=
\begin{bmatrix}
a_{11} & a_{12} \\
a_{21} & a_{22} \\
a_{31} & a_{32} \\
a_{41} & a_{42}
\end{bmatrix}
\begin{bmatrix}
b_{11} & b_{12} & b_{13} & b_{14} & b_{15} \\
b_{21} & b_{22} & b_{23} & b_{24} & b_{25}
\end{bmatrix}
$$

$$c_{24} = a_{21} b_{14} + a_{22} b_{24}$$

Another way to help remember the pattern for matrix multiplication is to write $\mathbf{B}$ above $\mathbf{C}$. This aligns the proper row from $\mathbf{A}$ with a column from $\mathbf{B}$ for each element in the result $\mathbf{C}$:

$$\begin{bmatrix} b_{11} & b_{12} & \boxed{b_{13}} & b_{14} & b_{15} \\ b_{21} & b_{22} & \boxed{b_{23}} & b_{24} & b_{25} \end{bmatrix}$$

$$\begin{bmatrix} a_{11} & a_{12} \\ a_{21} & a_{22} \\ a_{31} & a_{32} \\ \boxed{a_{41}} & \boxed{a_{42}} \end{bmatrix} \quad \begin{bmatrix} c_{11} & c_{12} & c_{13} & c_{14} & c_{15} \\ c_{21} & c_{22} & c_{23} & c_{24} & c_{25} \\ c_{31} & c_{32} & c_{33} & c_{34} & c_{35} \\ c_{41} & c_{42} & \boxed{c_{43}} & c_{44} & c_{45} \end{bmatrix}$$

$$c_{43} = a_{41}b_{13} + a_{42}b_{23}$$

For geometric applications, we are particularly interested in multiplying square matrices—the 2×2 and 3×3 cases being especially important. Equation (4.3) gives the complete equation for 2×2 matrix multiplication:

2×2 matrix multiplication

$$\mathbf{AB} = \begin{bmatrix} a_{11} & a_{12} \\ a_{21} & a_{22} \end{bmatrix} \begin{bmatrix} b_{11} & b_{12} \\ b_{21} & b_{22} \end{bmatrix}$$

$$= \begin{bmatrix} a_{11}b_{11} + a_{12}b_{21} & a_{11}b_{12} + a_{12}b_{22} \\ a_{21}b_{11} + a_{22}b_{21} & a_{21}b_{12} + a_{22}b_{22} \end{bmatrix}. \quad (4.3)$$

Let's look at a 2×2 example with some real numbers:

$$\mathbf{A} = \begin{bmatrix} -3 & 0 \\ 5 & 1/2 \end{bmatrix}, \quad \mathbf{B} = \begin{bmatrix} -7 & 2 \\ 4 & 6 \end{bmatrix},$$

$$\mathbf{AB} = \begin{bmatrix} -3 & 0 \\ 5 & 1/2 \end{bmatrix} \begin{bmatrix} -7 & 2 \\ 4 & 6 \end{bmatrix}$$

$$= \begin{bmatrix} (-3)(-7) + (0)(4) & (-3)(2) + (0)(6) \\ (5)(-7) + (1/2)(4) & (5)(2) + (1/2)(6) \end{bmatrix} = \begin{bmatrix} 21 & -6 \\ -33 & 13 \end{bmatrix}.$$

Applying the general matrix multiplication formula to the 3×3 case produces

3 × 3 matrix multiplication

$$\mathbf{AB} = \begin{bmatrix} a_{11} & a_{12} & a_{13} \\ a_{21} & a_{22} & a_{23} \\ a_{31} & a_{32} & a_{33} \end{bmatrix} \begin{bmatrix} b_{11} & b_{12} & b_{13} \\ b_{21} & b_{22} & b_{23} \\ b_{31} & b_{32} & b_{33} \end{bmatrix}$$

$$= \begin{bmatrix} a_{11}b_{11} + a_{12}b_{21} + a_{13}b_{31} & a_{11}b_{12} + a_{12}b_{22} + a_{13}b_{32} & a_{11}b_{13} + a_{12}b_{23} + a_{13}b_{33} \\ a_{21}b_{11} + a_{22}b_{21} + a_{23}b_{31} & a_{21}b_{12} + a_{22}b_{22} + a_{23}b_{32} & a_{21}b_{13} + a_{22}b_{23} + a_{23}b_{33} \\ a_{31}b_{11} + a_{32}b_{21} + a_{33}b_{31} & a_{31}b_{12} + a_{32}b_{22} + a_{33}b_{32} & a_{31}b_{13} + a_{32}b_{23} + a_{33}b_{33} \end{bmatrix} .$$

Here is a 3×3 example with some real numbers:

$$\mathbf{A} = \begin{bmatrix} 1 & -5 & 3 \\ 0 & -2 & 6 \\ 7 & 2 & -4 \end{bmatrix} , \quad \mathbf{B} = \begin{bmatrix} -8 & 6 & 1 \\ 7 & 0 & -3 \\ 2 & 4 & 5 \end{bmatrix} ;$$

$$\mathbf{AB} = \begin{bmatrix} 1 & -5 & 3 \\ 0 & -2 & 6 \\ 7 & 2 & -4 \end{bmatrix} \begin{bmatrix} -8 & 6 & 1 \\ 7 & 0 & -3 \\ 2 & 4 & 5 \end{bmatrix}$$

$$= \begin{bmatrix} 1 \cdot (-8) + (-5) \cdot 7 + 3 \cdot 2 & 1 \cdot 6 + (-5) \cdot 0 + 3 \cdot 4 & 1 \cdot 1 + (-5) \cdot (-3) + 3 \cdot 5 \\ 0 \cdot (-8) + (-2) \cdot 7 + 6 \cdot 2 & 0 \cdot 6 + (-2) \cdot 0 + 6 \cdot 4 & 0 \cdot 1 + (-2) \cdot (-3) + 6 \cdot 5 \\ 7 \cdot (-8) + 2 \cdot 7 + (-4) \cdot 2 & 7 \cdot 6 + 2 \cdot 0 + (-4) \cdot 4 & 7 \cdot 1 + 2 \cdot (-3) + (-4) \cdot 5 \end{bmatrix}$$

$$= \begin{bmatrix} -37 & 18 & 31 \\ -2 & 24 & 36 \\ -50 & 26 & -19 \end{bmatrix} .$$

Beginning in Section 6.4 we also use 4×4 matrices.

A few notes concerning matrix multiplication are of interest:

- Multiplying any matrix $\mathbf{M}$ by a square matrix $\mathbf{S}$ on either side results in a matrix of the same size as $\mathbf{M}$, provided that the sizes of the matrices are such that the multiplication is allowed. If $\mathbf{S}$ is the identity matrix $\mathbf{I}$, then the result is the original matrix $\mathbf{M}$:

$$\mathbf{MI} = \mathbf{IM} = \mathbf{M}.$$

(That's the reason it's called the *identity* matrix!)

- Matrix multiplication is not commutative. In general,

$$\mathbf{AB} \neq \mathbf{BA}.$$

- Matrix multiplication is associative:

$$(\mathbf{AB})\mathbf{C} = \mathbf{A}(\mathbf{BC}).$$

(Of course, this assumes that the sizes of $\mathbf{A}$, $\mathbf{B}$, and $\mathbf{C}$ are such that multiplication is allowed. Note that if $(\mathbf{A}\mathbf{B})\mathbf{C}$ is defined, then $\mathbf{A}(\mathbf{B}\mathbf{C})$ is always defined as well.) The associativity of matrix multiplication extends to multiple matrices. For example,

$$\begin{aligned} \mathbf{ABCDEF} &= ((((\mathbf{AB})\mathbf{C})\mathbf{D})\mathbf{E})\mathbf{F} \\ &= \mathbf{A}((((\mathbf{BC})\mathbf{D})\mathbf{E})\mathbf{F}) \\ &= (\mathbf{AB})(\mathbf{CD})(\mathbf{EF}). \end{aligned}$$

It is interesting to note that although all parenthesizations compute the correct result, some groupings require fewer scalar multiplications than others.[1]

- Matrix multiplication also associates with multiplication by a scalar or a vector:

$$\begin{aligned} (k\mathbf{A})\mathbf{B} &= k(\mathbf{AB}) = \mathbf{A}(k\mathbf{B}), \\ (\mathbf{v}\mathbf{A})\mathbf{B} &= \mathbf{v}(\mathbf{AB}). \end{aligned}$$

- Transposing the product of two matrices is the same as taking the product of their transposes in reverse order:

$$(\mathbf{AB})^T = \mathbf{B}^{\mathrm{T}}\mathbf{A}^{\mathrm{T}}.$$

This can be extended to more than two matrices:

$$(\mathbf{M}_1\mathbf{M}_2\cdots\mathbf{M}_{n-1}\mathbf{M}_n)^T = \mathbf{M}_n{}^{\mathrm{T}}\mathbf{M}_{n-1}{}^{\mathrm{T}}\cdots\mathbf{M}_2{}^{\mathrm{T}}\mathbf{M}_1{}^{\mathrm{T}}.$$

4.1.7 Multiplying a Vector and a Matrix

Since a vector can be considered a matrix with one row or one column, we can multiply a vector and a matrix by applying the rules discussed in the previous section. Now it becomes very important whether we are using row or column vectors. Equations (4.4)–(4.7) show how 3D row and column vectors may be pre- or post-multiplied by a 3×3 matrix:

---

[1]The problem of finding the parenthesization that minimizes the number of scalar multiplications is known as the *matrix chain* problem.

$$\begin{bmatrix} x & y & z \end{bmatrix} \begin{bmatrix} m_{11} & m_{12} & m_{13} \\ m_{21} & m_{22} & m_{23} \\ m_{31} & m_{32} & m_{33} \end{bmatrix} = \tag{4.4}$$

$$\begin{bmatrix} xm_{11} + ym_{21} + zm_{31} & xm_{12} + ym_{22} + zm_{32} & xm_{13} + ym_{23} + zm_{33} \end{bmatrix};$$

$$\begin{bmatrix} m_{11} & m_{12} & m_{13} \\ m_{21} & m_{22} & m_{23} \\ m_{31} & m_{32} & m_{33} \end{bmatrix} \begin{bmatrix} x \\ y \\ z \end{bmatrix} = \begin{bmatrix} xm_{11} + ym_{12} + zm_{13} \\ xm_{21} + ym_{22} + zm_{23} \\ xm_{31} + ym_{32} + zm_{33} \end{bmatrix}; \tag{4.5}$$

$$\begin{bmatrix} m_{11} & m_{12} & m_{13} \\ m_{21} & m_{22} & m_{23} \\ m_{31} & m_{32} & m_{33} \end{bmatrix} \begin{bmatrix} x & y & z \end{bmatrix} = \text{(undefined)}; \tag{4.6}$$

$$\begin{bmatrix} x \\ y \\ z \end{bmatrix} \begin{bmatrix} m_{11} & m_{12} & m_{13} \\ m_{21} & m_{22} & m_{23} \\ m_{31} & m_{32} & m_{33} \end{bmatrix} = \text{(undefined)}. \tag{4.7}$$

As you can see, when we multiply a row vector on the left by a matrix on the right, as in Equation (4.4), the result is a row vector. When we multiply a matrix on the left by a column vector on the right, as in Equation (4.5), the result is a column vector. (Please observe that this result *is* a column vector, even though it looks like a matrix.) The other two combinations are not allowed: you cannot multiply a matrix on the left by a row vector on the right, nor can you multiply a column vector on the left by a matrix on the right.

Let's make a few interesting observations regarding vector-times-matrix multiplication. First, each element in the resulting vector is the dot product of the original vector with a single row or column from the matrix.

Second, each element in the matrix determines how much "weight" a particular element in the input vector contributes to an element in the output vector. For example, in Equation (4.4) when row vectors are used, m_{12} controls how much of the input x value goes towards the output y value.

Third, vector-times-matrix multiplication distributes over vector addition, that is, for vectors $\mathbf{v}$, $\mathbf{w}$, and matrix $\mathbf{M}$,

$$(\mathbf{v} + \mathbf{w})\mathbf{M} = \mathbf{v}\mathbf{M} + \mathbf{w}\mathbf{M}.$$

Finally, and perhaps most important of all, the result of the multiplication is a linear combination of the rows or columns of the matrix. For example, in Equation (4.5), when column vectors are used, the resulting column vector can be interpreted as a linear combination of the columns of the matrix, where the coefficients come from the vector operand. This is a key fact, not just for our purposes but also for linear algebra in general, so bear it in mind. We will return to it shortly.

4.1.8 Row versus Column Vectors

This section explains why the distinction between row and column vectors is significant and gives our rationale for preferring row vectors. In Equation (4.4), when we multiply a row vector on the left with matrix on the right, we get the row vector

$$\begin{bmatrix} xm_{11} + ym_{21} + zm_{31} & xm_{12} + ym_{22} + zm_{32} & xm_{13} + ym_{23} + zm_{33} \end{bmatrix}.$$

Compare that with the result from Equation (4.5), when a column vector on the right is multiplied by a matrix on the left:

$$\begin{bmatrix} xm_{11} + ym_{12} + zm_{13} \\ xm_{21} + ym_{22} + zm_{23} \\ xm_{31} + ym_{32} + zm_{33} \end{bmatrix}.$$

Disregarding the fact that one is a row vector and the other is a column vector, the values for the components of the vector are *not* the same! This is why the distinction between row and column vectors is so important.

Although some matrices in video game programming do represent arbitrary systems of equations, a much larger majority are transformation matrices of the type we have been describing, which express relationships between coordinate spaces. For this purpose, we find row vectors to be preferable for the "eminently sensible" reason [1] that the order of transformations reads like a sentence from left to right. This is especially important when more than one transformation takes place. For example, if we wish to transform a vector $\mathbf{v}$ by the matrices $\mathbf{A}$, $\mathbf{B}$, and $\mathbf{C}$, in that order, we write $\mathbf{vABC}$, with the matrices listed in order of transformation from left to right. If column vectors are used, then the vector is on the right, and so the transformations will occur in order from right to left. In this case, we would write $\mathbf{CBAv}$. We discuss concatenation of multiple transformation matrices in detail in Section 5.6.

Unfortunately, row vectors lead to very "wide" equations; using column vectors on the right certainly makes things look better, especially as the dimension increases. (Compare the ungainliness of Equation (4.4) with the sleekness of Equation (4.5).) Perhaps this is why column vectors are the near universal standard in practically every other discipline. For most video game programming, however, readable computer code is more important than readable equations. For this reason, in this book we use row vectors in almost all cases where the distinction is relevant. Our limited use of column vectors is for aesthetic purposes, when either there are no matrices involved or those matrices are not transformation matrices and the left-to-right reading order is not helpful.

As expected, different authors use different conventions. Many graphics books and application programming interfaces (APIs), such as DirectX, use row vectors. But other APIs, such as OpenGL and the customized ports of OpenGL onto various consoles, use column vectors. And, as we have said, nearly every other science that uses linear algebra prefers column vectors. So be very careful when using someone else's equation or source code that you know whether it assumes row or column vectors.

If a book uses column vectors, its equations for matrices will be transposed compared to the equations we present in this book. Also, when column vectors are used, vectors are pre-multiplied by a matrix, as opposed to the convention chosen in this book, to multiply row vectors by a matrix on the right. This causes the order of multiplication to be reversed between the two styles when multiple matrices and vectors are multiplied together. For example, the multiplication $\mathbf{vABC}$ is valid only with row vectors. The corresponding multiplication would be written $\mathbf{CBAv}$ if column vectors were used. (Again, note that in this case $\mathbf{A}$, $\mathbf{B}$, and $\mathbf{C}$ would be transposed as compared to these matrices in the row vector case.)

Mistakes like this involving transposition can be a common source of frustration when programming 3D math. Luckily, with properly designed C++ classes, direct access to the individual matrix elements is seldom needed, and these types of errors can be minimized.

4.2   Geometric Interpretation of Matrix

In general, a square matrix can describe any *linear transformation*. In Section 5.7.1, we provide a complete definition of linear transformation, but for now, it suffices to say that a linear transformation preserves straight and parallel lines, and that there is no translation—that is, the origin does not move. However, other properties of the geometry, however, such as lengths, angles, areas, and volumes, *are* possibly altered by the transformation. In a nontechnical sense, a linear transformation may "stretch" the coordinate space, but it doesn't "curve" or "warp" it. This is a very useful set of transformations, including

- rotation

- scale

- orthographic projection

- reflection

- shearing

Chapter 5 derives matrices that perform all of these operations. For now, though, let's just attempt a general understanding of the relationship between a matrix and the transformation it represents.

The quotation at the beginning of this chapter is not only a line from a great movie, it's true for linear algebra matrices as well. Until you develop an ability to visualize a matrix, it will just be nine numbers in a box. We have stated that a matrix represents a coordinate space transformation. So when we visualize the matrix, we are visualizing the transformation, the new coordinate system. But what does this transformation look like? What is the relationship between a particular 3D transformation (i.e. rotation, shearing, etc.) and those nine numbers inside a 3×3 matrix? How can we construct a matrix to perform a given transform (other than just copying the equations blindly out of a book)?

To begin to answer this question, let's watch what happens when the standard basis vectors $\mathbf{i} = [1, 0, 0]$, $\mathbf{j} = [0, 1, 0]$, and $\mathbf{k} = [0, 0, 1]$ are multiplied by an arbitrary matrix $\mathbf{M}$:

$$\mathbf{iM} = \begin{bmatrix} 1 & 0 & 0 \end{bmatrix} \begin{bmatrix} m_{11} & m_{12} & m_{13} \\ m_{21} & m_{22} & m_{23} \\ m_{31} & m_{32} & m_{33} \end{bmatrix} = \begin{bmatrix} m_{11} & m_{12} & m_{13} \end{bmatrix};$$

$$\mathbf{jM} = \begin{bmatrix} 0 & 1 & 0 \end{bmatrix} \begin{bmatrix} m_{11} & m_{12} & m_{13} \\ m_{21} & m_{22} & m_{23} \\ m_{31} & m_{32} & m_{33} \end{bmatrix} = \begin{bmatrix} m_{21} & m_{22} & m_{23} \end{bmatrix};$$

$$\mathbf{kM} = \begin{bmatrix} 0 & 0 & 1 \end{bmatrix} \begin{bmatrix} m_{11} & m_{12} & m_{13} \\ m_{21} & m_{22} & m_{23} \\ m_{31} & m_{32} & m_{33} \end{bmatrix} = \begin{bmatrix} m_{31} & m_{32} & m_{33} \end{bmatrix}.$$

In other words, the first row of $\mathbf{M}$ contains the result of performing the transformation on $\mathbf{i}$, the second row is the result of transforming $\mathbf{j}$, and the last row is the result of transforming $\mathbf{k}$.

Once we know what happens to those basis vectors, we know everything about the transformation! This is because any vector can be written as a linear combination of the standard basis, as

$$\mathbf{v} = v_x \mathbf{i} + v_y \mathbf{j} + v_z \mathbf{k}.$$

Multiplying this expression by our matrix on the right, we have

$$
\begin{aligned}
\mathbf{v}\mathbf{M} &= \left(v_x\mathbf{i} + v_y\mathbf{j} + v_z\mathbf{k}\right)\mathbf{M} \\
&= (v_x\mathbf{i})\mathbf{M} + (v_y\mathbf{j})\mathbf{M} + (v_z\mathbf{k})\mathbf{M} \qquad\qquad (4.8)\\
&= v_x(\mathbf{i}\mathbf{M}) + v_y(\mathbf{j}\mathbf{M}) + v_z(\mathbf{k}\mathbf{M}) \\
&= v_x\begin{bmatrix} m_{11} & m_{12} & m_{13}\end{bmatrix} + v_y\begin{bmatrix} m_{21} & m_{22} & m_{23}\end{bmatrix} + v_z\begin{bmatrix} m_{31} & m_{32} & m_{33}\end{bmatrix}.
\end{aligned}
$$

Here we have confirmed an observation made in Section 4.1.7: the result of a vector × matrix multiplication is a linear combination of the rows of the matrix. The key is to *interpret those row vectors as basis vectors*. In this interpretation, matrix multiplication is simply a compact way to encode the operations for coordinate space transformations developed in Section 3.3.3. A small change of notation will make this connection more explicit. Remember that we introduced the convention to use the symbols **p**, **q**, and **r** to refer to a set of basis vectors. Putting these vectors as rows in our matrix **M**, we can rewrite the last line of Equation (4.8) as

$$
\mathbf{v}\mathbf{M} = \begin{bmatrix} v_x & v_y & v_z\end{bmatrix}\begin{bmatrix} -\mathbf{p}- \\ -\mathbf{q}- \\ -\mathbf{r}- \end{bmatrix} = v_x\mathbf{p} + v_y\mathbf{q} + v_z\mathbf{r}.
$$

Let's summarize what we have said.

By understanding how the matrix transforms the standard basis vectors, we know everything there is to know about the transformation. Since the results of transforming the standard basis are simply the rows[2] of the matrix, we interpret those rows as the basis vectors of a coordinate space.

We now have a simple way to take an arbitrary matrix and visualize what sort of transformation the matrix represents. Let's look at a couple of examples—first, a 2D example to get ourselves warmed up, and then a full-fledged 3D example. Examine the following 2 × 2 matrix:

$$
\mathbf{M} = \begin{bmatrix} 2 & 1 \\ -1 & 2 \end{bmatrix}.
$$

---

[2]It's rows in this book. If you're using column vectors, it's the columns of the matrix.

What sort of transformation does this matrix represent? First, let's extract
the basis vectors **p** and **q** from the rows of the matrix:

$$\mathbf{p} = \begin{bmatrix} 2 & 1 \end{bmatrix};$$
$$\mathbf{q} = \begin{bmatrix} -1 & 2 \end{bmatrix}.$$

Figure 4.1 shows these vectors in the Cartesian plane, along with the "original" basis vectors (the x-axis and y-axis), for reference.

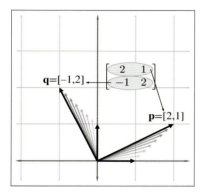

As Figure 4.1 illustrates, the $+x$ basis vector is transformed into the vector
labeled **p** above, and the y basis vector
is transformed into the vector labeled
q. So one way to visualize a matrix in
2D is to visualize the L shape formed
by the row vectors. In this example,
we can easily see that part of the transformation represented by matrix **M** is
a counterclockwise rotation of approximately 26.5°.

Of course, *all* vectors are affected
by a linear transformation, not just the
basis vectors. We can get a very good
idea what this transformation looks like
from the L, and we can gain further insight on the effect the transformation
has on the rest of the vectors by completing the 2D parallelogram formed
by the basis vectors, as shown in Figure 4.2.

Figure 4.1
Visualizing the row vectors of a 2D
transform matrix

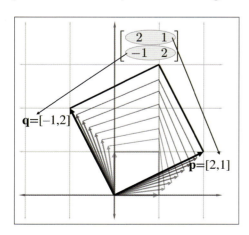

Figure 4.2
The 2D parallelogram formed by the
rows of a matrix

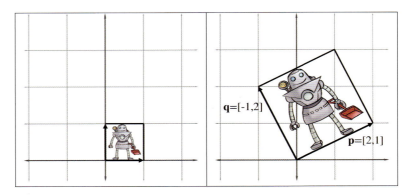

Figure 4.3. Drawing an object inside the box helps visualize the transformation

This parallelogram is also known as a "skew box." Drawing an object inside the box can also help, as illustrated in Figure 4.3.

Now it is clear that our example matrix **M** not only rotates the coordinate space, it also scales it.

We can extend the techniques we used to visualize 2D transformations into 3D. In 2D, we had two basis vectors that formed an L—in 3D, we have three basis vectors, and they form a "tripod." First, let's show an object before transformation. Figure 4.4 shows a teapot, a unit cube, and the basis vectors in the "identity" position.

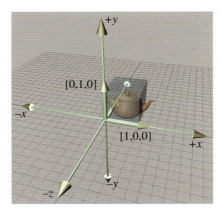

Figure 4.4
Teapot, unit cube, and basis vectors before transformation

(To avoid cluttering up the diagram, we have not labeled the $+z$ basis vector $[0, 0, 1]$, which is partially obscured by the teapot and cube.)

Now consider the 3D transformation matrix

$$\begin{bmatrix} 0.707 & -0.707 & 0 \\ 1.250 & 1.250 & 0 \\ 0 & 0 & 1 \end{bmatrix}.$$

Extracting the basis vectors from the rows of the matrix, we can visualize the transformation represented by this matrix. The transformed basis vectors, cube, and teapot are shown in Figure 4.5.

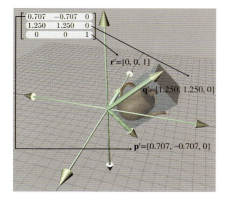

Figure 4.5
Teapot, unit cube, and basis vectors after transformation

As we can see, the transformation consists of a clockwise rotation of 45° about the z-axis as well as a nonuniform scale that makes the teapot "taller" than it was originally. Notice that the $+z$ basis vector is unaltered by the transformation, because the third row of the matrix is $[0, 0, 1]$.

By interpreting the rows of a matrix as basis vectors, we have a tool for deconstructing a matrix. But we also have a tool for *constructing* one! Given a desired transformation (i.e., rotation, scale, and so on), we can derive a matrix that represents that transformation. All we have to do is figure out what the transformation does to the basis vectors, and then place those transformed basis vectors into the rows of a matrix. We use this tool repeatedly in Chapter 5 to derive the matrices to perform basic linear transformations such as rotation, scale, shear, and reflection.

The bottom line about transformation matrices is this: there's nothing especially magical about matrices. Once we understand that coordinates are best understood as coefficients in a linear combination of the basis vectors (see Section 3.3.3), we really know all the math we need to know to do transformations. So from one perspective, matrices are just a compact way to write things down. A slightly less obvious but much more compelling reason to cast transformations in matrix notation is to take advantage of the large general-purpose toolset from linear algebra. For example, we can take simple transformations and derive more complicated transformations through matrix concatenation; more on this in Section 5.6.

Before we move on, let's review the key concepts of Section 4.2.

- The rows of a square matrix can be interpreted as the basis vectors of a coordinate space.

- To transform a vector from the original coordinate space to the new coordinate space, we multiply the vector by the matrix.

- The transformation from the original coordinate space to the coordinate space defined by these basis vectors is a linear transformation. A linear transformation preserves straight lines, and parallel lines remain parallel. However, angles, lengths, areas, and volumes may be altered after transformation.

- Multiplying the zero vector by any square matrix results in the zero vector. Therefore, the linear transformation represented by a square matrix has the same origin as the original coordinate space—the transformation does not contain translation.

- We can visualize a matrix by visualizing the basis vectors of the coordinate space after transformation. These basis vectors form an 'L' in 2D, and a tripod in 3D. Using a box or auxiliary object also helps in visualization.

4.3   The Bigger Picture of Linear Algebra

At the start of Chapter 2, we warned you that in this book we are focusing on just one small corner of the field of linear algebra—the geometric applications of vectors and matrices. Now that we've introduced the nuts and bolts, we'd like to say something about the bigger picture and how our part relates to it.

Linear algebra was invented to manipulate and solve systems of linear equations. For example, a typical introductory problem in a traditional course on linear algebra is to solve a system of equations such as

$$-5x_1 + x_2 + x_3 = -10,$$
$$2x_1 + 2x_2 + 4x_3 = 12,$$
$$x_1 - 3x_3 = 9,$$

which has the solution

$$x_1 = 3,$$
$$x_2 = 7,$$
$$x_3 = -2.$$

Matrix notation was invented to avoid the tedium involved in duplicating every x and $=$. For example, the system above can be more quickly written as

$$\begin{bmatrix} -5 & 1 & 1 \\ 2 & 2 & 4 \\ 1 & 0 & -3 \end{bmatrix} \begin{bmatrix} x_1 \\ x_2 \\ x_3 \end{bmatrix} = \begin{bmatrix} -10 \\ 12 \\ 9 \end{bmatrix}.$$

Perhaps the most direct and obvious place in a video game where a large system of equations must be solved is in the physics engine. The constraints to enforce nonpenetration and satisfy user-requested joints become a system of equations relating the velocities of the dynamic bodies. This large system[3] is then solved each and every simulation frame. Another common place for traditional linear algebra methods to appear is in least squares approximation and other data-fitting applications.

Systems of equations can appear where you don't expect them. Indeed, linear algebra has exploded in importance with the vast increase in computing power in the last half century because many difficult problems that were previously neither discrete nor linear are being approximated through methods that are both, such as the finite element method. The challenge begins with knowing how to transform the original problem into a matrix problem in the first place, but the resulting systems are often very large and can be difficult to solve quickly and accurately. Numeric stability becomes a factor in the choice of algorithms. The matrices that arise in practice are not boxes full of random numbers; rather, they express organized relationships and have a great deal of structure. Exploiting this structure artfully is the key to achieving speed and accuracy. The diversity of the types of structure that appear in applications explains why there is so very much to know about linear algebra, especially numerical linear algebra.

This book is intended to fill a gap by providing the geometric intuition that is the bread and butter of video game programming but is left out of

---

[3]It's a system of *inequalities*, but similar principles apply.

most linear algebra textbooks. However, we certainly know there is a larger world out there for you. Although traditional linear algebra and systems of equations do not play a prominent role for basic video game programming, they are essential for many advanced areas. Consider some of the technologies that are generating buzz today: fluid, cloth, and hair simulations (and rendering); more robust procedural animation of characters; real-time global illumination; machine vision; gesture recognition; and many more. What these seemingly diverse technologies all have in common is that they involve difficult linear algebra problems.

One excellent resource for learning the bigger picture of linear algebra and scientific computing is Professor Gilbert Strang's series of lectures, which can be downloaded free from MIT OpenCourseWare at ocw.mit.edu. He offers a basic undergraduate linear algebra course as well as graduate courses on computational science and engineering. The companion textbooks he writes for his classes [67, 68] are enjoyable books aimed at engineers (rather than math sticklers) and are recommended, but be warned that his writing style is a sort of shorthand that you might have trouble understanding without the lectures.

4.4   Exercises

(Answers on page 759.)

Use the following matrices for questions 1–3:

$$
\mathbf{A} = \begin{bmatrix} 13 & 4 & -8 \\ 12 & 0 & 6 \\ -3 & -1 & 5 \\ 10 & -2 & 5 \end{bmatrix} \quad
\mathbf{B} = \begin{bmatrix} k_x & 0 & 0 \\ 0 & k_y & 0 \\ 0 & 0 & k_z \end{bmatrix} \quad
\mathbf{C} = \begin{bmatrix} 15 & 8 \\ -7 & 3 \end{bmatrix}
$$

$$
\mathbf{D} = \begin{bmatrix} a & g \\ b & h \\ c & i \\ d & j \\ f & k \end{bmatrix} \quad
\mathbf{E} = \begin{bmatrix} 0 & 1 & 3 \end{bmatrix} \quad
\mathbf{F} = \begin{bmatrix} x \\ y \\ z \\ w \end{bmatrix}
$$

$$
\mathbf{G} = \begin{bmatrix} 10 & 20 & 30 & 1 \end{bmatrix} \quad
\mathbf{H} = \begin{bmatrix} \alpha \\ \beta \\ \gamma \end{bmatrix}
$$

1. For each matrix, give the dimensions of the matrix and identify whether it is square and/or diagonal.

2. Transpose each matrix.

3. Find all the possible pairs of matrices that can be legally multiplied, and give the dimensions of the resulting product. Include "pairs" in which a matrix is multiplied by itself. (Hint: there are 14 pairs.)

4. Compute the following matrix products. If the product is not possible, just say so.

(a) $\begin{bmatrix} 1 & -2 \\ 5 & 0 \end{bmatrix} \begin{bmatrix} -3 & 7 \\ 4 & 1/3 \end{bmatrix}$

(b) $\begin{bmatrix} 6 & -7 \\ -4 & 5 \end{bmatrix} \begin{bmatrix} 3 & 3 \end{bmatrix}$

(c) $\begin{bmatrix} 3 & -1 & 4 \end{bmatrix} \begin{bmatrix} -2 & 0 & 3 \\ 5 & 7 & -6 \\ 1 & -4 & 2 \end{bmatrix}$

(d) $\begin{bmatrix} x & y & z & w \end{bmatrix} \begin{bmatrix} 1 & 0 & 0 & 0 \\ 0 & 1 & 0 & 0 \\ 0 & 0 & 1 & 0 \\ 0 & 0 & 0 & 1 \end{bmatrix}$

(e) $\begin{bmatrix} 7 & -2 & 7 & 3 \end{bmatrix} \begin{bmatrix} -5 \\ 1 \end{bmatrix}$

(f) $\begin{bmatrix} 1 & 0 \\ 0 & 1 \end{bmatrix} \begin{bmatrix} m_{11} & m_{12} \\ m_{21} & m_{22} \end{bmatrix}$

(g) $\begin{bmatrix} 3 & 3 \end{bmatrix} \begin{bmatrix} 6 & -7 \\ -4 & 5 \end{bmatrix}$

(h) $\begin{bmatrix} a_{11} & a_{12} & a_{13} \\ a_{21} & a_{22} & a_{23} \\ a_{31} & a_{32} & a_{33} \end{bmatrix} \begin{bmatrix} b_{11} & b_{12} & b_{13} \\ b_{21} & b_{22} & b_{23} \end{bmatrix}$

5. For each of the following matrices, multiply on the left by the row vector $[5, -1, 2]$. Then consider whether multiplication on the right by the column vector $[5, -1, 2]^{\mathrm{T}}$ will give the same or a different result. Finally, perform this multiplication to confirm or correct your expectation.

(a) $\begin{bmatrix} 1 & 0 & 0 \\ 0 & 1 & 0 \\ 0 & 0 & 1 \end{bmatrix}$

(b) $\begin{bmatrix} 2 & 5 & -3 \\ 1 & 7 & 1 \\ -2 & -1 & 4 \end{bmatrix}$

(c) $\begin{bmatrix} 1 & 7 & 2 \\ 7 & 0 & -3 \\ 2 & -3 & -1 \end{bmatrix}$

This is an example of a *symmetric* matrix. A square matrix is symmetric if $\mathbf{A}^{\mathrm{T}} = \mathbf{A}$.

(d) $\begin{bmatrix} 0 & -4 & 3 \\ 4 & 0 & -1 \\ -3 & 1 & 0 \end{bmatrix}$

This is an example of a *skew symmetric* or *antisymmetric* matrix. A square matrix is skew symmetric if $\mathbf{A}^{\mathrm{T}} = -\mathbf{A}$. This implies that the diagonal elements of a skew symmetric matrix must be 0.

6. Manipulate the following matrix expressions to remove the parentheses.

   (a) $\left(\left(\mathbf{A}^{\mathrm{T}} \right)^{\mathrm{T}} \right)^{\mathrm{T}}$

   (b) $\left(\mathbf{B} \mathbf{A}^{\mathrm{T}} \right)^{\mathrm{T}} \left(\mathbf{C} \mathbf{D}^{\mathrm{T}} \right)$

   (c) $\left(\left(\mathbf{D}^{\mathrm{T}} \mathbf{C}^{\mathrm{T}} \right) \left(\mathbf{A} \mathbf{B} \right)^{\mathrm{T}} \right)^{\mathrm{T}}$

   (d) $\left(\left(\mathbf{A} \mathbf{B} \right)^{\mathrm{T}} \left(\mathbf{C} \mathbf{D} \mathbf{E} \right)^{\mathrm{T}} \right)^{\mathrm{T}}$

7. Describe the transformation $\mathbf{a}\mathbf{M} = \mathbf{b}$ represented by each of the following matrices.

   (a) $\mathbf{M} = \begin{bmatrix} 0 & -1 \\ 1 & 0 \end{bmatrix}$

   (b) $\mathbf{M} = \begin{bmatrix} \frac{\sqrt{2}}{2} & \frac{\sqrt{2}}{2} \\ -\frac{\sqrt{2}}{2} & \frac{\sqrt{2}}{2} \end{bmatrix}$

   (c) $\mathbf{M} = \begin{bmatrix} 2 & 0 \\ 0 & 2 \end{bmatrix}$

   (d) $\mathbf{M} = \begin{bmatrix} 4 & 0 \\ 0 & 7 \end{bmatrix}$

   (e) $\mathbf{M} = \begin{bmatrix} -1 & 0 \\ 0 & 1 \end{bmatrix}$

   (f) $\mathbf{M} = \begin{bmatrix} 0 & -2 \\ 2 & 0 \end{bmatrix}$

8. For 3D row vectors $\mathbf{a}$ and $\mathbf{b}$, construct a 3×3 matrix $\mathbf{M}$ such that $\mathbf{a} \times \mathbf{b} = \mathbf{a}\mathbf{M}$. That is, show that the cross product of $\mathbf{a}$ and $\mathbf{b}$ can be represented as the matrix product $\mathbf{a}\mathbf{M}$, for some matrix $\mathbf{M}$. (Hint: the matrix will be skew-symmetric.)

9. Match each of the following figures (1–4) with their corresponding transformations.

   (a) $\begin{bmatrix} 1 & 0 \\ 0 & -1 \end{bmatrix}$

   (b) $\begin{bmatrix} 2.5 & 0 \\ 0 & 2.5 \end{bmatrix}$

   (c) $\begin{bmatrix} -\frac{\sqrt{2}}{2} & -\frac{\sqrt{2}}{2} \\ -\frac{\sqrt{2}}{2} & \frac{\sqrt{2}}{2} \end{bmatrix}$

   (d) $\begin{bmatrix} 1.5 & 0 \\ 0 & 2.0 \end{bmatrix}$

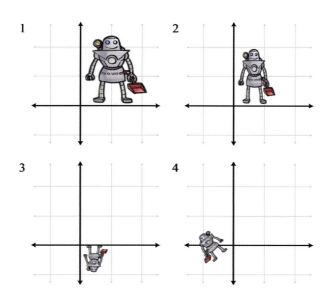

10. Given the 10×1 column vector $\mathbf{v}$, create a matrix $\mathbf{M}$ that, when multiplied
    by $\mathbf{v}$, produces a 10×1 column vector $\mathbf{w}$ such that

$$w_i = \begin{cases} v_1 & \text{if } i = 1, \\ v_i - v_{i-1} & \text{if } i > 1. \end{cases}$$

Matrices of this form arise when some continuous function is discretized.
Multiplication by this *first difference* matrix is the discrete equivalent of
continuous differentiation. (We'll learn about differentiation in Chapter 11
if you haven't already had calculus.)

11. Given the 10×1 column vector $\mathbf{v}$, create a matrix $\mathbf{N}$ that, when multiplied by $\mathbf{v}$, produces a 10×1 column vector $\mathbf{w}$ such that

$$w_i = \sum_{j=1}^{i} v_j.$$

In other words, each element becomes the sum of that element and all previous elements.

This matrix performs the discrete equivalent of *integration*, which as you might already know (but you certainly will know after reading Chapter 11) is the inverse operation of differentiation.

12. Consider $\mathbf{M}$ and $\mathbf{N}$, the matrices from Exercises 10 and 11.

   (a) Discuss your expectations of the product $\mathbf{MN}$.

   (b) Discuss your expectations of the product $\mathbf{NM}$.

   (c) Calculate both $\mathbf{MN}$ and $\mathbf{NM}$. Were your expectations correct?

To be civilized is to be potentially master of all possible ideas,
and that means that one has got beyond being shocked,
although one preserves one's own moral aesthetic preferences.

— Oliver Wendell Holmes (1809–1894)

Chapter 5

Matrices and Linear Transformations

It's time to transform!

— Super WHY!

Chapter 4 investigated some of the basic mathematical properties of matrices. It also developed a geometric understanding of matrices and their relationship to coordinate space transformations in general. This chapter continues our investigation of transformations.

To be more specific, this chapter is concerned with expressing *linear transformations* in 3D using 3×3 matrices. We give a more formal definition of linear transformations at the end of this chapter, but for now, recall from our informal introduction to linear transformations in Section 4.2 that one important property of linear transformations is that they do not contain translation. A transformation that contains translation is known as an *affine* transformation. Affine transformations in 3D cannot be implemented using 3×3 matrices. Section 5.7.2 gives a formal definition of affine transformations, and Section 6.4 shows how to use 4×4 matrices to represent affine transformations.

This chapter discusses the implementation of linear transformations via matrices. It is divided roughly into two parts. In the first part, Sections 5.1–5.5, we take the basic tools from previous chapters to derive matrices for primitive linear transformations of rotation, scaling, orthographic projection, reflection, and shearing. For each transformation, examples and equations in 2D and 3D are given. The same strategy will be used repeatedly: determine what happens to the standard basis vectors as a result of the transformation and then put those transformed basis vectors into the rows of our matrix. Note that these discussions assume an *active* transformation: the object is transformed while the coordinate space remains stationary. Remember from Section 3.3.1 that we can effectively perform a passive transformation (transform the coordinate space and keep the object still) by transforming the object by the opposite amount.

A lot of this chapter is filled with messy equations and details, so you might be tempted to skip over it—but don't! There are a lot of important, easily digested principles interlaced with the safely forgotten details. We think it's important to be able to understand how various transform matrices can be derived, so in principle you can derive them on your own from scratch. Commit the high-level principles in this chapter to memory, and don't get too bogged down in the details. This book will not self-destruct after you read it, so keep it on hand for reference when you need a particular equation.

The second part of this chapter returns to general principles of transformations. Section 5.6 shows how a sequence of primitive transformations may be combined by using matrix multiplication to form a more complicated transformation. Section 5.7 discusses various interesting categories of transformations, including linear, affine, invertible, angle-preserving, orthogonal, and rigid-body transforms.

5.1   Rotation

We have already seen general examples of rotation matrices. Now let's develop a more rigorous definition. First, Section 5.1.1 examines 2D rotation. Section 5.1.2 shows how to rotate about a cardinal axis. Finally, Section 5.1.3 tackles the most general case of rotation about an arbitrary axis in 3D.

5.1.1   Rotation in 2D

In 2D, there's really only one type of rotation that we can do: rotation about a point. This chapter is concerned with linear transformations, which do not contain translation, so we restrict our discussion even further to rotation about the origin. A 2D rotation about the origin has only one parameter, the angle θ, which defines the amount of rotation. The standard convention found in most math books is to consider counterclockwise rotation positive and clockwise rotation negative. (However, different conventions are more appropriate in different situations.) Figure 5.1 shows how the basis vectors $\mathbf{p}$ and $\mathbf{q}$ are rotated about the origin, resulting in the new basis vectors $\mathbf{p}'$ and $\mathbf{q}'$.

Now that we know the values of the basis vectors after rotation, we can build our matrix:

2D rotation matrix

$$\mathbf{R}(\theta) = \begin{bmatrix} -\mathbf{p}'- \\ -\mathbf{q}'- \end{bmatrix} = \begin{bmatrix} \cos\theta & \sin\theta \\ -\sin\theta & \cos\theta \end{bmatrix}.$$

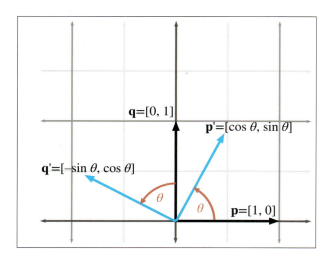

Figure 5.1
Rotation about the origin in 2D

5.1.2 3D Rotation about Cardinal Axes

In 3D, rotation occurs about an *axis* rather than a point, with the term *axis* taking on its more commonplace meaning of a line about which something rotates. An axis of rotation does not necessarily have to be one of the cardinal x, y, or z axes—but those special cases are the ones we consider

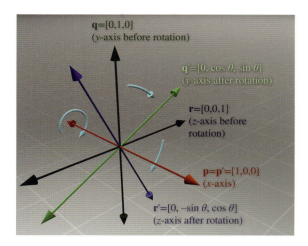

Figure 5.2
Rotating about the x-axis in 3D

in this section. Again, we are not considering translation in this chapter, so we will limit the discussion to rotation about an axis that passes through the origin. In any case, we'll need to establish which direction of rotation is considered "positive" and which is considered "negative." We're going to obey the left-hand rule for this. Review Section 1.3.3 if you've forgotten this rule.

Let's start with rotation about the x-axis, as shown in Figure 5.2. Constructing a matrix from the rotated basis vectors, we have

3D matrix to rotate about the x-axis

$$\mathbf{R}_x(\theta) = \begin{bmatrix} -\mathbf{p}'- \\ -\mathbf{q}'- \\ -\mathbf{r}'- \end{bmatrix} = \begin{bmatrix} 1 & 0 & 0 \\ 0 & \cos\theta & \sin\theta \\ 0 & -\sin\theta & \cos\theta \end{bmatrix}.$$

Rotation about the y-axis is similar (see Figure 5.3). The matrix to rotate about the y-axis is

3D matrix to rotate about the y-axis

$$\mathbf{R}_y(\theta) = \begin{bmatrix} -\mathbf{p}'- \\ -\mathbf{q}'- \\ -\mathbf{r}'- \end{bmatrix} = \begin{bmatrix} \cos\theta & 0 & -\sin\theta \\ 0 & 1 & 0 \\ \sin\theta & 0 & \cos\theta \end{bmatrix}.$$

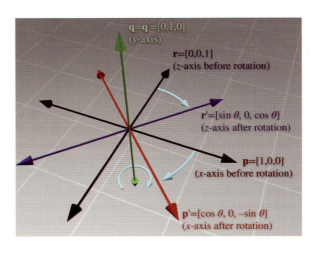

Figure 5.3
Rotating about the y-axis in 3D

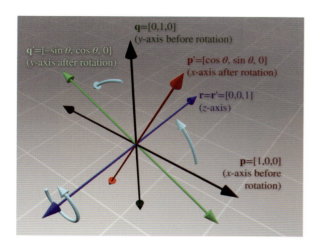

Figure 5.4
Rotating about the z-axis
in 3D

And finally, rotation about the z-axis (see Figure 5.4) is done with the matrix

$$\mathbf{R}_z(\theta) = \begin{bmatrix} -\mathbf{p}'- \\ -\mathbf{q}'- \\ -\mathbf{r}'- \end{bmatrix} = \begin{bmatrix} \cos\theta & \sin\theta & 0 \\ -\sin\theta & \cos\theta & 0 \\ 0 & 0 & 1 \end{bmatrix}.$$

3D matrix to rotate
about the z-axis

Please note that although the figures in this section use a left-handed convention, the matrices work in either left- or right-handed coordinate systems, due to the conventions used to define the direction of positive rotation. You can verify this visually by looking at the figures in a mirror.

5.1.3   3D Rotation about an Arbitrary Axis

We can also rotate about an arbitrary axis in 3D, provided, of course, that the axis passes through the origin, since we are not considering translation at the moment. This is more complicated and less common than rotating about a cardinal axis. As before, we define θ to be the amount of rotation about the axis. The axis will be defined by a unit vector $\hat{\mathbf{n}}$.

Let's derive a matrix to rotate about $\hat{\mathbf{n}}$ by the angle θ. In other words, we wish to derive the matrix $\mathbf{R}(\hat{\mathbf{n}}, \theta)$ such that when we multiply a vector $\mathbf{v}$ by $\mathbf{R}(\hat{\mathbf{n}}, \theta)$, the resulting vector $\mathbf{v}'$ is the result of rotating $\mathbf{v}$ about $\hat{\mathbf{n}}$ by the angle θ:

$$\mathbf{v}' = \mathbf{v}\,\mathbf{R}(\hat{\mathbf{n}}, \theta).$$

To derive the matrix $\mathbf{R}(\hat{\mathbf{n}}, \theta)$, let's first see if we can express $\mathbf{v}'$ in terms of $\mathbf{v}$, $\hat{\mathbf{n}}$, and θ. The basic idea is to solve the problem in the plane perpendicular to $\hat{\mathbf{n}}$, which is a much simpler 2D problem. To do this, we separate $\mathbf{v}$ into two vectors, $\mathbf{v}_{\parallel}$ and $\mathbf{v}_{\perp}$, which are parallel and perpendicular to $\hat{\mathbf{n}}$, respectively, such that $\mathbf{v} = \mathbf{v}_{\parallel} + \mathbf{v}_{\perp}$. (We learned how to do this with the dot product in Section 2.11.2.) By rotating each of these components individually, we can rotate the vector as a whole. In other words, $\mathbf{v}' = \mathbf{v}'_{\parallel} + \mathbf{v}'_{\perp}$. Since $\mathbf{v}_{\parallel}$ is parallel to $\hat{\mathbf{n}}$, it will not be affected by the rotation about $\hat{\mathbf{n}}$. In other words, $\mathbf{v}'_{\parallel} = \mathbf{v}_{\parallel}$. So all we need to do is compute $\mathbf{v}'_{\perp}$, and then we have $\mathbf{v}' = \mathbf{v}_{\parallel} + \mathbf{v}'_{\perp}$. To compute $\mathbf{v}'_{\perp}$, we construct the vectors $\mathbf{v}_{\parallel}$, $\mathbf{v}_{\perp}$, and an intermediate vector $\mathbf{w}$, as follows:

- The vector $\mathbf{v}_{\parallel}$ is the portion of $\mathbf{v}$ that is parallel to $\hat{\mathbf{n}}$. Another way of saying this is that $\mathbf{v}_{\parallel}$ is the value of $\mathbf{v}$ *projected onto* $\hat{\mathbf{n}}$. From Section 2.11.2, we know that $\mathbf{v}_{\parallel} = (\mathbf{v} \cdot \hat{\mathbf{n}})\hat{\mathbf{n}}$.

- The vector $\mathbf{v}_{\perp}$ is the portion of $\mathbf{v}$ that is perpendicular to $\hat{\mathbf{n}}$. Since $\mathbf{v} = \mathbf{v}_{\parallel} + \mathbf{v}_{\perp}$, $\mathbf{v}_{\perp}$ can be computed by $\mathbf{v} - \mathbf{v}_{\parallel}$. $\mathbf{v}_{\perp}$ is the result of projecting $\mathbf{v}$ onto the plane perpendicular to $\hat{\mathbf{n}}$.

- The vector $\mathbf{w}$ is mutually perpendicular to $\mathbf{v}_{\parallel}$ and $\mathbf{v}_{\perp}$ and has the same length as $\mathbf{v}_{\perp}$. It can be constructed by rotating $\mathbf{v}_{\perp}$ 90° about $\hat{\mathbf{n}}$; thus we see that its value is easily computed by $\mathbf{w} = \hat{\mathbf{n}} \times \mathbf{v}_{\perp}$.

These vectors are shown in Figure 5.5.

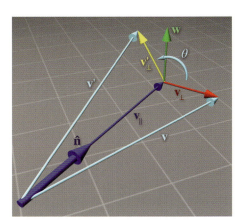

Figure 5.5
Rotating a vector about an arbitrary axis

How do these vectors help us compute $\mathbf{v}'_{\perp}$? Notice that $\mathbf{w}$ and $\mathbf{v}_{\perp}$ form a 2D coordinate space, with $\mathbf{v}_{\perp}$ as the "x-axis" and $\mathbf{w}$ as the "y-axis." (Note that the two vectors don't necessarily have unit length.) $\mathbf{v}'_{\perp}$ is the result of rotating $\mathbf{v}_{\perp}$ in this plane by the angle θ. Note that this is almost identical to rotating an angle into standard position. Section 1.4.4 showed that the endpoints of a unit ray rotated by an angle θ are $\cos \theta$ and $\sin \theta$. The only difference here is that our ray is not a unit ray, and we are using $\mathbf{v}_{\perp}$ and $\mathbf{w}$ as our basis vectors. Thus, $\mathbf{v}'_{\perp}$ can be computed as

$$\mathbf{v}'_{\perp} = \cos \theta \mathbf{v}_{\perp} + \sin \theta \mathbf{w}.$$

Let's summarize the vectors we have computed:

$$\mathbf{v}_\| = (\mathbf{v} \cdot \hat{\mathbf{n}}) \, \hat{\mathbf{n}},$$
$$\mathbf{v}_\perp = \mathbf{v} - \mathbf{v}_\|$$
$$= \mathbf{v} - (\mathbf{v} \cdot \hat{\mathbf{n}}) \, \hat{\mathbf{n}},$$
$$\mathbf{w} = \hat{\mathbf{n}} \times \mathbf{v}_\perp$$
$$= \hat{\mathbf{n}} \times (\mathbf{v} - \mathbf{v}_\|)$$
$$= \hat{\mathbf{n}} \times \mathbf{v} - \hat{\mathbf{n}} \times \mathbf{v}_\|$$
$$= \hat{\mathbf{n}} \times \mathbf{v} - \mathbf{0}$$
$$= \hat{\mathbf{n}} \times \mathbf{v},$$
$$\mathbf{v}'_\perp = \cos\theta \mathbf{v}_\perp + \sin\theta \mathbf{w}$$
$$= \cos\theta \left(\mathbf{v} - (\mathbf{v} \cdot \hat{\mathbf{n}}) , \hat{\mathbf{n}} \right) + \sin\theta \left(\hat{\mathbf{n}} \times \mathbf{v} \right).$$

Substituting for $\mathbf{v}'$, we have

$$\mathbf{v}' = \mathbf{v}'_\perp + \mathbf{v}_\|$$
$$= \cos\theta \left(\mathbf{v} - (\mathbf{v} \cdot \hat{\mathbf{n}}) \, \hat{\mathbf{n}} \right) + \sin\theta \left(\hat{\mathbf{n}} \times \mathbf{v} \right) + (\mathbf{v} \cdot \hat{\mathbf{n}}) \, \hat{\mathbf{n}}. \tag{5.1}$$

Equation (5.1) allows us to rotate any arbitrary vector about any arbitrary axis. We could perform arbitrary rotation transformations armed only with this equation, so in a sense we are done—the remaining arithmetic is essentially a notational change that expresses Equation (5.1) as a matrix multiplication.

Now that we have expressed $\mathbf{v}'$ in terms of $\mathbf{v}$, $\hat{\mathbf{n}}$, and θ, we can compute what the basis vectors are after transformation and construct our matrix. We're just presenting the results here; a reader interested in following each step can check out Exercise 2.24:

$$\mathbf{p} = \begin{bmatrix} 1 & 0 & 0 \end{bmatrix}, \qquad \mathbf{p}' = \begin{bmatrix} n_x{}^2 (1 - \cos\theta) + \cos\theta \\ n_x n_y (1 - \cos\theta) + n_z \sin\theta \\ n_x n_z (1 - \cos\theta) - n_y \sin\theta \end{bmatrix}^{\mathrm{T}},$$

$$\mathbf{q} = \begin{bmatrix} 0 & 1 & 0 \end{bmatrix}, \qquad \mathbf{q}' = \begin{bmatrix} n_x n_y (1 - \cos\theta) - n_z \sin\theta \\ n_y{}^2 (1 - \cos\theta) + \cos\theta \\ n_y n_z (1 - \cos\theta) + n_x \sin\theta \end{bmatrix}^{\mathrm{T}},$$

$$\mathbf{r} = \begin{bmatrix} 0 & 0 & 1 \end{bmatrix}, \qquad \mathbf{r}' = \begin{bmatrix} n_x n_z (1 - \cos\theta) + n_y \sin\theta \\ n_y n_z (1 - \cos\theta) - n_x \sin\theta \\ n_z{}^2 (1 - \cos\theta) + \cos\theta \end{bmatrix}^{\mathrm{T}}.$$

Note that $\mathbf{p}'$ and friends are actually row vectors, we are just writing them as transposed column vectors to fit on the page.

Constructing the matrix from these basis vectors, we get

3D matrix to rotate about an arbitrary axis

$$\mathbf{R}(\hat{\mathbf{n}}, \theta) = \begin{bmatrix} -\mathbf{p}'- \\ -\mathbf{q}'- \\ -\mathbf{r}'- \end{bmatrix}$$

$$= \begin{bmatrix} n_x^2 \left(1 - \cos\theta\right) + \cos\theta & n_x n_y \left(1 - \cos\theta\right) + n_z \sin\theta & n_x n_z \left(1 - \cos\theta\right) - n_y \sin\theta \\ n_x n_y \left(1 - \cos\theta\right) - n_z \sin\theta & n_y^2 \left(1 - \cos\theta\right) + \cos\theta & n_y n_z \left(1 - \cos\theta\right) + n_x \sin\theta \\ n_x n_z \left(1 - \cos\theta\right) + n_y \sin\theta & n_y n_z \left(1 - \cos\theta\right) - n_x \sin\theta & n_z^2 \left(1 - \cos\theta\right) + \cos\theta \end{bmatrix}.$$

5.2   Scale

We can scale an object to make it proportionally bigger or smaller by a factor of k. If we apply this scale to the entire object, thus "dilating" the object about the origin, we are performing a *uniform scale*. Uniform scale preserves angles and proportions. Lengths increase or decrease uniformly by a factor of k, areas by a factor of k^2, and volumes (in 3D) by a factor of k^3.

If we wish to "stretch" or "squash" the object, we can apply different scale factors in different directions, resulting in *nonuniform scale*. Nonuniform scale does not preserve angles. Lengths, areas, and volumes are adjusted by a factor that varies according to the orientation relative to the direction of scale.

If $|k| < 1$, then the object gets "shorter" in that direction. If $|k| > 1$, then the object gets "longer." If $k = 0$, then we have an *orthographic projection*, discussed in Section 5.3. If $k < 0$, then we have a *reflection*, covered in Section 5.4. For the remainder of this section, we will assume that $k > 0$.

Section 5.2.1 begins with the simple case of scaling along the cardinal axes. Then Section 5.2.2 examines the general case, scaling along an arbitrary axis.

5.2.1   Scaling along the Cardinal Axes

The simplest scale operation applies a separate scale factor along each cardinal axis. The scale *along* an axis is applied *about* the perpendicular axis (in 2D) or plane (in 3D). If the scale factors for all axes are equal, then the scale is uniform; otherwise, it is nonuniform.

In 2D, we have two scale factors, k_x and k_y. Figure 5.6 shows an object with various scale values for k_x and k_y.

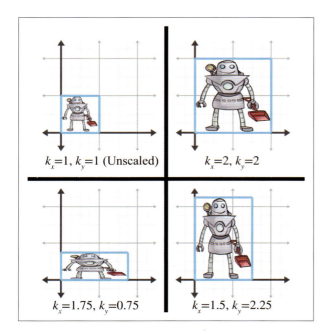

Figure 5.6
Scaling a 2D object
with various factors
for k_x and k_y

As is intuitively obvious, the basis vectors **p** and **q** are independently affected by the corresponding scale factors:

$$\begin{aligned}
\mathbf{p}' &= k_x\mathbf{p} = k_x\begin{bmatrix} 1 & 0 \end{bmatrix} = \begin{bmatrix} k_x & 0 \end{bmatrix}, \\
\mathbf{q}' &= k_y\mathbf{q} = k_y\begin{bmatrix} 0 & 1 \end{bmatrix} = \begin{bmatrix} 0 & k_y \end{bmatrix}.
\end{aligned}$$

Constructing the 2D scale matrix $\mathbf{S}(k_x, k_y)$ from these basis vectors, we get

$$\mathbf{S}(k_x, k_y) = \begin{bmatrix} -\mathbf{p}'- \\ -\mathbf{q}'- \end{bmatrix} = \begin{bmatrix} k_x & 0 \\ 0 & k_y \end{bmatrix}.$$

2D matrix to scale on cardinal axes

For 3D, we add a third scale factor k_z, and the 3D scale matrix is then given by

$$\mathbf{S}(k_x, k_y, k_z) = \begin{bmatrix} k_x & 0 & 0 \\ 0 & k_y & 0 \\ 0 & 0 & k_z \end{bmatrix}.$$

3D matrix to scale on cardinal axes

If we multiply any arbitrary vector by this matrix, then, as expected, each component is scaled by the appropriate scale factor:

$$\begin{bmatrix} x & y & z \end{bmatrix} \begin{bmatrix} k_x & 0 & 0 \\ 0 & k_y & 0 \\ 0 & 0 & k_z \end{bmatrix} = \begin{bmatrix} k_x x & k_y y & k_z z \end{bmatrix}.$$

5.2.2   Scaling in an Arbitrary Direction

We can apply scale independent of the coordinate system used by scaling in an arbitrary direction. We define $\hat{\mathbf{n}}$ to be the unit vector parallel to the direction of scale, and k to be the scale factor to be applied about the line (in 2D) or plane (in 3D) that passes through the origin and is perpendicular to $\hat{\mathbf{n}}$. We are scaling *along* $\hat{\mathbf{n}}$, not *about* $\hat{\mathbf{n}}$.

To derive a matrix that scales along an arbitrary axis, we'll use an approach similar to the one used in Section 5.1.3 for rotation about an arbitrary axis. Let's derive an expression that, given an arbitrary vector $\mathbf{v}$, computes $\mathbf{v}'$ in terms of $\mathbf{v}$, $\hat{\mathbf{n}}$, and k. As before, we separate $\mathbf{v}$ into two values, $\mathbf{v}_\parallel$ and $\mathbf{v}_\perp$, which are parallel and perpendicular to $\hat{\mathbf{n}}$, respectively, such that $\mathbf{v} = \mathbf{v}_\parallel + \mathbf{v}_\perp$. The parallel portion, $\mathbf{v}_\parallel$, is the projection of $\mathbf{v}$ onto $\hat{\mathbf{n}}$. From Section 2.11.2, we know that $\mathbf{v}_\parallel = (\mathbf{v} \cdot \hat{\mathbf{n}})\hat{\mathbf{n}}$. Since $\mathbf{v}_\perp$ is perpendicular to $\hat{\mathbf{n}}$, it will not be affected by the scale operation. Thus $\mathbf{v}' = \mathbf{v}'_\parallel + \mathbf{v}_\perp$, and all we have left to do is compute the value of $\mathbf{v}'_\parallel$. Since $\mathbf{v}_\parallel$ is parallel to the direction of scale, $\mathbf{v}'_\parallel$ is trivially given by $k\mathbf{v}_\parallel$. This is shown in Figure 5.7.

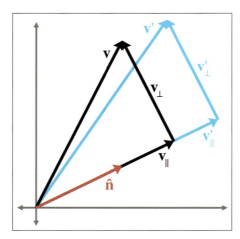

Figure 5.7
Scaling a vector along an arbitrary direction

Summarizing the known vectors and substituting gives us

$$\mathbf{v} = \mathbf{v}_\parallel + \mathbf{v}_\perp,$$

$$\mathbf{v}_\parallel = (\mathbf{v} \cdot \hat{\mathbf{n}})\,\hat{\mathbf{n}},$$

$$\mathbf{v}'_\perp = \mathbf{v}_\perp$$
$$= \mathbf{v} - \mathbf{v}_\parallel$$
$$= \mathbf{v} - (\mathbf{v} \cdot \hat{\mathbf{n}})\,\hat{\mathbf{n}},$$

$$\mathbf{v}'_\parallel = k\mathbf{v}_\parallel$$
$$= k\,(\mathbf{v} \cdot \hat{\mathbf{n}})\,\hat{\mathbf{n}},$$

$$\mathbf{v}' = \mathbf{v}'_\perp + \mathbf{v}'_\parallel$$
$$= \mathbf{v} - (\mathbf{v} \cdot \hat{\mathbf{n}})\,\hat{\mathbf{n}} + k\,(\mathbf{v} \cdot \hat{\mathbf{n}})\,\hat{\mathbf{n}}$$
$$= \mathbf{v} + (k - 1)\,(\mathbf{v} \cdot \hat{\mathbf{n}})\,\hat{\mathbf{n}}.$$

Now that we know how to scale an arbitrary vector, we can compute the value of the basis vectors after scale. We derive the first 2D basis vector; the other basis vector is similar, and so we merely present the results. (Note that column vectors are used in the equations below strictly to make the equations format nicely on the page.):

$$\mathbf{p} = \begin{bmatrix} 1 & 0 \end{bmatrix},$$

$$\mathbf{p}' = \mathbf{p} + (k-1)\,(\mathbf{p} \cdot \hat{\mathbf{n}})\,\hat{\mathbf{n}} = \begin{bmatrix} 1 \\ 0 \end{bmatrix} + (k-1)\left(\begin{bmatrix} 1 \\ 0 \end{bmatrix} \cdot \begin{bmatrix} n_x \\ n_y \end{bmatrix} \right) \begin{bmatrix} n_x \\ n_y \end{bmatrix}$$

$$= \begin{bmatrix} 1 \\ 0 \end{bmatrix} + (k-1)\,n_x \begin{bmatrix} n_x \\ n_y \end{bmatrix} = \begin{bmatrix} 1 \\ 0 \end{bmatrix} + \begin{bmatrix} (k-1)\,n_x{}^2 \\ (k-1)\,n_x n_y \end{bmatrix}$$

$$= \begin{bmatrix} 1 + (k-1)\,n_x{}^2 \\ (k-1)\,n_x n_y \end{bmatrix},$$

$$\mathbf{q} = \begin{bmatrix} 0 & 1 \end{bmatrix},$$

$$\mathbf{q}' = \begin{bmatrix} (k-1)\,n_x n_y \\ 1 + (k-1)\,n_y{}^2 \end{bmatrix}.$$

Forming a matrix from the basis vectors, we arrive at the 2D matrix to scale by a factor of k in an arbitrary direction specified by the unit vector $\hat{\mathbf{n}}$:

$$\mathbf{S}(\hat{\mathbf{n}}, k) = \begin{bmatrix} -\mathbf{p}'- \\ -\mathbf{q}'- \end{bmatrix} = \begin{bmatrix} 1 + (k-1)\,n_x{}^2 & (k-1)\,n_x n_y \\ (k-1)\,n_x n_y & 1 + (k-1)\,n_y{}^2 \end{bmatrix}.$$

2D matrix to scale in an arbitrary direction

In 3D, the values of the basis vectors are computed by

$$\mathbf{p} = \begin{bmatrix} 1 & 0 & 0 \end{bmatrix}, \qquad \mathbf{p}' = \begin{bmatrix} 1 + (k-1)\,n_x{}^2 \\ (k-1)\,n_x n_y \\ (k-1)\,n_x n_z \end{bmatrix}^{\mathrm{T}},$$

$$\mathbf{q} = \begin{bmatrix} 0 & 1 & 0 \end{bmatrix}, \qquad \mathbf{q}' = \begin{bmatrix} (k-1)\,n_x n_y \\ 1 + (k-1)\,n_y{}^2 \\ (k-1)\,n_y n_z \end{bmatrix}^{\mathrm{T}},$$

$$\mathbf{r} = \begin{bmatrix} 0 & 0 & 1 \end{bmatrix}, \qquad \mathbf{r}' = \begin{bmatrix} (k-1)\,n_x n_z \\ (k-1)\,n_y n_z \\ 1 + (k-1)\,n_z{}^2 \end{bmatrix}^{\mathrm{T}}.$$

A suspicious reader wondering if we just made that up can step through the derivation in Exercise 2.23.

Finally, the 3D matrix to scale by a factor of k in an arbitrary direction specified by the unit vector $\hat{\mathbf{n}}$ is

3D matrix to scale in an arbitrary direction

$$\mathbf{S}(\hat{\mathbf{n}}, k) = \begin{bmatrix} -\mathbf{p}'- \\ -\mathbf{q}'- \\ -\mathbf{r}'- \end{bmatrix}$$

$$= \begin{bmatrix} 1 + (k-1)\,n_x{}^2 & (k-1)\,n_x n_y & (k-1)\,n_x n_z \\ (k-1)\,n_x n_y & 1 + (k-1)\,n_y{}^2 & (k-1)\,n_y n_z \\ (k-1)\,n_x n_z & (k-1)\,n_y n_z & 1 + (k-1)\,n_z{}^2 \end{bmatrix}.$$

5.3   Orthographic Projection

In general, the term *projection* refers to any dimension-reducing operation. As we discussed in Section 5.2, one way we can achieve projection is to use a scale factor of zero in a direction. In this case, all the points are flattened or *projected* onto the perpendicular axis (in 2D) or plane (in 3D). This type of projection is an *orthographic projection*, also known as a *parallel projection*, since the lines from the original points to their projected counterparts are parallel. We present another type of projection, *perspective projection*, in Section 6.5.

First, Section 5.3.1 discusses orthographic projection onto a cardinal axis or plane, and then Section 5.3.2 examines the general case.

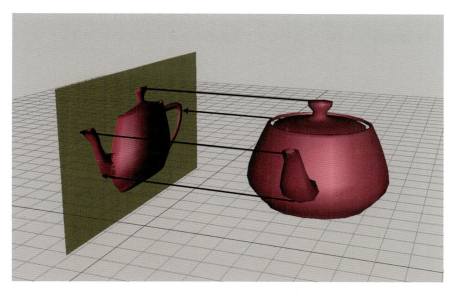

Figure 5.8. Projecting a 3D object onto a cardinal plane

5.3.1 Projecting onto a Cardinal Axis or Plane

The simplest type of projection occurs when we project onto a cardinal axis (in 2D) or plane (in 3D). This is illustrated in Figure 5.8.

Projection onto a cardinal axis or plane most frequently occurs not by actual transformation, but by simply discarding one of the coordinates while assigning the data into a variable of lesser dimension. For example, we may turn a 3D object into a 2D object by discarding the z components of the points and copying only x and y.

However, we can also project onto a cardinal axis or plane by using a scale value of zero on the perpendicular axis. For completeness, we present the matrices for these transformations:

$$\mathbf{P}_x = \mathbf{S}\left(\begin{bmatrix} 0 & 1 \end{bmatrix}, 0\right) = \begin{bmatrix} 1 & 0 \\ 0 & 0 \end{bmatrix},$$

Projecting onto a cardinal axis

$$\mathbf{P}_y = \mathbf{S}\left(\begin{bmatrix} 1 & 0 \end{bmatrix}, 0\right) = \begin{bmatrix} 0 & 0 \\ 0 & 1 \end{bmatrix},$$

Projecting onto a cardinal plane

$$\mathbf{P}_{xy} = \mathbf{S}\left(\begin{bmatrix} 0 & 0 & 1 \end{bmatrix}, 0\right) = \begin{bmatrix} 1 & 0 & 0 \\ 0 & 1 & 0 \\ 0 & 0 & 0 \end{bmatrix},$$

$$\mathbf{P}_{xz} = \mathbf{S}\left(\begin{bmatrix} 0 & 1 & 0 \end{bmatrix}, 0\right) = \begin{bmatrix} 1 & 0 & 0 \\ 0 & 0 & 0 \\ 0 & 0 & 1 \end{bmatrix},$$

$$\mathbf{P}_{yz} = \mathbf{S}\left(\begin{bmatrix} 1 & 0 & 0 \end{bmatrix}, 0\right) = \begin{bmatrix} 0 & 0 & 0 \\ 0 & 1 & 0 \\ 0 & 0 & 1 \end{bmatrix}.$$

5.3.2   Projecting onto an Arbitrary Line or Plane

We can also project onto any arbitrary line (in 2D) or plane (in 3D). As before, since we are not considering translation, the line or plane must pass through the origin. The projection will be defined by a unit vector $\hat{\mathbf{n}}$ that is perpendicular to the line or plane.

We can derive the matrix to project in an arbitrary direction by applying a zero scale factor along this direction, using the equations we developed in Section 5.2.2. In 2D, we have

2D matrix to project onto an arbitrary line

$$\mathbf{P}(\hat{\mathbf{n}}) = \mathbf{S}\left(\hat{\mathbf{n}}, 0\right) = \begin{bmatrix} 1 + (0-1)\,n_x{}^2 & (0-1)\,n_x n_y \\ (0-1)\,n_x n_y & 1 + (0-1)\,n_y{}^2 \end{bmatrix}$$

$$= \begin{bmatrix} 1 - n_x{}^2 & -n_x n_y \\ -n_x n_y & 1 - n_y{}^2 \end{bmatrix}.$$

Remember that $\hat{\mathbf{n}}$ is *perpendicular* to the line onto which we are projecting, not parallel to it. In 3D, we project onto the plane perpendicular to $\hat{\mathbf{n}}$:

3D matrix to project onto an arbitrary plane

$$\mathbf{P}(\hat{\mathbf{n}}) = \mathbf{S}\left(\hat{\mathbf{n}}, 0\right) = \begin{bmatrix} 1 + (0-1)\,n_x{}^2 & (0-1)\,n_x n_y & (0-1)\,n_x n_z \\ (0-1)\,n_x n_y & 1 + (0-1)\,n_y{}^2 & (0-1)\,n_y n_z \\ (0-1)\,n_x n_z & (0-1)\,n_y n_z & 1 + (0-1)\,n_z{}^2 \end{bmatrix}$$

$$= \begin{bmatrix} 1 - n_x{}^2 & -n_x n_y & -n_x n_z \\ -n_x n_y & 1 - n_y{}^2 & -n_y n_z \\ -n_x n_z & -n_y n_z & 1 - n_z{}^2 \end{bmatrix}.$$

5.4   Reflection

Reflection (also called *mirroring*) is a transformation that "flips" the object about a line (in 2D) or a plane (in 3D). Figure 5.9 shows the result of reflecting an object about the x- and y-axis.

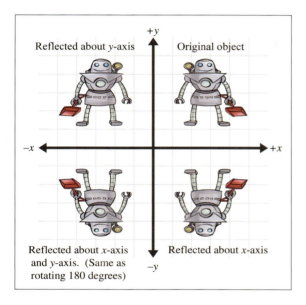

Figure 5.9
Reflecting an object about an axis in 2D

Reflection can be accomplished by applying a scale factor of -1. Let $\hat{\mathbf{n}}$ be a 2D unit vector. Then the matrix that performs a reflection about the axis of reflection that passes through the origin and is perpendicular to $\hat{\mathbf{n}}$ is given by

$$\mathbf{R}(\hat{\mathbf{n}}) = \mathbf{S}(\hat{\mathbf{n}}, -1) = \begin{bmatrix} 1 + (-1-1)\,n_x{}^2 & (-1-1)\,n_x n_y \\ (-1-1)\,n_x n_y & 1 + (-1-1)\,n_y{}^2 \end{bmatrix}$$

$$= \begin{bmatrix} 1 - 2n_x{}^2 & -2n_x n_y \\ -2n_x n_y & 1 - 2n_y{}^2 \end{bmatrix}.$$

2D matrix to reflect about an arbitrary axis

In 3D, we have a reflecting *plane* instead of axis. For the transformation to be linear, the plane must contain the origin, in which case the matrix to perform the reflection is

$$\mathbf{R}(\hat{\mathbf{n}}) = \mathbf{S}\left(\hat{\mathbf{n}}, -1\right)$$

3D matrix to reflect about an arbitrary plane

$$= \begin{bmatrix} 1 + (-1-1)\,{n_x}^2 & (-1-1)\,n_x n_y & (-1-1)\,n_x n_z \\ (-1-1)\,n_x n_y & 1 + (-1-1)\,{n_y}^2 & (-1-1)\,n_y n_z \\ (-1-1)\,n_x n_z & (-1-1)\,n_y n_z & 1 + (-1-1)\,{n_z}^2 \end{bmatrix}$$

$$= \begin{bmatrix} 1 - 2{n_x}^2 & -2n_x n_y & -2n_x n_z \\ -2n_x n_y & 1 - 2{n_y}^2 & -2n_y n_z \\ -2n_x n_z & -2n_y n_z & 1 - 2{n_z}^2 \end{bmatrix}.$$

Notice that an object can be "reflected" only once. If we reflect it again (even about a different axis or plane) then the object is flipped back to "right side out," and it is the same as if we had rotated the object from its initial position. An example of this is shown in the bottom-left corner of Figure 5.9.

5.5   Shearing

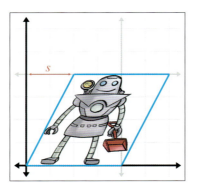

Figure 5.10
Shearing in 2D

Shearing is a transformation that "skews" the coordinate space, stretching it nonuniformly. Angles are not preserved; however, surprisingly, areas and volumes are. The basic idea is to add a multiple of one coordinate to the other. For example, in 2D, we might take a multiple of y and add it to x, so that $x' = x + sy$. This is shown in Figure 5.10.

The matrix that performs this shear is

$$\mathbf{H}_x(s) = \begin{bmatrix} 1 & 0 \\ s & 1 \end{bmatrix},$$

where the notation $\mathbf{H}_x$ denotes that the x-coordinate is sheared by the other coordinate, y. The parameter s controls the amount and direction of the shearing. The other 2D shear matrix, $\mathbf{H}_y$, is

$$\mathbf{H}_y(s) = \begin{bmatrix} 1 & s \\ 0 & 1 \end{bmatrix}.$$

In 3D, we can take one coordinate and add different multiples of that coordinate to the other two coordinates. The notation $\mathbf{H}_{xy}$ indicates that the x- and y-coordinates are shifted by the other coordinate, z. We present these matrices for completeness:

$$\mathbf{H}_{xy}(s,t) = \begin{bmatrix} 1 & 0 & 0 \\ 0 & 1 & 0 \\ s & t & 1 \end{bmatrix},$$

$$\mathbf{H}_{xz}(s,t) = \begin{bmatrix} 1 & 0 & 0 \\ s & 1 & t \\ 0 & 0 & 1 \end{bmatrix},$$

3D shear matrices

$$\mathbf{H}_{yz}(s,t) = \begin{bmatrix} 1 & s & t \\ 0 & 1 & 0 \\ 0 & 0 & 1 \end{bmatrix}.$$

Shearing is a seldom-used transform. It is also known as a *skew* transform. Combining shearing and scaling (uniform or nonuniform) creates a transformation that is indistinguishable from a transformation containing rotation and nonuniform scale.

5.6   Combining Transformations

This section shows how to take a sequence of transformation matrices and combine (or *concatenate*) them into one single transformation matrix. This new matrix represents the cumulative result of applying all of the original transformations in order. It's actually quite easy. The transformation that results from applying the transformation with matrix $\mathbf{A}$ followed by the transformation with matrix $\mathbf{B}$ has matrix $\mathbf{AB}$. That is, matrix multiplication is how we compose transformations represented as matrices.

One very common example of this is in rendering. Imagine there is an object at an arbitrary position and orientation in the world. We wish to render this object given a camera in any position and orientation. To do this, we must take the vertices of the object (assuming we are rendering some sort of triangle mesh) and transform them from object space into world space. This transform is known as the *model transform*, which we denote $\mathbf{M}_{\text{obj}\to\text{wld}}$. From there, we transform world-space vertices with the *view transform*, denoted $\mathbf{M}_{\text{wld}\to\text{cam}}$, into camera space. The math involved is summarized by

$$\mathbf{p}_{\text{wld}} = \mathbf{p}_{\text{obj}}\,\mathbf{M}_{\text{obj}\to\text{wld}},$$
$$\mathbf{p}_{\text{cam}} = \mathbf{p}_{\text{wld}}\,\mathbf{M}_{\text{wld}\to\text{cam}}$$
$$= (\mathbf{p}_{\text{obj}}\,\mathbf{M}_{\text{obj}\to\text{wld}})\,\mathbf{M}_{\text{wld}\to\text{cam}}.$$

From Section 4.1.6, we know that matrix multiplication is associative, and so we could compute one matrix to transform directly from object to camera space:

$$\mathbf{p}_{\text{cam}} = (\mathbf{p}_{\text{obj}} \, \mathbf{M}_{\text{obj}\to\text{wld}}) \, \mathbf{M}_{\text{wld}\to\text{cam}}$$
$$= \mathbf{p}_{\text{obj}} (\mathbf{M}_{\text{obj}\to\text{wld}} \, \mathbf{M}_{\text{wld}\to\text{cam}}).$$

Thus, we can concatenate the matrices outside the vertex loop, and have only one matrix multiplication inside the loop (remember there are many vertices):

$$\mathbf{M}_{\text{obj}\to\text{cam}} = \mathbf{M}_{\text{obj}\to\text{wld}} \, \mathbf{M}_{\text{wld}\to\text{cam}},$$
$$\mathbf{p}_{\text{cam}} = \mathbf{p}_{\text{obj}} \, \mathbf{M}_{\text{obj}\to\text{cam}}.$$

So we see that matrix concatenation works from an algebraic perspective by using the associative property of matrix multiplication. Let's see if we can get a more geometric interpretation of what's going on. Recall from Section 4.2, our breakthrough discovery, that the rows of a matrix contain the result of transforming the standard basis vectors. This is true even in the case of multiple transformations. Notice that in the matrix product $\mathbf{AB}$, each resulting row is the product of the corresponding row from $\mathbf{A}$ times the matrix $\mathbf{B}$. In other words, let the row vectors $\mathbf{a}_1$, $\mathbf{a}_2$, and $\mathbf{a}_3$ stand for the rows of $\mathbf{A}$. Then matrix multiplication can alternatively be written as

$$\mathbf{A} = \begin{bmatrix} -\mathbf{a}_1- \\ -\mathbf{a}_2- \\ -\mathbf{a}_3- \end{bmatrix}, \qquad \mathbf{AB} = \left(\begin{bmatrix} -\mathbf{a}_1- \\ -\mathbf{a}_2- \\ -\mathbf{a}_3- \end{bmatrix} \mathbf{B} \right) = \begin{bmatrix} -\mathbf{a}_1\mathbf{B}- \\ -\mathbf{a}_2\mathbf{B}- \\ -\mathbf{a}_3\mathbf{B}- \end{bmatrix}.$$

This explicitly shows that the rows of the product of $\mathbf{AB}$ are actually the result of transforming the basis vectors in $\mathbf{A}$ by $\mathbf{B}$.

5.7   Classes of Transformations

We can classify transformations according to several criteria. This section discuss classes of transformations. For each class, we describe the properties of the transformations that belong to that class and specify which of the primitive transformations from Sections 5.1 through 5.5 belong to that class. The classes of transformations are not mutually exclusive, nor do they necessarily follow an "order" or "hierarchy," with each one more or less restrictive than the next.

When we discuss transformations in general, we may make use of the synonymous terms *mapping* or *function*. In the most general sense, a mapping is simply a rule that takes an input and produces an output. We denote that the mapping F maps a to b by writing $F(a) = b$ (read "F of a equals b"). Of course, we are primarily interested in the transformations that can be expressed by using matrix multiplication, but it is important to note that other mappings are possible.

This section also mentions the *determinant* of a matrix. We're getting a bit ahead of ourselves here, since a full explanation of determinants isn't given until Section 6.1. For now, just know that the determinant of a matrix is a scalar quantity that is very useful for making certain high-level, shall we say, *determinations* about the matrix.

5.7.1   Linear Transformations

We met linear functions informally in Section 4.2. Mathematically, a mapping $F(\mathbf{a})$ is linear if

$$F(\mathbf{a} + \mathbf{b}) = F(\mathbf{a}) + F(\mathbf{b}) \tag{5.2}$$

and

$$F(k\mathbf{a}) = kF(\mathbf{a}). \tag{5.3}$$

Conditions satisfied if F is a linear mapping

This is a fancy way of stating that the mapping F is linear if it preserves the basic operations of addition and multiplication by a scalar. If we add two vectors, and then perform the transformation, we get the same result as if we perform the transformation on the two vectors individually and then add the transformed vectors. Likewise, if we scale a vector and then transform it, we should get the same resulting vector as when we transform the vector and then scale it.

There are two important implications of this definition of linear transformation. First, the mapping $F(\mathbf{a}) = \mathbf{aM}$, where $\mathbf{M}$ is any square matrix, is a linear transformation, because

$$F(\mathbf{a} + \mathbf{b}) = (\mathbf{a} + \mathbf{b})\mathbf{M} = \mathbf{aM} + \mathbf{bM} = F(\mathbf{a}) + F(\mathbf{b})$$

Matrix multiplication satisfies Equation (5.2)

and

$$F(k\mathbf{a}) = (k\mathbf{a})\mathbf{M} = k(\mathbf{aM}) = kF(\mathbf{a}).$$

Matrix multiplication satisfies Equation (5.3)

In other words:

Any transformation that can be accomplished with matrix multiplication is a linear transformation.

Second, any linear transformation will transform the zero vector into the zero vector. If $F(\mathbf{0}) = \mathbf{a}, \mathbf{a} \neq \mathbf{0}$, then F cannot be a linear mapping, since $F(k\mathbf{0}) = \mathbf{a}$ and therefore $F(k\mathbf{0}) \neq kF(\mathbf{0})$. Because of this,

Linear transformations do not contain translation.

Since all of the transformations we discussed in Sections 5.1 through 5.5 can be expressed using matrix multiplication, they are all linear transformations.

In some literature, a linear transformation is defined as one in which parallel lines remain parallel after transformation. This is almost completely accurate, with two exceptions. First, parallel lines remain parallel after translation, but translation is not a linear transformation. Second, what about projection? When a line is projected and becomes a single point, can we consider that point "parallel" to anything? Excluding these technicalities, the intuition is correct: a linear transformation may "stretch" things, but straight lines are not "warped" and parallel lines remain parallel.

5.7.2   Affine Transformations

An *affine* transformation is a linear transformation followed by translation. Thus, the set of affine transformations is a superset of the set of linear transformations: any linear transformation is an affine translation, but not all affine transformations are linear transformations.

Since all of the transformations discussed in this chapter are linear transformations, they are all also affine transformations (though none of them have a translation portion). Any transformation of the form $\mathbf{v}' = \mathbf{vM} + \mathbf{b}$ is an affine transformation.

5.7.3   Invertible Transformations

A transformation is *invertible* if there exists an opposite transformation, known as the *inverse* of F, that "undoes" the original transformation. In other words, a mapping $F(\mathbf{a})$ is invertible if there exists an inverse mapping F^{-1} such that

$$F^{-1}(F(\mathbf{a})) = F(F^{-1}(\mathbf{a})) = \mathbf{a}$$

for all $\mathbf{a}$. Notice that this implies that F^{-1} is also invertible.

There are nonaffine invertible transformations, but we will not consider them for the moment. For now, let's concentrate on determining if an affine transformation is invertible. As already stated, an affine transformation is a linear transformation followed by a translation. Obviously, we can always "undo" the translation portion by simply translating by the opposite amount. So the question becomes whether the linear transformation is invertible.

Intuitively, we know that all of the transformations other than projection can be "undone." If we rotate, scale, reflect, or skew, we can always "unrotate," "unscale," "unreflect," or "unskew." But when an object is projected, we effectively discard one or more dimensions' worth of information, and this information cannot be recovered. Thus, all of the primitive transformations other than projection are invertible.

Since any linear transformation can be expressed as multiplication by a matrix, finding the inverse of a linear transformation is equivalent to finding the inverse of a matrix. We discuss how to do this in Section 6.2. If the matrix has no inverse, we say that it is *singular*, and the transformation is noninvertible. The determinant of an invertible matrix is nonzero.

In a nonsingular matrix, the zero vector is the only input vector that is mapped to the zero vector in the output space; all other vectors are mapped to some other nonzero vector. In a singular matrix, however, there exists an entire subspace of the input vectors, known as the *null space* of the matrix, that is mapped to the zero vector. For example, consider a matrix that projects orthographically onto a plane containing the origin. The null space of this matrix consists of the line of vectors perpendicular to the plane, since they are all mapped to the origin.

When a square matrix is singular, its basis vectors are not linearly independent (see Section 3.3.3). If the basis vectors are linearly independent, then they have full rank, and coordinates of any given vector in the span are uniquely determined. If the vectors are linearly dependent, then there is a portion of the full n-dimensional space that is not in the span of the basis. Consider two vectors $\mathbf{a}$ and $\mathbf{b}$, which differ by a vector $\mathbf{n}$ that lies in the null space of a matrix $\mathbf{M}$, such that $\mathbf{b} = \mathbf{a} + \mathbf{n}$. Due to the linear nature of matrix multiplication, $\mathbf{M}$ maps $\mathbf{a}$ and $\mathbf{b}$ to the same output:

$$
\begin{aligned}
\mathbf{bM} &= (\mathbf{a} + \mathbf{n})\mathbf{M} \\
&= \mathbf{aM} + \mathbf{nM} \quad \text{(Matrix multiplication is linear and distributes)} \\
&= \mathbf{aM} + \mathbf{0} \quad \text{(}\mathbf{n}\text{ is in the null space of }\mathbf{M}\text{)} \\
&= \mathbf{aM}.
\end{aligned}
$$

5.7.4 Angle-Preserving Transformations

A transformation is *angle-preserving* if the angle between two vectors is not altered in either magnitude or direction after transformation. Only translation, rotation, and uniform scale are angle-preserving transformations. An angle-preserving matrix preserves proportions. We do not consider reflection an angle-preserving transformation because even though the magnitude of angle between two vectors is the same after transformation, the direction of angle may be inverted. All angle-preserving transformations are affine and invertible.

5.7.5   Orthogonal Transformations

Orthogonal is a term that is used to describe a matrix whose rows form an orthonormal basis. Remember from Section 3.3.3 that the basic idea is that the axes are perpendicular to each other and have unit length. Orthogonal transformations are interesting because it is easy to compute their inverse and they arise frequently in practice. We talk more about orthogonal matrices in Section 6.3.

Translation, rotation, and reflection are the only orthogonal transformations. All orthogonal transformations are affine and invertible. Lengths, angles, areas, and volumes are all preserved; however in saying this, we must be careful as to our precise definition of angle, area, and volume, since reflection is an orthogonal transformation and we just got through saying in the previous section that we didn't consider reflection to be an angle-preserving transformation. Perhaps we should be more precise and say that orthogonal matrices preserve the *magnitudes* of angles, areas, and volumes, but possibly not the signs.

As Chapter 6 shows, the determinant of an orthogonal matrix is ± 1.

5.7.6   Rigid Body Transformations

A *rigid body transformation* is one that changes the location and orientation of an object, but not its shape. All angles, lengths, areas, and volumes are preserved. Translation and rotation are the only rigid body transformations. Reflection is not considered a rigid body transformation.

Rigid body transformations are also known as *proper transformations*. All rigid body transformations are orthogonal, angle-preserving, invertible, and affine. Rigid body transforms are the most restrictive class of transforms discussed in this section, but they are also extremely common in practice.

The determinant of any rigid body transformation matrix is 1.

5.7.7   Summary of Types of Transformations

Table 5.1 summarizes the various classes of transformations. In this table, a Y means that the transformation in that row always has the property associated with that column. The absence of a Y does not mean "never"; rather, it means "not always."

Transform	Linear	Affine	Invertible	Angles preserved	Orthogonal	Rigid body	Lengths preserved	Areas/volumes preserved	Determinant
Linear transformations	Y	Y							
Affine transformations		Y							$\neq 0$
Invertible transformations			Y						
Angle-preserving transformations		Y	Y	Y					
Orthogonal transformations		Y	Y		Y				± 1
Rigid body transformations		Y	Y	Y	Y	Y	Y	Y	1
Translation		Y	Y	Y	Y	Y	Y	Y	1
Rotation[1]	Y	Y	Y	Y	Y	Y	Y	Y	1
Uniform scale[2]	Y	Y	Y	Y					k^{n} [3]
Non-uniform scale	Y	Y	Y						
Orthographic projection[4]	Y	Y							0
Reflection[5]	Y	Y	Y		Y		Y[6]	Y	-1
Shearing	Y	Y	Y					Y[7]	1

[1] About the origin in 2D or an axis passing through the origin in 3D.
[2] About the origin in 2D or an axis passing through the origin in 3D.
[3] The determinant is the square of the scale factor in 2D, and the cube of the scale factor in 3D.
[4] Onto a line (2D) or plane (3D) that passes through the origin.
[5] About a line (2D) or plane (3D) that passes through the origin.
[6] Not considering "negative" area or volume.
[7] Surprisingly!

Table 5.1. Types of transformations

5.8 Exercises

(Answers on page 763.)

1. Does the matrix below express a linear transformation? Affine?

$$\begin{bmatrix} 34 & 1.7 & \pi \\ \sqrt{2} & 0 & 18 \\ 4 & -9 & -1.3 \end{bmatrix}$$

2. Construct a matrix to rotate $-22°$ about the x-axis.

3. Construct a matrix to rotate $30°$ about the y-axis.

4. Construct a matrix to rotate $-15°$ about the axis $[0.267, -0.535, 0.802]$.

5. Construct a matrix that doubles the height, width, and length of an object in 3D.

6. Construct a matrix to scale by a factor of 5 about the plane through the origin perpendicular to the vector $[0.267, -0.535, 0.802]$.

7. Construct a matrix to orthographically project onto the plane through the origin perpendicular to the vector $[0.267, -0.535, 0.802]$.

8. Construct a matrix to reflect orthographically about the plane through the origin perpendicular to the vector $[0.267, -0.535, 0.802]$.

9. An object initially had its axes and origin coincident with the world axes and origin. It was then rotated $30°$ about the y-axis and then $-22°$ about the world x-axis.

   (a) What is the matrix that can be used to transform row vectors from object space to world space?

   (b) What about the matrix to transform vectors from world space to object space?

   (c) Express the object's z-axis using upright coordinates.

Chapter 6

More on Matrices

Man's mind stretched to a new idea
never goes back to its original dimensions.

— Oliver Wendell Holmes Jr. (1841–1935)

Chapter 4 presented a few of the most of the important properties and operations of matrices and discussed how matrices can be used to express geometric transformations in general. Chapter 5 considered matrices and geometric transforms in detail. This chapter completes our coverage of matrices by discussing a few more interesting and useful matrix operations.

- Section 6.1 covers the *determinant* of a matrix.

- Section 6.2 covers the *inverse* of a matrix.

- Section 6.3 discusses *orthogonal* matrices.

- Section 6.4 introduces *homogeneous* vectors and 4×4 matrices, and shows how they can be used to perform affine transformations in 3D.

- Section 6.5 discusses *perspective projection* and shows how to do it with a 4×4 matrix.

6.1   Determinant of a Matrix

For square matrices, there is a special scalar called the *determinant* of the matrix. The determinant has many useful properties in linear algebra, and it also has interesting geometric interpretations.

As is our custom, we first discuss some math, and then make some geometric interpretations. Section 6.1.1 introduces the notation for determinants and gives the linear algebra rules for computing the determinant of a 2×2 or 3×3 matrix. Section 6.1.2 discusses *minors* and *cofactors*. Then, Section 6.1.3 shows how to compute the determinant of an arbitrary $n \times n$ matrix, by using minors and cofactors. Finally, Section 6.1.4 interprets the determinant from a geometric perspective.

6.1.1   Determinants of 2×2 and 3×3 matrices

The determinant of a square matrix $\mathbf{M}$ is denoted $|\mathbf{M}|$ or, in some other books, as "det $\mathbf{M}$." The determinant of a nonsquare matrix is undefined. This section shows how to compute determinants of 2×2 and 3×3 matrices. The determinant of a general $n \times n$ matrix, which is fairly complicated, is discussed in Section 6.1.3

The determinant of a **2 $\times$ 2** matrix is given by

Determinant of a 2 $\times$ 2 matrix

$$|\mathbf{M}| = \begin{vmatrix} m_{11} & m_{12} \\ m_{21} & m_{22} \end{vmatrix} = m_{11}m_{22} - m_{12}m_{21}. \tag{6.1}$$

Notice that when we write the determinant of a matrix, we replace the brackets with vertical lines.

Equation (6.1) can be remembered easier with the following diagram. Simply multiply entries along the diagonal and back-diagonal, then subtract the back-diagonal term from the diagonal term.

$$|\mathbf{M}| = \begin{vmatrix} m_{11} & m_{12} \\ m_{21} & m_{22} \end{vmatrix} = m_{11}m_{22} - m_{12}m_{21}$$

Some examples help to clarify the simple calculation:

$$\begin{vmatrix} 2 & 1 \\ -1 & 2 \end{vmatrix} = (2)(2) - (1)(-1) = 4 + 1 = 5;$$

$$\begin{vmatrix} -3 & 4 \\ 2 & 5 \end{vmatrix} = (-3)(5) - (4)(2) = -15 - 8 = -23;$$

$$\begin{vmatrix} a & b \\ c & d \end{vmatrix} = ad - bc.$$

The determinant of a 3×3 matrix is given by

$$\begin{vmatrix} m_{11} & m_{12} & m_{13} \\ m_{21} & m_{22} & m_{23} \\ m_{31} & m_{32} & m_{33} \end{vmatrix}$$

Determinant of a 3×3 matrix

$$\begin{aligned} = \ & m_{11}m_{22}m_{33} + m_{12}m_{23}m_{31} + m_{13}m_{21}m_{32} \\ & - m_{13}m_{22}m_{31} - m_{12}m_{21}m_{33} - m_{11}m_{23}m_{32} \end{aligned} \quad (6.2)$$

$$\begin{aligned} = \ & m_{11}(m_{22}m_{33} - m_{23}m_{32}) \\ & + m_{12}(m_{23}m_{31} - m_{21}m_{33}) \\ & + m_{13}(m_{21}m_{32} - m_{22}m_{31}). \end{aligned}$$

A similar diagram can be used to memorize Equation (6.2). We write two copies of the matrix $\mathbf{M}$ side by side and multiply entries along the diagonals and back-diagonals, adding the diagonal terms and subtracting the back-diagonal terms.

$$m_{11}m_{22}m_{33} + m_{12}m_{23}m_{31} + m_{13}m_{21}m_{32}$$
$$- m_{13}m_{22}m_{31} - m_{12}m_{21}m_{33} - m_{11}m_{23}m_{32}$$

For example,

$$\begin{vmatrix} -4 & -3 & 3 \\ 0 & 2 & -2 \\ 1 & 4 & -1 \end{vmatrix} = \begin{array}{l} (-4)\big((\ 2)(-1) - (-2)(\ 4)\big) \\ +(-3)\big((-2)(\ 1) - (\ 0)(-1)\big) \\ +(\ 3)\big((\ 0)(\ 4) - (\ 2)(\ 1)\big) \end{array}$$

$$= \begin{array}{l} (-4)\big((-2) - (-8)\big) \\ +(-3)\big((-2) - (\ 0)\big) \\ +(\ 3)\big((\ 0) - (\ 2)\big) \end{array} = \begin{array}{l} (-4)(\ 6) \\ +(-3)(-2) \\ +(\ 3)(-2) \end{array} = \begin{array}{l} (-24) \\ +(\ 6) \\ +(\ -6) \end{array}$$

$$= -24. \quad (6.3)$$

If we interpret the rows of a 3×3 matrix as three vectors, then the determinant of the matrix is equivalent to the so-called "triple product" of the three vectors:

$$\begin{vmatrix} a_x & a_y & a_z \\ b_x & b_y & b_z \\ c_x & c_y & c_z \end{vmatrix} = \begin{array}{l} (a_y b_z - a_z b_y)\, c_x \\ + (a_z b_x - a_x b_z)\, c_y \\ + (a_x b_y - a_y b_x)\, c_z \end{array} = (\mathbf{a} \times \mathbf{b}) \cdot \mathbf{c}.$$

3×3 determinant vs. 3D vector triple product

6.1.2   Minors and Cofactors

Before we can look at determinants in the general case, we need to introduce some other constructs: *minors* and *cofactors*.

Assume $\mathbf{M}$ is a matrix with r rows and c columns. Consider the matrix obtained by deleting row i and column j from $\mathbf{M}$. This matrix will obviously have $r - 1$ rows and $c - 1$ columns. The determinant of this submatrix, denoted $M^{\{ij\}}$ is known as a *minor* of $\mathbf{M}$. For example, the minor $M^{\{12\}}$ is the determinant of the 2×2 matrix that is the result of deleting row 1 and column 2 from the 3×3 matrix $\mathbf{M}$:

A minor of a 3 × 3 matrix

$$\mathbf{M} = \begin{bmatrix} -4 & -3 & 3 \\ 0 & 2 & -2 \\ 1 & 4 & -1 \end{bmatrix} \implies M^{\{12\}} = \begin{vmatrix} 0 & -2 \\ 1 & -1 \end{vmatrix} = 2.$$

The *cofactor* of a square matrix $\mathbf{M}$ at a given row and column is the same as the corresponding minor, but with alternating minors negated:

Matrix cofactor

$$C^{\{ij\}} = (-1)^{i+j} M^{\{ij\}}. \tag{6.4}$$

As shown in Equation (6.4), we use the notation $C^{\{ij\}}$ to denote the cofactor of $\mathbf{M}$ in row i, column j. The $(-1)^{(i+j)}$ term has the effect of negating every other cofactor in a checkerboard pattern:

$$\begin{bmatrix} + & - & + & - & \cdots \\ - & + & - & + & \cdots \\ + & - & + & - & \cdots \\ - & + & - & + & \cdots \\ \vdots & \vdots & \vdots & \vdots & \ddots \end{bmatrix}.$$

In the next section, we use minors and cofactors to compute determinants of an arbitrary dimension $n \times n$, and again in Section 6.2 to compute the inverse of a matrix.

6.1.3   Determinants of Arbitrary $n \times n$ Matrices

Several equivalent definitions exist for the determinant of a matrix of arbitrary dimension $n \times n$. The definition we consider here expresses a determinant in terms of its cofactors. This definition is recursive, since cofactors are themselves signed determinants. First, we arbitrarily select a row or column from the matrix. Now, for each element in the row or column, we multiply this element by the corresponding cofactor. Summing these products yields the determinant of the matrix. For example, arbitrarily selecting row i, the determinant can be computed by

$$|\mathbf{M}| = \sum_{j=1}^{n} m_{ij} C^{\{ij\}} = \sum_{j=1}^{n} m_{ij} (-1)^{i+j} M^{\{ij\}}. \qquad (6.5)$$

Computing an $n \times n$ determinant by using cofactors of row i

As it turns out, it doesn't matter which row or column we choose; they all will produce the same result.

Let's look at an example. We'll rewrite the equation for 3×3 determinant using Equation (6.5):

$$\begin{vmatrix} m_{11} & m_{12} & m_{13} \\ m_{21} & m_{22} & m_{23} \\ m_{31} & m_{32} & m_{33} \end{vmatrix} = m_{11} \begin{vmatrix} m_{22} & m_{23} \\ m_{32} & m_{33} \end{vmatrix} - m_{12} \begin{vmatrix} m_{21} & m_{23} \\ m_{31} & m_{33} \end{vmatrix}$$

$$+ m_{13} \begin{vmatrix} m_{21} & m_{22} \\ m_{31} & m_{32} \end{vmatrix}.$$

Recursive definition of determinant applied to 3×3 case

Now, let's derive the 4×4 matrix determinant:

$$\begin{vmatrix} m_{11} & m_{12} & m_{13} & m_{14} \\ m_{21} & m_{22} & m_{23} & m_{24} \\ m_{31} & m_{32} & m_{33} & m_{34} \\ m_{41} & m_{42} & m_{43} & m_{44} \end{vmatrix} = m_{11} \begin{vmatrix} m_{22} & m_{23} & m_{24} \\ m_{32} & m_{33} & m_{34} \\ m_{42} & m_{43} & m_{44} \end{vmatrix}$$

Recursive definition of determinant applied to 4×4 case

$$- m_{12} \begin{vmatrix} m_{21} & m_{23} & m_{24} \\ m_{31} & m_{33} & m_{34} \\ m_{41} & m_{43} & m_{44} \end{vmatrix} + m_{13} \begin{vmatrix} m_{21} & m_{22} & m_{24} \\ m_{31} & m_{32} & m_{34} \\ m_{41} & m_{42} & m_{44} \end{vmatrix}$$

$$- m_{14} \begin{vmatrix} m_{21} & m_{22} & m_{23} \\ m_{31} & m_{32} & m_{33} \\ m_{41} & m_{42} & m_{43} \end{vmatrix}.$$

Expanding the cofactors, we have

$$m_{11} \left[m_{22}(m_{33}m_{44} - m_{34}m_{43}) + m_{23}(m_{34}m_{42} - m_{32}m_{44}) + m_{24}(m_{32}m_{43} - m_{33}m_{42}) \right]$$
$$- m_{12} \left[m_{21}(m_{33}m_{44} - m_{34}m_{43}) + m_{23}(m_{34}m_{41} - m_{31}m_{44}) + m_{24}(m_{31}m_{43} - m_{33}m_{41}) \right]$$
$$+ m_{13} \left[m_{21}(m_{32}m_{44} - m_{34}m_{42}) + m_{22}(m_{34}m_{41} - m_{31}m_{44}) + m_{24}(m_{31}m_{42} - m_{32}m_{41}) \right]$$
$$- m_{14} \left[m_{21}(m_{32}m_{43} - m_{33}m_{42}) + m_{22}(m_{33}m_{41} - m_{31}m_{43}) + m_{23}(m_{31}m_{42} - m_{32}m_{41}) \right].$$

Determinant of a 4×4 matrix in expanded form

As you can imagine, the complexity of explicit formulas for determinants of higher degree grows rapidly. Luckily, we can perform an operation known as "pivoting," which doesn't affect the value of the determinant, but causes a particular row or column to be filled with zeroes except for a single element (the "pivot" element). Then only one cofactor has to be evaluated. Since we won't need determinants of matrices higher than the 4×4 case, anyway, a complete discussion of pivoting is outside the scope of this book.

Let's briefly state some important characteristics concerning determinants.

- The determinant of an identity matrix of any dimension is 1:

Determinant of identity matrix

$$|\mathbf{I}| = 1.$$

- The determinant of a matrix product is equal to the product of the determinants:

Determinant of matrix product

$$|\mathbf{AB}| = |\mathbf{A}||\mathbf{B}|.$$

This extends to more than two matrices:

$$|\mathbf{M}_1\mathbf{M}_2\cdots\mathbf{M}_{n-1}\mathbf{M}_n| = |\mathbf{M}_1||\mathbf{M}_2|\cdots|\mathbf{M}_{n-1}||\mathbf{M}_n|.$$

- The determinant of the transpose of a matrix is equal to the original determinant:

Determinant of matrix transpose

$$\left|\mathbf{M}^{\mathrm{T}}\right| = |\mathbf{M}|.$$

- If any row or column in a matrix contains all 0s, then the determinant of that matrix is 0:

Determinant of matrix with a row/column full of 0s

$$
\begin{vmatrix}
? & ? & \cdots & ? \\
? & ? & \cdots & ? \\
\vdots & \vdots & & \vdots \\
0 & 0 & \cdots & 0 \\
\vdots & \vdots & & \vdots \\
? & ? & \cdots & ?
\end{vmatrix}
=
\begin{vmatrix}
? & ? & \cdots & 0 & \cdots & ? \\
? & ? & \cdots & 0 & \cdots & ? \\
\vdots & \vdots & & \vdots & & \vdots \\
? & ? & \cdots & 0 & \cdots & ?
\end{vmatrix}
= 0.
$$

- Exchanging any pair of rows negates the determinant:

Swapping rows negates the determinant

$$
\begin{vmatrix}
m_{11} & m_{12} & \cdots & m_{1n} \\
m_{21} & m_{22} & \cdots & m_{2n} \\
\vdots & \vdots & & \vdots \\
m_{i1} & m_{i2} & \cdots & m_{in} \\
\vdots & \vdots & & \vdots \\
m_{j1} & m_{j2} & \cdots & m_{jn} \\
\vdots & \vdots & & \vdots \\
m_{n1} & m_{n2} & \cdots & m_{nn}
\end{vmatrix}
= -
\begin{vmatrix}
m_{11} & m_{12} & \cdots & m_{1n} \\
m_{21} & m_{22} & \cdots & m_{2n} \\
\vdots & \vdots & & \vdots \\
m_{j1} & m_{j2} & \cdots & m_{jn} \\
\vdots & \vdots & & \vdots \\
m_{i1} & m_{i2} & \cdots & m_{in} \\
\vdots & \vdots & & \vdots \\
m_{n1} & m_{n2} & \cdots & m_{nn}
\end{vmatrix}.
$$

This same rule applies for exchanging a pair of columns.

- Adding any multiple of a row (column) to another row (column) *does not change* the value of the determinant!

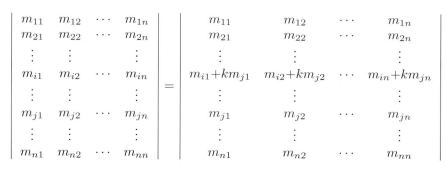

This explains why our shear matrices from Section 5.5 have a determinant of 1.

6.1.4 Geometric Interpretation of Determinant

The determinant of a matrix has an interesting geometric interpretation. In 2D, the determinant is equal to the signed area of the parallelogram or *skew box* that has the basis vectors as two sides (see Figure 6.1). (We discussed

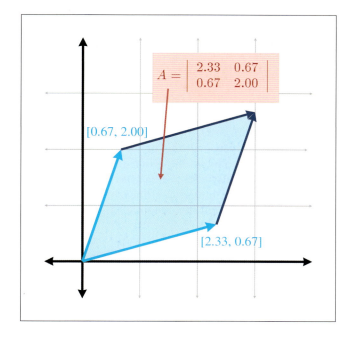

Figure 6.1
The determinant in 2D is the signed area of the skew box formed by the transformed basis vectors.

how we can use skew boxes to visualize coordinate space transformations in Section 4.2.) By *signed area*, we mean that the area is negative if the skew box is "flipped" relative to its original orientation.

In 3D, the determinant is the volume of the *parallelepiped* that has the transformed basis vectors as three edges. It will be negative if the object is reflected ("turned inside out") as a result of the transformation.

The determinant is related to the change in size that results from transforming by the matrix. The absolute value of the determinant is related to the change in area (in 2D) or volume (in 3D) that will occur as a result of transforming an object by the matrix, and the sign of the determinant indicates whether any reflection or projection is contained in the matrix.

The determinant of the matrix can also be used to help classify the type of transformation represented by a matrix. If the determinant of a matrix is zero, then the matrix contains a projection. If the determinant of a matrix is negative, then reflection is contained in the matrix. See Section 5.7 for more about different classes of transformations.

6.2   Inverse of a Matrix

Another important operation that applies only to square matrices is the *inverse* of a matrix. This section discusses the matrix inverse from a mathematical and geometric perspective.

The inverse of a square matrix $\mathbf{M}$, denoted $\mathbf{M}^{-1}$ is the matrix such that when we multiply $\mathbf{M}$ by $\mathbf{M}^{-1}$ on either side, the result is the identity matrix. In other words,

Matrix inverse
$$\mathbf{M}(\mathbf{M}^{-1}) = \mathbf{M}^{-1}\mathbf{M} = \mathbf{I}.$$

Not all matrices have an inverse. An obvious example is a matrix with a row or column filled with 0s—no matter what you multiply this matrix by, the corresponding row or column in the result will also be full of 0s. If a matrix has an inverse, it is said to be *invertible* or *nonsingular*. A matrix that does not have an inverse is said to be *noninvertible* or *singular*. For any invertible matrix $\mathbf{M}$, the vector equality $\mathbf{vM} = \mathbf{0}$ is true only when $\mathbf{v} = \mathbf{0}$. Furthermore, the rows of an invertible matrix are linearly independent, as are the columns. The rows (and columns) of a singular matrix are linearly dependent.

The determinant of a singular matrix is zero and the determinant of a nonsingular matrix is nonzero. Checking the magnitude of the determinant is the most commonly used test for invertibility because it's the easiest and quickest. In ordinary circumstances, this is OK, but please note that the method can break down. An example is an extreme shear matrix with basis vectors that form a very long, thin parallelepiped with unit volume. This *ill conditioned* matrix is nearly singular, even though its determinant is 1. The *condition number* is the proper tool for detecting such cases, but this is an advanced topic slightly beyond the scope of this book.

There are several ways to compute the inverse of a matrix. The one we use is based on the *classical adjoint*, which is the subject of the next section.

6.2.1 The Classical Adjoint

Our method for computing the inverse of a matrix is based on the *classical adjoint*. The classical adjoint of a matrix $\mathbf{M}$, denoted "adj $\mathbf{M}$," is defined as the transpose of the matrix of cofactors of $\mathbf{M}$.

Let's look at an example. Take the 3×3 matrix $\mathbf{M}$ given earlier:

$$\mathbf{M} = \begin{bmatrix} -4 & -3 & 3 \\ 0 & 2 & -2 \\ 1 & 4 & -1 \end{bmatrix}.$$

First, we compute the cofactors of $\mathbf{M}$, as discussed in Section 6.1.2:

$$C^{\{11\}} = + \begin{vmatrix} 2 & -2 \\ 4 & -1 \end{vmatrix} = 6, \quad C^{\{12\}} = - \begin{vmatrix} 0 & -2 \\ 1 & -1 \end{vmatrix} = -2, \quad C^{\{13\}} = + \begin{vmatrix} 0 & 2 \\ 1 & 4 \end{vmatrix} = -2,$$

$$C^{\{21\}} = - \begin{vmatrix} -3 & 3 \\ 4 & -1 \end{vmatrix} = 9, \quad C^{\{22\}} = + \begin{vmatrix} -4 & 3 \\ 1 & -1 \end{vmatrix} = 1, \quad C^{\{23\}} = - \begin{vmatrix} -4 & -3 \\ 1 & 4 \end{vmatrix} = 13,$$

$$C^{\{31\}} = + \begin{vmatrix} -3 & 3 \\ 2 & -2 \end{vmatrix} = 0, \quad C^{\{32\}} = - \begin{vmatrix} -4 & 3 \\ 0 & -2 \end{vmatrix} = -8, \quad C^{\{33\}} = + \begin{vmatrix} -4 & -3 \\ 0 & 2 \end{vmatrix} = -8.$$

The classical adjoint of $\mathbf{M}$ is the transpose of the matrix of cofactors:

$$\text{adj } \mathbf{M} = \begin{bmatrix} C^{\{11\}} & C^{\{12\}} & C^{\{13\}} \\ C^{\{21\}} & C^{\{22\}} & C^{\{23\}} \\ C^{\{31\}} & C^{\{32\}} & C^{\{33\}} \end{bmatrix}^{\text{T}} \qquad (6.6)$$

The classical adjoint

$$= \begin{bmatrix} 6 & -2 & -2 \\ 9 & 1 & 13 \\ 0 & -8 & -8 \end{bmatrix}^{\text{T}} = \begin{bmatrix} 6 & 9 & 0 \\ -2 & 1 & -8 \\ -2 & 13 & -8 \end{bmatrix}.$$

6.2.2 Matrix Inverse—Official Linear Algebra Rules

To compute the inverse of a matrix, we divide the classical adjoint by the determinant:

$$\mathbf{M}^{-1} = \frac{\text{adj } \mathbf{M}}{|\mathbf{M}|}.$$

Computing matrix inverse from classical adjoint and determinant

If the determinant is zero, the division is undefined, which jives with our earlier statement that matrices with a zero determinant are noninvertible.

Let's look at an example. In the previous section we calculated the classical adjoint of a matrix $\mathbf{M}$; now let's calculate its inverse:

$$\mathbf{M} = \begin{bmatrix} -4 & -3 & 3 \\ 0 & 2 & -2 \\ 1 & 4 & -1 \end{bmatrix};$$

$$\mathbf{M}^{-1} = \frac{\text{adj } \mathbf{M}}{|\mathbf{M}|} = \frac{1}{-24} \begin{bmatrix} 6 & 9 & 0 \\ -2 & 1 & -8 \\ -2 & 13 & -8 \end{bmatrix} = \begin{bmatrix} -1/4 & -3/8 & 0 \\ 1/12 & -1/24 & 1/3 \\ 1/12 & -13/24 & 1/3 \end{bmatrix}.$$

Here the value of adj $\mathbf{M}$ comes from Equation (6.6), and $|\mathbf{M}|$ is from Equation (6.3).

There are other techniques that can be used to compute the inverse of a matrix, such as Gaussian elimination. Many linear algebra textbooks assert that such techniques are better suited for implementation on a computer because they require fewer arithmetic operations, and this assertion is true for larger matrices and matrices with a structure that may be exploited. However, for arbitrary matrices of smaller order, such as the 2×2, 3×3, and 4×4 matrices encountered most often in geometric applications, the classical adjoint method is generally the method of choice. The reason is that the classical adjoint method provides for a branchless implementation, meaning there are no `if` statements, or loops that cannot be unrolled statically. On today's superscalar architectures and dedicated vector processors, this is a big win.

We close this section with a quick list of several important properties concerning matrix inverses.

- The inverse of the inverse of a matrix is the original matrix:

$$(\mathbf{M}^{-1})^{-1} = \mathbf{M}.$$

(Of course, this assumes that $\mathbf{M}$ is nonsingular.)

- The identity matrix is its own inverse:

$$\mathbf{I}^{-1} = \mathbf{I}.$$

Note that there are other matrices that are their own inverse. For example, consider any reflection matrix, or a matrix that rotates $180°$ about any axis.

- The inverse of the transpose of a matrix is the transpose of the inverse of the matrix:

$$(\mathbf{M}^{\mathrm{T}})^{-1} = (\mathbf{M}^{-1})^{\mathrm{T}}$$

- The inverse of a matrix product is equal to the product of the inverses of the matrices, taken in *reverse order*:

$$(\mathbf{AB})^{-1} = \mathbf{B}^{-1}\mathbf{A}^{-1}.$$

This extends to more than two matrices:

$$(\mathbf{M}_1\mathbf{M}_2\cdots\mathbf{M}_{n-1}\mathbf{M}_n)^{-1} = \mathbf{M}_n{}^{-1}\mathbf{M}_{n-1}{}^{-1}\cdots\mathbf{M}_2{}^{-1}\mathbf{M}_1{}^{-1}.$$

- The determinant of the inverse is the reciprocal of the determinant of the original matrix:

$$\left|\mathbf{M}^{-1}\right| = 1/|\mathbf{M}|.$$

6.2.3 Matrix Inverse—Geometric Interpretation

The inverse of a matrix is useful geometrically because it allows us to compute the "reverse" or "opposite" of a transformation—a transformation that "undoes" another transformation if they are performed in sequence. So, if we take a vector, transform it by a matrix $\mathbf{M}$, and then transform it by the inverse $\mathbf{M}^{-1}$, then we will get the original vector back. We can easily verify this algebraically:

$$(\mathbf{vM})\mathbf{M}^{-1} = \mathbf{v}(\mathbf{MM}^{-1}) = \mathbf{vI} = \mathbf{v}.$$

6.3 Orthogonal Matrices

Previously we made reference to a special class of square matrices known as *orthogonal matrices*. This section investigates orthogonal matrices a bit more closely. As usual, we first introduce some pure math (Section 6.3.1), and then give some geometric interpretations (Section 6.3.2). Finally, we discuss how to adjust an arbitrary matrix to make it orthogonal (Section 6.3.3).

6.3.1 Orthogonal Matrices—Official Linear Algebra Rules

A square matrix $\mathbf{M}$ is orthogonal if and only if[1] the product of the matrix and its transpose is the identity matrix:

---

[1] The notation "$P \Leftrightarrow Q$" should be read "P if and only if Q" and denotes that the statement P is true *if and only if* Q is also true. "If and only if" is sort of like an equals sign for Boolean values. In other words, if either P or Q are true, then both must be true, and if either P or Q are false, then both must be false. The $\Leftrightarrow$ notation is also like the standard "$=$" notation in that it is *reflexive*. This is a fancy way of saying that it doesn't matter which is on the left and which is on the right; $P \Leftrightarrow Q$ implies $Q \Leftrightarrow P$.

Definition of orthogonal matrix

$$\mathbf{M} \text{ is orthogonal} \quad \Longleftrightarrow \quad \mathbf{MM}^\mathrm{T} = \mathbf{I}. \qquad (6.7)$$

Recall from Section 6.2.2 that, by definition, a matrix times its inverse is the identity matrix ($\mathbf{MM}^{-1} = \mathbf{I}$). Thus, if a matrix is orthogonal, its transpose and inverse are equal:

Equivalent definition of orthogonal matrix

$$\mathbf{M} \text{ is orthogonal} \quad \Longleftrightarrow \quad \mathbf{M}^\mathrm{T} = \mathbf{M}^{-1}.$$

This is extremely powerful information, because the inverse of a matrix is often needed, and orthogonal matrices arise frequently in practice in 3D graphics. For example, as mentioned in Section 5.7.5, rotation and reflection matrices are orthogonal. If we know that our matrix is orthogonal, we can essentially avoid computing the inverse, which is a relatively costly computation.

6.3.2 Orthogonal Matrices—Geometric Interpretation

Orthogonal matrices are interesting to us primarily because their inverse is trivial to compute. But how do we know if a matrix is orthogonal in order to exploit its structure?

In many cases, we may have information about the way the matrix was constructed and therefore know *a priori* that the matrix contains only rotation and/or reflection. This is a very common situation, and it's very important to take advantage of this when using matrices to describe rotation. We return to this topic in Section 8.2.1.

But what if we don't know anything in advance about the matrix? In other words, how can we tell if an arbitrary matrix $\mathbf{M}$ is orthogonal? Let's look at the 3×3 case, which is the most interesting one for our purposes. The conclusions we draw in this section can be extended to matrices of any dimension.

Let $\mathbf{M}$ be an orthogonal 3×3 matrix. Expanding the definition of orthogonality given by Equation (6.7), we have

$$
\begin{matrix}
\mathbf{M} & \mathbf{M}^\mathrm{T} & = & \mathbf{I},
\end{matrix}
$$

$$
\begin{bmatrix}
m_{11} & m_{12} & m_{13} \\
m_{21} & m_{22} & m_{23} \\
m_{31} & m_{32} & m_{33}
\end{bmatrix}
\begin{bmatrix}
m_{11} & m_{21} & m_{31} \\
m_{12} & m_{22} & m_{32} \\
m_{13} & m_{23} & m_{33}
\end{bmatrix}
=
\begin{bmatrix}
1 & 0 & 0 \\
0 & 1 & 0 \\
0 & 0 & 1
\end{bmatrix}.
$$

This gives us nine equations, all of which must be true for $\mathbf{M}$ to be orthogonal:

$$m_{11}m_{11} + m_{12}m_{12} + m_{13}m_{13} = 1, \tag{6.8}$$

$$m_{11}m_{21} + m_{12}m_{22} + m_{13}m_{23} = 0,$$

$$m_{11}m_{31} + m_{12}m_{32} + m_{13}m_{33} = 0,$$

$$m_{21}m_{11} + m_{22}m_{12} + m_{23}m_{13} = 0,$$

$$m_{21}m_{21} + m_{22}m_{22} + m_{23}m_{23} = 1, \tag{6.9}$$

$$m_{21}m_{31} + m_{22}m_{32} + m_{23}m_{33} = 0,$$

$$m_{31}m_{11} + m_{32}m_{12} + m_{33}m_{13} = 0,$$

$$m_{31}m_{21} + m_{32}m_{22} + m_{33}m_{23} = 0,$$

$$m_{31}m_{31} + m_{32}m_{32} + m_{33}m_{33} = 1. \tag{6.10}$$

Conditions satisfied by an orthogonal matrix

Let the vectors $\mathbf{r}_1$, $\mathbf{r}_2$, and $\mathbf{r}_3$ stand for the rows of $\mathbf{M}$:

$$\mathbf{r}_1 = \begin{bmatrix} m_{11} & m_{12} & m_{13} \end{bmatrix},$$

$$\mathbf{r}_2 = \begin{bmatrix} m_{21} & m_{22} & m_{23} \end{bmatrix},$$

$$\mathbf{r}_3 = \begin{bmatrix} m_{31} & m_{32} & m_{33} \end{bmatrix},$$

$$\mathbf{M} = \begin{bmatrix} -\mathbf{r}_1- \\ -\mathbf{r}_2- \\ -\mathbf{r}_3- \end{bmatrix}.$$

Now we can rewrite the nine equations more compactly:

$\mathbf{r}_1 \cdot \mathbf{r}_1 = 1,$	$\mathbf{r}_1 \cdot \mathbf{r}_2 = 0,$	$\mathbf{r}_1 \cdot \mathbf{r}_3 = 0,$
$\mathbf{r}_2 \cdot \mathbf{r}_1 = 0,$	$\mathbf{r}_2 \cdot \mathbf{r}_2 = 1,$	$\mathbf{r}_2 \cdot \mathbf{r}_3 = 0,$
$\mathbf{r}_3 \cdot \mathbf{r}_1 = 0,$	$\mathbf{r}_3 \cdot \mathbf{r}_2 = 0,$	$\mathbf{r}_3 \cdot \mathbf{r}_3 = 1.$

Conditions satisfied by an orthogonal matrix

This notational changes makes it easier for us to make some interpretations.

- First, the dot product of a vector with itself is 1 if and only if the vector is a unit vector. Therefore, the equations with a 1 on the right-hand side of the equals sign (Equations (6.8), (6.9), and (6.10)) will be true only when $\mathbf{r}_1$, $\mathbf{r}_2$, and $\mathbf{r}_3$ are unit vectors.

- Second, recall from Section 2.11.2 that the dot product of two vectors is 0 if and only if they are perpendicular. Therefore, the other six equations (with 0 on the right-hand side of the equals sign) are true when $\mathbf{r}_1$, $\mathbf{r}_2$, and $\mathbf{r}_3$ are mutually perpendicular.

So, for a matrix to be orthogonal, the following must be true:

- Each row of the matrix must be a unit vector.

- The rows of the matrix must be mutually perpendicular.

Similar statements can be made regarding the *columns* of the matrix, since if $\mathbf{M}$ is orthogonal, then $\mathbf{M}^T$ must be orthogonal as well.

Notice that these criteria are precisely those that we said in Section 3.3.3 were satisfied by an orthonormal set of basis vectors. In that section, we also noted that an orthonormal basis was particularly useful because we could perform, by using the dot product, the "opposite" coordinate transform from the one that is always available. When we say that the transpose of an orthogonal matrix is equal to its inverse, we are just restating this fact in the formal language of linear algebra.

Also notice that three of the orthogonality equations are duplicates, because the dot product is commutative. Thus, these nine equations actually express only six constraints. In an arbitrary 3×3 matrix there are nine elements and thus nine degrees of freedom, but in an orthogonal matrix, six degrees of freedom are removed by the constraints, leaving three degrees of freedom. It is significant that three is also the number of degrees of freedom inherent in 3D rotation. (However, rotation matrices cannot contain a reflection, so there is "slightly more freedom" in the set of orthogonal matrices than in the set of orientations in 3D.)

When computing a matrix inverse, we will usually only take advantage of orthogonality if we know *a priori* that a matrix is orthogonal. If we don't know in advance, it's probably a waste of time to check. In the best case, we check for orthogonality and find that the matrix is indeed orthogonal, and then we transpose the matrix. But this may take almost as much time as doing the inversion. In the worst case, the matrix is not orthogonal, and any time we spent checking was definitely wasted. Finally, even matrices that are orthogonal in the abstract may not be exactly orthogonal when represented in floating point, and so we must use tolerances, which have to be tuned.

One important note is needed here on terminology that can be slightly confusing. In linear algebra, we describe a set of basis vectors as *orthogonal* if they are mutually perpendicular. It is not required that they have unit length. If they do have unit length, they are an *orthonormal basis*. Thus the rows and columns of an *orthogonal matrix* are *orthonormal basis vectors*. However, constructing a matrix from a set of orthogonal basis vectors does not necessarily result in an orthogonal matrix (unless the basis vectors are also orthonormal).

6.3.3 Orthogonalizing a Matrix

It is sometimes the case that we encounter a matrix that is slightly out of orthogonality. We may have acquired bad data from an external source, or we may have accumulated floating point error (which is called *matrix creep*). For basis vectors used for bump mapping (see Section 10.9), we will often adjust the basis to be orthogonal, even if the texture mapping gradients aren't quite perpendicular. In these situations, we would like to *orthogonalize* the matrix, resulting in a matrix that has mutually perpendicular unit vector axes and is (hopefully) as close to the original matrix as possible.

The standard algorithm for constructing a set of orthogonal basis vectors (which is what the rows of an orthogonal matrix are) is *Gram-Schmidt* orthogonalization. The basic idea is to go through the basis vectors in order. For each basis vector, we subtract off the portion of that vector that is parallel to the proceeding basis vectors, which must result in a perpendicular vector.

Let's look at the 3×3 case as an example. As before, let $\mathbf{r}_1$, $\mathbf{r}_2$, and $\mathbf{r}_3$ stand for the rows of a 3×3 matrix $\mathbf{M}$. (Remember, you can also think of these as the x-, y-, and z-axes of a coordinate space.) Then an orthogonal set of row vectors, $\mathbf{r}_1'$, $\mathbf{r}_2'$, and $\mathbf{r}_3'$, can be computed according to the following algorithm:

$$\mathbf{r}_1' \Leftarrow \mathbf{r}_1,$$

$$\mathbf{r}_2' \Leftarrow \mathbf{r}_2 - \frac{\mathbf{r}_2 \cdot \mathbf{r}_1'}{\mathbf{r}_1' \cdot \mathbf{r}_1'}\mathbf{r}_1',$$

$$\mathbf{r}_3' \Leftarrow \mathbf{r}_3 - \frac{\mathbf{r}_3 \cdot \mathbf{r}_1'}{\mathbf{r}_1' \cdot \mathbf{r}_1'}\mathbf{r}_1' - \frac{\mathbf{r}_3 \cdot \mathbf{r}_2'}{\mathbf{r}_2' \cdot \mathbf{r}_2'}\mathbf{r}_2'.$$

Gram-Schmidt orthogonalization of 3D basis vectors

After applying these steps, the vectors $\mathbf{r}_1$, $\mathbf{r}_2$, and $\mathbf{r}_3$ are guaranteed to be mutually perpendicular, and thus will form an orthogonal basis. However, they may not necessarily be unit vectors. We need an orthonormal basis to form an orthogonal matrix, and so we must normalize the vectors. (Again, the terminology can be confusing, see the note at the end of the previous section.) Notice that if we normalize the vectors as we go, rather than in a second pass, then we can avoid all of the divisions. Also, a trick that works in 3D (but not in higher dimensions) is to compute the third basis vector using the cross product:

$$\mathbf{r}_3' \Leftarrow \mathbf{r}_1' \times \mathbf{r}_2'.$$

The Gram-Schmidt algorithm is biased, depending on the order in which the basis vectors are listed. For instance, $\mathbf{r}_1$ never changes, and $\mathbf{r}_3$ is likely to change the most. A variation on the algorithm that is not biased towards any particular axis is to abandon the attempt to completely orthogonalize the entire matrix in one pass. We select some fraction k, and instead of subtracting off all of the projection, we subtract off only k of it. We also subtract the projection onto the original axis, not the adjusted one. In this way, the order in which we perform the operations does not matter and we have no dimensional bias. This algorithm is summarized by

Nonbiased incremental orthogonalization algorithm

$$\mathbf{r}_1' \Leftarrow \mathbf{r}_1 - k\frac{\mathbf{r}_1 \cdot \mathbf{r}_2}{\mathbf{r}_2 \cdot \mathbf{r}_2}\mathbf{r}_2 - k\frac{\mathbf{r}_1 \cdot \mathbf{r}_3}{\mathbf{r}_3 \cdot \mathbf{r}_3}\mathbf{r}_3,$$

$$\mathbf{r}_2' \Leftarrow \mathbf{r}_2 - k\frac{\mathbf{r}_2 \cdot \mathbf{r}_1}{\mathbf{r}_1 \cdot \mathbf{r}_1}\mathbf{r}_1 - k\frac{\mathbf{r}_2 \cdot \mathbf{r}_3}{\mathbf{r}_3 \cdot \mathbf{r}_3}\mathbf{r}_3,$$

$$\mathbf{r}_3' \Leftarrow \mathbf{r}_3 - k\frac{\mathbf{r}_3 \cdot \mathbf{r}_1}{\mathbf{r}_1 \cdot \mathbf{r}_1}\mathbf{r}_1 - k\frac{\mathbf{r}_3 \cdot \mathbf{r}_2}{\mathbf{r}_2 \cdot \mathbf{r}_2}\mathbf{r}_2.$$

One iteration of this algorithm results in a set of basis vectors that are slightly "more orthogonal" than the original vectors, but possibly not completely orthogonal. By repeating this procedure multiple times, we can eventually converge on an orthogonal basis. Selecting an appropriately small value for k (say, 1/4) and iterating a sufficient number of times (say, ten) gets us fairly close. Then, we can use the standard Gram-Schmidt algorithm to guarantee a perfectly orthogonal basis.

6.4 4 × 4 Homogeneous Matrices

Up until now, we have used only 2D and 3D vectors. In this section, we introduce 4D vectors and the so-called "homogeneous" coordinate. There is nothing magical about 4D vectors and matrices (and no, the fourth coordinate in this case isn't "time"). As we will see, 4D vectors and 4×4 matrices are nothing more than a notational convenience for what are simple 3D operations.

This section introduces 4D homogeneous space and 4×4 transformation matrices and their application to affine 3D geometry. Section 6.4.1 discusses the nature of 4D homogeneous space and how it is related to physical 3D space. Section 6.4.2 explains how 4×4 transformation matrices can be used to express translations. Section 6.4.3 explains how 4×4 transformation matrices can be used to express affine transformations.

6.4.1 4D Homogeneous Space

As was mentioned in Section 2.1, 4D vectors have four components, with the first three components being the standard x, y, and z components. The fourth component in a 4D vector is w, sometimes referred to as the *homogeneous coordinate*.

To understand how the standard physical 3D space is extended into 4D, let's first examine homogeneous coordinates in 2D, which are of the form (x, y, w). Imagine the standard 2D plane as existing in 3D at the plane $w = 1$, such that physical 2D point (x, y) is represented in homogeneous space $(x, y, 1)$. For all points that are not in the plane $w = 1$, we can compute the corresponding 2D point by projecting the point onto the plane $w = 1$, by dividing by w. So the homogeneous coordinate (x, y, w) is mapped to the physical 2D point $(x/w, y/w)$. This is shown in Figure 6.2.

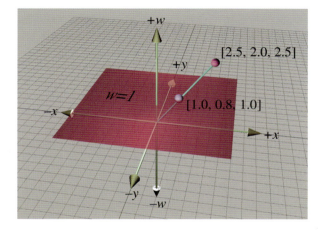

Figure 6.2
Projecting homogeneous coordinates onto the plane $w = 1$ in 2D

For any given physical 2D point (x, y) there are an infinite number of corresponding points in homogeneous space, all of the form (kx, ky, k), provided that $k \neq 0$. These points form a line through the (homogeneous) origin.

When $w = 0$, the division is undefined and there is no corresponding physical point in 2D space. However, we can interpret a 2D homogeneous point of the form $(x, y, 0)$ as a "point at infinity," which defines a *direction* rather than a *location*. When we make the conceptual distinction between "points" and "vectors" (see Section 2.4), then the "locations" where $w \neq 0$ are "points" and the "directions" with $w = 0$ are are "vectors." There is more on this in the next section.

The same basic idea applies when extending physical 3D space to 4D homogeneous space (although it's a lot harder to visualize). The physical 3D points can be thought of as living in the hyperplane in 4D at $w = 1$. A 4D point is of the form (x, y, z, w), and we project a 4D point onto this hyperplane to yield the corresponding physical 3D point $(x/w, y/w, z/w)$. When $w = 0$, the 4D point represents a "point at infinity," which defines a direction rather than a location.

Homogeneous coordinates and projection by division by w are interesting, but why on earth would we want to use 4D space? There are two primary reasons for using 4D vectors and 4×4 matrices. The first reason, which we discuss in the next section, is actually nothing more than a notational convenience. The second reason is that if we put the proper value into w, the homogenous division will result in a perspective projection, as we discuss in Section 6.5.

6.4.2   4 × 4 Translation Matrices

Recall from Section 4.2 that a 3×3 transformation matrix represents a *linear transformation*, which does not contain translation. Due to the nature of matrix multiplication, the zero vector is always transformed into the zero vector, and therefore any transformation that can be represented by a matrix multiplication cannot contain translation. This is unfortunate, because matrix multiplication and inversion are very convenient tools for composing complicated transformations out of simple ones and manipulating nested coordinate space relationships. It would be nice if we could find a way to somehow extend the standard 3×3 transformation matrix to be able to handle transformations with translation; 4×4 matrices provide a mathematical "kludge" that allows us to do this.

Assume for the moment that w is always 1. Thus, the standard 3D vector $[x, y, z]$ will always be represented in 4D as $[x, y, z, 1]$. Any 3×3 transformation matrix can by represented in 4D by using the conversion

Extending a 3 × 3 transform matrix into 4D

$$\begin{bmatrix} m_{11} & m_{12} & m_{13} \\ m_{21} & m_{22} & m_{23} \\ m_{31} & m_{32} & m_{33} \end{bmatrix} \Longrightarrow \begin{bmatrix} m_{11} & m_{12} & m_{13} & 0 \\ m_{21} & m_{22} & m_{23} & 0 \\ m_{31} & m_{32} & m_{33} & 0 \\ 0 & 0 & 0 & 1 \end{bmatrix}.$$

When we multiply a 4D vector of the form $[x, y, z, 1]$ by a 4×4 matrix of this form, we get the same result as the standard 3×3 case, the only difference being the additional coordinate $w = 1$:

$$\begin{bmatrix} x & y & z \end{bmatrix} \begin{bmatrix} m_{11} & m_{12} & m_{13} \\ m_{21} & m_{22} & m_{23} \\ m_{31} & m_{32} & m_{33} \end{bmatrix}$$

$$= \begin{bmatrix} xm_{11}+ym_{21}+zm_{31} & xm_{12}+ym_{22}+zm_{32} & xm_{13}+ym_{23}+zm_{33} \end{bmatrix};$$

$$\begin{bmatrix} x & y & z & 1 \end{bmatrix} \begin{bmatrix} m_{11} & m_{12} & m_{13} & 0 \\ m_{21} & m_{22} & m_{23} & 0 \\ m_{31} & m_{32} & m_{33} & 0 \\ 0 & 0 & 0 & 1 \end{bmatrix}$$

$$= \begin{bmatrix} xm_{11}+ym_{21}+zm_{31} & xm_{12}+ym_{22}+zm_{32} & xm_{13}+ym_{23}+zm_{33} & 1 \end{bmatrix}.$$

Now for the interesting part. In 4D, we can also express translation as a matrix multiplication, something we were not able to do in 3D:

$$
\begin{bmatrix} x & y & z & 1 \end{bmatrix}
\begin{bmatrix}
1 & 0 & 0 & 0 \\
0 & 1 & 0 & 0 \\
0 & 0 & 1 & 0 \\
\Delta x & \Delta y & \Delta z & 1
\end{bmatrix}
= \begin{bmatrix} x+\Delta x & y+\Delta y & z+\Delta z & 1 \end{bmatrix}.
$$

Using a 4 × 4 matrix to perform translation in 3D

$$(6.11)$$

It is important to understand that this matrix multiplication is still a *linear transformation*. Matrix multiplication cannot represent "translation" in 4D, and the 4D zero vector will always be transformed back into the 4D zero vector. The reason this trick works to transform points in 3D is that we are actually *shearing* 4D space. (Compare Equation (6.11) with the shear matrices from Section 5.5.) The 4D hyperplane that corresponds to physical 3D space does not pass through the origin in 4D. Thus, when we shear 4D space, we are able to translate in 3D.

Let's examine what happens when we perform a transformation without translation followed by a transformation with only translation. Let $\mathbf{R}$ be a rotation matrix. (In fact, $\mathbf{R}$ could possibly contain other 3D linear transformations, but for now, let's assume $\mathbf{R}$ only contains rotation.) Let $\mathbf{T}$ be a translation matrix of the form in Equation (6.11):

$$
\mathbf{R} = \begin{bmatrix}
r_{11} & r_{12} & r_{13} & 0 \\
r_{21} & r_{22} & r_{23} & 0 \\
r_{31} & r_{32} & r_{33} & 0 \\
0 & 0 & 0 & 1
\end{bmatrix}, \qquad
\mathbf{T} = \begin{bmatrix}
1 & 0 & 0 & 0 \\
0 & 1 & 0 & 0 \\
0 & 0 & 1 & 0 \\
\Delta x & \Delta y & \Delta z & 1
\end{bmatrix}.
$$

Then we could rotate and then translate a point $\mathbf{v}$ to compute a new point $\mathbf{v}'$ by

$$\mathbf{v}' = \mathbf{vRT}.$$

Remember that the order of transformations is important, and since we have chosen to use row vectors, the order of transformations coincides with the order that the matrices are multiplied, from left to right. We are rotating *first* and then translating.

Just as with 3×3 matrices, we can concatenate the two matrices into a single transformation matrix, which we assign to the matrix $\mathbf{M}$:

$$\mathbf{M} = \mathbf{RT},$$
$$\mathbf{v}' = \mathbf{vRT} = \mathbf{v}(\mathbf{RT}) = \mathbf{vM}.$$

Let's now examine the contents of $\mathbf{M}$:

$$\mathbf{M} = \mathbf{RT} = \begin{bmatrix} r_{11} & r_{12} & r_{13} & 0 \\ r_{21} & r_{22} & r_{23} & 0 \\ r_{31} & r_{32} & r_{33} & 0 \\ 0 & 0 & 0 & 1 \end{bmatrix} \begin{bmatrix} 1 & 0 & 0 & 0 \\ 0 & 1 & 0 & 0 \\ 0 & 0 & 1 & 0 \\ \Delta x & \Delta y & \Delta z & 1 \end{bmatrix}$$

$$= \begin{bmatrix} r_{11} & r_{12} & r_{13} & 0 \\ r_{21} & r_{22} & r_{23} & 0 \\ r_{31} & r_{32} & r_{33} & 0 \\ \Delta x & \Delta y & \Delta z & 1 \end{bmatrix}.$$

Notice that the upper 3×3 portion of $\mathbf{M}$ contains the rotation portion, and the bottom row contains the translation portion. The rightmost column (for now) will be $[0, 0, 0, 1]^{\mathrm{T}}$.

Applying this information in reverse, we can take any 4×4 matrix and separate it into a linear transformation portion, and a translation portion. We can express this succinctly with block matrix notation, by assigning the translation vector $[\Delta x, \Delta y, \Delta z]$ to the vector $\mathbf{t}$:

$$\mathbf{M} = \begin{bmatrix} \mathbf{R} & \mathbf{0} \\ \mathbf{t} & 1 \end{bmatrix}.$$

For the moment, we are assuming that the rightmost column of a 4×4 transformation matrix is always $[0, 0, 0, 1]^{\mathrm{T}}$. We will begin to encounter situations where this is not be the case in Section 6.5.

Let's see what happens with the so-called "points at infinity" (those vectors with $w = 0$). Multiplying by a "standard" 3×3 linear transformation matrix extended into 4D (a transformation that does not contain translation), we get

Multiplying a "point at infinity" by a 4×4 matrix *without* translation

$$\begin{bmatrix} x & y & z & 0 \end{bmatrix} \begin{bmatrix} r_{11} & r_{12} & r_{13} & 0 \\ r_{21} & r_{22} & r_{23} & 0 \\ r_{31} & r_{32} & r_{33} & 0 \\ 0 & 0 & 0 & 1 \end{bmatrix}$$

$$= \begin{bmatrix} xr_{11}{+}yr_{21}{+}zr_{31} & xr_{12}{+}yr_{22}{+}zr_{32} & xr_{13}{+}yr_{23}{+}zr_{33} & 0 \end{bmatrix}.$$

In other words, when we transform a point-at-infinity vector of the form $[x, y, z, 0]$ by a transformation matrix containing rotation, scale, etc., the expected transformation occurs, and the result is another point-at-infinity vector of the form $[x', y', z', 0]$.

When we transform a point-at-infinity vector by a transformation that *does* contain translation, we get the following result:

$$\begin{bmatrix} x & y & z & 0 \end{bmatrix} \begin{bmatrix} r_{11} & r_{12} & r_{13} & 0 \\ r_{21} & r_{22} & r_{23} & 0 \\ r_{31} & r_{32} & r_{33} & 0 \\ \Delta x & \Delta y & \Delta z & 1 \end{bmatrix}$$

Multiplying a "point at infinity" by a 4 × 4 matrix *with* translation

$$= \begin{bmatrix} xr_{11}+yr_{21}+zr_{31} & xr_{12}+yr_{22}+zr_{32} & xr_{13}+yr_{23}+zr_{33} & 0 \end{bmatrix}.$$

Notice that the result is the same—that is, no translation occurs.

In other words, the w component of a 4D vector can be used to selectively "switch off" the translation portion of a 4×4 matrix. This is useful because some vectors represent "locations" and should be translated, and other vectors represent "directions," such as surface normals, and should not be translated. In a geometric sense, we can think of the first type of data, with $w = 1$, as "points," and the second type of data, the "points at infinity" with $w = 0$, as "vectors."

So, one reason why 4×4 matrices are useful is that a 4×4 transformation matrix can contain translation. When we use 4×4 matrices solely for this purpose, the right-most column of the matrix will always be $[0, 0, 0, 1]^{\mathrm{T}}$. Since this is the case, why don't we just drop the column and use a 4×3 matrix? According to linear algebra rules, 4×3 matrices are undesirable for several reasons:

- We cannot multiply a 4×3 matrix by another 4×3 matrix.

- We cannot invert a 4×3 matrix, since the matrix is not square.

- When we multiply a 4D vector by a 4×3 matrix, the result is a 3D vector.

Strict adherence to linear algebra rules forces us to add the fourth column. Of course, in our code, we are not bound by linear algebra rules. It is a common technique to write a 4×3 matrix class that is useful for representing transformations that contain translation. Basically, such a matrix is a 4×4 matrix, where the right-most column is assumed to be $[0, 0, 0, 1]^{\mathrm{T}}$ and therefore isn't explicitly stored.

6.4.3   General Affine Transformations

Chapter 5 presented 3×3 matrices for many primitive transformations. Because a 3×3 matrix can represent only *linear* transformations in 3D, translation was not considered. Armed with 4×4 transform matrices, though, we can now create more general *affine* transformations that contain translation, such as:

- rotation about an axis that does not pass through the origin,

- scale about a plane that does not pass through the origin,

- reflection about a plane that does not pass through the origin, and

- orthographic projection onto a plane that does not pass through the origin.

The basic idea is to translate the "center" of the transformation to the origin, perform the linear transformation by using the techniques developed in Chapter 5, and then transform the center back to its original location. We start with a translation matrix $\mathbf{T}$ that translates the point $\mathbf{p}$ to the origin, and a linear transform matrix $\mathbf{R}$ from Chapter 5 that performs the linear transformation. The final affine transformation matrix $\mathbf{A}$ will be equal to the matrix product $\mathbf{T}\mathbf{R}(\mathbf{T}^{-1})$, where $\mathbf{T}^{-1}$ is the translation matrix with the opposite translation amount as $\mathbf{T}$.

It is interesting to observe the general form of such a matrix. Let's first write $\mathbf{T}$, $\mathbf{R}$, and $\mathbf{T}^{-1}$ in the partitioned form we used earlier:

$$
\mathbf{T} = \begin{bmatrix} 1 & 0 & 0 & 0 \\ 0 & 1 & 0 & 0 \\ 0 & 0 & 1 & 0 \\ -p_x & -p_y & -p_z & 1 \end{bmatrix} = \begin{bmatrix} \mathbf{I} & \mathbf{0} \\ -\mathbf{p} & 1 \end{bmatrix};
$$

$$
\mathbf{R}_{4\times4} = \begin{bmatrix} r_{11} & r_{12} & r_{13} & 0 \\ r_{21} & r_{22} & r_{23} & 0 \\ r_{31} & r_{32} & r_{33} & 0 \\ 0 & 0 & 0 & 1 \end{bmatrix} = \begin{bmatrix} \mathbf{R}_{3\times3} & \mathbf{0} \\ \mathbf{0} & 1 \end{bmatrix};
$$

$$
\mathbf{T}^{-1} = \begin{bmatrix} 1 & 0 & 0 & 0 \\ 0 & 1 & 0 & 0 \\ 0 & 0 & 1 & 0 \\ p_x & p_y & p_z & 1 \end{bmatrix} = \begin{bmatrix} \mathbf{I} & \mathbf{0} \\ \mathbf{p} & 1 \end{bmatrix}.
$$

Evaluating the matrix multiplication, we get

$$
\mathbf{T}\mathbf{R}_{4\times4}\mathbf{T}^{-1} = \begin{bmatrix} \mathbf{I} & \mathbf{0} \\ -\mathbf{p} & 1 \end{bmatrix} \begin{bmatrix} \mathbf{R}_{3\times3} & \mathbf{0} \\ \mathbf{0} & 1 \end{bmatrix} \begin{bmatrix} \mathbf{I} & \mathbf{0} \\ \mathbf{p} & 1 \end{bmatrix} = \begin{bmatrix} \mathbf{R}_{3\times3} & \mathbf{0} \\ -\mathbf{p}\,(\mathbf{R}_{3\times3}) + \mathbf{p} & 1 \end{bmatrix}.
$$

Thus, the extra translation in an affine transformation changes only the last row of the 4×4 matrix. The upper 3×3 portion, which contains the linear transformation, is not affected.

Our use of "homogeneous" coordinates so far has really been nothing more than a mathematical kludge to allow us to include translation in our transformations. We use quotations around "homogeneous" because the w value was always 1 (or 0, in the case of points at infinity). In the next section, we will remove the quotations, and discuss meaningful ways to use 4D coordinates with other w values.

6.5 4×4 Matrices and Perspective Projection

Section 6.4.1 showed that when we interpret a 4D homogeneous vector in 3D, we divide by w. This division is a mathematical tool that we did not really take advantage of in the previous section, since w was always 1 or 0. However, if we play our cards right, we can use the division by w to encapsulate very succinctly the important geometric operation of *perspective projection*.

We can learn a lot about perspective projection by comparing it to another type of projection we have already discussed, *orthographic projection*. Section 5.3 showed how to project 3D space onto a 2D plane, known as the *projection plane*, by using orthographic projection. Orthographic projection is also known as *parallel projection*, because the projectors are parallel. (A *projector* is a line from the original point to the resulting projected point on the plane). The parallel projectors used in orthographic projection are shown in Figure 6.3.

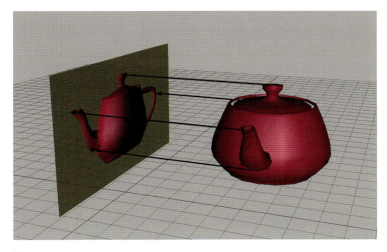

Figure 6.3
Orthographic projection uses parallel projectors.

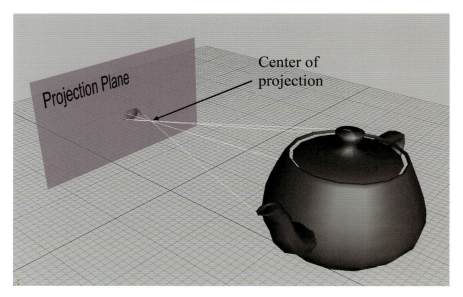

Figure 6.4. With perspective projection, the projectors intersect at the center of projection.

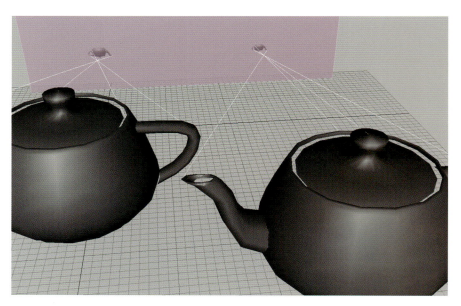

Figure 6.5. Due to perspective foreshortening, the projection of the teapot on the left is larger than the projection of the teapot on the right. The left-hand teapot is closer to the projection plane.

Perspective projection in 3D also projects onto a 2D plane. However, the projectors are not parallel. In fact, they intersect at a point, known as the *center of projection*. This is shown in Figure 6.4.

Because the center of projection is in front of the projection plane, the projectors cross before striking the plane, and thus the image is inverted. As we move an object farther away from the center of projection, its orthographic projection remains constant, but the perspective projection gets smaller, as illustrated in Figure 6.5. The teapot on the right is further from the projection plane, and the projection is (slightly) smaller than the closer teapot. This is a very important visual cue known as *perspective foreshortening*.

6.5.1   A Pinhole Camera

Perspective projection is important in graphics because it models the way the human visual system works. Actually, the human visual system is more complicated because we have two eyes, and for each eye, the projection surface (our retina) is not flat; so let's look at the simpler example of a *pinhole camera*. A pinhole camera is a box with a tiny hole on one end. Rays of light enter the pinhole (thus converging at a point), and then strike the opposite end of the box, which is the projection plane. This is shown in Figure 6.6.

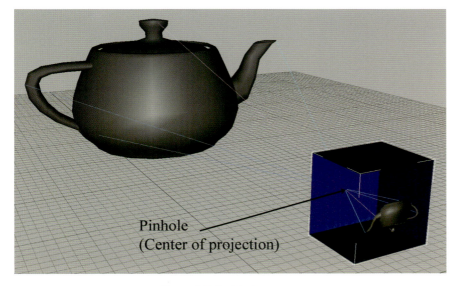

Figure 6.6. A pinhole camera.

In this view, the left and back sides of the box have been removed so you can see the inside. Notice that the image projected onto the back of the box is inverted. This is because the rays of light (the projectors) cross as they meet at the pinhole (the center of projection).

Let's examine the geometry behind the perspective projection of a pinhole camera. Consider a 3D coordinate space with the origin at the pinhole, the z-axis perpendicular to the projection plane, and the x- and y-axes parallel to the plane of projection, as shown in Figure 6.7.

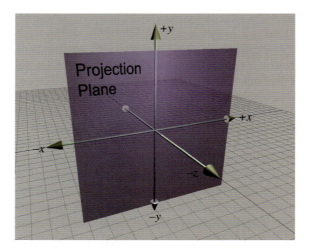

Figure 6.7
A projection plane parallel to the xy-plane

Let's see if we can't compute, for an arbitrary point $\mathbf{p}$, the 3D coordinates of $\mathbf{p}'$, which is $\mathbf{p}$ projected through the pinhole onto the projection plane. First, we need to know the distance from the pinhole to the projection plane. We assign this distance to the variable d. Thus, the plane is defined by the equation $z = -d$. Now let's view things from the side and solve for y (see Figure 6.8).

By similar triangles, we can see that

$$\frac{-p'_y}{d} = \frac{p_y}{z} \quad \Longrightarrow \quad p'_y = \frac{-dp_y}{z}.$$

Notice that since a pinhole camera flips the image upside down, the signs of p_y and p'_y are opposite. The value of p'_x is computed in a similar manner:

$$p'_x = \frac{-dp_x}{z}.$$

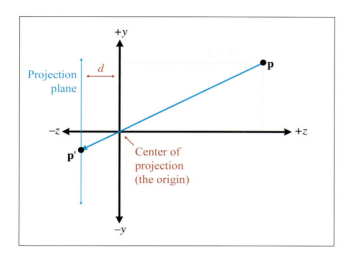

Figure 6.8
Viewing the
projection plane
from the side

The z values of all the projected points are the same: $-d$. Thus, the result of projecting a point $\mathbf{p}$ through the origin onto a plane at $z = -d$ is

$$\mathbf{p} = \begin{bmatrix} x & y & z \end{bmatrix} \implies \mathbf{p}' = \begin{bmatrix} x' & y' & z' \end{bmatrix} = \begin{bmatrix} -dx/z & -dy/z & -d \end{bmatrix}.$$

Projecting onto the plane $z = -d$

In practice, the extra minus signs create unnecessary complexities, and so we move the plane of projection to $z = d$, which is in *front* of the center of projection, as shown in Figure 6.9. Of course, this would never work for a real pinhole camera, since the purpose of the pinhole in the first place is

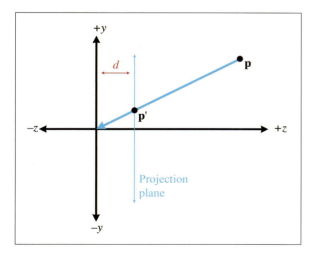

Figure 6.9
Projection plane in
front of the center of
projection

to allow in only light that passes through a single point. However, in the mathematical universe inside a computer, it works just fine.

As expected, moving the plane of projection in front of the center of projection removes the annoying minus signs:

Projecting a point onto the plane $z = d$

$$\mathbf{p}' = \begin{bmatrix} x' & y' & z' \end{bmatrix} = \begin{bmatrix} dx/z & dy/z & d \end{bmatrix}. \tag{6.12}$$

6.5.2 Perspective Projection Matrices

Because the conversion from 4D to 3D space implies a division, we can encode a perspective projection in a 4×4 matrix. The basic idea is to come up with an equation for $\mathbf{p}'$ with a common denominator for x, y, and z, and then set up a 4×4 matrix that will set w equal to this denominator. We assume that the original points have $w = 1$.

First, we manipulate Equation (6.12) to have a common denominator:

$$\mathbf{p}' = \begin{bmatrix} dx/z & dy/z & d \end{bmatrix} = \begin{bmatrix} dx/z & dy/z & dz/z \end{bmatrix} = \frac{\begin{bmatrix} x & y & z \end{bmatrix}}{z/d}.$$

To divide by this denominator, we put the denominator into w, so the 4D point will be of the form

$$\begin{bmatrix} x & y & z & z/d \end{bmatrix}.$$

So we need a 4×4 matrix that multiplies a homogeneous vector $[x, y, z, 1]$ to produce $[x, y, z, z/d]$. The matrix that does this is

Projecting onto the plane $z = d$ using a 4×4 matrix

$$\begin{bmatrix} x & y & z & 1 \end{bmatrix} \begin{bmatrix} 1 & 0 & 0 & 0 \\ 0 & 1 & 0 & 0 \\ 0 & 0 & 1 & 1/d \\ 0 & 0 & 0 & 0 \end{bmatrix} = \begin{bmatrix} x & y & z & z/d \end{bmatrix}.$$

Thus, we have derived a 4×4 projection matrix.

There are several important points to be made here:

- Multiplication by this matrix doesn't *actually* perform the perspective transform, it just computes the proper denominator into w. Remember that the perspective division actually occurs when we convert from 4D to 3D by dividing by w.

- There are many variations. For example, we can place the plane of projection at $z = 0$, and the center of projection at $[0, 0, -d]$. This results in a slightly different equation.

- This seems overly complicated. It seems like it would be simpler to just divide by z, rather than bothering with matrices. So why is homogeneous space interesting? First, 4×4 matrices provide a way to express projection as a transformation that can be concatenated with other transformations. Second, projection onto nonaxially aligned planes is possible. Basically, we don't *need* homogeneous coordinates, but 4×4 matrices provide a compact way to represent and manipulate projection transformations.

- The projection matrix in a real graphics geometry pipeline (perhaps more accurately known as the "clip matrix") does more than just copy z into w. It differs from the one we derived in two important respects:

    ○ Most graphics systems apply a normalizing scale factor such that $w = 1$ at the far clip plane. This ensures that the values used for depth buffering are distributed appropriately for the scene being rendered, to maximize precision of depth buffering.

    ○ The projection matrix in most graphics systems also scales the x and y values according to the field of view of the camera.

    We'll get into these details in Section 10.3.2, when we show what a projection matrix looks like in practice, using both DirectX and OpenGL as examples.

6.6 Exercises

(Answers on page 765.)

1. Compute the determinant of the following matrix:

$$\begin{bmatrix} 3 & -2 \\ 1 & 4 \end{bmatrix}$$

2. Compute the determinant, adjoint, and inverse of the following matrix:

$$\begin{bmatrix} 3 & -2 & 0 \\ 1 & 4 & 0 \\ 0 & 0 & 2 \end{bmatrix}$$

3. Is the following matrix orthogonal?

$$\begin{bmatrix} -0.1495 & -0.1986 & -0.9685 \\ -0.8256 & 0.5640 & 0.0117 \\ -0.5439 & -0.8015 & 0.2484 \end{bmatrix}$$

4. Invert the matrix from the previous exercise.

5. Invert the 4×4 matrix

$$\begin{bmatrix} -0.1495 & -0.1986 & -0.9685 & 0 \\ -0.8256 & 0.5640 & 0.0117 & 0 \\ -0.5439 & -0.8015 & 0.2484 & 0 \\ 1.7928 & -5.3116 & 8.0151 & 1 \end{bmatrix}.$$

6. Construct a 4×4 matrix to translate by $[4, 2, 3]$.

7. Construct a 4×4 matrix to rotate $20°$ about the x-axis and then translate by $[4, 2, 3]$.

8. Construct a 4×4 matrix to translate by $[4, 2, 3]$ and then rotate $20°$ about the x-axis.

9. Construct a 4×4 matrix to perform a perspective projection onto the plane $x = 5$. (Assume the origin is the center of projection.)

10. Use the matrix from the previous exercise to compute the 3D coordinates of the projection of the point $(105, -243, 89)$ onto the plane $x = 5$.

An attempt at visualizing the Fourth Dimension:
Take a point, stretch it into a line, curl it into a circle,
twist it into a sphere, and punch through the sphere.

— Albert Einstein (1879–1955)

Chapter 7

Polar Coordinate Systems

First of all, we must note that the universe is spherical.

— Nicolaus Copernicus (1473–1543)

The Cartesian coordinate system isn't the only system for mapping out space and defining locations precisely. An alternative to the Cartesian system is the *polar* coordinate system, which is the subject of this chapter. If you're not very familiar with polar coordinates, it might seem like an esoteric or advanced topic (especially because of the trig), and you might be tempted to gloss over. Please don't make this mistake. There are many very practical problems in areas such as AI and camera control whose solutions (and inherent difficulties!) can be readily understood in the framework of polar coordinates.

This chapter is organized into the following sections:

- Section 7.1 describes 2D polar coordinates.

- Section 7.2 gives some examples where polar coordinates are preferable to Cartesian coordinates.

- Section 7.3 shows how polar space works in 3D and introduces *cylindrical* and *spherical* coordinates.

- Finally, Section 7.4 makes it clear that polar space can be used to describe vectors as well as positions.

7.1   2D Polar Space

This section introduces the basic idea behind polar coordinates, using two dimensions to get us warmed up. Section 7.1.1 shows how to use polar coordinates to describe position. Section 7.1.2 discusses *aliasing* of polar coordinates. Section 7.1.3 shows how to convert between polar and Cartesian coordinates in 2D.

7.1.1 Locating Points by Using 2D Polar Coordinates

Remember that a 2D Cartesian coordinate space has an origin, which establishes the position of the coordinate space, and two axes that pass through the origin, which establish the orientation of the space. A 2D polar coordinate space also has an origin (known as the *pole*), which has the same basic purpose—it defines the "center" of the coordinate space. A polar coordinate space has only *one* axis, however, sometimes called the *polar axis*, which is usually depicted as a ray from the origin. It is customary in math literature for the polar axis to point to the right in diagrams, and thus it corresponds to the $+x$ axis in a Cartesian system, as shown in Figure 7.1.

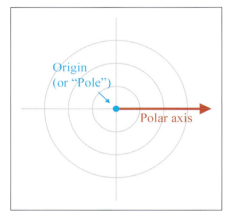

Figure 7.1
A 2D polar coordinate space

It's often convenient to use different conventions than this, as shown in Section 7.3.3. Until then, our discussion adopts the traditional conventions of the math literature.

In the Cartesian coordinate system, we described a 2D point using two signed distances, x and y. The polar coordinate system uses one distance and one *angle*. By convention, the distance is usually assigned to the variable r (which is short for "radius") and the angle is usually called θ. The polar coordinate pair (r, θ) specifies a point in 2D space as follows:

Locating the point described by 2D polar coordinates (r, θ)

Step 1. Start at the origin, facing in the direction of the polar axis, and rotate by the angle θ. Positive values of θ are usually interpreted to mean counterclockwise rotation, negative values mean clockwise rotation.

Step 2. Now move forward from the origin a distance of r units. You have arrived at the point described by the polar coordinates (r, θ).

This process is shown in Figure 7.2.

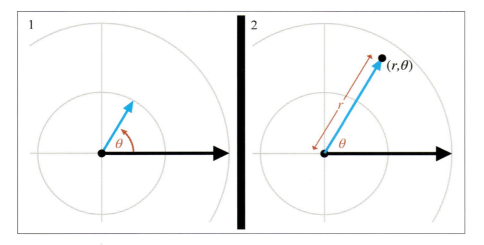

Figure 7.2. Locating a point using 2D polar coordinates

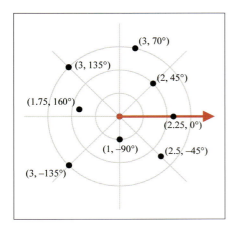

Figure 7.3
Example points labeled with 2D polar coordinates

In summary, r defines the distance from the point to the origin, and θ defines the direction of the point from the origin. Figure 7.3 shows several points and their polar coordinates. You should study this figure until you are convinced that you know how it works.

You might have noticed that the diagrams of polar coordinate spaces contain grid lines, but that these grid lines are slightly different from the grid lines used in diagrams of Cartesian coordinate systems. Each grid line in a Cartesian coordinate system is composed of points with the same value for one of the coordinates. A vertical line is composed of points that all have the same x-coordinate, and a horizontal line is composed of points that all have the same y-coordinate. The grid lines in a polar coordinate system are similar:

- The "grid circles" show lines of constant r. This makes sense; after all, the definition of a circle is the set of all points equidistant from its center. That's why the letter r is the customary variable to hold this distance, because it is a *radial* distance.

- The straight grid lines that pass through the origin show lines of constant θ, consisting of points that are the same direction from the origin.

One note regarding angle measurements. With Cartesian coordinates, the unit of measure wasn't really significant. We could interpret diagrams using feet, meters, miles, yards, light-years, beard-seconds, or picas, and it didn't really matter.[1] If you take some Cartesian coordinate data, interpreting that data using different physical units just makes whatever you're looking at get bigger or smaller, but it's proportionally the same shape. However, interpreting the angular component of polar coordinates using different angular units can produce drastically distorted results.

It really doesn't matter whether you use degrees or radians (or grads, mils, minutes, signs, sextants, or Furmans), as long as you keep it straight. In the text of this book, we almost always give specific angular measurements in degrees and use the ° symbol after the number. We do this because we are human beings, and most humans who are not math professors find it easier to deal with whole numbers rather than fractions of π. Indeed, the choice of the number 360 was specifically designed to make fractions avoidable in many common cases. However, computing machines[2] prefer to work with angles expressed using radians, and so the code snippets in this book use radians rather than degrees.

7.1.2 Aliasing

Hopefully you're starting to get a good feel for how polar coordinates work and what polar coordinate space looks like. But there may be some nagging thoughts in the back of your head. Consciously or subconsciously, you may have noticed a fundamental difference between Cartesian and polar space. Perhaps you imagined a 2D Cartesian space as a perfectly even continuum of space, like a flawless sheet of Jell-O, spanning infinitely in all directions, each infinitely thin bite identical to all the others. Sure, there

---

[1] There might be some employees at NASA who feel otherwise, since the $125 million Mars Climate Orbiter went astray due to a bug involving confusion between metric and English units. Perhaps we should say that knowing the specific units of measurement isn't necessary to understand the *concepts* of Cartesian coordinates.

[2] Such as math professors.

are some "special" places, like the origin, and the axes, but those are just like marks on the bottom of the pan—the Jell-O itself is the same there as everywhere else. But when you imagined the fabric of polar coordinate space, something was different. Polar coordinate space has some "seams" in it, some discontinuities where things are a bit "patched together." In the infinitely large circular pan of Jell-O, there are multiple sheets of Jell-O stacked on top of each other. When you put your spoon down a particular place to get a bite, you often end up with multiple bites! There's a piece of hair in the block of Jell-O, a singularity that requires special precautions.

Whether your mental image of polar space was of Jell-O, or some other yummy dessert, you were probably pondering some of these questions:

1. Can the radial distance r ever be negative?

2. Can θ ever go outside the interval $[-180°, +180°]$?

3. The value of the angle θ directly "west" of the origin (i.e., for points where $x < 0$ and $y = 0$ using Cartesian coordinates) is ambiguous. You may have noticed that none of these points are labeled in Figure 7.3. Is θ equal to $+180°$ or $-180°$ for these points?

4. The polar coordinates for the origin itself are also ambiguous. Clearly $r = 0$, but what value of θ should we use? Wouldn't *any* value work?

The answer to all of these questions is "yes."[3] In fact, we must face a rather harsh reality about polar space.

For any given point, there are infinitely many polar coordinate pairs that can be used to describe that point.

This phenomenon is known as *aliasing*. Two coordinate pairs are said to be *aliases* of each other if they have different numeric values but refer to the same point in space. Notice that aliasing doesn't happen in Cartesian space—each point in space is assigned exactly one (x, y) coordinate pair; the mapping of points to coordinate pairs is one-to-one.

Before we discuss some of the difficulties created by aliasing, let's be clear about one task for which aliasing does *not* pose any problems: interpreting a particular polar coordinate pair (r, θ) and locating the point in space referred to by those coordinates. No matter what the values of r and θ, we can come up with a sensible interpretation.

---

[3]Even question 3.

When $r < 0$, it is interpreted as "backward" movement—displacement in the opposite direction that we would move if r were positive. If θ is outside the range $[-180°, +180°]$, that's not a cause for panic; we can still determine the resulting direction.[4] In other words, although there may be some "unusual" polar coordinates, there's no such thing as "invalid" polar coordinates. A given point in space corresponds to many coordinate pairs, but a coordinate pair unambiguously designates exactly one point in space.

One way to create an alias for a point (r, θ) is to add a multiple of $360°$ to θ. This adds one or more whole "revolutions," but doesn't change the resulting direction defined by θ. Thus (r, θ) and $(r, \theta + k360°)$ describe the same point, where k is an integer. We can also generate an alias by adding $180°$ to θ and negating r; which means we face the other direction, but we displace by the opposite amount.

In general, for any point (r, θ) other than the origin, all of the polar coordinates that are aliases for (r, θ) can be expressed as

$$\left((-1)^k r, \theta + k180°\right),$$

where k is any integer.

So, in spite of aliasing, we can all agree what point is described by the polar coordinates (r, θ), no matter what values of r and θ are used. But what about the reverse problem? Given an arbitrary point $\mathbf{p}$ in space, can we all agree what polar coordinates (r, θ) should be used to describe $\mathbf{p}$? We've just said that there are an infinite number of polar coordinate pairs that could be used to describe the location $\mathbf{p}$. Which do we use? The short answer is: "Any one that works is *OK*, but only one is the *preferred* one to use."

It's like reducing fractions. We all agree that $13/26$ is a perfectly valid fraction, and there's no dispute as to what the value of this fraction is. Even so, $13/26$ is an "unusual" fraction; most of us would prefer that this value be expressed as $1/2$, which is simpler and easier to understand. A fraction is in the "preferred" format when it's expressed in lowest terms, meaning there isn't an integer greater than 1 that evenly divides both the numerator and denominator. We don't *have* to reduce $13/26$ to $1/2$, but by convention we normally do. A person's level of commitment to this convention is usually based on how many points their math teacher counted off on their homework for not reducing fractions to lowest terms.[5]

---

[4] Warning: extremely large values of θ may cause dizziness if step 1 in Figure 7.2 is followed literally.

[5] Speaking of math teachers and reduced fractions, one author remembers his middle school math teacher engaged in a fierce debate about whether a mixed fraction such as 2 3/5 is "simpler" than the corresponding improper fraction 13/5. Luckily, the answer to this profound mystery isn't necessary in the context of polar coordinate aliasing.

For polar coordinates, the "preferred" way to describe any given point is known as the *canonical* coordinates for that point. A 2D polar coordinate pair (r, θ) is in the canonical set if r is nonnegative and θ is in the interval $(-180°, 180°]$. Notice that the interval is half open: for points directly "west" of the origin $(x < 0, y = 0)$, we will use $\theta = +180°$. Also, if $r = 0$ (which is only true at the origin), then we usually assign $\theta = 0$. If you apply all these rules, then for any given point in 2D space, there is exactly *one* way to represent that point using canonical polar coordinates. We can summmarize this succintly with some math notation. A polar coordinate pair (r, θ) is in the canonical set if all of the following are true:

$r \geq 0$ We don't measure distances "backwards."

$-180° < \theta \leq 180°$ The angle is limited to 1/2 revolution. We use $+180°$ for "west."

$r = 0 \;\Rightarrow\; \theta = 0$ At the origin, set the angle to zero.

Conditions satisfied by canonical coordinates

The following algorithm can be used to convert a polar coordinate pair into its canonical form:

1. If $r = 0$, then assign $\theta = 0$.

2. If $r < 0$, then negate r, and add $180°$ to θ.

3. If $\theta \leq -180°$, then add $360°$ to θ until $\theta > -180°$.

4. If $\theta > 180°$, then subtract $360°$ from θ until $\theta \leq 180°$.

Converting a polar coordinate pair (r, θ) to canonical form

Listing 7.1 shows how it could be done in C. As discussed in Section 7.1.1, our computer code will normally store angles using radians.

```
// Radial distance
float r;

// Angle in RADIANS
float theta;

// Declare a constant for 2*pi (360 degrees)
const float TWOPI = 2.0f*PI;

// Check if we are exactly at the origin
if (r == 0.0f) {

    // At the origin — slam theta to zero
    theta = 0.0f;
} else {

    // Handle negative distance
    if (r < 0.0f) {
```

```
        r = −r;
        theta += PI;
    }

    // Theta out of range?  Note that this if() check is not
    // strictly necessary, but we try to avoid doing floating
    // point operations if they aren't necessary.  Why
    // incur floating point precision loss if we don't
    // need to?
    if (fabs(theta) > PI) {

        // Offset by PI
        theta += PI;

        // Wrap in range 0...TWOPI
        theta −= floor(theta / TWOPI) * TWOPI;

        // Undo offset, shifting angle back in range −PI...PI
        theta −= PI;
    }
}
```

Listing 7.1
Converting polar coordinates to canonical form

Picky readers may notice that while this code ensures that θ is in the *closed* interval $[-\pi, +\pi]$, it does not explicitly avoid the case where $\theta = -\pi$. The value of π is not exactly representable in floating point. In fact, because π is an irrational number, it can never be represented exactly in floating point, or with any finite number of digits in any base, for that matter! The value of the constant `PI` in our code is not exactly equal to π, it's the closest number to π that is representable by a `float`. Using double-precision arithmetic can get us closer to the exact value, but it is still not exact. So you can think of this function as returning a value from the *open* interval $(-\pi, +\pi)$.

7.1.3 Converting between Cartesian and Polar Coordinates in 2D

This section describes how to convert between the Cartesian and polar coordinate systems in 2D. By the way, if you were wondering when we were going to make use of the trigonometry that we reviewed in Section 1.4.5, this is it.

Figure 7.4 shows the geometry involved in converting between polar and Cartesian coordinates in 2D.

Converting polar coordinates (r, θ) to the corresponding Cartesian coordinates follows almost immediately from the definitions of sine and cosine:

Converting 2D polar coordinates to Cartesian

$$x = r \cos\theta; \qquad\qquad y = r \sin\theta. \qquad (7.1)$$

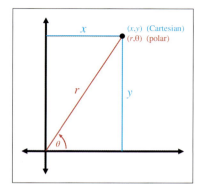

Figure 7.4
Converting between Cartesian and polar coordinates

Notice that aliasing is a nonissue; Equation (7.1) works even for "weird" values of r and θ.

Computing the polar coordinates (r, θ) from the Cartesian coordinates (x, y) is the tricky part. Due to aliasing, there isn't only one right answer; there are infinitely many (r, θ) pairs that describe the point (x, y). Usually, we want the canonical coordinates.

We can easily compute r by using the Pythagorean theorem,

$$r = \sqrt{x^2 + y^2}.$$

Since the square root function always returns the positive root, we don't have to worry about r causing our computed polar coordinates to be outside the canonical set.

Computing r was pretty easy, so now let's solve for θ:

$$\frac{y}{x} = \frac{r \sin \theta}{r \cos \theta},$$

$$\frac{y}{x} = \frac{\sin \theta}{\cos \theta},$$

$$y/x = \tan \theta,$$

$$\theta = \arctan(y/x).$$

Unfortunately, there are two problems with this approach. The first is that if $x = 0$, the division is undefined. The second is that the arctan function has a range of only $[-90°, +90°]$. The basic problem is that the division y/x effectively discards some useful information. Both x and y can either be positive or negative, resulting in four different possibilities, corresponding to the four different quadrants that may contain the point. But the division y/x results in a single value. If we negate both x and y, we move to a different quadrant in the plane, but the ratio y/x doesn't change.

Because of these problems, the complete equation for conversion from Cartesian to polar coordinates requires some "if statements" to handle each quadrant, and is a bit of a mess for "math people." Luckily, "computer people" have the `atan2` function, which properly computes the angle θ for all x and y, except for the pesky case at the origin. Borrowing this notation, let's define an atan2 function we can use in this book in our math notation:

The atan2 function used in this book

$$\text{atan2}(y, x) = \begin{cases} 0, & x = 0, y = 0, \\ +90°, & x = 0, y > 0, \\ -90°, & x = 0, y < 0, \\ \arctan(y/x), & x > 0, \\ \arctan(y/x) + 180°, & x < 0, y \geq 0, \\ \arctan(y/x) - 180°, & x < 0, y < 0. \end{cases} \tag{7.2}$$

Let's make two key observations about Equation (7.2). First, following the convention of the `atan2` function found in the standard libraries of most computer languages, the arguments are in the "reverse" order: y, x. You can either just remember that it's reversed, or you might find it handy to remember the lexical similarity between $\text{atan2}(y, x)$ and $\arctan(y/x)$. Or remember that $\tan \theta = \sin \theta / \cos \theta$, and $\theta = \text{atan2}(\sin \theta, \cos \theta)$.

Second, in many software libraries, the `atan2` function is *undefined* at the origin, when $x = y = 0$. The atan2 function we are defining for use in our equations in the text of this book is defined such that $\text{atan2}(0, 0) = 0$. In our code snippets, we use the library function `atan2` and explicitly handle the origin as a special case, but in our equations, we use the abstract function atan2, which is defined at the origin. (Note the difference in typeface.)

Back to the task at hand: computing the polar angle θ from a set of 2D Cartesian coordinates. Armed with the atan2 function, we can easily convert 2D Cartesian coordinates to polar form:

2D Cartesian to polar coordinate conversion

$$r = \sqrt{x^2 + y^2}; \qquad\qquad \theta = \text{atan2}(y, x).$$

The C code in Listing 7.2 shows how to convert a Cartesian (x, y) coordinate pair to the corresponding canonical polar (r, θ) coordinates.

```
// Input: Cartesian coordinates
float x,y;

// Output: polar radial distance, and angle in RADIANS
float r, theta;

// Check if we are at the origin
```

```
if (x == 0.0f && y == 0.0f) {

    // At the origin — slam both polar coordinates to zero
    r = 0.0f;
    theta = 0.0f;
} else {

    // Compute values.  Isn't the atan2 function great?
    r = sqrt(x*x + y*y);
    theta = atan2(y,x);
}
```

Listing 7.2
Converting 2D Cartesian coordinates to polar form

7.2   Why Would Anybody Use Polar Coordinates?

With all of the complications with aliasing, degrees and radians, and trig, why would anybody use polar coordinates when Cartesian coordinates work just fine, without any hairs in the Jell-O? Actually, you probably use polar coordinates more often than you do Cartesian coordinates. They arise frequently in informal conversation.

For example, one author is from Alvarado, Texas. When people ask where Alvarado, Texas, is, he tells them, "About 15 miles southeast of Burleson." He's describing where Alvarado is by using polar coordinates, specifying an origin (Burleson), a distance (15 miles), and an angle (southeast). Of course, most people who aren't from Texas (and many people who are) don't know where Burleson is, either, so it's more natural to switch to a different polar coordinate system and say, "About 50 miles southwest of Dallas." Luckily, even people from outside the United States usually know where Dallas is.[6] By the way, everyone in Texas does *not* wear a cowboy hat and boots. We *do* use the words "y'all" and "fixin'," however.[7]

In short, polar coordinates often arise because people naturally think about locations in terms of distance and direction. (Of course, we often aren't very *precise* when using polar coordinates, but precision is not really one of the brain's strong suits.) Cartesian coordinates are just not our native language. The opposite is true of computers—in general, when using a computer to solve geometric problems, it's easier to use Cartesian coordinates than polar coordinates. We discuss this difference between humans and computers again in Chapter 8 when we compare different methods for describing orientation in 3D.

---

[6]This is due to Dallas's two rather unfortunate claims to fame: the assassination of President Kennedy and a soap opera named after the city, which inexplicably had international appeal.

[7]These two facts have nothing to do with math, but everything to do with correcting misconceptions.

Perhaps the reason for our affinity for polar coordinates is that each polar coordinate has concrete meaning all by itself. One fighter pilot may say to another "Bogey, six o'clock!"[8] In the midst of a dogfight, these brave fighter pilots are actually using polar coordinates. "Six o'clock" means "behind you" and is basically the angle θ that we've been studying. Notice that the pilot didn't need to specify a distance, presumably because the other pilot could turn around and see for himself faster than the other pilot could tell him. So one polar coordinate (in this case, a direction) is useful information by itself. The same types of examples can be made for the other polar coordinate, distance (r). Contrast that with the usefulness of a lone Cartesian coordinate. Imagine a fighter pilot saying, "Bogey, $x = 1000$ ft!" This information is more difficult to process, and isn't as useful.

In video games, one of the most common times that polar coordinates arise is when we want to aim a camera, weapon, or something else at some target. This problem is easily handled by using a Cartesian-to-polar coordinate conversion, since it's usually the angles we need. Even when angular data can be avoided for such purposes (we might be able to completely use vector operations, for example, if the orientation of the object is specified using a matrix), polar coordinates are still useful. Usually, cameras and turrets and assassins' arms cannot move instantaneously (no matter how good the assassin), but targets *do* move. In this situation, we usually "chase" the target in some manner. This chasing (whatever type of control system is used, whether a simple velocity limiter, a lag, or a second-order system) is usually best done in polar space, rather than, say, interpolating a target position in 3D space.

Polar coordinates are also often encountered with physical data acquisition systems that provide basic raw measurements in terms of distance and direction.

One final occasion worth mentioning when polar coordinates are more natural to use than Cartesian coordinates is moving around on the surface of a sphere. When would anybody do that? You're probably doing it right now. The latitude/longitude coordinates used to precisely describe geographic locations are really not Cartesian coordinates, they are polar coordinates. (To be more precise, they are a type of 3D polar coordinates known as *spherical* coordinates, which we'll discuss in Section 7.3.2.) Of course, if you are looking at a relatively small area compared to the size of the planet and you're not too far away from the equator, you can use latitude and longitude as Cartesian coordinates without too many problems. We do it all the time in Dallas.

---

[8]The authors have never actually heard anything like this first-hand. However, they have seen it in movies.

7.3   3D Polar Space

Polar coordinates can be used in 3D as well as 2D. As you probably have already guessed, 3D polar coordinates have *three* values. But is the third coordinate another linear distance (like r) or is it another angle (like θ)? Actually, we can choose to do either; there are *two* different types of 3D polar coordinates. If we add a linear distance, we have *cylindrical coordinates*, which are the subject of the next section. If we add another angle instead, we have *spherical coordinates*, which are covered in the later sections. Although cylindrical coordinates are less commonly used than spherical coordinates, we describe them first because they are easier to understand.

Section 7.3.1 discusses one kind of 3D polar coordinates, *cylindrical coordinates*, and Section 7.3.2 discusses the other kind of 3D polar coordinates, *spherical coordinates*. Section 7.3.3 presents some alternative polar coordinate conventions that are often more streamlined for use in video game code. Section 7.3.4 describes the special types of aliasing that can occur in spherical coordinate space. Section 7.3.5 shows how to convert between spherical coordinates and 3D Cartesian coordinates.

7.3.1   Cylindrical Coordinates

To extend Cartesian coordinates into 3D, we start with the 2D system, used for working in the plane, and add a third axis perpendicular to this plane. This is basically how cylindrical coordinates work to extend polar coordinates into 3D. Let's call the third axis the z-axis, as we do with Cartesian coordinates. To locate the point described by the cylindrical coordinates (r, θ, z), we start by processing r and θ just like we would for 2D polar coordinates, and then move "up" or "down" according to the z coordinate. Figure 7.5 shows how to locate a point (r, θ, z) by using cylindrical coordinates.

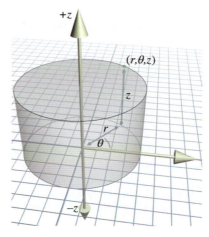

Figure 7.5
Cylindrical coordinates

Conversion between 3D Cartesian coordinates and cylindrical coordinates is straightforward. The z coordinate is the same in either representation, and we convert between (x, y) and (r, θ) via the 2D techniques from Section 7.1.3.

We don't use cylindrical coordinates much in this book, but they are useful in some situations when working in a cylinder-shaped environment or describing a cylinder-shaped object. In the same way that people often use polar coordinates without knowing it (see Section 7.2), people who don't know the term "cylindrical coordinates" may still use them. Be aware that even when people do acknowledge that they are using cylindrical coordinates, notation and conventions vary widely. For example, some people use the notation (ρ, ϕ, z). Also, the orientation of the axes and definition of positive rotation are set according to whatever is most convenient for a given situation.

7.3.2   Spherical Coordinates

The more common kind of 3D polar coordinate system is a *spherical coordinate system*. Whereas a set of cylindrical coordinates has two distances and one angle, a set of spherical coordinates has two angles and one distance.

Let's review the essence of how polar coordinates work in 2D. A point is specified by giving a direction (θ) and a distance (r). Spherical coordinates also work by defining a direction and distance; the only difference is that in 3D it takes *two* angles to define a direction. There are also two polar axes in a 3D spherical space. The first axis is "horizontal" and corresponds to the polar axis in 2D polar coordinates or $+x$ in our 3D Cartesian conventions. The other axis is vertical, corresponding to $+y$ in our 3D Cartesian conventions.

Different people use different conventions and notation for spherical coordinates, but most math people have agreed that the two angles are named θ and ϕ.[9] Math people also are in general agreement about how these two angles are to be interpreted to define a direction. The entire process works like this:

Locating points in 3D using polar coordinates

Step 1. Begin by standing at the origin, facing the direction of the horizontal polar axis. The vertical axis points from your feet to your head. Point your right[10] arm straight up, in the direction of the vertical polar axis.

Step 2. Rotate counterclockwise by the angle θ (the same way that we did for 2D polar coordinates).

---

[9] ϕ is the Greek letter phi, which is pronounced "fee" by most people. Some people prefer to make it rhyme with "fly."

[10] We mean no prejudice against our left-handed readers; you may imagine using your left arm if you wish. However, this is a right-handed coordinate system, so you may feel more official using your imaginary right arm. Save your left arm for later, when we discuss some left-handed conventions.

Step 3. Rotate your arm downward by the angle ϕ. Your arm now points in the direction specified by the polar angles θ and ϕ.

Step 4. Displace from the origin along this direction by the distance r. You've arrived at the point described by the spherical coordinates (r, θ, ϕ).

Figure 7.6 shows how this works.

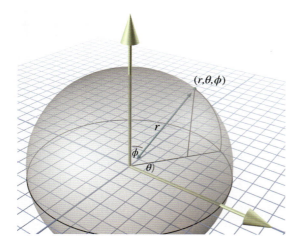

(r,θ,ϕ)

r

ϕ

$\theta)$

Figure 7.6
Spherical coordinates used by math people

Other people use different notation. The convention in which the symbols θ and ϕ are reversed is frequently used, especially in physics. Other authors, perhaps intent on replacing all Roman letters with Greek, use ρ instead of r as the name of the radial distance. We present some conventions that are a bit more practical for video game purposes in Section 7.3.3.

The horizontal angle θ is known as the *azimuth*, and ϕ is the *zenith*. Other terms that you've probably heard are *longitude* and *latitude*. *Longitude* is basically the same as θ, and *latitude* is the angle of inclination, $90° - \phi$. So, you see, the latitude/longitude system for describing locations on planet Earth is actually a type of spherical coordinate system. We're often interested only in describing points on the planet's surface, and so the radial distance r, which would measure the distance to the center of the Earth, isn't necessary. We can think of r as being roughly equivalent to *altitude*, although the value is offset by Earth's radius[11] in order to make either ground level or sea level equal to zero, depending on exactly what is meant by "altitude."

7.3.3   Some Polar Conventions Useful in 3D Virtual Worlds

The spherical coordinate system described in the previous section is the traditional right-handed system used by math people, and the formulas for converting between Cartesian and spherical coordinates are rather elegant under these assumptions. However, for most people in the video game industry, this elegance is only a minor benefit to be weighed against the following irritating disadvantages of the traditional conventions:

- The default horizontal direction at $\theta = 0$ points in the direction of $+x$. This is unfortunate, since for us, $+x$ points "to the right" or "east," neither of which are the "default" directions in most people's mind. Similar to the way that numbers on a clock start at the top, it would be nicer for us if the horizontal polar axis pointed towards $+z$, which is "forward" or "north."

- The conventions for the angle ϕ are unfortunate in several respects. It would be nicer if the 2D polar coordinates (r, θ) were extended into 3D simply by adding a third coordinate of zero, similar to how we extend the Cartesian system from 2D to 3D. But the spherical coordinates $(r, \theta, 0)$ don't correspond to the 2D polar coordinates (r, θ) as we'd like. In fact, assigning $\phi = 0$ puts us in the awkward situation of *Gimbal lock*, a singularity we describe in Section 7.3.4. Instead, the points in the 2D plane are represented as $(r, \theta, 90°)$. It might have been more intuitive to measure latitude, rather than zenith. Most people think of the default as "horizontal," and "up" as the extreme case.

- No offense to the Greeks, but θ and ϕ take a little while to get used to. The symbol r isn't so bad because at least it stands for something meaningful: radial distance or radius. Wouldn't it be great if the symbols we used to denote the angles were similarly short for English words, rather than completely arbitrary Greek symbols?

- It would be nice if the two angles for spherical coordinates were the same as the first two angles we use for *Euler angles*,[12] which are used to describe orientation in 3D. We're not going to discuss Euler angles

---

[11] Earth's radius is about 6,371 km (3,959 miles), on average.

[12] It's been said that the name *Euler* is a one-word math test: if you know how to pronounce it, then you've learned some math. Please make the authors of this book proud by passing this test, and pronouncing it "oiler," not "yooler."

until Section 8.3, so for now let us disagree with Descartes twice-over by saying "It'd be nice because we told you so."[13]

- It's a right-handed system, and we use a left-handed system (in this book at least).

Let's describe some spherical coordinate conventions that are better suited for our purposes. We have no complaints against the standard conventions for the radial distance r, and so we preserve both the name and semantics of this coordinate. Our grievances are primarily concerning the two angles, both of which we rename and repurpose.

The horizontal angle θ is renamed h, which is short for *heading* and is similar to a compass heading. A heading of zero indicates a direction of "forward" or "to the north," depending on the context. This matches standard aviation conventions. If we assume our 3D Cartesian conventions described in Section 1.3.4, then a heading of zero (and thus our primary polar axis) corresponds to $+z$. Also, since we prefer a left-handed coordinate system, positive rotation will rotate *clockwise* when viewed from above.

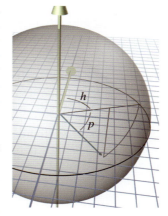

Figure 7.7
Heading and pitch angles used in this book

The vertical angle ϕ is renamed p, which is short for *pitch* and measures how much we are looking up or down. The default pitch value of zero indicates a horizontal direction, which is what most of us intuitively expect. Perhaps not so intuitively, positive pitch rotates *downward*, which means that pitch actually measures the *angle of declination*. This might seem to be a bad choice, but it is consistent with the left-hand rule (see Figure 1.14). Later we see how consistency with the left-hand rule bears fruit worth suffering this small measure of counterintuitiveness.

Figure 7.7 shows how heading and pitch conspire to define a direction.

7.3.4   Aliasing of Spherical Coordinates

Section 7.1.2 examined the bothersome phenomenon of *aliasing* of 2D polar coordinates: different numerical coordinate pairs are *aliases* of each other when they refer to the same point in space. Three basic types of aliasing were presented, which we review here because they are also present in the 3D spherical coordinate system.

---

[13] You read the first part of Chapter 1, right?

The first sure-fire way to generate an alias is to add a multiple of $360°$ to either angle. This is really the most trivial form of aliasing and is caused by the cyclic nature of angular measurements.

The other two forms of aliasing are a bit more interesting because they are caused by the interdependence of the coordinates. In other words, the meaning of one coordinate, r, depends on the values of the other coordinate(s), the angles. This dependency creates a form of aliasing and a singularity:

- The *aliasing* in 2D polar space can be triggered by negating the radial distance r and adjusting the angle so that the opposite direction is indicated. We can do the same with spherical coordinates. Using the heading and pitch conventions described in Section 7.3.3, all we need to do is flip the heading by adding an odd multiple of $180°$, and then negate the pitch.

- The *singularity* in 2D polar space occurs at the origin, because the angular coordinate is irrelevant when $r = 0$. With spherical coordinates, both angles are irrelevant at the origin.

So spherical coordinates exhibit similar aliasing behavior because the meaning of r changes depending on the values of the angles. However, spherical coordinates also suffer additional forms of aliasing because the pitch angle rotates about an axis that varies depending on the heading angle. This creates an additional form of aliasing and an additional singularity, which are analogous to those caused by the dependence of r on the direction.

- Different heading and pitch values can result in the same direction, even excluding trivial aliasing of each individual angle. An *alias* of (h, p) can be generated by $(h \pm 180°, 180° - p)$. For example, instead of turning right $90°$(facing "east") and pitching down $45°$, we could turn left $90°$(facing "west") and then pitch down $135°$. Although we would be upside down, we would still be looking in the same direction.

- A *singularity* occurs when the pitch angle is set to $\pm 90°$ (or any alias of these values). In this situation, known as *Gimbal lock*, the direction indicated is purely vertical (straight up or straight down), and the heading angle is irrelevant. We have a great deal more to say about Gimbal lock when we discuss Euler angles in Section 8.3.

Just as we did in 2D, we can define a set of canonical spherical coordinates such that any given point in 3D space maps unambiguously to exactly one coordinate triple within the canonical set. We place similar restrictions on r and h as we did for polar coordinates. Two additional constraints are added related to the pitch angle. First, pitch is restricted to be on the interval $[-90°, +90°]$. Second, since the heading value is irrelevant when pitch reaches the extreme values in the case of Gimbal lock, we force $h = 0$ in that case. The conditions that are satisfied by the points in the canonical set are summarized by the criteria below. (Note that these criteria assume our heading and pitch conventions, not the traditional math conventions with θ and ϕ.)

$r \geq 0$ — We don't measure distances "backwards."

Conditions satisfied by canonical spherical coordinates, assuming the conventions for spherical coordinates in this book

$-180° < h \leq 180°$ — Heading is limited to 1/2 revolution. We use $+180°$ for "south."

$-90° \leq p \leq 90°$ — Pitch limits are straight up and down. We can't "pitch over backwards."

$r = 0 \implies h = p = 0$ — At the origin, we set the angles to zero.

$|p| = 90° \implies h = 0$ — When looking directly up or down, we set the heading to zero.

The following algorithm can be used to convert a spherical coordinate triple into its canonical form:

1. If $r = 0$, then assign $h = p = 0$.

Converting a spherical coordinate triple (r, h, p) to canonical form

2. If $r < 0$, then negate r, add $180°$ to h, and negate p.

3. If $p < -90°$, then add $360°$ to p until $p \geq -90°$.

4. If $p > 270°$, then subtract $360°$ from p until $p \leq 270°$.

5. If $p > 90°$, then add $180°$ to h and set $p = 180° - p$.

6. If $h \leq -180°$, then add $360°$ to h until $h > -180°$.

7. If $h > 180°$, then subtract $360°$ from h until $h \leq 180°$.

Listing 7.3 shows how it could be done in C. Remember that computers like radians.

```
// Radial distance
float r;

// Angles in radians
float heading, pitch;

// Declare a few constants
const float TWOPI = 2.0f*PI; // 360 degrees
const float PIOVERTWO = PI/2.0f; // 90 degrees

// Check if we are exactly at the origin
if (r == 0.0f) {
```

```
        // At the origin - slam angles to zero
        heading = pitch = 0.0f;
} else {

        // Handle negative distance
        if (r < 0.0f) {
            r = -r;
            heading += PI;
            pitch = -pitch;
        }

        // Pitch out of range?
        if (fabs(pitch) > PIOVERTWO) {

            // Offset by 90 degrees
            pitch += PIOVERTWO;

            // Wrap in range 0...TWOPI
            pitch -= floor(pitch / TWOPI) * TWOPI;

            // Out of range?
            if (pitch > PI) {

                // Flip heading
                heading += PI;

                // Undo offset and also set pitch = 180-pitch
                pitch = 3.0f*PI/2.0f - pitch; // p = 270 degrees - p

            } else {

                // Undo offset, shifting pitch in range
                // -90 degrees ... +90 degrees
                pitch -= PIOVERTWO;
            }
        }

        // Gimbal lock?  Test using a relatively small tolerance
        // here, close to the limits of single precision.
        if (fabs(pitch) >= PIOVERTWO*0.9999) {
            heading = 0.0f;
        } else {

            // Wrap heading, avoiding math when possible
            // to preserve precision
            if (fabs(heading) > PI) {

                // Offset by PI
                heading += PI;

                // Wrap in range 0...TWOPI
                heading -= floor(heading / TWOPI) * TWOPI;

                // Undo offset, shifting angle back in range -PI...PI
                heading -= PI;
            }
        }
    }
}
```

Listing 7.3
Converting spherical coordinates to canonical form

7.3.5   Converting between Spherical and Cartesian Coordinates

Let's see if we can convert spherical coordinates to 3D Cartesian coordinates. Examine Figure 7.8, which shows both spherical and Cartesian coordinates. We first develop the conversions using the traditional right-handed conventions for both Cartesian and spherical spaces, and then we show conversions applicable to our left-handed conventions.

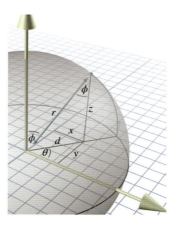

Notice in Figure 7.8 that we've introduced a new variable d, which is the horizontal distance between the point and the vertical axis. From the right triangle with hypotenuse r and legs d and z, we get

$$z/r = \cos\phi,$$
$$z = r\cos\phi.$$

Figure 7.8
Spherical and Cartesian coordinates for math people

and so we're left to compute x and y.

Consider that if $\phi = 90°$, we basically have 2D polar coordinates. Let's assign x' and y' to stand for the x and y coordinates that would result if $\phi = 90°$. From Section 7.1.3, we have

$$x' = r\cos\theta, \qquad\qquad y' = r\sin\theta.$$

Notice that when $\phi = 90°$, $d = r$. As ϕ decreases, d decreases, and by the properties of similar triangles, $x/x' = y/y' = d/r$. Looking at $\triangle drz$ again, we observe that $d/r = \sin\phi$. Putting all this together, we have

$$x = r\sin\phi\,\cos\theta, \qquad y = r\sin\phi\,\sin\theta, \qquad z = r\cos\phi.$$

Converting spherical coordinates used by math people to 3D Cartesian coordinates

These equations are applicable for right-handed math people. If we adopt our conventions for both the Cartesian (see Section 1.3.4) and spherical (see Section 7.3.3) spaces, the following formulas should be used:

$$x = r\cos p\,\sin h, \qquad y = -r\sin p, \qquad z = r\cos p\,\cos h. \qquad (7.3)$$

Spherical-to-Cartesian conversion for the conventions used in this book

Converting from Cartesian coordinates to spherical coordinates is more complicated, due to aliasing. We know that there are multiple sets of spherical coordinates that map to any given 3D position; we want the canonical coordinates. The derivation that follows uses our preferred aviation-inspired conventions in Equation (7.3) because those conventions are the ones most commonly used in video games.

As with 2D polar coordinates, computing r is a straightforward application of the distance formula:

$$r = \sqrt{x^2 + y^2 + z^2}.$$

As before, the singularity at the origin, where $r = 0$, is handled as a special case.

The heading angle is surprisingly simple to compute using our atan2 function:

$$h = \text{atan2}(x, z).$$

The trick works because atan2 uses only the ratio of its arguments and their signs. By examining Equation (7.3), we notice that the scale factor of $r \cos p$ is common to both x and z. Furthermore, by using canonical coordinates, we are assuming $r > 0$ and $-90° \leq p \leq 90°$; thus, $\cos p \geq 0$ and the common scale factor is always nonnegative. The Gimbal lock case is dealt with by our definition of atan2.

Finally, once we know r, we can solve for p from y:

$$y = -r \sin p,$$

$$-y/r = \sin p,$$

$$p = \arcsin(-y/r).$$

The arcsin function has a range of $[-90°, 90°]$, which fortunately coincides with the range for p within the canonical set.

Listing 7.4 illustrates the entire procedure.

```
// Input Cartesian coordinates
float x,y,z;

// Output radial distance
float r;

// Output angles in radians
float heading, pitch;

// Declare a few constants
const float TWOPI = 2.0f*PI; // 360 degrees
const float PIOVERTWO = PI/2.0f; // 90 degrees

// Compute radial distance
r = sqrt(x*x + y*y + z*z);

// Check if we are exactly at the origin
if (r > 0.0f) {

    // Compute pitch
    pitch = asin(-y/r);
```

```
    // Check for gimbal lock, since the library atan2
    // function is undefined at the (2D) origin
    if (fabs(pitch) >= PIOVERTWO*0.9999) {
        heading = 0.0f;
    } else {
        heading = atan2(x,z);
    }
} else {

    // At the origin — slam angles to zero
    heading = pitch = 0.0f;
}
```

Listing 7.4
Cartesian to spherical coordinate conversion

7.4 Using Polar Coordinates to Specify Vectors

We've seen how to describe a point by using polar coordinates, and how to describe a vector by using Cartesian coordinates. It's also possible to use polar form to describe vectors. Actually, to say that we can "also" use polar form is sort of like saying that a computer is controlled with a keyboard but it can "also" be controlled with the mouse. Polar coordinates directly describe the two key properties of a vector—its direction and length. In Cartesian form, these values are stored indirectly and obtained only through some computations that essentially boil down to a conversion to polar form. This is why, as we discussed in Section 7.2, polar coordinates are the local currency in everyday conversation.

But it isn't just laymen who prefer polar form. It's interesting to notice that most physics textbooks contain a brief introduction to vectors, and this introduction is carried out using a framework of polar coordinates. This is done despite the fact that it makes the math significantly more complicated.

As for the details of how polar vectors work, we've actually already covered them. Consider our "algorithm" for locating a point described by 2D polar coordinates on page 192. If you take out the phrase "start at the origin" and leave the rest intact, the instructions describe how to visualize the displacement (vector) described by any given polar coordinates. This is the same idea from Section 2.4: a vector is related to the point with the same coordinates because it gives us the displacement from the origin to that point.

We've also already learned the math for converting vectors between Cartesian and polar form. The methods discussed in Section 7.1.3 were presented in terms of points, but they are equally valid for vectors.

7.5   Exercises

(Answers on page 767.)

1. Plot and label the points with the following polar coordinates:

$$\mathbf{a} = (2, 60°) \qquad\qquad \mathbf{b} = (5, 195°)$$
$$\mathbf{c} = (3, -45°) \qquad\qquad \mathbf{d} = (-2.75, 300°)$$
$$\mathbf{e} = (4, \pi/6 \text{ rad}) \qquad\qquad \mathbf{f} = (1, 4\pi/3 \text{ rad})$$
$$\mathbf{g} = (-5/2, -\pi/2 \text{ rad})$$

2. Convert the following 2D polar coordinates to canonical form:

   (a) $(4, 207°)$

   (b) $(-5, -720°)$

   (c) $(0, 45.2°)$

   (d) $(12.6, 11\pi/4 \text{ rad})$

3. Convert the following 2D polar coordinates to Cartesian form:

   (a) $(1, 45°)$

   (b) $(3, 0°)$

   (c) $(4, 90°)$

   (d) $(10, -30°)$

   (e) $(5.5, \pi \text{ rad})$

4. Convert the polar coordinates in Exercise 2 to Cartesian form.

5. Convert the following 2D Cartesian coordinates to (canonical) polar form:

   (a) $(10, 20)$

   (b) $(-12, -5)$

   (c) $(0, 4.5)$

   (d) $(-3, 4)$

   (e) $(0, 0)$

   (f) $(-5280, 0)$

6. Convert the following cylindrical coordinates to Cartesian form:

   (a) $(4, 120°, 5)$

   (b) $(2, 45°, -1)$

   (c) $(6, -\pi/6, -3)$

   (d) $(3, 3\pi, 1)$

7. Convert the following 3D Cartesian coordinates to (canonical) cylindrical form:

   (a) $(1, 1, 1)$

   (b) $(0, -5, 2)$

   (c) $(-3, 4, -7)$

   (d) $(0, 0, -3)$

8. Convert the following spherical coordinates (r, θ, ϕ) to Cartesian form according to the standard mathematical convention:

   (a) $(4, \pi/3, 3\pi/4)$

   (b) $(5, -5\pi/6, \pi/3)$

   (c) $(2, -\pi/6, \pi)$

   (d) $(8, 9\pi/4, \pi/6)$

9. Interpret the spherical coordinates (a)–(d) from the previous exercise as (r, h, p) triples, switching to our video game conventions.

   1. Convert to canonical (r, h, p) coordinates.

   2. Use the canonical coordinates to convert to Cartesian form (using the video game conventions).

10. Convert the following 3D Cartesian coordinates to (canonical) spherical form using our modified convention:

    (a) $(\sqrt{2}, 2\sqrt{3}, -\sqrt{2})$

    (b) $(2\sqrt{3}, 6, -4)$

    (c) $(-1, -1, -1)$

    (d) $(2, -2\sqrt{3}, 4)$

    (e) $(-\sqrt{3}, -\sqrt{3}, 2\sqrt{2})$

    (f) $(3, 4, 12)$

11. What do the "grid lines" look like in spherical space? Assuming the spherical conventions used in this book, describe the shape defined by the set of all points that meet the following criteria. Do not restrict the coordinates to the canonical set.

    (a) A fixed radius $r = r_0$, but any arbitrary values for h and p.

    (b) A fixed heading $h = h_0$, but any arbitrary values for r and p.

    (c) A fixed pitch $p = p_0$, but any arbitrary values for r and h.

12. During crunch time one evening, a game developer decided to get some fresh air and go for a walk. The developer left the studio walking south and walked for 5 km. She then turned east and walked another 5 km. Realizing that all the fresh air was making her light-headed, she decided to return to the studio. She turned north, walked 5 km and was back at the

studio, ready to squash the few remaining programming bugs left on her list. Unfortunately, waiting for her at the door was a hungry bear, and she was eaten alive.[14] What color was the bear?

For the execution of the voyage to the Indies,
I did not make use of intelligence, mathematics or maps.

— Christopher Columbus (1451-1506)

---

[14] We know this scenario is totally impossible. I mean, a game developer taking a walk during crunch time?!

Chapter 8

Rotation in Three Dimensions

If you do not change direction,
you may end up where you are heading.

— Lao Tzu (600–531 BCE)

This chapter tackles the difficult problem of describing the orientation of an object in 3D. It also discusses the closely related concepts of rotation and angular displacement. There are several different ways we can express orientation and angular displacement in 3D. Here we discuss the three most important methods—matrices, Euler angles, and quaternions—as well as two lesser known forms—axis-angle and exponential map. For each method, we define precisely how the representation method works, and discuss the peculiarities, advantages, and disadvantages of the method.

Different techniques are needed in different circumstances, and each technique has its advantages and disadvantages. It is important to know not only how each method works, but also which technique is most appropriate for a particular situation and how to convert between representations.

The discussion of orientation in 3D is divided into the following sections:

- Section 8.1 discusses the subtle differences between terms like "orientation," "direction," and "angular displacement."

- Section 8.2 describes how to express orientation using a matrix.

- Section 8.3 describes how to express angular displacement using Euler angles.

- Section 8.4 describes the axis-angle and exponential map forms.

- Section 8.5 describes how to express angular displacement using a quaternion.

- Section 8.6 compares and contrasts the different methods.

- Section 8.7 explains how to convert an orientation from one form to another.

This chapter makes extensive use of the terms *object space* and *upright space*. If you aren't familiar with these terms, you should flip back to Section 3.2, where the terms were first introduced.

8.1   What Exactly is "Orientation"?

Before we can begin to discuss how to describe orientation in 3D, let us first define exactly what it is that we are attempting to describe. The term *orientation* is related to other similar terms, such as

- direction

- angular displacement

- rotation.

Intuitively, we know that the "orientation" of an object basically tells us what direction the object is facing. However, "orientation" is not exactly the same as "direction."

For example, a vector has a direction, but not an orientation. The difference is that when a vector points in a certain direction, you can twist the vector along its length (see Figure 8.1), and there is no real change to the vector, since a vector has no thickness or dimension other than its length.

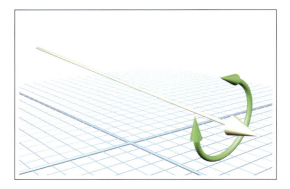

Figure 8.1
Twisting a vector results in no appreciable change to the vector

In contrast to a simple vector, consider an object, such as a jet, facing a certain direction. If we twist the jet (see Figure 8.2) in the same way that we twisted the vector, we *will* change the *orientation* of the jet. In Section 8.3, we refer to this twisting component of an object's orientation as *bank*.

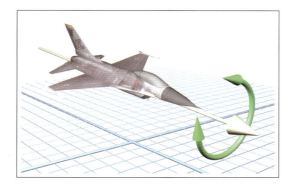

Figure 8.2
Twisting an object changes its orientation

The fundamental difference between direction and orientation is seen concretely by the fact that we can parameterize a direction in 3D with just two numbers (the spherical coordinate angles—see Section 7.3.2), whereas an orientation requires a minimum of three numbers (Euler angles—see Section 8.3).

Section 2.4.1 discussed that it's impossible to describe the position of an object in absolute terms—we must always do so within the context of a specific reference frame. When we investigated the relationship between "points" and "vectors," we noticed that specifying a position is actually the same as specifying an amount of translation from some other given reference point (usually the origin of some coordinate system).

In the same way, orientation cannot be described in absolute terms. Just as a position is given by a translation from some known point, an orientation is given by a *rotation* from some known reference orientation (often called the "identity" or "home" orientation). The amount of rotation is known as an *angular displacement*. In other words, describing an orientation is mathematically equivalent to describing an angular displacement.

We say "mathematically equivalent" because in this book, we make a subtle distinction between "orientation" and terms such as "angular displacement" and "rotation." It is helpful to think of an "angular displacement" as an operator that accepts an input and produces an output. A particular direction of transformation is implied; for example, the angular displacement *from* the old orientation *to* the new orientation, or *from* upright space *to* object space. An example of an angular displacement is, "Rotate 90° about the z-axis." It's an action that we can perform on a vector.

However, we frequently encounter state variables and other situations in which this operator framework of input/output is not helpful and a parent/ child relationship is more natural. We tend to use the word "orientation" in those situations. An example of an orientation is, "Standing upright and facing east." It describes a state of affairs.

Of course, we can describe the orientation "standing upright and facing east" as an angular displacement by saying, "Stand upright, facing north, and then rotate 90° about the z-axis." This distinction between orientation and angular displacement is similar to the distinction between points and vectors, which are two other terms that are equivalent mathematically but not identical conceptually. In both cases, the first term is used primarily to describe a single state, and the second term primarily used to describe a difference between two states. Of course, these conventions are purely a matter of preference, but they can be helpful.

You might also hear the word "attitude" used to refer the orientation of an object, especially if that object is an aircraft.

8.2   Matrix Form

One way to describe the orientation of a coordinate space in 3D is to tell which way the basis vectors of that coordinate space (the $+x$, $+y$, and $+z$ axes) point. Of course, we don't measure these vectors in the coordinate space we are attempting to describe—by definition, they are $[1, 0, 0]$, $[0, 1, 0]$, and $[0, 0, 1]$ no matter what orientation the coordinate space is in. We must describe the basis vectors using some *other* coordinate space. By doing so, we've established the relative orientation of the two coordinate spaces.

When these basis vectors are used to form the rows of a 3×3 matrix, we have expressed the orientation in *matrix form*.[1] Another way of saying all this is that we can express the relative orientation of two coordinate spaces by giving a rotation matrix that can be used to transform vectors from one coordinate space to the other.

8.2.1   Which Matrix?

We have already seen how a matrix can be used to transform points from one coordinate space to another. In Figure 8.3, the matrix in the upper right-hand corner can be used to rotate points from the object space of the jet into upright space. We've pulled out the rows of this matrix to

---

[1] Actually, we can put the vectors into the columns of a matrix, too. Certainly this is true if we are using column vectors—but it turns out to work even if our preference is to use row vectors. This is because rotation matrices are orthonormal, which means we can invert them by taking their transpose. We discuss this in Section 8.2.1.

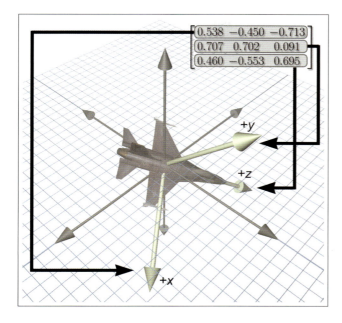

$$\begin{bmatrix} 0.538 & -0.450 & -0.713 \\ 0.707 & 0.702 & 0.091 \\ 0.460 & -0.553 & 0.695 \end{bmatrix}$$

Figure 8.3
Defining an
orientation using
a matrix

emphasize their direct relationship to the coordinates for the jet's body axes. The rotation matrix contains the object axes, expressed in upright space. Simultaneously, it is a rotation matrix: we can multiply row vectors by this matrix to transform those vectors from object-space coordinates to upright-space coordinates.

Legitimate question to ask are: Why does the matrix contain the body axes expressed using upright-space coordinates? Why not the upright axes expressed in object-space coordinates? Another way to phrase this is, Why did we choose to give a rotation matrix that transforms vectors from object space to upright space? Why not from upright space to object space?

From a mathematical perspective, this question is a bit ridiculous. Because rotation matrices are orthogonal, their inverse is the same as their transpose (see Section 6.3.2). Thus, the decision is entirely a cosmetic one.

But practically speaking, in our opinion, it is quite important. At issue is whether you can write code that is intuitive to read and works the first time, or whether it requires a lot of work to decipher, or a knowledge of conventions that are not stated because they are "obvious" to everyone but you. So please allow us a brief digression to continue a line of thought begun when we introduced the term "upright space" in Section 3.2.4 concerning the practical aspects of what happens when the math of coordinate space transformations gets translated into code. Also please allow some latitude to express some opinions based on our observations watching programmers grapple with rotation matrices. We don't expect that everyone will agree with our assertions, but we hope that every reader will at least appreciate the value in considering these issues.

Certainly every good math library will have a 3×3 matrix class that can represent any arbitrary transformation, which is to say that it makes no assumptions about the value of the matrix elements. (Or perhaps it is a 4×4 matrix that can do projection, or a 4×3 which can do translation but not projection—those distinctions are not important here.) For a matrix like this, the operations inherently are in terms of some input coordinate space and an output coordinate space. This is just implicit in the idea of matrix multiplication. If you need to go from output to input, then you must obtain the inverse of the matrix.

It is common practice to use the generic transform matrix class to describe the orientation of an object. In this case, rotation is treated just like any other transformation. The interface remains in terms of a source and destination space. Unfortunately, it is our experience that the following two matrix operations are by far the most commonly used:[2]

- Take an object-space vector and express it in upright coordinates.

- Take an upright-space vector and express it in object coordinates.

Notice that we need to be able to go in both directions. We have no experience or evidence that either direction is significantly more common than the other. But more important, the very nature of the operations and the way programmers think about the operations is in terms of "object space" and "upright space" (or some other equivalent terminology, such as "parent space" and "child space"). We do *not* think of them in terms of a source space and a destination space. It is in this context that we wish to consider the question posed at the beginning of this section: Which matrix should we use?

First, we should back up a bit and remind ourselves of the mathematically moot but yet conceptually important distinction between orientation and angular displacement. (See the notes on terminology at the end of Section 8.1.) If your purpose is to create a matrix that performs a specific angular displacement (for example, "rotate 30 degrees about the x-axis"), then the two operations above are not really the ones you probably have in your head, and using a generic transform matrix with its implied direction of transformation is no problem, and so this discussion does not apply.

---

[2]We are measuring frequency of use based on how many times the operation is coded, not how often it is executed at run time. For example, transforming vertices through the graphics pipeline is certainly an extremely commonly used matrix operation, but there are relatively few lines of code that do this operation. This has proven to be true in a wide variety of game genres, such as racing, combat, 3D board games, and shooters, although, of course, we cannot speak for everyone's work environment.

Right now, we are focusing on the situation in which the orientation of some object is stored as a state variable.

Let's assume that we adopt the common policy and store orientation using the generic transformation matrix. We are forced to arbitrarily pick a convention, so let's decide that multiplication by this matrix will transform from object to upright space. If we have a vector in upright space and we need to express it in object-space coordinates, we must multiply this vector by the inverse[3] of the matrix.

Now let's see how our policy affects the code that is written and read hundreds of times by average game programmers.

- *Rotate some vector from object space to upright space* is translated into code as multiplication by the matrix.

- *Rotate a vector from upright space to object space* is translated into code as multiplication by the inverse (or transpose) of the matrix.

Notice that the code does not match one-to-one with the high-level intentions of the programmer. It forces every user to remember what the conventions are every time they use the matrix. It is our experience that this coding style is a contributing factor to the difficulty that beginning programmers have in learning how to use matrices; they often end up transposing and negating things randomly when things don't look right.

We have found it helpful to have a special 3×3 matrix class that is used exclusively for storing the orientation of an object, not for arbitrary transforms. The class assumes, as an invariant, that the matrix is orthogonal, meaning it contains only rotation. (We also would probably assume that the matrix does not contain a reflection, even though that is possible in an orthogonal matrix.) With these assumptions in place, we are now free to perform rotations using the matrix at a higher level of abstraction. Our interface functions match exactly the high-level intentions of the programmer. Furthermore, we have removed the confusing linear algebra details having to do with row vectors versus column vectors, which space is on the left or right, which way is the regular way and which is the inverse, and so forth. Or rather, we have confined such details to the class internals—the person implementing the class certainly needs to pick a convention (and hopefully document it). In fact, in this specialized matrix class, the operations of "multiply a vector" and "invert this matrix" really are not that useful. We would advocate keeping this dedicated matrix class confined to operations in terms of upright space and object space, rather than multiply a vector.

---

[3]Actually, probably multiplication by the transpose, since rotation matrices are orthogonal—but that is not the point here.

So, back to the question posed at the start of this section: Which matrix should we use? Our answer is, "It shouldn't matter." By that we mean there is a way to design your matrix code in such a way that it can be used without knowing what choice was made. As far as the C++ code goes, this is purely a cosmetic change. For example, perhaps we just replace the function name `multiply()` with `objectToUpright()`, and likewise we replace `multiplyByTranspose()` with `uprightToObject()`. The version of the code with descriptive, named coordinate spaces is easier to read and write.

8.2.2   Direction Cosines Matrix

You might come across the (very old school) term *direction cosines* in the context of using a matrix to describe orientation. A direction cosines matrix is the same thing as a rotation matrix; the term just refers to a special way to interpret (or construct) the matrix, and this interpretation is interesting and educational, so let's pause for a moment to take a closer look. Each element in a rotation matrix is equal to the dot product of a cardinal axis in one space with a cardinal axis in the other space. For example, the center element m_{22} in a 3×3 matrix gives the dot product that the y-axis in one space makes with the y-axis in the other space.

More generally, let's say that the basis vectors of a coordinate space are the mutually orthogonal unit vectors $\mathbf{p}$, $\mathbf{q}$, and $\mathbf{r}$, while a second coordinate space with the same origin has as its basis a different (but also orthonormal) basis $\mathbf{p}'$, $\mathbf{q}'$, and $\mathbf{r}'$. (Please allow us to break from convention by dropping all the hats from the unit vectors in this section, to avoid distracting clutter in the equations.) The rotation matrix that rotates row vectors from the first space to the second can be constructed from the cosines of the angles between each pair of basis vectors. Of course, the dot product of two unit vectors is exactly equal to the cosine of the angle between them, so the matrix product is

$$\mathbf{v} \begin{bmatrix} \mathbf{p} \cdot \mathbf{p}' & \mathbf{q} \cdot \mathbf{p}' & \mathbf{r} \cdot \mathbf{p}' \\ \mathbf{p} \cdot \mathbf{q}' & \mathbf{q} \cdot \mathbf{q}' & \mathbf{r} \cdot \mathbf{q}' \\ \mathbf{p} \cdot \mathbf{r}' & \mathbf{q} \cdot \mathbf{r}' & \mathbf{r} \cdot \mathbf{r}' \end{bmatrix} = \mathbf{v}'. \tag{8.1}$$

These axes can be interpreted as geometric rather than numeric entities, so it really does not matter what coordinates are used to describe the axes (provided we use the same coordinate space to describe all of them), the rotation matrix will be the same.

For example, let's say that our axes are described using coordinates relative to the first basis. Then $\mathbf{p}$, $\mathbf{q}$, and $\mathbf{r}$ have the trivial forms $[1, 0, 0]$, $[0, 1, 0]$ and $[0, 0, 1]$, respectively. The basis vectors of the second space, $\mathbf{p}'$, $\mathbf{q}'$, and $\mathbf{r}'$ have arbitrary coordinates. When we substitute the trivial vectors $\mathbf{p}$, $\mathbf{q}$, and $\mathbf{r}$ into the matrix in Equation (8.1) and expand the dot products, we get

$$\begin{bmatrix} [1,0,0] \cdot \mathbf{p}' & [0,1,0] \cdot \mathbf{p}' & [0,0,1] \cdot \mathbf{p}' \\ [1,0,0] \cdot \mathbf{q}' & [0,1,0] \cdot \mathbf{q}' & [0,0,1] \cdot \mathbf{q}' \\ [1,0,0] \cdot \mathbf{r}' & [0,1,0] \cdot \mathbf{r}' & [0,0,1] \cdot \mathbf{r}' \end{bmatrix} = \begin{bmatrix} p'_x & p'_y & p'_z \\ q'_x & q'_y & q'_z \\ r'_x & r'_y & r'_z \end{bmatrix} = \begin{bmatrix} -\mathbf{p}'- \\ -\mathbf{q}'- \\ -\mathbf{r}'- \end{bmatrix}.$$

In other words, the rows of the rotation matrix are the basis vectors of the output coordinate space, expressed by using the coordinates of the input coordinate space. Of course, this fact is not just true for rotation matrices, it's true for *all* transformation matrices. This is the central idea of why a transformation matrix works, which was developed in Section 4.2.

Now let's look at the other case. Instead of using coordinates relative to the first basis, we'll measure everything using the second coordinate space (the output space). This time, $\mathbf{p}'$, $\mathbf{q}'$, and $\mathbf{r}'$ have trivial forms, and $\mathbf{p}$, $\mathbf{q}$, and $\mathbf{r}$ are arbitrary. Putting these into the direction cosines matrix produces

$$\begin{bmatrix} \mathbf{p} \cdot [1,0,0] & \mathbf{q} \cdot [1,0,0] & \mathbf{r} \cdot [1,0,0] \\ \mathbf{p} \cdot [0,1,0] & \mathbf{q} \cdot [0,1,0] & \mathbf{r} \cdot [0,1,0] \\ \mathbf{p} \cdot [0,0,1] & \mathbf{q} \cdot [0,0,1] & \mathbf{r} \cdot [0,0,1] \end{bmatrix} = \begin{bmatrix} p_x & q_x & r_x \\ p_y & q_y & r_y \\ p_z & q_z & r_z \end{bmatrix} = \begin{bmatrix} | & | & | \\ \mathbf{p}^{\mathrm{T}} & \mathbf{q}^{\mathrm{T}} & \mathbf{r}^{\mathrm{T}} \\ | & | & | \end{bmatrix}.$$

This says that the *columns* of the rotation matrix are formed from the basis vectors of the input space, expressed using the coordinates of the output space. This is *not* true of transformation matrices in general; it applies only to orthogonal matrices such as rotation matrices.

Also, remember that our convention is to use row vectors on the left. If you are using column vectors on the right, things will be transposed.

8.2.3   Advantages of Matrix Form

Matrix form is a very explicit form of representing orientation. This explicit nature provides some benefits.

- *Rotation of vectors is immediately available.* The most important property of matrix form is that you can use a matrix to rotate vectors between object and upright space. No other representation of orientation allows this[4]—to rotate vectors, we must convert the orientation to matrix form.

---

[4]It is an often-touted advantage of quaternions that they can be used to perform rotations through quaternion multiplication (see Section 8.5.7). However, if we examine the math, we see that this "shortcut" amounts to multiplication by the corresponding rotation matrix.

- *Format used by graphics APIs.* Partly due to reasons in the previous item, graphics APIs use matrices to express orientation. (API stands for Application Programming Interface. Basically, this is the code we use to communicate with the graphics hardware.) When we are communicating with the API, we are going to have to express our transformations as matrices. How we store transformations internally in our program is up to us, but if we choose another representation, we are going to have to convert them into matrices at some point in the graphics pipeline.

- *Concatenation of multiple angular displacements.* A third advantage of matrices is that it is possible to "collapse" nested coordinate space relationships. For example, if we know the orientation of object A relative to object B, and we know the orientation of object B relative to object C, then by using matrices, we can determine the orientation of object A relative to object C. We encountered these concepts before when we discussed nested coordinate spaces in Chapter 3, and then we discussed how matrices could be concatenated in Section 5.6.

- *Matrix inversion.* When an angular displacement is represented in matrix form, it is possible to compute the "opposite" angular displacement by using matrix inversion. What's more, since rotation matrices are orthogonal, this computation is a trivial matter of transposing the matrix.

8.2.4   Disadvantages of Matrix Form

The explicit nature of a matrix provides some advantages, as we have just discussed. However, a matrix uses nine numbers to store an orientation, and it is possible to parameterize orientation with only three numbers. The "extra" numbers in a matrix can cause some problems.

- *Matrices take more memory.* If we need to store many orientations (for example, keyframes in an animation sequence), that extra space for nine numbers instead of three can really add up. Let's take a modest example. Let's say we are animating a model of a human that is broken up into 15 pieces for different body parts. Animation is accomplished strictly by controlling the orientation of each part relative to its parent part. Assume we are storing one orientation for each part, per frame, and our animation data is stored at a modest rate, say, 15 Hz. This means we will have 225 orientations per second. Using matrices and 32-bit floating point numbers, each frame will take 8,100 bytes. Using Euler angles (which we will meet next in Section 8.3), the same data would take only 2,700 bytes. For a mere

30 seconds of animation data, matrices would take 162K more than the same data stored using Euler angles!

- *Difficult for humans to use.* Matrices are not intuitive for humans to work with directly. There are just too many numbers, and they are all between -1 and $+1$. What's more, humans naturally think about orientation in terms of angles, but a matrix is expressed in terms of vectors. With practice, we can learn how to decipher the orientation from a given matrix. (The techniques from Section 4.2 for visualizing a matrix help a lot for this.) But still, this is much more difficult than Euler angles. And going the other way is *much* more difficult—it would take forever to construct the matrix for a nontrivial orientation by hand. In general, matrices just aren't the way people naturally think about orientation.

- *Matrices can be ill-formed.* As we have said, a matrix uses nine numbers, when only three are necessary. In other words, a matrix contains six degrees of redundancy. There are six constraints that must be satisfied for a matrix to be "valid" for representing an orientation. The rows must be unit vectors, and they must be mutually perpendicular (see Section 6.3.2).

Let's consider this last point in more detail. If we take any nine numbers at random and create a 3×3 matrix, it is very unlikely that these six constraints will be satisfied, and thus the nine numbers will not form a valid rotation matrix. In other words, matrices can be ill-formed, at least for purposes of representing an orientation. Ill-formed matrices can be a problem because they can lead to numerical exceptions, weird stretched graphics, and other unexpected behavior.

How could we ever end up with a bad matrix? There are several ways:

- We may have a matrix that contains scale, skew, reflection, or projection. What is the "orientation" of an object that has been affected by such operations? There really isn't a clear definition for this. Any nonorthogonal matrix is not a well-defined rotation matrix. (See Section 6.3 for a complete discussion on orthogonal matrices.) And reflection matrices (which are orthogonal) are not valid rotation matrices, either.

- We may just get bad data from an external source. For example, if we are using a physical data acquisition system, such as motion capture, there could be errors due to the capturing process. Many modeling packages are notorious for producing ill-formed matrices.

- We can actually *create* bad data due to floating point round off error. For example, suppose we apply a large number of incremental changes to an orientation, which could routinely happen in a game or simulation that allows a human to interactively control the orientation of an object. The large number of matrix multiplications, which are subject to limited floating point precision, can result in an ill-formed matrix. This phenomenon is known as *matrix creep*. We can combat matrix creep by *orthogonalizing* the matrix, as we already discussed in Section 6.3.3.

8.2.5   Summary of Matrix Form

Let's summarize what Section 8.2 has said about matrices.

- Matrices are a "brute force" method of expressing orientation: we explicitly list the basis vectors of one space in the coordinates of some different space.

- The term *direction cosines matrix* alludes to the fact that each element in a rotation matrix is equal to the dot product of one input basis vector with one output basis vector. Like all transformation matrices, the rows of the matrix are the output-space coordinates of the input-space basis vectors. Furthermore, the columns of a rotation matrix are the input-space coordinates of the output-space basis vectors, a fact that is only true by virtue of the orthogonality of a rotation matrix.

- The matrix form of representing orientation is useful primarily because it allows us to rotate vectors between coordinate spaces.

- Modern graphics APIs express orientation by using matrices.

- We can use matrix multiplication to collapse matrices for nested coordinate spaces into a single matrix.

- Matrix inversion provides a mechanism for determining the "opposite" angular displacement.

- Matrices can take two to three times as much memory as other techniques. This can become significant when storing large numbers of orientations, such as animation data.

- The numbers in a matrix aren't intuitive for humans to work with.

- Not all matrices are valid for describing an orientation. Some matrices contain mirroring or skew. We can end up with a ill-formed matrix either by getting bad data from an external source or through matrix creep.

8.3   Euler Angles

Another common method of representing orientation is known as *Euler angles*. (Remember, Euler is pronounced "oiler," not "yoolur.") The technique is named after the famous mathematician who developed them, Leonhard Euler (1707–1783). Section 8.3.1 describes how Euler angles work and discusses the most common conventions used for Euler angles. Section 8.3.2 discusses other conventions for Euler angles, including the important *fixed axis* system. We consider the advantages and disadvantages of Euler angles in Section 8.3.3 and Section 8.3.4. Section 8.3.5 summarizes the most important concepts concerning of Euler angles.

This section utilizes many ideas, terms, and conventions from Section 7.3.2 concerning spherical coordinates.

8.3.1   What Are Euler Angles?

The basic idea behind Euler angles is to define an angular displacement as a sequence of three rotations about three mutually perpendicular axes. This sounds complicated, but actually it is quite intuitive. (In fact, its ease of use by humans is one of its primary advantages.)

So Euler angles describe orientation as three rotations about three mutually perpendicular axes. But which axes? And in what order? As it turns out, any three axes in any order will work, but most people have found it practical to use the cardinal axes in a particular order. The most common convention, and the one we use in this book, is the so-called "heading-pitch-bank" convention for Euler angles. In this system, an orientation is defined by a *heading* angle, a *pitch* angle, and a *bank* angle.

Before we define the terms heading, pitch, and bank precisely, let us briefly review the coordinate space conventions we use in this book. We use a left-handed system, where $+x$ is to the right, $+y$ is up, and $+z$ is forward. (Check out Figure 1.15 on page 19 for an illustration.) Also, if you have forgotten how positive rotation is defined according to the left-hand rule, you might want to flip back to Figure 1.14 on page 17 to refresh your memory.

Given heading, pitch, and bank angles, we can determine the orientation described by these Euler angles using a simple four-step process.

Step 1. Begin in the "identity" orientation—that is, with the object-space axes aligned with the upright axes.

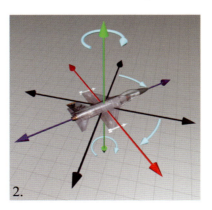

Figure 8.4
Step 1: An object in its identity orientation

Step 2. Perform the *heading* rotation, rotating about the y-axis, as shown in Figure 8.5. Positive rotation rotates to the right (clockwise when viewed from above).

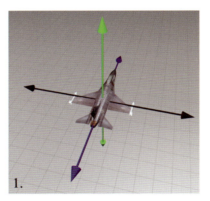

Figure 8.5
Step 2: Heading is the first rotation and rotates about the vertical axis

Step 3. After heading has been applied, *pitch* measures the amount of rotation about the x-axis. This is the object-space x-axis, not the upright x-axis. Staying consistent with the left-hand rule, positive rotation rotates *downward*. In other words, pitch actually measures the angle of *declination*. This is illustrated in Figure 8.6.

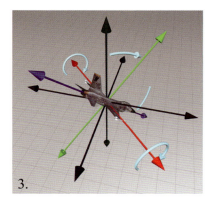

3.

Figure 8.6
Step 3: Pitch is the second rotation and rotates about the object lateral axis

Step 4. After heading and pitch angles have been applied, *bank* measures the amount of rotation about the z-axis. Again, this is the object-space z-axis, not the original upright-space z-axis. The left-hand rule dictates that positive bank rotates counterclockwise when viewed from the origin looking towards $+z$. This is illustrated in Figure 8.7.

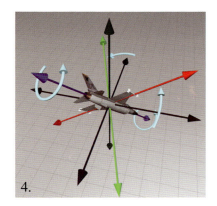

4.

Figure 8.7
Step 4: Bank is the third and final rotation and rotates about the object longitudinal axis

It may seem contradictory that positive bank is counterclockwise, since positive heading is clockwise. But notice that positive heading is clockwise when viewed from the positive end of the axis towards the origin, the opposite perspective from the one used when judging clockwise/counterclockwise for bank. If we look from the origin to the positive end of the y-axis, then positive heading does rotate counterclockwise. Or if we look from the positive end of the z-axis towards the origin (looking backward from in front of the object), then positive bank appears to rotate the object clockwise. In either case, the left-hand rule prevails.

Now you have reached the orientation described by the Euler angles. Notice the similarity of Steps 1–3 to the procedure used in Section 7.3.2 to locate the direction described by the spherical coordinate angles. In other words, we can think of heading and pitch as defining the basic direction that the object is facing, and bank defining the amount of twist.

8.3.2   Other Euler Angle Conventions

The heading-pitch-bank system described in the previous section isn't the only way to define a rotation using three angles about mutually perpendicular axes. There are many variations on this theme. Some of these differences turn out to be purely nomenclature; others are more meaningful. Even if you like our conventions, we encourage you to not skip this section, as some very important concepts are discussed; these topics are the source of much confusion, which we hope to dispel.

First of all, there is the trivial issue of naming. The most common variation you will find was made popular by the field of aerospace, the *yaw-pitch-roll* method.[5] The term "roll" is completely synonymous with bank, and for all purposes they are identical. Similarly, within the limited context of yaw-pitch-roll, the term "yaw" is practically identical to the term heading. (However, in a broader sense, the word "yaw" actually has a subtly different meaning, and it is this subtle difference that drives our preference for the term *heading*. We discuss this rather nit-picky distinction in just a moment, but for the moment yaw and heading are the same.) So essentially yaw-pitch-roll is the same system as heading-pitch-bank.

Other less common terms are often used. Heading also goes by the name *azimuth*. The vertical angle that we call pitch is also called attitude or *elevation*. The final angle of rotation, which we call "bank," is sometimes called *tilt* or *twist*.

---

[5] It's probably a bit presumptuous of us to refer to yaw-pitch-roll as a variation. After all, there is a Wikipedia article for yaw-pitch-roll, but none for heading-pitch-bank, so who's to say our preference isn't the variation? We admit this presumption and are prepared to defend our preference shortly.

And, of course, there are those perverse mathematicians who (motivated by the need to save space when writing on a chalkboard?) insist on assaulting your eyeballs with a slew of Greek letters. You may see any of the following:

$$(\phi, \theta, \psi) \quad (\psi, \theta, \phi)$$
$$(\Omega, i, \omega) \quad (\alpha, \beta \, \gamma).$$

It's all Greek to us

Of course, these are cosmetic differences. Perhaps more interesting is that fact that you will often hear these same three words listed in the opposite order: roll-pitch-yaw. (A quick Google search for "roll pitch yaw" or "yaw pitch roll" yields plenty of results for both forms, with neither appearing more predominant.) Considering how the order of rotations is so critical, are people really that perverse that they choose to list them in the reverse order? We're not just dwelling on terminology here; the distinctions in thinking hinted at by the differences in terminology will actually become useful when we consider how to convert Euler angles to a rotation matrix. As it turns out, there is a perfectly reasonable explanation for this "backwards" convention: it's the order in which we actually do the rotations inside a computer!

The *fixed-axis* system is very closely related to the Euler angle system. In an Euler angle system, the rotation occurs about the *body* axes, which change after each rotation. Thus, for example, the physical axis for the bank angle is always the longitudinal body space axis, but in general it is arbitrarily oriented in upright space. In a fixed-axis system, in contrast, the axes of rotation are always the fixed, *upright* axes. But as it turns out, the fixed-axis system and the Euler angle system are actually equivalent, *provided that we take the rotations in the opposite order.*

You should visualize the following example to convince yourself this is true. Let's say we have a heading (yaw) of h and a pitch of p. (We'll ignore bank/roll for the moment.) According to the Euler angle convention, we first do the heading axis and rotate about the vertical axis (the y-axis) by h. Then we rotate about the *object-space* lateral axis (the x-axis) by the angle p. Using a fixed-axis scheme, we arrive at this same ending orientation by doing the rotations in the opposite order. First, we do the pitch, rotating about the upright x-axis by p. Then, we perform the heading rotation, rotating about the upright y-axis by h. Although we might visualize Euler angles, inside a computer when rotating vectors from upright space to object space, we actually use a fixed-axis system. We discuss this in greater detail in Section 8.7.1, when we show how to convert Euler angles to a rotation matrix. The fixed-axis conventions are also called *extrinsic*, the typical Euler angle conventions being referred to as *intrinsic*.

Euler angles rotate about the body axes, so the axis of rotation for a given step depends on the angles used in prior rotations. In the fixed-axis system, the axes of rotation are always the same—the upright axes. The two systems are equivalent, provided that the rotations are performed in the opposite order.

Now we'd like to make a brief but humble campaign for a more precise use of the term "yaw." A lot of aeronautical terminology is inherited nautical terminology.[6] In a nautical context, the original meaning of the word "yaw" was essentially the same thing as heading, both in terms of absolute angle and also a change in that angle. In the context of airplanes and other freely rotating bodies, however, we don't feel that yaw and heading are the same thing. A yawing motion produces a rotation about the *object* y-axis, whereas a change in heading produces a rotation about the *upright* y-axis. For example, when the pilot of an airplane uses the pedals to control the rudder, he is performing a *yaw* rotation, because the rotation caused by the rudder is always about the object-space y-axis of the plane. Imagine a plane diving straight down. If the pilot performs a 90° yaw, the plane will end up "on its ear," no longer looking downward, but looking towards the horizon, banked 90°. This is illustrated in Figure 8.8.

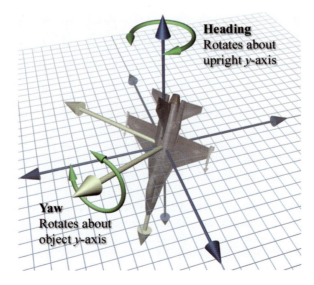

Figure 8.8
Heading versus yaw

---

[6]The word "aeronautical" is a prime example.

In contrast, when players navigating a first-person shooter move the mouse from left to right, they are performing a *heading* rotation. The rotation is always about the vertical axis (the upright y-axis). If players are looking downward and move the mouse horizontally to perform a heading rotation, they continue to look downward and spin in place. The point is certainly not that heading is better than yaw because that's what we do in first-person shooters. The point is that a yawing motion cannot be accomplished by adjusting a single Euler angle, but a heading motion can. That's why we think "heading" is a better term: it's the action that results when you make an incremental change to the first Euler angle.

Alas, the same argument can be leveled against the term "pitch." If bank is nonzero, an incremental change to the middle Euler angle does not produce a rotation about the object's lateral axis. But then, there isn't really a simple, good word to describe the angle that the object's longitudinal axis makes with the horizontal, which is what the middle Euler angle really specifies. ("Inclination" is no good as it is specific to the right-handed conventions.)

We hope you have read our opinions with the humility we intended, and also have received the more important message: investigating (seemingly cosmetic) differences in convention can sometimes lead us to a deeper understanding of the finer points. And then sometimes it's just plain nit-picking. Generations of aerospace engineers have been putting men on the moon and robots on Mars, and building airplanes that safely shuttle the authors to and from distant cities, all the while using the terms yaw and roll. Would you believe that some of these guys don't even know who we are!? Given the choice to pick your own terminology, we say to use the word "heading" when you can, but if you hear the word "yaw," then for goodness sake don't make as big of a deal out of it as we have in these pages, especially if the person you are talking to is smarter than you.

Although in this book we do not follow the right-handed aerospace coordinate conventions (and we have a minor quibble about terminology), when it comes to the basic strategy of Euler angles, in a physical sense, we believe complete compliance with the wisdom of the aerospace forefathers is the only way to go, at least if your universe has some notion of "ground." Remember that, in theory, any three axes can be used as the axes of rotation, in any order. But really, the conventions they chose are the only ones that make any practical sense, if you want the individual angles to be useful and meaningful. No matter how you label your axes, the first angle needs to rotate about the vertical, the second about the body lateral axis, and the third about the body longitudinal axis.

As if these weren't enough complications, let us throw in a few more. In the system we have been describing, each rotation occurs about a different body axes. However, Euler's own original system was a "symmetric" system in which the first and last rotations are performed around the *same* axis. These methods are more convenient in certain situations, such as describing the motion of a top, where the three angles correspond to precession, nutation, and spin. You may encounter some purists who object to the name "Euler angles" being attached to an asymmetric system, but this usage is widespread in many fields, so rest assured that you outnumber them. To distinguish between the two systems, the symmetric Euler angles are sometimes called "proper" Euler angles, with the more common conventions being called *Tait-Bryan* angles, first documented by the aerospace forefathers we mentioned [10]. O'Reilly [51] discusses proper Euler angles, even *more* methods of describing rotation, such as the Rodrigues vector, Cayley-Klein parameters, and interesting historical remarks. James Diebel's summary [13] compares different Euler angle conventions and the other major methods for describing rotation, much as this chapter does, but assumes a higher level of mathematical sophistication.

If you have to deal with Euler angles that use a different convention from the one you prefer, we offer two pieces of advice:

- First, make sure you understand exactly how the other Euler angle system works. Little details such as the definition of positive rotation and order of rotations make a big difference.

- Second, the easiest way to convert the Euler angles to your format is to convert them to matrix form and then convert the matrix back to your style of Euler angles. We will learn how to perform these conversions in Section 8.7. Fiddling with the angles directly is much more difficult than it would seem. See [63] for more information.

8.3.3 Advantages of Euler Angles

Euler angles parameterize orientation using only three numbers, and these numbers are angles. These two characteristics of Euler angles provide certain advantages over other forms of representing orientation.

- *Euler angles are easy for humans to use*—considerably easier than matrices or quaternions. Perhaps this is because the numbers in an Euler angle triple are *angles*, which is naturally how people think about orientation. If the conventions most appropriate for the situation are chosen, then the *most practical* angles can be expressed directly. For example, the angle of declination is expressed directly by the heading-pitch-bank system. This ease of use is a serious advantage. When an orientation needs to be displayed numerically or entered at the keyboard, Euler angles are really the only choice.

- *Euler angles use the smallest possible representation.* Euler angles use three numbers to describe an orientation. No system can parameterize 3D orientation using fewer than three numbers. If memory is at a premium, then Euler angles are the most economical way to represent an orientation.

  Another reason to choose Euler angles when you need to save space is that the numbers you are storing are more easily compressed. It's relatively easy to pack Euler angles into a smaller number of bits using a trivial fixed-precision system. Because Euler angles are *angles*, the data loss due to quantization is spread evenly. Matrices and quaternions require using very small numbers, because the values stored are sines and cosines of the angles. The absolute numeric difference between two values is not proportionate to the perceived difference, however, as it is with Euler angles. In general, matrices and quaternions don't pack into a fixed-point system easily.

  Bottom line: if you need to store a lot of 3D rotational data in as little memory as possible, as is very common when handling animation data, Euler angles (or the exponential map format—to be discussed in Section 8.4) are the best choices.

- *Any set of three numbers is valid.* If we take any three numbers at random, they form a valid set of Euler angles that we can interpret as an expression of an orientation. In other words, there is no such thing as an invalid set of Euler angles. Of course, the numbers may not be *correct* but at least they are *valid*. This is not the case with matrices and quaternions.

8.3.4   Disadvantages of Euler Angles

This section discusses some disadvantages of the Euler angle method of representing orientation; primarily,

- The representation for a given orientation is not unique.

- Interpolating between two orientations is problematic.

Let's address these points in detail. First, we have the problem that for a given orientation, there are many different Euler angle triples that can be used to describe that orientation. This is known as *aliasing* and can be somewhat of an inconvenience. These irritating problems are very similar to those we met dealing with spherical coordinates in Section 7.3.4. Basic questions such as "Do two Euler angle triples represent the same angular displacement?" are difficult to answer due to aliasing.

We've seen one trivial type of aliasing before with polar coordinates: adding a multiple of 360° does not change the orientation expressed, even though the numbers are different.

A second and more troublesome form of aliasing occurs because the three angles are not completely independent of each other. For example, pitching down 135° is the same as heading 180°, pitching down 45°, and then banking 180°.

To deal with aliasing of spherical coordinates, we found it useful to establish a *canonical set*; any given point has a unique representation in the canonical set that is the "official" way to describe that point using polar coordinates. We use a similar technique for Euler angles. In order to guarantee a unique Euler angle representation for any given orientation, we restrict the ranges of the angles. One common technique is to limit heading and bank to $(-180°, +180°]$ and to limit pitch to $[-90°, +90°]$. For any orientation, there is only one Euler angle triple in the canonical set that represents that orientation. (Actually, there is one more irritating singularity that must be handled, which we describe in just a moment.) Using canonical Euler angles simplifies many basic tests such as "am I facing approximately east?"

The most famous (and irritating) type of aliasing problem suffered by Euler angles is illustrated by this example: if we head right 45° and then pitch down 90°, this is the same as pitching down 90°and then banking 45°. In fact, once we chose ±90° as the pitch angle, we are restricted to rotating about the vertical axis. This phenomenon, in which an angle of ±90° for the second rotation can cause the first and third rotations to rotate about the same axis, is known as *Gimbal lock*. To remove this aliasing from the canonical set of Euler angle triples, we assign all rotation about the vertical axis to *heading* in the Gimbal lock case. In other words, in the canonical set, if pitch is ±90°, then bank is zero.

This last rule for Gimbal lock completes the rules for the canonical set of Euler angles:

Conditions satisfied by Euler angles in the canonical set

$$-180° < h \leq 180°$$
$$-90° \leq p \leq 90°$$
$$-180° < b \leq 180°$$
$$p = \pm 90° \quad \Rightarrow \quad b = 0.$$

When writing C++ that accepts Euler angle arguments, it's usually best to ensure that they work given Euler angles in any range. Luckily this is usually easy; things frequently just work without taking any extra precaution, especially if the angles are fed into trig functions. However, when writing code that computes or *returns* Euler angles, it's good practice to return the canonical Euler angle triple. The conversion methods shown in Section 8.7 demonstrate these principles.

A common misconception is that, because of Gimbal lock, certain orientations cannot be described using Euler angles. Actually, for the purposes of *describing* an orientation, aliasing doesn't pose any problems. To be clear, *any* orientation in 3D can be described by using Euler angles, and that representation is unique within the canonical set. Also, as we mentioned in the previous section, there is no such thing as an "invalid" set of Euler angles. Even if the angles are outside the usual range, we can always agree on what orientation is described by the Euler angles.

So for purposes of simply *describing* orientation, aliasing isn't a huge problem, especially when canonical Euler angles are used. So what's so bad about aliasing and Gimbal lock? Let's say we wish to interpolate between two orientations $\mathbf{R}_0$ and $\mathbf{R}_1$. In other words, for a given parameter t, $0 \le t \le 1$, we wish to compute an intermediate orientation $\mathbf{R}(t)$ that interpolates smoothly from $\mathbf{R}_0$ to $\mathbf{R}_1$ as t varies from 0 to 1. This is an extremely useful operation for character animation and camera control, for example.

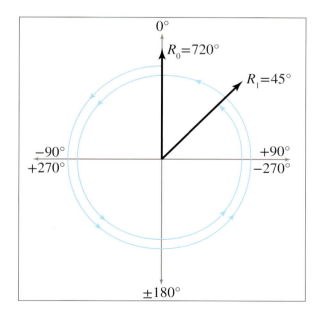

Figure 8.9
Naïve interpolation can cause excessive rotation

The naïve approach to this problem is to apply the standard linear interpolation formula ("lerp") to each of the three angles independently:

Simple linear
interpolation between
two angles

$$\Delta\theta = \theta_1 - \theta_0,$$
$$\theta_t = \theta_0 + t\,\Delta\theta.$$

This is fraught with problems.

First, if canonical Euler angles are not used, we may have large angle values. For example, imagine the heading of $\mathbf{R}_0$, denoted h_0, is 720°. Assume $h_1 = 45°$. Now, $720° = 2 \times 360°$, which is the same as 0°, so basically the h_1 and h_2 are only 45° apart. However, naïve interpolation will spin around nearly twice in the wrong direction, as shown in Figure 8.9.

Of course, the solution to this problem is to use canonical Euler angles. We could assume that we will always be interpolating between two sets of canonical Euler angles. Or we could attempt to enforce this by converting to canonical values inside our interpolation routine. (Simply wrapping angles within the $(-180°, +180°]$ range is easy, but dealing with pitch values outside the $[-90°, +90°]$ range is more challenging.)

However, even using canonical angles doesn't completely solve the problem. A second type of interpolation problem can occur because of the cyclic nature of rotation angles. Suppose $h_0 = -170°$ and $h_1 = 170°$. Notice that these are canonical values for heading, both in the range $(-180°, +180°]$.

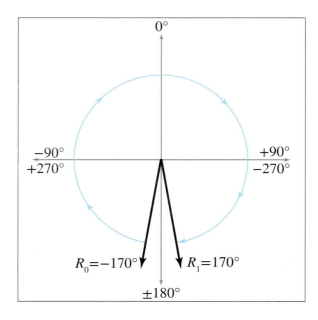

Figure 8.10
Naïve interpolation can rotate the long way around.

```
float wrapPi(float theta) {

    // Check if already in range. This is not strictly necessary,
    // but it will be a very common situation. We don't want to
    // incur a speed hit and perhaps floating precision loss if
    // it's not necessary
    if (fabs(theta) <= PI) {

        // One revolution is 2 PI.
        const float TWOPPI = 2.0f*PI;

        // Out of range. Determine how many "revolutions"
        // we need to add.
        float revolutions = floor((theta + PI) * (1.0f/TWOPPI));

        // Subtract it off
        theta -= revolutions*TWOPPI;
    }

    return theta;
}
```

Listing 8.1
Wrapping an angle between $\pm 180°$

The two heading values are only $20°$ apart, but again, naïve interpolation will not behave correctly, rotating the long way around by a clockwise rotation of $340°$ instead of taking the shorter counterclockwise path of $20°$, as shown in Figure 8.10.

The solution to this second type of problem is to wrap the differences between angles used in the interpolation equation in the range $(-180°, +180°]$ in order to find the shortest arc. To do this, we introduce the notation

$$\text{wrapPi}(x) = x - 360° \lfloor (x + 180°)/360° \rfloor,$$

Wrapping an angle
between $\pm 180°$

where $\lfloor \cdot \rfloor$ denotes the floor function.

The wrapPi function is a small, sharp tool that every game programmer should have in their toolbox. It elegantly handles common situations in which we must account for the cyclic nature of angles. It works by adding or subtracting the appropriate multiple of $360°$. Listing 8.1 shows how it would be implemented in C.

Let's go back to Euler angles. As expected, using wrapPi() makes it easy to take the shortest arc when interpolating between two angles:

$$\Delta\theta = \text{wrapPi}(\theta_1 - \theta_0),$$
$$\theta_t = \theta_0 + t\,\Delta\theta.$$

Taking the shortest arc
when interpolating
between two angles

But even with these two Band-Aids, Euler angle interpolation still suffers from Gimbal lock, which in many situations causes a jerky, unnatural course. The object whips around suddenly and appears to be hung somewhere. The basic problem is that the angular velocity is not constant during the interpolation. If you have never experienced what Gimbal lock looks like, you may be wondering what all the fuss is about. Unfortunately, it is very difficult to fully appreciate the problem from illustrations in a book—you need to experience it in real time. Fortunately, though, it's easy to find an animation demonstrating the problem: just do a youtube.com search for "gimbal lock."

The first two problems with Euler angle interpolation were irritating, but certainly not insurmountable. Canonical Euler angles and wrapPi provide relatively simple workarounds. Gimbal lock, unfortunately, is more than a minor nuisance; it's a fundamental problem. Could we perhaps reformulate our rotations and devise a system that does not suffer from these problems? Unfortunately, this is not possible. There is simply an inherent problem with using three numbers to describe 3D orientation. We could change our problems, but not eliminate them. Any system that parameterizes 3-space orientation using three numbers is guaranteed to have singularities in the parameterization space and therefore be subject to problems such as Gimbal lock. The exponential map form (see Section 8.4), a different scheme for parameterizing 3D rotation with three numbers, manages to consolidate the singularities to a single point: the antipodes. This behavior is more benign for certain practical situations, but it does not remove the singularities completely. To do that, we must use quaternions, which are discussed in Section 8.5.

8.3.5   Summary of Euler Angles

Let's summarize our findings from Section 8.3 about Euler angles.

- Euler angles store orientation by using three angles. These angles are ordered rotations about the three object-space axes.

- The most common system of Euler angles is the heading-pitch-bank system. Heading and pitch tell which way the object is facing—heading gives a "compass reading" and pitch measures the angle of declination. Bank measures the amount of "twist."

- In a fixed-axis system, the rotations occur about the upright axes rather than the moving body axes. This system is equivalent to Euler angles, provided that we perform the rotations in the opposite order.

- Lots of smart people use lots of different terms for Euler angles, and they can have good reasons for using different conventions.[7] It's best

---

[7] There are also lots of stupid people who do *not* have good reasons for the choices they make—but in the end, the result is the same.

not to rely on terminology when using Euler angles. Always make sure you get a precise working definition, or you're likely to get very confused.

- In most situations, Euler angles are more intuitive for humans to work with compared to other methods of representing orientation.

- When memory is at a premium, Euler angles use the minimum amount of data possible for storing an orientation in 3D, and Euler angles are more easily compressed than quaternions.

- There is no such thing as an invalid set of Euler angles. Any three numbers have a meaningful interpretation.

- Euler angles suffer from aliasing problems due to the cyclic nature of rotation angles and because the rotations are not completely independent of one another.

- Using canonical Euler angles can simplify many basic queries on Euler angles. An Euler angle triple is in the canonical set if heading and bank are in the range $(-180°, +180°]$ and pitch is in the range $[-90°, +90°]$. What's more, if pitch is $\pm 90°$, then bank is zero.

- Gimbal lock occurs when pitch is $\pm 90°$. In this case, one degree of freedom is lost because heading and bank both rotate about the vertical axis.

- Contrary to popular myth, *any* orientation in 3D can be represented by using Euler angles, and we can agree on a unique representation for that orientation within the canonical set.

- The wrapPi function is a very handy tool that simplifies situations in which we have to deal with the cyclic nature of angles. Such situations arise frequently in practice, especially in the context of Euler angles, but at other times as well.

- Simple forms of aliasing are irritating, but there are workarounds. Gimbal lock is a more fundamental problem with no easy solution. Gimbal lock is a problem because the parameter space of orientation has a discontinuity. This means small changes in orientation can result in large changes in the individual angles. Interpolation between orientations using Euler angles can freak out or take a wobbly path.

8.4   Axis-Angle and Exponential Map Representations

Euler's name is attached to all sorts of stuff related to rotation (we just discussed Euler angles in Section 8.3). His name is also attached to *Euler's rotation theorem*, which basically says that any 3D angular displacement can be accomplished via a *single* rotation about a carefully chosen axis. To be more precise, given any two orientations $\mathbf{R}_1$ and $\mathbf{R}_2$, there exists an axis $\hat{\mathbf{n}}$ such that we can get from $\mathbf{R}_1$ to $\mathbf{R}_2$ by performing *one* rotation about $\hat{\mathbf{n}}$. With Euler angles, we need three rotations to describe any orientation, since we are restricted to rotate about the cardinal axis. However, when we are free to choose the axis of rotation, its possible to find one such that only one rotation is needed. Furthermore, as we will show in this section, except for a few minor details, this axis of rotation is uniquely determined.

Euler's rotation theorem leads to two closely related methods for describing orientation. Let's begin with some notation. Assume we have chosen a rotation angle θ and an axis of rotation that passes through the origin and is parallel to the unit vector $\hat{\mathbf{n}}$. (In this book, positive rotation is defined according to the left-hand rule; see Section 1.3.3.)

Taking the two values $\hat{\mathbf{n}}$ and θ as is, we have described an angular displacement in the *axis-angle* form. Alternatively, since $\hat{\mathbf{n}}$ has unit length, we can multiply it by θ without loss of information, yielding the single vector $\mathbf{e} = \theta\hat{\mathbf{n}}$. This scheme for describing rotation goes by the rather intimidating and obscure name of *exponential map*.[8] The rotation angle can be deduced from the length of $\mathbf{e}$; in other words, $\theta = \|\mathbf{e}\|$, and the axis is obtained by normalizing $\mathbf{e}$. The exponential map is not only more compact than the axis-angle (three numbers instead of four), it elegantly avoids certain singularities and has better interpolation and differentiation properties.

We're not going to discuss the axis-angle and exponential map forms in quite as much detail as the other methods of representing orientation because in practice their use is a bit specialized. The axis-angle format is primarily a conceptual tool. It's important to understand, but the method gets relatively little direct use compared to the other formats. It's one notable capability is that we can directly obtain an arbitrary multiple of the displacement. For example, given a rotation in axis-angle form, we can obtain a rotation that represents one third of the rotation or 2.65 times the rotation, simply by multiplying θ by the appropriate amount. Of course,

---

[8]The reason for this is that it comes from the equally intimidating and obscure branch of mathematics known as Lie algebra. (Lie is pronounced "lee," since it's named after a person.) The exponential map has a broader definition in this context, and the space of 3D rotations (sometimes denoted as $SO(3)$) is just one type of Lie group. More regretful comments about terminology to come.

we can do this same operation with the exponential map just as easily. Quaternions can do this through exponentiation, but an inspection of the math reveals that it's really using the axis-angle format under the hood. (Even though quaternions claim to be using the exponential map under the hood!) Quaternions can also do a similar operation using slerp, but in a more roundabout way and without the ability for intermediate results to store rotations beyond 180 degrees. We look at quaternions in Section 8.5.

The exponential map gets more use than the axis-angle. First of all, its interpolation properties are nicer than Euler angles. Although it does have singularities (discussed next), they are not as troublesome as for Euler angles. Usually, when one thinks of interpolating rotations, one immediately thinks of quaternions; however, for some applications, such as storage of animation data, the underappreciated exponential map can be a viable alternative [27]. But the most important and frequent use of the exponential map is to store not angular displacement, but rather angular *velocity*. This is because the exponential map differentiates nicely (which is somewhat related to its nicer interpolation properties) and can represent multiple rotations easily.

Like Euler angles, the axis-angle and exponential map forms exhibit aliasing and singularities, although of a slightly more restricted and benign manner. There is an obvious singularity at the identity orientation, or the quantity "no angular displacement." In this case, $\theta = 0$, and our choice of axis is irrelevant—any axis may be used. Notice, however, that the exponential map nicely tucks this singularity away, since multiplication by θ causes $\mathbf{e}$ to vanish, no matter what rotation axis $\hat{\mathbf{n}}$ is chosen. Another trivial form of aliasing in axis-angle space can be produced by negating both θ and $\hat{\mathbf{n}}$. However, the exponential map dodges this issue as well, since negating both θ and $\hat{\mathbf{n}}$ leaves $\mathbf{e} = \theta\hat{\mathbf{n}}$ unchanged!

The other aliases cannot be dispatched so easily. As with Euler angles, adding a multiple of $360°$ to θ produces an angular displacement that results in the same ending orientation, and this form of aliasing affects both the axis-angle and exponential map. However, this is not always a shortcoming—for describing angular velocity, this ability to represent such "extra" rotation is an important and useful property. For example, it's quite important to be able to distinguish between rotation about the x-axis at a rate of $720°$ per second versus rotation about the same axis at a rate of $1080°$ per second, even though these displacements result in the same ending orientation if applied for an integral number of seconds. It is not possible to capture this distinction in quaternion format.

As it turns out, given any angular displacement that can be described by a rotation matrix, the exponential map representation is uniquely determined. Although more than one exponential map may produce the same rotation matrix, it is possible to take a subset of the exponential maps (those for which $\|\mathbf{e}\| < 2\pi$) and form a one-to-one correspondence with the rotation matrices. This is the essence of Euler's rotation theorem.

Now let's consider concatenating multiple rotations. Let's say $\mathbf{e}_1$ and $\mathbf{e}_2$ are two rotations in exponential map format. The result of performing the rotations in sequence, for example, $\mathbf{e}_1$ and then $\mathbf{e}_2$, is *not* the same as performing a single rotation $\mathbf{e}_1 + \mathbf{e}_2$. We know this cannot be true, because ordinary vector addition is commutative, but three-space rotations are not. Assume that $\mathbf{e}_1 = [90°, 0, 0]$, and $\mathbf{e}_2 = [0, 90°, 0]$. With our conventions, this is a 90° downward pitch rotation, and a 90° heading rotation to the east. Performing $\mathbf{e}_1$ followed by $\mathbf{e}_2$, we would end up looking downward with our head pointing east, but doing them in the opposite order, we end up "on our ear" facing east. But what if the angles were much smaller, say 2° instead of 90°? Now the ending rotations are more similar. As we take the magnitude of the rotation angles down, the importance of the order decreases, and at the extreme, for "infinitesimal" rotations, the order is completely irrelevant. In other words, for infinitesimal rotations, exponential maps *can* be added vectorially. Infinitesimals are important topics from calculus, and they are at the heart of defining rate of change. We look at these topics in Chapter 11, but for now, the basic idea is that exponential maps do not add vectorially when used to define an amount of rotation (an angular displacement or an orientation), but they do properly add vectorially when they describe a *rate* of rotation. This is why exponential maps are perfectly suited for describing angular velocity.

Before we leave this topic, a regretful word of warning regarding terminology. Alternative names for these two simple concepts abound. We have tried to choose the most standard names possible, but it was difficult to find strong consensus. Some authors use the term "axis-angle" to describe *both* of these (closely related) methods and don't really distinguish between them. Even more confusing is the use of the term "Euler axis" to refer to either form (but not to Euler angles!). "Rotation vector" is another term you might see attached to what we are calling exponential map. Finally, the term "exponential map," in the broader context of Lie algebra, from whence the term originates, actually refers to an *operation* (a "map") rather than a quantity. We apologize for the confusion, but it's not our fault.

8.5 Quaternions

The term *quaternion* is somewhat of a buzzword in 3D math. Quaternions carry a certain mystique—which is a euphemismistic way of saying that many people find quaternions complicated and confusing. We think the way quaternions are presented in most texts contributes to their confusion, and we hope that our slightly different approach will help dispel quaternions' "mystique."

There is a mathematical reason why using only three numbers to represent a 3-space orientation is guaranteed to cause the problems we discussed with Euler angles, such as Gimbal lock. It has something to do with some fairly advanced[9] math terms such as "manifolds." A quaternion avoids these problems by using *four* numbers to express an orientation (hence the name *quat*ernion).

This section describes how to use a quaternion to define an angular displacement. We're going to deviate somewhat from the traditional presentation, which emphasizes the interesting (but, in our opinion, nonessential) interpretation of quaternions as complex numbers. Instead, we will be developing quaternions from a primarily geometric perspective. Here's what's in store: First, Section 8.5.1 introduces some basic notation. Section 8.5.2 is probably the most important section—it explains how a quaternion may be interpreted geometrically. Sections 8.5.3 through Section 8.5.11 review the basic quaternion properties and operations, examining each from a geometric perspective. Section 8.5.12 discusses the important slerp operation, which is used to interpolate between two quaternions and is one of the primary advantages of quaternions. Section 8.5.13 discusses the advantages and disadvantages of quaternions. Section 8.5.14 is an optional digression into how quaternions may be interpreted as 4D complex numbers. Section 8.5.15 summaries the properties of quaternions.

Don't be scared off by what seems like a lot of hairy math in this section. The most important things to remember about quaternions are the high-level concepts that are summarized in Section 8.5.15. The nuts and bolts of quaternions are given here to show that everything about quaternions can be derived, and you don't have to take our word on faith. A detailed understanding of quaternions is not really needed in order to use them,[10] but you need to understand what quaternions can do.

8.5.1 Quaternion Notation

A quaternion contains a scalar component and a 3D vector component. We usually refer to the scalar component as w. We may refer to the vector

---

[9]In this case, the word "advanced" means "outside the authors' expertise."
[10]That is, if your class libraries are designed well.

component as a single entity $\mathbf{v}$ or as individual components x, y, and z. Here are examples of both notations:

Two types of quaternion notation

$$\begin{bmatrix} w & \mathbf{v} \end{bmatrix}, \qquad \begin{bmatrix} w & (x & y & z) \end{bmatrix}.$$

In some cases it will be convenient to use the shorter notation, using $\mathbf{v}$, and in some cases the "expanded" version is clearer. This chapter presents most equations in both forms.

We also may write expanded quaternions vertically:

$$\begin{bmatrix} w \\ \begin{pmatrix} x \\ y \\ z \end{pmatrix} \end{bmatrix}.$$

Unlike regular vectors, there is no significant distinction between "row" and "column" quaternions. We are free to make the choice strictly for aesthetic purposes.

We denote quaternion variables with the same typeface conventions used for vectors: lowercase letters in bold (e.g., $\mathbf{q}$). When vectors and quaternions appear together, the context (and the letters chosen for the variables!) usually make clear which are which.

8.5.2   What Do Those Four Numbers Mean?

The quaternion form is closely related to the axis-angle and exponential map forms from Section 8.4. Let's briefly review the notation from that section, as the same notation will be used here. The unit vector $\hat{\mathbf{n}}$ defines an axis of rotation, and the scalar θ is the amount of rotation about this axis. Thus, the pair $(\theta, \hat{\mathbf{n}})$ define an angular displacement using the axis-angle system. You need a left or right hand[11] to determine which way is positive rotation.

A quaternion also contains an axis and angle, but $\hat{\mathbf{n}}$ and θ aren't simply stored in the four numbers of the quaternion directly, as they are in axis angle (that would be too easy!). Instead, they are encoded in a way that at first might seem weird, but turns out to be highly practical. Equation (8.2) shows how the values of a quaternion are related to θ and $\hat{\mathbf{n}}$, using both forms of quaternion notation:

Geometric meaning of the four values of a quaternion

$$\begin{bmatrix} w & \mathbf{v} \end{bmatrix} = \begin{bmatrix} \cos(\theta/2) & \sin(\theta/2)\hat{\mathbf{n}} \end{bmatrix}, \tag{8.2}$$
$$\begin{bmatrix} w & (x & y & z) \end{bmatrix} = \begin{bmatrix} \cos(\theta/2) & (\sin(\theta/2)n_x & \sin(\theta/2)n_y & \sin(\theta/2)n_z) \end{bmatrix}.$$

Keep in mind that w is related to θ, but they are not the same thing. Likewise, $\mathbf{v}$ and $\hat{\mathbf{n}}$ are related, but not identical.

---

[11]With apologies to our readers in India, we prefer the left.

The next several sections discuss a number of quaternion operations from mathematical and geometric perspectives.

8.5.3 Quaternion Negation

Quaternions can be negated. This is done in the obvious way of negating each component:

$$-\mathbf{q} = -\begin{bmatrix} w & (x & y & z) \end{bmatrix} = \begin{bmatrix} -w & (-x & -y & -z) \end{bmatrix}$$
$$= -\begin{bmatrix} w & \mathbf{v} \end{bmatrix} = \begin{bmatrix} -w & -\mathbf{v} \end{bmatrix}.$$

(8.3)

Quaternion negation

The surprising fact about negating a quaternion is that it really doesn't do anything, at least in the context of angular displacement.

The quaternions $\mathbf{q}$ and $-\mathbf{q}$ describe the *same* angular displacement. Any angular displacement in 3D has exactly *two* distinct representations in quaternion format, and they are negatives of each other.

It's not too difficult to see why this is true. If we add $360°$ to θ, it doesn't change the angular displacement represented by $\mathbf{q}$, but it negates all four components of $\mathbf{q}$.

8.5.4 Identity Quaternion(s)

Geometrically, there are two "identity" quaternions that represent "no angular displacement." They are

$$\begin{bmatrix} 1 & \mathbf{0} \end{bmatrix} \quad \text{and} \quad \begin{bmatrix} -1 & \mathbf{0} \end{bmatrix}.$$

Identity quaternions

(Note the boldface zero, which indicates the zero vector.) When θ is an even multiple of $360°$, then $\cos(\theta/2) = 1$, and we have the first form. If θ is an odd multiple of $360°$, then $\cos(\theta/2) = -1$, and we have the second form. In both cases, $\sin(\theta/2) = 0$, so the value of $\hat{\mathbf{n}}$ is irrelevant. This makes intuitive sense; if the rotation angle θ is a whole number of complete revolutions about *any* axis, then no real change is made to the orientation.

Algebraically, there is really only one identity quaternion: $[1, \mathbf{0}]$. When we multiply any quaternion $\mathbf{q}$ by the identity quaternion, the result is $\mathbf{q}$. (We present quaternion multiplication in Section 8.5.7.) When we multiply a quaternion $\mathbf{q}$ by the other "geometric identity" quaternion $[-1, \mathbf{0}]$, we get $-\mathbf{q}$. Geometrically, this results in the same quaternion, since $\mathbf{q}$ and $-\mathbf{q}$ represent the same angular displacement. Mathematically, however, $\mathbf{q}$ and $-\mathbf{q}$ are not equal, so $[-1, \mathbf{0}]$ is not a "true" identity quaternion.

8.5.5 Quaternion Magnitude

We can compute the magnitude of a quaternion, just as we can for vectors and complex numbers. The notation and formula shown in Equation (8.4) are similar to those used for vectors:

Quaternion magnitude

$$\|\mathbf{q}\| = \left\|\begin{bmatrix} w & (x & y & z) \end{bmatrix}\right\| = \sqrt{w^2 + x^2 + y^2 + z^2}$$
$$= \left\|\begin{bmatrix} w & \mathbf{v} \end{bmatrix}\right\| = \sqrt{w^2 + \|\mathbf{v}\|^2}. \tag{8.4}$$

Let's see what this means geometrically for a rotation quaternion:

Rotation quaternions have unit magnitude

$$\|\mathbf{q}\| = \left\|\begin{bmatrix} w & \mathbf{v} \end{bmatrix}\right\| = \sqrt{w^2 + \|\mathbf{v}\|^2}$$
$$= \sqrt{\cos^2(\theta/2) + (\sin(\theta/2)\|\hat{\mathbf{n}}\|)^2} \quad \text{(substituting using } \theta \text{ and } \hat{\mathbf{n}})$$
$$= \sqrt{\cos^2(\theta/2) + \sin^2(\theta/2)\|\hat{\mathbf{n}}\|^2}$$
$$= \sqrt{\cos^2(\theta/2) + \sin^2(\theta/2)(1)} \quad (\hat{\mathbf{n}} \text{ is a unit vector})$$
$$= \sqrt{1} \quad (\sin^2 x + \cos^2 x = 1)$$
$$= 1.$$

This is an important observation.

For our purposes of using quaternions to represent orientation, all quaternions are so-called *unit quaternions*, which have a magnitude equal to unity.

For information concerning nonnormalized quaternions, see the technical report by Dam et al. [11].

8.5.6 Quaternion Conjugate and Inverse

The *conjugate* of a quaternion, denoted $\mathbf{q}^*$, is obtained by negating the vector portion of the quaternion:

Quaternion conjugate

$$\mathbf{q}^* = \begin{bmatrix} w & \mathbf{v} \end{bmatrix}^* = \begin{bmatrix} w & -\mathbf{v} \end{bmatrix} \tag{8.5}$$
$$= \begin{bmatrix} w & (x & y & z) \end{bmatrix}^* = \begin{bmatrix} w & (-x & -y & -z) \end{bmatrix}.$$

The term "conjugate" is inherited from the interpretation of a quaternion as a complex number. We look at this interpretation in more detail in Section 8.5.14.

The *inverse* of a quaternion, denoted $\mathbf{q}^{-1}$, is defined as the conjugate of a quaternion divided by its magnitude:

$$\mathbf{q}^{-1} = \frac{\mathbf{q}^*}{\|\mathbf{q}\|}.$$

(8.6)

Quaternion inverse

The quaternion inverse has an interesting correspondence with the multiplicative inverse for real numbers (scalars). For real numbers, the multiplicative inverse a^{-1} is $1/a$. In other words, $a(a^{-1}) = a^{-1}a = 1$. The same applies to quaternions. When we multiply a quaternion $\mathbf{q}$ by its inverse $\mathbf{q}^{-1}$, we get the identity quaternion $[1, \mathbf{0}]$. (We discuss quaternion multiplication in Section 8.5.7.)

Equation (8.6) is the *official* definition of quaternion inverse. However, if you are interested only in quaternions that represent pure rotations, like we are in this book, then all the quaternions are unit quaternions and so the conjugate and inverse are equivalent.

The conjugate (inverse) is interesting because $\mathbf{q}$ and $\mathbf{q}^*$ represent opposite angular displacements. It is easy to see why this is the case. By negating $\mathbf{v}$, we are negating the axis of rotation $\hat{\mathbf{n}}$. This doesn't change the axis in the physical sense, since $\hat{\mathbf{n}}$ and $-\hat{\mathbf{n}}$ are parallel. However, it does flip the direction that we consider to be positive rotation. Thus, $\mathbf{q}$ rotates about an axis by an amount θ, and $\mathbf{q}^*$ rotates in the opposite direction by the same amount.

For our purposes, an alternative definition of quaternion conjugate could have been to negate w, leaving $\mathbf{v}$ (and thus $\hat{\mathbf{n}}$) unchanged. This would negate the amount of rotation θ, rather than reversing what is considered positive rotation by flipping the axis of rotation. This would have been equivalent to the definition given in Equation (8.5) (for our geometric purposes, at least) and provided for a slightly more intuitive geometric interpretation. However, the term *conjugate* has a special significance in the context of complex numbers, so let's stick with the original definition.

8.5.7 Quaternion Multiplication

Quaternions can be multiplied. The result is similar to the cross product for vectors, in that it yields another quaternion (not a scalar), and it is not commutative. However, the notation is different: we denote quaternion multiplication simply by placing the two operands side-by-side. The formula for quaternion multiplication can be easily derived based upon the definition of quaternions as complex numbers (see Exercise 6), but we state it here without development, using both quaternion notations:

Quaternion product

$$\mathbf{q}_1\mathbf{q}_2 = \begin{bmatrix} w_1 & \begin{pmatrix} x_1 & y_1 & z_1 \end{pmatrix} \end{bmatrix}\begin{bmatrix} w_2 & \begin{pmatrix} x_2 & y_2 & z_2 \end{pmatrix} \end{bmatrix}$$

$$= \begin{bmatrix} \begin{pmatrix} w_1w_2 - x_1x_2 - y_1y_2 - z_1z_2 \\ w_1x_2 + x_1w_2 + y_1z_2 - z_1y_2 \\ w_1y_2 + y_1w_2 + z_1x_2 - x_1z_2 \\ w_1z_2 + z_1w_2 + x_1y_2 - y_1x_2 \end{pmatrix} \end{bmatrix}$$

$$= \begin{bmatrix} w_1 & \mathbf{v}_1 \end{bmatrix}\begin{bmatrix} w_2 & \mathbf{v}_2 \end{bmatrix}$$

$$= \begin{bmatrix} w_1w_2 - \mathbf{v}_1 \cdot \mathbf{v}_2 & w_1\mathbf{v}_2 + w_2\mathbf{v}_1 + \mathbf{v}_1 \times \mathbf{v}_2 \end{bmatrix}.$$

The quaternion product is also known as the *Hamilton product*; you'll understand why after reading about the history of quaternions in Section 8.5.14.

Let's quickly mention three properties of quaternion multiplication, all of which can be easily shown by using the definition given above. First, quaternion multiplication is associative, but not commutative:

Quaternion multiplication is associative, but not commutative

$$(\mathbf{ab})\mathbf{c} = \mathbf{a}(\mathbf{bc}),$$
$$\mathbf{ab} \neq \mathbf{ba}.$$

Second, the magnitude of a quaternion product is equal to the product of the magnitudes (see Exercise 9):

Magnitude of quaternion product

$$\|\mathbf{q}_1\mathbf{q}_2\| = \|\mathbf{q}_1\|\|\mathbf{q}_2\|.$$

This is very significant because it guarantees us that when we multiply two unit quaternions, the result is a unit quaternion.

Finally, the inverse of a quaternion product is equal to the product of the inverses taken in reverse order:

Inverse of quaternion product

$$(\mathbf{ab})^{-1} = \mathbf{b}^{-1}\mathbf{a}^{-1},$$
$$(\mathbf{q}_1\mathbf{q}_2\cdots\mathbf{q}_{n-1}\mathbf{q}_n)^{-1} = \mathbf{q}_n{}^{-1}\mathbf{q}_{n-1}{}^{-1}\cdots\mathbf{q}_2{}^{-1}\mathbf{q}_1{}^{-1}.$$

Now that we know some basic properties of quaternion multiplication, let's talk about why the operation is actually useful. Let us "extend" a standard 3D point (x, y, z) into quaternion space by defining the quaternion $\mathbf{p} = [0, (x, y, z)]$. In general, $\mathbf{p}$ is not a valid rotation quaternion, since it can have any magnitude. Let $\mathbf{q}$ be a rotation quaternion in the form we have been discussing, $[\cos\theta/2, \hat{\mathbf{n}}\sin\theta/2]$, where $\hat{\mathbf{n}}$ is a unit vector axis of rotation, and θ is the rotation angle. It is surprising to realize that we can rotate the 3D point $\mathbf{p}$ about $\hat{\mathbf{n}}$ by performing the rather odd-looking quaternion multiplication

Using quaternion
multiplication to rotate a
3D vector

$$\mathbf{p}' = \mathbf{q}\mathbf{p}\mathbf{q}^{-1}. \tag{8.7}$$

We could prove this by expanding the multiplication, substituting in $\hat{\mathbf{n}}$ and θ, and comparing the result to the matrix we derived to rotate about an arbitrary axis (Equation (5.1.3), page 144), and indeed this is the approach taken in most texts on quaternions. While this certainly is an effective way to verify that the trick works, it leaves us wondering how the heck somebody could have ever stumbled upon it in the first place. In Section 8.7.3, we derive the conversion from quaternion to matrix form in a straightforward way, solely from the geometry of the rotations and without referring to $\mathbf{q}\mathbf{p}\mathbf{q}^{-1}$. As for how the association was discovered, we cannot say for sure, but we will offer a train of thought that can lead a person to discover the connection between this strange product and rotations in Section 8.5.14. This discussion also explains how a person might have discovered that it would be fruitful to use *half* of the rotation angle for the components.

As it turns out, the correspondence between quaternion multiplication and 3D vector rotations is more of a theoretical interest than a practical one. Some people ("quaternio-philes?") like to attribute quaternions with the useful property that vector rotations are immediately accessible by using Equation (8.7). To the quaternion lovers, we admit that this compact notation is an advantage of sorts, but its practical benefit in computations is dubious. If you actually work through this math, you will find that it is just about the same number of operations involved as converting the quaternion to the equivalent rotation matrix (by using Equation (8.20), which is developed in Section 8.7.3) and then multiplying the vector by this matrix. Because of this, we don't consider quaternions to possess any direct ability to rotate vectors, at least for practical purposes in a computer.

Although the correspondence between $\mathbf{q}\mathbf{p}\mathbf{q}^{-1}$ and rotation is not of direct practical importance, it is of supreme theoretical importance. It leads us to a slightly different use of quaternion multiplication, and this use *is* highly practical in programming. Examine what happens when multiple rotations are applied to a vector. We'll rotate the vector $\mathbf{p}$ by the quaternion $\mathbf{a}$, and then rotate that result by another quaternion $\mathbf{b}$:

$$\begin{aligned} \mathbf{p}' &= \mathbf{b}(\mathbf{a}\mathbf{p}\mathbf{a}^{-1})\mathbf{b}^{-1} \\ &= (\mathbf{b}\mathbf{a})\mathbf{p}(\mathbf{a}^{-1}\mathbf{b}^{-1}) \\ &= (\mathbf{b}\mathbf{a})\mathbf{p}(\mathbf{b}\mathbf{a})^{-1}. \end{aligned}$$

Concatenating multiple
rotations with
quaternion algebra

Notice that rotating by $\mathbf{a}$ and then by $\mathbf{b}$ is equivalent to performing a *single* rotation by the quaternion product $\mathbf{b}\mathbf{a}$. This is a key observation.

Quaternion multiplication can be used to concatenate multiple rotations, just like matrix multiplication.

We say "just like matrix multiplication," but in fact there is a slightly irritating difference. With matrix multiplication, our preference to use row vectors puts the vectors on the left, resulting in the nice property that concatenated rotations read left-to-right in the order of transformation. With quaternions, we don't have this flexibility: concatenation of multiple rotations will always read "inside out" from right to left.[12]

8.5.8   Quaternion "Difference"

Using the quaternion multiplication and inverse, we can compute the difference between two quaternions, with "difference" meaning the angular displacement from one orientation to another. In other words, given orientations $\mathbf{a}$ and $\mathbf{b}$, we can compute the angular displacement $\mathbf{d}$ that rotates from $\mathbf{a}$ to $\mathbf{b}$. This can be expressed compactly as

$$\mathbf{da} = \mathbf{b}.$$

(Remember that quaternion multiplication performs the rotations from right-to-left.)

Let's solve for $\mathbf{d}$. If the variables in the equation represented scalars, we could simply divide by $\mathbf{a}$. However, we can't divide quaternions; we can only multiply them. Perhaps multiplication by the inverse will achieve the desired effect? Multiplying both sides by $\mathbf{a}^{-1}$ on the right (we have to be careful since quaternion multiplication is not commutative) gives us

The quaternion "difference"

$$(\mathbf{da})\mathbf{a}^{-1} = \mathbf{ba}^{-1},$$
$$\mathbf{d}(\mathbf{aa}^{-1}) = \mathbf{ba}^{-1},$$
$$\mathbf{d}\begin{bmatrix}1 & \mathbf{0}\end{bmatrix} = \mathbf{ba}^{-1},$$
$$\mathbf{d} = \mathbf{ba}^{-1}.$$

Now we have a way to generate a quaternion that represents the angular displacement from one orientation to another. We use this in Section 8.5.12, when we explore slerp.

---

[12]Actually, you *do* have some flexibility if you're willing to buck the system. Some crazy authors [16] have gone so far as to provide an alternative definition of the quaternion product with the operands reversed. This can lead to code that is easier to understand, and this option might be worth considering in your own code. However, we'll stick with the standard in this book.

Mathematically, the angular difference between two quaternions is actually more similar to a division than a true difference (subtraction).

8.5.9 Quaternion Dot Product

The dot product operation is defined for quaternions. The notation and definition for this operation is very similar to the vector dot product:

$$\mathbf{q}_1 \cdot \mathbf{q}_2 = \begin{bmatrix} w_1 & \mathbf{v}_1 \end{bmatrix} \cdot \begin{bmatrix} w_2 & \mathbf{v}_2 \end{bmatrix}$$
$$= w_1 w_2 + \mathbf{v}_1 \cdot \mathbf{v}_2$$
$$= \begin{bmatrix} w_1 & (x_1 \quad y_1 \quad z_1) \end{bmatrix} \cdot \begin{bmatrix} w_2 & (x_2 \quad y_2 \quad z_2) \end{bmatrix}$$
$$= w_1 w_2 + x_1 x_2 + y_1 y_2 + z_1 z_2.$$

Quaternion dot product

Like the vector dot product, the result is a scalar. For unit quaternions $\mathbf{a}$ and $\mathbf{b}$, $-1 \le \mathbf{a} \cdot \mathbf{b} \le 1$.

The dot product is perhaps not one of the most frequently used quaternion operators, at least in video game programming, but it does have an interesting geometric interpretation. In Section 8.5.8, we considered the difference quaternion $\mathbf{d} = \mathbf{ba}^*$, which describes the angular displacement from orientation $\mathbf{a}$ to orientation $\mathbf{b}$. (We assume unit quaternions and replace the quaternion inverse with the conjugate.) If we expand the product and examine the contents of $\mathbf{d}$, we find that the w component is equal to the dot product $\mathbf{a} \cdot \mathbf{b}$!

What does this mean geometrically? Remember Euler's rotation theorem: we can rotate from the orientation $\mathbf{a}$ into the orientation $\mathbf{b}$ via a single rotation about a carefully chosen axis. This uniquely determined (up to a reversal of sign) axis and angle are precisely the ones encoded in $\mathbf{d}$. Remembering the relationship between the w component and the rotation angle θ, we see that $\mathbf{a} \cdot \mathbf{b} = \cos(\theta/2)$, where θ is the amount of rotation needed to go from the orientation $\mathbf{a}$ to the orientation $\mathbf{b}$.

In summary, the quaternion dot product has an interpretation similar to the vector dot product. The larger the absolute value of the quaternion dot product $\mathbf{a} \cdot \mathbf{b}$, the more "similar" are the angular displacements represented by $\mathbf{a}$ and $\mathbf{b}$. While the vector dot product gives the cosine of the angle between vectors, the quaternion dot product gives the cosine of *half* of the angle needed to rotate one quaternion into the other. For the purpose of measuring similarity, usually we are interested only in the absolute value of $\mathbf{a} \cdot \mathbf{b}$, since $\mathbf{a} \cdot \mathbf{b} = -(\mathbf{a} \cdot -\mathbf{b})$, even though $\mathbf{b}$ and $-\mathbf{b}$ represent the same angular displacement.

Although direct use of the dot product is infrequent in most video game code, the dot product is the first step in the calculation of the slerp function, which we discuss in Section 8.5.12.

8.5.10   Quaternion log, exp, and Multiplication by a Scalar

This section discusses three operations on quaternions that, although they are seldom used directly, are the basis for several important quaternion operations. These operations are the quaternion logarithm, exponential, and multiplication by a scalar.

First, let us reformulate our definition of a quaternion by introducing a variable α to equal the half-angle, $\theta/2$:

Defining a quaternion in terms of the half-angle α

$$\alpha = \theta/2, \qquad\qquad \mathbf{q} = \begin{bmatrix} \cos\alpha & \hat{\mathbf{n}}\sin\alpha \end{bmatrix}.$$

The logarithm of $\mathbf{q}$ is defined as

The logarithm of a quaternion

$$\log\mathbf{q} = \log\left(\begin{bmatrix} \cos\alpha & \hat{\mathbf{n}}\sin\alpha \end{bmatrix}\right) \equiv \begin{bmatrix} 0 & \alpha\hat{\mathbf{n}} \end{bmatrix}.$$

We use the notation $\equiv$ to mean equal by definition. In general, $\log\mathbf{q}$ is not a unit quaternion. Note the similarity between taking the logarithm of a quaternion, and the exponential map format (see Section 8.4).

The exponential function is defined in the exact opposite manner. First we define the quaternion $\mathbf{p}$ to be of the form $[0, \alpha\hat{\mathbf{n}}]$, with $\hat{\mathbf{n}}$ a unit vector:

$$\mathbf{p} = \begin{bmatrix} 0 & \alpha\hat{\mathbf{n}} \end{bmatrix}, \qquad\qquad (\|\hat{\mathbf{n}}\| = 1).$$

Then the exponential function is defined as

The exponential function of a quaternion

$$\exp\mathbf{p} = \exp\left(\begin{bmatrix} 0 & \alpha\hat{\mathbf{n}} \end{bmatrix}\right) \equiv \begin{bmatrix} \cos\alpha & \hat{\mathbf{n}}\sin\alpha \end{bmatrix}.$$

Note that, by definition, $\exp\mathbf{p}$ always returns a unit quaternion.

The quaternion logarithm and exponential are related to their scalar analogs. For any scalar a,

$$e^{\ln a} = a.$$

In the same way, the quaternion exp function is defined to be the inverse of the quaternion log function:

$$\exp(\log\mathbf{q}) = \mathbf{q}.$$

Finally, quaternions can be multiplied by a scalar, with the result computed in the obvious way of multiplying each component by the scalar. Given a scalar k and a quaternion $\mathbf{q}$,

Multiplying a quaternion by a scalar

$$k\mathbf{q} = k\begin{bmatrix} w & \mathbf{v} \end{bmatrix} = \begin{bmatrix} kw & k\mathbf{v} \end{bmatrix}.$$

This will not usually result in a unit quaternion, which is why multiplication by a scalar is not a very useful operation in the context of representing angular displacement. (But we will find a use for it in Section 8.5.11.)

8.5.11   Quaternion Exponentiation

Quaternions can be *exponentiated*, which means that we can raise a quaternion to a scalar power. Quaternion exponentiation, denoted $\mathbf{q}^t$, should not be confused with the exponential function $\exp \mathbf{q}$. The exponential function accepts only one argument: a quaternion. Quaternion exponentiation has two arguments: the quaternion $\mathbf{q}$ and the scalar exponent t.

The meaning of quaternion exponentiation is similar to that of real numbers. Recall that for any scalar a, besides zero, $a^0 = 1$ and $a^1 = a$. As the exponent t varies from 0 to 1 the value of a^t varies from 1 to a. A similar statement holds for quaternion exponentiation: as t varies from 0 to 1 the quaternion exponentiation $\mathbf{q}^t$ varies from $[1, \mathbf{0}]$ to $\mathbf{q}$.

Quaternion exponentiation is useful because it allows us to extract a "fraction" of an angular displacement. For example, to compute a quaternion that represents one third of the angular displacement represented by the quaternion $\mathbf{q}$, we would compute $\mathbf{q}^{1/3}$.

Exponents outside the $[0, 1]$ range behave mostly as expected—with one major caveat. For example, $\mathbf{q}^2$ represents twice the angular displacement as $\mathbf{q}$. If $\mathbf{q}$ represents a clockwise rotation of $30°$ about the x-axis, then $\mathbf{q}^2$ represents a clockwise rotation of $60°$ about the x-axis, and $\mathbf{q}^{-1/3}$ represents a counterclockwise rotation of $10°$ about the x-axis. Notice in particular that the inverse notation $\mathbf{q}^{-1}$ can also be interpreted in this context and the result is the same: the quaternion that performs the opposite rotation.

The caveat we mentioned is this: a quaternion represents angular displacements using the shortest arc. Multiple spins cannot be represented. Continuing our example above, $\mathbf{q}^8$ is not a $240°$ clockwise rotation about the x-axis as expected; it is a $120°$ counterclockwise rotation. Of course, rotating $240°$ in one direction produces the same end result as rotating $120°$ in the opposite direction, and this is the point: quaternions really capture only the end result. In general, many of the algebraic identities concerning exponentiation of scalars, such as $(a^s)^t = a^{st}$, do not apply to quaternions.

In some situations, we *do* care about the total amount of rotation, not just the end result. (The most important example is that of angular velocity.) In these situations, quaternions are not the correct tool for the job; use the exponential map (or its cousin, the axis-angle format) instead.

Now that we understand what quaternion exponentiation is used for, let's see how it is mathematically defined. Quaternion exponentiation is defined in terms of the "utility" operations we learned in the previous section. The definition is given by

$$\mathbf{q}^t = \exp\left(t \log \mathbf{q}\right). \tag{8.8}$$

Raising a quaternion to a power

Notice that a similar statement is true regarding exponentiation of a scalar:

$$a^t = e^{(t \ln a)}.$$

It is not too difficult to understand why $\mathbf{q}^t$ interpolates from identity to $\mathbf{q}$ as t varies from 0 to 1. Notice that the log operation essentially converts the quaternion to exponential map format (except for a factor of 2). Then, when we perform the scalar multiplication by the exponent t, the effect is to multiply the angle by t. Finally, the exp "undoes" what the log operation did, recalculating the new w and $\mathbf{v}$ from the exponential vector. At least this is how it works academically in an equation. Although Equation (8.8) is the official mathematical definition and works elegantly in theory, direct translation into code is more complicated than necessary. Listing 8.2 shows how we could compute the value of $\mathbf{q}^t$ in C. Essentially, instead of working with a single exponential-map-like quantity as the formula tells us to, we break out the axis and half-angle separately.

```c
// Quaternion (input and output)
float w,x,y,z;

// Input exponent
float exponent;

// Check for the case of an identity quaternion.
// This will protect against divide by zero
if (fabs(w) < .9999f) {

    // Extract the half angle alpha (alpha = theta/2)
    float alpha = acos(w);

    // Compute new alpha value
    float newAlpha = alpha * exponent;

    // Compute new w value
    w = cos(newAlpha);

    // Compute new xyz values
    float mult = sin(newAlpha) / sin(alpha);
    x *= mult;
    y *= mult;
    z *= mult;
}
```

Listing 8.2
Raising a quaternion to an exponent

There are a few points to notice about this code. First, the check for the identity quaternion is necessary since a value of $w = \pm1$ would cause the computation of `mult` to divide by zero. Raising an identity quaternion to any power results in the identity quaternion, so if we detect an identity quaternion on input, we simply ignore the exponent and return the original quaternion.

Second, when we compute `alpha`, we use the arccos function, which always returns a positive angle. This does not create a loss of generality. Any quaternion can be interpreted as having a positive angle of rotation, since negative rotation about an axis is the same as positive rotation about the axis pointing in the opposite direction.

8.5.12   Quaternion Interpolation, a.k.a. Slerp

The *raison d'être* of quaternions in games and graphics today is an operation known as *slerp*, which stands for **S**pherical **L**inear int**erp**olation. The slerp operation is useful because it allows us to smoothly interpolate between two orientations. Slerp avoids all the problems that plagued interpolation of Euler angles (see Section 8.3.4).

Slerp is a ternary operator, meaning it accepts three operands. The first two operands to slerp are the two quaternions between which we wish to interpolate. We'll assign these starting and ending orientations to the variables $\mathbf{q}_0$ and $\mathbf{q}_1$, respectively. The interpolation parameter will be assigned to the variable t, and as t varies from 0 to 1, the slerp function $\text{slerp}(\mathbf{q}_0, \mathbf{q}_1, t)$ returns an orientation that interpolates from $\mathbf{q}_0$ to $\mathbf{q}_1$.

Let's see if we can't derive the slerp formula by using the tools we have so far. If we were interpolating between two scalar values a_0 and a_1, we could use the standard linear interpolation (lerp) formula:

$$\Delta a = a_1 - a_0,$$
$$\text{lerp}(a_0, a_1, t) = a_0 + t\,\Delta a.$$

Simple linear interpolation

The standard linear interpolation formula works by starting at a_0 and adding the fraction t of the difference between a_1 and a_0. This requires three basic steps:

1. Compute the difference between the two values.

2. Take a fraction of this difference.

3. Take the original value and adjust it by this fraction of the difference.

We can use the same basic idea to interpolate between orientations. (Again, remember that quaternion multiplication reads right-to-left.)

1. *Compute the difference between the two values.* We showed how to do this in Section 8.5.8. The angular displacement from $\mathbf{q}_0$ to $\mathbf{q}_1$ is given by
$$\Delta \mathbf{q} = \mathbf{q}_1 \mathbf{q}_0{}^{-1}.$$

2. *Take a fraction of this difference.* To do this, we use quaternion exponentiation, which we discussed in Section 8.5.11. The fraction of the difference is given by
$$(\Delta \mathbf{q})^t.$$

3. *Take the original value and adjust it by this fraction of the difference.* We "adjust" the initial value by composing the angular displacements via quaternion multiplication:
$$(\Delta \mathbf{q})^t \mathbf{q}_0.$$

Thus, the equation for slerp is given by

$$\text{slerp}(\mathbf{q}_0, \mathbf{q}_1, t) = (\mathbf{q}_1 \mathbf{q}_0{}^{-1})^t \mathbf{q}_0.$$

This algebraic form is how slerp is computed in theory. In practice, we use a formulation that is mathematically equivalent, but computationally more efficient. To derive this alternative formula, we start by interpreting the quaternions as existing in a 4D Euclidian space. Since all of the quaternions of interest are unit quaternions, they "live" on the surface of a 4D hypersphere. The basic idea is to interpolate around the arc that connects the two quaternions, along the surface of the 4D hypersphere. (Hence the name *spherical* linear interpolation.)

We can visualize this in the plane (see Figure 8.11). Imagine two 2D vectors $\mathbf{v}_0$ and $\mathbf{v}_1$, both of unit length. We wish to compute the value of $\mathbf{v}_t$, which is the result of smoothly interpolating around the arc by a fraction t of the distance from $\mathbf{v}_0$ to $\mathbf{v}_1$. If we let ω[13] be the angle intercepted by the arc from $\mathbf{v}_0$ to $\mathbf{v}_1$, then $\mathbf{v}_t$ is the result of rotating $\mathbf{v}_0$ around this arc by an angle of $t\omega$.

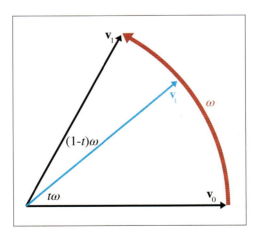

Figure 8.11
Interpolating a rotation

We can express $\mathbf{v}_t$ as a linear combination of $\mathbf{v}_0$ and $\mathbf{v}_1$. In other words, there exist nonnegative constants k_0 and k_1 such that $\mathbf{v}_t = k_0 \mathbf{v}_0 + k_1 \mathbf{v}_1$. We can use elementary geometry to determine the values of k_0 and k_1. Figure 8.12 shows how this can be done.

---

[13]This is the Greek letter omega, pronounced "oh-MAY-guh."

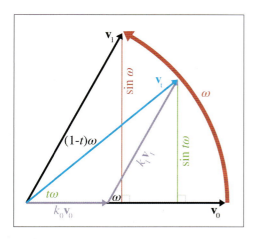

Figure 8.12
Interpolating a vector about an arc

Applying some trig to the right triangle with $k_1\mathbf{v}_1$ as the hypotenuse (and recalling that $\mathbf{v}_1$ is a unit vector), we see that

$$\sin \omega = \frac{\sin t\omega}{k_1},$$

$$k_1 = \frac{\sin t\omega}{\sin \omega}.$$

A similar technique to solve for k_0 yields the following result:

$$k_0 = \frac{\sin(1-t)\omega}{\sin \omega}.$$

Thus, $\mathbf{v}_t$ can be expressed as

$$\mathbf{v}_t = k_0\mathbf{v}_0 + k_1\mathbf{v}_1 = \frac{\sin(1-t)\omega}{\sin \omega}\mathbf{v}_0 + \frac{\sin t\omega}{\sin \omega}\mathbf{v}_1.$$

The same basic idea can be extended into quaternion space, and we can reformulate the slerp as

$$\text{slerp}(\mathbf{q}_0, \mathbf{q}_1, t) = \frac{\sin(1-t)\omega}{\sin \omega}\mathbf{q}_0 + \frac{\sin t\omega}{\sin \omega}\mathbf{q}_1.$$

Quaternion slerp in practice

We just need a way to compute ω, the "angle" between the two quaternions. As it turns out, an analogy from 2D vector math can be carried into quaternion space; we can think of the quaternion dot product as returning $\cos \omega$.

There are two slight complications. First, the two quaternions $\mathbf{q}$ and $-\mathbf{q}$ represent the same orientation, but may produce different results when used as an argument to slerp. This problem doesn't happen in 2D or 3D, but the surface of a 4D hypersphere has a different topology than Euclidian space. The solution is to choose the signs of $\mathbf{q}_0$ and $\mathbf{q}_1$ such that the dot product $\mathbf{q}_0 \cdot \mathbf{q}_1$ is nonnegative. This has the effect of always selecting the shortest rotational arc from $\mathbf{q}_0$ to $\mathbf{q}_1$. The second complication is that if $\mathbf{q}_0$ and $\mathbf{q}_1$ are very close, then ω is very small, and thus $\sin \omega$ is also very small, which will cause problems with the division. To avoid this, if $\sin \omega$ is very small, we will use simple linear interpolation. The code snippet in Listing 8.3 applies all of this advice to compute the quaternion slerp.

```
// The two input quaternions
float w0,x0,y0,z0;
float w1,x1,y1,z1;

// The interpolation parameter
float t;

// The output quaternion will be computed here
float w,x,y,z;

// Compute the "cosine of the angle" between the
// quaternions, using the dot product
float cosOmega = w0*w1 + x0*x1 + y0*y1 + z0*z1;

// If negative dot, negate one of the input
// quaternions, to take the shorter 4D "arc"
if (cosOmega < 0.0f) {
    w1 = -w1;
    x1 = -x1;
    y1 = -y1;
    z1 = -z1;
    cosOmega = -cosOmega;
}

// Check if they are very close together, to protect
// against divide-by-zero
float k0, k1;
if (cosOmega > 0.9999f) {

    // Very close - just use linear interpolation
    k0 = 1.0f-t;
    k1 = t;

} else {

    // Compute the sin of the angle using the
    // trig identity sin^2(omega) + cos^2(omega) = 1
    float sinOmega = sqrt(1.0f - cosOmega*cosOmega);

    // Compute the angle from its sine and cosine
    float omega = atan2(sinOmega, cosOmega);

    // Compute inverse of denominator, so we only have
    // to divide once
    float oneOverSinOmega = 1.0f / sinOmega;

    // Compute interpolation parameters
    k0 = sin((1.0f - t) * omega) * oneOverSinOmega;
    k1 = sin(t * omega) * oneOverSinOmega;
```

```
}

// Interpolate
w = w0*k0 + w1*k1;
x = x0*k0 + x1*k1;
y = y0*k0 + y1*k1;
z = z0*k0 + z1*k1;
```

Listing 8.3
Quaternion slerp

8.5.13 Advantages and Disadvantages of Quaternions

Quaternions offer a number of advantages over other methods of representing angular displacement:

- *Smooth interpolation.* The interpolation provided by slerp provides smooth interpolation between orientations. No other representation method provides for smooth interpolation.

- *Fast concatenation and inversion of angular displacements.* We can concatenate a sequence of angular displacements into a single angular displacement by using the quaternion cross product. The same operation using matrices involves more scalar operations, although which one is actually faster on a given architectures is not so clean-cut: single instruction multiple data (SIMD) vector operations can make very quick work of matrix multiplication. The quaternion conjugate provides a way to compute the opposite angular displacement very efficiently. This can be done by transposing a rotation matrix, but is not easy with Euler angles.

- *Fast conversion to and from matrix form.* As we see in Section 8.7, quaternions can be converted to and from matrix form a bit faster than Euler angles.

- *Only four numbers.* Since a quaternion contains four scalar values, it is considerably more economical than a matrix, which uses nine numbers. (However, it still is 33% larger than Euler angles.)

These advantages do come at some cost, however. Quaternions suffer from a few of the problems that affect matrices, only to a lesser degree:

- *Slightly bigger than Euler angles.* That one additional number may not seem like much, but an extra 33% can make a difference when large amounts of angular displacements are needed, for example, when storing animation data. And the values inside a quaternion are not "evenly spaced" along the $[-1, +1]$ interval; the component values do not interpolate smoothly, even if the orientation does. This makes quaternions more difficult to pack into a fixed-point number than Euler angles or an exponential map.

- *Can become invalid.* This can happen either through bad input data, or from accumulated floating point roundoff error. (We can address this problem by normalizing the quaternion to ensure that it has unit magnitude.)

- *Difficult for humans to work with.* Of the three representation methods, quaternions are the most difficult for humans to work with directly.

8.5.14 Quaternions as Complex Numbers

We end our discussion on quaternions in the place that most texts begin: a discussion of their interpretation as complex numbers. If you are interested in quaternions solely for rotations, you can safely skip this section. If you want a bit deeper understanding or are interested in the mathematical heritage of quaternions and the circumstances that surrounded their invention, this section will be interesting. We will be following an approach due to John McDonald of DePaul University [45]. Among other things, this method is able to explain two peculiarities of quaternions: the appearance of $\theta/2$ rather than θ and the unusual mathematical form $\mathbf{qvq}^{-1}$:

We begin by considering how we can embed the set of real numbers in the set of 2×2 matrices. For any given scalar a, we associate it with exactly one 2×2 matrix, namely the matrix that has a on both of the diagonal elements:

Each real scalar maps to a 2×2 matrix

$$a \equiv \begin{bmatrix} a & 0 \\ 0 & a \end{bmatrix}.$$

We have chosen a subset of the 2×2 matrices, and established a one-to-one correspondence between this smaller set of matrices and the set of all real numbers. We could have established this one-to-one relationship in other ways, but this particular way of doing it is important because it preserves all the ordinary algebra laws of addition, subtraction, and multiplication: the associative property, distributive property, nonfactorability of zero, and so on. (We can even include division if we treat division as multiplication by the inverse.) For example,

Addition, subtraction, and multiplication work the same

$$\begin{bmatrix} a & 0 \\ 0 & a \end{bmatrix} + \begin{bmatrix} b & 0 \\ 0 & b \end{bmatrix} = \begin{bmatrix} a+b & 0 \\ 0 & a+b \end{bmatrix},$$

$$\begin{bmatrix} a & 0 \\ 0 & a \end{bmatrix} - \begin{bmatrix} b & 0 \\ 0 & b \end{bmatrix} = \begin{bmatrix} a-b & 0 \\ 0 & a-b \end{bmatrix},$$

$$\begin{bmatrix} a & 0 \\ 0 & a \end{bmatrix} \begin{bmatrix} b & 0 \\ 0 & b \end{bmatrix} = \begin{bmatrix} ab & 0 \\ 0 & ab \end{bmatrix}.$$

Now let's see if we can create a similar mapping for the set of complex numbers. You probably already have been introduced to complex numbers; if so, you should remember that the complex pair (a, b) defines the number $a + bi$. The number i is a special number such that $i^2 = -1$. It's often called the *imaginary number* because no ordinary scalar (a "real" number) can have this property. The word "imaginary" gives one the impression that the number doesn't really exist; we're going avoid this term and instead stick with the more descriptive one: "complex."

Complex numbers can be added, subtracted, and multiplied. All we need to do is follow the ordinary rules for arithmetic, and replace i^2 with -1 whenever it appears. This results in the following identities:

$$\begin{aligned}
(a + bi) + (c + di) &= (a + c) + (b + d)i, \\
(a + bi) - (c + di) &= (a - c) + (b - d)i, \\
(a + bi)(c + di) &= ac + adi + bci + bdi^2 \\
&= ac + (ad + bc)i + bd(-1) \\
&= (ac - bd) + (ad + bc)i.
\end{aligned}$$

Adding, subtracting, and multiplying complex numbers

Now, how can we extend our system of embedding numbers in the space of 2×2 matrices to include complex numbers? Before, we only had one degree of freedom, a, and now we have two, a and b. The mapping we use is

$$a + bi \equiv \begin{bmatrix} a & -b \\ b & a \end{bmatrix}. \tag{8.9}$$

Mapping each complex number to a 2×2 matrix

We can easily verify that the complex number on the left behaves exactly the same as the matrix on the right. In a certain sense, they are just two notations for writing the same quantity:

$$\begin{aligned}
(a + bi) + (c + di) &\equiv \begin{bmatrix} a & -b \\ b & a \end{bmatrix} + \begin{bmatrix} c & -d \\ d & c \end{bmatrix} = \begin{bmatrix} a + c & -(b + d) \\ b + d & a + c \end{bmatrix} \\
&\equiv (a + c) + (b + d)i, \\
(a + bi) - (c + di) &\equiv \begin{bmatrix} a & -b \\ b & a \end{bmatrix} - \begin{bmatrix} c & -d \\ d & c \end{bmatrix} = \begin{bmatrix} a - c & -(b - d) \\ b - d & a - c \end{bmatrix} \\
&\equiv (a - c) + (b - d)i, \\
(a + bi)(c + di) &\equiv \begin{bmatrix} a & -b \\ b & a \end{bmatrix} \begin{bmatrix} c & -d \\ d & c \end{bmatrix} = \begin{bmatrix} ac - bd & -(ad + bc) \\ ad + bc & ac - bd \end{bmatrix} \\
&\equiv (ac - bd) + (ad + bc)i.
\end{aligned}$$

Addition, subtraction, and multiplication in standard notation and our 2×2 form

We also verify that the equation $i^2 = -1$ still holds:

i doesn't seem quite as
"imaginary" in 2 × 2
form

$$i^2 \equiv \begin{bmatrix} 0 & -1 \\ 1 & 0 \end{bmatrix}^2 = \begin{bmatrix} 0 & -1 \\ 1 & 0 \end{bmatrix}\begin{bmatrix} 0 & -1 \\ 1 & 0 \end{bmatrix} = \begin{bmatrix} -1 & 0 \\ 0 & -1 \end{bmatrix} \equiv -1.$$

Let's apply the geometric perspective from Chapter 5. Interpreting the columns[14] $[0, 1]$ and $[-1, 0]$ as the basis vectors of a coordinate space, we see that this matrix performs a 90° rotation.

We can interpret multiplication by i as a 90° rotation.[15]

There's nothing "imaginary" about this. Instead of thinking of i as the square root of -1, think instead of the complex number $a + bi$ as a mathematical entity with two degrees of freedom that behaves in particular ways when multiplied. The part we usually call the "real" part, a, is the main degree of freedom, and b measures some secondary degree of freedom. The two degrees of freedom are in some sense "orthogonal" to one another.

Continuing this further, we see that we can represent rotations by any arbitrary angle θ using this scheme. The basic 2×2 rotation matrix derived in Section 5.1.1 happens to be in this special set of matrices that we are mapping to the complex numbers. It maps to the complex number $\cos\theta + i\sin\theta$:

Unit complex numbers
as rotations

$$\cos\theta + i\sin\theta \equiv \begin{bmatrix} \cos\theta & -\sin\theta \\ \sin\theta & \cos\theta \end{bmatrix}.$$

Notice how complex conjugation (negating the complex part) corresponds to matrix transposition. This is particularly pleasing. Remember that the conjugate of a quaternion expresses the inverse angular displacement. A corresponding fact is true for transposing rotation matrices: since they are orthogonal, their transpose is equal to their inverse.

How do ordinary 2D vectors fit into this scheme? We interpret the vector $[x, y]$ as the complex number $x + iy$, and then we can interpret the

---

[14]Our usual convention is to use row vectors, but we are going to use column vectors here because the right-to-left order of rotations matches quaternions more closely.

[15]Whether the rotation is clockwise or counterclockwise is a matter of convention and not inherent to complex numbers. In fact, you probably already noticed that we could have negated the other b in the matrix from Equation (8.9) and still had a valid way to map the set of complex numbers to 2×2 matrices. Our arbitrary choice will become useful in just a moment.

multiplication of two complex numbers

$$(\cos\theta + i\sin\theta)(x + iy) = x\cos\theta + iy\cos\theta + ix\sin\theta + i^2 y\sin\theta$$
$$= (x\cos\theta - y\sin\theta) + i(x\sin\theta + y\cos\theta)$$

as performing a rotation. This is equivalent to the matrix multiplication

$$\begin{bmatrix} \cos\theta & -\sin\theta \\ \sin\theta & \cos\theta \end{bmatrix} \begin{bmatrix} x \\ y \end{bmatrix} = \begin{bmatrix} x\cos\theta - y\sin\theta \\ x\sin\theta + y\cos\theta \end{bmatrix}.$$

While this a not much more that mathematical trivia so far, our goal is to build up some parallels that we can carry forward to quaternions, so let's repeat the key result.

In 2D, we can interpret the vector $[x, y]$ as a complex number $x + yi$ and rotate it by using the complex multiplication $(\cos\theta + i\sin\theta)(x + iy)$.

A similar conversion from ordinary vectors to complex numbers is necessary in order to multiply quaternions and 3D vectors.

Before we leave 2D, let's summarize what we've learned so far. Complex numbers are mathematical objects with two degrees of freedom that obey certain rules when we multiply them. These objects are usually written as $a + bi$, but can equivalently be written as a 2×2 matrix. When we write complex numbers as matrices, it begs the geometric interpretation of multiplication by i as a 90° rotation. The rule $i^2 = -1$ has the interpretation that combining two 90° rotations yields a 180° rotation, and that leaves us with a warm fuzzy feeling. More generally, any complex number with unit length can be written as $\cos\theta + i\sin\theta$ and interpreted as a rotation by the angle θ. If we convert a 2D vector into complex form and multiply it by $\cos\theta + i\sin\theta$, it has the effect of performing the rotation.

It's very tempting to extend this trick from 2D into 3D. Tempting, but alas not possible in the straightforward way. The Irish mathematician William Hamilton (1805–1865) apparently fell victim to just this temptation, and had looked for a way to extend complex numbers from 2D to 3D for years. This new type of complex number, he thought, would have one real part and two imaginary parts. However, Hamilton was unable to create a useful type of complex number with two imaginary parts. Then, as the story goes, in 1843, on his way to a speech at the Royal Irish Academy, he suddenly realized that *three* imaginary parts were needed rather than two. He carved the equations that define the properties of this new type of complex number on the Broom Bridge. His original marks have faded into legend, but a commemorative plaque holds their place. Thus quaternions were invented.

Since we weren't on the Broom Bridge in 1843, we can't say for certain what made Hamilton realize that a 3D system of complex numbers was no good, but we can show how such a set could not be easily mapped to 3×3 matrices and rotations. A 3D complex number would have two complex parts, i and j, with the properties $i^2 = j^2 = -1$. (We would also need to define the values of the products ij and ji. Exactly what these rules should be, we're not sure; perhaps Hamilton realized this was a dead end. In any case, it doesn't matter for the present discussion.) Now, a straightforward extension of the ideas from 2D would mean that we could somehow associate the numbers 1, i, and j with the set of 3×3 matrices, such that all the usual algebra laws hold. The number 1 must obviously map to the 3D identity matrix $\mathbf{I}_3$. The number -1 should map to its negative, $-\mathbf{I}_3$, which has -1s on the diagonal. But now we run into a problem trying to a find matrices for i and j whose square is $-\mathbf{I}_3$. We can quickly see that this is not possible because the determinant of $-\mathbf{I}_3$ is -1. To be a root of this matrix, i or j must have a determinant that is the square root of -1 because the determinant of a matrix product is the product of the determinants. The only way this can work is for i and j to contain entries that are complex. In short, there doesn't seem to be a coherent system of 3D complex numbers; certainly there isn't one that maps elegantly to rotations analogously to standard complex numbers and 2D rotations. For that, we need quaternions.

Quaternions extend the complex number system by having three imaginary numbers, i, j, and k, which are related by Hamilton's famous equations:

The rules for 4D complex numbers that Hamilton wrote on the Broom Bridge

$$\begin{aligned}
i^2 = j^2 = k^2 &= -1 \\
ij = k, \quad ji &= -k, \\
jk = i, \quad kj &= -i, \\
ki = j, \quad ik &= -j.
\end{aligned} \tag{8.10}$$

The quaternion we have been denoting $[w, (x, y, z)]$ corresponds to the complex number $w + xi + yj + zk$. The definition of the quaternion product given in Section 8.5.7 follows from these rules. (Also see Exercise 6.) The dot product, however, basically ignores all of this complex i, j, k business and treats the operands as simple 4D vectors.

Now we return to matrices. Can we embed the set of quaternions into the set of matrices such that Hamilton's rules in Equation (8.10) still hold? Yes, we can, although, as you might expect, we map them to 4×4 matrices. Real numbers are mapped to a matrix with the number on each entry of the diagonal as before,

$$a \equiv \begin{bmatrix} a & 0 & 0 & 0 \\ 0 & a & 0 & 0 \\ 0 & 0 & a & 0 \\ 0 & 0 & 0 & a \end{bmatrix},$$

and the complex quantities are mapped to the matrices

$$i \equiv \begin{bmatrix} 0 & 0 & 0 & 1 \\ 0 & 0 & -1 & 0 \\ 0 & 1 & 0 & 0 \\ -1 & 0 & 0 & 0 \end{bmatrix}, \quad j \equiv \begin{bmatrix} 0 & 0 & 1 & 0 \\ 0 & 0 & 0 & 1 \\ -1 & 0 & 0 & 0 \\ 0 & -1 & 0 & 0 \end{bmatrix}, \quad k \equiv \begin{bmatrix} 0 & -1 & 0 & 0 \\ 1 & 0 & 0 & 0 \\ 0 & 0 & 0 & 1 \\ 0 & 0 & -1 & 0 \end{bmatrix}.$$

(8.11)

Mapping the three complex quantities to 4 × 4 matrices

We encourage you to convince yourself that these mappings do preserve all of Hamilton's rules before moving on.

Combining the above associations, we can map an arbitrary quaternion to a 4×4 matrix as

$$w + xi + yj + zk \equiv \begin{bmatrix} w & -z & y & x \\ z & w & -x & y \\ -y & x & w & z \\ -x & -y & -z & w \end{bmatrix}.$$

(8.12)

Encoding quaternions as 4 × 4 matrices

Once again, we notice how complex conjugation (negating x, y, and z) corresponds to matrix transposition.

Everything we've said so far applies to quaternions of any length. Now let's get back to rotations. We can see that the i, j, k matrices in Equation (8.11) permute and negate axes, so they bear some similarity to 90° rotations or reflections. Let's see if we can carry the simple ideas from 2D forward with these matrices. Notice how the upper left 2×2 portion of the k matrix is the same as the very first 2×2 matrix for i; in other words, part of k is a 90° rotation about z. By analogy with the 2D case, we might reasonably expect the quaternion $\cos\theta + k\sin\theta$ to represent a rotation about the z-axis by an arbitrary angle θ. Let's multiply it by the vector $[1,0,0]$ and see what happens. As in the 2D case, we need to "promote" the vector into the complex domain; what's different here is that quaternions have an extra number. We'll map $[x,y,z]$ to the complex number $0 + xi + yj + zk$, so the vector $[1,0,0]$ is simply i. Expanding the multiplication, we have

$$(\cos\theta + k\sin\theta)i = i\cos\theta + ki\sin\theta,$$
$$= i\cos\theta + j\sin\theta,$$

which corresponds to $[\cos\theta, \sin\theta, 0]$, exactly what we would expect when rotating the x-axis about the z-axis. So far, all is good. Let's try a slightly more general vector $[1, 0, 1]$, which is represented in the complex domain as $i + k$:

$$(\cos\theta + k\sin\theta)(i + k) = i\cos\theta + k\cos\theta + ki\sin\theta + k^2\sin\theta$$
$$= i\cos\theta + j\sin\theta + k\cos\theta - \sin\theta. \qquad (8.13)$$

This result does not correspond to a vector at all, since it has a nonzero value for w. The rotation in the xy-plane worked as expected, but unfortunately, the z component did not come out right. There is unwanted rotation in the zw-hyperplane. This is made perfectly clear by looking at how $(\cos\theta + k\sin\theta)$ is represented as a 4×4 matrix:

$$\cos\theta + k\sin\theta \equiv \begin{bmatrix} \cos\theta & -\sin\theta & 0 & 0 \\ \sin\theta & \cos\theta & 0 & 0 \\ 0 & 0 & \cos\theta & \sin\theta \\ 0 & 0 & -\sin\theta & \cos\theta \end{bmatrix}.$$

The upper-left 2×2 rotation matrix is the one we want; the lower-right 2×2 rotation matrix is not wanted.

Now we are left wondering if maybe we did something wrong. Perhaps there are other 4×4 roots of -1 we could use for i, j, and k—alternative ways that we could embed the quaternion set within the set of 4×4 matrices. In fact, there are other alternatives, and this is a hint that something is a bit different from the 2D case. Unfortunately, all of these alternatives exhibit variations of what is essentially the same behavior we are seeing here. Perhaps, instead, our problem is that we did the multiplication in the wrong order. (After all, multiplication of i, j, and k is not commutative.) Let's try putting the vector on the left and the rotation quaternion on the right:

$$(i + k)(\cos\theta + k\sin\theta) = i\cos\theta + ik\sin\theta + k\cos\theta + k^2\sin\theta$$
$$= i\cos\theta - j\sin\theta + k\cos\theta - \sin\theta.$$

Comparing this to Equation (8.13), when the operands were in the opposite order, we see that the only difference is the sign of the y-coordinate. At first glance, it looks like this is actually worse. The rotation in the xz-plane that we want got inverted; now we have a rotation by $-\theta$. Meanwhile, the extra rotation we didn't want is exactly the same as before. But perhaps you can already see the solution. If we use the opposite rotation, which corresponds to using the conjugate of the quaternion, we fix both problems:

$$(i + k)(\cos\theta - k\sin\theta) = i\cos\theta + j\sin\theta - k\cos\theta + \sin\theta.$$

So, multiplying on the left by $(\cos\theta + k\sin\theta)$ produced the rotation we wanted, plus some extra rotation we didn't want, and multiplication on the right by the conjugate achieved the same rotation that was desired, with the opposite undesired rotation. If we combine these two steps, the unwanted rotation is canceled out, and we are left with only the rotation we want. Well, not quite, we are left with *twice* the rotation we want, but this is easily fixed by using $\theta/2$ instead of θ. Of course, we knew that $\theta/2$ would appear somewhere, but now we see the reason. Let's summarize our findings from the preceding paragraphs.

To extend the ideas about complex numbers and rotations from 2D to quaternions, we first convert the vector $[x, y, z]$ to quaternion form as $\mathbf{v} = [0, (x, y, z)]$. A straightforward approach to rotate the vector by the angle θ about the axis $\hat{\mathbf{n}}$ would be to create the quaternion $\mathbf{q} = [\cos\theta, \sin\theta\,\hat{\mathbf{n}}]$ and then perform the multiplication $\mathbf{qv}$. This, however, does not work; while the result contains the rotation we want, it also contains an unwanted rotation into w. The multiplication $\mathbf{vq}^*$ also produces the rotation we want plus some unwanted rotation, but in this case the unwanted rotation is exactly opposite of that produced by $\mathbf{qv}$. The solution is to use the half angle and set $\mathbf{q} = [\cos(\theta/2), \sin(\theta/2)\,\hat{\mathbf{n}}]$, and the rotation is accomplished by performing both multiplications: $\mathbf{qvq}^*$. The first rotation rotates halfway to the goal, plus some unwanted rotation involving w. The second rotation completes the desired rotation while also canceling the unwanted rotation.

Before we leave this section, let us go back and clear up one last finer point. We mentioned that there are other ways we could embed the set of quaternions within the set of 4×4 matrices. (Equations (8.11) and (8.12) aren't the only way to do it.) McDonald [45] explores this idea in more detail; here we merely want to note that this is another underlying cause of the need for $\mathbf{qvq}^{-1}$. Using just a single multiplication, the variations in the embedding would produce variations in the rotated result. When both multiplications are present, the change from one style to another produces a change on the left that is exactly canceled by the matching change on the right.

8.5.15   Summary of Quaternions

Section 8.5 has covered a lot of math, and most of it isn't important to remember. The facts that *are* important to remember about quaternions are summarized here.

- Conceptually, a quaternion expresses angular displacement by using an axis of rotation and an amount of rotation about that axis.

- A quaternion contains a scalar component w and a vector component $\mathbf{v}$. They are related to the angle of rotation θ and the axis of rotation $\hat{\mathbf{n}}$ by

$$w = \cos(\theta/2), \qquad\qquad \mathbf{v} = \hat{\mathbf{n}}\sin(\theta/2).$$

- Every angular displacement in 3D has exactly two different representations in quaternion space, and they are negatives of each other.

- The identity quaternion, which represents "no angular displacement," is $[1, \mathbf{0}]$.

- All quaternions that represent angular displacement are "unit quaternions" with magnitude equal to 1.

- The conjugate of a quaternion expresses the opposite angular displacement and is computed by negating the vector portion $\mathbf{v}$. The inverse of a quaternion is the conjugate divided by the magnitude. If you use quaternions only to describe angular displacement (as we do in this book), then the conjugate and inverse are equivalent.

- Quaternion multiplication can be used to concatenate multiple rotations into a single angular displacement. In theory, quaternion multiplication can also be used to perform 3D vector rotations, but this is of little practical value.

- Quaternion exponentiation can be used to calculate a multiple of an angular displacement. This always captures the correct end result; however, since quaternions always take the shortest arc, multiple revolutions cannot be represented.

- Quaternions can be interpreted as 4D complex numbers, which creates interesting and elegant parallels between mathematics and geometry.

A lot more has been written about quaternions than we have had the space to discuss here. The technical report by Dam et al [11] is a good mathematical summary. Kuiper's book [41] is written from an aerospace perspective and also does a good job of connecting quaternions and Euler angles. Hanson's modestly titled *Visualizing Quaternions* [30] analyzes quaternions using tools from several different disciplines (Riemannian Geometry, complex numbers, lie algebra, moving frames) and is sprinkled with interesting engineering and mathematical lore; it also discusses how to visualize quaternions. A shorter presentation on visualizing quaternions is given by Hart et al. [31].

8.6   Comparison of Methods

Let's review the most important discoveries from the previous sections. Table 8.1 summarizes the differences among the three representation methods.

Some situations are better suited for one orientation format or another. The following advice should aid you in selecting the best format:

- Euler angles are easiest for humans to work with. Using Euler angles greatly simplifies human interaction when specifying the orientation of objects in the world. This includes direct keyboard entry of an orientation, specifying orientations directly in the code (i.e., positioning the camera for rendering), and examination in the debugger. This advantage should not be underestimated. Certainly don't sacrifice ease of use in the name of "optimization" until you are certain that it will make a difference.

- Matrix form must eventually be used if vector coordinate space transformations are needed. However, this doesn't mean you can't store the orientation in another format and then generate a rotation matrix when you need it. A common strategy is to store the "main copy" of an orientation in Euler angle or quaternion form, but also to maintain a matrix for rotations, recomputing this matrix any time the Euler angles or quaternion change.

- For storage of large numbers of orientations (e.g., animation data), Euler angles, exponential maps, and quaternions offer various trade-offs. In general, the components of Euler angles and exponential maps quantize better than quaternions. It is possible to store a rotation quaternion in only three numbers. Before discarding the fourth component, we check its sign; if it's negative, we negate the quaternion. Then the discarded component can be recovered by assuming the quaternion has unit length.

- Reliable quality interpolation can be accomplished only by using quaternions. Even if you are using a different form, you can always convert to quaternions, perform the interpolation, and then convert

- Rotating points between coordinate spaces (object and upright)

  - **Matrix**: Possible; can often by highly optimized by SIMD instructions.
  - **Euler Angles**: Impossible (must convert to rotation matrix).
  - **Exponential Map**: Impossible (must convert to rotation matrix).
  - **Quaternion**: On a chalkboard, yes. Practically, in a computer, not really. You might as well convert to rotation matrix.

- Concatenation of multiple rotations

  - **Matrix**: Possible; can often be highly optimized by SIMD instructions. Watch out for matrix creep.
  - **Euler Angles**: Impossible.
  - **Exponential Map**: Impossible.
  - **Quaternion**: Possible. Fewer scalar operations than matrix multiplication, but maybe not as easy to take advantage of SIMD instructions. Watch out for error creep.

- Inversion of rotations

  - **Matrix**: Easy and fast, using matrix transpose.
  - **Euler Angles**: Not easy.
  - **Exponential Map**: Easy and fast, using vector negation.
  - **Quaternion**: Easy and fast, using quaternion conjugate.

- Interpolation

  - **Matrix**: Extremely problematic.
  - **Euler Angles**: Possible, but Gimbal lock causes quirkiness.
  - **Exponential Map**: Possible, with some singularities, but not as troublesome as Euler angles.
  - **Quaternion**: Slerp provides smooth interpolation.

- Direct human interpretation

  - **Matrix**: Difficult.
  - **Euler Angles**: Easiest.
  - **Exponential Map**: Very difficult.
  - **Quaternion**: Very difficult.

- Storage efficiency in memory or in a file

  - **Matrix**: Nine numbers.
  - **Euler Angles**: Three numbers that can be easily quantized.
  - **Exponential Map**: Three numbers that can be easily quantized.
  - **Quaternion**: Four numbers that do not quantize well; can be reduced to three by assuming fourth component is always nonnegative and quaternion has unit length.

- Unique representation for a given rotation

  - **Matrix**: Yes.
  - **Euler Angles**: No, due to aliasing.
  - **Exponential Map**: No, due to aliasing, but not as complicated as Euler angles.
  - **Quaternion**: Exactly two distinct representations for any angular displacement, and they are negatives of each other.

- Possible to become invalid

  - **Matrix**: Six degrees of redundancy inherent in orthogonal matrix. Matrix creep can occur.
  - **Euler Angles**: Any three numbers can be interpreted unambiguously.
  - **Exponential Map**: Any three numbers can be interpreted unambiguously.
  - **Quaternion**: Error creep can occur.

Table 8.1. Comparison of matrices, Euler angles, exponential maps, and quaternions

back to the original form. Direct interpolation using exponential maps might be a viable alternative in some cases, as the points of singularity are at very extreme orientations and in practice are often easily avoided.

- For angular velocity or any other situation where "extra spins" need to be represented, use the exponential map or axis-angle.

8.7 Converting between Representations

We have established that different methods of representing orientation are appropriate in different situations, and have also provided some guidelines for choosing the most appropriate method. This section discusses how to convert an angular displacement from one format to another. It is divided into six subsections:

- Section 8.7.1 shows how to convert Euler angles to a matrix.

- Section 8.7.2 shows how to convert a matrix to Euler angles.

- Section 8.7.3 shows how to convert a quaternion to a matrix.

- Section 8.7.4 shows how to convert a matrix to a quaternion.

- Section 8.7.5 shows how to convert Euler angles to a quaternion.

- Section 8.7.6 shows how to convert a quaternion to Euler angles.

For more on converting between representation forms, see the paper by James Diebel [13].

8.7.1 Converting Euler Angles to a Matrix

Euler angles define a sequence of three rotations. Each of these three rotations is a simple rotation about a cardinal axis, so each is easy to convert to matrix form individually. We can compute the matrix that defines the total angular displacement by concatenating the matrices for each individual rotation. This exercise is carried out in numerous books and websites. If you've ever tried to use one of these references, you may have been left wondering, "Exactly what happens if I multiply a vector by this matrix?" The reason for your confusion is because they forgot to mention whether the matrix rotates from object space to upright space or from upright space to object space. In other words, there are actually *two* different matrices, not just one. (Of course, they are transposes of each other, so, in a sense, there really is only one matrix.) This section shows how to compute both.

Some readers[16] might think that we are belaboring the point. Maybe you figured out how to use that rotation matrix from that book or website, and now it's totally obvious to you. But we've seen this be a stumbling block for too many programmers, so we have chosen to dwell on this point. One common example will illustrate the confusion we've seen.

Consider a typical real-time situation with objects moving around. Assume that the orientation of each object is maintained in Euler angle format as a state variable. One of these objects is the camera, and naturally we use the same system of Euler angles to describe the orientation of the camera as we do for any other object. Now, at some point we are going to need to communicate those reference frames to the graphics API. This is where the confusion occurs: the matrix we use to describe the orientation of the objects is *not* the same matrix we use to describe the orientation of the camera! The graphics API needs two types of matrices. (We'll discuss them in more detail in Section 10.3.1.) The *model transform* is a matrix that transforms vectors from object space to world space. The *view transform* transforms vectors from world space to the camera's object space. The rotation portion of the model transform matrix is an object-to-upright matrix, but the rotation portion of the view transform matrix is an upright-to-object matrix. So to speak of *the* Euler rotation matrix leaves out some important practical details.

Now to derive the matrices. We start by deriving the object-to-upright matrix, which rotates points from object space to upright space. We will be using the simple rotation matrices developed in Section 5.1.2, and those were developed using the perspective of active transformation (see Section 3.3.1 if you don't remember the difference between activate and passive transformations). Thus, to visualize the task at hand, imagine an arbitrary point on our object. The object starts out in the "identity" orientation, and the body coordinates for our point, which are the coordinates we know, also happen to be the upright coordinates at this time because the two spaces are aligned. We perform the sequence of Euler rotations on the object, and the point moves in space until, after the third rotation, the object has arrived in the orientation described by the Euler angles. All the while, our upright coordinate space used to measure the coordinates remains fixed. So the final result of these calculations is the upright coordinates for the point in its arbitrary orientation.

There's one last catch. The elementary rotation matrices that we wish to use as building blocks each rotate about a cardinal axis. With Euler angles, the axes of rotation are the body axes, which (after the first rotation) will be arbitrarily oriented. So instead of doing the Euler rotations about the

---

[16]Probably those who already know this stuff. Hey, what are you doing in the kiddie pool—don't you have some backflips off the diving board you should be doing right about now?

body axes, we do fixed-axis rotations, where the rotations are about the upright axes. This means we actually do the rotations in the reverse order: first bank, then pitch, and finally heading. See Section 8.3.2 if you don't remember what fixed-axis rotations are.

In summary, the generation of the object-to-upright rotation matrix is a straightforward concatenation of three simple rotation matrices,

$$\mathbf{M}_{object \rightarrow upright} = \mathbf{BPH},$$

where $\mathbf{B}$, $\mathbf{P}$, and $\mathbf{H}$ are the rotation matrices for bank, pitch, and heading, which rotate about the z-, x-, and y-axes, respectively. We learned how to compute these elementary rotation matrices in Section 5.1.2.

$$\mathbf{B} = \mathbf{R}_z(b) = \begin{bmatrix} \cos b & \sin b & 0 \\ -\sin b & \cos b & 0 \\ 0 & 0 & 1 \end{bmatrix},$$

Elementary rotation matrices for bank, pitch, and heading

$$\mathbf{P} = \mathbf{R}_x(p) = \begin{bmatrix} 1 & 0 & 0 \\ 0 & \cos p & \sin p \\ 0 & -\sin p & \cos p \end{bmatrix},$$

$$\mathbf{H} = \mathbf{R}_y(h) = \begin{bmatrix} \cos h & 0 & -\sin h \\ 0 & 1 & 0 \\ \sin h & 0 & \cos h \end{bmatrix}.$$

Putting it all together (and leaving out the messy math to actually do the matrix multiplications), we have

$$\mathbf{M}_{object \rightarrow upright} = \mathbf{BPH}$$
$$= \begin{bmatrix} ch\,cb + sh\,sp\,sb & sb\,cp & -sh\,cb + ch\,sp\,sb \\ -ch\,sb + sh\,sp\,cb & cb\,cp & sb\,sh + ch\,sp\,cb \\ sh\,cp & -sp & ch\,cp \end{bmatrix}, \quad (8.14)$$

Object-to-upright rotation matrix from Euler angles

where we have introduced the shorthand notation

$$\begin{array}{lll} ch = \cos h, & cp = \cos p, & cb = \cos b, \\ sh = \sin h, & sp = \sin p, & sb = \sin b. \end{array}$$

To rotate vectors from upright space to object space, we will use the inverse of this object-to-upright matrix. We know that since a rotation matrix is orthogonal, the inverse is simply the transpose. However, let's verify this.

To visualize the upright-to-object transform, we imagine the undoing the fixed-axis rotations. We first undo the heading, and then the pitch, and finally the bank. As before, the object (and its points) are moving in space, and we are using upright coordinates to measure everything. The only difference is that we are starting with upright coordinates this time. At the end of these rotations, the objects' body axes are aligned with the upright axes, and the resulting coordinates are the object-space coordinates:

Upright-to-object rotation matrix from Euler angles

$$\mathbf{M}_{upright \to object} = \mathbf{H}^{-1}\mathbf{P}^{-1}\mathbf{B}^{-1} = \mathbf{R}_y(-h)\,\mathbf{R}_x(-p)\,\mathbf{R}_z(-b)$$

$$= \begin{bmatrix} ch\,cb + sh\,sp\,sb & -ch\,sb + sh\,sp\,cb & sh\,cp \\ sb\,cp & cb\,cp & -sp \\ -sh\,cb + ch\,sp\,sb & sb\,sh + ch\,sp\,cb & ch\,cp \end{bmatrix}.$$

$$(8.15)$$

When we compare Equations (8.14) and (8.15), we see that the object-to-upright matrix is indeed the transpose of the upright-to-object matrix, as expected.

Also notice that we can think of the rotation matrices $\mathbf{H}^{-1}$, $\mathbf{P}^{-1}$, and $\mathbf{B}^{-1}$ either as the inverse matrices of their counterparts or as regular rotation matrices using the opposite rotation angles.

8.7.2   Converting a Matrix to Euler angles

Converting an angular displacement from matrix form to Euler angle representation entails several considerations:

- We must know which rotation the matrix performs: either object-to-upright, or upright-to-object. This section develops a technique using the object-to-upright matrix. The process of converting an upright-to-object matrix to Euler angles is very similar, since the two matrices are transposes of each other.

- For any given angular displacement, there are an infinite number of Euler angle representations, due to Euler angle aliasing (see Section 8.3.4). The technique we present here always returns canonical Euler angles, with heading and bank in the range ±180° and pitch in the range ±90°.

- Some matrices may be ill-formed, and so we must be tolerant of floating point precision errors. Some matrices contain transformations other than rotation, such as scale, mirroring, or skew. The technique described here works only on proper rotation matrices, perhaps with the usual floating point imprecision but nothing grossly out of orthogonality. If this technique is used on a non-orthogonal matrix, the results are unpredictable.

With those considerations in mind, we set out to solve for the Euler angles from the rotation matrix (Equation (8.14)) directly. For your convenience, the matrix is expanded below:

$$\begin{bmatrix} \cos h \cos b + \sin h \sin p \sin b & \sin b \cos p & -\sin h \cos b + \cos h \sin p \sin b \\ -\cos h \sin b + \sin h \sin p \cos b & \cos b \cos p & \sin b \sin h + \cos h \sin p \cos b \\ \sin h \cos p & -\sin p & \cos h \cos p \end{bmatrix}.$$

We can solve for p immediately from m_{32}:

$$m_{32} = -\sin p,$$
$$-m_{32} = \sin p,$$
$$\arcsin(-m_{32}) = p.$$

The C standard library function `asin()` returns a value in the range $[-\pi/2, +\pi/2]$ radians, which is $[-90°, +90°]$, exactly the range of values for pitch allowed in the canonical set.

Now that we know p, we also know $\cos p$. Let us first assume that $\cos p \neq 0$. Since $-90° \leq p \leq +90°$, this means that $\cos p > 0$. We can determine $\sin h$ and $\cos h$ by dividing m_{13} and m_{33} by $\cos p$:

$$m_{31} = \sin h \cos p, \qquad\qquad m_{33} = \cos h \cos p,$$
$$m_{31}/\cos p = \sin h, \qquad\qquad m_{33}/\cos p = \cos h. \qquad (8.16)$$

Once we know the sine and cosine of an angle, we can compute the value of the angle with the C standard library function `atan2()`. This function returns an angle in the range $[-\pi, +\pi]$ radians ($[-180°, +180°]$), which is again our desired output range. Knowing just the sine or cosine of angle isn't enough to uniquely identify an angle that is allowed to take on any value in this range, which is why we cannot just use `asin()` or `acos()`.

Substituting the results from Equation (8.16) yields

$$h = \text{atan2}(\sin h, \cos h) = \text{atan2}(m_{31}/\cos p, m_{33}/\cos p).$$

However, we can actually simplify this because `atan2(y,x)` works by taking the arctangent of the quotient y/x, using the signs of the two arguments to place the angle in the correct quadrant. Since $\cos p > 0$, the divisions do not affect the signs of x or y, nor do they change the quotient y/x. Omitting the unnecessary divisions by $\cos p$, heading can be computed more simply by

$$h = \text{atan2}(m_{31}, m_{33}).$$

Bank is computed in a similar manner from m_{12} and m_{22}:

$$m_{12} = \sin b \cos p,$$
$$m_{12}/\cos p = \sin b;$$
$$m_{22} = \cos b \cos p,$$
$$m_{22}/\cos p = \cos b;$$
$$b = \text{atan2}(\sin b, \cos b) = \text{atan2}(m_{12}/\cos p, m_{22}/\cos p)$$
$$= \text{atan2}(m_{12}, m_{22}).$$

Now we've got all three angles. However, if $\cos p = 0$, then we cannot use the above trick since it would result in division by zero. But notice that when $\cos p = 0$, then $p = \pm 90°$, which means we are either looking straight up or straight down. This is the Gimbal lock situation, where heading and bank effectively rotate about the same physical axis (the vertical axis). In other words, the mathematical and geometric singularities occur at the same time. In this case, we will arbitrarily assign all rotation about the vertical axis to heading, and set bank equal to zero. This means that we know values of pitch and bank, and all we have left is to solve for heading. If we take the simplifying assumptions

$$\cos p = 0, \qquad b = 0, \qquad \sin b = 0, \qquad \cos b = 1,$$

and plug these assumptions into Equation (8.14), we get

$$\begin{bmatrix} \cos h \cos b + \sin h \sin p \sin b & \sin b \cos p & -\sin h \cos b + \cos h \sin p \sin b \\ -\cos h \sin b + \sin h \sin p \cos b & \cos b \cos p & \sin b \sin h + \cos h \sin p \cos b \\ \sin h \cos p & -\sin p & \cos h \cos p \end{bmatrix}$$

$$= \begin{bmatrix} \cos h\,(1) + \sin h \sin p\,(0) & (0)(0) & -\sin h\,(1) + \cos h \sin p\,(0) \\ -\cos h\,(0) + \sin h \sin p\,(1) & (1)(0) & (0)\sin h + \cos h \sin p\,(1) \\ \sin h\,(0) & -\sin p & \cos h\,(0) \end{bmatrix}$$

$$= \begin{bmatrix} \cos h & 0 & -\sin h \\ \sin h \sin p & 0 & \cos h \sin p \\ 0 & -\sin p & 0 \end{bmatrix}.$$

Now we can compute h from $-m_{13}$ and m_{11}, which contain the sine and cosine of heading, respectively.

Listing 8.4 is C code that extracts the Euler angles from an object-to-upright rotation matrix, using the technique developed above.

```
// Assume the matrix is stored in these variables:
float m11,m12,m13;
float m21,m22,m23;
float m31,m32,m33;

// We will compute the Euler angle values in radians
// and store them here:
float h,p,b;

// Extract pitch from m32, being careful for domain errors with
// asin(). We could have values slightly out of range due to
// floating point arithmetic.
float sp = −m32;
if (sp <= −1.0f) {
    p = −1.570796f; // −pi/2
} else if (sp >= 1.0f) {
    p = 1.570796f; // pi/2
} else {
    p = asin(sp);
}

// Check for the Gimbal lock case, giving a slight tolerance
// for numerical imprecision
if (fabs(sp) > 0.9999f) {

    // We are looking straight up or down.
    // Slam bank to zero and just set heading
    b = 0.0f;
    h = atan2(−m13, m11);

} else {

    // Compute heading from m13 and m33
    h = atan2(m31, m33);

    // Compute bank from m21 and m22
    b = atan2(m12, m22);
}
```

Listing 8.4
Extracting Euler angles from an object-to-upright matrix

8.7.3 Converting a Quaternion to a Matrix

We have a few options for converting a quaternion to a rotation matrix. The more common way is to expand the quaternion multiplication $\mathbf{qvq}^{-1}$. This produces the correct matrix, but we are left with no real confidence why the matrix is correct. (We are, however, left with some experience manipulating quaternions; see Exercise 10.) We take a different option and stick purely to the geometric interpretation of the components of the quaternion. Since a quaternion is essentially an encoded version of an axis-angle rotation, we attempt to construct the matrix from Section 5.1.3, which rotates about an arbitrary axis:

$$\begin{bmatrix} n_x{}^2 \left(1 - \cos\theta\right) + \cos\theta & n_x n_y \left(1 - \cos\theta\right) + n_z \sin\theta & n_x n_z \left(1 - \cos\theta\right) - n_y \sin\theta \\ n_x n_y \left(1 - \cos\theta\right) - n_z \sin\theta & n_y{}^2 \left(1 - \cos\theta\right) + \cos\theta & n_y n_z \left(1 - \cos\theta\right) + n_x \sin\theta \\ n_x n_z \left(1 - \cos\theta\right) + n_y \sin\theta & n_y n_z \left(1 - \cos\theta\right) - n_x \sin\theta & n_z{}^2 \left(1 - \cos\theta\right) + \cos\theta \end{bmatrix}.$$

Unfortunately, this matrix is in terms of $\hat{\mathbf{n}}$ and θ, but the components of a quaternion are

$$w = \cos(\theta/2),$$
$$x = n_x \sin(\theta/2),$$
$$y = n_y \sin(\theta/2),$$
$$z = n_z \sin(\theta/2).$$

Let's see if we can't manipulate the matrix into a form that we can substitute in w, x, y, and z. We need to do this for all nine elements of the matrix. Luckily, the matrix has a great deal of structure, and there are really only two major cases to handle: the diagonal elements, and the nondiagonal elements.

This is a tricky derivation, and it is not necessary to understand how the matrix is derived in order to *use* the matrix. If you're not interested in the math, skip to Equation (8.20).

Let's start with the diagonal elements of the matrix. We work through m_{11} here; m_{22} and m_{33} can be solved similarly:

$$m_{11} = n_x{}^2 \left(1 - \cos\theta\right) + \cos\theta.$$

We first perform some manipulations that may seem to be a detour. The purpose of these steps will become apparent in just a moment:

$$\begin{aligned}
m_{11} &= n_x{}^2 \left(1 - \cos\theta\right) + \cos\theta \\
&= n_x{}^2 - n_x{}^2 \cos\theta + \cos\theta \\
&= 1 - 1 + n_x{}^2 - n_x{}^2 \cos\theta + \cos\theta \\
&= 1 - \left(1 - n_x{}^2 + n_x{}^2 \cos\theta - \cos\theta\right) \\
&= 1 - \left(1 - \cos\theta - n_x{}^2 + n_x{}^2 \cos\theta\right) \\
&= 1 - \left(1 - n_x{}^2\right)\left(1 - \cos\theta\right).
\end{aligned}$$

Now we need to get rid of the $\cos\theta$ term; and we'd like to replace it with something that contains $\cos\theta/2$ or $\sin\theta/2$, since the components of a quaternion contain those terms. As we have done before, let $\alpha = \theta/2$. We write one of the double-angle formulas for cosine from Section 1.4.5 in terms of α, and then substitute in θ:

$$\cos 2\alpha = 1 - 2\sin^2 \alpha,$$
$$\cos\theta = 1 - 2\sin^2(\theta/2). \qquad (8.17)$$

Substituting for $\cos\theta$ in Equation (8.17), we have

$$\begin{aligned}
m_{11} &= 1 - (1 - n_x{}^2)(1 - \cos\theta) \\
&= 1 - (1 - n_x{}^2)\left(1 - \left(1 - 2\sin^2(\theta/2)\right)\right) \\
&= 1 - (1 - n_x{}^2)\left(2\sin^2(\theta/2)\right).
\end{aligned}$$

Now we use the fact that since $\hat{\mathbf{n}}$ is a unit vector, $n_x{}^2 + n_y{}^2 + n_z{}^2 = 1$, and therefore $1 - n_x{}^2 = n_y{}^2 + n_z{}^2$:

$$\begin{aligned}
m_{11} &= 1 - (1 - n_x{}^2)\left(2\sin^2(\theta/2)\right) \\
&= 1 - (n_y{}^2 + n_z{}^2)\left(2\sin^2(\theta/2)\right) \\
&= 1 - 2n_y{}^2\sin^2(\theta/2) - 2n_z{}^2\sin^2(\theta/2) \\
&= 1 - 2y^2 - 2z^2.
\end{aligned}$$

Elements m_{22} and m_{33} are derived in a similar fashion. The results are presented at the end of this section when we give the complete matrix in Equation (8.20).

Now let's look at the nondiagonal elements of the matrix; they are easier than the diagonal elements. We'll use m_{12} as an example:

$$m_{12} = n_x n_y \left(1 - \cos\theta\right) + n_z \sin\theta. \qquad (8.18)$$

We use the reverse of the double-angle formula for sine (see Section 1.4.5):

$$\sin 2\alpha = 2\sin\alpha\cos\alpha,$$
$$\sin\theta = 2\sin(\theta/2)\cos(\theta/2). \qquad (8.19)$$

Now we substitute Equations (8.17) and (8.19) into Equation (8.18) and simplify:

$$\begin{aligned}
m_{12} &= n_x n_y \left(1 - \cos\theta\right) + n_z \sin\theta \\
&= n_x n_y \left(1 - \left(1 - 2\sin^2(\theta/2)\right)\right) + n_z \left(2\sin(\theta/2)\cos(\theta/2)\right) \\
&= n_x n_y \left(2\sin^2(\theta/2)\right) + 2n_z \sin(\theta/2)\cos(\theta/2) \\
&= 2\left(n_x \sin(\theta/2)\right)\left(n_y \sin(\theta/2)\right) + 2\cos(\theta/2)\left(n_z \sin(\theta/2)\right) \\
&= 2xy + 2wz.
\end{aligned}$$

The other nondiagonal elements are derived in a similar fashion.

Finally, we present the complete rotation matrix constructed from a quaternion:

Converting a quaternion to a 3 × 3 rotation matrix

$$\begin{bmatrix} 1 - 2y^2 - 2z^2 & 2xy + 2wz & 2xz - 2wy \\ 2xy - 2wz & 1 - 2x^2 - 2z^2 & 2yz + 2wx \\ 2xz + 2wy & 2yz - 2wx & 1 - 2x^2 - 2y^2 \end{bmatrix}. \tag{8.20}$$

Other variations can be found in other sources.[17] For example $m_{11} = -1 + 2w^2 + 2z^2$ also works, since $w^2 + x^2 + y^2 + z^2 = 1$. In deference to Shoemake, who brought quaternions to the attention of the computer graphics community, we've tailored our derivation to produce the version from his early and authoritative source [62].

8.7.4 Converting a Matrix to a Quaternion

To extract a quaternion from the corresponding rotation matrix, we reverse engineer Equation (8.20). Examining the sum of the diagonal elements (known as the *trace* of the matrix) we get

$$\begin{aligned} \text{tr}(\mathbf{M}) &= m_{11} + m_{22} + m_{33} \\ &= (1 - 2y^2 - 2z^2) + (1 - 2x^2 - 2z^2) + (1 - 2x^2 - 2y^2) \\ &= 3 - 4(x^2 + y^2 + z^2) \\ &= 3 - 4(1 - w^2) \\ &= 4w^2 - 1, \end{aligned}$$

and therefore we can compute w by

$$w = \frac{\sqrt{m_{11} + m_{22} + m_{33} + 1}}{2}.$$

The other three elements can be computed in a similar way, by negating two of the three elements in the trace:

$$\begin{aligned} m_{11} - m_{22} - m_{33} &= (1 - 2y^2 - 2z^2) - (1 - 2x^2 - 2z^2) - (1 - 2x^2 - 2y^2) \\ &= 4x^2 - 1, \end{aligned} \tag{8.21}$$

$$\begin{aligned} -m_{11} + m_{22} - m_{33} &= -(1 - 2y^2 - 2z^2) + (1 - 2x^2 - 2z^2) - (1 - 2x^2 - 2y^2) \\ &= 4y^2 - 1, \end{aligned} \tag{8.22}$$

---

[17]Including the first edition of this book [16].

$$-m_{11} - m_{22} + m_{33} = -(1-2y^2-2z^2) - (1-2x^2-2z^2) + (1-2x^2-2y^2)$$

$$= 4z^2 - 1, \tag{8.23}$$

$$x = \frac{\sqrt{m_{11} - m_{22} - m_{33} + 1}}{2}, \tag{8.24}$$

$$y = \frac{\sqrt{-m_{11} + m_{22} - m_{33} + 1}}{2}, \tag{8.25}$$

$$z = \frac{\sqrt{-m_{11} - m_{22} + m_{33} + 1}}{2}. \tag{8.26}$$

Unfortunately, we cannot use this trick for all four components, since the square root will always yield positive results. (More accurately, we have no basis for choosing the positive or negative root.) However, since $\mathbf{q}$ and $-\mathbf{q}$ represent the same orientation, we can arbitrarily choose to use the nonnegative root for one of the four components and still always return a correct quaternion. We just can't use the above technique for all four values of the quaternion.

Another line of attack is to examine the sum and difference of diagonally opposite matrix elements:

$$m_{12} + m_{21} = (2xy + 2wz) + (2xy - 2wz) = 4xy, \tag{8.27}$$

$$m_{12} - m_{21} = (2xy + 2wz) - (2xy - 2wz) = 4wz, \tag{8.28}$$

$$m_{31} + m_{13} = (2xz + 2wy) + (2xz - 2wy) = 4xz, \tag{8.29}$$

$$m_{31} - m_{13} = (2xz + 2wy) - (2xz - 2wy) = 4wy, \tag{8.30}$$

$$m_{23} + m_{32} = (2yz + 2wx) + (2yz - 2wx) = 4yz, \tag{8.31}$$

$$m_{23} - m_{32} = (2yz + 2wx) - (2yz - 2wx) = 4wx. \tag{8.32}$$

Armed with these formulas, we develop a two-step strategy. We first solve for one of the components from the trace, using one of Equations (8.21)–(8.26). Then we plug that known value into Equations (8.27)–(8.32) to solve for the other three. Essentially, this strategy boils down to selecting a row from Table 8.2 and then solving the equations in that row from left to right.

The only question is, "Which row should we use?" In other words, which component should we solve for first? The simplest strategy would be to just pick one arbitrarily and always use the same procedure, but this is no good. Let's say we choose to always use the top row, meaning we solve for w from the trace, and then for x, y, and z with the equations on the right side of the arrow. But if $w = 0$, the divisions to follow will be undefined. Even if $w > 0$, a small w will produce numeric instability. Shoemake [62] suggests the strategy of first determining which of w, x, y, and z has the largest absolute value (which we can do without performing any square roots), computing that component using the diagonal of the matrix, and then using it to compute the other three according to Table 8.2.

$$w = \frac{\sqrt{m_{11}+m_{22}+m_{33}+1}}{2} \implies x = \frac{m_{23}-m_{32}}{4w} \quad y = \frac{m_{31}-m_{13}}{4w} \quad z = \frac{m_{12}-m_{21}}{4w}$$

$$x = \frac{\sqrt{m_{11}-m_{22}-m_{33}+1}}{2} \implies w = \frac{m_{23}-m_{32}}{4x} \quad y = \frac{m_{12}+m_{21}}{4x} \quad z = \frac{m_{31}+m_{13}}{4x}$$

$$y = \frac{\sqrt{-m_{11}+m_{22}-m_{33}+1}}{2} \implies w = \frac{m_{31}-m_{13}}{4y} \quad x = \frac{m_{12}+m_{21}}{4y} \quad z = \frac{m_{23}+m_{32}}{4y}$$

$$z = \frac{\sqrt{-m_{11}-m_{22}+m_{33}+1}}{2} \implies w = \frac{m_{12}-m_{21}}{4z} \quad x = \frac{m_{31}+m_{13}}{4z} \quad y = \frac{m_{23}+m_{32}}{4z}$$

Table 8.2. Extracting a quaternion from a rotation matrix

Listing 8.5 implements this strategy in a straightforward manner.

```
// Input matrix:
float m11,m12,m13;
float m21,m22,m23;
float m31,m32,m33;

// Output quaternion
float w,x,y,z;

// Determine which of w, x, y, or z has the largest absolute value
float fourWSquaredMinus1 = m11 + m22 + m33;
float fourXSquaredMinus1 = m11 − m22 − m33;
float fourYSquaredMinus1 = m22 − m11 − m33;
float fourZSquaredMinus1 = m33 − m11 − m22;

int biggestIndex = 0;
float fourBiggestSquaredMinus1 = fourWSquaredMinus1;
if (fourXSquaredMinus1 > fourBiggestSquaredMinus1) {
    fourBiggestSquaredMinus1 = fourXSquaredMinus1;
    biggestIndex = 1;
}
if (fourYSquaredMinus1 > fourBiggestSquaredMinus1) {
    fourBiggestSquaredMinus1 = fourYSquaredMinus1;
    biggestIndex = 2;
}
if (fourZSquaredMinus1 > fourBiggestSquaredMinus1) {
    fourBiggestSquaredMinus1 = fourZSquaredMinus1;
    biggestIndex = 3;
}

// Perform square root and division
float biggestVal = sqrt(fourBiggestSquaredMinus1 + 1.0f) * 0.5f;
float mult = 0.25f / biggestVal;

// Apply table to compute quaternion values
switch (biggestIndex) {
    case 0:
        w = biggestVal;
        x = (m23 − m32) * mult;
```

```
        y = (m31 - m13) * mult;
        z = (m12 - m21) * mult;
        break;

    case 1:
        x = biggestVal;
        w = (m23 - m32) * mult;
        y = (m12 + m21) * mult;
        z = (m31 + m13) * mult;
        break;

    case 2:
        y = biggestVal;
        w = (m31 - m13) * mult;
        x = (m12 + m21) * mult;
        z = (m23 + m32) * mult;
        break;

    case 3:
        z = biggestVal;
        w = (m12 - m21) * mult;
        x = (m31 + m13) * mult;
        y = (m23 + m32) * mult;
        break;
}
```

Listing 8.5
Converting a rotation matrix to a quaternion

8.7.5   Converting Euler Angles to a Quaternion

To convert an angular displacement from Euler angle form to quaternion, we use a technique similar to the one used in Section 8.7.1 to generate a rotation matrix from Euler angles. We first convert the three rotations to quaternion format individually, which is a trivial operation. Then we concatenate these three quaternions in the proper order. Just as with matrices, there are two cases to consider: one when we wish to generate an object-to-upright quaternion, and a second when we want the upright-to-object quaternion. Since the two are conjugates of each other, we walk through the derivation only for the object-to-upright quaternion.

As we did in Section 8.7.1, we assign the Euler angles to the variables h, p, and b. Let $\mathbf{h}$, $\mathbf{p}$, and $\mathbf{b}$ be quaternions that perform the rotations about the y, x, and z-axes, respectively:

$$\mathbf{h} = \left[\begin{array}{c} \cos(h/2) \\ \begin{pmatrix} 0 \\ \sin(h/2) \\ 0 \end{pmatrix} \end{array} \right], \quad \mathbf{p} = \left[\begin{array}{c} \cos(p/2) \\ \begin{pmatrix} \sin(p/2) \\ 0 \\ 0 \end{pmatrix} \end{array} \right], \quad \mathbf{b} = \left[\begin{array}{c} \cos(b/2) \\ \begin{pmatrix} 0 \\ 0 \\ \sin(b/2) \end{pmatrix} \end{array} \right].$$

Now we concatenate these in the correct order. We have two sources of "backwardness," which cancel each other out. We are using fixed-axis rotations, so the order of rotations actually is bank, pitch, and lastly heading.

However, quaternion multiplication performs the rotations from right-to-left (see Section 8.7.1 if you don't understand the former source of backwardness, and Section 8.5.7 for the latter):

Computing the object-to-upright quaternion from a set of Euler angles

$$\mathbf{q}_{object \rightarrow upright}(h, p, b) = \mathbf{hpb}$$

$$= \left[\begin{pmatrix} \cos(h/2) \\ 0 \\ \sin(h/2) \\ 0 \end{pmatrix} \right] \left[\begin{pmatrix} \cos(p/2) \\ \sin(p/2) \\ 0 \\ 0 \end{pmatrix} \right] \left[\begin{pmatrix} \cos(b/2) \\ 0 \\ 0 \\ \sin(b/2) \end{pmatrix} \right]$$

$$= \left[\begin{pmatrix} \cos(h/2)\cos(p/2) \\ \cos(h/2)\sin(p/2) \\ \sin(h/2)\cos(p/2) \\ -\sin(h/2)\sin(p/2) \end{pmatrix} \right] \left[\begin{pmatrix} \cos(b/2) \\ 0 \\ 0 \\ \sin(b/2) \end{pmatrix} \right]$$

$$= \left[\begin{pmatrix} \cos(h/2)\cos(p/2)\cos(b/2) + \sin(h/2)\sin(p/2)\sin(b/2) \\ \cos(h/2)\sin(p/2)\cos(b/2) + \sin(h/2)\cos(p/2)\sin(b/2) \\ \sin(h/2)\cos(p/2)\cos(b/2) - \cos(h/2)\sin(p/2)\sin(b/2) \\ \cos(h/2)\cos(p/2)\sin(b/2) - \sin(h/2)\sin(p/2)\cos(b/2) \end{pmatrix} \right].$$

The upright-to-object quaternion is simply the conjugate:

The upright-to-object quaternion from a set of Euler angles

$$\mathbf{q}_{upright \rightarrow object}(h, p, b) = \mathbf{q}_{object \rightarrow upright}(h, p, b)^*$$

$$= \left[\begin{pmatrix} \cos(h/2)\cos(p/2)\cos(b/2) + \sin(h/2)\sin(p/2)\sin(b/2) \\ -\cos(h/2)\sin(p/2)\cos(b/2) - \sin(h/2)\cos(p/2)\sin(b/2) \\ \cos(h/2)\sin(p/2)\sin(b/2) - \sin(h/2)\cos(p/2)\cos(b/2) \\ \sin(h/2)\sin(p/2)\cos(b/2) - \cos(h/2)\cos(p/2)\sin(b/2) \end{pmatrix} \right]. \quad (8.33)$$

8.7.6   Converting a Quaternion to Euler Angles

To extract Euler angles from a quaternion, we could solve for the Euler angles from Equation (8.33) directly. However, let's see if we can take advantage of our work in previous sections and arrive at the answer without going through so much effort. We've already come up with a technique to extract Euler angles from a matrix in Section 8.7.2. And we showed how to convert a quaternion to a matrix. So let's just take our technique for converting a matrix to Euler angles and plug in our results from Equation (8.20).

Our method from Section 8.7.2 for extracting Euler angles from an object-to-upright matrix is summarized below.

$$p = \arcsin(-m_{32}) \tag{8.34}$$

$$h = \begin{cases} \text{atan2}(m_{31}, m_{33}) & \text{if } \cos p \neq 0, \\ \text{atan2}(-m_{13}, m_{11}) & \text{otherwise.} \end{cases} \tag{8.35}$$

$$b = \begin{cases} \text{atan2}(m_{12}, m_{22}) & \text{if } \cos p \neq 0, \\ 0 & \text{otherwise.} \end{cases} \tag{8.36}$$

For convenience, we repeat the needed matrix elements from Equation (8.20):

$$m_{11} = 1 - 2y^2 - 2z^2, \quad m_{12} = 2xy + 2wz, \quad m_{13} = 2xz - 2wy, \tag{8.37}$$

$$m_{22} = 1 - 2x^2 - 2z^2, \tag{8.38}$$

$$m_{31} = 2xz + 2wy, \quad m_{32} = 2yz - 2wx, \quad m_{33} = 1 - 2x^2 - 2y^2. \tag{8.39}$$

Substituting Equations (8.37)–(8.39) into Equations (8.34)–(8.36) and simplifying, we have

$$\begin{aligned} p &= \arcsin(-m_{32}) \\ &= \arcsin\left(-2(yz - wx)\right) \end{aligned}$$

$$h = \begin{cases} \begin{aligned} &\text{atan2}(m_{31}, m_{33}) \\ &= \text{atan2}(2xz + 2wy, 1 - 2x^2 - 2y^2) \\ &= \text{atan2}(xz + wy, 1/2 - x^2 - y^2) \end{aligned} & \text{if } \cos p \neq 0, \\[2em] \begin{aligned} &\text{atan2}(-m_{13}, m_{11}) \\ &= \text{atan2}(-2xz + 2wy, 1 - 2y^2 - 2z^2) \\ &= \text{atan2}(-xz + wy, 1/2 - y^2 - z^2) \end{aligned} & \text{otherwise.} \end{cases}$$

$$b = \begin{cases} \begin{aligned} &\text{atan2}(m_{12}, m_{22}) \\ &= \text{atan2}(2xy + 2wz, 1 - 2x^2 - 2z^2) \\ &= \text{atan2}(xy + wz, 1/2 - x^2 - z^2) \end{aligned} & \text{if } \cos p \neq 0, \\[2em] 0 & \text{otherwise.} \end{cases}$$

We can translate this directly into code, as shown in Listing 8.6, which converts an object-to-upright quaternion into Euler angles.

```
// Input quaternion
float w,x,y,z;

// Output Euler angles (radians)
float h,p,b;

// Extract sin(pitch)
float sp = -2.0f * (y*z - w*x);

// Check for Gimbal lock, giving slight tolerance
// for numerical imprecision
if (fabs(sp) > 0.9999f) {

    // Looking straight up or down
    p = 1.570796f * sp; // pi/2

    // Compute heading, slam bank to zero
    h = atan2(-x*z + w*y, 0.5f - y*y - z*z);
    b = 0.0f;

} else {

    // Compute angles
    p = asin(sp);
    h = atan2(x*z + w*y, 0.5f - x*x - y*y);
    b = atan2(x*y + w*z, 0.5f - x*x - z*z);
}
```

Listing 8.6
Converting an object-to-upright quaternion to Euler angles

To convert an upright-to-object quaternion to Euler angle format, we use nearly identical code, only with the x, y, and z values negated, since we assume the upright-to-object quaternion is the conjugate of the object-to-upright quaternion.

```
// Extract sin(pitch)
float sp = -2.0f * (y*z + w*x);

// Check for Gimbal lock, giving slight tolerance
// for numerical imprecision
if (fabs(sp) > 0.9999f) {

    // Looking straight up or down
    p = 1.570796f * sp; // pi/2

    // Compute heading, slam bank to zero
    h = atan2(-x*z - w*y, 0.5f - y*y - z*z);
    b = 0.0f;

} else {

    // Compute angles
    p = asin(sp);
    h = atan2(x*z - w*y, 0.5f - x*x - y*y);
    b = atan2(x*y - w*z, 0.5f - x*x - z*z);
}
```

Listing 8.7
Converting an upright-to-object quaternion to Euler angles

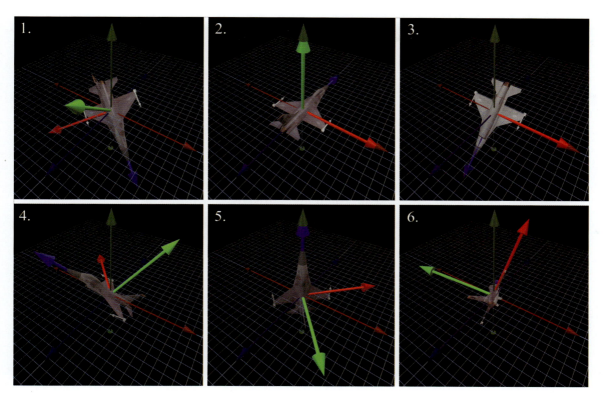

Figure 8.13
Sample orientations used for Exercises 1, 2, 4, and 5.

8.8   Exercises

(Answers on page 772.)

1. Match each of the rotation matrices below with the corresponding orientation from Figure 8.13. These matrices transform row vectors on the left from object space to upright space.

(a) $\begin{bmatrix} 0.707 & 0.000 & 0.707 \\ 0.707 & 0.000 & -0.707 \\ 0.000 & 1.000 & 0.000 \end{bmatrix}$

(b) $\begin{bmatrix} 1.000 & 0.000 & 0.000 \\ 0.000 & -0.707 & 0.707 \\ 0.000 & -0.707 & -0.707 \end{bmatrix}$

$$
\text{(c)} \begin{bmatrix} 0.061 & 0.814 & 0.578 \\ -0.900 & 0.296 & -0.322 \\ -0.433 & -0.500 & 0.750 \end{bmatrix}
$$

$$
\text{(d)} \begin{bmatrix} -0.713 & -0.450 & -0.538 \\ 0.091 & 0.702 & -0.706 \\ 0.696 & -0.552 & -0.460 \end{bmatrix}
$$

$$
\text{(e)} \begin{bmatrix} 1.000 & 0.000 & 0.000 \\ 0.000 & 1.000 & 0.000 \\ 0.000 & 0.000 & 1.000 \end{bmatrix}
$$

$$
\text{(f)} \begin{bmatrix} -0.707 & 0.000 & 0.707 \\ 0.500 & 0.707 & 0.500 \\ -0.500 & 0.707 & -0.500 \end{bmatrix}
$$

2. Match each of the following Euler angle triples with the corresponding orientation from Figure 8.13, and determine whether the orientation is in the canonical set of Euler angles. If not, say why not.

   (a) $h = 180°$, $p = 45°$, $b = 180°$

   (b) $h = -135°$, $p = -45°$, $b = 0°$

   (c) $h = 0°$, $p = -90°$, $b = -45°$

   (d) $h = 123°$, $p = 33.5°$, $b = -32.7°$

   (e) $h = 0°$, $p = 0°$, $b = 0°$

   (f) $h = 0°$, $p = 135°$, $b = 0°$

   (g) $h = -45°$, $p = -90°$, $b = 0°$

   (h) $h = 180°$, $p = -180°$, $b = 180°$

   (i) $h = -30°$, $p = 30°$, $b = 70°$

3. (a) Construct a quaternion to rotate $30°$ about the x-axis.

   (b) What is the magnitude of this quaternion?

   (c) What is its conjugate?

   (d) Assume the quaternion is used to rotate points from object space to upright space of some object. What is the orientation, in Euler angles, of this object?

4. Match each of the following quaternions with the corresponding orientation from Figure 8.13. These quaternions transform vectors from object space to upright space. (We told you that quaternions are harder for humans to use! Try converting these to matrix or Euler angle form and take advantage of your previous work.)

   (a) $\begin{bmatrix} -1.000 & (0.000 & 0.000 & 0.000) \end{bmatrix}$

   (b) $\begin{bmatrix} 0.653 & (-0.653 & -0.271 & -0.271) \end{bmatrix}$

(c) $\begin{bmatrix} 0.364 & (-0.106 & 0.848 & -0.372) \end{bmatrix}$

(d) $\begin{bmatrix} 0.383 & (0.924 & 0.000 & 0.000) \end{bmatrix}$

(e) $\begin{bmatrix} 1.000 & (0.000 & 0.000 & 0.000) \end{bmatrix}$

(f) $\begin{bmatrix} -0.364 & (0.106 & -0.848 & 0.372) \end{bmatrix}$

(g) $\begin{bmatrix} 0.354 & (-0.146 & -0.853 & -0.354) \end{bmatrix}$

(h) $\begin{bmatrix} 0.726 & (0.061 & -0.348 & 0.590) \end{bmatrix}$

(i) $\begin{bmatrix} -0.383 & (-0.924 & 0.000 & 0.000) \end{bmatrix}$

5. Match each of the following axis-angle orientations with the corresponding orientation from Figure 8.13. (Hint: this can be quite difficult to visualize. Try converting the axis and angle to quaternion and then using the results of your previous results. Then see if you can visualize it.)

   (a) $98.4°, [-0.863, -0.357, -0.357]$

   (b) $0°, [0.707, -0.707, 0.000]$

   (c) $87.0°, [0.089, -0.506, 0.857]$

   (d) $137°, [-0.114, 0.910, 0.399]$

   (e) $135°, [1.000, 0.000, 0.000]$

   (f) $261.6°, [0.863, 0.357, 0.357]$

   (g) $139°, [-0.156, -0.912, -0.378]$

   (h) $7200°, [0.000, -1.000, 0.000]$

   (i) $-135°, [-1.000, 0.000, 0.000]$

6. Derive the quaternion multiplication formula, by interpreting quaternions as 4D complex numbers and applying the rules from Equation (8.10).

7. Compute a quaternion that performs twice the rotation of the quaternion $\begin{bmatrix} 0.965 & (0.149 & -0.149 & 0.149) \end{bmatrix}$.

8. Consider the quaternions:

$$\mathbf{a} = \begin{bmatrix} 0.233 & (0.060 & -0.257 & -0.935) \end{bmatrix}$$
$$\mathbf{b} = \begin{bmatrix} -0.752 & (0.286 & 0.374 & 0.459) \end{bmatrix}$$

   (a) Compute the dot product $\mathbf{a} \cdot \mathbf{b}$.

   (b) Compute the quaternion product $\mathbf{ab}$.

   (c) Compute the difference from $\mathbf{a}$ to $\mathbf{b}$.

9. Prove the statement made in Section 8.5.7: the magnitude of the product of two quaternions is the product of the magnitudes.

10. Expand the multiplication $\mathbf{q}\mathbf{v}\mathbf{q}^{-1}$, and verify that the matrix Equation (8.20) is correct.

11. Make a survey of some game engines and open source code and figure out the conventions regarding row/column vectors, rotation matrices, and Euler angles.

I wear a necklace, cause I wanna know when I'm upside down.
— Mitch Hedberg (1968–2005)

Chapter 9

Geometric Primitives

Triangle man, triangle man.
Triangle man hates particle man.
They have a fight, triangle wins.
Triangle man.

— *Particle Man* (1990) by They Might Be Giants

This chapter is about geometric primitives in general and in specific.

- Section 9.1 discusses some general principles related to representing geometric primitives.

- Sections 9.2–9.7 cover a number of specific important geometric primitives, including methods for representing those primitives and some classic properties and operations. Along the way, we'll present a few C++ snippets.

9.1 Representation Techniques

Let's begin with a brief overview of the major strategies for describing geometric shapes. For any given primitive, one or more of these techniques may be applicable, and different techniques are useful in different situations.

We can describe an object in *implicit form* by defining a Boolean function $f(x, y, z)$ that is true for all points of the primitive and false for all other points. For example, the equation

$$x^2 + y^2 + z^2 = 1$$

Unit sphere in implicit form

is true for all points on the surface of a unit sphere centered at the origin. The *conic sections* are classic examples of implicit representations of geometric shapes that you may already know. A conic section is a 2D shape formed by the intersection of a cone with a plane. The conic sections are the circle, ellipse, parabola, and hyperbola, all of which can be described in the standard implicit form $Ax^2 + Bxy + Cy^2 + D = 0$.

Metaballs [7] is an implicit method for representing fluid and organic shapes. The volume is defined by a collection of fuzzy "balls." Each ball defines a three-dimensional scalar density function based on the distance from the center of the ball, with zero distance being the maximal value and greater distances having lower values. We can define an aggregate density function for any arbitrary point in space by taking the sum of the density of all the balls at that point. The twist with metaballs is that the volume of the fluid or organic object is defined to be the region where the density exceeds some *nonzero* threshold. In other words, the balls have a "fuzzy" region around them that extends outside of the volume when the ball is in isolation. When two or more balls come together, the fuzzy regions interfere constructively, causing a graceful "bond" of solid volume to materialize in the region in between the balls, where no such solid would exist if either ball were in isolation. The *marching cubes* algorithm [43] is a classic technique for converting an arbitrary implicit form into a surface description (such as a polygon mesh).

Another general strategy for describing shapes is the *parametric form*. Once again, the primitive is defined by a function, but instead of the spatial coordinates being the input to the function, they are the output. Let's begin with a simple 2D example. We define the following two functions of t:

Unit circle in parametric form

$$x(t) = \cos 2\pi t, \qquad\qquad y(t) = \sin 2\pi t.$$

The argument t is known as the *parameter* and is independent of the coordinate system used. As t varies from 0 to 1, the point $(x(t), y(t))$ traces out

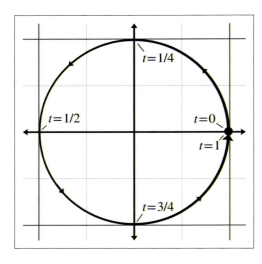

Figure 9.1
Parametric circle

the outline of the shape we are describing—in this example, a unit circle centered at the origin (see Figure 9.1).

It is often convenient to normalize the parameter to be in the range $[0, 1]$, although we may allow t to assume any range of values we wish. Another common choice is $[0, l]$, where l is some measure of the length of the primitive.

When our functions are in terms of one parameter, we say that the functions are *univariate*. Univariate functions trace out a 1D shape: a curve. (Chapter 13 presents more about parametric curves.) It's also possible to use more than one parameter. A *bivariate function* accepts two parameters, usually assigned to the variables s and t. Bivariate functions trace out a surface rather than a line.

We have dubbed the final method for representing primitives, for lack of a better term, *straightforward forms*. By this we mean all the ad-hoc methods that capture the most important and obvious information directly. For example, to describe a line segment, we could name the two endpoints. A sphere is described most simply by giving its center and radius. The straightforward forms are the easiest for humans to work with directly.

Regardless of the method of representation, each geometric primitive has an inherent number of degrees of freedom. This is the minimum number of "pieces of information" that are required to describe the entity unambiguously. It is interesting to notice that for the same geometric primitive, some representation forms use more numbers than others. However, we find that any "extra" numbers are always due to a redundancy in the parameterization of the primitive, which could be eliminated by assuming the appropriate constraint, such as a vector having unit length. For example, a circle in the plane has three degrees of freedom: two for the position of the center (x_c, y_c) and one for the radius r. In parametric form these variables appear directly:

$$x(t) = x_c + r \cos 2\pi t, \qquad y(t) = y_c + r \sin 2\pi t.$$

Parametric circle with arbitrary center and radius

However, the general conic section equation (the implicit form) is $Ax^2 + Bxy + Cy^2 + D = 0$, which has four coefficients. A general conic section can be recognized as a circle if it can be manipulated into the form

$$(x - x_c)^2 + (y - y_c)^2 = r^2.$$

Implicit circle with arbitrary center and radius

9.2 Lines and Rays

Now for some specific types of primitives. We begin with what is perhaps the most basic and important one of all: the linear segment. Let's meet the three basic types of linear segments, and also clarify some terminology. In classical geometry, the following definitions are used:

- A *line* extends infinitely in two directions.

- A *line segment* is a finite portion of a line that has two endpoints.

- A *ray* is "half" of a line that has an origin and extends infinitely in one direction.

In computer science and computational geometry, there are variations on these definitions. This book uses the classical definitions for line and line segment. However, the definition of "ray" is altered slightly:

- A *ray* is a directed line segment.

So to us, a ray will have an origin and an endpoint. Thus a ray defines a position, a *finite* length, and (unless the ray has zero length) a direction. Since a ray is just a line segment where we have differentiated between the two ends, and a ray also can be used to define an infinite line, rays are of fundamental importance in computational geometry and graphics and will be the focus of this section. A ray can be imagined as the result of sweeping a point through space over time; rays are everywhere in video games. An obvious example is the rendering strategy known as *raytracing*, which uses eponymous rays representing the paths of photons. For AI, we trace "line of sight" rays through the environment to detect whether an enemy can see the player. Many user interface tools use raytracing to determine what object is under the mouse cursor. Bullets and lasers are always whizzing through the air in video games, and we need rays to determine what they hit. Figure 9.2 compares the line, line segment, and ray.

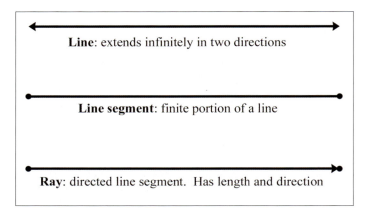

Line: extends infinitely in two directions

Line segment: finite portion of a line

Ray: directed line segment. Has length and direction

Figure 9.2
Line, line segment, and ray

The remainder of this section surveys different methods for representing lines and rays in 2D and 3D. Section 9.2.1 discusses some simple ways to represent a ray, including the all-important parametric form. Section 9.2.2 discusses some special ways to define an infinite line in 2D. Section 9.2.3 gives some examples of converting from one representation to another.

9.2.1 Rays

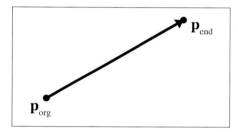

Figure 9.3
Defining a ray using the starting and ending points

The most obvious way to define a ray (the "straightforward form") is by two points, the ray origin and the ray endpoint, which we will denote as $\mathbf{p}_{org}$ and $\mathbf{p}_{end}$ (see Figure 9.3).

The parametric form of the ray is only slightly different, and is quite important:

$$\mathbf{p}(t) = \mathbf{p}_0 + t\mathbf{d}. \qquad (9.1)$$

Parametric definition of a ray using vector notation

The ray starts at the point $\mathbf{p}(0) = \mathbf{p}_0$. Thus $\mathbf{p}_0$ contains information about the *position* of the ray, while the "delta vector" $\mathbf{d}$ contains its *length and direction*. We restrict the parameter t to the normalized range $[0, 1]$, and so the ray ends at the point $\mathbf{p}(1) = \mathbf{p}_0 + \mathbf{d}$, as shown in Figure 9.4.

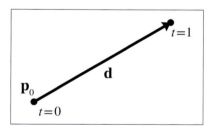

Figure 9.4
Defining a ray parametrically

We can also write out a separate scalar function for each coordinate, although the vector format is more compact and also has the nice property that it makes the equations the same in any dimension. For example, a 2D ray is defined parametrically by using the two scalar functions,

$$x(t) = x_0 + t\,\Delta x, \qquad\qquad y(t) = y_0 + t\,\Delta y.$$

Parametric definition of a 2D ray

A slight variation on Equation (9.1) that we use in some of the intersection tests is to use a unit vector $\hat{\mathbf{d}}$ and change the domain of the parameter t to $[0, l]$, where l is the length of the ray.

9.2.2   Special 2D Representations of Lines

Now let's look a bit closer at some special ways of describing (infinite) lines. These methods are applicable only in 2D; in 3D, techniques similar to these are used to define a plane, as we show in Section 9.5. A 2D ray inherently has four degrees of freedom (x_0, y_0, Δx, and Δy), but an infinite line has only two degrees of freedom.

Most readers are probably familiar with the *slope-intercept* form, which is an implicit method for representing an infinite line in 2D:

Slope-intercept form

$$y = mx + y_0. \tag{9.2}$$

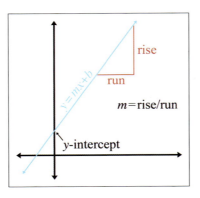

Figure 9.5
The slope and *y*-intercept of a line

The symbol m is the traditional one used to denote the slope of the line, expressed as a ratio of *rise* over *run*: for every *rise* units that we move up, we will move *run* units to the right (see Figure 9.5). The *y-intercept* is where the line crosses the *y*-axis, and is the value that we have denoted y_0 in Equation (9.2). (We're bucking tradition and not using the traditional variable, b, in order to avoid some confusion later on.) Substituting $x = 0$ clearly shows that the line crosses the *y*-axis at $y = y_0$.

The slope-intercept makes it easy to verify that an infinite line does, in fact, have two degrees of freedom: one degree for rotation and another for translation. Unfortunately, a vertical line has infinite slope and cannot be represented in slope-intercept form, since the implicit form of a vertical line is $x = k$. (Horizontal lines are no problem, their slope is zero.)

We can work around this singularity by using the slightly different implicit form

Implicit definition of infinite line in 2D

$$ax + by = d. \tag{9.3}$$

 Most sources use the form $ax + by + d = 0$. This flips the sign of d compared to our equations. We will use the form in Equation (9.3) because it has fewer terms, and we also feel that d has a more intuitive meaning geometrically in this form.

The remainder of this section surveys different methods for representing lines and rays in 2D and 3D. Section 9.2.1 discusses some simple ways to represent a ray, including the all-important parametric form. Section 9.2.2 discusses some special ways to define an infinite line in 2D. Section 9.2.3 gives some examples of converting from one representation to another.

9.2.1 Rays

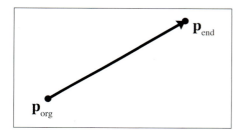

Figure 9.3
Defining a ray using the starting and ending points

The most obvious way to define a ray (the "straightforward form") is by two points, the ray origin and the ray endpoint, which we will denote as $\mathbf{p}_{org}$ and $\mathbf{p}_{end}$ (see Figure 9.3).

The parametric form of the ray is only slightly different, and is quite important:

$$\mathbf{p}(t) = \mathbf{p}_0 + t\mathbf{d}. \qquad (9.1)$$

The ray starts at the point $\mathbf{p}(0) = \mathbf{p}_0$. Thus $\mathbf{p}_0$ contains information about the *position* of the ray, while the "delta vector" $\mathbf{d}$ contains its *length and direction*. We restrict the parameter t to the normalized range $[0, 1]$, and so the ray ends at the point $\mathbf{p}(1) = \mathbf{p}_0 + \mathbf{d}$, as shown in Figure 9.4.

Parametric definition of a ray using vector notation

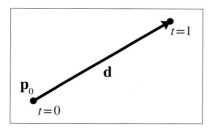

Figure 9.4
Defining a ray parametrically

We can also write out a separate scalar function for each coordinate, although the vector format is more compact and also has the nice property that it makes the equations the same in any dimension. For example, a 2D ray is defined parametrically by using the two scalar functions,

$$x(t) = x_0 + t\,\Delta x, \qquad\qquad y(t) = y_0 + t\,\Delta y.$$

Parametric definition of a 2D ray

A slight variation on Equation (9.1) that we use in some of the intersection tests is to use a unit vector $\hat{\mathbf{d}}$ and change the domain of the parameter t to $[0, l]$, where l is the length of the ray.

9.2.2   Special 2D Representations of Lines

Now let's look a bit closer at some special ways of describing (infinite) lines. These methods are applicable only in 2D; in 3D, techniques similar to these are used to define a plane, as we show in Section 9.5. A 2D ray inherently has four degrees of freedom (x_0, y_0, Δx, and Δy), but an infinite line has only two degrees of freedom.

Most readers are probably familiar with the *slope-intercept* form, which is an implicit method for representing an infinite line in 2D:

$$y = mx + y_0. \tag{9.2}$$

Slope-intercept form

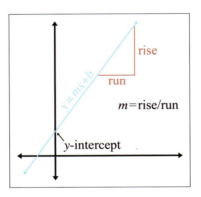

Figure 9.5
The slope and y-intercept of a line

The symbol m is the traditional one used to denote the slope of the line, expressed as a ratio of *rise* over *run*: for every *rise* units that we move up, we will move *run* units to the right (see Figure 9.5). The *y-intercept* is where the line crosses the y-axis, and is the value that we have denoted y_0 in Equation (9.2). (We're bucking tradition and not using the traditional variable, b, in order to avoid some confusion later on.) Substituting $x = 0$ clearly shows that the line crosses the y-axis at $y = y_0$.

The slope-intercept makes it easy to verify that an infinite line does, in fact, have two degrees of freedom: one degree for rotation and another for translation. Unfortunately, a vertical line has infinite slope and cannot be represented in slope-intercept form, since the implicit form of a vertical line is $x = k$. (Horizontal lines are no problem, their slope is zero.)

We can work around this singularity by using the slightly different implicit form

Implicit definition of infinite line in 2D

$$ax + by = d. \tag{9.3}$$

Most sources use the form $ax + by + d = 0$. This flips the sign of d compared to our equations. We will use the form in Equation (9.3) because it has fewer terms, and we also feel that d has a more intuitive meaning geometrically in this form.

If we assign the vector $\mathbf{n} = [a, b]$, we can write Equation (9.3) using vector notation as

$$\mathbf{p} \cdot \mathbf{n} = d. \tag{9.4}$$

Implicit definition of
infinite 2D line using
vector notation

Since this form has three degrees of freedom, and we said that an infinite line in 2D has only two, we know there is some redundancy. Note that we can multiply both sides of the equation by any constant; by so doing, we are free to choose the length of $\mathbf{n}$ without loss of generality. It is often convenient for $\mathbf{n}$ to be a unit vector. This gives $\mathbf{n}$ and d interesting geometric interpretations, as shown in Figure 9.6.

The vector $\mathbf{n}$ is the unit vector orthogonal to the line, and d gives the signed distance from the origin to the line. This distance is measured perpendicular to the line (parallel to $\mathbf{n}$). By *signed distance*, we mean that d is positive if the line is on the same side of the origin as the normal points. As d increases, the line moves in the direction of $\mathbf{n}$. At least, this is the case when we put d on the right side of the equals sign, as in Equation (9.4). If we move d to the left side of the equals sign and put zero on the right side, as in the traditional form, then the sign of d is flipped and these statements are reversed.

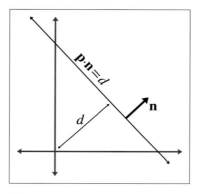

Figure 9.6
Defining a line using a perpendicular vector and distance to the origin

Notice that $\mathbf{n}$ describes the "orientation" of the line, while d describes its position. Another way to describe the position of the line is to give a point $\mathbf{q}$ that is on the line. Of course there are infinitely many points on the line, so any point will do (see Figure 9.7).

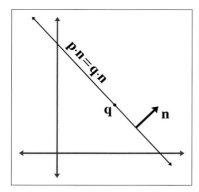

Figure 9.7
Defining a line using a perpendicular vector and a point on the line

One final way to define a line is as the perpendicular bisector of two points, to which we assign the variables $\mathbf{q}$ and $\mathbf{r}$ (see Figure 9.8). This is actually one of the earliest definitions of a line: the set of all points equidistant from two given points.

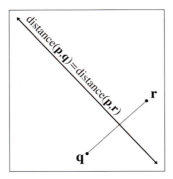

Figure 9.8
Defining a line as the perpendicular bisector of a line segment

9.2.3   Converting between Representations

Now let's give a few examples of how to convert a ray or line between the various representation techniques. We will not cover all of the combinations. Remember that the techniques we learned for infinite lines are applicable only in 2D.

To convert a ray defined using two points to parametric form:

$$\mathbf{p}_0 = \mathbf{p}_{\text{org}}, \qquad\qquad \mathbf{d} = \mathbf{p}_{\text{end}} - \mathbf{p}_{\text{org}}.$$

The opposite conversion, from parametric form to two-points form, is

$$\mathbf{p}_{\text{org}} = \mathbf{p}_0, \qquad\qquad \mathbf{p}_{\text{end}} = \mathbf{p}_0 + \mathbf{d}.$$

Given a parametric ray, we can compute the implicit line that contains this ray:

$$a = d_y, \qquad b = -d_x, \qquad d = x_{\text{org}} d_y - y_{\text{org}} d_x. \qquad (9.5)$$

To convert a line expressed implicitly to slope-intercept form:

$$m = -a/b, \qquad\qquad y_0 = d/b. \qquad (9.6)$$

Converting a line expressed implicitly to "normal and distance" form:

$$\hat{\mathbf{n}} = \begin{bmatrix} a & b \end{bmatrix} / \sqrt{a^2 + b^2}, \qquad\qquad \text{distance} = d/\sqrt{a^2 + b^2}.$$

Converting a normal and a point on the line to normal and distance form:

$$\text{distance} = \hat{\mathbf{n}} \cdot \mathbf{q}.$$

(This assumes that $\hat{\mathbf{n}}$ is a unit vector.)

Finally, to convert perpendicular bisector form to implicit form, we use

$$a = q_y - r_y,$$
$$b = r_x - q_x,$$
$$d = \frac{\mathbf{q} + \mathbf{r}}{2} \cdot \begin{bmatrix} a & b \end{bmatrix} = \frac{\mathbf{q} + \mathbf{r}}{2} \cdot \begin{bmatrix} q_y - r_y & r_x - q_x \end{bmatrix}$$
$$= \frac{(q_x + r_x)(q_y - r_y) + (q_y + r_y)(r_x - q_x)}{2}$$
$$= \frac{(q_x q_y - q_x r_y + r_x q_y - r_x r_y) + (q_y r_x - q_y q_x + r_y r_x - r_y q_x)}{2}$$
$$= r_x q_y - q_x r_y.$$

9.3   Spheres and Circles

A sphere is a 3D object defined as the set of all points that are a fixed distance from a given point. The distance from the center of the sphere to a point is known as the *radius* of the sphere. The straightforward representation of a sphere is to describe its center $\mathbf{c}$ and radius r.

Spheres appear often in computational geometry and graphics because of their simplicity. A bounding sphere is often used for trivial rejection because the equations for intersection with a sphere are simple. Also important is that rotating a sphere does not change its extents. Thus, when a bounding sphere is used for trivial rejection, if the center of the sphere is the origin of the object, then the orientation of the object can be ignored. A bounding box (see Section 9.4) does not have this property.

The implicit form of a sphere comes directly from its definition: the set of all points that are a given distance from the center. The implicit form of a sphere with center $\mathbf{c}$ and radius r is

$$\|\mathbf{p} - \mathbf{c}\| = r, \tag{9.7}$$

Implicit definition of a sphere using vector notation

where $\mathbf{p}$ is any point on the *surface* of the sphere. For a point $\mathbf{p}$ inside the sphere to satisfy the equation, we must change the "=" to a "≤". Since Equation (9.7) uses vector notation, it also works in 2D, as the implicit definition of a circle. Another more common form is to expand the vector notation and square both sides:

$$(x - c_x)^2 + (y - c_y)^2 = r^2 \qquad \text{(2D circle)} \qquad (9.8)$$

$$(x - c_x)^2 + (y - c_y)^2 + (z - c_z)^2 = r^2 \qquad \text{(3D sphere)} \qquad (9.9)$$

We might be interested in the diameter (distance from one point to a point on the exact opposite side), and circumference (the distance all the way around the circle) of a circle or sphere. Elementary geometry provides formulas for those quantities, as well as for the area of a circle, surface area of a sphere, and volume of a sphere:

$$D = 2r \qquad \text{(diameter)}$$

$$C = 2\pi r = \pi D \qquad \text{(circumference)}$$

$$A = \pi r^2 \qquad \text{(area of circle)}$$

$$S = 4\pi r^2 \qquad \text{(surface area of sphere)}$$

$$V = \frac{4}{3}\pi r^3 \qquad \text{(volume of sphere)}$$

For the calculus buffs, it is interesting to notice that the derivative of the area of a circle with respect to r is the circumference, and the derivative for the volume of a sphere is the surface area.

9.4   Bounding Boxes

Another simple geometric primitive commonly used as a bounding volume is the *bounding box*. Bounding boxes may be either axially aligned, or arbitrarily oriented. Axially aligned bounding boxes have the restriction that their sides be perpendicular to principal axes. The acronym AABB is often used for **a**xially **a**ligned **b**ounding **b**ox.

A 3D AABB is a simple 6-sided box with each side parallel to one of the cardinal planes. The box is *not* necessarily a cube—the length, width, and height of the box may each be different. Figure 9.9 shows a few simple 3D objects and their axially aligned bounding boxes.

Another frequently used acronym is OBB, which stands for **o**riented **b**ounding **b**ox. We don't discuss OBBs much in this section, for two reasons. First, axially aligned bounding boxes are simpler to create and use. But more important, you can think about an OBB as simply an AABB with an orientation; every bounding box is an AABB in *some* coordinate space; in fact any one with axes perpendicular to the sides of the box will do. In other words, the difference between an AABB and an OBB is not in the box itself, but in whether you are performing calculations in a coordinate space aligned with the bounding box.

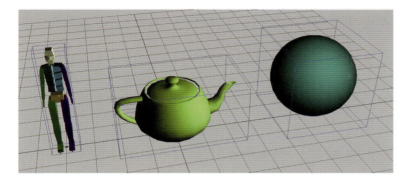

Figure 9.9. 3D objects and their AABB's

As an example, let's say that for objects in our world, we store the AABB of the object in the objects' object space. When performing operations in object space, this bounding box is an AABB. But when performing calculations in world (or upright) space, then this same bounding box is an OBB, since it may be "at an angle" relative to the world axes.

Although this section focuses on 3D AABBs, most of the information can be applied in a straightforward manner in 2D by simply dropping the third dimension.

The next four sections cover the basic properties of AABBs. Section 9.4.1 introduces the notation we use and describes the options we have for representing an AABB. Section 9.4.2 shows how to compute the AABB for a set of points. Section 9.4.3 compares AABBs to bounding spheres. Section 9.4.4 shows how to construct an AABB for a transformed AABB.

9.4.1 Representing AABBs

Let us introduce several important properties of an AABB, and the notation we use when referring to these values. The points inside an AABB satisfy the inequalities

$$x_{\min} \le x \le x_{\max}, \qquad y_{\min} \le y \le y_{\max}, \qquad z_{\min} \le z \le z_{\max}.$$

Two corner points of special significance are

$$\mathbf{p}_{\min} = \begin{bmatrix} x_{\min} & y_{\min} & z_{\min} \end{bmatrix}, \qquad \mathbf{p}_{\max} = \begin{bmatrix} x_{\max} & y_{\max} & z_{\max} \end{bmatrix}.$$

The center point $\mathbf{c}$ is given by

$$\mathbf{c} = (\mathbf{p}_{\min} + \mathbf{p}_{\max})/2.$$

The "size vector" $\mathbf{s}$ is the vector from $\mathbf{p}_{min}$ to $\mathbf{p}_{max}$ and contains the width, height, and length of the box:

$$\mathbf{s} = \mathbf{p}_{max} - \mathbf{p}_{min}.$$

We can also refer to the "radius vector" $\mathbf{r}$ of the box, which is half of the size vector $\mathbf{s}$, and can be interpreted as the vector from $\mathbf{c}$ to $\mathbf{p}_{max}$:

$$\mathbf{r} = \mathbf{p}_{max} - \mathbf{c} = \mathbf{s}/2.$$

To unambiguously define an AABB requires only two of the five vectors $\mathbf{p}_{min}$, $\mathbf{p}_{max}$, $\mathbf{c}$, $\mathbf{s}$, and $\mathbf{r}$. Other than the pair $\mathbf{s}$ and $\mathbf{r}$, any pair may be used. Some representation forms are more useful in particular situations than others. We advise representing a bounding box by using $\mathbf{p}_{min}$ and $\mathbf{p}_{max}$, since in practice these values are needed far more frequently than $\mathbf{s}$, $\mathbf{c}$, and $\mathbf{r}$. And, of course, computing any of these three vectors from $\mathbf{p}_{min}$ and $\mathbf{p}_{max}$ is very fast. In C, an AABB might be represented by using a `struct` like in Listing 9.1.

```
struct AABB3 {
    Vector3 min;
    Vector3 max;
};
```

Listing 9.1
The most straightforward way to describe an AABB

9.4.2   Computing AABBs

Computing an AABB for a set of points is a simple process. We first reset the minimum and maximum values to "infinity," or what is effectively bigger than any number we will encounter in practice. Then, we pass through the list of points, expanding our box as necessary to contain each point.

```
void AABB3::empty() {
    min.x = min.y = min.z = FLT_MAX;
    max.x = max.y = max.z = -FLT_MAX;
}

void AABB3::add(const Vector3 &p) {
    if (p.x < min.x) min.x = p.x;
    if (p.x > max.x) max.x = p.x;
    if (p.y < min.x) min.y = p.y;
    if (p.y > max.x) max.y = p.y;
    if (p.z < min.x) min.z = p.z;
    if (p.z > max.x) max.z = p.z;
}
```

Listing 9.2
Two helpful AABB functions

An AABB class often defines two functions to help with this. The first function "empties" the AABB. The other function adds a single point into the AABB by expanding the AABB if necessary to contain the point. Listing 9.2 shows such code.

Now, to create a bounding box from a set of points, we could use the following code:

```
// Our list of points
const int N;
Vector3 list[N];

// First , empty the box
AABB3 box;
box.empty();

// Add each point into the box
for (int i = 0 ; i < N ; ++i) {
    box.add(list[i]);
}
```

Listing 9.3
Computing the AABB for a set of points

9.4.3 AABBs versus Bounding Spheres

In many cases, we have a choice between using an AABB or a bounding sphere. AABBs offer two main advantages over bounding spheres.

The first advantage of AABBs over bounding spheres is that computing the optimal AABB for a set of points is easy to program and can be run in linear time. Computing the optimal bounding sphere is a much more difficult problem. (O'Rourke [52] and Lengyel [42] describe algorithms for computing bounding spheres.)

Second, for many objects that arise in practice, an AABB provides a tighter bounding volume, and thus better trivial rejection. Of course, for some objects, the bounding sphere is better. (Imagine an object that is itself a sphere!) In the worst case, an AABB will have a volume of just under twice the volume of the sphere, but when a sphere is bad, it can be *really* bad. Consider the bounding sphere and AABB of a telephone pole, for example.

The basic problem with a sphere is that there is only one degree of freedom to its shape—the radius of the sphere. An AABB has three degrees of freedom—the length, width, and height. Thus, it can usually adapt to differently shaped objects better. For most of the objects in Figure 9.10, the AABB is smaller than the bounding sphere. The exception is the star in the upper right-hand corner, where the bounding sphere is slightly smaller than the AABB. Notice that the AABB is highly sensitive to the orientation of the object, as shown by the AABBs for the two rifles on the bottom. In each

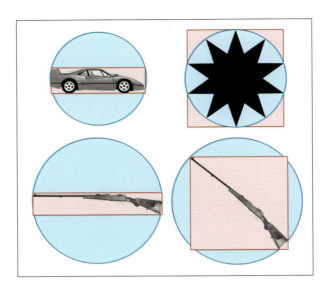

Figure 9.10
The AABB and
bounding sphere for
various objects

case, the size of the rifle is the same, and only the orientation is different. Also notice that the bounding spheres are the same size since bounding spheres are not sensitive to the orientation of the object. When the objects are free to rotate, some of the advantage of AABBs can be eroded. There is an inherent trade-off between a tighter volume (OBB) and a compact, fast representation (bounding spheres). Which bounding primitive is best will depend highly on the application.

9.4.4 Transforming AABBs

Sometimes we need to transform an AABB from one coordinate space to another. For example, let's say that we have the AABB in object space (which, from the perspective of world space, is basically the same thing as an OBB; see Section 9.4) and we want to get an AABB in world space. Of course, in theory, we could compute a world-space AABB of the object itself. However, we assume that the description of the object shape (perhaps a triangle mesh with a thousand vertices) is more complicated than the AABB that we already have computed in object space. So to get an AABB in world space, we will transform the object-space AABB.

What we get as a result is not necessarily axially aligned (if the object is rotated), and is not necessarily a box (if the object is skewed). However, computing an AABB for the "transformed AABB" (we should perhaps call it a NNAABNNB—a "not-necessarily axially aligned bounding not-necessarily box") is faster than computing a new AABB for all but the most simple transformed objects because AABBs have only eight vertices.

To compute an AABB for a transformed AABB it is *not* enough to simply transform the original $\mathbf{p}_{\min}$ and $\mathbf{p}_{\max}$. This could result in a bogus bounding box, for example, if $x_{\min} > x_{\max}$. To compute a new AABB, we must transform the eight corner points, and then form an AABB from these eight transformed points.

Depending on the transformation, this usually results in a bounding box that is larger than the original bounding box. For example, in 2D, a rotation of 45 degrees will increase the size of the bounding box significantly (see Figure 9.11).

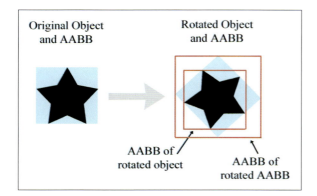

Original Object and AABB

Rotated Object and AABB

AABB of rotated object

AABB of rotated AABB

Figure 9.11
The AABB of a transformed box

Compare the size of the original AABB in Figure 9.11 (the blue box), with the new AABB (the largest red box on the right) which was computed solely from the rotated AABB. The new AABB is almost twice as big. Notice that if we were able to compute an AABB from the rotated object rather than the rotated AABB, it is about the same size as the original AABB.

As it turns out, the structure of an AABB can be exploited to speed up the generation of the new AABB, so it is not necessary to actually transform all eight corner points and build a new AABB from these points.

Let's quickly review what happens when we transform a 3D point by a 3×3 matrix (see Section 4.1.7 if you have forgotten how to multiply a vector by a matrix):

$$\begin{bmatrix} x' & y' & z' \end{bmatrix} = \begin{bmatrix} x & y & z \end{bmatrix} \begin{bmatrix} m_{11} & m_{12} & m_{13} \\ m_{21} & m_{22} & m_{23} \\ m_{31} & m_{32} & m_{33} \end{bmatrix};$$

$$x' = m_{11}x + m_{21}y + m_{31}z,$$
$$y' = m_{12}x + m_{22}y + m_{32}z,$$
$$z' = m_{13}x + m_{23}y + m_{33}z.$$

Assume the original bounding box is in x_{min}, x_{max}, y_{min}, etc., and the new bounding box will be computed into x'_{min}, x'_{max}, y'_{min}, etc. Let's examine how we might more quickly compute x'_{min} as an example. In other words, we wish to find the minimum value of

$$m_{11}x + m_{21}y + m_{31}z,$$

where $[x, y, z]$ is any of the original eight corner points. Our job is to figure out which of these corner points would have the smallest x value after transformation. The trick to minimizing the entire sum is to minimize each of the three products individually. Let's look at the first product, $m_{11}x$. We must decide which of x_{min} or x_{max} to substitute for x in order to minimize the product. Obviously, if $m_{11} > 0$, then the smaller of the two, x_{min}, will result in the smaller product. Conversely, if $m_{11} < 0$, then x_{max} gives smaller product. Conveniently, whichever of x_{min} or x_{max} we use for computing x'_{min}, we use the other value for computing x'_{max}. We then apply this process for each of the nine elements in the matrix.

This technique is illustrated in Listing 9.4. The class **Matrix4x3** is a 4×3 transform matrix, which can represent any affine transform. (It's a 4×4 matrix that acts on row vectors, where the right-most column is assumed to be $[0, 0, 0, 1]^T$.)

```
void AABB3::setToTransformedBox(const AABB3 &box, const Matrix4x3 &m) {

    // Start with the last row of the matrix, which is the translation
    // portion, i.e. the location of the origin after transformation.
    min = max = getTranslation(m);

    //
    // Examine each of the 9 matrix elements
    // and compute the new AABB
    //

    if (m.m11 > 0.0f) {
        min.x += m.m11 * box.min.x;  max.x += m.m11 * box.max.x;
    } else {
        min.x += m.m11 * box.max.x;  max.x += m.m11 * box.min.x;
    }

    if (m.m12 > 0.0f) {
        min.y += m.m12 * box.min.x;  max.y += m.m12 * box.max.x;
    } else {
        min.y += m.m12 * box.max.x;  max.y += m.m12 * box.min.x;
    }

    if (m.m13 > 0.0f) {
        min.z += m.m13 * box.min.x;  max.z += m.m13 * box.max.x;
    } else {
        min.z += m.m13 * box.max.x;  max.z += m.m13 * box.min.x;
    }

    if (m.m21 > 0.0f) {
        min.x += m.m21 * box.min.y;  max.x += m.m21 * box.max.y;
    } else {
```

```
        min.x += m.m21 * box.max.y; max.x += m.m21 * box.min.y;
    }

    if (m.m22 > 0.0f) {
        min.y += m.m22 * box.min.y; max.y += m.m22 * box.max.y;
    } else {
        min.y += m.m22 * box.max.y; max.y += m.m22 * box.min.y;
    }

    if (m.m23 > 0.0f) {
        min.z += m.m23 * box.min.y; max.z += m.m23 * box.max.y;
    } else {
        min.z += m.m23 * box.max.y; max.z += m.m23 * box.min.y;
    }

    if (m.m31 > 0.0f) {
        min.x += m.m31 * box.min.z; max.x += m.m31 * box.max.z;
    } else {
        min.x += m.m31 * box.max.z; max.x += m.m31 * box.min.z;
    }

    if (m.m32 > 0.0f) {
        min.y += m.m32 * box.min.z; max.y += m.m32 * box.max.z;
    } else {
        min.y += m.m32 * box.max.z; max.y += m.m32 * box.min.z;
    }

    if (m.m33 > 0.0f) {
        min.z += m.m33 * box.min.z; max.z += m.m33 * box.max.z;
    } else {
        min.z += m.m33 * box.max.z; max.z += m.m33 * box.min.z;
    }
}
```

Listing 9.4
Computing a transformed AABB

9.5   Planes

A *plane* is a flat, 2D subspace of 3D. Planes are extremely common tools
in video games, and the concepts in this section are especially useful. The
definition of a plane that Euclid would probably recognize is similar to
the perpendicular bisector definition of an infinite line in 2D: the set of
all points that are equidistant from two given points. This similarity in
definitions hints at the fact that planes in 3D share many properties with
infinite lines in 2D. For example, they both subdivide the space into two
"half-spaces."

This section covers the fundamental properties of planes. Section 9.5.1
shows how to define a plane implicitly with the plane equation. Section 9.5.2
shows how three points may be used to define a plane. Section 9.5.3 describes how to find the "best-fit" plane for a set of points that may not be
exactly planar. Section 9.5.4 describes how to compute the distance from
a point to a plane.

9.5.1   The Plane Equation: An Implicit Definition of a Plane

We can represent planes using techniques similar to the ones we used to describe infinite 2D lines in Section 9.2.2. The implicit form of a plane is given by all points $\mathbf{p} = (x, y, z)$ that satisfy the *plane equation*:

The plane equation

$$ax + by + cz = d \qquad \text{(scalar notation)},$$

$$\mathbf{p} \cdot \mathbf{n} = d \qquad \text{(vector notation)}. \qquad (9.10)$$

Note that in the vector form, $\mathbf{n} = [a, b, c]$. Once we know $\mathbf{n}$, we can compute d from any point known to be in the plane.

Most sources give the plane equation as $ax + by + cz + d = 0$. This has the effect of flipping the sign of d. Our comments in Section 9.2.2 explaining our preference to put d on the left side of the equals sign also apply here: our experience is that this form results in fewer terms and minus signs and a more intuitive geometric interpretation for d.

The vector $\mathbf{n}$ is called the plane *normal* because it is perpendicular (normal) to the plane. Although $\mathbf{n}$ is often normalized to unit length, this is not strictly necessary. We use a hat ($\hat{\mathbf{n}}$) when we are assuming unit length. The normal determines the orientation of the plane; d defines its position. More specifically, it determines the signed distance to the plane from the origin, measured in the direction of the normal. Increasing d slides the plane forward, in the direction of the normal. If $d > 0$, the origin is on the back side of the plane, and if $d < 0$, the origin is on the front side. (This assumes we put d on the right-hand side of the equals sign, as in Equation (9.10). The standard homogenous form with d on the left has the opposite sign conventions.)

Let's verify that $\mathbf{n}$ is perpendicular to the plane. Assume $\mathbf{p}$ and $\mathbf{q}$ are arbitrary points in the plane, and therefore satisfy the plane equation. Substituting $\mathbf{p}$ and $\mathbf{q}$ into Equation (9.10), we get

$$\mathbf{n} \cdot \mathbf{p} = d,$$

$$\mathbf{n} \cdot \mathbf{q} = d,$$

$$\mathbf{n} \cdot \mathbf{p} = \mathbf{n} \cdot \mathbf{q},$$

$$\mathbf{n} \cdot \mathbf{p} - \mathbf{n} \cdot \mathbf{q} = 0,$$

$$\mathbf{n} \cdot (\mathbf{p} - \mathbf{q}) = 0. \qquad (9.11)$$

The geometric implication of Equation (9.11) is that $\mathbf{n}$ is perpendicular to the vector from $\mathbf{q}$ to $\mathbf{p}$ (see Section 2.11). This is true for any points $\mathbf{p}$ and $\mathbf{q}$ in the plane, and therefore $\mathbf{n}$ is perpendicular to every vector in the plane.

We often consider a plane as having a "front" side and a "back" side. Usually, the front side of the plane is the direction that $\mathbf{n}$ points; i.e. when looking from the head of $\mathbf{n}$ towards the tail, we are looking at the front side of the plane (see Figure 9.12).

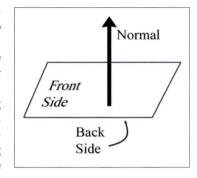

Figure 9.12
The front and back sides of a plane

As mentioned previously, it's often useful to restrict $\mathbf{n}$ to have unit length. We can do this without loss of generality, since we can multiply the entire plane equation by any constant.

9.5.2   Defining a Plane by Using Three Points

Another way we can define a plane is to give three noncollinear points that lie in the plane. Collinear points (points in a straight line) won't work because there would be an infinite number of planes that contain that line, and there would be no way of telling which plane we meant.

Let's compute $\mathbf{n}$ and d from three points $\mathbf{p}_1$, $\mathbf{p}_2$, and $\mathbf{p}_3$ known to be in the plane. First, we must compute $\mathbf{n}$. Which way will $\mathbf{n}$ point? The standard way to do this in a left-handed coordinate system is to assume that $\mathbf{p}_1$, $\mathbf{p}_2$, and $\mathbf{p}_3$ are listed in clockwise order, when viewed from the front side of the plane, as illustrated in Figure 9.13. (In a right-handed coordinate system, we usually assume the points are listed in counterclockwise order. Under these conventions, the equations are the same no matter what coordinate system is used.)

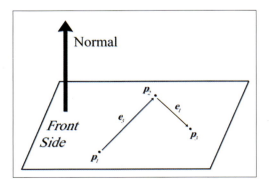

Figure 9.13
Computing a plane normal from three points in the plane

We will construct two vectors according to the clockwise ordering. The notation $\mathbf{e}$ stands for "edge" vector, since these equations commonly arise when computing the plane equation for a triangle. (The indexing may seem weird, but bear with us—this is the indexing we'll use in Section 9.6.1 where triangles are discussed in more detail.) The cross product of these two vectors yields the perpendicular vector $\mathbf{n}$, but this vector is not necessarily of unit length. As mentioned earlier, we usually normalize $\mathbf{n}$. All of this is summarized succinctly by

The normal of a plane
containing three points

$$\mathbf{e}_3 = \mathbf{p}_2 - \mathbf{p}_1, \qquad \mathbf{e}_1 = \mathbf{p}_3 - \mathbf{p}_2, \qquad \hat{\mathbf{n}} = \frac{\mathbf{e}_3 \times \mathbf{e}_1}{\|\mathbf{e}_3 \times \mathbf{e}_1\|}. \qquad (9.12)$$

Notice that if the points are collinear, then $\mathbf{e}_3$ and $\mathbf{e}_1$ will be parallel, and thus the cross product will be $\mathbf{0}$, which cannot be normalized. This mathematical singularity coincides with the physical singularity that collinear points do not unambiguously define a plane.

Now that we know $\hat{\mathbf{n}}$, all that is left to do is compute d. This is easily done by taking the dot product of one of the points and $\hat{\mathbf{n}}$.

9.5.3 "Best Fit" Plane for More than Three Points

Occasionally, we may wish to compute the plane equation for a set of more than three points. The most common example of such a set of points is the vertices of a polygon. In this case, the vertices are assumed to be enumerated in a clockwise fashion around the polygon. (The ordering matters because it is how we decide which side is the front and which is the back, which in turn determines which direction our normal will point.)

One naïve solution is to arbitrarily select three consecutive points and compute the plane equation from those three points. However, the three points we choose might be collinear, or nearly collinear, which is almost as bad because it is numerically inaccurate. Or perhaps the polygon is concave and the three points we have chosen are a point of concavity and therefore form a counterclockwise turn (which would result in a normal that points in the wrong direction). Or the vertices of the polygon may not be coplanar, which can happen due to numeric imprecision or the method used to generate the polygon. What we really want is a way to compute the "best fit" plane for a set of points that takes into account *all* of the points. Given n points,

$$\mathbf{p}_1 = \begin{bmatrix} x_1 & y_1 & z_1 \end{bmatrix},$$
$$\mathbf{p}_2 = \begin{bmatrix} x_2 & y_2 & z_2 \end{bmatrix},$$
$$\vdots$$

$$\mathbf{p}_{n-1} = \begin{bmatrix} x_{n-1} & y_{n-1} & z_{n-1} \end{bmatrix},$$
$$\mathbf{p}_n = \begin{bmatrix} x_n & y_n & z_n \end{bmatrix},$$

the best fit perpendicular vector $\mathbf{n}$ is given by

$$
\begin{aligned}
n_x =& (z_1 + z_2)(y_1 - y_2) + (z_2 + z_3)(y_2 - y_3) + \cdots \\
&\cdots + (z_{n-1} + z_n)(y_{n-1} - y_n) + (z_n + z_1)(y_n - y_1), \\
n_y =& (x_1 + x_2)(z_1 - z_2) + (x_2 + x_3)(z_2 - z_3) + \cdots \\
&\cdots + (x_{n-1} + x_n)(z_{n-1} - z_n) + (x_n + x_1)(z_n - z_1), \\
n_z =& (y_1 + y_2)(x_1 - x_2) + (y_2 + y_3)(x_2 - x_3) + \cdots \\
&\cdots + (y_{n-1} + y_n)(x_{n-1} - x_n) + (y_n + y_1)(x_n - x_1).
\end{aligned}
\tag{9.13}
$$

Computing the best-fit plane normal from n points

This vector must then be normalized if we wish to enforce the restriction that $\mathbf{n}$ be of unit length.

We can express Equation (9.13) succinctly by using summation notation. Adopting a circular indexing scheme such that $\mathbf{p}_{n+1} \equiv \mathbf{p}_1$, we can write

$$n_x = \sum_{i=1}^{n} (z_i + z_{i+1})(y_i - y_{i+1}),$$

$$n_y = \sum_{i=1}^{n} (x_i + x_{i+1})(z_i - z_{i+1}),$$

$$n_z = \sum_{i=1}^{n} (y_i + y_{i+1})(x_i - x_{i+1}).$$

Listing 9.5 illustrates how we might compute a best-fit plane normal for a set of points.

```
Vector3 computeBestFitNormal(const Vector3 v[], int n) {

    // Zero out sum
    Vector3 result = kZeroVector;

    // Start with the ''previous'' vertex as the last one.
    // This avoids an if-statement in the loop
    const Vector3 *p = &v[n-1];

    // Iterate through the vertices
    for (int i = 0 ; i < n ; ++i) {

        // Get shortcut to the ''current'' vertex
        const Vector3 *c = &v[i];

        // Add in edge vector products appropriately
        result.x += (p->z + c->z) * (p->y - c->y);
        result.y += (p->x + c->x) * (p->z - c->z);
        result.z += (p->y + c->y) * (p->x - c->x);
```

```
        // Next vertex, please
        p = c;
    }

    // Normalize the result and return it
    result.normalize();
    return result;
}
```

Listing 9.5
Computing the best-fit plane normal for a set of points

The best-fit d value can be computed as the average of the d values for each point:

Computing the best-fit plane d value

$$d = \frac{1}{n}\sum_{i=1}^{n}(\mathbf{p}_i \cdot \mathbf{n}) = \frac{1}{n}\left(\sum_{i=1}^{n}\mathbf{p}_i\right) \cdot \mathbf{n}.$$

9.5.4   Distance from Point to Plane

It is often the case that we have a plane and a point $\mathbf{q}$ that is not in the plane, and we want to calculate the distance from the plane to $\mathbf{q}$, or at least classify $\mathbf{q}$ as being on the front or back side of the plane. To do this, we imagine the point $\mathbf{p}$ that lies in the plane and is the closest point in the plane to $\mathbf{q}$. Clearly, the vector from $\mathbf{p}$ to $\mathbf{q}$ is perpendicular to the plane, and thus is of the form $a\mathbf{n}$, as shown in Figure 9.14.

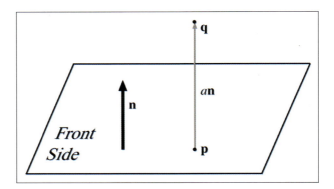

Figure 9.14
Computing the distance between a point and a plane

If we assume the plane normal $\hat{\mathbf{n}}$ is a unit vector, then the distance from $\mathbf{p}$ to $\mathbf{q}$ (and thus the distance from the plane to $\mathbf{q}$) is simply a. This is a signed distance, which means that it will be negative when $\mathbf{q}$ is on the back side of the plane. What's surprising is that we can compute a without knowing the location of $\mathbf{p}$. We go back to our original definition of $\mathbf{q}$ and then perform some vector algebra to eliminate $\mathbf{p}$:

$$\mathbf{p} + a\hat{\mathbf{n}} = \mathbf{q},$$
$$(\mathbf{p} + a\hat{\mathbf{n}}) \cdot \hat{\mathbf{n}} = \mathbf{q} \cdot \hat{\mathbf{n}},$$
$$\mathbf{p} \cdot \hat{\mathbf{n}} + (a\hat{\mathbf{n}}) \cdot \hat{\mathbf{n}} = \mathbf{q} \cdot \hat{\mathbf{n}},$$
$$d + a = \mathbf{q} \cdot \hat{\mathbf{n}},$$
$$a = \mathbf{q} \cdot \hat{\mathbf{n}} - d. \tag{9.14}$$

Computing the signed distance from a plane to an arbitrary 3D point

9.6 Triangles

Triangles are of fundamental importance in modeling and graphics. The surface of a complex 3D object, such as a car or a human body, is approximated with many triangles. Such a group of connected triangles forms a *triangle mesh*, which is the topic of Section 10.4. But before we can learn how to manipulate many triangles, we must first learn how to manipulate one triangle.

This section covers the fundamental properties of triangles. Section 9.6.1 introduces some notation and basic properties of triangles. Section 9.6.2 lists several methods for computing the area of a triangle in 2D or 3D. Section 9.6.3 discusses barycentric space. Section 9.6.5 discusses a few points on a triangle that are of special geometric significance.

9.6.1 Notation

A triangle is defined by listing its three vertices. The order that these points are listed is significant. In a left-handed coordinate system, we typically enumerate the points in clockwise order when viewed from the front side of the triangle. We will refer to the three vertices as $\mathbf{v}_1$, $\mathbf{v}_2$, and $\mathbf{v}_3$.

A triangle lies in a plane, and the equation of this plane (the normal $\mathbf{n}$ and distance to origin d) is important in a number of applications. We just discussed planes, including how to compute the plane equation given three points, in Section 9.5.2.

Let us label the interior angles, clockwise edge vectors, and side lengths as shown in Figure 9.15.

Let l_i denote the length of $\mathbf{e}_i$. Notice that $\mathbf{e}_i$ and l_i are opposite $\mathbf{v}_i$, the vertex with the corresponding index, and are given by

$$\mathbf{e}_1 = \mathbf{v}_3 - \mathbf{v}_2, \qquad \mathbf{e}_2 = \mathbf{v}_1 - \mathbf{v}_3, \qquad \mathbf{e}_3 = \mathbf{v}_2 - \mathbf{v}_1,$$
$$l_1 = \|\mathbf{e}_1\|, \qquad l_2 = \|\mathbf{e}_2\|, \qquad l_3 = \|\mathbf{e}_3\|.$$

Notation for edge vectors and lengths

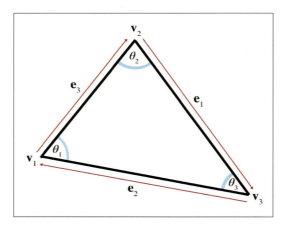

Figure 9.15
Labeling triangles

For example, let's write the law of sines and law of cosines using this notation:

$$\frac{\sin \theta_1}{l_1} = \frac{\sin \theta_2}{l_2} = \frac{\sin \theta_3}{l_3},$$

Law of sines

Low of cosines

$$l_1{}^2 = l_2{}^2 + l_3{}^2 - 2l_2l_3 \cos \theta_1,$$
$$l_2{}^2 = l_1{}^2 + l_3{}^2 - 2l_1l_3 \cos \theta_2,$$
$$l_3{}^2 = l_1{}^2 + l_2{}^2 - 2l_1l_2 \cos \theta_3.$$

The perimeter of the triangle is often an important value, and is computed trivially by summing the three sides:

Perimeter of a triangle

$$p = l_1 + l_2 + l_3.$$

9.6.2   Area of a Triangle

This section investigates several techniques for computing the area of a triangle. The most well known method is to compute the area from the *base* and *height* (also known as the *altitude*). Examine the parallelogram and enclosed triangle in Figure 9.16.

From classical geometry, we know that the area of a parallelogram is equal to the product of the base and height. (See Section 2.12.2 for an explanation of why this is true.) Since the triangle occupies exactly one half of this area, the area of a triangle, is

Area of a triangle

$$A = bh/2.$$

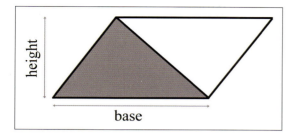

Figure 9.16
A triangle enclosed in a parallelogram

If the altitude is not known, then *Heron's formula* can be used, which requires only the lengths of the three sides. Let s equal one half the perimeter (also known as the *semiperimeter*). Then the area is given by

$$s = \frac{l_1 + l_2 + l_3}{2} = \frac{p}{2},$$

$$A = \sqrt{s(s - l_1)(s - l_2)(s - l_3)}.$$

Heron's formula for the area of a triangle

Heron's formula is particularly interesting because of the ease with which it can be applied in 3D.

Often the altitude or lengths of the sides are not readily available and all we have are the Cartesian coordinates of the vertices. (Of course, we could always compute the side lengths from the coordinates, but there are situations for which we wish to avoid this relatively costly computation.) Let's see if we can compute the area of a triangle from the vertex coordinates alone.

Let's first tackle this problem in 2D. The basic idea is to compute, for each of the three edges of the triangle, the signed area of the trapezoid

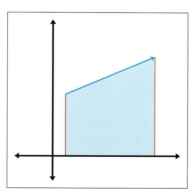

Figure 9.17
The area "beneath" an edge vector

bounded above by the edge and below by the x-axis, as shown in Figure 9.17. By "signed area," we mean that the area is positive if the edge points from left to right, and negative if the edge points from right to left. Notice that no matter how the triangle is oriented, there will always be at least one positive edge and at least one negative edge. A vertical edge will have zero area. The formulas for the areas under each edge are

$$A(\mathbf{e}_1) = \frac{(y_3 + y_2)(x_3 - x_2)}{2},$$

$$A(\mathbf{e}_2) = \frac{(y_1 + y_3)(x_1 - x_3)}{2},$$

$$A(\mathbf{e}_3) = \frac{(y_2 + y_1)(x_2 - x_1)}{2}.$$

By summing the signed areas of the three trapezoids, we arrive at the area of the triangle itself. In fact, the same idea can be used to compute the area of a polygon with any number of sides.

We assume a clockwise ordering of the vertices around the triangle. Enumerating the vertices in the opposite order flips the sign of the area. With these considerations in mind, we sum the areas of the trapezoids to compute the signed area of the triangle:

$$A = A(\mathbf{e}_1) + A(\mathbf{e}_2) + A(\mathbf{e}_3)$$

$$= \frac{(y_3 + y_2)(x_3 - x_2) + (y_1 + y_3)(x_1 - x_3) + (y_2 + y_1)(x_2 - x_1)}{2}$$

$$= \frac{\left(\begin{array}{c}(y_3x_3 - y_3x_2 + y_2x_3 - y_2x_2) \\ + (y_1x_1 - y_1x_3 + y_3x_1 - y_3x_3) \\ + (y_2x_2 - y_2x_1 + y_1x_2 - y_1x_1)\end{array}\right)}{2}$$

$$= \frac{-y_3x_2 + y_2x_3 - y_1x_3 + y_3x_1 - y_2x_1 + y_1x_2}{2}$$

$$= \frac{y_1(x_2 - x_3) + y_2(x_3 - x_1) + y_3(x_1 - x_2)}{2}.$$

We can actually simplify this just a bit further. The basic idea is to realize that we can translate the triangle without affecting the area. We make an arbitrary choice to shift the triangle vertically, subtracting y_3 from each of the y coordinates (in case you were wondering whether the trapezoid summing trick works if some of the triangle extends below the x-axis, this shifting properly shows that it does):

Computing the area of a 2D triangle from the coordinates of the vertices

$$A = \frac{y_1(x_2 - x_3) + y_2(x_3 - x_1) + y_3(x_1 - x_2)}{2}$$

$$= \frac{(y_1 - y_3)(x_2 - x_3) + (y_2 - y_3)(x_3 - x_1) + (y_3 - y_3)(x_1 - x_2)}{2} \quad (9.15)$$

$$= \frac{(y_1 - y_3)(x_2 - x_3) + (y_2 - y_3)(x_3 - x_1)}{2}.$$

In 3D, we can use the cross product to compute the area of a triangle. Recall from Section 2.12.2 that the magnitude of the cross product of two vectors $\mathbf{a}$ and $\mathbf{b}$ is equal to the area of the parallelogram formed on two sides by $\mathbf{a}$ and $\mathbf{b}$. Since the area of a triangle is half the area of the enclosing parallelogram, we have a simple way to calculate the area of the triangle. Given two edge vectors from the triangle, $\mathbf{e}_1$ and $\mathbf{e}_2$, the area of the triangle is given by

$$A = \frac{\|\mathbf{e}_1 \times \mathbf{e}_2\|}{2}. \tag{9.16}$$

Notice that if we extend a 2D triangle into 3D by assuming $z = 0$, then Equation (9.15) and Equation (9.16) are equivalent.

9.6.3 Barycentric Space

Even though we certainly use triangles in 3D, the surface of a triangle lies in a plane and is inherently a 2D object. Moving around on the surface of a triangle that is arbitrarily oriented in 3D is somewhat awkward. It would be nice to have a coordinate space that is related to the surface of the triangle and is independent of the 3D space in which the triangle "lives." *Barycentric space* is just such a coordinate space. Many practical problems that arise when making video games, such as interpolation and intersection, can be solved by using barycentric coordinates. We are introducing barycentric coordinates in the context of triangles here, but they have wide applicability. In fact, we meet them again in a slightly more general form in the context of 3D curves in Chapter 13.

Any point in the plane of a triangle can be expressed as a *weighted average* of the vertices. These weights are known as *barycentric coordinates*. The conversion from barycentric coordinates (b_1, b_2, b_3) to standard 3D space is defined by

$$(b_1, b_2, b_3) \equiv b_1 \mathbf{v}_1 + b_2 \mathbf{v}_2 + b_3 \mathbf{v}_3. \tag{9.17}$$

Computing a 3D point from barycentric coordinates

Of course, this is simply a linear combination of some vectors. Section 3.3.3 showed how ordinary Cartesian coordinates can also be interpreted as a linear combination of the basis vectors, but the subtle distinction between barycentric coordinates and ordinary Cartesian coordinates is that for barycentric coordinates the sum of the coordinates is restricted to be unity:

$$b_1 + b_2 + b_3 = 1.$$

This normalization constraint removes one degree of freedom, which is why even though there are three coordinates, it is still a 2D space.

The values b_1, b_2, b_3 are the "contributions" or "weights" that each vertex contributes to the point. Figure 9.18 shows some examples of points and their barycentric coordinates.

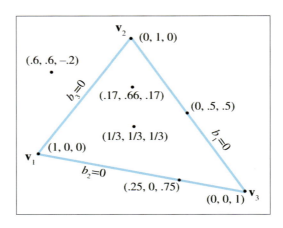

Figure 9.18
Examples of barycentric
coordinates

Let's make a few observations here. First, notice that the three vertices of the triangle have a trivial form in barycentric space:

$$(1, 0, 0) \equiv \mathbf{v}_1, \qquad (0, 1, 0) \equiv \mathbf{v}_2, \qquad (0, 0, 1) \equiv \mathbf{v}_3.$$

Second, all points on the side opposite a vertex will have a zero for the barycentric coordinate corresponding to that vertex. For example, $b_1 = 0$ for all points on the line containing $\mathbf{e}_1$ (which is opposite $\mathbf{v}_1$).

Finally, *any* point in the plane can be described in barycentric coordinates, not just the points inside the triangle. The barycentric coordinates of a point inside the triangle will all be in the range $[0, 1]$. Any point outside the triangle will have at least one negative coordinate. Barycentric space tessellates the plane into triangles of the same size as the original triangle, as shown in Figure 9.19.

There's another way to think about barycentric coordinates. Discarding b_3, we can interpret (b_1, b_2) as regular (x, y) 2D coordinates, where the origin is at $\mathbf{v}_3$, the x-axis is $\mathbf{v}_1 - \mathbf{v}_3$, and the y-axis is $\mathbf{v}_1 - \mathbf{v}_2$. This can be made more explicit by rearranging Equation (9.17):

Interpreting (b_1, b_2) as ordinary 2D coordinates

$$
\begin{aligned}
(b_1, b_2, b_3) &\equiv b_1\mathbf{v}_1 + b_2\mathbf{v}_2 + b_3\mathbf{v}_3 \\
&\equiv b_1\mathbf{v}_1 + b_2\mathbf{v}_2 + (1 - b_1 - b_2)\mathbf{v}_3 \\
&\equiv b_1\mathbf{v}_1 + b_2\mathbf{v}_2 + \mathbf{v}_3 - b_1\mathbf{v}_3 - b_2\mathbf{v}_3 \\
&\equiv \mathbf{v}_3 + b_1(\mathbf{v}_1 - \mathbf{v}_3) + b_2(\mathbf{v}_2 - \mathbf{v}_3).
\end{aligned}
$$

This makes it very clear that, due to the normalization constraint, although there are three coordinates, there are only two degrees of freedom. We could completely describe a point in barycentric space using only two of the coordinates. In fact, the rank of the space described by the coordinates

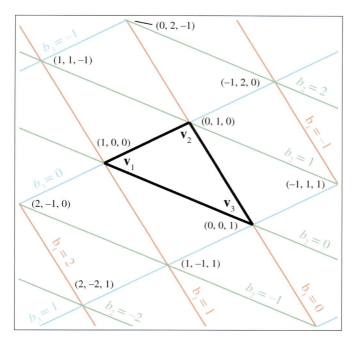

Figure 9.19
Barycentric coordinates tessellate the plane

doesn't depend on the dimension of the "sample points" but rather on the *number* of sample points. The number of degrees of freedom is one less than the number of barycentric coordinates, since we have the constraint that the coordinates sum to one. For example, if we have two sample points, the dimension of the barycentric coordinates is two, and the space that can be described using these coordinates is a line, which is a 1D space. Notice that the line may be a 1D line (i.e., interpolation of a scalar), a 2D line, a 3D line, or a line in some higher dimensional space. In this section, we've had three sample points (the vertices of our triangle) and three barycentric coordinates, resulting in a 2D space—a plane. If we had four sample points in 3D, then we could use barycentric coordinates to locate points in 3D. Four barycentric coordinates induce a "tetrahedronal" space, rather than the "triangular" space we get when there are three coordinates.

To convert a point from barycentric coordinates to standard Cartesian coordinates, we simply compute the weighted average of the vertices by applying Equation (9.17). The opposite conversion—computing the barycentric coordinates from Cartesian coordinates—is slightly more difficult, and is discussed in Section 9.6.4. However, before we get too far into the details (which might be skipped over by a casual reader), now that you have the basic idea behind barycentric coordinates, let us take this opportunity to mention a few places where barycentric coordinates are useful.

In graphics, it is common for parameters to be edited (or computed) per vertex, such as texture coordinates, colors, surface normals, lighting values, and so forth. We often then need to determine the interpolated value of one of those parameters at an arbitrary location within the triangle. Barycentric coordinates make this task easy. We first determine the barycentric coordinates of the interior point in question, and then take the weighted average of the values at the vertices for the parameter we seek.

Another important example is intersection testing. One simple way to perform ray-triangle testing is to determine the point where the ray intersects the infinite plane containing the triangle, and then to decide whether this point lies within the triangle. An easy way to make this decision is to calculate the barycentric coordinates of the point, using the techniques described here. If all of the coordinates lie in the $[0, 1]$ range, then the point is inside the triangle; otherwise at least one coordinate lies outside this range and the point is outside the triangle. It is common for the calculated barycentric coordinates to be further used to fetch some interpolated surface property. For example, let's say we are casting a ray to determine whether a light is visible to some point or if the point is in shadow. We strike a triangle on some model at an arbitrary location. If the model is opaque, the light is not visible. However, if the model uses transparency, we may need to determine the opacity at that location to determine what fraction of the light is blocked. Typically, this transparency is in a texture map, which is indexed using UV coordinates. (More about texture mapping is presented in Section 10.5.) To fetch the transparency at the location of ray intersection, we use the barycentric coordinates at the point to interpolate the UVs from the vertices. Then we use these UVs to fetch the texel from the texture map, and determine the transparency of that particular location on the surface.

9.6.4   Calculating Barycentric Coordinates

Now let's see how to determine barycentric coordinates from Cartesian coordinates. We start in 2D with Figure 9.20, which shows the three vertices $\mathbf{v}_1$, $\mathbf{v}_2$, and $\mathbf{v}_3$ and the point $\mathbf{p}$. We have also labeled the three "subtriangles" T_1, T_2, T_3, which are opposite the vertex of the same index. These will become useful in just a moment.

We know the Cartesian coordinates of the three vertices and the point $\mathbf{p}$. Our task is to compute the barycentric coordinates b_1, b_2, and b_3. This gives

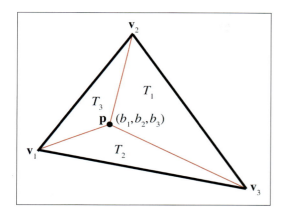

Figure 9.20
Computing the barycentric
coordinates for an arbitrary
point **p**

us three equations and three unknowns:

$$b_1 x_1 + b_2 x_2 + b_3 x_3 = p_x,$$
$$b_1 y_1 + b_2 y_2 + b_3 y_3 = p_y,$$
$$b_1 + b_2 + b_3 = 1.$$

Solving this system of equations yields

$$b_1 = \frac{(p_y - y_3)(x_2 - x_3) + (y_2 - y_3)(x_3 - p_x)}{(y_1 - y_3)(x_2 - x_3) + (y_2 - y_3)(x_3 - x_1)},$$

$$b_2 = \frac{(p_y - y_1)(x_3 - x_1) + (y_3 - y_1)(x_1 - p_x)}{(y_1 - y_3)(x_2 - x_3) + (y_2 - y_3)(x_3 - x_1)}, \qquad (9.18)$$

$$b_3 = \frac{(p_y - y_2)(x_1 - x_2) + (y_1 - y_2)(x_2 - p_x)}{(y_1 - y_3)(x_2 - x_3) + (y_2 - y_3)(x_3 - x_1)}.$$

Computing barycentric coordinates for a 2D point

Examining Equation (9.18) closely, we see that the denominator is the same in each expression—it is equal to twice the area of the triangle, according to Equation (9.15). What's more, for each barycentric coordinate b_i, the numerator is equal to twice the area of the "subtriangle" T_i. In other words,

$$b_1 = A(T_1)/A(T), \qquad b_2 = A(T_2)/A(T), \qquad b_3 = A(T_3)/A(T).$$

Interpreting barycentric coordinates as ratios of areas

Note that this interpretation applies even if **p** is outside the triangle, since our equation for computing area yields a negative result if the vertices are enumerated in a counterclockwise order. If the three vertices of the triangle are collinear, then the triangle is degenerate and the area in the denominator will be zero, and thus the barycentric coordinates cannot be computed.

Computing barycentric coordinates for an arbitrary point **p** in 3D is more complicated than in 2D. We cannot solve a system of equations as we did before, since we have three unknowns and four equations (one equation for each coordinate of **p**, plus the normalization constraint on the barycentric coordinates). Another complication is that **p** may not lie in the plane that contains the triangle, in which case the barycentric coordinates are undefined. For now, let's assume that **p** lies in the plane containing the triangle.

One trick that works is to turn the 3D problem into a 2D problem simply by discarding one of x, y, or z. This has the effect of projecting the triangle onto one of the three cardinal planes. Intuitively, this works because the projected areas are proportional to the original areas.

But which coordinate should we discard? We can't just always discard the same one, since the projected points will be collinear if the triangle is perpendicular to the projection plane. If our triangle is nearly perpendicular to the plane of projection, we will have problems with floating point accuracy. A solution to this dilemma is to choose the plane of projection so as to maximize the area of the projected triangle. This can be done by examining the plane normal, and whichever coordinate has the largest absolute value is the coordinate that we will discard. For example, if the normal is $[0.267, -0.802, 0.535]$ then we would discard the y values of the vertices and **p**, projecting onto the xz-plane. The code snippet in Listing 9.6 shows how to compute the barycentric coordinates for an arbitrary 3D point.

```
bool computeBarycentricCoords3d(
    const Vector3 v[3],  // vertices of the triangle
    const Vector3 &p,    // point that we wish to compute coords for
    float b[3]           // barycentric coords returned here
) {

    // First, compute two clockwise edge vectors
    Vector3 d1 = v[1] - v[0];
    Vector3 d2 = v[2] - v[1];

    // Compute surface normal using cross product. In many cases
    // this step could be skipped, since we would have the surface
    // normal precomputed. We do not need to normalize it, although
    // if a precomputed normal was normalized, it would be OK.
    Vector3 n = crossProduct(d1, d2);

    // Locate dominant axis of normal, and select plane of projection
    float u1, u2, u3, u4;
    float v1, v2, v3, v4;
    if ((fabs(n.x) >= fabs(n.y)) && (fabs(n.x) >= fabs(n.z))) {

        // Discard x, project onto yz plane
        u1 = v[0].y - v[2].y;
        u2 = v[1].y - v[2].y;
        u3 = p.y - v[0].y;
        u4 = p.y - v[2].y;
```

```
         v1 = v[0].z − v[2].z;
         v2 = v[1].z − v[2].z;
         v3 = p.z − v[0].z;
         v4 = p.z − v[2].z;

    } else if (fabs(n.y) >= fabs(n.z)) {

         // Discard y, project onto xz plane
         u1 = v[0].z − v[2].z;
         u2 = v[1].z − v[2].z;
         u3 = p.z − v[0].z;
         u4 = p.z − v[2].z;

         v1 = v[0].x − v[2].x;
         v2 = v[1].x − v[2].x;
         v3 = p.x − v[0].x;
         v4 = p.x − v[2].x;

    } else {

         // Discard z, project onto xy plane
         u1 = v[0].x − v[2].x;
         u2 = v[1].x − v[2].x;
         u3 = p.x − v[0].x;
         u4 = p.x − v[2].x;

         v1 = v[0].y − v[2].y;
         v2 = v[1].y − v[2].y;
         v3 = p.y − v[0].y;
         v4 = p.y − v[2].y;
    }

    // Compute denominator, check for invalid
    float denom = v1*u2 − v2*u1;
    if (denom == 0.0f) {

         // Bogus triangle − probably triangle has zero area
         return false;
    }

    // Compute barycentric coordinates
    float oneOverDenom = 1.0f / denom;
    b[0] = (v4*u2 − v2*u4) * oneOverDenom;
    b[1] = (v1*u3 − v3*u1) * oneOverDenom;
    b[2] = 1.0f − b[0] − b[1];

    // OK
    return true;
}
```

Listing 9.6
Computing barycentric coordinates in 3D

Another technique for computing barycentric coordinates in 3D is based on the method for computing the area of a 3D triangle using the cross product, which is discussed in Section 9.6.2. Recall that given two edge vectors $\mathbf{e}_1$ and $\mathbf{e}_2$ of a triangle, we can compute the area of the triangle as $\|\mathbf{e}_1 \times \mathbf{e}_2\|/2$. Once we have the area of the entire triangle and the areas of the three subtriangles, we can compute the barycentric coordinates.

There is one slight problem with this: the magnitude of the cross product is not sensitive to the ordering of the vertices—magnitude is by definition always positive. This will not work for points outside the triangle, since these points must always have at least one negative barycentric coordinate.

Let's see if we can find a way to work around this problem. It seems like what we really need is a way to calculate the length of the cross product vector that would yield a negative value if the vertices were enumerated in the "incorrect" order. As it turns out, there is a very simple way to do this with the dot product.

Let's assign $\mathbf{c}$ to be the cross product of two edge vectors of a triangle. Remember that the magnitude of $\mathbf{c}$ will equal twice the area of the triangle. Assume we have a normal $\hat{\mathbf{n}}$ of unit length. Now, $\hat{\mathbf{n}}$ and $\mathbf{c}$ are parallel, since they are both perpendicular to the plane containing the triangle. However, they may point in opposite directions. Recall from Section 2.11.2 that the dot product of two vectors is equal to the product of their magnitudes times the cosine of the angle between them. Since we know that $\hat{\mathbf{n}}$ is a unit vector, and the vectors are either pointing in the exact same or the exact opposite direction, we have

$$\begin{aligned} \mathbf{c} \cdot \hat{\mathbf{n}} &= \|\mathbf{c}\| \|\hat{\mathbf{n}}\| \cos \theta \\ &= \|\mathbf{c}\|(1)(\pm 1) \\ &= \pm \|\mathbf{c}\|. \end{aligned}$$

Dividing this result by two, we have a way to compute the "signed area" of a triangle in 3D. Armed with this trick, we can now apply the observation from the previous section, that each barycentric coordinate b_i is proportional to the area of the subtriangle T_i. Let us first label all of the vectors involved, as shown in Figure 9.21.

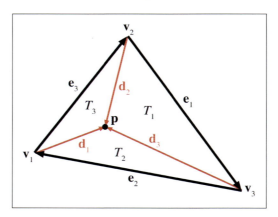

Figure 9.21
Computing barycentric coordinates in 3D

As Figure 9.21 shows, each vertex has a vector from $\mathbf{v}_i$ to $\mathbf{p}$, named $\mathbf{d}_i$. Summarizing the equations for the vectors, we have

$$\mathbf{e}_1 = \mathbf{v}_3 - \mathbf{v}_2, \qquad \mathbf{e}_2 = \mathbf{v}_1 - \mathbf{v}_3, \qquad \mathbf{e}_3 = \mathbf{v}_2 - \mathbf{v}_1,$$
$$\mathbf{d}_1 = \mathbf{p} - \mathbf{v}_1, \qquad \mathbf{d}_2 = \mathbf{p} - \mathbf{v}_2, \qquad \mathbf{d}_3 = \mathbf{p} - \mathbf{v}_3.$$

We'll also need a surface normal, which can be computed by

$$\hat{\mathbf{n}} = \frac{\mathbf{e}_1 \times \mathbf{e}_2}{\|\mathbf{e}_1 \times \mathbf{e}_2\|}.$$

Now the areas for the entire triangle (which we'll simply call T) and the three subtriangles are given by

$$A(T) = ((\mathbf{e}_1 \times \mathbf{e}_2) \cdot \hat{\mathbf{n}})/2,$$
$$A(T_1) = ((\mathbf{e}_1 \times \mathbf{d}_3) \cdot \hat{\mathbf{n}})/2,$$
$$A(T_2) = ((\mathbf{e}_2 \times \mathbf{d}_1) \cdot \hat{\mathbf{n}})/2,$$
$$A(T_3) = ((\mathbf{e}_3 \times \mathbf{d}_2) \cdot \hat{\mathbf{n}})/2.$$

Each barycentric coordinate b_i is given by $A(T_i)/A(T)$:

$$b_1 = A(T_1)/A(T) = \frac{(\mathbf{e}_1 \times \mathbf{d}_3) \cdot \hat{\mathbf{n}}}{(\mathbf{e}_1 \times \mathbf{e}_2) \cdot \hat{\mathbf{n}}},$$

Computing barycentric coordinates in 3D

$$b_2 = A(T_2)/A(T) = \frac{(\mathbf{e}_2 \times \mathbf{d}_1) \cdot \hat{\mathbf{n}}}{(\mathbf{e}_1 \times \mathbf{e}_2) \cdot \hat{\mathbf{n}}},$$

$$b_3 = A(T_3)/A(T) = \frac{(\mathbf{e}_3 \times \mathbf{d}_2) \cdot \hat{\mathbf{n}}}{(\mathbf{e}_1 \times \mathbf{e}_2) \cdot \hat{\mathbf{n}}}.$$

Notice that $\hat{\mathbf{n}}$ is used in all of the numerators and all of the denominators, and so it is doesn't necessarily have to be a unit vector.

This technique for computing barycentric coordinates involves more scalar math operations than the method of projection into 2D. However, it is branchless and offers better SIMD optimization.

9.6.5 Special Points

In this section we discuss three points on a triangle that have special geometric significance:

- center of gravity

- incenter

- circumcenter.

To present these classic calculations, we follow Goldman's article [25] from *Graphics Gems*. For each point, we discuss its geometric significance and construction and give its barycentric coordinates.

The *center of gravity* is the point on which the triangle would balance perfectly. It is the intersection of the medians. (A *median* is a line from one vertex to the midpoint of the opposite side.) Figure 9.22 shows the center of gravity of a triangle.

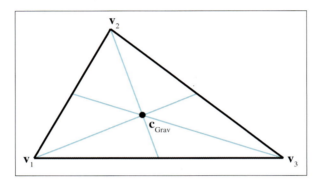

Figure 9.22
The center of gravity
of a triangle

The center of gravity is the geometric average of the three vertices:

$$\mathbf{c}_{\text{Grav}} = \frac{\mathbf{v}_1 + \mathbf{v}_2 + \mathbf{v}_3}{3}.$$

The barycentric coordinates are

$$\left(\frac{1}{3}, \frac{1}{3}, \frac{1}{3} \right).$$

The center of gravity is also known as the *centroid*.

The *incenter* is the point in the triangle that is equidistant from the sides. It is called the incenter because it is the center of the circle inscribed in the triangle. The incenter is constructed as the intersection of the angle bisectors, as shown in Figure 9.23.

The incenter is computed by

$$\mathbf{c}_{\text{In}} = \frac{l_1 \mathbf{v}_1 + l_2 \mathbf{v}_2 + l_3 \mathbf{v}_3}{p},$$

where $p = l_1 + l_2 + l_3$ is the perimeter of the triangle. Thus the barycentric coordinates of the incenter are

$$\left(\frac{l_1}{p}, \frac{l_2}{p}, \frac{l_3}{p} \right).$$

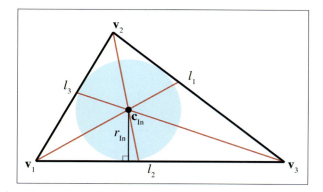

Figure 9.23
The incenter of a triangle

The radius of the inscribed circle can be computed by dividing the area of the triangle by its perimeter:

$$r_{\text{In}} = \frac{A}{p}.$$

The inscribed circle solves the problem of finding a circle tangent to three lines.

The *circumcenter* is the point in the triangle that is equidistant from the vertices. It is the center of the circle that circumscribes the triangle. The circumcenter is constructed as the intersection of the perpendicular bisectors of the sides. Figure 9.24 shows the circumcenter of a triangle.

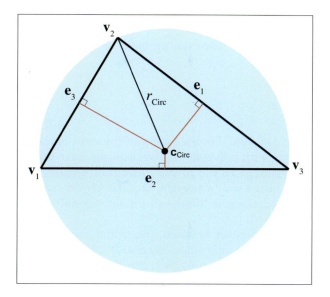

Figure 9.24
The circumcenter of a triangle

To compute the circumcenter, we first define the following intermediate values:

$$d_1 = -\mathbf{e}_2 \cdot \mathbf{e}_3,$$
$$d_2 = -\mathbf{e}_3 \cdot \mathbf{e}_1,$$
$$d_3 = -\mathbf{e}_1 \cdot \mathbf{e}_2,$$
$$c_1 = d_2 d_3,$$
$$c_2 = d_3 d_1,$$
$$c_3 = d_1 d_2,$$
$$c = c_1 + c_2 + c_3.$$

With those intermediate values, the barycentric coordinates for the circumcenter are given by

$$\left(\frac{c_2 + c_3}{2c}, \frac{c_3 + c_1}{2c}, \frac{c_1 + c_2}{2c} \right);$$

thus, the circumcenter is given by

$$\mathbf{c}_{\text{Circ}} = \frac{(c_2 + c_3)\mathbf{v}_1 + (c_3 + c_1)\mathbf{v}_2 + (c_1 + c_2)\mathbf{v}_3}{2c}.$$

The circumradius is given by

$$r_{\text{Circ}} = \frac{\sqrt{(d_1 + d_2)(d_2 + d_3)(d_3 + d_1)/c}}{2}.$$

The circumradius and circumcenter solve the problem of finding a circle that passes through three points.

9.7   Polygons

This section introduces polygons and discusses a few of the most important issues that arise when dealing with polygons. It is difficult to come up with a simple definition for polygon, since the precise definition usually varies depending on the context. In general, a polygon is a flat object made up of vertices and edges. The next few sections will discuss several ways in which polygons may be classified.

Section 9.7.1 presents the difference between simple and complex polygons and mentions self-intersecting polygons. Section 9.7.2 discusses the difference between convex and concave polygons. Section 9.7.3 describes how any polygon may be turned into connected triangles.

9.7.1 Simple versus Complex Polygons

A *simple* polygon does not have any "holes," whereas a *complex* polygon may have holes (see Figure 9.25.) A simple polygon can be described by enumerating the vertices in order around the polygon. (Recall that in a left-handed world, we usually enumerate them in clockwise order when viewed from the "front" side of the polygon.) Simple polygons are used much more frequently than complex polygons.

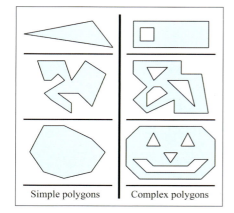

Simple polygons | Complex polygons

Figure 9.25
Simple versus complex polygons

We can turn any complex polygon into a simple one by adding pairs of "seam" edges, as shown in Figure 9.26. As the close-up on the right shows, we add two edges per seam. The edges are actually coincident, although in the close-up they have been separated so you could see them. When we think about the edges being ordered around the polygon, the two seam edges point in opposite directions.

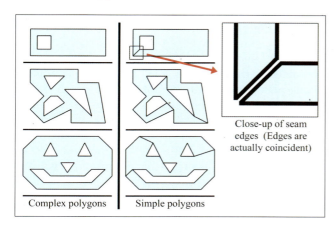

Complex polygons | Simple polygons

Close-up of seam
edges (Edges are
actually coincident)

Figure 9.26
Turning complex
polygons into
simple ones by
adding pairs of
seam edges

The edges of most simple polygons do not intersect each other. If the edges do intersect, the polygon is considered a *self-intersecting* polygon. An example of a self-intersecting polygon is shown in Figure 9.27.

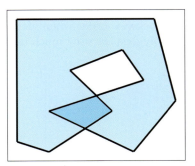

Figure 9.27
A self-intersecting polygon

Most people usually find it easier to arrange things so that self-intersecting polygons are either avoided, or simply rejected. In most situations, this is not a huge burden on the user.

9.7.2   Convex versus Concave Polygons

Non-self-intersecting simple polygons may be further classified as either *convex* or *concave*. Giving a precise definition for "convex" is actually somewhat tricky because there are many sticky degenerate cases. For most polygons, the following commonly used definitions are equivalent, although some degenerate polygons may be classified as convex according to one definition and concave according to another.

- Intuitively, a convex polygon doesn't have any "dents." A concave polygon has at least one vertex that is a "dent," which is called a *point of concavity* (see Figure 9.28).

- In a convex polygon, the line between any two points in the polygon is completely contained within the polygon. In a concave polygon, there is at least one pair of points in the polygon for which the line between the points is partially outside the polygon.

- As we move around the perimeter of a convex polygon, at each vertex we will turn in the same direction. In a concave polygon, we will make some left-hand turns and some right-hand turns. We will turn the opposite direction at the point(s) of concavity. (Note that this applies to non-self-intersecting polygons only.)

As we mentioned, degenerate cases can make even these relatively clear-cut definitions blurry. For example, what about a polygon with two consecutive coincident vertices, or an edge that doubles back on itself? Are those polygons considered convex? In practice, the following "definitions" for convexity are often used:

- If my code, which is supposed to work only for convex polygons, can deal with it, then it's convex. (This is the "if it ain't broke, don't fix it" definition.)

- If my algorithm that tests for convexity decides it's convex, then it's convex. (This is an "algorithm as definition" explanation.)

For now, let's ignore the pathological cases, and give some examples of polygons that we can all agree are definitely convex or definitely concave. The top concave polygon in Figure 9.28 has one point of concavity. The bottom concave polygon has five points of concavity.

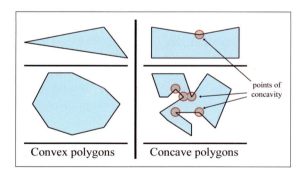

points of
concavity

Figure 9.28
Convex vs. concave
polygons

| Convex polygons | Concave polygons |

Any concave polygon may be divided into convex pieces. The basic idea is to locate the points of concavity (called "reflex vertices") and systematically remove them by adding diagonals. O'Rourke [52] provides an algorithm that works for simple polygons, and de Berg et al. [12] show a more complicated method that works for complex polygons as well.

How can we know if a polygon is convex or concave? One method is to examine the sum of the angles at the vertices. Consider a convex polygon with n vertices. The sum of interior angles in a convex polygon is $(n-2)180°$. We have two different ways to show this to be true.

First, let θ_i measure the interior angle at vertex i. Clearly, if the polygon is convex, then $\theta_i \leq 180°$. The amount of "turn" that occurs at each vertex will be $180° - \theta_i$. A closed polygon will of course turn one complete revolution, or 360°. Therefore,

$$\sum_{i=1}^{n}(180° - \theta_i) = 360°,$$

$$n180° - \sum_{i=1}^{n}\theta_i = 360°,$$

$$-\sum_{i=1}^{n}\theta_i = 360° - n180°,$$

$$\sum_{i=1}^{n}\theta_i = n180° - 360°,$$

$$\sum_{i=1}^{n}\theta_i = (n - 2)180°.$$

Second, as we will show in Section 9.7.3, any convex polygon with n vertices can be triangulated into $n - 2$ triangles. From classical geometry, the sum of the interior angles of a triangle is 180°. The sum of the interior angles of all of the triangles is $(n - 2)180°$, and we can see that this sum must also be equal to the sum of the interior angles of the polygon itself.

Unfortunately, the sum of the interior angles is $(n-2)180°$ for concave as well as convex polygons. So how does this get us any closer to determining whether or not a polygon is convex? As shown in (see Figure 9.29), the dot product can be used to measure the *smaller* of the exterior and interior angles. The exterior angle of a polygon vertex is the compliment of the

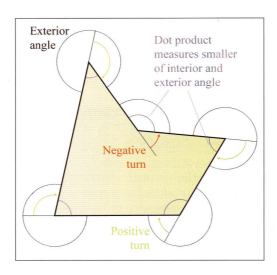

Figure 9.29
Using the dot product to determine whether a polygon is convex or concave

interior angle, meaning they sum to 360°. (It's not the same as the "turn angle," and you might notice that the definition of exterior angle for polygon vertices is different from the classic one used for triangle vertices.) So, if we take the sum of the smaller angle (interior or exterior) at each vertex, then the sum will be $(n-2)180°$ for convex polygons, and less than that if the polygon is concave.

Listing 9.7 shows how to determine if a polygon is convex by summing the angles.

```cpp
bool isPolygonConvex(
    int n,              // Number of vertices
    const Vector3 vl[], // pointer to array of of vertices
) {

    // Initialize sum to 0 radians
    float angleSum = 0.0f;

    // Go around the polygon and sum the angle at each vertex
    for (int i = 0 ; i < n ; ++i) {

        // Get edge vectors.  We have to be careful on
        // the first and last vertices.  Also, note that
        // this could be optimized considerably
        Vector3 e1;
        if (i == 0) {
            e1 = vl[n-1] - vl[i];
        } else {
            e1 = vl[i-1] - vl[i];
        }

        Vector3 e2;
        if (i == n-1) {
            e2 = vl[0] - vl[i];
        } else {
            e2 = vl[i+1] - vl[i];
        }

        // Normalize and compute dot product
        e1.normalize();
        e2.normalize();
        float dot = e1 * e2;

        // Compute smaller angle using ''safe'' function that protects
        // against range errors which could be caused by
        // numerical imprecision
        float theta = safeAcos(dot);

        // Sum it up
        angleSum += theta;
    }

    // Figure out what the sum of the angles should be, assuming
    // we are convex.  Remember that pi rad = 180 degrees
    float convexAngleSum = (float)(n - 2) * kPi;

    // Now, check if the sum of the angles is less than it should be,
    // then we're concave.  We give a slight tolerance for
    // numerical imprecision
    if (angleSum < convexAngleSum - (float)n * 0.0001f) {
```

```
        // We're concave
        return false;
    }

    // We're convex, within tolerance
    return true;
}
```

Listing 9.7
3D polygon convexity test using angle sum

Another method for determining convexity is to search for vertices that are points of concavity. If none are found, then the polygon is convex. The basic idea is that each vertex should turn in the same direction. Any vertex that turns in the opposite direction is a point of concavity. We can determine which way a vertex turns using by the cross product on the edge vectors. Recall from Section 2.12.2 that in a left-handed coordinate system, the cross product will point towards you if the vectors form a clockwise turn. By "towards you," we assume you are viewing the polygon from the front, as determined by the polygon normal. If this normal is not available to us initially, care must be exercised in computing it; because we do not know if the polygon is convex or not, we cannot simply choose any three vertices to compute the normal from. The techniques in Section 9.5.3 for computing the best fit normal from a set of points can be used in this case.

Once we have a normal, we check each vertex of the polygon, computing a normal at that vertex using the adjacent clockwise edge vectors. We take the dot product of the polygon normal with the normal computed at that vertex to determine if they point in opposite directions. If so (the dot product is negative), then we have located a point of concavity.

In 2D, we can simply act as if the polygon were in 3D at the plane $z = 0$, and assume the normal is $[0, 0, -1]$. There are subtle difficulties with any method for determining convexity. Schorn and Fisher [60] discuss the topic in greater detail.

9.7.3   Triangulation and Fanning

Any polygon can be divided into triangles. Thus, all of the operations and calculations for triangles can be piecewise applied to polygons. Triangulating complex, self-intersecting, or even simple concave polygons is no trivial task [12, 52] and is slightly out of the scope of this book.

Luckily, triangulating simple convex polygons *is* a trivial matter. One obvious triangulation technique is to pick one vertex (say, the first one) and "fan" the polygon around this vertex. Given a polygon with n vertices, enumerated $\mathbf{v}_1 \dots \mathbf{v}_n$ around the polygon, we can easily form $n-2$ triangles, each of the form $\{\mathbf{v}_1, \mathbf{v}_{i-1}, \mathbf{v}_i\}$ with the index i going from 3 to n, as shown in Figure 9.30.

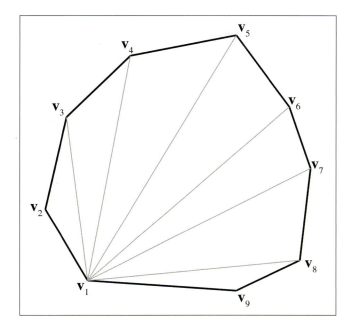

Figure 9.30
Triangulating a
convex polygon
by fanning

Fanning tends to create many long, thin sliver triangles, which can be troublesome in some situations, such as computing a surface normal. Certain consumer hardware can run into precision problems when clipping very long edges to the view frustum. Smarter techniques exist that attempt to minimize this problem. One idea is to triangulate as follows: Consider that we can divide a polygon into two pieces with a diagonal between two vertices. When this occurs, the two interior angles at the vertices of the diagonal are each divided into two new interior angles. Thus, a total of four new interior angles are created. To subdivide a polygon, select the diagonal that maximizes the smallest of these four new interior angles. Divide the polygon in two using this diagonal. Recursively apply the procedure to each half, until only triangles remain. This algorithm results in a triangulation with fewer slivers.

9.8   Exercises

(Answers on page 774.)

1. Given the 2D ray in parametric form

$$\mathbf{p}(t) = \begin{bmatrix} 5 \\ 3 \end{bmatrix} + t \begin{bmatrix} -7 \\ 5 \end{bmatrix},$$

determine the line that contains this ray, in slope-intercept form.

2. Give the slope and y-intercept of the 2D line defined implicitly by $4x+7y = 42$.

3. Consider the set of five points:

$$\mathbf{v}_1 = (7, 11, -5), \qquad \mathbf{v}_2 = (2, 3, 8), \qquad \mathbf{v}_3 = (-3, 3, 1),$$
$$\mathbf{v}_4 = (-5, -7, 0), \qquad \mathbf{v}_5 = (6, 3, 4).$$

(a) Determine the AABB of this box. What are $\mathbf{p}_{\min}$ and $\mathbf{p}_{\max}$?

(b) List all eight corner points.

(c) Determine the center point $\mathbf{c}$ and size vector $\mathbf{s}$.

(d) Multiply the five points by the following matrix, which we hope you recognize as a $45°$ rotation about the z-axis:

$$\mathbf{M} = \begin{bmatrix} 0.707 & 0.707 & 0 \\ -0.707 & 0.707 & 0 \\ 0 & 0 & 1 \end{bmatrix}.$$

(e) What is the AABB of these transformed points?

(f) What is the AABB we get by transforming the original AABB? (The bounding box of the transformed corner points.)

4. Consider a triangle defined by the clockwise enumeration of the vertices $(6, 10, -2)$, $(3, -1, 17)$, $(-9, 8, 0)$.

(a) What is the plane equation of the plane containing this triangle?

(b) Is the point $(3, 4, 5)$ on the front or back side of this plane? How far is this point from the plane?

(c) Compute the barycentric coordinates of the point $(13.60, -0.46, 17.11)$.

(d) What is the center of gravity?

(e) What is the incenter?

(f) What is the circumcenter?

5. What is the best-fit plane equation for the following points, which are not quite coplanar?

$$\mathbf{p}_1 = (-29.74, 13.90, 12.70) \qquad \mathbf{p}_4 = (14.62, 10.64, -7.09)$$
$$\mathbf{p}_2 = (11.53, 12.77, -9.22) \qquad \mathbf{p}_5 = (-3.31, 3.16, 18.68)$$
$$\mathbf{p}_3 = (9.16, 2.34, 12.67)$$

6. Consider a convex polygon P that has seven vertices numbered $\mathbf{v}_1 \ldots \mathbf{v}_7$. Show how to fan this polygon.

A square was sitting quietly
Outside his rectangular shack
When a triangle came down–keerplunk!–
"I must go to the hospital,"
Cried the wounded square,
So a passing rolling circle
Picked him up and took him there.

— *Shapes* (1981) by Shel Silverstein

Chapter 10

Mathematical Topics
from 3D Graphics

I don't think there's anything wrong with pretty graphics.

— Shigeru Miyamoto (1952–)

This chapter discusses a number of mathematical issues that arise when creating 3D graphics on a computer. Of course, we cannot hope to cover the vast subject of computer graphics in any amount of detail in a single chapter. Entire books are written that merely survey the topic. This chapter is to graphics what this entire book is to interactive 3D applications: it presents an extremely brief and high level overview of the subject matter, focusing on topics for which mathematics plays a critical role. Just like the rest of this book, we try to pay special attention to those topics that, from our experience, are glossed over in other sources or are a source of confusion in beginners.

To be a bit more direct: this chapter alone is not enough to teach you how to get some pretty pictures on the screen. However, it should be used parallel with (or preceding!) some other course, book, or self-study on graphics, and we hope that it will help you breeze past a few traditional sticky points. Although we present some example snippets in High Level Shading Language (HLSL) at the end of this chapter, you will not find much else to help you figure out which DirectX or OpenGL function calls to make to achieve some desired effect. These issues are certainly of supreme practical importance, but alas, they are also in the category of knowledge that Robert Maynard Hutchins dubbed "rapidly aging facts," and we have tried to avoid writing a book that requires an update every other year when ATI releases a new card or Microsoft a new version of DirectX. Luckily, up-to-date API references and examples abound on the Internet, which is a much more appropriate place to get that sort of thing. (API stands for application programming interface. In this chapter, API will mean the software that we use to communicate with the rendering subsystem.)

One final caveat is that since this is a book on math for video games, we will have a real-time bias. This is not to say that the book cannot be used if you are interested in learning how to write a raytracer; only that our expertise and focus is in real-time graphics.

This chapter proceeds roughly in order from ivory tower theory to down-and-dirty code snippets.

- Section 10.1 gives a very high-level (and high-brow) theoretical approach to graphics, culminating in the *rendering equation*.

- We then lower our brows somewhat to focus attention on matters of more direct practical application, while still maintaining our platform independence and attempt to be relevant ten years from now.

  ○ Section 10.2 discusses some basic mathematics related to *viewing* in 3D.

  ○ Section 10.3 introduces some important coordinate spaces and transformations.

  ○ Section 10.4 looks at how to represent the surfaces of the geometry in our scene using a *polygon mesh*.

  ○ Section 10.5 shows how to control material properties (such as the "color" of the object) using *texture maps*.

- The next sections are about lighting.

  ○ Section 10.6 defines the ubiquitous *Blinn-Phong lighting model*.

  ○ Section 10.7 discusses some common methods for representing light sources.

- With a little nudge further away from timeless theory, the next sections discuss two issues of particular contemporary interest.

  ○ Section 10.8 is about *skeletal animation*.

  ○ Section 10.9 tells how *bump mapping* works.

- The last third of this chapter is the most in danger of becoming irrelevant in coming years, because it is the most immediately practical.

  ○ Section 10.10 gives an overview of a simple real-time graphics pipeline, and then descends that pipeline and talks about some mathematical issues along the way.

  ○ Section 10.11 concludes the chapter squarely in the "rapidly aging facts" territory with several HLSL examples demonstrating some of the techniques covered earlier.

10.1   How Graphics Works

We begin our discussion of graphics by telling you how things *really* work, or perhaps more accurately, how they really *should* work, if we had enough knowledge and processing power to make things work the right way. The beginner student is to be warned that much introductory material (especially tutorials on the Internet) and API documentation suffers from a great lack of perspective. You might get the impression from reading these sources that diffuse maps, Blinn-Phong shading, and ambient occlusion are "The way images in the real world work," when in fact you are probably reading a description of how one particular lighting model was implemented in one particular language on one particular piece of hardware through one particular API. Ultimately, any down-to-the-details tutorial must choose a lighting model, language, platform, color representation, performance goals, etc.—as we will have to do later in this chapter. (This lack of perspective is usually purposeful and warranted.) However, we think it's important to know which are the fundamental and timeless principles, and which are arbitrary choices based on approximations and trade-offs, guided by technological limitations that might by applicable only to real-time rendering, or are likely to change in the near future. So before we get too far into the details of the particular type of rendering most useful for introductory real-time graphics, we want to take our stab at describing how rendering *really* works.

We also hasten to add that this discussion assumes that the goal is photorealism, simulating how things work in nature. In fact, this is often not the goal, and it certainly is never the only goal. Understanding how nature works is a very important starting place, but artistic and practical factors often dictate a different strategy than just simulating nature.

10.1.1   The Two Major Approaches to Rendering

We begin with the end in mind. The end goal of rendering is a bitmap, or perhaps a sequence of bitmaps if we are producing an animation. You almost certainly already know that a bitmap is a rectangular array of colors, and each grid entry is known as *pixel*, which is short for "picture element." At the time we are producing the image, this bitmap is also known as the *frame buffer*, and often there is additional post-processing or conversion that happens when we copy the frame buffer to the final bitmap output.

How do we determine the color of each pixel? That is the fundamental question of rendering. Like so many challenges in computer science, a great place to start is by investigating how nature works.

We see light. The image that we perceive is the result of light that bounces around the environment and finally enters the eye. This process is complicated, to say the least. Not only is the physics[1] of the light bouncing around very complicated, but so are the physiology of the sensing equipment in our eyes[2] and the interpreting mechanisms in our minds. Thus, ignoring a great number of details and variations (as any introductory book must do), the basic question that any rendering system must answer for each pixel is "What color of light is approaching the camera from the direction corresponding to this pixel?"

There are basically two cases to consider. Either we are looking directly at a light source and light traveled directly from the light source to our eye, or (more commonly) light departed from a light source in some other direction, bounced one or more times, and then entered our eye. We can decompose the key question asked previously into two tasks. This book calls these two tasks *the rendering algorithm*, although these two highly abstracted procedures obviously conceal a great deal of complexity about the actual algorithms used in practice to implement it.

The rendering algorithm

- *Visible surface determination.* Find the surface that is closest to the eye, in the direction corresponding to the current pixel.

- *Lighting.* Determine what light is emitted and/or reflected off this surface in the direction of the eye.

At this point it appears that we have made some gross simplifications, and many of you no doubt are raising your metaphorical hands to ask "What about translucency?" "What about reflections?" "What about refraction?" "What about atmospheric effects?" Please hold all questions until the end of the presentation.

The first step in the rendering algorithm is known as *visible surface determination.* There are two common solutions to this problem. The first is known as *raytracing.* Rather than following light rays in the direction that they travel from the emissive surfaces, we trace the rays backward, so that we can deal only with the light rays that matter: the ones that enter our eye from the given direction. We send a ray out from the eye in the direction through the center of each pixel[3] to see the first object in the scene this ray strikes. Then we compute the color that is being emitted

---

[1] Actually, almost everybody approximates the true physics of light by using simpler *geometric optics.*

[2] Speaking of equipment, there are also many phenomena that occur in a camera but not the eye, or as a result of the storage of an image on film. These effects, too, are often simulated to make it look as if the animation was filmed.

[3] Actually, it's probably not a good idea to think of pixels as having a "center," as they are not really rectangular blobs of color, but rather are best interpreted as infinitely small point samples in a continuous signal. The question of which mental model is best is incredibly important [33, 66], and is intimately related to the process by which the pixels are combined to reconstruct an image. On CRTs, pixels were definitely not little rectangles, but on modern display devices such as LCD monitors, "rectangular blob of

```
for (each x,y screen pixel) {

    // Select a ray for this pixel
    Ray ray = getRayForPixel(x,y);

    // Intersect the ray against the geometry.  This will
    // not just return the point of intersection, but also
    // a surface normal and some other information needed
    // to shade the point, such as an object reference,
    // material information, local S,T coordinates, etc.
    // Don't take this pseudocode too literally.
    Vector3 pos, normal;
    Object *obj; Material *mtl;
    if (rayIntersectScene(ray, pos, normal, obj, mtl)) {

        // Shade the intersection point.  (What light is
        // emitted/reflected from this point towards the camera?)
        Color c = shadePoint(ray, pos, normal, obj, mtl);

        // Put it into the frame buffer
        writeFrameBuffer(x,y, c);

    } else {

        // Ray missed the entire scene.  Just use a generic
        // background color at this pixel
        writeFrameBuffer(x,y, backgroundColor);
    }
}
```

Listing 10.1
Pseudocode for the raytracing algorithm

or reflected from that surface back in the direction of the ray. A highly simplified summary of this algorithm is illustrated by Listing 10.1.

The other major strategy for visible surface determination, the one used for real-time rendering at the time of this writing, is known as *depth buffering*. The basic plan is that at each pixel we store not only a color value, but also a depth value. This depth buffer value records the distance from the eye to the surface that is reflecting or emitting the light used to determine the color for that pixel. As illustrated in Listing 10.1, the "outer loop" of a raytracer is the screen-space pixels, but in real-time graphics, the "outer loop" is the geometric elements that make up the surface of the scene.

The different methods for describing surfaces are not important here. What *is* important is that we can project the surface onto screen-space and map them to screen-space pixels through a process known as *rasterization*. For each pixel of the surface, known as the *source fragment*, we compute the depth of the surface at that pixel and compare it to the existing value

color" is not too bad of a description of the reconstruction process. Nonetheless, whether pixels are rectangles or point samples, we still might not send a single ray through the center of each pixel, but rather we might send several rays ("samples") in a smart pattern, and average them together them in a smart way.

```
// Clear the frame and depth buffers
fillFrameBuffer(backgroundColor);
fillDepthBuffer(infinity);

// Outer loop iterates over all the primitives (usually triangles)
for (each geometric primitive) {

    // Rasterize the primitive
    for (each pixel x,y in the projection of the primitive) {

        // Test the depth buffer, to see if a closer pixel has
        // already been written.
        float primDepth = getDepthOfPrimitiveAtPixel(x,y);
        if (primDepth > readDepthBuffer(x,y)) {

            // Pixel of this primitive is obscured, discard it
            continue;
        }

        // Determine primitive color at this pixel.
        Color c = getColorOfPrimitiveAtPixel(x,y);

        // Update the color and depth buffers
        writeFrameBuffer(x,y, c);
        writeDepthBuffer(x,y, primDepth);
    }
}
```

Listing 10.2
Pseudocode for forward rendering using the depth buffer

in the depth buffer, sometimes known as the *destination fragment*. If the source fragment we are currently rendering is farther away from the camera than the existing value in the buffer, then whatever we rendered before this is obscuring the surface we are now rendering (at least at this one pixel), and we move on to the next pixel. However, if our depth value is closer than the existing value in the depth buffer, then we know this is the closest surface to the eye (at least of those rendered so far) and so we update the depth buffer with this new, closer depth value. At this point we might also proceed to step 2 of the rendering algorithm (at least for this pixel) and update the frame buffer with the color of the light being emitted or reflected from the surface that point. This is known as *forward rendering*, and the basic idea is illustrated by Listing 10.2.

Opposed to forward rendering is *deferred rendering*, an old technique that is becoming popular again due to the current location of bottlenecks in the types of images we are producing and the hardware we are using to produce them. A deferred renderer uses, in addition to the frame buffer and the depth buffer, additional buffers, collectively known as the *G-buffer* (short for "geometry" buffer), which holds extra information about the surface closest to the eye at that location, such as the 3D location of the surface, the surface normal, and material properties needed for lighting calculations, such as the "color" of the object and how "shiny" it is at that particular location. (Later, we see how those intuitive terms in quotes are a bit too vague for rendering purposes.) Compared to a forward renderer, a deferred renderer follows our two-step rendering algorithm a bit more literally. First we "render" the scene into the G-buffer, essentially performing only visibility determination—fetching the material properties of the point that is "seen" by each pixel but not yet performing lighting calculations. The second pass actually performs the lighting calculations. Listing 10.3 explains deferred rendering in pseudocode.

```
// Clear the geometry and depth buffers
clearGeometryBuffer();
fillDepthBuffer(infinity);

// Rasterize all primitives into the G-buffer
for (each geometric primitive) {
    for (each pixel x,y in the projection of the primitive) {

        // Test the depth buffer, to see if a closer pixel has
        // already been written.
        float primDepth = getDepthOfPrimitiveAtPixel(x,y);
        if (primDepth > readDepthBuffer(x,y)) {

            // Pixel of this primitive is obscured, discard it
            continue;
        }

        // Fetch information needed for shading in the next pass.
        MaterialInfo mtlInfo;
        Vector3 pos, normal;
        getPrimitiveShadingInfo(mtlInfo, pos, normal);

        // Save it off into the G-buffer and depth buffer
        writeGeometryBuffer(x,y, mtlInfo, pos, normal);
        writeDepthBuffer(x,y, primDepth);
    }
}

// Now perform shading in a 2nd pass, in screen space
for (each x,y screen pixel) {
    if (readDepthBuffer(x,y) == infinity) {

        // No geometry here.  Just write a background color
        writeFrameBuffer(x,y, backgroundColor);

    } else {

        // Fetch shading info back from the geometry buffer
        MaterialInfo mtlInfo;
        Vector3 pos, normal;
        readGeometryBuffer(x,y, mtlInfo, pos, normal);

        // Shade the point
        Color c = shadePoint(pos, normal, mtlInfo);

        // Put it into the frame buffer
        writeFrameBuffer(x,y, c);
    }
}
```

Listing 10.3
Pseudocode for deferred rendering using the depth buffer

Before moving on, we must mention one important point about why deferred rendering is popular. When multiple light sources illuminate the same surface point, hardware limitations or performance factors may prevent us from computing the final color of a pixel in a single calculation, as was shown in the pseudocode listings for both forward and deferred rendering. Instead, we must using multiple passes, one pass for each light, and *accumulate* the reflected light from each light source into the frame buffer. In forward rendering, these extra passes involve rerendering the primitives. Under deferred rendering, however, extra passes are in image space, and thus depend on the 2D size of the light in screen space, not on the complexity of the scene! It is in this situation that deferred rendering really begins to have large performance advantages over forward rendering.

10.1.2   Describing Surface Properties: The BRDF

Now let's talk about the second step in the rendering algorithm: lighting. Once we have located the surface closest to the eye, we must determine the amount of light emitted directly from that surface, or emitted from some other source and reflected off the surface in the direction of the eye. The light directly transmitted from a surface to the eye—for example, when looking directly at a light bulb or the sun—is the simplest case. These *emissive* surfaces are a small minority in most scenes; most surfaces do not emit their own light, but rather they only reflect light that was emitted from somewhere else. We will focus the bulk of our attention on the nonemissive surfaces.

Although we often speak informally about the "color" of an object, we know that the perceived color of an object is actually the light that is entering our eye, and thus can depend on many different factors. Important questions to ask are: What colors of light are incident on the surface, and from what directions? From which direction are we viewing the surface? How "shiny" is the object?[4] So a description of a surface suitable for use in rendering doesn't answer the question "What color is this surface?" This question is sometimes meaningless—what color is a mirror, for example? Instead, the salient question is a bit more complicated, and it goes something like, "When light of a given color strikes the surface from a given incident direction, how much of that light is reflected in some other particular direction?" The answer to this question is given by the *bidirectional reflectance distribution function*, or BRDF for short. So rather than "What color is the object?" we ask, "What is the distribution of reflected light?"

---

[4]Further relevant questions that should influence what color we write into the frame buffer could be asked concerning the general viewing conditions, but these issues have no bearing on the light coming into our eye; rather, they affect our perception of that light.

Symbolically, we write the BRDF as the function $f(\mathbf{x}, \hat{\boldsymbol{\omega}}_{\text{in}}, \hat{\boldsymbol{\omega}}_{\text{out}}, \lambda)$.[5]
The value of this function is a scalar that describes the relatively likelihood
that light incident at the point $\mathbf{x}$ from direction $\hat{\boldsymbol{\omega}}_{\text{in}}$ will be reflected in
the outgoing direction $\hat{\boldsymbol{\omega}}_{\text{out}}$ rather than some other outgoing direction. As
indicated by the boldface type and hat, $\hat{\boldsymbol{\omega}}$ might be a unit vector, but
more generally it can be any way of specifying a direction; polar angles are
another obvious choice and are commonly used. Different colors of light are
usually reflected differently; hence the dependence on λ, which is the color
(actually, the wavelength) of the light.

Although we are particularly interested in the incident directions that
come from emissive surfaces and the outgoing directions that point towards
our eye, in general, the entire distribution is relevant. First of all, lights,
eyes, and surfaces can move around, so in the context of creating a surface
description (for example, "red leather"), we don't know which directions
will be important. But even in a particular scene with all the surfaces,
lights, and eyes fixed, light can bounce around multiple times, so we need
to measure light reflections for arbitrary pairs of directions.

Before moving on, it's highly instructive to see how the two intuitive
material properties that were earlier disparaged, color and shininess, can
be expressed precisely in the framework of a BRDF. Consider a green ball.
A green object is green and not blue because it reflects incident light that
is green more strongly than incident light of any other color.[6] For example,
perhaps green light is almost all reflected, with only a small fraction absorbed, while 95% of the blue and red light is absorbed and only 5% of light
at those wavelengths is reflected in various directions. White light actually
consists of all the different colors of light, so a green object essentially filters
out colors other than green. If a different object responded to green and
red light in the same manner as our green ball, but absorbed 50% of the
blue light and reflected the other 50%, we might perceive the object as teal.
Or if most of the light at all wavelengths was absorbed, except for a small
amount of green light, then we would perceive it as a dark shade of green.
To summarize, a BRDF accounts for the difference in color between two
objects through the dependence on λ: any given wavelength of light has its
own reflectance distribution.

Next, consider the difference between shiny red plastic and diffuse red
construction paper. A shiny surface reflects incident light much more
strongly in one particular direction compared to others, whereas a diffuse
surface scatters light more evenly across all outgoing directions. A perfect
reflector, such as a mirror, would reflect all the light from one incoming

---

[5]Remember that ω and λ are the lowercase Greek letters omega and lambda, respectively.

[6]Here and elsewhere, we use the word "color" in a way that's technically a bit dodgy,
but is OK under the assumptions about light and color made in most graphics systems.

direction in a single outgoing direction, whereas a perfectly diffuse surface would reflect light equally in all outgoing directions, regardless of the direction of incidence. In summary, a BRDF accounts for the difference in "shininess" of two objects through its dependence on $\hat{\boldsymbol{\omega}}_{\text{in}}$ and $\hat{\boldsymbol{\omega}}_{\text{out}}$.

More complicated phenomena can be expressed by generalizing the BRDF. Translucence and light refraction can be easily incorporated by allowing the direction vectors to point back into the surface. We might call this mathematical generalization a *bidirectional surface scattering distribution function* (BSSDF). Sometimes light strikes an object, bounces around inside of it, and then exits at a different point. This phenomenon is known as *subsurface scattering* and is an important aspect of the appearances of many common substances, such as skin and milk. This requires splitting the single reflection point $\mathbf{x}$ into $\mathbf{x}_{\text{in}}$ and $\mathbf{x}_{\text{out}}$, which is used by the *bidirectional surface scattering distribution function* (BSSDF). Even volumetric effects, such as fog and subsurface scattering, can be expressed, by dropping the words "surface" and defining a *bidirectional scattering distribution function* (BSDF) at any point in space, not just on the "surfaces." Taken at face value, these might seem like impractical abstractions, but they can be useful in understanding how to design practical tools.

By the way, there are certain criteria that a BRDF must satisfy in order to be physically plausible. First, it doesn't make sense for a negative amount of light to be reflected in any direction. Second, it's not possible for the total reflected light to be more than the light that was incident, although the surface may absorb some energy so the reflected light can be less than the incident light. This rule is usually called the *normalization constraint*. A final, less obvious principle obeyed by physical surfaces is *Helmholtz reciprocity*: if we pick two arbitrary directions, the same fraction of light should be reflected, no matter which is the incident direction and which is the outgoing direction. In other words,

Helmholtz reciprocity
$$f(\mathbf{x}, \hat{\boldsymbol{\omega}}_1, \hat{\boldsymbol{\omega}}_2, \lambda) = f(\mathbf{x}, \hat{\boldsymbol{\omega}}_2, \hat{\boldsymbol{\omega}}_1, \lambda).$$

Due to Helmholtz reciprocity, some authors don't label the two directions in the BRDF as "in" and "out" because to be physically plausible the computation must be symmetric.

The BRDF contains the complete description of an object's appearance at a given point, since it describes how the surface will reflect light at that point. Clearly, a great deal of thought must be put into the design of this function. Numerous lighting models have been proposed over the last several decades, and what is surprising is that one of the earliest models, Blinn-Phong, is still in widespread use in real-time graphics today. Although it is not physically accurate (nor plausible: it violates the normalization constraint), we study it because it is a good educational stepping stone and an important bit of graphics history. Actually, describing Blinn-Phong as "history" is wishful thinking—perhaps the most important reason to study this model is that it still is in such widespread use! In fact, it's the best example of the phenomena we mentioned at the start of this chapter: particular methods being presented as if they are "the way graphics work."

Different lighting models have different goals. Some are better at simulating rough surfaces, others at surfaces with multiple strata. Some focus on providing intuitive "dials" for artists to control, without concern for whether those dials have any physical significance at all. Others are based on taking real-world surfaces and measuring them with special cameras called goniophotometers, essentially sampling the BRDF and then using interpolation to reconstruct the function from the tabulated data. The notable Blinn-Phong model discussed in Section 10.6 is useful because it is simple, inexpensive, and well understood by artists. Consult the sources in the suggested reading for a survey of lighting models.

10.1.3   A Very Brief Introduction to Colorimetry and Radiometry

Graphics is all about measuring light, and you should be aware of some important subtleties, even though we won't have time to go into complete detail here. The first is how to measure the color of light, and the second is how to measure its brightness.

In your middle school science classes you might have learned that every color of light is some mixture of red, green, and blue (RGB) light. This is the popular conception of light, but it's not quite correct. Light can take on any single frequency in the visible band, or it might be a combination of any number of frequencies. Color is a phenomena of *human perception* and is not quite the same thing as frequency. Indeed different combinations of frequencies of light can be perceived as the same color—these are known as *metamers*. The infinite combinations of frequencies of light are sort of like all the different chords that can be played on a piano (and also tones between the keys). In this metaphor our color perception is unable to pick out all the different individual notes, but instead, any given chord sounds to us like some combination of middle C, F, and G. Three color channels is not a magic number as far as physics is concerned, it's peculiar to human vision. Most other mammals have only two different types of receptors (we would call them "color blind"), and fish, reptiles, and birds have *four* types of color receptors (they would call *us* color blind).

However, even very advanced rendering systems project the continuous spectrum of visible light onto some discrete basis, most commonly, the RGB basis. This is a ubiquitous simplification, but we still wanted to let you know that it is a simplification, as it doesn't account for certain phenomena. The RGB basis is not the only color space, nor is it necessarily the best one for many purposes, but it is a very convenient basis because it is the one used by most display devices. In turn, the reason that this basis is used by so many display devices is due to the similarity to our own visual system. Hall [29] does a good job of describing the shortcomings of the RGB system.

Since the visible portion of the electromagnetic spectrum is continuous, an expression such as $f(\mathbf{x}, \hat{\boldsymbol{\omega}}_{\text{in}}, \hat{\boldsymbol{\omega}}_{\text{out}}, \lambda)$ is continuous in terms of λ. At least it should be in theory. In practice, because we are producing images for human consumption, we reduce the infinite number of different λs down to three particular wavelengths. Usually, we choose the three wavelengths to be those perceived as the colors red, green, and blue. In practice, you can think of the presence of λ in an equation as an integer that selects which of the three discrete "color channels" is being operated on.

Key Points about Color

- To describe the spectral distribution of light requires a continuous function, not just three numbers. However, to describe the human perception of that light, three numbers are essentially sufficient.

- The RGB system is a convenient color space, but it's not the only one, and not even the best one for many practical purposes. In practice, we usually treat light as being a combination of red, green, and blue because we are making images for human consumption.

You should also be aware of the different ways that we can measure the intensity of light. If we take a viewpoint from physics, we consider light as energy in the form of electromagnetic radiation, and we use units of measurement from the field of *radiometry*. The most basic quantity is *radiant energy*, which in the SI system is measured in the standard unit of energy, the *joule* (J). Just like any other type of energy, we are often interested in the rate of energy flow per unit time, which is known as *power*. In the SI system power is measured using the *watt* (W), which is one joule per second (1 W = 1 J/s). Power in the form of electromagnetic radiation is called *radiant power* or *radiant flux*. The term "flux," which comes from the Latin *fluxus* for "flow," refers to some quantity flowing across some cross-sectional area. Thus, radiant flux measures the total amount of energy that is arriving, leaving, or flowing across some area per unit time.

Imagine that a certain amount of radiant flux is emitted from a 1 m^2 surface, while that same amount of power is emitted from a different surface that is 100 m^2. Clearly, the smaller surface is "brighter" than the larger surface; more precisely, it has a greater flux per unit area, also known as *flux density*. The radiometric term for flux density, the radiant flux per unit area, is called *radiosity*, and in the SI system it is measured in watts per meter. The relationship between flux and radiosity is analogous to the relationship between force and pressure; confusing the two will lead to similar sorts of conceptual errors.

Several equivalent terms exist for radiosity. First, note that we can measure the flux density (or total flux, for that matter) across any cross-sectional area. We might be measuring the radiant power emitted from some surface with a finite area, or the surface through which the light flows might be an imaginary boundary that exists only mathematically (for example, the surface of some imaginary sphere that surrounds a light source). Although in all cases we are measuring flux density, and thus the term "radiosity" is perfectly valid, we might also use more specific terms, depending on whether the light being measured is coming or going. If the area is a surface and the light is arriving on the surface, then the term *irradiance* is used. If light is being emitted from a surface, the term *radiant exitance* or *radiant emittance* is used. In digital image synthesis, the word "radiosity" is most often used to refer to light that is leaving a surface, having been either reflected or emitted.

When we are talking about the brightness at a particular point, we cannot use plain old radiant power because the area of that point is infinitesimal (essentially zero). We can speak of the flux *density* at a single point, but to measure flux, we need a finite area over which to measure. For a surface of finite area, if we have a single number that characterizes the total for the entire surface area, it will be measured in flux, but to capture the fact that different locations within that area might be brighter than others, we use a function that varies over the surface that will measure the flux density.

Now we are ready to consider what is perhaps the most central quantity we need to measure in graphics: the intensity of a "ray" of light. We can see why the radiosity is not the unit for the job by an extension of the ideas from the previous paragraph. Imagine a surface point surrounded by an emissive dome and receiving a certain amount of irradiance coming from all directions in the hemisphere centered on the local surface normal. Now imagine a second surface point experiencing the same amount of irradiance, only all of the illumination is coming from a single direction, in a very thin beam. Intuitively, we can see that a ray along this beam is somehow "brighter" than any one ray that is illuminating the first surface point. The irradiance is somehow "denser." It is denser *per unit solid angle*.

The idea of a solid angle is probably new to some readers, but we can easily understand the idea by comparing it to angles in the plane. A "regular" angle is measured (in radians) based on the length of its projection onto the unit circle. In the same way, a solid angle measures the *area* as projected onto the unit sphere surrounding the point. The SI unit for solid angle is the *steradian*, abbreviated "sr." The complete sphere has 4π sr; a hemisphere encompasses 2π sr.

By measuring the radiance per unit solid angle, we can express the intensity of light at a certain point as a function that varies based upon the direction of incidence. We are very close to having the unit of measurement that describes the intensity of a ray. There is just one slight catch, illustrated by Figure 10.1, which is a close-up of a very thin pencil of light rays striking a surface. On the top, the rays strike the surface perpendicularly, and on the bottom, light rays of the same strength strike a different surface at an angle. The key point is that the area of the top surface is smaller than the area of the bottom surface; therefore, the irradiance on the top surface is larger than the irradiance on the bottom surface, despite the fact that the two surfaces are being illuminated by the "same number" of identical light rays. This basic phenomenon, that the angle of the surface causes incident light rays to be spread out and thus contribute less irradiance, is known as *Lambert's law*. We have more to say about Lambert's law in Section 10.6.3, but for now, the key idea is that the contribution of a bundle of light to the irradiance at a surface depends on the angle of that surface.

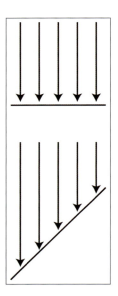

Figure 10.1
The two surfaces are receiving identical bundles of light, but the surface on the bottom has a larger area, and thus has a lower irradiance.

Due to Lambert's law, the unit we use in graphics to measure the strength of a ray, *radiance*, is defined as the radiant flux per unit *projected area*, per unit solid angle. To measure a projected area, we take the actual surface area and project it onto the plane perpendicular to the ray. (In Figure 10.1, imagine taking the bottom surface and projecting it upwards onto the top surface). Essentially this counteracts Lambert's law.

Table 10.1 summarizes the most important radiometric terms.

Whereas radiometry takes the perspective of physics by measuring the raw energy of the light, the field of *photometry* weighs that same light using the human eye. For each of the corresponding radiometric terms, there is

Quantity	Units	SI unit	Rough translation
Radiant energy	Energy	J	Total illumination during an interval of time
Radiant flux	Power	W	Brightness of a finite area from all directions
Radiant flux density	Power per unit area	W/m^2	Brightness of a single point from all directions
Irradiance	Power per unit area	W/m^2	Radiant flux density of incident light
Radiant exitance	Power per unit area	W/m^2	Radiant flux density of emitted light
Radiosity	Power per unit area	W/m^2	Radiant flux density of emitted or reflected light
Radiance	Power per unit projected area, per unit solid angle	$W/(m^2 \cdot sr)$	Brightness of a ray

Table 10.1. Common radiometric terms

a similar term from photometry (Table 10.2). The only real difference is a nonlinear conversion from raw energy to perceived brightness.

Throughout the remainder of this chapter, we try to use the proper radiometric units when possible. However, the practical realities of graphics make using proper units confusing, for two particular reasons. It is common in graphics to need to take some integral over a "signal"—for example, the color of some surface. In practice we cannot do the integral analytically, and so we must integrate numerically, which boils down to taking a weighted average of many samples. Although mathematically we are taking a weighted average (which ordinarily would not cause the units to change), in fact what we are doing is *integrating*, and that means each sample is really being multiplied by some differential quantity, such as a differential area or differential solid angle, which causes the physical units to change. A second cause of confusion is that, although many signals have a finite nonzero domain in the real world, they are represented in a computer by signals that are nonzero at a single point. (Mathematically, we say that

Radiometric term	Photometric term	SI Photometric unit
Radiant energy	Luminous energy	talbot, or lumen second $(lm \cdot s)$
Radiant flux	Luminous flux, luminous power	lumen (lm)
Irradiance	Illuminance	lux $(lx = lm/m^2)$
Radiant exitance	Luminous emittance	lux $(lx = lm/m^2)$
Radiance	Luminance	$lm/(m^2 \cdot sr)$

Table 10.2. Units of measurement from radiometry and photometry

the signal is a multiple of a Direc delta; see Section 12.4.3.) For example, a real-world light source has a finite area, and we would be interested in the radiance of the light at a given point on the emissive surface, in a given direction. In practice, we imagine shrinking the area of this light down to zero while holding the radiant flux constant. The flux density becomes infinite in theory. Thus, for a real area light we would need a signal to describe the flux density, whereas for a point light, the flux density becomes infinite and we instead describe the brightness of the light by its total flux. We'll repeat this information when we talk about point lights.

Key Points about Radiometry

- Vague words such as "intensity" and "brightness" are best avoided when the more specific radiometric terms can be used. The scale of our numbers is not that important and we don't need to use real world SI units, but it is helpful to understand what the different radiometric quantities measure to avoid mixing quantities together inappropriately.

- Use radiant flux to measure the total brightness of a finite area, in all directions.

- Use radiant flux density to measure the brightness at a single point, in all directions. Irradiance and radiant exitance refer to radiant flux density of light that is incident and emitted, respectively. Radiosity is the radiant flux density of light that is leaving a surface, whether the light was reflected or emitted.

- Due to Lambert's law, a given ray contributes more differential irradiance when it strikes a surface at a perpendicular angle compared to a glancing angle.

- Use radiance to measure the brightness of a ray. More specifically, radiance is the flux per unit projected area, per solid angle. We use projected area so that the value for a given ray is a property of a ray alone and does not depend on the orientation of the surface used to measure the flux density.

- Practical realities thwart our best intentions of doing things "the right way" when it comes to using proper units. Numerical integration is a lot like taking a weighted average, which hides the change of units that really occurs. Point lights and other Dirac deltas add further confusion.

10.1.4 The Rendering Equation

Now let's fit the BRDF into the rendering algorithm. In step 2 of our rendering algorithm (Section 10.1), we're trying to determine the radiance leaving a particular surface in the direction of our eye. The only way this can happen is for light to arrive from some direction onto the surface and get reflected in our direction. With the BRDF, we now have a way to measure this. Consider all the potential directions that light might be incident upon the surface, which form a hemisphere centered on $\mathbf{x}$, oriented according to the local surface normal $\hat{\mathbf{n}}$. For each potential direction $\hat{\boldsymbol{\omega}}_{\text{in}}$, we measure the color of light incident from that direction. The BRDF tells us how much of the radiance from $\hat{\boldsymbol{\omega}}_{\text{in}}$ is reflected in the direction $\hat{\boldsymbol{\omega}}_{\text{out}}$ towards our eye (as opposed to scattered in some other direction or absorbed). By summing up the radiance reflected towards $\hat{\boldsymbol{\omega}}_{\text{out}}$ over all possible incident directions, we obtain the total radiance reflected along $\hat{\boldsymbol{\omega}}_{\text{out}}$ into our eye. We add the reflected light to any light that is being *emitted* from the surface in our direction (which is zero for most surfaces), and voila, we have the total radiance. Writing this in math notation, we have the *rendering equation*.

The Rendering Equation

$$L_{\text{out}}(\mathbf{x}, \hat{\boldsymbol{\omega}}_{\text{out}}, \lambda) = L_{\text{emis}}(\mathbf{x}, \hat{\boldsymbol{\omega}}_{\text{out}}, \lambda)$$
$$+ \int_{\Omega} L_{\text{in}}(\mathbf{x}, \hat{\boldsymbol{\omega}}_{\text{in}}, \lambda) f(\mathbf{x}, \hat{\boldsymbol{\omega}}_{\text{in}}, \hat{\boldsymbol{\omega}}_{\text{out}}, \lambda)(-\hat{\boldsymbol{\omega}}_{\text{in}} \cdot \hat{\mathbf{n}}) \, d\hat{\boldsymbol{\omega}}_{\text{in}}. \quad (10.1)$$

As fundamental as Equation (10.1) may be, its development is relatively recent, having been published in SIGGRAPH in 1986 by Kajiya [37]. Furthermore, it was the *result* of, rather than the cause of, numerous strategies for producing realistic images. Graphics researchers pursued the creation of images through different techniques that seemed to make sense to them before having a framework to describe the problem they were trying to solve. And for many years after that, most of us in the video game industry were unaware that the problem we were trying to solve had finally been clearly defined. (Many still are.)

Now let's convert this equation into English and see what the heck it means. First of all, notice that $\mathbf{x}$ and λ appear in each function. The whole equation governs a balance of radiance at a single surface point $\mathbf{x}$ for a single wavelength ("color channel") λ. So this balance equation applies to each color channel individually, at all surface points simultaneously.

The term $L_{\text{out}}(\mathbf{x}, \hat{\boldsymbol{\omega}}_{\text{out}}, \lambda)$ on the left side of the equals sign is simply "The radiance leaving the point in the direction $\hat{\boldsymbol{\omega}}_{\text{out}}$." Of course, if $\mathbf{x}$ is the visible surface at a given pixel, and $\hat{\boldsymbol{\omega}}_{\text{out}}$ is the direction from $\mathbf{x}$ to the eye, then this quantity is exactly what we need to determine the pixel color. But note that the equation is more general, allowing us to compute the outgoing radiance in any arbitrary direction $\hat{\boldsymbol{\omega}}_{\text{out}}$ and for any given point $\mathbf{x}$, whether or not $\hat{\boldsymbol{\omega}}_{\text{out}}$ points towards our eye.

On the right-hand side, we have a sum. The first term in the sum $L_{\text{emis}}(\mathbf{x}, \hat{\boldsymbol{\omega}}_{\text{out}}, \lambda)$, is "the radiance emitted from $\mathbf{x}$ in the direction $\hat{\boldsymbol{\omega}}_{\text{out}}$" and will be nonzero only for special emissive surfaces. The second term, the integral, is "the light reflected from $\mathbf{x}$ in the direction of $\hat{\boldsymbol{\omega}}_{\text{out}}$." Thus, from a high level the rendering equation would seem to state the rather obvious relation

$$\begin{pmatrix} \text{Total radiance} \\ \text{towards } \hat{\boldsymbol{\omega}}_{\text{out}} \end{pmatrix} = \begin{pmatrix} \text{Radiance emitted} \\ \text{towards } \hat{\boldsymbol{\omega}}_{\text{out}} \end{pmatrix} + \begin{pmatrix} \text{Radiance reflected} \\ \text{towards } \hat{\boldsymbol{\omega}}_{\text{out}} \end{pmatrix}.$$

Now let's dig into that intimidating integral. (By the way, if you haven't had calculus and haven't read Chapter 11 yet, just replace the word "integral" with "sum," and you won't miss any of the main point of this section.) We've actually already discussed how it works when we talked about the BRDF, but let's repeat it with different words. We might rewrite the integral as

$$\begin{pmatrix} \text{Radiance reflected} \\ \text{towards } \hat{\boldsymbol{\omega}}_{\text{out}} \end{pmatrix} = \int_{\Omega} \begin{pmatrix} \text{Radiance incident from } \hat{\boldsymbol{\omega}}_{\text{in}} \\ \text{and reflected towards } \hat{\boldsymbol{\omega}}_{\text{out}} \end{pmatrix} d\hat{\boldsymbol{\omega}}_{\text{in}}.$$

Note that symbol Ω (uppercase Greek omega) appears where we normally would write the limits of integration. This is intended to mean "sum over the hemisphere of possible incoming directions." For each incoming direction $\hat{\boldsymbol{\omega}}_{\text{in}}$, we determine how much radiance was incident in this incoming direction and got scattered in the outgoing direction $\hat{\boldsymbol{\omega}}_{\text{out}}$. The sum of all these contributions from all the different incident directions gives the total radiance reflected in the direction $\hat{\boldsymbol{\omega}}_{\text{out}}$. Of course, there are an infinite number of incident directions, which is why this is an integral. In practice, we cannot evaluate the integral analytically, and we must sample a discrete number of directions, turning the "$\int$" into a "$\sum$."

Now all that is left is to dissect the integrand. It's a product of three factors:

$$\begin{pmatrix} \text{Radiance incident from } \hat{\boldsymbol{\omega}}_{\text{in}} \\ \text{and reflected towards } \hat{\boldsymbol{\omega}}_{\text{out}} \end{pmatrix} = L_{\text{in}}(\mathbf{x}, \hat{\boldsymbol{\omega}}_{\text{in}}, \lambda)\, f(\mathbf{x}, \hat{\boldsymbol{\omega}}_{\text{in}}, \hat{\boldsymbol{\omega}}_{\text{out}}, \lambda)\, (-\hat{\boldsymbol{\omega}}_{\text{in}} \cdot \hat{\mathbf{n}}).$$

The first factor denotes the radiance incident from the direction of $\hat{\boldsymbol{\omega}}_{\text{in}}$. The next factor is simply the BRDF, which tells us how much of the radiance incident from this particular direction will be reflected in the outgoing direction we care about. Finally, we have the *Lambert factor*. As discussed in Section 10.1.2, this accounts for the fact that more incident light is available to be reflected, per unit surface area, when $\hat{\omega}_{in}$ is perpendicular to the surface than when at a glancing angle to the surface. The vector $\hat{\mathbf{n}}$ is the outward-facing surface normal; the dot product $-\hat{\omega}_{in} \cdot \hat{\mathbf{n}}$ peaks at 1 in the perpendicular direction and trails off to zero as the angle of incidence becomes more glancing. We discuss the Lambert factor once more in Section 10.6.3.

In purely mathematical terms, the rendering equation is an *integral equation*: it states a relationship between some unknown function $L_{out}(\mathbf{x}, \hat{\omega}_{out}, \lambda)$, the distribution of light on the surfaces in the scene, in terms of its own integral. It might not be apparent that the rendering equation is recursive, but L_{out} actually appears on both sides of the equals sign. It appears in the evaluation of $L_{in}(\mathbf{x}, \hat{\omega}_{in}, \lambda)$, which is precisely the expression we set out to solve for each pixel: what is the radiance incident on a point from a given direction? Thus to find the radiance exiting a point $\mathbf{x}$, we need to know all the radiance incident at $\mathbf{x}$ from all directions. But the radiance incident on $\mathbf{x}$ is the same as the radiance leaving from *all other surfaces visible to* $\mathbf{x}$, in the direction pointing from the other surface towards $\mathbf{x}$.

To render a scene realistically, we must solve the rendering equation, which requires us to know (in theory) not only the radiance arriving at the camera, but also the entire distribution of radiance in the scene in every direction at every point. Clearly, this is too much to ask for a finite, digital computer, since both the set of surface locations and the set of potential incident/exiting directions are infinite. The real art in creating software for digital image synthesis is to allocate the limited processor time and memory most efficiently, to make the best possible approximation.

The simple rendering pipeline we present in Section 10.10 accounts only for direct light. It doesn't account for indirect light that bounced off of one surface and arrived at another. In other words, it only does "one recursion level" in the rendering equation. A huge component of realistic images is accounting for the indirect light—solving the rendering equation more completely. The various methods for accomplishing this are known as *global illumination* techniques.

This concludes our high-level presentation of how graphics works. Although we admit we have not yet presented a single practical idea, we believe it's very important to understand what you are trying to approximate before you start to approximate it. Even though the compromises we are forced to make for the sake of real-time are quite severe, the available computing power is growing. A video game programmer whose only exposure to graphics has been OpenGL tutorials or demos made by video card manufacturers or books that focused exclusively on real-time rendering will have a much more difficult time understanding even the global illumination techniques of today, much less those of tomorrow.

10.2   Viewing in 3D

Before we render a scene, we must pick a camera and a window. That is, we must decide where to render it *from* (the view position, orientation, and zoom) and where to render it *to* (the rectangle on the screen). The output window is the simpler of the two, and so we will discuss it first.

Section 10.2.1 describes how to specify the output window. Section 10.2.2 discusses the pixel aspect ratio. Section 10.2.3 introduces the view frustum. Section 10.2.4 describes field of view angles and zoom.

10.2.1   Specifying the Output Window

We don't have to render our image to the entire screen. For example, in split-screen multiplayer games, each player is given a portion of the screen. The output window refers to the portion of the output device where our image will be rendered. This is shown in Figure 10.2.

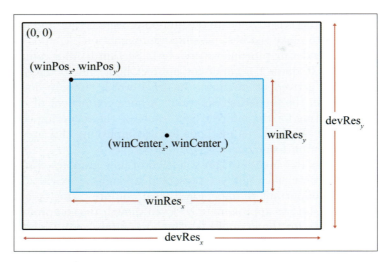

Figure 10.2. Specifying the output window

The position of the window is specified by the coordinates of the upper left-hand pixel (winPos_x, winPos_y). The integers winRes_x and winRes_y are the dimensions of the window in pixels. Defining it this way, using the size of the window rather than the coordinates of the lower right-hand corner, avoids some sticky issues caused by integer pixel coordinates. We are also careful to distinguish between the size of the window in pixels, and the physical size of the window. This distinction will become important in Section 10.2.2.

With that said, it is important to realize that we do not necessarily have to be rendering to the screen at all. We could be rendering into a buffer to be saved into a .TGA file or as a frame in an .AVI, or we may be rendering into a texture as a subprocess of the "main" render, to produce a shadow map, or a reflection, or the image on a monitor in the virtual world. For these reasons, the term *render target* is often used to refer to the current destination of rendering output.

10.2.2   Pixel Aspect Ratio

Regardless of whether we are rendering to the screen or an off-screen buffer, we must know the aspect ratio of the pixels, which is the ratio of a pixel's height to its width. This ratio is often 1:1—that is, we have "square" pixels—but *this is not always the case!* We give some examples below, but it is common for this assumption to go unquestioned and become the source of complicated kludges applied in the wrong place, to fix up stretched or squashed images.

The formula for computing the aspect ratio is

$$\frac{\text{pixPhys}_x}{\text{pixPhys}_y} = \frac{\text{devPhys}_x}{\text{devPhys}_y} \cdot \frac{\text{devRes}_y}{\text{devRes}_x}. \qquad (10.2)$$

Computing the pixel aspect ratio

The notation pixPhys refers to the physical size of a pixel, and devPhys is the physical height and width of the device on which the image is displayed. For both quantities, the individual measurements may be unknown, but that's OK because the ratio is all we need, and this usually is known. For example, standard desktop monitors come in all different sizes, but the viewable area on many older monitors has a ratio of 4:3, meaning it is 33% wider than it is tall. Another common ratio is 16:9 or wider[7] on high-definition televisions. The integers devRes_x and devRes_y are the number

---

[7]Monitor manufacturers must have been overjoyed to find that people perceived a premium quality to these "widescreen" monitors. Monitor sizes are typically measured by the diagonal, but costs are more directly tied to number of pixels, which is proportional to area, not diagonal length. Thus, a 16:9 monitor with the same number of pixels as a 4:3 will have a longer diagonal measurement, which is perceived as a "bigger" monitor. We're not sure if the proliferation of monitors with even wider aspect ratios is fueled more by market forces or market*ing* forces.

of pixels in the x and y dimensions. For example, a resolution of 1280×720 means that $\text{devRes}_x = 1280$ and $\text{devRes}_y = 720$.

But, as mentioned already, we often deal with square pixels with an aspect ratio of 1:1. For example, on a desktop monitor with a physical width:height ratio of 4:3, some common resolutions resulting in square pixel ratios are 640×480, 800×600, 1024×768, and 1600×1200. On 16:9 monitors, common resolutions are 1280×720, 1600×900, 1920×1080. The aspect ratio 8:5 (more commonly known as 16:10) is also very common, for desktop monitor sizes and televisions. Some common display resolutions that are 16:10 are 1153×720, 1280×800, 1440×900, 1680×1050, and 1920×1200. In fact, on the PC, it's common to just assume a 1:1 pixel ratio, since obtaining the dimensions of the display device might be impossible. Console games have it easier in this respect.

Notice that nowhere in these calculations is the size or location of the *window* used; the location and size of the rendering window has no bearing on the physical proportions of a pixel. However, the size of the window will become important when we discuss field of view in Section 10.2.4, and the position is important when we map from camera space to screen space Section 10.3.5.

At this point, some readers may be wondering how this discussion makes sense in the context of rendering to a bitmap, where the word "physical" implied by the variable names pixPhys and devPhys doesn't apply. In most of these situations, it's appropriate simply to act as if the pixel aspect ratio is 1:1. In some special circumstances, however, you may wish to render *anamorphically*, producing a squashed image in the bitmap that will later be stretched out when the bitmap is used.

10.2.3   The View Frustum

The *view frustum* is the volume of space that is potentially visible to the camera. It is shaped like a pyramid with the tip snipped off. An example of a view frustum is shown in Figure 10.3.

The view frustum is bounded by six planes, known as the *clip planes*. The first four of the planes form the sides of the pyramid and are called the top, left, bottom, and right planes, for obvious reasons. They correspond to the sides of the output window. The near and far clip planes, which correspond to certain camera-space values of z, require a bit more explanation.

The reason for the far clip plane is perhaps easier to understand. It prevents rendering of objects beyond a certain distance. There are two practical reasons why a far clip plane is needed. The first is relatively easy to understand: a far clip plane can limit the number of objects that need to be rendered in an outdoor environment. The second reason is slightly

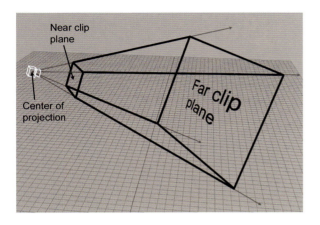

Figure 10.3
The 3D view frustum

more complicated, but essentially it has to do with how the depth buffer values are assigned. As an example, if the depth buffer entries are 16-bit fixed point, then the largest depth value that can be stored is 65,535. The far clip establishes what (floating point) z value in camera space will correspond to the maximum value that can be stored in the depth buffer. The motivation for the near clip plane will have to wait until we discuss *clip space* in Section 10.3.2.

Notice that each of the clipping planes are *planes*, with emphasis on the fact that they extend infinitely. The view volume is the intersection of the six half-spaces defined by the clip planes.

10.2.4 Field of View and Zoom

A camera has position and orientation, just like any other object in the world. However, it also has an additional property known as *field of view*. Another term you probably know is *zoom*. Intuitively, you already know what it means to "zoom in" and "zoom out." When you zoom in, the object you are looking at appears bigger on screen, and when you zoom out, the apparent size of the object is smaller. Let's see if we can develop this intuition into a more precise definition.

The *field of view* (FOV) is the angle that is intercepted by the view frustum. We actually need two angles: a horizontal field of view, and a vertical field of view. Let's drop back to 2D briefly and consider just one of these angles. Figure 10.4 shows the view frustum from above, illustrating precisely the angle that the horizontal field of view measures. The labeling of the axes is illustrative of camera space, which is discussed in Section 10.3.

Zoom measures the ratio of the apparent size of the object relative to a 90° field of view. For example, a zoom of 2.0 means that object will appear

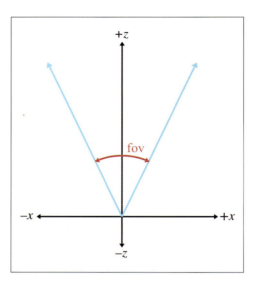

Figure 10.4
Horizontal field of view

twice as big on screen as it would if we were using a 90° field of view. So larger zoom values cause the image on screen to become larger ("zoom in"), and smaller values for zoom cause the images on screen to become smaller ("zoom out").

Zoom can be interpreted geometrically as shown in Figure 10.5. Using some basic trig, we can derive the conversion between zoom and field of

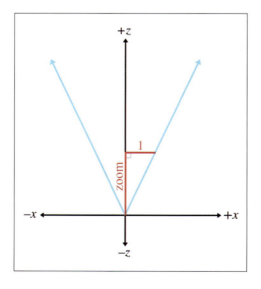

Figure 10.5
Geometric interpretation of zoom

view:

$$\text{zoom} = \frac{1}{\tan{(\text{fov}/2)}}, \qquad \text{fov} = 2\ \arctan{(1/\text{zoom})}. \qquad (10.3)$$

Converting between zoom and field of view

Notice the inverse relationship between zoom and field of view. As zoom gets larger, the field of view gets smaller, causing the view frustum to narrow. It might not seem intuitive at first, but when the view frustum gets more narrow, the perceived size of visible objects increases.

Field of view is a convenient measurement for humans to use, but as we discover in Section 10.3.4, zoom is the measurement that we need to feed into the graphics pipeline.

We need *two* different field of view angles (or zoom values), one horizontal and one vertical. We are certainly free to choose any two arbitrary values we fancy, but if we do not maintain a proper relationship between these values, then the rendered image will appear stretched. If you've ever watched a movie intended for the wide screen that was simply squashed anamorphically to fit on a regular TV, or watched content with a 4:3 aspect on a 16:9 TV in "full"[8] mode, then you have seen this effect.

In order to maintain proper proportions, the zoom values must be inversely proportional to the physical dimensions of the output window:

$$\frac{\text{zoom}_y}{\text{zoom}_x} = \frac{\text{winPhys}_x}{\text{winPhys}_y} = \text{window aspect ratio.} \qquad (10.4)$$

The usual relationship between vertical and horizontal zoom

The variable winPhys refers to the physical size of the output window. As indicated in Equation (10.4), even though we don't usually know the actual size of the render window, we can determine its aspect ratio. But how do we do this? Usually, all we know is the *resolution* (number of pixels) of the output window. Here's where the pixel aspect ratio calculations from Section 10.2.2 come in:

$$
\begin{aligned}
\frac{\text{zoom}_y}{\text{zoom}_x} &= \frac{\text{winPhys}_x}{\text{winPhys}_y} = \frac{\text{winRes}_x}{\text{winRes}_y} \cdot \frac{\text{pixPhys}_x}{\text{pixPhys}_y} \\
&= \frac{\text{winRes}_x}{\text{winRes}_y} \cdot \frac{\text{devPhys}_x}{\text{devPhys}_y} \cdot \frac{\text{devRes}_y}{\text{devRes}_x}.
\end{aligned}
\qquad (10.5)
$$

In this formula,

- zoom refers to the camera's zoom values,

- winPhys refers to the physical window size,

---

[8]While it causes videophiles extreme stress to see an image manhandled this way, apparently some TV owners prefer a stretched image to the black bars, which give them the feeling that they are not getting all their money's worth out of their expensive new TV.

- winRes refers to the resolution of the window, in pixels,

- pixPhys refers to the physical dimensions of a pixel,

- devPhys refers to the physical dimensions of the output device. Remember that we usually don't know the individual sizes, but we do know the ratio,

- devRes refers to the resolution of the output device.

Many rendering packages allow you to specify only one field of view angle (or zoom value). When you do this, they automatically compute the other value for you, assuming you want uniform display proportions. For example, you may specify the horizontal field of view, and they compute the vertical field of view for you.

Now that we know how to describe zoom in a manner suitable for consumption by a computer, what do we do with these zoom values? They go into the *clip matrix*, which is described in Section 10.3.4.

10.2.5   Orthographic Projection

The discussion so far has centered on perspective projection, which is the most commonly used type of projection, since that's how our eyes perceive the world. However, in many situations *orthographic projection* is also useful. We introduced orthographic projection in Section 5.3; to briefly review, in orthographic projection, the lines of projection (the lines that connect all the points in space that project onto the same screen coordinates) are parallel, rather than intersecting at a single point. There is no perspective foreshortening in orthographic projection; an object will appear the same size on the screen no matter how far away it is, and moving the camera forward or backward along the viewing direction has *no apparent effect* so long as the objects remain in front of the near clip plane.

Figure 10.6 shows a scene rendered from the same position and orientation, comparing perspective and orthographic projection. On the left, notice that with perspective projection, parallel lines do not remain parallel, and the closer grid squares are larger than the ones in the distance. Under orthographic projection, the grid squares are all the same size and the grid lines remain parallel.

Orthographic views are very useful for "schematic" views and other situations where distances and angles need to be measured precisely. Every modeling tool will support such a view. In a video game, you might use an orthographic view to render a map or some other HUD element.

For an orthographic projection, it makes no sense to speak of the "field of view" as an angle, since the view frustum is shaped like a box, not a pyramid. Rather than defining the x and y dimensions of the view frustum

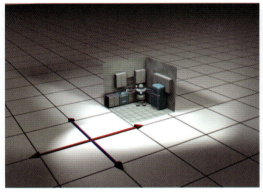 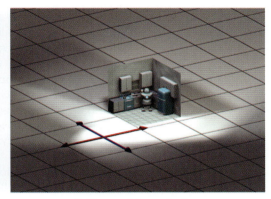

Perspective projection   Orthographic projection

Figure 10.6
Perspective versus orthographic projection

in terms of two angles, we give two sizes: the physical width and height of
the box.

The zoom value has a different meaning in orthographic projection compared to perspective. It is related to the physical size of the frustum box:

$$\text{zoom} = 2/\text{size}, \qquad \text{size} = 2/\text{zoom}.$$

**Converting between
zoom and frustum size in
orthographic projection**

As with perspective projections, there are *two* different zoom values, one
for x and one for y, and their ratio must be coordinated with the aspect ratio
of the rendering window in order to avoid producing a "squashed" image.
We developed Equation (10.5) with perspective projection in mind, but this
formula also governs the proper relationship for orthographic projection.

10.3 Coordinate Spaces

This section reviews several important coordinate spaces related to 3D viewing. Unfortunately, terminology is not consistent in the literature on the
subject, even though the concepts are. Here, we discuss the coordinate
spaces in the order they are encountered as geometry flows through the
graphics pipeline.

10.3.1 Model, World, and Camera Space

The geometry of an object is initially described in object space, which is
a coordinate space local to the object being described (see Section 3.2.2).

The information described usually consists of vertex positions and surface normals. Object space is also known as *local space* and, especially in the context of graphics, *model space*.

From model space, the vertices are transformed into world space (see Section 3.2.1). The transformation from modeling space to world space is often called the *model transform*. Typically, lighting for the scene is specified in world space, although, as we see in Section 10.11, it doesn't really matter what coordinate space is used to perform the lighting calculations provided that the geometry and the lights can be expressed in the same space.

From world space, vertices are transformed by the *view transform* into *camera space* (see Section 3.2.3), also known as *eye space* and *view space* (not to be confused with canonical view volume space, discussed later). Camera space is a 3D coordinate space in which the origin is at the center of projection, one is axis parallel to the direction the camera is facing (perpendicular to the projection plane), one axis is the intersection of the top and bottom clip planes, and the other axis is the intersection of the left and right clip planes. If we assume the perspective of the camera, then one axis will be "horizontal" and one will be "vertical."

In a left-handed world, the most common convention is to point $+z$ in the direction that the camera is facing, with $+x$ and $+y$ pointing "right" and "up" (again, from the perspective from the camera). This is fairly intuitive, as shown in Figure 10.7. The typical right-handed convention is to have $-z$ point in the direction that the camera is facing. We assume the left-handed conventions for the remainder of this chapter

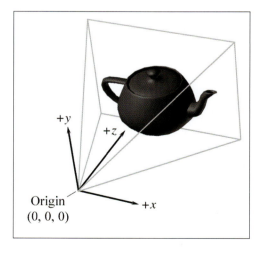

Figure 10.7
Typical camera-space conventions for left-handed coordinate systems

10.3.2   Clip Space and the Clip Matrix

From camera space, vertices are transformed once again into *clip space*, also known as the *canonical view volume space*. The matrix that transforms vertices from camera space into clip space is called the *clip matrix*, also known as the *projection matrix*.

Up until now, our vertex positions have been "pure" 3D vectors—that is, they only had three coordinates, or if they have a fourth coordinate, then w was always equal to 1 for position vectors and 0 for direction vectors such as surface normals. (In some special situations, we might use more exotic transforms, but most basic transforms are 3D affine transformations.) The clip matrix, however, puts meaningful information into w. The clip matrix serves two primary functions:

- *Prepare for projection.* We put the proper value into w so that the homogeneous division produces the desired projection. For the typical perspective projection, this means we copy z into w. We talk about this in Section 10.3.3.

- *Apply zoom and prepare for clipping.* We scale x, y, and z so that they can be compared against w for clipping. This scaling takes the camera's zoom values into consideration, since those zoom values affect the shape of the view frustum against which clipping occurs. This is discussed in Section 10.3.4.

10.3.3   The Clip Matrix: Preparing for Projection

Recall from Section 6.4.1 that a 4D homogeneous vector is mapped to the corresponding physical 3D vector by dividing by w:

$$\begin{bmatrix} x \\ y \\ z \\ w \end{bmatrix} \implies \begin{bmatrix} x/w \\ y/w \\ z/w \end{bmatrix}.$$

Converting 4D homogeneous coordinates to 3D

The first goal of the clip matrix is to get the correct value into w such that this division causes the desired projection (perspective or orthographic). That's the reason this matrix is sometimes called the projection matrix, although this term is a bit misleading—the projection doesn't take place during the multiplication by this matrix, it happens when we divide x, y, and z by w.

If this was the only purpose of the clip matrix, to place the correct value into w, the clip matrix for perspective projection would simply be

$$\begin{bmatrix} 1 & 0 & 0 & 0 \\ 0 & 1 & 0 & 0 \\ 0 & 0 & 1 & 1 \\ 0 & 0 & 0 & 0 \end{bmatrix}.$$

Multiplying a vector of the form $[x, y, z, 1]$ by this matrix, and then performing the homogeneous division by w, we get

$$\begin{bmatrix} x & y & z & 1 \end{bmatrix} \begin{bmatrix} 1 & 0 & 0 & 0 \\ 0 & 1 & 0 & 0 \\ 0 & 0 & 1 & 1 \\ 0 & 0 & 0 & 0 \end{bmatrix} = \begin{bmatrix} x & y & z & z \end{bmatrix} \implies \begin{bmatrix} x/z & y/z & 1 \end{bmatrix}.$$

At this point, many readers might very reasonably ask two questions. The first question might be, "Why is this so complicated? This seems like a lot of work to accomplish what basically amounts to just dividing by z." You're right. In many old school software rasterizers, where the projection math was hand-coded, w didn't appear anywhere, and there was just an explicit divide by z. So why do we tolerate all this complication? One reason for homogeneous coordinates is that they can represent a wider range of camera specifications naturally. At the end of this section we'll see how orthographic projections can be handled easily, without the "if statement" that was necessary in the old hand-coded systems. But there are other types of projections that are also useful and are handled naturally in this framework. For example, the frustum planes do not need to be symmetric about the viewing direction, which corresponds to the situation where your view direction does not look through the center of the window. This is useful, for example, when rendering a very high resolution image in smaller blocks, or for seamless dynamic splitting and merging of split screen views. Another advantage of using homogeneous coordinates is that they make z-clipping (against the near and far clipping planes) identical to x- and y-clipping. This similarity makes things nice and tidy, but, more important, on some hardware the vector unit can be exploited to perform clipping comparison tests in parallel. In general, the use of homogeneous coordinates and 4×4 matrices makes things more compact and general purpose, and (in some peoples' minds) more elegant. But regardless of whether the use of 4×4 matrices improves the process, it's the way most APIs want things delivered, so that's the way it works, for better or worse.

The second question a reader might have is, "What happened to d?" Remember that d is the *focal distance*, the distance from the projection plane to the center of projection (the "focal point"). Our discussion of perspective projection via homogeneous division in Section 6.5 described

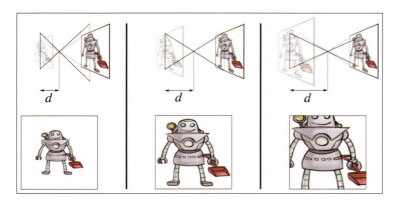

Figure 10.8. In a physical camera, increasing the focal distance d while keeping the size of the "film" the same has the effect of zooming in.

how to project onto a plane perpendicular to the z-axis and d units away from the origin. (The plane is of the form $z = d$.) But we didn't use d anywhere in the above discussion. As it turns out, the value we use for d isn't important, and so we choose the most convenient value possible for d, which is 1.

To understand why d doesn't matter, let's compare the projection that occurs in a computer to the projection that occurs in a physical camera. Inside a real camera, increasing this distance causes the camera to zoom in (objects appear bigger), and decreasing it zooms out (objects appear smaller). This is shown in Figure 10.8.

The vertical line on the left side of each diagram represents the film (or, for modern cameras, the sensing element), which lies in the infinite plane of projection. Importantly, notice that the film is the same height in each diagram. As we increase d, the film moves further away from the focal plane, and the field of view angle intercepted by the view frustum decreases. As the view frustum gets smaller, an object inside this frustum takes a larger proportion of the visible volume, and thus appears larger in the projected image. The perceived result is that we are zooming in. The key point here is that changing the focal length causes an object to appear bigger because the projected image is larger *relative to the size of the film.*

Now let's look at what happens inside a computer. The "film" inside a computer is the rectangular portion of the projection plane that intersects the view frustum.[9] Notice that if we increase the focal distance,

---

[9] The "film" is in front of the focal point rather than behind the focal point like in a real camera, but that fact is not significant to this discussion.

the size of the projected image increases, just like it did in a real camera. However, inside a computer, the film actually increases by this same proportion, rather than the view frustum changing in size. Because the projected image and the film increase by the same proportion, there is no change to the rendered image or the apparent size of objects within this image.

In summary, zoom is always accomplished by changing the shape of the view frustum, whether we're talking about a real camera or inside a computer. In a real camera, changing the focal length changes the shape of the view frustum because the film stays the same size. However, in a computer, adjusting the focal distance d does not affect the rendered image, since the "film" increases in size and the shape of the view frustum does not change.

Some software allow the user to specify the field of view by giving a focal length measured in millimeters. These numbers are in reference to some standard film size, almost always 35 mm film.

What about orthographic projection? In this case, we do *not* want to divide by z, so our clip matrix will have a right-hand column of $[0, 0, 0, 1]^\mathrm{T}$, the same as the identity matrix. When multiplied by a vector of the form $[x, y, z, 1]$, this will result in a vector with $w = 1$, rather than $w = z$. The homogeneous division still occurs, but this time we are dividing by 1:

$$\begin{bmatrix} x & y & z & 1 \end{bmatrix} \begin{bmatrix} 1 & 0 & 0 & 0 \\ 0 & 1 & 0 & 0 \\ 0 & 0 & 1 & 0 \\ 0 & 0 & 0 & 1 \end{bmatrix} = \begin{bmatrix} x & y & z & 1 \end{bmatrix} \quad \implies \quad \begin{bmatrix} x & y & z \end{bmatrix}.$$

The next section fills in the rest of the clip matrix. But for now, the key point is that a perspective projection matrix will always have a right-hand column of $[0, 0, 1, 0]$, and a orthographic projection matrix will always have a right-hand column of $[0, 0, 0, 1]$. Here, the word "always" means "we've never seen anything else." You might come across some obscure case on some particular hardware for which other values are needed, and it is important to understand that 1 isn't a magic number here, it is just the simplest number. Since the homogeneous conversion is a division, what is important is the *ratio* of the coordinates, not their magnitude.

Notice that multiplying the entire matrix by a constant factor doesn't have any effect on the projected values x/w, y/w, and z/w, but it will adjust the value of w, which is used for perspective correct rasterization. So a different value might be necessary for some reason. Then again, certain hardware (such as the Wii) assume that these are the only two cases, and no other right-hand column is allowed.

10.3.4 The Clip Matrix: Applying Zoom and Preparing for Clipping

The second goal of the clip matrix is to scale the x, y, and z components such that the six clip planes have a trivial form. Points are outside the view frustum if they satisfy at least one of the inequalities:

$$\begin{array}{ll} \text{Bottom} & y < -w, \\ \text{Top} & y > w, \\ \text{Left} & x < -w, \\ \text{Right} & x > w, \\ \text{Near} & z < -w, \\ \text{Far} & z > w. \end{array}$$

The six planes of the view frustum in clip space

So the points inside the view volume satisfy

$$\begin{array}{ccc} -w & \leq x & \leq w, \\ -w & \leq y & \leq w, \\ -w & \leq z & \leq w. \end{array}$$

Any geometry that does not satisfy these equalities must be *clipped* to the view frustum. Clipping is discussed in Section 10.10.4.

To stretch things to put the top, left, right, and bottom clip planes in place, we scale the x and y values by the zoom values of the camera. We discussed how to compute these values in Section 10.2.4. For the near and far clip planes, the z-coordinate is biased and scaled such that at the near clip plane, $z/w = -1$, and at the far clip plane, $z/w = 1$.

Let zoom_x and zoom_y be the horizontal and vertical zoom values, and let n and f be the distances to the near and far clipping planes. Then the matrix that scales x, y, and z appropriately, while simultaneously outputting the z-coordinate into w, is

$$\begin{bmatrix} \text{zoom}_x & 0 & 0 & 0 \\ 0 & \text{zoom}_y & 0 & 0 \\ 0 & 0 & \frac{f+n}{f-n} & 1 \\ 0 & 0 & \frac{-2nf}{f-n} & 0 \end{bmatrix}. \tag{10.6}$$

Clip matrix for perspective projection with $z = -w$ at the near clip plane

This clip matrix assumes a coordinate system with z pointing into the screen (the usual left-handed convention), row vectors on the left, and z values in the range $[-w, w]$ from the near to far clip plane. This last detail is yet another place where conventions can vary. Other APIs, (notably, DirectX) want the projection matrix such that z is in the range $[0, w]$. In other words, a point in clip space is outside the clip plane if

$$\begin{array}{ll} \text{near} & z < 0, \\ \text{far} & z > w. \end{array}$$

Near and far clip planes in DirectX-style clip space

Under these DirectX-style conventions, the points inside the view frustum satisfy the inequality $0 \leq z \leq w$. A slightly different clip matrix is used in this case:

Clip matrix for perspective projection with $z = 0$ at the near clip plane

$$\begin{bmatrix} \text{zoom}_x & 0 & 0 & 0 \\ 0 & \text{zoom}_y & 0 & 0 \\ 0 & 0 & \frac{f}{f-n} & 1 \\ 0 & 0 & \frac{-nf}{f-n} & 0 \end{bmatrix}. \qquad (10.7)$$

We can easily tell that the two matrices in Equations (10.6) and (10.7) are *perspective* projection matrices because the right-hand column is $[0, 0, 1, 0]^T$. (OK, the caption in the margin is a bit of a hint, too.)

What about orthographic projection? The first and second columns of the projection matrix don't change, and we know the fourth column will become $[0, 0, 0, 1]^T$. The third column, which controls the output z value, must change. We start by assuming the first set of conventions for z, that is the output z value will be scaled such that z/w takes on the values -1 and $+1$ at the near and far clip planes, respectively. The matrix that does this is

Clip matrix for orthographic projection with $z = -w$ at the near clip plane

$$\begin{bmatrix} \text{zoom}_x & 0 & 0 & 0 \\ 0 & \text{zoom}_y & 0 & 0 \\ 0 & 0 & \frac{2}{f-n} & 0 \\ 0 & 0 & -\frac{f+n}{f-n} & 1 \end{bmatrix}.$$

Alternatively, if we are using a DirectX-style range for the clip space z values, then the matrix we use is

Clip matrix for orthographic projection with $z = 0$ at the near clip plane

$$\begin{bmatrix} \text{zoom}_x & 0 & 0 & 0 \\ 0 & \text{zoom}_y & 0 & 0 \\ 0 & 0 & \frac{1}{f-n} & 0 \\ 0 & 0 & \frac{n}{n-f} & 1 \end{bmatrix}.$$

In this book, we prefer a left-handed convention and row vectors on the left, and all the projection matrices so far assume those conventions. However, both of these choices differ from the OpenGL convention, and we know that many readers may be working in environments that are similar to OpenGL. Since this can be very confusing, let's repeat these matrices, but with the right-handed, column-vector OpenGL conventions. We'll only discuss the $[-1, +1]$ range for clip-space z values, because that's what OpenGL uses.

It's instructive to consider how to convert these matrices from one set of conventions to the other. Because OpenGL uses column vectors, the first thing we need to do is transpose our matrix. Second, the right-handed conventions have $-z$ pointing into the screen in camera space ("eye space" in the OpenGL vocabulary), but the clip-space $+z$ axis points into the screen just like the left-handed conventions assumed earlier. (In OpenGL, clip space is actually a left-handed coordinate space!) This means we need to negate our *incoming* z values, or alternatively, negate the third column (after we've transposed the matrix), which is the column that is multiplied by z.

The above procedure results in the following perspective projection matrix

$$\begin{bmatrix} \text{zoom}_x & 0 & 0 & 0 \\ 0 & \text{zoom}_y & 0 & 0 \\ 0 & 0 & -\frac{f+n}{f-n} & \frac{-2nf}{f-n} \\ 0 & 0 & -1 & 0 \end{bmatrix},$$

Clip matrix for perspective projection assuming OpenGL conventions

and the orthographic projection matrix is

$$\begin{bmatrix} \text{zoom}_x & 0 & 0 & 0 \\ 0 & \text{zoom}_y & 0 & 0 \\ 0 & 0 & \frac{-2}{f-n} & -\frac{f+n}{f-n} \\ 0 & 0 & 0 & 1 \end{bmatrix}.$$

Clip matrix for orthographic projection assuming OpenGL conventions

So, for OpenGL conventions, you can tell whether a projection matrix is perspective or orthographic based on the bottom *row*. It will be $[0, 0, -1, 0]$ for perspective, and $[0, 0, 0, 1]$ for orthographic.

Now that we know a bit about clip space, we can understand the need for the near clip plane. Obviously, there is a singularity precisely at the origin, where a perspective projection is not defined. (This corresponds to a perspective division by zero.) In practice, this singularity would be extremely rare, and however we wanted to handle it—say, by arbitrarily projecting the point to the center of the screen—would be OK, since putting the camera directly in a polygon isn't often needed in practice.

But projecting polygons onto pixels isn't the only issue. Allowing for arbitrarily small (but positive) values of z will result in arbitrarily large values for w. Depending on the hardware, this can cause problems with perspective-correct rasterization. Another potential problem area is depth buffering. Suffice it to say that for practical reasons it is often necessary to restrict the range of the z values so that there is a known minimum value, and we must accept the rather unpleasant necessity of a near clip plane. We say "unpleasant" because the near clip plane is an artifact of implementation, not an inherent part of a 3D world. (Raytracers don't necessarily have this issue.) It cuts off objects when you get too close to them, when in reality you should be able to get arbitrarily close. Many readers are probably familiar with the phenomena where a camera is placed in the middle of a very large ground polygon, just a small distance above it, and a gap opens up at the bottom of the screen, allowing the camera to see through the ground. A similar situation exists if you get very close to practically any object other than a wall. A hole will appear in the middle of the object, and this hole will expand as you move closer.

10.3.5   Screen Space

Once we have clipped the geometry to the view frustum, it is projected into screen space, which corresponds to actual pixels in the frame buffer. Remember that we are rendering into an output window that does not necessarily occupy the entire display device. However, we usually want our screen-space coordinates to be specified using coordinates that are absolute to the rendering device (Figure 10.9).

Screen space is a 2D space, of course. Thus, we must project the points from clip space to screen space to generate the correct 2D coordinates. The first thing that happens is the standard homogeneous division by w. (OpenGL calls the result of this division the *normalized device coordinates*.) Then, the x- and y-coordinates must be scaled to map into the output window. This is summarized by

Projecting and mapping to screen space

$$\text{screen}_x = \frac{\text{clip}_x \cdot \text{winRes}_x}{2 \cdot \text{clip}_w} + \text{winCenter}_x, \qquad (10.8)$$

$$\text{screen}_y = -\frac{\text{clip}_y \cdot \text{winRes}_y}{2 \cdot \text{clip}_w} + \text{winCenter}_y. \qquad (10.9)$$

A quick comment is warranted about the negation of the y component in the math above. This reflects DirectX-style coordinate conventions where

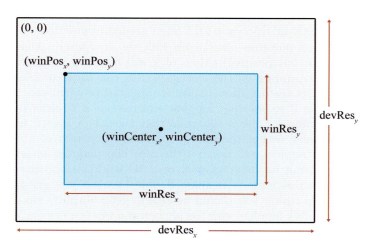

Figure 10.9. The output window in screen space

(0,0) is in the upper-left corner. Under these conventions, $+y$ points up in clip space, but down in screen space. In fact, if we continue to think about $+z$ pointing into the screen, then screen space actually becomes a right-handed coordinate space, even though it's left-handed everywhere else in DirectX. In OpenGL, the origin is in the *lower* left corner, and the negation of the y-coordinate does not occur. (As already discussed, in OpenGL, they choose a different place to introduce confusion, by flipping the z-axis between eye space, where $-z$ points into the screen, to clip space, where $+z$ points into the screen.)

Speaking of z, what happens to clip_z? In general it's used in some way for depth buffering. A traditional method is to take the normalized depth value $\text{clip}_z/\text{clip}_w$ and store this value in the depth buffer. The precise details depend on exactly what sort of clip values are used for clipping, and what sort of depth values go into the depth buffer. For example, in OpenGL, the conceptual convention is for the view frustum to contain $-1 \leq \text{clip}_z/\text{clip}_w \leq +1$, but this might not be optimal for depth buffering. Driver vendors must convert from the API's conceptual conventions to whatever is optimal for the hardware.

An alternative strategy, known as *w-buffering*, is to use clip_w as the depth value. In most situations clip_w is simply a scaled version of the camera-space z value; thus by using clip_w in the depth buffer, each value has a linear relationship to the viewing depth of the corresponding pixel. This method can be attractive, especially if the depth buffer is fixed-point with limited precision, because it spreads out the available precision more evenly. The traditional method of storing $\text{clip}_z/\text{clip}_w$ in the depth buffer results in greatly increased precision up close, but at the expense of (sometimes drastically) reduced precision near the far clip plane. If the depth buffer values are stored in floating-point, this issue is much less important. Also note that w-buffering doesn't work for orthographic projection, since an orthographic projection matrix always outputs $w = 1$.

The clip_w value is also not discarded. As we've said, it serves the important purpose as the denominator in the homogeneous division to normalized device coordinates. But this value is also usually needed for proper perspective-correct interpolation of texture coordinates, colors, and other vertex-level values during rasterization.

On modern graphics APIs at the time of this writing, the conversion of vertex coordinates from clip space to screen space is done for you. Your vertex shader outputs coordinates in clip space. The API clips the triangles to the view frustum and then projects the coordinates to screen space. But that doesn't mean that you will never use the equations in this section in your code. Quite often, we need to perform these calculations in software for visibility testing, level-of-detail selection, and so forth.

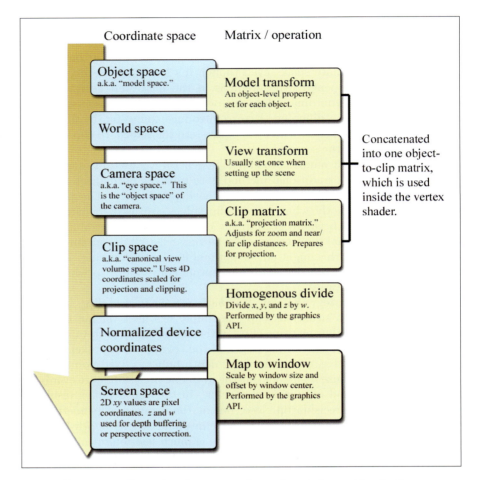

Figure 10.10. Conversion of vertex coordinates through the graphics pipeline

10.3.6   Summary of Coordinate Spaces

Figure 10.10 summarizes the coordinate spaces and matrices discussed in this section, showing the data flow from object space to screen space.

The coordinate spaces we've mentioned are the most important and common ones, but other coordinate spaces are used in computer graphics. For example, a projected light might have its own space, which is essentially the same as camera space, only it is from the perspective that the light "looks" onto the scene. This space is important when the light projects an image (sometimes called a *gobo*) and also for shadow mapping to determine whether a light can "see" a given point.

Another space that has become very important is *tangent space*, which is a local space on the surface of an object. One basis vector is the surface normal and the other two basis vectors are locally tangent to the surface, essentially establishing a 2D coordinate space that is "flat" on the surface at that spot. There are many different ways we could determine these basis vectors, but by far the most common reason to establish such a coordinate space is for bump mapping and related techniques. A more complete discussion of tangent space will need to wait until after we discuss texture mapping in Section 10.5, so we'll come back to this subject in Section 10.9.1. Tangent space is also sometimes called *surface-local* space.

10.4   Polygon Meshes

To render a scene, we need a mathematical description of the geometry in that scene. Several different methods are available to us. This section focuses on the one most important for real-time rendering: the *triangle mesh*. But first, let's mention a few alternatives to get some context. *Constructive solid geometry* (CSG) is a system for describing an object's shape using Boolean operators (union, intersection, subtraction) on primitives. Within video games, CSG can be especially useful for rapid prototyping tools, with the *Unreal* engine being a notable example. Another technique that works by modeling volumes rather than their surfaces is *metaballs*, sometimes used to model organic shapes and fluids, as was discussed in Section 9.1. CSG, metaballs, and other volumetric descriptions are very useful in particular realms, but for rendering (especially real-time rendering) we are interested in a description of the *surface* of the object, and seldom need to determine whether a given point is inside or outside this surface. Indeed, the surface need not be closed or even define a coherent volume.

The most common surface description is the *polygon mesh*, of which you are probably already aware. In certain circumstances, it's useful to allow the polygons that form the surface of the object to have an arbitrary number of vertices; this is often the case in importing and editing tools. For real-time rendering, however, modern hardware is optimized for *triangle meshes*, which are polygon meshes in which every polygon is a triangle. Any given polygon mesh can be converted into an equivalent triangle mesh by decomposing each polygon into triangles individually, as was discussed briefly in Section 9.7.3. Please note that many important concepts introduced in the context of a single triangle or polygon were covered in

Section 9.6 and Section 9.7, respectively. Here, our focus is on how more than one triangle can be connected in a mesh.

One very straightforward way to store a triangle mesh would be to use an array of triangles, as shown in Listing 10.4.

```
struct Triangle {
    Vector3 vertPos[3];    // vertex positions
};

struct TriangleMesh {
    int        triCount; // number of triangles
    Triangle *triList; // array of triangles
};
```

Listing 10.4
A trivial representation of a triangle mesh

For some applications this trivial representation might be adequate. However, the term "mesh" implies a degree of connectivity between adjacent triangles, and this connectivity is not expressed in our trivial representation. There are three basic types of information in a triangle mesh:

- *Vertices.* Each triangle has exactly three vertices. Each vertex may be shared by multiple triangles. The *valence* of a vertex refers to how many faces are connected to the vertex.

- *Edges.* An edge connects two vertices. Each triangle has three edges. In many cases, each edge is shared by exactly two faces, but there are certainly exceptions. If the object is not closed, an *open edge* with only one neighboring face can exist.

- *Faces.* These are the surfaces of the triangles. We can store a face as either a list of three vertices, or a list of three edges.

A variety of methods exist to represent this information efficiently, depending on the operations to be performed most often on the mesh. Here we will focus on a standard storage format known as an *indexed triangle mesh.*

10.4.1   Indexed Triangle Mesh

An indexed triangle mesh consists of two lists: a list of vertices, and a list of triangles.

- Each *vertex* contains a position in 3D. We may also store other information at the vertex level, such as texture-mapping coordinates, surface normals, or lighting values.

- A *triangle* is represented by three integers that index into the vertex list. Usually, the order in which these vertices are listed is significant, since we may consider faces to have "front" and "back" sides. We adopt the left-handed convention that the vertices are listed in clockwise order when viewed from the front side. Other information may also be stored at the triangle level, such as a precomputed normal of the plane containing the triangle, surface properties (such as a texture map), and so forth.

Listing 10.5 shows a highly simplified example of how an indexed triangle mesh might be stored in C.

```
// struct Vertex is the information we store at the vertex level
struct Vertex {

    // 3D position of the vertex
    Vector3 pos;

    // Other information could include
    // texture mapping coordinates, a
    // surface normal, lighting values, etc.
};

// struct Triangle is the information we store at the triangle level
struct Triangle {

    // Indices into the vertex list. In practice, 16-bit indices are
    // almost always used rather than 32-bit, to save memory and bandwidth.
    int vertexIndex[3];

    // Other information could include
    // a normal, material information, etc
};

// struct TriangleMesh stores an indexed triangle mesh
struct TriangleMesh {

    // The vertices
    int     vertexCount;
    Vertex *vertexList;

    // The triangles
    int      triangleCount;
    Triangle *triangleList;
};
```

Listing 10.5
Indexed triangle mesh

Figure 10.11 shows how a cube and a pyramid might be represented as a polygon mesh or a triangle mesh. Note that both objects are part of a *single* mesh with 13 vertices. The lighter, thicker wires show the outlines of polygons, and the thinner, dark green wires show one way to add edges to triangulate the polygon mesh.

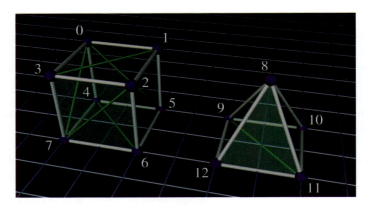

Figure 10.11. A simple mesh containing a cube and a pyramid

Assuming the origin is on the "ground" directly between the two objects, the vertex coordinates might be as shown in Table 10.3.

0	$(-3, 2, 1)$	4	$(-3, 0, 1)$	8	$(2, 2, 0)$	12	$(1, 0, -1)$
1	$(-1, 2, 1)$	5	$(-1, 0, 1)$	9	$(1, 0, 1)$		
2	$(-1, 2, -1)$	6	$(-1, 0, -1)$	10	$(3, 0, 1)$		
3	$(-3, 2, -1)$	7	$(-3, 0, -1)$	11	$(3, 0, -1)$		

Table 10.3. Vertex positions in our sample mesh

Table 10.4 shows the vertex indices that would form faces of this mesh, either as a polygon mesh or as a triangle mesh. Remember that the order of the vertices is significant; they are listed in clockwise order when viewed from the outside. You should study these figures until you are sure you understand them.

Description	Vertex indices (Polygon mesh)	Vertex indices (Triangle mesh)
Cube top	$\{0, 1, 2, 3\}$	$\{1, 2, 3\}, \{1, 3, 0\}$
Cube front	$\{2, 6, 7, 3\}$	$\{2, 6, 7\}, \{2, 7, 3\}$
Cube right	$\{2, 1, 5, 6\}$	$\{2, 1, 5\}, \{2, 5, 6\}$
Cube left	$\{0, 3, 7, 4\}$	$\{0, 3, 7\}, \{0, 7, 4\}$
Cube back	$\{0, 4, 5, 1\}$	$\{0, 4, 5\}, \{0, 5, 1\}$
Cube bottom	$\{4, 7, 6, 5\}$	$\{4, 7, 6\}, \{4, 6, 5\}$
Pyramid front	$\{12, 8, 11\}$	$\{12, 8, 11\}$
Pyramid left	$\{9, 8, 12\}$	$\{9, 8, 12\}$
Pyramid right	$\{8, 10, 11\}$	$\{8, 10, 11\}$
Pyramid back	$\{8, 9, 10\}$	$\{8, 9, 10\}$
Pyramid bottom	$\{9, 12, 11, 10\}$	$\{9, 12, 11\}, \{9, 11, 10\}$

Table 10.4. The vertex indices that form the faces of our sample mesh, either as a polygon mesh or a triangle mesh

The vertices must be listed in clockwise order around a face, but it doesn't matter which one is considered the "first" vertex; they can be cycled without changing the logical structure of the mesh. For example, the quad forming the cube top could equivalently have been given as $\{1, 2, 3, 0\}$, $\{2, 3, 0, 1\}$, or $\{3, 0, 1, 2\}$.

As indicated by the comments in Listing 10.5, additional data are almost always stored per vertex, such as texture coordinates, surface normals, basis vectors, colors, skinning data, and so on. Each of these is discussed in later sections in the context of the techniques that make use of the data. Additional data can also be stored at the triangle level, such as an index that tells which material to use for that face, or the plane equation (part of which is the surface normal—see Section 9.5) for the face. This is highly useful for editing purposes or in other tools that perform mesh manipulations in software. For real-time rendering, however, we seldom store data at the triangle level beyond the three vertex indices. In fact, the most common method is to not have a `struct Triangle` at all, and to represent the entire list of triangles simply as an array (e.g. `unsigned short triList[]`), where the length of the array is the number of triangles times 3. Triangles with identical properties are grouped into batches so that an entire batch can be fed to the GPU in this optimal format. After we review many of the concepts that give rise to the need to store additional data per vertex, Section 10.10.2 looks at several more specific examples of how we might feed that data to the graphics API. By the way, as a general rule, things are a lot easier if you do not try to use the same mesh class for both rendering and editing. The requirements are very different, and a bulkier data structure with more flexibility is best for use in tools, importers, and the like.

Note that in an indexed triangle mesh, the edges are not stored explicitly, but rather the adjacency information contained in an indexed triangle list is stored implicitly: to locate shared edges between triangles, we must search the triangle list. Our original trivial "array of triangles" format in Listing 10.4 did not have *any* logical connectivity information (although we could have attempted to detect whether the vertices on an edge were identical by comparing the vertex positions or other properties). What's surprising is that the "extra" connectivity information contained in the indexed representation actually results in a reduction of memory usage in most cases, compared to the flat method. The reason for this is that the information stored at the vertex level, which is duplicated in the trivial flat format, is relatively large compared to a single integer index. (At a minimum, we must store a 3D vector position.) In meshes that arise in practice, a typical vertex has a valence of around 3–6, which means that the flat format duplicates quite a lot of data.

The simple indexed triangle mesh scheme is appropriate for many applications, including the very important one of rendering. However, some operations on triangle meshes require a more advanced data structure in order to be implemented more efficiently. The basic problem is that the adjacency between triangles is not expressed explicitly and must be extracted by searching the triangle list. Other representation techniques exist that make this information available in constant time. One idea is to maintain an edge list explicitly. Each edge is defined by listing the two vertices on the ends. We also maintain a list of triangles that share the edge. Then the triangles can be viewed as a list of three edges rather than a list of three vertices, so they are stored as three indices into the edge list rather than the vertex list. An extension of this idea is known as the *winged-edge* model [22], which also stores, for each vertex, a reference to one edge that uses the vertex. The edges and triangles may be traversed intelligently to quickly locate all edges and triangles that use the vertex.

10.4.2   Surface Normals

Surface normals are used for several different purposes in graphics; for example, to compute proper lighting (Section 10.6), and for backface culling (Section 10.10.5). In general, a surface normal is a unit[10] vector that is perpendicular to a surface. We might be interested in the normal of a given face, in which case the surface of interest is the plane that contains the face. The surface normals for polygons can be computed easily by using the techniques from Section 9.5.

Vertex-level normals are a bit trickier. First, it should be noted that, strictly speaking, there is not a true surface normal at a vertex (or an edge for that matter), since these locations mark discontinuities in the surface of the polygon mesh. Rather, for rendering purposes, we typically interpret a polygon mesh as an approximation to some *smooth* surface. So we don't want a normal to the piecewise linear surface defined by the polygon mesh; rather, we want (an approximation of) the surface normal of the smooth surface.

The primary purpose of vertex normals is lighting. Practically every lighting model takes a surface normal at the spot being lit as an input. Indeed, the surface normal is part of the rendering equation itself (in the Lambert factor), so it is always an input, even if the BRDF does not depend on it. We have normals available only at the vertices, but yet we need to compute lighting values over the entire surface. What to do? If hardware resources permit (as they usually do nowadays), then we can approximate the normal of the continuous surface corresponding to any point on a given face by interpolating vertex normals and renormalizing the result. This technique is illustrated in Figure 10.12, which shows a cross section of a

---

[10]This is not strictly necessary in some cases, but in practice we almost always use unit vectors.

cylinder (black circle) that is being approximated by a hexagonal prism (blue outline). Black normals at the vertices are the true surface normals, whereas the interior normals are being approximated through interpolation. (The actual normals used would be the result of stretching these out to unit length.)

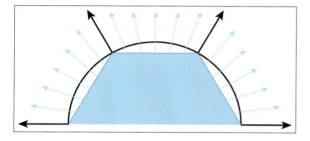

Figure 10.12
A cylinder approximated with a hexagonal prism.

Once we have a normal at a given point, we can perform the full lighting equation per pixel. This is known as *per-pixel* shading.[11] An alternative strategy to per-pixel shading, known as *Gouraud*[12] shading [26], is to perform lighting calculations only at the vertex level, and then interpolate the results themselves, rather than the normal, across the face. This requires less computation, and is still done on some systems, such as the Nintendo Wii.

Figure 10.13 shows per-pixel lighting of cylinders with a different number of sides. Although the illusion breaks down on the ends of the cylinder, where the silhouette edge gives away the low-poly nature of the geometry, this method of approximating a smooth surface can indeed make even a very low-resolution mesh look "smooth." Cover up the ends of the cylinder, and even the 5-sided cylinder is remarkably convincing.

Now that we understand how normals are interpolated in order to approximately reconstruct a curved surface, let's talk about how to obtain vertex normals. This information may not be readily available, depending on how the triangle mesh was generated. If the mesh is generated procedurally, for example, from a parametric curved surface, then the vertex normals can be supplied at that time. Or you may simply be handed the vertex normals from the modeling package as part of the mesh. However, sometimes the surface normals are not provided, and we must approximate them by interpreting the only information available to us: the vertex positions and the triangles. One trick that works is to average the normals of the adjacent triangles, and then renormalize the result. This classic technique is demonstrated in Listing 10.6.

---

[11] This technique of interpolating the vertex normals is also sometimes confusingly known as *Phong* shading, not to be confused with the Phong model for specular reflection.

[12] Pronounced "guh-ROH."

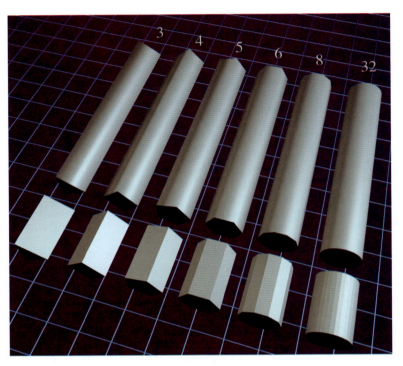

Figure 10.13. Approximating cylinders with prisms of varying number of sides.

```
struct Vertex {
    Vector3 pos;
    Vector3 normal;
};
struct Triangle {
    int      vertexIndex[3];
    Vector3 normal;
};
struct TriangleMesh {
    int      vertexCount;
    Vertex  *vertexList;
    int      triangleCount;
    Triangle *triangleList;

    void computeVertexNormals() {

        // First clear out the vertex normals
        for (int i = 0 ; i < vertexCount ; ++i) {
            vertexList[i].normal.zero();
        }

        // Now add in the face normals into the
        // normals of the adjacent vertices
        for (int i = 0 ; i < triangleCount ; ++i) {
```

```
        // Get shortcut
        Triangle &tri = triangleList[i];

        // Compute triangle normal.
        Vector3 v0 = vertexList[tri.vertexIndex[0]].pos;
        Vector3 v1 = vertexList[tri.vertexIndex[1]].pos;
        Vector3 v2 = vertexList[tri.vertexIndex[2]].pos;
        tri.normal = cross(v1-v0, v2-v1);
        tri.normal.normalize();

        // Sum it into the adjacent vertices
        for (int j = 0 ; j < 3 ; ++j) {
            vertexList[tri.vertexIndex[j]].normal += tri.normal;
        }
    }

    // Finally, average and normalize the results.
    // Note that this can blow up if a vertex is isolated
    // (not used by any triangles), and in some other cases.
    for (int i = 0 ; i < vertexCount ; ++i) {
        vertexList[i].normal.normalize();
    }
  }
};
```

Listing 10.6
Simple method for calculating vertex normals as the average of adjacent face normals

Averaging face normals to compute vertex normals is a tried-and-true technique that works well in most cases. However, there are a few things to watch out for. The first is that sometimes the mesh is *supposed* to have a discontinuity, and if we're not careful, this discontinuity will get "smoothed out." Take the very simple example of a box. There should be a sharp lighting discontinuity at its edges. However, if we use vertex normals computed from the average of the surface normals, then there is no lighting discontinuity, as shown in Figure 10.14.

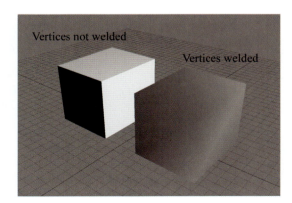

Figure 10.14
On the right, the box edges are not visible because there is only one normal at each corner

	Vertices		Faces	
#	Position	Normal	Description	Indices
0	$(-1, +1, +1)$	$[-0.577, +0.577, +0.577]$	Top	$\{0, 1, 2, 3\}$
1	$(+1, +1, +1)$	$[+0.577, +0.577, +0.577]$	Front	$\{2, 6, 7, 3\}$
2	$(+1, +1, -1)$	$[+0.577, +0.577, -0.577]$	Right	$\{2, 1, 5, 6\}$
3	$(-1, +1, -1)$	$[-0.577, +0.577, -0.577]$	Left	$\{0, 3, 7, 4\}$
4	$(-1, -1, +1)$	$[-0.577, -0.577, +0.577]$	Back	$\{0, 4, 5, 1\}$
5	$(+1, -1, +1)$	$[+0.577, -0.577, +0.577]$	Bottom	$\{4, 7, 6, 5\}$
6	$(+1, -1, -1)$	$[+0.577, -0.577, -0.577]$		
7	$(-1, -1, -1)$	$[-0.577, -0.577, -0.577]$		

Table 10.5. Polygon mesh of a box with welded vertices and smoothed edges

The basic problem is that the surface discontinuity at the box edges cannot be properly represented because *there is only one normal stored per vertex*. The solution to this problem is to "detach" the faces; in other words, duplicate the vertices along the edge where there is a true geometric discontinuity, creating a topological discontinuity to prevent the vertex normals from being averaged. After doing so, the faces are no longer logically connected, but this seam in the topology of the mesh doesn't cause a problem for many important tasks, such as rendering and raytracing. Table 10.5 shows a smoothed box mesh with eight vertices. Compare that mesh to the one in Table 10.6, in which the faces have been detached, resulting in 24 vertices.

	Vertices		Faces	
#	Position	Normal	Description	Indices
0	$(-1, +1, +1)$	$[0, +1, 0]$	Top	$\{0, 1, 2, 3\}$
1	$(+1, +1, +1)$	$[0, +1, 0]$	Front	$\{4, 5, 6, 7\}$
2	$(+1, +1, -1)$	$[0, +1, 0]$	Right	$\{8, 9, 10, 11\}$
3	$(-1, +1, -1)$	$[0, +1, 0]$	Left	$\{12, 13, 14, 15\}$
4	$(-1, +1, -1)$	$[0, 0, -1]$	Back	$\{16, 17, 18, 19\}$
5	$(+1, +1, -1)$	$[0, 0, -1]$	Bottom	$\{20, 21, 22, 23\}$
6	$(+1, -1, -1)$	$[0, 0, -1]$		
7	$(-1, -1, -1)$	$[0, 0, -1]$		
8	$(+1, +1, -1)$	$[+1, 0, 0]$		
9	$(+1, +1, +1)$	$[+1, 0, 0]$		
10	$(+1, -1, +1)$	$[+1, 0, 0]$		
11	$(+1, -1, -1)$	$[+1, 0, 0]$		
12	$(-1, +1, +1)$	$[-1, 0, 0]$		
13	$(-1, +1, -1)$	$[-1, 0, 0]$		
14	$(-1, -1, -1)$	$[-1, 0, 0]$		
15	$(-1, -1, +1)$	$[-1, 0, 0]$		
16	$(+1, +1, +1)$	$[0, 0, +1]$		
17	$(-1, +1, +1)$	$[0, 0, +1]$		
18	$(-1, -1, +1)$	$[0, 0, +1]$		
19	$(+1, -1, +1)$	$[0, 0, +1]$		
20	$(+1, -1, -1)$	$[0, -1, 0]$		
21	$(-1, -1, -1)$	$[0, -1, 0]$		
22	$(-1, -1, +1)$	$[0, -1, 0]$		
23	$(+1, -1, +1)$	$[0, -1, 0]$		

Table 10.6. Polygon mesh of a box with detached faces and lighting discontinuities at the edges

An extreme version of this situation occurs when two faces are placed back-to-back. Such infinitely thin double-sided geometry can arise with foliage, cloth, billboards, and the like. In this case, since the normals are exactly opposite, averaging them produces the zero vector, which cannot be normalized. The simplest solution is to detach the faces so that the vertex normals will not average together. Or if the front and back sides are mirror images, the two "single-sided" polygons can be replaced by one "double-sided" one. This requires special treatment during rendering to disable backface culling (Section 10.10.5) and intelligently dealing with the normal in the lighting equation.

A more subtle problem is that the averaging is biased towards large numbers of triangles with the same normal. For example, consider the vertex at index 1 in Figure 10.11. This vertex is adjacent to two triangles on the top of the cube, but only one triangle on the right side and one triangle on the back side. The vertex normal computed by averaging the triangle normals is biased because the top face normal essentially gets twice as many "votes" as each of the side face normals. But this topology is the result of an arbitrary decision as to where to draw the edges to triangulate the faces of the cube. For example, if we were to triangulate the top face by drawing an edge between vertices 0 and 2 (this is known as "turning" the edge), all of the normals on the top face would change.

Techniques exist to deal with this problem, such as weighing the contribution from each adjacent face based on the interior angle adjacent to the vertex, but it's often ignored in practice. Most of the really terrible examples are contrived ones like this, where the faces should be detached anyway. Furthermore, the normals are an approximation to begin with, and having a slightly perturbed normal is often difficult to tell visually.

Although some modeling packages can deliver vertex normals for you, fewer provide the basis vectors needed for bump mapping. As we see in Section 10.9, techniques used to synthesize vertex basis vectors are similar to those described here.

Before we go on, there is one very important fact about surface normals that we must mention. In certain circumstances, they cannot be transformed by the same matrix that is used to transform positions. (This is an entirely separate issue from the fact that normals should not be translated like positions.) The reason for this is that normals are *covariant* vectors. "Regular" vectors, such as position and velocity, are said to be *contravariant*: if we scale the coordinate space used to describe the vector, the coordinates will respond in the *opposite* direction. If we use a coordinate space with a larger scale (for example, using meters instead of feet) the coordinates of a contravariant vector respond to the contrary, by becoming smaller. Notice that this is all about scale; translation and rotation are not

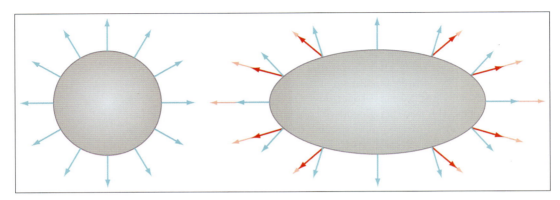

Figure 10.15
Transforming normals with nonuniform scale. The light red vectors show the normals multiplied by the same transform matrix used to transform the object; the dark red vectors are their normalized versions. The light blue vectors show the correct normals.

part of the discussion. Normals and other types of gradients, known as *dual vectors*, do not behave like this.

Imagine that we stretch a 2D object, such as a circle, horizontally, as shown in Figure 10.15. Notice that the normals (shown in light blue in the right figure) begin to turn to point more *vertically*—the horizontal coordinates of the normals are decreasing in absolute value while the horizontal coordinates of the positions are increasing. A stretching of the object (object getting bigger while coordinate space stays the same) has the same effect as scaling down the coordinate space while holding the object at the same size. The coordinates of the normal change in the same direction as the scale of the coordinate space, which is why they are called covariant vectors.

To properly transform surface normals, we must use the *inverse transpose* of the matrix used to transform positions; that is, the result of transposing and inverting the matrix. This is sometimes denoted $\mathbf{M}^{-T}$, since it doesn't matter if we transpose first, or invert first: $(\mathbf{M}^{-1})^{T} = (\mathbf{M}^{T})^{-1}$. If the transform matrix doesn't contain any scale (or skew), then the matrix is orthonormal, and thus the inverse transpose is simply the same as the original matrix, and we can safely transform normals with this transform. If the matrix contains uniform scale, then we can still ignore this, but we must renormalize the normals after transforming them. If the matrix contains nonuniform scale (or skew, which is indistinguishable from nonuniform scale combined with rotation), then to properly transform the normals, we must use the inverse transpose transform matrix, and then re-normalize the resulting transformed normals.

In general, normals must be transformed with the inverse transpose of the matrix used to transform positions. This can safely be ignored if the transform matrix is without scale. If the matrix contains uniform scale, then all that is required is to renormalize the normals after transformation. If the matrix contains nonuniform scale, then we must use the inverse transpose transform and renormalize after transforming.

10.5 Texture Mapping

There is much more to the appearance of an object than its shape. Different objects are different colors and have different patterns on their surface. One simple yet powerful way to capture these qualities is through *texture mapping*. A texture map is a bitmap image that is "pasted" to the surface of an object. Rather than controlling the color of an object per triangle or per vertex, with texture mapping we can control the color at a much finer level—per *texel*. (A texel is a single pixel in a texture map. This is a handy word to know, since in graphics contexts, there are lots of different bitmaps being accessed, and it's nice to have a short way to differentiate between a pixel in the frame buffer and a pixel in a texture.)

So a texture map is just a regular bitmap that is applied onto the surface of a model. Exactly how does this work? Actually, there are many different ways to apply a texture map onto a mesh. *Planar mapping* projects the texture orthographically onto the mesh. *Spherical*, *cylindrical*, and *cubic* mapping are various methods of "wrapping" the texture around the object. The details of each of these techniques are not important to us at the moment, since modeling packages such as *3DS Max* deal with these user interface issues. The key idea is that, at each point on the surface of the mesh, we can obtain *texture-mapping coordinates*, which define the 2D location in the texture map that corresponds to this 3D location. Traditionally, these coordinates are assigned the variables (u, v), where u is the horizontal coordinate and v is the vertical coordinate; thus, texture-mapping coordinates are often called *UV coordinates* or simply *UV*s.

Although bitmaps come in different sizes, UV coordinates are normalized such that the mapping space ranges from 0 to 1 over the entire width (u) or height (v) of the image, rather than depending on the image dimensions. The origin of this space is either in the upper left-hand corner of the image, which is the DirectX-style convention, or in the lower left-hand corner, the OpenGL conventions. We use the DirectX conventions in this book. Figure 10.16 shows the texture map that we use in several examples and the DirectX-style coordinate conventions.

In principle, it doesn't matter how we determine the UV coordinates for a given point on the surface. However, even when UV coordinates are calculated dynamically, rather than edited by an artist, we typically compute or assign UV coordinates only at the *vertex* level, and the UV coordinates at an arbitrary interior position on a face are obtained through interpolation. If you imagine the texture map as a stretchy cloth, then when we assign texture-mapping coordinates to a vertex, it's like sticking a pin through the cloth at those UV coordinates, and then pinning the cloth onto the surface at that vertex. There is one pin per vertex, so the whole surface is covered.

Figure 10.16
An example texture map, with labeled UV coordinates according to the DirectX convention, which places the origin in the upper-left corner.

Figure 10.17
A texture-mapped quad, with different UV coordinates assigned to the vertices

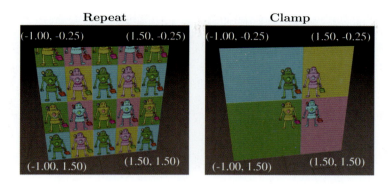

Figure 10.18. Comparing repeating and clamping texture addressing modes

Let's look at some examples. Figure 10.17 shows a single texture-mapped quad, with different UV values assigned to the vertices. The bottom of each diagram shows the UV space of the texture. You should study these examples until you are sure you understand them.

UV coordinates outside of the range $[0, 1]$ are allowed, and in fact are quite useful. Such coordinates are interpreted in a variety of ways. The most common addressing modes are *repeat* (also known as *tile* or *wrap*) and *clamp*. When repeating is used, the integer portion is discarded and only the fractional portion is used, causing the texture to repeat, as shown in the left side of Figure 10.18. Under clamping, when a coordinate outside the range $[0, 1]$ is used to access a bitmap, it is clamped in range. This has the effect of streaking the edge pixels of the bitmap outwards, as depicted on the right side of Figure 10.18. The mesh in both cases is identical: a single polygon with four vertices. And the meshes have identical UV coordinates. The only difference is how coordinates outside the $[0, 1]$ range are interpreted.

There are other options supported on some hardware, such as mirror, which is similar to repeat except that every other tile is mirrored. (This can be beneficial because it guarantees that no "seam" will exist between adjacent tiles.) On most hardware, the addressing mode can be set for the u- and v-coordinates independently. It's important to understand that these rules are applied at the last moment, when the coordinates are used to index into the texture. The coordinates at the vertex are not limited or processed in any way; otherwise, they could not be interpolated properly across the face.

Figure 10.19 shows one last instructive example: the same mesh is texture mapped two different ways.

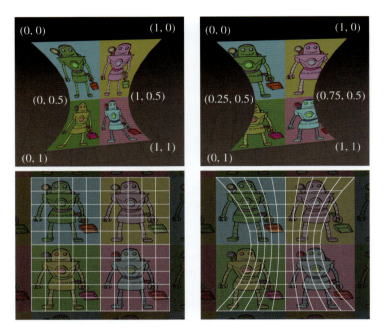

Figure 10.19. Texture mapping works on stuff that's not just a single quad

10.6   The Standard Local Lighting Model

In the rendering equation, the BRDF describes the scattering distribution for light of a given frequency and direction of incidence. The differences in distributions between different surfaces is precisely what causes those surfaces (or even different surface points on the same object) to look different from one another. Most BRDFs are expressed in a computer by some sort of formula, where certain numbers in the formula are adjusted to match the desired material properties. The formula itself is often called a *lighting model*, and the particular values going into the formula come from the *material* assigned to the surface. It is common for a game engine to use only a handful of lighting models, even though the materials in the scene may be quite diverse and there may be thousands of different BRDFs. Indeed, just a few years ago, almost all real-time rendering was done with a *single* lighting model. In fact, the practice is not uncommon today.

This lighting model was so ubiquitous that it was hardwired into the very rendering APIs of OpenGL and DirectX. Although these older parts of the API have effectively become legacy features on hardware with programmable shaders, the standard model is still commonly used in the more general framework of shaders and generic constants and interpolants. The great diversity and flexibility available is usually used to determine the best way to feed the parameters into the model (for example, by doing multiple lights at once, or doing all the lighting at the end with deferred shading), rather than using different models. But even ignoring programmable shaders, at the time of this writing, the most popular video game console is the Nintendo Wii,[13] which has hardwired support for this standard model.

The venerable standard lighting model is the subject of this section. Since its development precedes the framework of the BRDF and the rendering equation by at least a decade, we first present this model in the simplified context that surrounded its creation. This notation and perspective are still predominant in the literature today, which is why we think we should present the idea in its own terms. Along the way, we show how one component of the model (the diffuse component) is modeled as a BRDF. The standard model is important in the present, but you must understand the rendering equation if you want to be prepared for the future.

10.6.1 The Standard Lighting Equation: Overview

Bui Tuong Phong [54] introduced the basic concepts behind the standard lighting model in 1975. Back then, the focus was on a fast way to model *direct* reflection. While certainly researchers understood the importance of indirect light, it was a luxury that could not yet be afforded. Thus while the rendering equation (which, as we noted previously, came into focus a decade or so after the proposal of the standard model) is an equation for the radiance outgoing from a point in any particular direction, the only outgoing direction that mattered in those days were the directions that pointed to the eye. Similarly, while the rendering equation considers incident light from the entire hemisphere surrounding the surface normal, if we ignore indirect light, then we need not cast about in all incident directions. We need to consider only those directions that aim at a light source. We examine some different ways that light sources are modeled in real-time graphics in more detail in Section 10.7, but for now an important point is that the light sources are not emissive *surfaces* in the scene, as they are in the rendering equation and in the real world. Instead, lights are special entities without any corresponding geometry, and are simulated as if the light were emitting from a single point. Thus, rather than including a solid angle of directions corresponding to the projection of the emissive surface of each light source onto the hemisphere surrounding $\mathbf{x}$, we only care

---

[13]This is a very important lesson. Realistic graphics might be important to hardcore gamers, but for a more general audience they are not nearly as important as we once believed. The recent surge in popularity of facebook games further underscores this point.

about a *single* incident direction for the light. To summarize, the original goal of the standard model was to determine the light reflected back in the direction of the camera, only considering direct reflections, incident from a finite number of directions, one direction for each light source.

Now for the model. The basic idea is to classify light coming into the eye into four distinct categories, each of which has a unique method for calculating its contribution. The four categories are

- The *emissive* contribution, denoted $\mathbf{c}_{\text{emis}}$, is the same as the rendering equation. It tells the amount of radiance emitted directly from the surface in the given direction. Note that without global illumination techniques, these surfaces do not actually light up anything (except themselves).

- The *specular* contribution, denoted $\mathbf{c}_{\text{spec}}$, accounts for light incident directly from a light source that is scattered preferentially in the direction of a perfect "mirror bounce."

- The *diffuse* contribution, denoted $\mathbf{c}_{\text{diff}}$, accounts for light incident directly from a light source that is scattered in every direction evenly.

- The *ambient* contribution, denoted $\mathbf{c}_{\text{amb}}$, is a fudge factor to account for all indirect light.

The letter $\mathbf{c}$ is intended to be short for "contribution." Note the bold typeface, indicating that these contributions are not scalar quantities representing the amount of light of a particular wavelength, but rather they are vectors representing *colors* in some basis with a discrete number of components ("channels"). As stated before, due to the tri-stimulus human vision system, the number of channels is almost always chosen to be three. A less fundamental choice is which three basis functions to use, but in real-time graphics, by far the most common choice is to make one channel for red, one channel for blue, and one channel for green. These details are surprisingly irrelevant from a high-level discussion (they will not appear anywhere in the equations), but, of course, they are important practical considerations.

The emissive term is the same as in the rendering equation, so there's not much more detail to say about it. In practice, the emissive contribution is simply a constant color at any given surface point $\mathbf{x}$. The specular, diffuse, and ambient terms are more involved, so we discuss each in more detail in the next three sections.

10.6.2   The Specular Component

The *specular component* of the standard lighting model accounts for the light that is reflected (mostly) in a "perfect mirror bounce" off the surface. The specular component is what gives surfaces a "shiny" appearance.

Rougher surfaces tend to scatter the light in a much broader pattern of
directions, which is modeled by the diffuse component described in Section 10.6.3.

Now let's see how the standard model calculates the specular contribution. The important vectors are labeled in Figure 10.20.

- **n** is a the local outward-pointing surface normal.

- **v** points towards the viewer. (The symbol **e**, for "eye," is also sometimes used to name this vector.)

- **l** points towards the light source.

- **r** is the *reflection* vector, which is the direction of a "perfect mirror
  bounce." It's the result of reflecting **l** about **n**.

- θ is the angle between **r** and **v**.

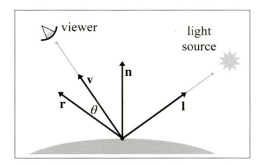

Figure 10.20
Phong model for specular
reflection

For convenience, we assume that all of these vectors are unit vectors.
Our convention in this book is to denote unit vectors with hats, but we'll
drop the hats to avoid decorating the equations excessively. Many texts
on the subject use these standard variable names and, especially in the
video game community, they are effectively part of the vernacular. It is
not uncommon for job interview questions to be posed in such a way that
assumes the applicant is familiar with this framework.

One note about the **l** vector before we continue. Since lights are abstract
entities, they need not necessarily have a "position." Directional lights and
Doom-style volumetric lights (see Section 10.7) are examples for which the
position of the light might not be obvious. The key point is that the *position*
of the light isn't important, but the abstraction being used for the light must
facilitate the computation of a *direction of incidence* at any given shading
point. (It must also provide the color and intensity of incident light.)

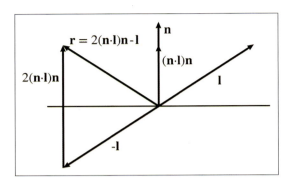

Figure 10.21
Constructing the reflection vector $\mathbf{r}$

Of the four vectors, the first three are inherent degrees of freedom of the problem, and the reflection vector $\mathbf{r}$ is a derived quantity and must be computed. The geometry is shown in Figure 10.21.

As you can see, the reflection vector can be computed by

Computing the reflection vector is a popular job interview question

$$\mathbf{r} = 2(\mathbf{n} \cdot \mathbf{l})\mathbf{n} - \mathbf{l}. \tag{10.10}$$

There are many interviewers for whom this equation is a favorite topic, which is why we have displayed it on a line by itself, despite the fact that it would have fit perfectly fine inline in the paragraph. A reader seeking a job in the video game industry is advised to fully digest Figure 10.21, to be able to produce Equation (10.10) under pressure. Notice that if we assume $\mathbf{n}$ and $\mathbf{l}$ are unit vectors, then $\mathbf{r}$ will be as well.

Now that we know $\mathbf{r}$, we can compute the specular contribution by using the *Phong model* for specular reflection (Equation (10.11)).

The Phong Model for Specular Reflection

$$\mathbf{c}_{\text{spec}} = (\mathbf{s}_{\text{spec}} \otimes \mathbf{m}_{\text{spec}}) (\cos \theta)^{m_{\text{gls}}} = (\mathbf{s}_{\text{spec}} \otimes \mathbf{m}_{\text{spec}}) (\mathbf{v} \cdot \mathbf{r})^{m_{\text{gls}}}. \tag{10.11}$$

In this formula and elsewhere in this book, the symbol $\otimes$ denotes componentwise multiplication of colors. Let's look at the inputs to this formula in more detail.

First, let's consider m_{gls}, which is the *glossiness* of the material, also known as the *Phong exponent*, *specular exponent*, or just as the material *shininess*. This controls how wide the "hotspot" is—a smaller m_{gls} produces a larger, more gradual falloff from the hotspot, and a larger m_{gls} produces a very tight hotspot with sharp falloff. (Here we are talking about the hotspot of a reflection, not to be confused with the hotspot of a spot light.) Perfectly reflective surfaces, such as chrome, would have an extremely high value for m_{gls}. When rays of light strike the surface from the incident direction $\mathbf{l}$, there is very little variation in the reflected directions. They are reflected in a very narrow solid angle ("cone") surrounding the direction described by $\mathbf{r}$, with very little scattering. Shiny surfaces that are not perfect reflectors—for example, the surface of an apple—have lower specular exponents, resulting in a larger hotspot. Lower specular exponents model a less perfect reflection of light rays. When rays of light strike the surface at the same incident direction given by $\mathbf{l}$, there is more variation in the reflected directions. The distribution clusters about the bounce direction $\mathbf{r}$, but the falloff in intensity as we move away from $\mathbf{r}$ is more gradual. We'll show this difference visually in just a moment.

Like all of the material properties that are input to the lighting equation, the value for m_{gls} can vary over the surface, and the specific value for any given location on that surface may be determined in any way you wish, for example with a texture map (see Section 10.5). However, compared to the other material properties, this is relatively rare; in fact it is quite common in real-time graphics for the glossiness value to be a constant for an entire material and not vary over the surface.

Another value in Equation (10.11) related to "shininess" is the material's *specular color*, denoted $\mathbf{m}_{\text{spec}}$. While m_{gls} controls the size of the hotspot, $\mathbf{m}_{\text{spec}}$ controls its intensity and color. Highly reflective surfaces will have a higher value for $\mathbf{m}_{\text{spec}}$, and more matte surfaces will have a lower value. If desired, a *specular map*[14] may be used to control the color of the hotspot using a bitmap, much as a texture map controls the color of an object.

The *light specular color*, denoted $\mathbf{s}_{\text{spec}}$, is essentially the "color" of the light, which contains both its color and intensity. Although many lights will have a single constant color, the strength of this color will attenuate with distance (Section 10.7.2), and this attenuation is contained in $\mathbf{s}_{\text{spec}}$ in our formulation. Furthermore, even ignoring attenuation, the same light source may shine light of different colors in different directions. For rectangular spot lights, we might determine the color from a *gobo*, which is a projected bitmap image. A colored gobo might be used to simulate a light shining through a stained glass window, or an animated gobo could be used to fake shadows of spinning ceiling fans or trees blowing in the wind. We use the letter $\mathbf{s}$ to stand for "source." The subscript "spec" indicates that this color is used for specular calculations. A different light color can be used for

---

[14]Unfortunately, some people refer to this map as the *gloss map*, creating confusion as to exactly which material property is being specified on a per-texel basis.

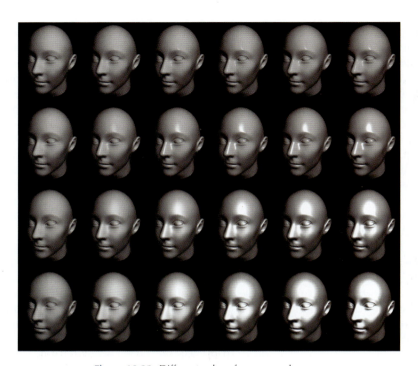

Figure 10.22. Different values for m_{gls} and $\mathbf{m}_{spec}$

diffuse calculations—this is a feature of the lighting model used to achieve special effects in certain circumstances, but it doesn't have any real-world meaning. In practice, $\mathbf{s}_{spec}$ is almost always equal to the light color used for diffuse lighting, which, not surprisingly, is denoted in this book as $\mathbf{s}_{diff}$.

Figure 10.22 shows how different values of m_{gls} and $\mathbf{m}_{spec}$ affect the appearance of an object with specular reflection. The material specular color $\mathbf{m}_{spec}$ goes from black on the leftmost column to white on the right-most column. The specular exponent m_{gls} is large on the top row and decreases with each subsequent row. Notice that the heads in the left-most column all look the same; since the specular strength is zero, the specular exponent is irrelevant and there is no specular contribution in any case. (The lighting comes from the diffuse and ambient components, which are discussed in Sections 10.6.3 and 10.6.4, respectively.)

Blinn [6] popularized a slight modification to the Phong model that produces very similar visual results, but at the time was a significant optimization. In many cases, it is still faster to compute today, but beware that vector operations (which are reduced with this model) are not always the performance bottleneck. The basic idea is this: if the distance to the viewer is large relative to the size of an object, then $\mathbf{v}$ may be computed once and then considered constant for an entire object. Likewise for a light source and the vector $\mathbf{l}$. (In fact, for directional lights, $\mathbf{l}$ is always constant.) However, since the surface normal $\mathbf{n}$ is not constant, we must still compute the reflection vector $\mathbf{r}$, a computation that we would like to avoid, if possible. The Blinn model introduces a new vector $\mathbf{h}$, which stands for "halfway" vector and is the result of averaging $\mathbf{v}$ and $\mathbf{l}$ and then normalizing the result:

$$\mathbf{h} = \frac{\mathbf{v} + \mathbf{l}}{\|\mathbf{v} + \mathbf{l}\|}.$$

The halfway vector h, used in the Blinn specular model

Then, rather than using the angle between $\mathbf{v}$ and $\mathbf{r}$, as the Phong model does, the cosine of the angle between $\mathbf{n}$ and $\mathbf{h}$ is used. The situation is shown in Figure 10.23.

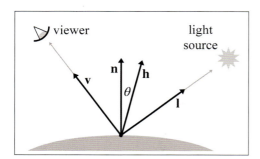

Figure 10.23
Blinn model for specular reflection

The formula for the Blinn model is quite similar to the original Phong model. Only the dot product portion is changed.

The Blinn Model for Specular Reflection

$$\mathbf{c}_{\text{spec}} = (\mathbf{s}_{\text{spec}} \otimes \mathbf{m}_{\text{spec}}) \, (\cos\theta)^{m_{\text{gls}}} = (\mathbf{s}_{\text{spec}} \otimes \mathbf{m}_{\text{spec}}) \, (\mathbf{n} \cdot \mathbf{h})^{m_{\text{gls}}}.$$

The Blinn model can be faster to implement in hardware than the Phong model, if the viewer and light source are far enough away from the object to be considered a constant, since then $\mathbf{h}$ is a constant and only needs to be computed once. But when $\mathbf{v}$ or $\mathbf{l}$ may not be considered constant, the Phong calculation might be faster. As we've said, the two models produce similar, but not identical, results (see Fisher and Woo [21] for a comparison). Both are empirical models, and the Blinn model should not be considered an "approximation" to the "correct" Phong model. In fact, Ngan et al. [48] have demonstrated that the Blinn model has some objective advantages and more closely matches experimental data for certain surfaces.

One detail we have omitted is that in either model, $\cos \theta$ may be less than zero. In this case, we usually clamp the specular contribution to zero.

10.6.3   The Diffuse Component

The next component in the standard lighting model is the *diffuse* component. Like the specular component, the diffuse component also models light that traveled directly from the light source to the shading point. However, whereas specular light accounts for light that reflects preferentially in a particular direction, diffuse light models light that is reflected randomly in all directions due to the rough nature of the surface material. Figure 10.24 compares how rays of light reflect on a perfectly reflective surface and on a rough surface.

To compute specular lighting, we needed to know the location of the viewer, to see how close the eye is to the direction of the perfect mirror bounce. For diffuse lighting, in contrast, the location of the viewer is *not relevant*, since the reflections are scattered randomly, and no matter where we position the camera, it is equally likely that a ray will be sent our way. However, the direction if incidence **l**, which is dictated by the position of the light source relative to the surface, *is* important. We've mentioned Lambert's law previously, but let's review it here, since the diffuse portion of Blinn-Phong is the most important place in real-time graphics that it comes into play. If we imagine counting the photons that hit the surface of the object and have a chance of reflecting into the eye, a surface that is perpendicular to the rays of light receives more photons per unit area than a surface oriented at a more glancing angle, as shown in Figure 10.25.

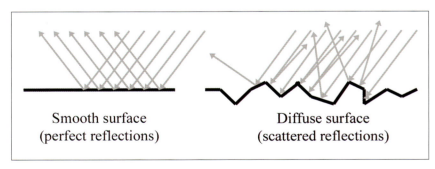

Smooth surface
(perfect reflections)

Diffuse surface
(scattered reflections)

Figure 10.24. Diffuse lighting models scattered reflections

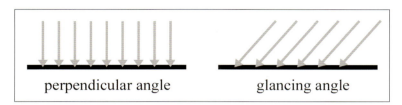

Figure 10.25. Surfaces more perpendicular to the light rays receive more light per unit area

Notice that, in both cases, the perpendicular distance between the rays is the same. (Due to an optical illusion in the diagram, the rays on the right may appear to be farther apart, but they are not.) So, the perpendicular distance between the rays is the same, but notice that on the right side of Figure 10.25, they strike the object at points that are farther apart. The surface on the left receives nine light rays, and the surface on the right receives only six, even though the "area" of both surfaces is the same. Thus the number of photons per unit area[15] is higher on the left, and it will appear brighter, all other factors being equal. This same phenomenon is responsible for the fact that the climate near the equator is warmer than near the poles. Since Earth is round, the light from the sun strikes Earth at a more perpendicular angle near the equator.

Diffuse lighting obeys *Lambert's law*: the intensity of the reflected light is proportional to the cosine of the angle between the surface normal and the rays of light. We will compute this cosine with the dot product.

Calculating the Diffuse Component according to Lambert's Law

$$\mathbf{c}_{\text{diff}} = (\mathbf{s}_{\text{diff}} \otimes \mathbf{m}_{\text{diff}}) (\mathbf{n} \cdot \mathbf{l}). \qquad (10.12)$$

As before, $\mathbf{n}$ is the surface normal and $\mathbf{l}$ is a unit vector that points towards the light source. The factor $\mathbf{m}_{\text{diff}}$ is the material's diffuse color, which is the value that most people think of when they think of the "color" of an object. The diffuse material color often comes from a texture map. The diffuse color of the light source is $\mathbf{s}_{\text{diff}}$; this is usually equal to the light's specular color, $\mathbf{s}_{\text{spec}}$.

---

[15]The proper radiometric term is *irradiance*, which measures the radiant power arriving per unit area.

Just as with specular lighting, we must prevent the dot product from going negative by clamping it to zero. This prevents objects from being lit from behind.

It's very instructive to see how diffuse surfaces are implemented in the framework of the rendering equation.

Diffuse reflection models light that is scattered completely randomly, and any given outgoing direction is equally likely, no matter what the incoming light direction. Thus, the BRDF for a perfectly diffuse surface is a constant.

Note the similarity of Equation (10.12) with the contents of the integral from the rendering equation,

$$L_{\text{in}}(\mathbf{x}, \hat{\boldsymbol{\omega}}_{\text{in}}, \lambda) f(\mathbf{x}, \hat{\boldsymbol{\omega}}_{\text{in}}, \hat{\boldsymbol{\omega}}_{\text{out}}, \lambda)(-\hat{\boldsymbol{\omega}}_{\text{in}} \cdot \hat{\mathbf{n}}).$$

The first factor is the incident light color. The material color $\mathbf{m}_{\text{diff}}$ is the constant value of the BRDF, which comes next. Finally, we have the Lambert factor.

10.6.4   The Ambient and Emmissive Components

Specular and diffuse lighting both account for light rays that travel directly from the light source to the surface of the object, "bounce" one time, and then arrive in the eye. However, in the real world, light often bounces off one or more intermediate objects before hitting an object and reflecting to the eye. When you open the refrigerator door in the middle of the night, the entire kitchen will get just a bit brighter, even though the refrigerator door blocks most of the direct light.

To model light that is reflected more than one time before it enters the eye, we can use a very crude approximation known as "ambient light." The ambient portion of the lighting equation depends only on the properties of the material and an ambient lighting value, which is often a global value used for the entire scene. None of the light sources are involved in the computation. (In fact, a light source is not even necessary.) Equation (10.13) is used to compute the ambient component:

Ambient contribution to the lighting equation

$$\mathbf{c}_{\text{amb}} = \mathbf{g}_{\text{amb}} \otimes \mathbf{m}_{\text{amb}}. \tag{10.13}$$

The factor $\mathbf{m}_{\text{amb}}$ is the material's "ambient color." This is almost always the same as the diffuse color (which is often defined using a texture map). The other factor, $\mathbf{g}_{\text{amb}}$, is the ambient light value. We use the notation $\mathbf{g}$ for "global," because often one global ambient value is used for the entire scene. However, some techniques, such as lighting probes, attempt to provide more localized and direction-dependent indirect lighting.

Sometimes a ray of light travels directly from the light source to the eye, without striking any surface in between. The standard lighting equation accounts for such rays by assigning a material an *emissive* color. For example, when we render the surface of a light bulb, this surface will probably appear very bright, even if there are no other light sources in the scene, because the light bulb is emitting light.

In many situations, the emissive contribution doesn't depend on environmental factors; it is simply the emissive color of the material:

$$\mathbf{c}_{\text{emis}} = \mathbf{m}_{\text{emis}}.$$

The emissive contribution depends only on the material

Most surfaces don't emit light, so their emissive component is $\mathbf{0}$. Surfaces that have a nonzero emissive component are called "self-illuminated."

It's important to understand that in real-time graphics, a self-illuminated surface does *not* light the other surfaces—you need a light source for that. In other words, we don't actually render light sources, we only render the effects that those light sources have on the surfaces in the scene. We *do* render self-illuminated surfaces, but those surfaces don't interact with the other surfaces in the scene. When using the rendering equation properly, however, emissive surfaces do light up their surroundings.

We may choose to attenuate the emissive contribution due to atmospheric conditions, such as fog, and of course there may be performance reasons to have objects fade out and disappear in the distance. However, as explained in Section 10.7.2, in general the emissive contribution should *not* be attenuated due to distance in the same way that light sources are.

10.6.5 The Lighting Equation: Putting It All Together

We have discussed the individual components of the lighting equation in detail. Now it's time to give the complete equation for the standard lighting model.

$$\mathbf{c}_{\text{lit}} = \begin{matrix} \mathbf{c}_{\text{spec}} \\ + \mathbf{c}_{\text{diff}} \\ + \mathbf{c}_{\text{amb}} \\ + \mathbf{c}_{\text{emis}} \end{matrix} = \begin{matrix} (\mathbf{s}_{\text{spec}} \otimes \mathbf{m}_{\text{spec}}) \max{(\mathbf{n} \cdot \mathbf{h}, 0)}^{m_{\text{gls}}} \\ + (\mathbf{s}_{\text{diff}} \otimes \mathbf{m}_{\text{diff}}) \max{(\mathbf{n} \cdot \mathbf{l}, 0)} \\ + \mathbf{g}_{\text{amb}} \otimes \mathbf{m}_{\text{amb}} \\ + \mathbf{m}_{\text{emis}} \end{matrix}$$

The standard lighting equation for one light source

Figure 10.26 shows what the ambient, diffuse, and specular lighting components actually look like in isolation from the others. (We are ignoring the emissive component, assuming that this particular floating head doesn't emit light.) There are several interesting points to be noted:

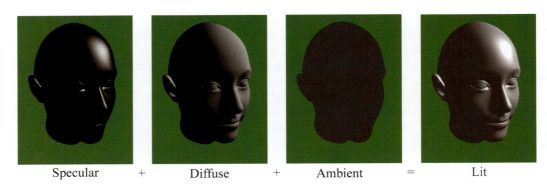

Specular   +   Diffuse   +   Ambient   =   Lit

Figure 10.26
The visual contribution of each of the components of the lighting equation

- The ear is lit just as bright as the nose, even though it is actually in the shadow of the head. For shadows, we must determine whether the light can actually "see" the point being shaded, using techniques such as shadow mapping.

- In the first two images, without ambient light, the side of the head that is facing away from the light is completely black. In order to light the "back side" of objects, you must use ambient light. Placing enough lights in your scene so that every surface is lit directly is the best situation, but it's not always possible. One common hack, which Mitchell et al. [47] dubbed "Half Lambert" lighting, is to bias the Lambert term, allowing diffuse lighting to "wrap around" to the back side of the model to prevent it from ever being flattened out and lit only by ambient light. This can easily be done by replacing the standard $\mathbf{n} \cdot \mathbf{l}$ term with $\alpha + (1 - \alpha)(\mathbf{n} \cdot \mathbf{l})$, where α is a tunable parameter that specifies the extra wraparound effect. (Mitchell et al. suggest using $\alpha = 1/2$, and they also square the result.) Although this adjustment has little physical basis, it has a very high perceptual benefit, especially considering the small computational cost.

- With only ambient lighting, just the silhouette is visible. Lighting is an extremely powerful visual cue that makes the object appear "3D." The solution to this "cartoon" effect is to place a sufficient number of lights in the scene so that every surface is lit directly.

Speaking of multiple lights, how do multiple light sources work with the lighting equation? We must sum up the lighting values for all the lights. To simplify the notation, we'll go ahead and make the almost universal assumption that $\mathbf{s}_{\mathrm{spec}} = \mathbf{s}_{\mathrm{diff}}$. Then we can let $\mathbf{s}_j$ denote the color of the

jth light source, including the attenuation factor. The index j goes from 1 to n, where n is the number of lights. Now the lighting equation becomes

$$\mathbf{c}_{\text{lit}} = \sum_{j=1}^{n} \left[(\mathbf{s}_j \otimes \mathbf{m}_{\text{spec}}) \max (\mathbf{n} \cdot \mathbf{h}_j, 0)^{m_{\text{gls}}} + (\mathbf{s}_j \otimes \mathbf{m}_{\text{diff}}) \max (\mathbf{n} \cdot \mathbf{l}_j, 0) \right]$$

$$+ \, \mathbf{g}_{\text{amb}} \otimes \mathbf{m}_{\text{amb}} + \mathbf{m}_{\text{emis}}. \tag{10.14}$$

The standard lighting equation for multiple lights

Since there is only one ambient light value and one emissive component for any given surface, these components are not summed per light source.

10.6.6   Limitations of the Standard Model

Nowadays we have the freedom of programmable shaders and can choose any lighting model we wish. Since the standard model has some fairly serious shortcomings, you might very well ask, "Why learn about this ancient history?" First, it isn't exactly ancient history; it is alive and well. The reasons that made it a good compromise between realism, usability, and speed of calculation still apply. Yes, we have more processing power; but we also want to render more pixels and more lights, and it currently is very common for the standard lighting model to be the winner when programmers are deciding whether to spend the cycles on more pixels (higher resolution) or more accurate pixels (a more accurate lighting model). Second, the current local lighting model is one that content creators can understand and use. This advantage is not to be taken lightly. Artists have *decades* of experience with diffuse and specular maps. Switching to a lighting model that replaces those familiar inputs with different ones such as "metalness" (from Strauss's model [69]) for which artists do not have an intuitive grasp is a big price to pay. A final reason to learn the standard lighting model is because many newer models bear similarities to the standard model, and you cannot know when to use more advanced lighting models without understanding the old standard.

If you have read the OpenGL or DirectX documentation for setting material parameters, you are forgiven for thinking that ambient, diffuse, and specular are "how light works" (remember our warning at the beginning of this chapter) as opposed being arbitrary practical constructs peculiar to a particular lighting model. The dichotomy between diffuse and specular is *not* an inherent physical reality; rather, it arose (and continues to be used) due to practical considerations. These are descriptive terms for two extreme scattering patterns, and by taking arbitrary combinations of these two patterns, many phenomena are able to be approximated to a decent degree.

Because of the near unanimous adoption of this model, it is often used without giving it a name, and in fact there is still some confusion as to exactly what to call it. You might call it the *Phong* lighting model, because Phong introduced the basic idea of modeling reflection as the sum of diffuse and specular contributions, and also provided a useful empirically based calculation for specular reflection. (The Lambert model for diffuse reflection was already known.) We saw that Blinn's computation for specular reflection is similar but sometimes faster. Because this is the specific calculation most often used, perhaps we should call it the Blinn model? But Blinn's name is also attached to a different microfacet model in which diffuse and specular are at different ends of a continuous spectrum, rather than independent "orthogonal" components being mixed together. Since most implementations use Blinn's optimization for Phong's basic idea, the name *Blinn-Phong* is the one most often used for this model, and that's the name we use.

A huge part of realistic lighting is, of course, realistic shadows. Although the techniques for producing shadows are interesting and important, alas we will not have time to discuss them here. In the theory of the rendering equation, shadows are accounted for when we determine the radiance incident in a given direction. If a light (more accurately, an emissive surface) exists in a particular direction, and the point can "see" that surface, then its light will be incident upon the point. If, however, there is some other surface that obscures the light source when looking in that direction, then the point is in shadow with respect to that light source. More generally, shadows can be cast not just due to the light from emissive surfaces; the light bouncing off reflective surfaces can cause shadows. In all cases, shadows are an issue of light visibility, not reflectance model.

Finally, we would like to mention several important physical phenomena not properly captured by the Blinn-Phong model. The first is *Fresnel*[16] *reflectance*, which predicts that the reflectance of nonmetals is strongest when the light is incident at a glancing angle, and least when incident from the normal angle. Some surfaces, such as velvet, exhibit *retroreflection*; you might guess this means that the surface looks like Madonna's earrings, but it actually means that the primary direction of reflection is not the "mirror bounce" as predicted by Blinn-Phong, but rather back towards the light source. Finally, Blinn-Phong is *isotropic*, which means that if we rotate the surface while keeping the viewer and light source stationary, the reflectance will not change. Some surfaces have *anisotropic* reflection, due to grooves or other patterns in the surface. This means that the strength of the reflection varies, based on the direction of incidence relative to the direction of the grooves, which is sometimes called the *scratch direction*. Classic examples

---

[16]Pronounced "fre-NELL."

of anisotropic materials are brushed metal, hair, and those little Christmas ornaments made of shiny fibers.

10.6.7 Flat and Gouraud Shading

On modern shader-based hardware, lighting calculations are usually done on a per-pixel basis. By this we mean that for each pixel, we determine a surface normal (whether by interpolating the vertex normal across the face or by fetching it from a bump map), and then we perform the full lighting equation using this surface normal. This is *per-pixel* lighting, and the technique of interpolating vertex normals across the face is sometimes called *Phong shading*, not to be confused with the Phong calculation for specular reflection. The alternative to Phong shading is to perform the lighting equation less frequently (per face, or per vertex). These two techniques are known as *flat shading* and *Gouraud shading*, respectively. Flat shading is almost never used in practice except in software rendering. This is because most modern methods of sending geometry efficiently to the hardware do not provide any face-level data whatsoever. Gouraud shading, in contrast, still has some limited use on some platforms. Some important general principles can be gleaned from studying these methods, so let's examine their results.

When using flat shading, we compute a single lighting value for the entire triangle. Usually the "position" used in lighting computations is the centroid of the triangle, and the surface normal is the normal of the triangle. As you can see in Figure 10.27, when an object is lit using flat shading, the faceted nature of the object becomes painfully apparent, and any illusion of smoothness is lost.

Figure 10.27
A flat shaded teapot

Gouraud shading, also known as *vertex shading*, *vertex lighting*, or *interpolated shading*, is a trick whereby values for lighting, fog, and so forth are computed at the vertex level. These values are then linearly interpolated across the face of the polygon. Figure 10.28 shows the same teapot rendered with Gouraud shading.

As you can see, Gouraud shading does a relatively good job at restoring the smooth nature of the object. When the values being approximated are basically linear across the triangle, then, of course, the linear interpolation used by Gouraud shading works well. Gouraud shading breaks down when the values are not linear, as in the case of specular highlights.

Figure 10.28
A Gouraud shaded teapot

Compare the specular highlights in the Gouraud shaded teapot with the highlights in a Phong (per-pixel) shaded teapot, shown in Figure 10.29. Notice how much smoother the highlights are. Except for the silhouette and areas of extreme geometric discontinuities, such as the handle and spout, the illusion of smoothness is very convincing. With Gouraud shading, the in-

Figure 10.29
A Phong shaded teapot

dividual facets are detectable due to the specular highlights.

The basic problem with interpolated shading is that no value in the middle of the triangle can be larger than the largest value at a vertex; highlights can occur only at a vertex. Sufficient tessellation can overcome this problem. Despite its limitations, Gouraud shading is still in use on some limited hardware, such as hand-held platforms and the Nintendo Wii.

One question that you should be asking is how the lighting can be computed at the vertex level if any maps are used to control inputs to the lighting equation. We can't use the lighting equation as given in Equation (10.14) directly. Most notably, the diffuse color $\mathbf{m}_{\text{diff}}$ is not usually a vertex-level material property; this value is typically defined by a texture map. In order to make Equation (10.14) more suitable for use in an interpolated lighting scheme, it must be manipulated to isolate $\mathbf{m}_{\text{diff}}$. We first split the sum and move the constant material colors outside:

$$
\begin{aligned}
\mathbf{c}_{\text{lit}} &= \sum_{j=1}^{n} \left[(\mathbf{s}_j \otimes \mathbf{m}_{\text{spec}}) \max (\mathbf{n} \cdot \mathbf{h}_j, 0)^{m_{\text{gls}}} + (\mathbf{s}_j \otimes \mathbf{m}_{\text{diff}}) \max (\mathbf{n} \cdot \mathbf{l}_j, 0) \right] \\
&\quad + \mathbf{g}_{\text{amb}} \otimes \mathbf{m}_{\text{amb}} + \mathbf{m}_{\text{emis}} \\
&= \sum_{j=1}^{n} (\mathbf{s}_j \otimes \mathbf{m}_{\text{spec}}) \max (\mathbf{n} \cdot \mathbf{h}_j, 0)^{m_{\text{gls}}} + \sum_{j=1}^{n} (\mathbf{s}_j \otimes \mathbf{m}_{\text{diff}}) \max (\mathbf{n} \cdot \mathbf{l}_j, 0) \\
&\quad + \mathbf{g}_{\text{amb}} \otimes \mathbf{m}_{\text{amb}} + \mathbf{m}_{\text{emis}}
\end{aligned}
$$

$$= \left[\sum_{j=1}^{n} \mathbf{s}_j \max \left(\mathbf{n} \cdot \mathbf{h}_j, 0 \right)^{m_{\text{gls}}} \right] \otimes \mathbf{m}_{\text{spec}} + \left[\sum_{j=1}^{n} \mathbf{s}_j \max \left(\mathbf{n} \cdot \mathbf{l}_j, 0 \right) \right] \otimes \mathbf{m}_{\text{diff}}$$

$$+ \mathbf{g}_{\text{amb}} \otimes \mathbf{m}_{\text{amb}} + \mathbf{m}_{\text{emis}}.$$

Finally, we make the very reasonable assumption that $\mathbf{m}_{\text{amb}} = \mathbf{m}_{\text{diff}}$:

$$\mathbf{c}_{\text{lit}} = \left[\sum_{j=1}^{n} \mathbf{s}_j \max \left(\mathbf{n} \cdot \mathbf{h}_j, 0 \right)^{m_{\text{gls}}} \right] \otimes \mathbf{m}_{\text{spec}}$$

$$+ \left[\mathbf{g}_{\text{amb}} + \sum_{j=1}^{n} \mathbf{s}_j \max \left(\mathbf{n} \cdot \mathbf{l}_j, 0 \right) \right] \otimes \mathbf{m}_{\text{diff}} \qquad (10.15)$$

$$+ \mathbf{m}_{\text{emis}}.$$

A version of the standard lighting equation more suitable for vertex-level lighting computations

With the lighting equation in the format of Equation (10.15), we can see how to use interpolated lighting values computed at the vertex level. At each vertex, we will compute two values: $\mathbf{v}_{\text{spec}}$ contains the specular portion of Equation (10.15) and $\mathbf{v}_{\text{diff}}$ contains the ambient and diffuse terms:

$$\mathbf{v}_{\text{spec}} = \sum_{j=1}^{n} \mathbf{s}_j \max \left(\mathbf{n} \cdot \mathbf{h}_j, 0 \right)^{m_{\text{gls}}} \qquad \mathbf{v}_{\text{diff}} = \mathbf{g}_{\text{amb}} + \sum_{j=1}^{n} \mathbf{s}_j \max \left(\mathbf{n} \cdot \mathbf{l}_j, 0 \right).$$

Vertex-level diffuse and specular lighting values

Each of these values is computed per vertex and interpolated across the face of the triangle. Then, per pixel, the light contributions are multiplied by the corresponding material colors and summed:

$$\mathbf{c}_{\text{lit}} = \mathbf{v}_{\text{spec}} \otimes \mathbf{m}_{\text{spec}} + \mathbf{v}_{\text{diff}} \otimes \mathbf{m}_{\text{diff}} + \mathbf{m}_{\text{emis}}.$$

Shading pixels using interpolated lighting values

As mentioned earlier, $\mathbf{m}_{\text{spec}}$ is sometimes a constant color, in which case we could move this multiplication into the vertex shader. But it also can come from a specular map.

What coordinate space should be used for lighting computations? We could perform the lighting computations in world space. Vertex positions and normals would be transformed into world space, lighting would be performed, and then the vertex positions would be transformed into clip space. Or we may transform the lights into modeling space, and perform lighting computations in modeling space. Since there are usually fewer lights than there are vertices, this results in fewer overall vector-matrix multiplications. A third possibility is to perform the lighting computations in camera space.

10.7   Light Sources

In the rendering equation, light sources produce their effect when we factor in the emissive component of a surface. As mentioned earlier, in real-time graphics, doing this "properly" with emissive surfaces is usually a luxury we cannot afford. Even in offline situations where it can be afforded, we might have reasons to just emit light out of nowhere, to make it easier to get control of the look of the scene for dramatic lighting, or to simulate the light that would be reflecting from a surface for which we're not wasting time to model geometry since it's off camera. Thus we usually have light sources that are abstract entities within the rendering framework with no surface geometry to call their own. This section discusses some of the most common types of light sources.

Section 10.7.1 covers the classic point, directional, and spot lights. Section 10.7.2 considers how light attenuates in the real world and how deviations from this reality are common for practical reasons. The next two sections move away from the theoretically pure territory and into the messy domain of ad-hoc lighting techniques in use in real-time graphics today. Section 10.7.3 presents the subject of *Doom*-style volumetric lights. Finally, Section 10.7.4 discusses how lighting calculations can be done offline and then used at runtime, especially for the purpose of incorporating indirect lighting effects.

10.7.1   Standard Abstract Light Types

This section lists some of the most basic light types that are supported by most rendering systems, even older or limited platforms, such as the OpenGL and DirectX fixed-function lighting pipelines or the Nintendo Wii. Of course, systems with programmable shaders often use these light types, too. Even when completely different methods, such as spherical harmonics, are used at runtime, standard light types are usually used as an offline editing interface.

A *point light* source represents light that emanates from a single point outward in all directions. Point lights are also called *omni* lights (short for "omnidirectional") or *spherical* lights. A point light has a position and color, which controls not only the hue of the light, but also its intensity. Figure 10.30 shows how *3DS Max* represents point lights visually.

As Figure 10.30 illustrates, a point light may have a *falloff radius*, which controls the size of the sphere that is illuminated by the light. The intensity of the light usually decreases the farther away we are from the center of the light. Although not realistic, it is desirable for many reasons that the intensity drop to zero at the falloff distance, so that the volume of the

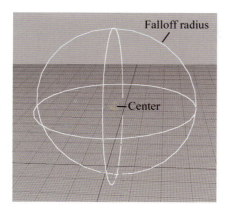

Figure 10.30
A point light

effect of the light can be bounded. Section 10.7.2 compares real-world attenuation with the simplified models commonly used. Point lights can be used to represent many common light sources, such as light bulbs, lamps, fires, and so forth.

A *spot light* is used to represent light from a specific location in a specific direction. These are used for lights such as flashlights, headlights, and of course, spot lights! A spot light has a position and an orientation, and optionally a falloff distance. The shape of the lit area is either a cone or a pyramid.

A *conical spot* light has a circular "bottom." The width of the cone is defined by a *falloff angle* (not to be confused with the falloff distance). Also, there is an inner angle that measures the size of the hotspot. A conical spot light is shown in Figure 10.31.

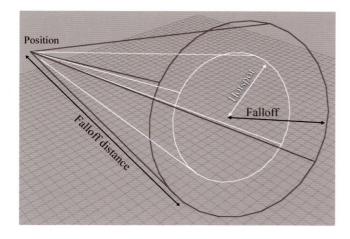

Figure 10.31
A conical spot light

A *rectangular spot light* forms a pyramid rather than a cone. Rectangular spot lights are especially interesting because they are used to project an image. For example, imagine walking in front of a movie screen while a movie is being shown. This projected image goes by many names, including *projected light map*, *gobo*, and even *cookie*.[17] The term *gobo* originated from the world of theater, where it refers to a mask or filter placed over a spot light used to create a colored light or special effect, and it's the term we use in this book. Gobos are very useful for faking shadows and other lighting effects. If conical spot lights are not directly supported, they can be implemented with an appropriately designed circular gobo.

A *directional light* represents light emanating from a point in space sufficiently far away that all the rays of light involved in lighting the scene (or at least the object we are currently considering) can be considered as parallel. The sun and moon are the most obvious examples of directional lights, and certainly we wouldn't try to specify the actual position of the sun in world space in order to properly light the scene. Thus directional lights usually do not have a position, at least as far as lighting calculations are concerned, and they usually do not attenuate. For editing purposes, however, it's often useful to create a "box" of directional light that can be moved around and placed strategically, and we might include additional attenuation factors to cause the light to drop off at the edge of the box. Directional lights are sometimes called *parallel* lights. We might also use a gobo on a directional light, in which case the projection of the image is orthographic rather than perspective, as it is with rectangular spot lights.

As we've said, in the rendering equation and in the real world, lights are emissive surfaces with finite surface areas. Abstract light types do not have any surface area, and thus require special handling during integration. Typically in a Monte Carlo integrator, a sample is specifically chosen to be in the direction of the light source, and the multiplication by $d\hat{\omega}_{\text{in}}$ is ignored. Imagine if, rather than the light coming from a single point, it comes instead from a disk of some nonzero surface area that is facing the point being illuminated. Now imagine that we shrink the area of the disk down to zero, all the while increasing the radiosity (energy flow per unit area) from the disk such that radiant flux (total energy flow) remains constant. An abstract light can be considered the result of this limiting process in a manner very similar to a Dirac delta (see Section 12.4.3). The radiosity is infinite, but the flux is finite.

While the light types discussed so far are the classic ones supported by fixed-function real-time pipelines, we certainly are free to define light volumes in any way we find useful. The volumetric lights discussed in

---

[17] "Gobo" is short for "go between," and "cookie" is short for "cucoloris." The subtle technical difference between these terms from the world of theater is not relevant for computer-generated imagery.

Section 10.7.3 are an alternative system that is flexible and also amenable to real-time rendering. Warn [71] and Barzel [5] discuss more flexible systems for shaping lights in greater detail.

10.7.2   Light Attenuation

Light *attenuates* with distance. That is, objects receive less illumination from a light as the distance between the light and the object increases. In the real world, the intensity of a light is inversely proportional to the square of the distance between the light and the object, as

$$\frac{i_1}{i_2} = \left(\frac{d_2}{d_1}\right)^2,$$

(10.16)

Real-world light attenuation

where i is the radiant flux (the radiant power per unit area) and d is the distance. To understand the squaring in real-world attenuation, consider the sphere formed by all the photons emitted from a point light at the same instant. As these photons move outward, a larger and larger sphere is formed by the same number of photons. The density of this photon flow per unit area (the radiant flux) is inversely proportional to the surface area of the sphere, which is proportional to the square of the radius (see Section 9.3).

Let's pause here to discuss a finer point: the perceived brightness of an object (or light source) *does not* decrease with increased distance from the viewer, ignoring atmospheric effects. As a light or object recedes from the viewer, the irradiance on our eye decreases for the reasons just described. However, perceived brightness is related to *radiance*, not irradiance. Remember that radiance measures power per unit projected area *per unit solid angle*, and as the object recedes from view, the decrease in irradiance is compensated for by the decrease in solid angle subtended by the object. It's particularly educational to understand how the rendering equation naturally accounts for light attenuation. Inside the integral, for each direction on the hemisphere surrounding the shading point $\mathbf{x}$, we measure the incident radiance from an emissive surface in that direction. We've just said that this radiance does not attenuate with distance. However, as the light source moves away from $\mathbf{x}$, it occupies a smaller solid angle on this hemisphere. Thus, attenuation happens automatically in the rendering equation if our light sources have finite area. However, for abstract light sources emanating from a single point (Dirac delta), attenuation must be manually factored in. Because this is a bit confusing, let's summarize the general rule for real-time rendering. Emissive surfaces, which are rendered and have finite area, typically are *not* attenuated due to distance—but they might be affected by atmospheric effects such as fog. For purposes of calculating the effective light color when shading a particular spot, the standard abstract light types *are* attenuated.

In practice, Equation (10.16) can be unwieldy for two reasons. First, the light intensity theoretically increases to infinity at $d = 0$. (This is a result of the light being a Dirac delta, as mentioned previously.) Barzel [5] describes a simple adjustment to smoothly transition from the inverse square curve near the light origin, to limit the maximum intensity near the center. Second, the light intensity never falls off completely to zero.

Instead of the real-world model, a simpler model based on *falloff distance* is often used. Section 10.7 mentioned that the falloff distance controls the distance beyond which the light has no effect. It's common to use a simple linear interpolation formula such that the light gradually fades with the distance d:

$$i(d) = \begin{cases} 1 & \text{if } d \leq d_{\min}, \\ \dfrac{d_{\max} - d}{d_{\max} - d_{\min}} & \text{if } d_{\min} < d < d_{\max}, \\ 0 & \text{if } d \geq d_{\max}. \end{cases} \quad (10.17)$$

As Equation (10.17) shows, there are actually two distances used to control the attenuation. Within $d_{\min}$, the light is at full intensity (100%). As the distance goes from $d_{\min}$ to $d_{\max}$, the intensity varies linearly from 100% down to 0%. At $d_{\max}$ and beyond, the light intensity is 0%. So basically, $d_{\min}$ controls the distance at which the light begins to fall off; it is frequently zero, which means that the light begins falling off immediately. The quantity $d_{\max}$ is the actual falloff distance—the distance where the light has fallen off completely and no longer has any effect. Figure 10.32 compares real-world light attenuation to the simple linear attenuation model.

Distance attenuation can be applied to point and spot lights; directional lights are usually not attenuated. An additional attenuation factor is used for spot lights. *Hotspot falloff* attenuates light as we move closer to the edge of the cone.

10.7.3   *Doom*-style Volumetric Lights

In the theoretical framework of the rendering equation as well as HLSL shaders doing lighting equations using the standard Blinn-Phong model, all that is required of a light source for it to be used in shading calculations at a particular point $\mathbf{x}$ is a light color (intensity) and direction of incidence. This section discusses a type of volumetric light, popularized by the *Doom 3* engine (also known as *id Tech 4*) around 2003, which specifies these values in a novel way. Not only are these types of lights interesting to understand from a practical standpoint (they are still useful today), they are interesting from a theoretical perspective because they illustrate an

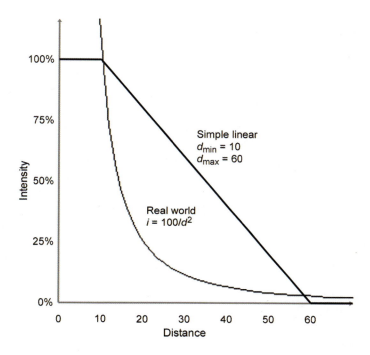

Figure 10.32. Real-world light attenuation vs. simple linear attenuation

elegant, fast approximation. Such approximations are the essence of the art of real-time rendering.

The most creative aspect of *Doom*-style volumetric lights is how they determine the intensity at a given point. It is controlled through two texture maps. One map is essentially a gobo, which can be projected by either orthographic or perspective projection, similar to a spot or directional light. The other map is a *one-dimensional* map, known as the *falloff map*, which controls the falloff. The procedure for determining the light intensity at point $\mathbf{x}$ is as follows: $\mathbf{x}$ is multiplied by a 4×4 matrix, and the resulting coordinates are used to index into the two maps. The 2D gobo is indexed using $(x/w, y/w)$, and the 1D falloff map is indexed with z. The product of these two texels defines the light intensity at $\mathbf{x}$.

The examples in Figure 10.33 will make this clear. Let's look at each of the examples in more detail. The omni light projects the circular gobo orthographically across the box, and places the "position" of the light (which is used to compute the $\mathbf{l}$ vector) in the center of the box. The 4×4 matrix

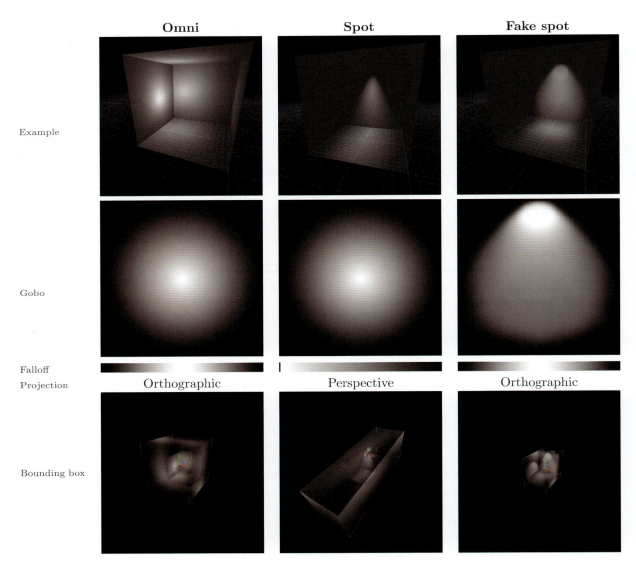

Figure 10.33
Examples of *Doom*-style volumetric lights

used to generate the texture coordinates in this case is

$$\begin{bmatrix} 1/s_x & 0 & 0 & 0 \\ 0 & -1/s_y & 0 & 0 \\ 0 & 0 & 1/s_z & 0 \\ 1/2 & 1/2 & 1/2 & 1 \end{bmatrix},$$

Texture coordinate generation matrix for a *Doom*-style omni light

where s_x, s_y, and s_z are the dimensions of the box on each axis. This matrix operates on points in the object space of the light, where the position of the light is in the center of the box, so for the matrix that operates on world-space coordinates, we would need to multiply this matrix by a 4×4 world-to-object matrix on the left. Note the right-most column is $[0, 0, 0, 1]^T$, since we use an orthographic projection onto the gobo. The translation of $1/2$ is to adjust the coordinates from the $[-1/2, +1/2]$ range into the $[0, 1]$ range of the texture. Also, note the flipping of the y-axis, since $+y$ points up in our 3D conventions, but $+v$ points down in the texture.

Next, let's look at the spot light. It uses a perspective projection, where the center of projection is at one end of the box. The position of the light used for calculating the **l** vector is at this same location, but that isn't always the case! Note that the same circular gobo is used as for the omni, but due to the perspective projection, it forms a cone shape. The falloff map is brightest at the end of the box nearest the center of projection and falls off linearly along the $+z$ axis, which is the direction of projection of the gobo in all cases. Notice that the very first pixel of the spot light falloff map is black, to prevent objects "behind" the light from getting lit; in fact, all of the gobos and falloff maps have black pixels at their edges, since these pixels will be used for any geometry outside the box. (The addressing mode must be set to clamp to avoid the gobo and falloff map tiling across 3D space.) The texture generation matrix for perspective spots is

$$\begin{bmatrix} s_z/s_x & 0 & 0 & 0 \\ 0 & -s_z/s_y & 0 & 0 \\ 1/2 & 1/2 & 1/s_z & 1 \\ 0 & 0 & 0 & 0 \end{bmatrix}.$$

Texture coordinate generation matrix for a *Doom*-style spot light

The "fake spot" on the right is perhaps the most interesting. Here, projection is orthographic, and it is *sideways*. The conical nature of the light as well as its falloff (what we ordinarily think of as the falloff, that is) are both encoded in the gobo. The falloff map used for this light is the same as for the omni light: it is brightest in the center of the box, and causes the light to fade out as we approach the $-z$ and $+z$ faces of the box. The texture coordinate matrix in this case is actually the same as that for the omni. The entire change comes from using a different gobo, and orienting the light properly!

You should study these examples until you are sure you know how they work.

Doom-style volumetric lights can be attractive for real-time graphics for several reasons:

- They are simple and efficient, requiring only the basic functionality of texture coordinate generation, and two texture lookups. These are flexible operations that are easily hardwired into fixed-function hardware such as the Nintendo Wii.

- Many different light types and effects can be represented in the same framework. This can be helpful to limit the number of different shaders that are needed. Lighting models, light types, material properties, and lighting passes can all be dimensions in the matrix of shaders, and the size of this matrix can grow quite quickly. It can also be useful to reduce the amount of switching of render states.

- Arbitrary falloff curves can be encoded in the gobo and falloff maps. We are not restricted to linear or real-world inverse squared attenuation.

- Due to the ability to control the falloff, the bounding box that contains the lighting volume can usually be relatively tight compared to traditional spot and omni lights. In other words, a large percentage of the volume within the box is receiving significant lighting, and the light falls off more rapidly than for traditional models, so the volume is as small and as tight as possible. Looking at the bottom row of Figure 10.33, compare the size of the box needed to contain the true spot light, versus the fake spot light.

  This is perhaps the most important feature behind the introduction of these sorts of lights in *Doom 3*, which used an accumulated rendering technique with no lightmaps or precomputed lighting; every object was fully lit in real time. Each light was *added* into the scene by rerendering the geometry within the volume of the light and adding the light's contribution into the frame buffer. Limiting the amount of geometry that had to be redrawn (as well as the geometry that had to be processed for purposes of the stencil shadows that were used) was a huge performance win.

10.7.4   Precalculated Lighting

One of the greatest sources of error in the images produced in real time (those positive thinkers among you might say the greatest opportunity for improvement) is indirect lighting: light that has "bounced" at least one time before illuminating the pixel being rendered. This is an extremely difficult problem. A first important step to making it tractable is to break up the surfaces in the scene into discrete patches or sample points. But even with a relatively modest number of patches, we still have to determine which patches can "see" each other and have a conduit of radiance, and which cannot see each other and do not exchange radiance. Then we must solve for the balance of light in the rendering equation. Furthermore, when any object moves, it can potentially alter which patches can see which. In other words, practically *any* change will alter the distribution of light in the *entire scene.*

However, it is usually the case that certain lights and geometry in the scene are *not* moving. In this case, we can perform more detailed lighting calculations (solve the rendering equation more fully), and then use those results, ignoring any error that results due to the difference in the current lighting configuration and the one that was used during the offline calculations. Let's consider several examples of this basic principle.

One technique is *lightmapping.* In this case, an extra UV channel is used to arrange the polygons of the scene into a special texture map that contains precalculated lighting information. This process of finding a good way to arrange the polygons within the texture map is often called *atlasing.* In this case, the discrete "patches" that we mentioned earlier are the lightmap texels. Lightmapping works well on large flat surfaces, such as floors and ceilings, which are relatively easy to arrange within the lightmap effectively. But more dense meshes, such as staircases, statues, machinery, and trees, which have much more complicated topology, are not so easily atlased. Luckily, we can just as easily store precomputed lighting values in the vertices, which often works better for relatively dense meshes.

What exactly is the precomputed information that is stored in lightmaps (or vertices)? Essentially, we store incident illumination, but there are many options. One option is the number of samples per patch. If we have only a single lightmap or vertex color, then we cannot account for the directional distribution of this incident illumination and must simply use the sum over the entire hemisphere. (As we have shown in Section 10.1.3, this "directionless" quantity, the incident radiant power per unit area, is properly known as *radiosity,* and for historical reasons algorithms for calculating lightmaps are sometimes confusingly known as radiosity techniques, even if the lightmaps include a directional component.) If we can afford more than one lightmap or vertex color, then we can more accurately capture the distribution. This directional information is then projected onto a particular basis. We might have each basis correspond to a single direction. A technique known as *spherical harmonics* [44, 64] uses sinusoidal basis functions similar to 2D Fourier techniques. The point in any case is that the directional distribution of incident light does matter, but when saving precomputed incident light information, we are usually forced to discard or compress this information.

Another option is whether the precalculated illumination includes direct lighting, indirect light, or both. This decision can often be made on a per-light basis. The earliest examples of lightmapping simply calculated the direct light from each light in the scene for each patch. The primary advantage of this was that it allowed for shadows, which at the time were prohibitively expensive to produce in real time. (The same basic idea is still useful today, only now the goal is usually to reduce the total number of real-time shadows that must be generated.) Then the view could be moved around in real time, but obviously, any lights that were burned into the lightmaps could not move, and if any geometry moved, the shadows would be "stuck" to them and the illusion would break down. An identical runtime system can be used to render lightmaps that also include indirect lighting, although the offline calculations require much more finesse. It is possible for certain lights to have both their direct and indirect lighting baked into the lightmaps, while other lights have just the indirect portion included in the precalculated lighting and direct lighting done at runtime. This might offer advantages, such as shadows with higher precision than the lightmap texel density, improved specular highlights due to the correct modeling of the direction of incidence (which is lost when the light is burned into the lightmaps), or some limited ability to dynamically adjust the intensity of the light or turn it off or change its position. Of course, the presence of precalculated lighting for some lights doesn't preclude the use of completely dynamic techniques for other lights.

The lightmapping techniques just discussed work fine for static geometry, but what about dynamic objects such as characters, vehicles, platforms, and items? These must be lit dynamically, which makes the inclusion of indirect lighting challenging. One technique, popularized by Valve's *Half Life 2* [28,47], is to strategically place *light probes* at various locations in the scene. At each probe, we render a cubic environment map offline. When rendering a dynamic object, we locate the closest nearby probe and use this probe to get localized indirect lighting. There are many variations on this technique—for example, we might use one environment map for diffuse reflection of indirect light, where each sample is prefiltered to contain the entire cosine-weighted hemisphere surrounding this direction, and a different cubic map for specular reflection of indirect light, which does not have this filtering.

10.8   Skeletal Animation

The animation of human creatures is certainly of great importance in video games and in computer graphics in general. One of the most important techniques for animating characters is *skeletal animation*, although it is certainly not limited to this purpose. The easiest way to appreciate skeletal animation is to compare it to other alternatives, so let's review those first.

Let's say we have created a model of a humanoid creature such as a robot. How do we animate it? Certainly, we could treat it like a chess piece and move it around just like a box of microwavable herring sandwiches or any other solid object—this is obviously not very convincing. Creatures are *articulated*, meaning they are composed of connected, movable parts. The simplest method of animating an articulated creature is to break the model up into a hierarchy of connected parts—left forearm, left upper arm, left thigh, left shin, left foot, torso, head, and so on—and animate this hierarchy. An early example of this was Dire Straits' *Money for Nothing* music video. Newer examples include practically every PlayStation 1 game, such as the first *Tomb Raider*. The common feature here is that each part is still rigid; it does not bend or flex. Hence, no matter how skillfully the character is animated, it still looks like a robot.

The idea behind skeletal animation is to replace the hierarchy of parts with an imaginary hierarchy of *bones*. Then each vertex of the model is associated with one or more bones, each of which exert influence over the vertex but do not totally determine its position. If a vertex is associated with a single bone, then it will maintain a fixed offset relative to this bone. Such a vertex is known as a *rigid* vertex, and that vertex behaves exactly like any vertex from the first Laura Croft model. However, more generally, a vertex will receive influence from more than one bone. An artist needs to specify which bones influence which vertices. This process is known as *skinning*,[18] and a model thus annotated is known as a *skinned* model. When more than one bone influences a vertex, the animator can distribute, per vertex, differing amounts of influence to each bone. As you can imagine, this can be very labor intensive. Automated tools exist that can provide a quick first pass at the skin weights, but a well-skinned character requires expertise and time.

To determine the animated position of a vertex, we iterate over all the bones that exert some influence over the vertex, and compute the position that the vertex would have if it were rigid relative to that bone. The final vertex position is then taken as the weighted average of those positions.

---

[18]You might also hear the term *rigging*, but this term can imply a wider range of tasks. For example, often a rigger creates an extra apparatus that assists with animation but is not used directly for rendering.

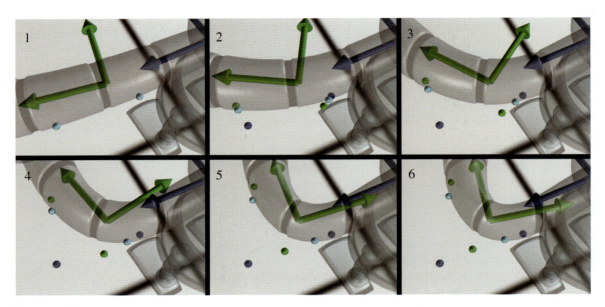

Figure 10.34
Two skinned vertices

Let's look at an example. Figure 10.34 shows two example skinned vertices near the elbow of a robot. The blue and green dots show what a vertex would look like if it were rigid to the corresponding bone, and the cyan dot is the skinned vertex; notice that it stays attached to the surface of the mesh.

The vertex on the right, closer to the shoulder, is influenced approximately 60% by the upper arm bone and 40% by the forearm bone. You can see that as the arm bends, this vertex stays closer to the blue rigid vertex. In contrast, the vertex closer to the hand appears to be influenced approximately 80% by the forearm bone and only 20% by the upper arm bone, and thus it stays closer to its green rigid vertex.

So a simple strategy for implementing skeletal animation might be as follows. For each vertex, we keep a list of bones that influence the vertex. Typically we set a limit on the number of bones that may influence any one vertex (four is a common number). For each bone, we know the position of the vertex relative to the bone's local axes, and we have a weight for that bone. To compute the skinned vertex positions for a model in an arbitrary pose, we need a transform matrix for each bone that tells how to convert from bone coordinate space into modeling coordinate space. Varying these transform matrices over time is what makes the character appear to animate.

Listing 10.7 illustrates this basic technique. Note that we are also planning ahead by including vertex normals. These are handled in the same way as vertex positions, only we discard the translation portion of the matrix. In theory, the same matrix should not be used to transform positions and normals. Remember that if nonuniform scale or skew is included in the matrix, we really should use the inverse transpose matrix, as was discussed in Section 10.4.2. In practice, however, computing and sending two sets of matrices to the GPU is too expensive, so for the sake of efficiency this error is ignored, or nonuniform scale is simply avoided. (Uniform scale is typically OK because the normals have to be renormalized anyway.) Basis vectors for bump mapping are also commonly part of the process, but they are handled in a manner very similar to normals, so we will leave those out for now.

```
// Set a limit on the max number of bones that can influence one vertex
const int kMaxBonesPerVertex = 4;

// Describes a vertex in a skeletal model
struct SkinnedVertex {

    // Number of bones that influence this vertex
    int boneCount;

    // Which bones influence the vertex? These are indices
    // into a list of bones.
    int boneIndex[kMaxBonesPerVertex];

    // Bone weights. These must sum to 1
    float boneWeight[kMaxBonesPerVertex];

    // Vertex position and normal, in bone space
    Vector3 posInBoneSpace[kMaxBonesPerVertex];
    Vector3 normalInBoneSpace[kMaxBonesPerVertex];
};

// Describes a vertex as we will use it for rendering
struct Vertex {
    Vector3 pos;
    Vector3 normal;
};

// Compute skinned vertex positions and normals.
void computeSkinnedVertices(
    int vertexCount,                        // number of verts to skin
    const SkinnedVertex *inSkinVertList,    // input vert list
    const Matrix4x3 *boneToModelList,       // Pos/orient of each bone
    Vertex *outVertList                     // output goes here
) {

    // Iterate over all the vertices
    for (int i = 0 ; i < vertexCount ; ++i) {
        const SkinnedVertex &s = inSkinVertList[i];
        Vertex &d = outVertList[i];

        // Loop over all bones that influence this vertex, and
        // compute weighted average
```

```
d.pos.zero();
d.normal.zero();
for (int j = 0 ; j < s.boneCount ; ++j) {

    // Locate the transform matrix
    const Matrix4x3 &boneToModel
        = boneToModelList[s.boneIndex[j]];

    // Transform from bone to model space (using
    // overloaded vector * matrix operator which does
    // matrix multiplication), and sum in this bone's
    // contribution
    d.pos += s.posInBoneSpace[j] * boneToModel
        * s.boneWeight[j];

    // *Rotate* the vertex into body space, ignoring the
    // translation portion of the affine transform.  The
    // normal is a "vector" and not a "point", so it is not
    // translated.
    d.normal += boneToModel.rotate(s.normalInBoneSpace[j])
        * s.boneWeight[j];
}

// Make sure the normal is normalized
d.normal.normalize();
}
}
```

Listing 10.7
A simple strategy for skinning vertices

Like all of the code snippets in this book, the purpose of this code is to explain principles, not to show how things are optimized in practice. In reality, the skinning computations shown here are usually done in hardware in a vertex shader; we'll show how this is done in Section 10.11.5. But there's plenty more theory to talk about, so let's stay at a high level. As it turns out, the technique just presented is easy to understand, but there's an important high-level optimization. In practice, a slightly different technique is used.

We'll get to the optimization in just a moment, but for now, let's back up and ask ourselves where the bone space coordinates (the member variables named `posInBoneSpace` and `normalInBoneSpace` in Listing 10.7) came from in the first place. "That's easy," you might say, "we just export them directly from Maya!" But how did Maya determine them? The answer is they come from the *binding pose*. The binding pose (sometimes called the *home pose*) describes an orientation of the bones in some default position. When an artist creates a character mesh, he starts by building a mesh without any bones or skinning data, just like any other model. During this process, he builds the character posed in the binding pose. Figure 10.35 shows our skinned model in her binding pose, along with the skeleton that is used to animate her. Remember that bones are really just coordinate spaces and don't have any actual geometry. The geometry you see exists only as an aid to visualization

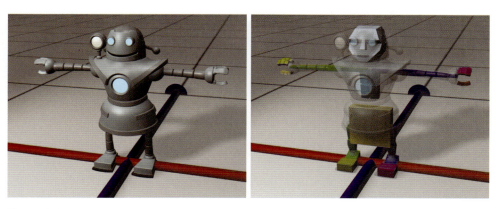

Figure 10.35
The robot model in the binding pose (left), and the bones used to animate the model (right)

When the mesh is done,[19] it is *rigged*, which means a hierarchy of bones (a skeleton) is created and the skinning data is edited to associate vertices with the appropriate bones. During this process, the rigger will bend bones to various extreme angles to preview how well the model reacts to these contortions. Is the weighting done properly so that joints don't collapse? This is where the skill and experience of both the character modeler and the rigger come into play. The point for us is that although Maya is constantly calculating new vertex positions in response to the manipulation of the bones, it has saved the original modeling space coordinates of each vertex at the location it had in the binding pose, before it got attached to a skeleton. Everything starts with that original vertex position.

So, to compute the bone-space coordinates of a vertex, we start with the modeling-space coordinates of that vertex in the binding pose. We also know the position and orientation of each bone in the binding pose. We simply transform the vertex positions from modeling space into bone space based on those positions and orientations.

That's the big picture of mesh skinning, in principle. Now let's get to the optimization. The basic idea is to store the position of each vertex only in the binding pose, rather than storing it relative to each bone that exerts influence. Then, when rendering a mesh, rather than having a bone-to-model transform for each bone, we have a matrix that transforms coordinates from the original binding space to modeling space in the cur-

---

[19]Well, this is the ideal process. In reality, changes to the mesh are often necessary after the mesh is rigged. The mesh might require adjustments in order to make it bend better, although an experienced character modeler can anticipate the needs of rigging. Of course, changes are often needed for aesthetic purposes having nothing to do with rigging—especially if executives or focus groups are involved.

rent pose. In other words, this matrix describes the difference between the bone's orientation in the binding pose and the bone's current orientation in the current pose. This is shown in Listing 10.8.

```cpp
// Set a limit on the max number of bones that can influence one vertex
const int kMaxBonesPerVertex = 4;

// Describes a vertex in a skeletal model
struct SkinnedVertex {

    // Number of bones that influence this vertex
    int boneCount;

    // Which bones influence the vertex?  These are indices
    // into a list of bones.
    int boneIndex[kMaxBonesPerVertex];
    // Bone weights.  These must sun to 1
    float boneWeight[kMaxBonesPerVertex];

    // Vertex position and normal in the binding pose,
    // in model space
    Vector3 pos;
    Vector3 normal;
};

// Describes a vertex as we will use it for rendering
struct Vertex {
    Vector3 pos;
    Vector3 normal;
};

// Compute skinned vertex positions and normals.
void computeSkinnedVertices(
    int vertexCount,                    // number of verts to skin
    const SkinnedVertex *inSkinVertList, // input vert list
    const Matrix4x3 *boneTransformList,  // From binding to current pose
    Vertex *outVertList                  // output goes here
) {

    // Iterate over all the vertices
    for (int i = 0 ; i < vertexCount ; ++i) {
        const SkinnedVertex &s = inSkinVertList[i];
        Vertex &d = outVertList[i];

        // Loop over all bones that influence this vertex, and compute
        // a blended *matrix* for this vertex
        Matrix4x3 blendedMat;
        blendedMat.zero();
        for (int j = 0 ; j < s.boneCount ; ++j) {
            blendedMat += boneTransformList[s.boneIndex[j]]
                * s.boneWeight[j];
        }

        // Transform position and normal using blended matrix
        d.pos = s.pos * blendedMat;
        d.normal = blendedMat.rotate(s.normal);
        // Make sure the normal is normalized
        d.normal.normalize();
    }
}
```

Listing 10.8
A more optimized strategy for skinning vertices

This produces a significant reduction in bandwidth to the GPU (due to the decrease in `sizeof(SkinnedVertex)`), as well as a reduction in per-vertex computation, especially when basis vectors are present. It just requires a bit more manipulation of the matrices before handing them to the GPU.

We have presented the basic idea behind simple skinning. Certainly, in situations where computing resources (and human resources!) are available and worth expending to produce the highest fidelity characters possible, such as in fighting games or sports games, more advanced techniques can be employed. For example, we might want to make the bicep bulge as the arm bends up, or squish out the flesh of a dinosaur foot as weight is transferred and the foot is pressed harder into the ground.

10.9   Bump Mapping

The first use of texture mapping in computer graphics was to define the *color* of an object. But texture mapping can be used when we want to specify *any* surface property with more granularity than at the vertex level. The particular surface property that perhaps is closest to controlling its "texture," in the sense most laypersons would understand it, is actually the surface normal.

Bump mapping is a general term that can refer to at least two different methods of controlling the surface normal per texel. A *height map* is a grayscale map, in which the intensity indicates the local "elevation" of the surface. Lighter colors indicate portions of the surface that are "bumped out," and darker colors are areas where the surface is "bumped in." Height maps are attractive because they are very easy to author, but they are not ideal for real-time purposes because the normal is not directly available; instead, it must be calculated from the intensity gradient. We focus here on the technique of *normal mapping*, which is very common nowadays and what most people usually mean when they say "bump map."

In a normal map, the coordinates of the surface normal are directly encoded in the map. The most basic way is to encode x, y, and z in the red, green, and blue channels, respectively, although certain hardware supports more optimized formats. The values are usually scaled, biased, and quantized such that a coordinate value of -1 is encoded as a 0, and $+1$ is encoded using the maximum color value (usually 255). Now, in principle, using a normal map is simple. In our lighting calculation, rather than using the result of interpolating the vertex normals, we fetch a normal out of the normal map and use it instead. Voila! Ah, if only it were that easy...

Complications arise for two primary reasons. First, normal maps are not intuitive to edit. While a height map (or true displacement map) can be easily painted in *Photoshop*, normal maps are not so easily visualized and edited. Cut-and-paste operations on normal maps are usually safe, but for the normal map to be valid, each pixel should encode a vector that is normalized. The usual technique for making a normal map is for an artist to actually model a low- and high-res version of the mesh. The low-res mesh is the one actually used at runtime, and the high-res mesh is solely to create the bump map,[20] using an automated tool that raytraces against the higher resolution mesh to determine the surface normal for every texel in the normal map.

The trickier issue is that texture memory is a precious resource.[21] In some simple cases, every texel in the normal map is used at most once on the surface of the mesh. In this case, we could simply encode the normal in object space, and our earlier description would work just fine. But real-world objects exhibit a great deal of symmetry and self-similarity, and patterns are often repeated. For example, a box often has similar bumps and notches on more than one side. Because of this, it is currently a more efficient use of the same amount of memory (and artist time) to increase the resolution of the map and reuse the same normal map (or perhaps just portions of it) on multiple models (or perhaps just on multiple places in the same model). Of course, the same principle applies to any sort of texture map, not just normal maps. But normal maps are different in that they cannot be arbitrarily rotated or mirrored because they encode a vector. Imagine using the same normal map on all six sides of a cube. While shading a point on the surface of the cube, we will fetch a texel from the map and decode it into a 3D vector. A particular normal map texel on the top will produce a surface normal that points in the same direction as that same texel on the bottom of the cube, when they should be opposites! We need some other kind of information to tell us how to interpret the normal we get from the texture, and this extra bit of information is stored in the *basis vectors*.

10.9.1   Tangent Space

The most common technique these days is for the normal encoded in the map to use coordinates in *tangent space*. In tangent space, $+z$ points out from the surface; the $+z$ basis vector is actually just the surface normal $\hat{\mathbf{n}}$. The x basis vector is known as the *tangent* vector, which we'll denote $\hat{\mathbf{u}}$, and it points in the direction of increasing u in texture space. In other words, when we move in the direction of the tangent vector in 3D, this corresponds

---

[20]And for high-res renders for the box art. Some people also use high-res models to make disingenuous screen shots of "in-game" footage, sort of like how the hamburger you get at a fast-food restaurant doesn't look like the one in the picture on the menu.

[21]We don't all have *id Tech 5*'s MegaTexturing yet.

to moving to the right in 2D in the normal map. (Often, the bump map shares the same UV coordinates as other maps, but if they differ, it's the coordinates used for bump mapping that count.) Similarly, the y basis vector, known as the *binormal*[22] and denoted here as $\hat{\mathbf{v}}$, corresponds to the direction of increasing v, although whether this motion is "up" or "down" in the texture space depends on the conventions for the origin in (u, v) space, which can differ, as we discussed earlier. Of course, the coordinates for the tangent and binormal are given in model space, just like the surface normal. As implied by the hats over the variables, basis vectors are usually stored as unit vectors.

For example, assume that a certain texel in a normal map has the RGB triple $[37, 128, 218]$, which is decoded to the unit vector $[-0.707, 0, 0.707]$. We interpret this to mean that the local surface normal is pointing at about a $45°$ angle from a "flat" surface normal defined by the interpolated vertex normal. It points "to the left," where "left" is meaningful in the image space of the normal map and really means "in the direction of decreasing u."

In summary, the tangent, binormal, and normal are the axes of a coordinate space known as tangent space, and the coordinates of the per-texel normal are interpreted by using this coordinate space. To obtain the model-space normal from a tangent-space normal, we first decode the normal from the map and then transform it into model space just like any other vector. Let $\mathbf{s}^t = [s_x^t, s_y^t, s_z^t]$ denote the tangent-space surface normal and $\mathbf{s}^m = [s_x^m, s_y^m, s_z^m]$ denote the model-space surface normal. We can determine $\mathbf{s}^m$ simply by taking the linear combination of the basis vectors

$$\mathbf{s}^m = s_x^t \hat{\mathbf{u}} + s_y^t \hat{\mathbf{v}} + s_z^t \hat{\mathbf{n}}.$$

By now, we know that this is the same thing as multiplying $\mathbf{s}^t$ by a matrix whose rows are the basis vectors:

$$\mathbf{s}^m = \mathbf{s}^t \begin{bmatrix} -\hat{\mathbf{u}}- \\ -\hat{\mathbf{v}}- \\ -\hat{\mathbf{n}}- \end{bmatrix}. \tag{10.18}$$

Remember that the polygon mesh is just an approximation for a potentially curved surface, so the surface normal we use for lighting varies continuously over each face in order to approximate the true surface normal. In the same way, the tangent and binormal basis vectors also vary continuously over the mesh, since they should be perpendicular to the surface normal and tangent to the surface being approximated. But even on a flat surface, the basis vectors can change over the surface if a texture is

---

[22]The term "bitangent" is probably more correct; but it's less commonly used.

squeezed, squashed, or twisted. Two instructive examples can be found in Figure 10.19 on page 396. The left side shows an example of "squishing." In this case, the tangent vector $\hat{\mathbf{u}}$ would be pointing to the right, parallel to the horizontal polygon edges, while the binormal $\hat{\mathbf{v}}$ would be locally parallel to the vertical (curving) polygon edges at each vertex. To determine the basis vectors at any given point in the interior of the face, we interpolate the basis vectors from the vertices, just like we do with the surface normal. Compare this with the texture mapping on the right side, where the texture mapping is planar. In this example, the binormal at every vertex (and every interior point) points directly down.

Notice that in the texture mapping used in the left side of the figure, the tangent and binormal vectors are not perpendicular. Despite this possibility, it's common to assume the basis vectors form an orthonormal basis (or to adjust them so that they do), even if the texture is being manhandled. We make this assumption in order to facilitate two optimizations. The first optimization is that we can perform our lighting calculations in tangent space rather than in model space. If we do the lighting in model space, we must interpolate the three basis vectors across the face, and then in the pixel shader we must transform our tangent-space normal into model space. When we do the lighting in tangent space, however, we can instead transform the vectors needed for lighting ($\mathbf{l}$ and $\mathbf{h}$) into tangent space once in the vertex shader, and then during rasterization the interpolation is done in tangent space. In many circumstances, this is faster. If we have an orthonormal basis, then the inverse of the transform matrix is simply its transpose, and we can transform from model space to tangent space just by using the dot product. (If this isn't making sense, see Section 3.3.3 and Section 6.3.) Of course, we are free to rotate vectors into tangent space by using the dot product even if our basis isn't orthonormal; in fact, after interpolating basis vectors and renormalizing them, it's likely that it will be slightly out of orthogonality. In this case, our transform is not completely correct, but usually this doesn't cause any problems. It's important to remember that the whole idea of interpolating surface normals and basis vectors is an approximation to begin with.

The second optimization we can make by assuming perpendicular basis vectors is to completely avoid storing one of the two basis vectors (usually we drop the binormal) and compute it on the fly. This can be faster when the performance bottleneck is the shuffling around of memory rather than per-vertex calculations. There's just one complication: mirrored bump maps. It is very common on symmetrical objects for texture maps, including the bump map, to be used twice; on one side the "regular" way, and mirrored on the other side. Essentially, we need to know whether the texture is applied in its regular orientation or mirrored. This is done by storing a flag that indicates whether the texture is mirrored. A value of $+1$

indicates the usual orientation, and -1 indicates the mirrored state. It's common to conveniently tuck away this flag in the w component of the one basis vector we are keeping. Now when we need to compute the dropped basis vector, we take the appropriate cross product (for example $\hat{\mathbf{v}} = \hat{\mathbf{n}} \times \hat{\mathbf{u}}$), and then multiply by our flag to flip the basis vector if necessary. This flag is calculated by the triple product $\hat{\mathbf{n}} \times \hat{\mathbf{u}} \cdot \hat{\mathbf{v}}$, which is the same thing as the determinant of the transform matrix in Equation (10.18).

10.9.2 Calculating Tangent Space Basis Vectors

Finally, let's talk about how to compute basis vectors. Our development follows Lengyel [42]. We are given a triangle with vertex positions $\mathbf{p}_0 = (x_0, y_0, z_0)$, $\mathbf{p}_1 = (x_1, y_1, z_1)$, and $\mathbf{p}_2 = (x_2, y_2, z_2)$, and at those vertices we have the UV coordinates (u_0, v_0), (u_1, v_1), and (u_2, v_2). Under these circumstances, it is always possible to find a planar mapping, meaning the mapping gradient is constant over the entire triangle.

Looking ahead, the math will be simplified if we shift the origin to $\mathbf{p}_0$ by introducing

$$\mathbf{q}_1 = \mathbf{p}_1 - \mathbf{p}_0, \qquad s_1 = u_1 - u_0, \qquad t_1 = v_1 - v_0,$$
$$\mathbf{q}_2 = \mathbf{p}_2 - \mathbf{p}_0, \qquad s_2 = u_2 - u_0, \qquad t_2 = v_2 - v_0.$$

We seek basis vectors that lie in the plane of the triangle, and thus we can express the triangle edge vectors $\mathbf{q}_1$ and $\mathbf{q}_2$ as a linear combination of the basis vectors, where the known u and v displacements on those edges are the coordinates:

$$\mathbf{u}s_1 + \mathbf{v}t_1 = \mathbf{q}_1$$
$$\mathbf{u}s_2 + \mathbf{v}t_2 = \mathbf{q}_2.$$

Normalizing $\mathbf{u}$ and $\mathbf{v}$ produces the unit vectors we seek. We can write these equations more compactly in matrix notation as

$$\begin{bmatrix} s_1 & t_1 \\ s_2 & t_2 \end{bmatrix} \begin{bmatrix} -\mathbf{u}- \\ -\mathbf{v}- \end{bmatrix} = \begin{bmatrix} -\mathbf{q}_1- \\ -\mathbf{q}_2- \end{bmatrix},$$

whence an elegant solution presents itself. By multiplying both sides by the inverse of the s, t matrix on the left, we have

$$\begin{bmatrix} -\mathbf{u}- \\ -\mathbf{v}- \end{bmatrix} = \begin{bmatrix} s_1 & t_1 \\ s_2 & t_2 \end{bmatrix}^{-1} \begin{bmatrix} -\mathbf{q}_1- \\ -\mathbf{q}_2- \end{bmatrix}$$

$$= \frac{1}{s_1 t_2 - s_2 t_1} \begin{bmatrix} t_2 & -t_1 \\ -s_2 & s_1 \end{bmatrix} \begin{bmatrix} -\mathbf{q}_1- \\ -\mathbf{q}_2- \end{bmatrix}.$$

Since we are planning on normalizing our basis vectors, we can drop the leading constant fraction, and we are left with

$$\mathbf{u} = t_2\mathbf{q}_1 - t_1\mathbf{q}_2,$$
$$\mathbf{v} = -s_2\mathbf{q}_1 + s_1\mathbf{q}_2.$$

This gives us basis vectors for each triangle. They are not guaranteed to be perpendicular, but they are usable for our main purpose: determining basis vectors at the vertex level. These can be calculated by using a trick similar to computing vertex normals: for each vertex we take the average of the basis vectors of the adjacent triangles. We also usually enforce an orthonormal basis. This is done most simply via Gram-Schmidt orthogonalization (Section 6.3.3). Also, if we are dropping one of the basis vectors, then this is where we need to save the determinant of the basis. Listing 10.9 shows how we might compute vertex basis vectors.

```
struct Vertex {
    Vector3  pos;
    float    u,v;
    Vector3  normal;
    Vector3  tangent;
    float    det;       // determinant of tangent transform.  (-1 if mirrored)
};
struct Triangle {
    int  vertexIndex[3];
};
struct TriangleMesh {
    int       vertexCount;
    Vertex    *vertexList;
    int       triangleCount;
    Triangle *triangleList;

    void computeBasisVectors() {

        // Note: we assume vertex normals are valid
        Vector3 *tempTangent = new Vector3[vertexCount];
        Vector3 *tempBinormal = new Vector3[vertexCount];

        // First clear out the accumulators
        for (int i = 0 ; i < vertexCount ; ++i) {
            tempTangent[i].zero();
            tempBinormal[i].zero();
        }

        // Average in the basis vectors for each face
        // into its neighboring vertices
        for (int i = 0 ; i < triangleCount ; ++i) {

            // Get shortcuts
            const Triangle &tri = triangleList[i];
            const Vertex &v0 = vertexList[tri.vertexIndex[0]];
            const Vertex &v1 = vertexList[tri.vertexIndex[1]];
            const Vertex &v2 = vertexList[tri.vertexIndex[2]];

            // Compute intermediate values
            Vector3 q1 = v1.pos - v0.pos;
            Vector3 q2 = v2.pos - v0.pos;
```

```
            float s1 = v1.u − v0.u;
            float s2 = v2.u − v0.u;
            float t1 = v1.v − v0.v;
            float t2 = v2.v − v0.v;

            // Compute basis vectors for this triangle
            Vector3 tangent = t2*q1 − t1*q2; tangent.normalize();
            Vector3 binormal = −s2*q1 + s1*q2; binormal.normalize();

            // Add them into the running totals for neighboring verts
            for (int j = 0 ; j < 3 ; ++j) {
                tempTangent[tri.vertexIndex[j]] += tangent;
                tempBinormal[tri.vertexIndex[j]] += binormal;
            }
        }

        // Now fill in the values into the vertices
        for (int i = 0 ; i < vertexCount ; ++i) {
            Vertex &v = vertexList[i];
            Vector3 t = tempTangent[i];

            // Ensure tangent is perpendicular to the normal.
            // (Gram–Schmit), then keep normalized version
            t −= v.normal * dot(t, v.normal);
            t.normalize();
            v.tangent = t;

            // Figure out if we're mirrored
            if (dot(cross(v.normal, t), tempBinormal[i]) < 0.0f) {
                v.det = −1.0f; // we're mirrored
            } else {
                v.det = +1.0f; // not mirrored
            }
        }

        // Clean up
        delete[] tempTangent;
        delete[] tempBinormal;
    }
};
```

Listing 10.9
Simple method for calculating basis vectors as the average of adjacent triangle normals

One irritating complication that Listing 10.9 doesn't address is that there may be a discontinuity in the mapping, where the basis vectors *should not* be averaged together, and the basis vectors must be different across a shared edge. Most of the time, the faces will have already be detached from each other (the vertices will be duplicated) along such an edge, since the UV coordinates or normals will not match. Unfortunately, there is one particularly common case where this is not true: mirrored textures on symmetric objects. For example, it is common for character models and other symmetric meshes to have a line down their center, across which the texture has been mirrored. The vertices along this seam very often require identical UVs but an opposite $\hat{\mathbf{u}}$ or $\hat{\mathbf{v}}$. These vertices must be detached in order to avoid producing invalid basis vectors along this seam.

Section 10.11.4 shows some sample shader code that actually uses the basis vectors to perform bump mapping. The runtime code is surprisingly simple, once all the data has been munged into the right format. This illustrates a common theme of contemporary real-time graphics: at least 75% of the code is in the *tools* that manipulate the data—optimizing, packing, and otherwise manipulating it into just the right format—so that the runtime code (the other 25%) can run as fast as possible.

10.10   The Real-Time Graphics Pipeline

The rendering equation is the correct way to produce images, assuming you have an infinite amount of computing power. But if you want to produce images in the real world on a real computer, you need to understand the contemporary trade-offs that are being made. The remainder of this chapter is more focused on those techniques, by attempting to describe a typical simple real-time graphics pipeline, circa 2010. After giving an overview of the graphics pipeline, we then descend that pipeline and discuss each section in more detail, stopping along the way to focus on some key mathematical ideas. The reader of this section should be aware of several serious flaws in this discussion:

- There is no such thing as the "typical" modern graphics pipeline. The number of different rendering strategies is equal to the number of graphics programmers. Everyone has his or her own preferences, tricks, and optimizations. Graphics hardware continues to evolve *rapidly*. As evidence, the use of shader programs is now in widespread use in consumer hardware such as gaming consoles, and this technology was in its infancy at the time of the writing of the first edition of this book. Still, although there is great variance in graphics systems and graphics programmers, most systems do have a great deal in common.[23] We'd like to reiterate that our goal in this chapter (indeed, this entire book!) is to give you a solid overview, especially where the mathematics is involved, from which you can expand your knowledge. This is not a survey of the latest cutting-edge techniques. (*Real-Time Rendering* [1] is the best such survey at the time of this writing.)

- We attempt to describe the basic procedure for generating a single rendered image with very basic lighting. We do not consider animation, and we only briefly mention techniques for global illumination in passing.

---

[23]And most programmers have a lot in common, too, even though we might hate to admit it.

- Our description is of the *conceptual* flow of data through the graphics pipeline. In practice, tasks are often performed in parallel or out of sequence for performance reasons.

- We are interested in real-time rendering systems which, at the time of this writing, are primarily geared for rendering triangle meshes. Other means of producing an image, such as raytracing, have a very different high-level structure than that discussed here. A reader is warned that in the future, techniques for real-time and offline rendering could converge if parallel raytracing becomes a more economical way to keep up with the march of Moore's law.

With the above simplifications in mind, the following is a rough outline of the flow of data through the graphics pipeline.

- *Setting up the scene.* Before we can begin rendering, we must set several options that apply to the entire scene. For example, we need to set up the camera, or more specifically, pick a point of view in the scene from which to render it, and choose where on the screen to render it. We discussed the math involved in this process in Section 10.2. We also need to select lighting and fog options, and prepare the depth buffer.

- *Visibility determination.* Once we have a camera in place, we must then decide which objects in the scene are visible. This is extremely important for real-time rendering, since we don't want to waste time rendering anything that isn't actually visible. This high-level culling is very important for real games, but is usually ignored for simple applications when you're getting started, and is not covered here.

- *Setting object-level rendering states.* Once we know that an object is potentially visible, it's time to actually draw the object. Each object may have its own rendering options. We must install these options into the rendering context before rendering any primitives associated with the object. Perhaps the most basic property associated with an object is a *material* that describes the surface properties of the object. One of the most common material properties is the diffuse color of the object, which is usually controlled by using a *texture map*, as we discussed in Section 10.5.

- *Geometry generation/delivery.* Next, the geometry is actually submitted to the rendering API. Typically, the data is delivered in the form of triangles; either as individual triangles, or an indexed triangle mesh, triangle strip, or some other form. At this stage, we may also

perform level of detail (LOD) selection or generate geometry procedurally. We discuss a number of issues related to delivering geometry
to the rendering API in Section 10.10.2.

- *Vertex-level operations.* Once the rendering API has the geometry
  in some triangulated format, a number of various operations are performed at the vertex level. Perhaps the most important such operation is the transformation of vertex positions from modeling space
  into camera space. Other vertex level operations might include skinning for animation of skeletal models, vertex lighting, and texture
  coordinate generation. In consumer graphics systems at the time of
  this writing, these operations are performed by a user-supplied micro-
  program called a *vertex shader*. We give several examples of vertex
  and pixel shaders at the end of this chapter, in Section 10.11.

- *Culling, clipping, and projection.* Next, we must perform three operations to get triangles in 3D onto the screen in 2D. The exact order in
  which these steps are taken can vary. First, any portion of a triangle
  outside the view frustum is removed, by a process known as *clipping*,
  which is discussed in Section 10.10.4. Once we have a clipped polygon in 3D clip space, we then project the vertices of that polygon,
  mapping them to the 2D screen-space coordinates of the output window, as was explained in Section 10.3.5. Finally, individual triangles
  that face away from the camera are removed ("culled"), based on the
  clockwise or counterclockwise ordering of their vertices, as we discuss
  in Section 10.10.5.

- *Rasterization.* Once we have a clipped polygon in screen space, it is
  rasterized. Rasterization refers to the process of selecting which pixels
  on the screen should be drawn for a particular triangle; interpolating
  texture coordinates, colors, and lighting values that were computed
  at the vertex level across the face for each pixel; and passing these
  down to the next stage for pixel shading. Since this operation is
  usually performed at the hardware level, we will only briefly mention
  rasterization in Section 10.10.6.

- *Pixel shading.* Next we compute a color for the pixel, a process known
  as *shading*. Of course, the innocuous phrase "compute a color" is the
  heart of computer graphics! Once we have picked a color, we then
  write that color to the frame buffer, possibly subject to alpha blending
  and z-buffering. We discuss this process in Section 10.10.6. In today's
  consumer hardware, pixel shading is done by a *pixel shader*, which is a
  small piece of code you can write that takes the values from the vertex
  shader (which are interpolated across the face and supplied per-pixel),
  and then outputs the color value to the final step: blending.

- *Blending and output.* Finally, at the very bottom of the render pipeline, we have produced a color, opacity, and depth value. The depth value is tested against the depth buffer for per-pixel visibility determination to ensure that an object farther away from the camera doesn't obscure one closer to the camera. Pixels with an opacity that is too low are rejected, and the output color is then combined with the previous color in the frame buffer in a process known as *alpha blending*.

The pseudocode in Listing 10.10 summarizes the simplified rendering pipeline outlined above.

```
// First, figure how to view the scene
setupTheCamera();

// Clear the zbuffer
clearZBuffer();

// Setup environmental lighting and fog
setGlobalLightingAndFog();

// get a list of objects that are potentially visible
potentiallyVisibleObjectList = highLevelVisibilityDetermination(scene);

// Render everything we found to be potentially visible
for (all objects in potentiallyVisibleObjectList) {

    // Perform lower-level VSD using bounding volume test
    if (!object.isBoundingVolumeVisible()) continue;

    // Fetch or procedurally generate the geometry
    triMesh = object.getGeometry()

    // Clip and render the faces
    for (each triangle in the geometry) {

        // Transform the vertices to clip space, and perform
        // vertex-level calculations (run the vertex shader)
        clipSpaceTriangle = transformAndLighting(triangle);

        // Clip the triangle to the view volume
        clippedTriangle = clipToViewVolume(clipSpaceTriangle);
        if (clippedTriangle.isEmpty()) continue;

        // Project the triangle onto screen space
        screenSpaceTriangle = clippedTriangle.projectToScreenSpace();

        // Is the triangle backfacing?
        if (screenSpaceTriangle.isBackFacing()) continue;

        // Rasterize the triangle
        for (each pixel in the triangle) {

            // Scissor the pixel here (if triangle was
            // not completely clipped to the frustum)
            if (pixel is off-screen) continue;
```

```
                   // Interpolate color, zbuffer value,
                   // and texture mapping coords

                   // The pixel shader takes interpolated values
                   // and computes a color and alpha value
                   color = shadePixel();

                   // Perform zbuffering
                   if (!zbufferTest()) continue;

                   // Alpha test to ignore pixels that are "too
                   // transparent"
                   if (!alphaTest()) continue;

                   // Write to the frame buffer and zbuffer
                   writePixel(color, interpolatedZ);

                   // Move on to the next pixel in this triangle
                }

             // Move on to the next triangle in this object
          }

       // Move on to the next potentially visible object
    }
```

Listing 10.10
Pseudocode for the graphics pipeline

It wasn't too long ago that a graphics programmer would be responsible for writing code to do all of the steps shown Listing 10.10 in software. Nowadays, we delegate many tasks to a graphics API such as DirectX or OpenGL. The API may perform some of these tasks in software on the main CPU, and other tasks (ideally, as many as possible) are dispatched to specialized graphics hardware. Modern graphics hardware allows the graphics programmer (that's us) very low level control through *vertex shaders* and *pixel shaders*, which are basically microprograms we write that are executed on the hardware for each vertex and pixel that is processed. While performance concerns in the old single processor software rendering days were addressed with hand-tuned assembly, nowadays the concerns are more about using the GPU as efficiently as possible, and ensuring that it is never idle, waiting on the CPU to do anything. Of course, both now and then the simplest way to speed up rendering something is to simply avoid rendering it at all (if it isn't visible) or to render a cheaper approximation of it (if it's not large on the screen).

In summary, a modern graphics pipeline involves close cooperation of our code and the rendering API. When we say "rendering API," we mean the API software *and* the graphics hardware. On PC platforms the API software layer is necessarily very "thick," due to the wide variety of underlying hardware that must be supported. On console platforms where the

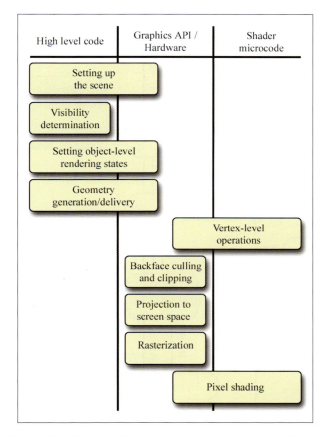

Figure 10.36. Division of labor between our code and the graphics API

hardware is standardized, the layer can be significantly leaner. A notable example of this was the PlayStation 2, which allowed programmers direct access to hardware registers and very low level control over direct memory access (DMA). Figure 10.36 illustrates the division of labor involved in this cooperation.

A slightly different summary of the real-time graphics pipeline is illustrated in Figure 10.37, this time focusing more on the lower end of the pipeline and the conceptual flow of data. The blue boxes represent data that we provide, and blue ovals are our shaders that we write. The yellow ovals are operations that are performed by the API.

The remainder of this chapter discusses a number of various topics in computer graphics. We proceed roughly in the order that these topics are encountered in the graphics pipeline.

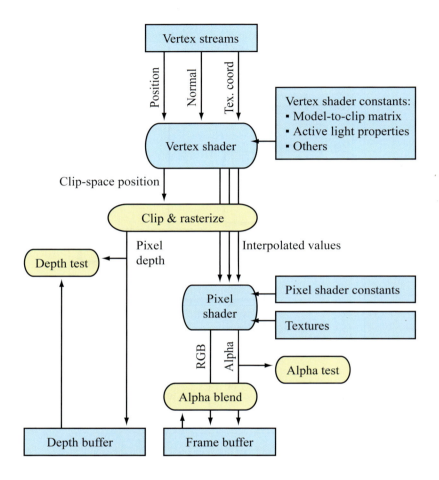

Figure 10.37. Data flow through the graphics pipeline.

10.10.1   Buffers

Rendering involves many buffers. In this context, a buffer is simply a rectangular region of memory that stores some sort of data per pixel. The most important buffers are the *frame buffer* and the *depth buffer*.

The frame buffer stores one color per pixel—it holds the rendered image. The color for a single pixel may be stored in a variety of formats; the variations are not significant for the current discussion. If we're rendering a single image, the frame buffer may be in regular RAM, to be saved to disk.

A more interesting situation arises in real-time animation. In this case, the frame buffer is normally located in video RAM. The video card is constantly reading this area of video RAM, converting the binary data into the appropriate signal to be sent to the display device. But how can the monitor read this memory while we're trying to render to it? A technique known as *double buffering* is used to prevent an image from being displayed before it is completely rendered. Under double buffering, there are actually two frame buffers. One frame buffer, the *front buffer*, holds the image currently displayed on the monitor. The *back buffer* is the off-screen buffer, which holds the image currently being rendered.

When we have finished rendering an image and are ready for it to be displayed, we "flip" the buffers. We can do this in one of two ways. If we use page flipping, then we instruct the video hardware to begin reading from the buffer that was the off-screen buffer. We then swap the roles of the two buffers; the buffer that was being displayed now becomes the off-screen buffer. Or we may blit (copy) the off-screen buffer over the display buffer. Double buffering is shown in Figure 10.38.

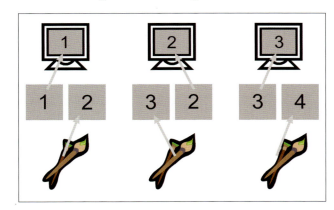

Figure 10.38
Double buffering

The more modern terminology for making visible the image that was rendered into the back buffer is *presenting* the image.

The second important buffer used for rendering is the *depth buffer*, also known as the *z-buffer*. Rather than storing a color at each pixel, the depth buffer stores a depth value per pixel. There are many variations in the specifics of exactly what value goes into the depth buffer, but the basic idea is that it is related to the distance from the camera. Often the clip-space z-coordinate is used as a depth value, which is why the depth buffer is also known as the z-buffer.

The depth buffer is used to determine which objects occlude which objects, as follows. As we are rasterizing a triangle, we compute an interpolated depth value per pixel. Before rendering a pixel, we compare this depth value with the value already in the depth buffer for this pixel. If the new depth is farther from the camera than the value currently in the depth buffer, then the pixel is discarded. Otherwise, the pixel color is written to the frame buffer, and the depth buffer is updated with the new, closer depth value.

Before we can begin rendering an image, we must clear the depth buffer to a value that means "very far from the camera." (In clip space, this value is 1.0). Then, the first pixels to be rendered are guaranteed to pass the depth buffer test. There's normally no need to double buffer the depth buffer like we do the frame buffer.

10.10.2   Delivering the Geometry

After deciding which objects to render, we need to actually render them. This is actually a two-step process. First, we must set up the *render context*. This involves telling the renderer what vertex and pixel shaders to use, what textures to use, and setting any other constants needed by the shaders, such as the transform matrices, lighting positions, colors, fog settings, and so forth. The details of this process depend greatly on your high-level rendering strategy and target platform, so there isn't much more specific we can say here, although we give several examples in Section 10.11. Instead, we would like to focus on the second step, which is essentially the top box in Figure 10.37, where vertex data is delivered to the API for rendering. Nowadays a programmer has quite a bit of flexibility in what data to send, how to pack and format each data element, and how to arrange the bits in memory for maximum efficiency.

What values might we need to supply per vertex? Basically, the answer is, "whatever properties you want to use to render the triangles." Ultimately, there are only two required outputs of the vertex and pixel shader. First, the vertex shader must output a position for each vertex so that the hardware can perform rasterization. This position is typically specified in clip space, which means the hardware will do the perspective divide and conversion to screen space coordinates (see Section 10.3.5) for you. The pixel shader really has only one required output: a color value (which typically includes an alpha channel). Those two outputs are the only things that are *required*. Of course, to properly determine the proper clip-space coordinates, we probably need the matrix that transforms from model space to clip space. We can pass parameters like this that apply to all the vertices or pixels in a given batch of triangles by setting *shader constants*. This is conceptually just a large table of vector values that is part of the render context and for us to use as needed. (Actually, there is usually one set of registers assigned for use in the vertex shader and a different set of registers that can be accessed in the pixel shader.)

Some typical pieces of information that are stored at the vertex level include

- *Position.* This describes the location of the vertex. This can be a 3D vector or a 2D screen-space position, or it could be a position already transformed into clip space that is simply passed directly through the vertex shader. If a 3D vector is used, the position must be transformed into clip space by the current model, view, and projection transforms. If 2D window coordinates (ranging according to the resolution of the screen, not normalized) are used, then they must be converted back into clip space in the vertex shader. (Some hardware allows your shader to output coordinates that are already projected to screen space.)

  If the model is a skinned model (see Section 10.8), then the positional data must also include the indices and weights of the bones that influence the vertex. The animated matrices can be delivered in a variety of ways. A standard technique is to pass them as vertex shader constants. A newer technique that works on some hardware is to deliver them in a separate vertex stream, which must be accessed through special instructions since the access pattern is random rather than streaming.

- *Texture-mapping coordinates.* If we are using texture-mapped triangles, then each vertex must be assigned a set of mapping coordinates. In this simplest case, this is a 2D location into the texture map. We usually denote the coordinates (u, v). If we are using multitexturing, then we might need one set of mapping coordinates per texture map. Optionally, we can generate one or more sets of texture-mapping coordinates procedurally (for example, if we are projecting a gobo onto a surface).

- *Surface normal.* Most lighting calculations need the surface normal. Even though these lighting equations are often done per-pixel, with the surface normal being determined from a normal map, we still often store a normal at the vertex level, in order to establish the basis for tangent space.

- *Color.* Sometimes it's useful to assign a color input to each vertex. For example, if we are rendering particles, the color of the particle may change over time. Or we may use one channel (such as alpha) to control the blending between two texture layers. An artist can edit the vertex alpha to control this blending. We might also have per-vertex lighting calculations that were done offline.

- *Basis vectors.* As discussed in Section 10.9, for tangent-space normal maps (and a few other similar techniques) we need basis vectors in order to define the local tangent space. The basis vectors and surface normal establish this coordinate space at each vertex. These vectors are then interpolated across the triangle during rasterization, to provide an approximate tangent space per pixel.

With all that in mind, let's give a few examples of C `structs` that could be used to deliver vertex data in some situations that might arise in practice.

One of the most basic vertex formats contains a 3D position, surface normal, and mapping coordinates. A basic triangle mesh with a simple diffuse map is stored using this vertex type. We can't use tangent space normal maps with this vertex format, since there are no basis vectors:

```
// Untransformed, unlit vertex
struct RenderVertex {
    Vector3 p;    // position
    float   u,v;  // texture mapping coordinates
    Vector3 n;    // normal
};
```

If we want to use a tangent-space normal map, we'll need to include basis vectors:

```
// Untransformed, unlit vertex with basis vectors
struct RenderVertexBasis {
    Vector3 p;        // position
    Vector3 n;        // normal
    Vector3 tangent;  // 1st basis vector
    float   det;      // Determinant of tangent space
                      // transform (mirror flag)
    float   u,v;      // texture mapping coordinates
};
```

Another common format, used for heads-up displays, text rendering, and other 2D items, is a vertex with screen space coordinates and pre-lit vertices (no normal needs to be supplied since no lighting calculations will take place):

```
// 2D screen-space pre-lit.
struct RenderVertex2D {
    float    x,y;  // 2D screen-space position
    unsigned argb; // prelit color (0xAARRGGBB)
    float    u,v;  // texture mapping coordinates
};
```

The following vertex is expressed in 3D, but does not need to be lit by the graphics API's lighting engine. This format is often useful for particle effects, such as explosions, flames, and self-illuminated objects, and for rendering debugging objects such as bounding boxes, waypoints, markers, and the like:

```
// Untransformed, lit vertex
struct RenderVertexL {
    Vector3  p;      // 3D position
    unsigned argb;   // prelit color (0xAARRGGBB)
    float    u,v;    // texture mapping coordinates
};
```

The next example is a vertex used for lightmapped, bump-mapped geometry. It has basis vectors for lightmapping, and two sets of UVs, one for the regular diffuse texture, and another for the lightmap, which stores baked-in lighting that was calculated offline:

```
// Lightmapped, bump mapped vertex
struct RenderVertexLtMapBump {
    Vector3 p;          // position
    Vector3 n;          // normal
    Vector3 tangent;    // 1st basis vector
    float   det;        // Determinant of tangent space
                        // transform (mirror flag)
    float   u,v;        // regular coordinates for diffuse and bump map
    float   lmu,lmv;    // texture coords into lightmap
};
```

Finally, here's a vertex that might be used for skeletal rendering. The indices are stored in four 8-bit values, and the weights are stored as four floats:

```
// Lightmapped, bump mapped vertex
struct RenderVertexSkinned {
    Vector3 p;              // position
    Vector3 n;              // normal
    Vector3 tangent;        // 1st basis vector
    float   det;            // Determinant of tangent space
                            // transform (mirror flag)
    float   u,v;            // regular coordinates for diffuse and bump map
    unsigned boneIndices;   // bone indices for up to 4 bones
                            // (8-bit values)
    Vector4  boneWeights;   // weights for up to 4 bones
};
```

The preceding examples were all declared as `struct`s. As you can see, the combinations can grow quite quickly. Dealing with this simply but efficiently is a challenge. One idea is to allocate the fields as a structure of arrays (SOA) rather than array of structures (AOS):

```
struct VertexListSOA {
    Vector3 *p;             // positions
    Vector3 *n;             // normals
    Vector4 *tangentDet;    // xyz tangent + det in w
```

```
    Vector2   *uv0;            // first channel mapping coords
    Vector2   *uv1;            // second channel mapping
    Vector2   *ltMap;          // lightmap coords
    unsigned  *boneIndices;    // bone indices for up to 4 bones
                               // (8-bit values)
    Vector4   *boneWeights;    // weights for up to 4 bones
    unsigned  *argb;           // vertex color
};
```

In this case, if a value was not present, the array pointer would simply be NULL.

Another idea is to use a raw block of memory, but declare a vertex format class with accessor functions that do the address arithmetic to locate a vertex by index, based on the variable stride, and access a member based on its variable offset within the structure.

10.10.3   Vertex-Level Operations

After mesh data has been submitted to the API, a wide range of vertex-level computations are performed. In a shader-based renderer (as opposed to a fixed-function pipeline), this happens in our vertex shader. The input to a vertex shader is essentially one of the structs that we described in the previous section. As discussed earlier, a vertex shader can produce many different types of output, but there are two basic responsibilities it *must* fulfill. The first is that it must output, at the very minimum, a clip-space (or in some circumstances screen-space) position. The second responsibility is to provide to the pixel shader any inputs that are needed for the pixel shader to perform the shading calculations. In many cases, we can simply pass through vertex values received from the input streams, but other times, we must perform calculations, such as transforming raw vertex values from modeling space to some other coordinate space in which we are performing lighting or generating texture coordinates.

Some of the most common operations that are done in a vertex shader are

- Transforming model-space vertex positions into clip space.

- Performing skinning for skeletal models.

- Transforming normals and basis vectors into the appropriate space for lighting.

- Calculating vectors needed for lighting ($\mathbf{l}$ and $\mathbf{h}$) and transforming them into the appropriate coordinate space.

- Computing fog density values from the vertex position.

- Generating texture mapping coordinates procedurally. Examples include projected spot lights, *Doom*-style volumetric light, reflecting a view vector about the normal for environment mapping, various fake reflection techniques, scrolling or otherwise animated textures, and so on.

- Passing through raw vertex inputs without modification, if they are already in the correct format and coordinate space.

If we are using Gouraud shading, we might actually perform the lighting calculations here, and interpolate the lighting results. We'll show some examples of this later in the chapter.

The transformation from modeling to clip space is the most common operation, so let's review the process. We do it with matrix multiplication. Conceptually, the vertices undergo a sequence of transformations as follows:

- The model transform transforms from modeling space to world space.

- The view transform transforms from world space to camera space.

- The clip matrix is used to transform from camera space to clip space.

Conceptually, the matrix math is

$$\mathbf{v}_{\text{clip}} = (\mathbf{v}_{\text{model}})(\mathbf{M}_{\text{model}\to\text{world}})(\mathbf{M}_{\text{world}\to\text{camera}})(\mathbf{M}_{\text{camera}\to\text{clip}}).$$

In practice, we don't actually perform three separate matrix multiplications. We have one matrix that transforms from object space to clip space, and inside the vertex shader we perform one matrix multiplication using this matrix.

10.10.4 Clipping

After vertices have been transformed into clip space, two important tests are performed on the triangle: clipping and culling. Both operations are usually performed by the rendering API, so although you won't usually have to perform these operations yourself, it's important to know how they work. The order in which we discuss these tests is not necessarily the order in which they will occur on a particular piece of hardware. Most hardware culls in screen space, whereas older software renderers did it earlier, in 3D, in order to reduce the number of triangles that had to be clipped.

Before we can project the vertices onto screen space, we must ensure that they are completely inside the view frustum. This process is known as *clipping*. Since clipping is normally performed by the hardware, we will describe the process with only cursory detail.

The standard algorithm for clipping polygons is the *Sutherland-Hodgman* algorithm. This algorithm tackles the difficult problem of polygon clipping by breaking it down into a sequence of easy problems. The input polygon is clipped against one plane at a time.

To clip a polygon against one plane, we iterate around the polygon, clipping each edge against the plane in sequence. Each of the two vertices of the edge may be inside or outside the plane; thus, there are four cases. Each case may generate zero, one, or two output vertices, as shown in Figure 10.39.

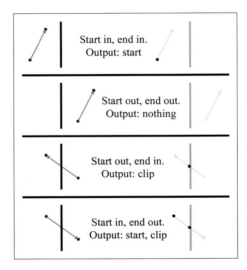

Figure 10.39
Clipping a single edge—the four cases

Figure 10.40 shows an example of how we can apply these rules to clip a polygon against the right clip plane. Remember that the clipper outputs vertices, not edges. In Figure 10.40, the edges are drawn only for illustration. In particular, the final clip step appears to output two edges when actually only one vertex was output—the last edge is implicit to complete the polygon.

At the end of each stage, if there are fewer than three vertices remaining, then the polygon is rejected as being invisible. (Notice that it is impossible to output only one or two vertices. The number of vertices output by any one pass will either be zero, or at least three.)

Some graphics hardware does not clip polygons to all six planes in 3D (or 4D). Instead, only the near clip is performed, and then *scissoring* is done in 2D to clip to the window. This can be a performance win because clipping is slow on certain hardware. A variation on this technique is to employ a *guard band*. Polygons completely outside the screen are rejected, polygons completely inside the guard band are scissored rather than clipped

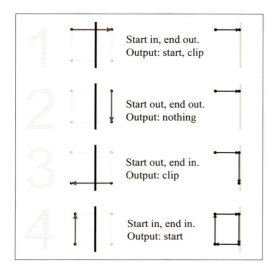

Figure 10.40
Clipping a polygon against the right clip plane

in 3D, and polygons that are partially on screen but outside of the guard band are clipped in 3D.

10.10.5   Backface Culling

The second test used to reject hidden surfaces is known as *backface culling*, and the purpose of this test is to reject triangles that don't face the camera. In standard closed meshes, we should never see the back side of a triangle unless we are allowed to go inside the mesh. Removal of the backfacing triangles is not strictly necessary in an opaque mesh—we could draw them and still generate a correct image, since they will be covered up by a closer, front-facing triangle. However, we don't want to waste time drawing anything that isn't visible, so we usually want to cull backfaces. In theory, about half of the triangles will be backfacing. In practice, less than half of the triangles can be culled, especially in static scenery, which in many cases is created without backfaces in the first place. One obvious example is a terrain system. Certainly we may be able to eliminate some backfacing triangles, for example, on the backside of a hill, but in general most triangles will be frontfacing because we are usually above the ground. However, for dynamic objects that move around in the world freely, roughly half of the faces will be backfacing.

Backfacing triangles can be detected in 3D (before projection) or 2D (after projection). On modern graphics hardware, backface culling is performed in 2D based on clockwise or counterclockwise enumeration of vertices in screen space. In a left-handed coordinate system like we use in this book, the convention is to order the vertices in a clockwise fashion around the triangle when viewed from the front side. Thus, as shown in Figure 10.41, we will normally remove any triangle whose vertices are ordered in a counterclockwise fashion on the screen. (Right-handers usually employ the opposite conventions.)

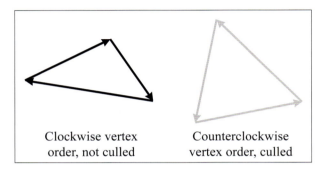

Clockwise vertex
order, not culled

Counterclockwise
vertex order, culled

Figure 10.41
Backface culling of triangles with vertices enumerated counterclockwise in screen space

The API will let you control backface culling. You may want to turn backface culling off while rendering certain geometry. Or, if geometry has been reflected, you may need to invert the culling, since reflection flips the vertex order around the faces. Rendering using stencil shadows requires rendering the frontfaces in one pass and the backfaces in another pass.

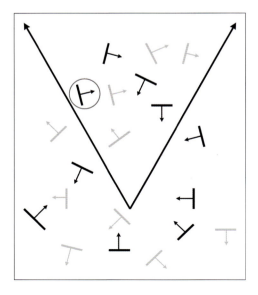

Figure 10.42
Backface culling in 3D

The bottlenecks are different in software rendering compared to hardware rendering (notably, the bandwidth required for raw delivery of data to the hardware), and in software, backface culling is usually done in 3D. The basic idea with the 3D backfacing test is to determine whether the camera position is on the front side of the triangle's plane. To make this determination quickly, we store a precomputed triangle normal. This is shown in Figure 10.42, in which the backfacing triangles that could be culled are drawn in gray. Notice that backface culling doesn't depend on whether a triangle is inside or outside the view frustum. In fact, it doesn't depend on the orientation of the camera at all—only the position of the camera relative to the triangle is relevant.

To detect backfacing triangles in 3D, we need the normal of the plane containing the triangle, and a vector from the eye to the triangle (any point on the triangle will do—usually we just pick one vertex arbitrarily). If these two vectors point in basically the same direction (their dot product is greater than zero), then the triangle is backfacing. A variation on this theme is to also precompute and store the d value of the plane equation (see Section 9.5.1). Then the backfacing check can be done with one dot product and scalar comparison. One quick note about a tempting optimization trick that doesn't work: you might try to only use the z-component of the normal of the triangle in camera (or clip) space. Although it might seem like if the z value is positive, then the triangle faces away from the camera and could be culled, an example where this isn't true is circled in Figure 10.42.

10.10.6   Rasterization, Shading, and Output

After clipping, the vertices are projected and mapped into the screen coordinates of the output window, according to Equations (10.8)–(10.9). Of course, these coordinates are floating-point coordinates, which are "continuous" (see Section 1.1). But we typically render pixels, which are discrete. So how do we know which pixels actually get drawn? Devising an algorithm to answer this question is surprisingly complicated. If we answer wrong, then gaps can appear between triangles. Rendering a pixel more than once can be bad, too, if we are using alpha blending. In other words, we must make sure that when we render a surface represented as triangles, every pixel is rendered *exactly once*. Luckily, the graphics hardware takes care of this for us and we don't have to sweat the details.

During rasterization, the rendering system may perform *scissoring*, which rejects pixels that are outside of the rendering window. This is impossible if the polygon is clipped to the edge of the screen, but it might be advantageous for performance reasons to skip that step. The guard band is a technique that can be used to tune performance trade-offs between clipping and scissoring (see Section 10.10.4).

Even though we don't necessarily have to understand exactly how the graphics hardware decides which pixels to render for a given triangle, we do need to understand how it determines what to do with a single pixel. Conceptually, five basic steps are performed:

1. *Interpolate.* Any quantities computed at the vertex level, such as texture coordinates, colors, and normals, are interpolated across the face. The interpolated values for each quantity must be computed for the pixel before it can be shaded.

2. *Depth test.* We reject pixels by using the depth buffer (see Section 10.10.1) if the pixel we are about to shade would be obscured by a closer pixel. Note that, in some circumstances, the pixel shader is allowed to modify the depth value, in which case this test must be deferred until after shading.

3. *Shade.* *Pixel shading* refers to the process of computing a color for a pixel. On shader-based hardware, this is where your pixel shader is executed. In a basic forward renderer pass, where we are actually rendering objects into the frame buffer (as opposed to writing into a shadow map, or doing some other lighting pass), the pixel is typically first lit and then fogged if fogging is being used. The output of a pixel shader consists of not only an RGB color, but also an alpha value, which is usually interpreted as the "opacity" of the pixel, used for blending. The next section shows several examples of pixel shaders.

4. *Alpha test.* This rejects pixels based on the alpha value of the pixel. All sorts of different alpha tests can be used, but the most common one is to reject pixels that are "too transparent." Although such invisible pixels may not cause any change the frame buffer were we to write them, we do need to reject them so that they do not cause us to write into the depth buffer.

5. *Write.* If the pixel passes the depth and alpha tests, then the frame buffer and depth buffers are updated.

   - The depth buffer is updated simply by replacing the old depth value with the new one.
   - The frame buffer update is more complicated. If blending is not used, then the new pixel color replaces the old one. Otherwise, the new pixel color is blended with the old one, with the relative contributions of the old and new colors controlled by the alpha value. Other mathematical operations, such as addition, subtraction, and multiplication, are also often available, depending on the graphics hardware.

10.11 Some HLSL Examples

In this section we will present some examples of HLSL vertex and pixel shaders that demonstrate many of the techniques discussed in previous sections. These code snippets are well commented, as we intend for this code to be *read*. We are giving examples in HLSL for the same reason that we show code snippets in C: we expect that it will be applicable to a large number of our readers, and although we know not every reader will be using this specific language, we think the language is sufficiently high level that many of the basic principles can be conveyed and appreciated by almost everyone.

HLSL is essentially the same language as the shader language developed by NVIDIA known as "Cg." HLSL is also very similar, although not identical, to GLSL, the shading language used in OpenGL.

One aspect of HLSL that we realize introduces an unwanted impediment to those not interested in real-time rendering is the division of labor between vertex and pixel shaders.[24] Unfortunately, this is where some of the messy guts cannot be fully hidden. This book is *not* a book on HLSL, so we don't fully explain these details, and some exposure to HLSL is helpful. However, since the language uses the C syntax, it is relatively approachable, and our examples should be readable. For those who are unfamiliar with HLSL, the comments in the examples introduce the HLSL specifics as they come up.

Because these examples are all very basic, they were written targeting shader model 2.0.

10.11.1 Decal Shading and HLSL Basics

We'll start with a very simple example to get warmed up and demonstrate the basic mechanisms of HLSL for declaring constants and passing interpolated arguments. Perhaps the simplest type of shading is to just output the color from a texture map directly, without any lighting at all. This is sometimes called *decal* shading. The vertex shader in Listing 10.11 illustrates several of the basic mechanisms of HLSL, as explained by the source comments.

---

[24]For example, the *RenderMan* shading language does not have this property.

```
// This struct declares the inputs we receive from the mesh.
// Note that the order here is not important.  Inputs are identified
// by their "semantic", which is the thing on the right after the
// colon.  When sending the vertex list to the renderer, we must
// specify the "semantic" of each vertex element.
struct Input {
  float4 pos : POSITION;   // position in modeling space
  float2 uv  : TEXCOORD0;  // texture coordinates
};

// This is the data we will output from our vertex shader.  These
// are matched up with the pixel shader inputs based on the semantic
struct Output {
  float4 pos : POSITION;   // position in CLIP space
  float2 uv  : TEXCOORD0;  // texture coordinate
};

// Here we declare a global variable, which is a shader constant
// that holds the model->clip transform matrix.
uniform float4x4 modelToClip;

// The body of our vertex shader
Output main(Input input) {
  Output output;

  // Transform vertex position to clip space.  The mul() intrinsic
  // function performs matrix multiplication.  Note that mul()
  // treats any vector passed as the first operand as a row vector.
  output.pos = mul(input.pos, modelToClip);

  // Pass through the supplied UV coordinates without modification
  output.uv = input.uv;

  return output;
}
```

Listing 10.11
Vertex shader for decal rendering

A vertex shader like this could be be used with the pixel shader of
Listing 10.12, which actually does the decal shading. However, to make
things interesting and demonstrate that pixel shader constants work the
same as vertex shader constants, we've added a global constant color, which
we consider to be part of the global render context. We have found it very
useful to have a constant such as this, which modulates the color and opacity
of every rendered primitive.

```
// This struct declares the interpolated inputs we receive from the
// rasterizer.  They will usually match the vertex outputs exactly,
// except that we often leave off the clip-space position.
struct Input {
  float2 uv : TEXCOORD0;  // texture coordinates
};

// Here, just to show how a pixel shader constant works, we declare
// a global constant color.  The output of our shader is multiplied
// by this RGBA value.  One of the most common reasons to have such a
// constant is to add an opacity setting into the render context,
// but it's very handy to have a full RGBA constant color.
```

```
uniform float4 constantColor;

// We are going to do a texture lookup.  Here we declare a "variable"
// to refer to this texture, and annotate it with sufficient information
// so that our rendering code can select the appropriate texture into
// the rendering context before drawing primitives.
sampler2D diffuseMap;

// The body of our pixel shader.  It only has one output, which falls
// under the semantic "COLOR"
float4 main(Input input): COLOR {

    // Fetch the texel
    float4 texel = tex2D(diffuseMap, input.uv);

    // Modulate it by the constant color and output. Note that
    // operator* performs component-wise multiplication.
    return texel*constantColor;
}
```

Listing 10.12
Pixel shader for decal rendering

Clearly, the higher-level code must supply the shader constants and the primitive data properly. The simplest way to match up a shader constant with the higher-level code is to specifically assign a register number to a constant by using special HLSL variable declaration syntax, but there are subtler techniques, such as locating constants by name. These practical details are certainly important, but they are don't belong in this book.

10.11.2 Basic Per-Pixel Blinn-Phong Lighting

Now let's look at a simple example that actually does some lighting calculations. We start with basic per-pixel lighting, although we don't use a bump map just yet. This example simply illustrates the Phong shading technique of interpolating the normal across the face and evaluating the full lighting equation per-pixel. We compare Phong shading with Gouraud shading in Section 10.11.3, and we show an example of normal mapping in Section 10.11.4.

All of our lighting examples use the standard Blinn-Phong lighting equation. In this example and most of the examples to follow, the lighting environment consists of a single linearly-attenuated omni light plus a constant ambient.

For the first example (Listings 10.13 and 10.14), we do all of the work in the pixel shader. In this case, the vertex shader is fairly trivial; it just needs to pass through the inputs to the pixel shader.

```
// Mesh inputs.
struct Input {
    float4 pos    : POSITION;  // position in model space
    float3 normal : NORMAL;    // vertex normal in model space
    float2 uv     : TEXCOORD0; // texture coords for diffuse, spec maps
};
```

```
// Vertex shader output.  Note that with the exception of the output
// position, which is output under the POSITION sematic, all others go
// under the TEXCOORDx semantic.  Despite its name, this semantic is
// actually used for pretty much ANY interpolated vector value up to 4D
// that we want to pass to the pixel shader, not just texture coords.
struct Output {
    float4 clipPos  : POSITION;   // clip-space position
    float2 uv       : TEXCOORD0;  // texture coords for diffuse, spec maps
    float3 normal   : TEXCOORD1;  // vertex normal in model space
    float3 modelPos : TEXCOORD2;  // position in model space
};

// Model->clip transform matrix.
uniform float4x4 modelToClip;

// The body of our vertex shader
Output main(Input input) {
    Output output;

    // Transform vertex position to clip space.
    output.clipPos = mul(input.pos, modelToClip);

    // Pass through vertex inputs without modification
    output.normal = input.normal;
    output.uv = input.uv;
    output.modelPos = input.pos;

    return output;
}
```

Listing 10.13
Vertex shader for per-pixel lighting of a single omni plus ambient

Listing 10.14 is the corresponding pixel shader, where all the action happens. Notice that we are using two different texture maps, one for the diffuse color and another for the specular color. We assume that the two maps use the same texture-mapping coordinates.

```
// Interpolated inputs from the vertex shader.
struct Input {
    float2 uv       : TEXCOORD0;  // texture coords for diffuse, spec maps
    float3 normal   : TEXCOORD1;  // vertex normal in model space
    float3 modelPos : TEXCOORD2;  // model space position (for lighting)
};

// A global constant RGB and opacity
uniform float4 constantColor;

// Omni light position, in MODEL space,
uniform float3 omniPos;

// Reciprocal of omni light radius.  (The light will falloff
// linearly to zero at this radius).  Note that it's common to tuck
// this into the w component of the position, to reduce the number of
// constants, since each  constant usually takes a full 4D vector slot.
uniform float invOmniRad;
```

```
// Unattenuated omni light color
uniform float3 omniColor;

// View position, in MODEL space
uniform float3 viewPos;

// Constant ambient light color
uniform float3 ambientLightColor;

// Material glossiness (phong exponent)
uniform float specExponent;

// Diffuse and specular map samplers.  Note we assume that diffuse
// and spec maps use the same UV coords
sampler2D diffuseMap;
sampler2D specularMap;

// Pixel shader body
float4 main(Input input): COLOR {

  // Fetch the texels to get the material colors
  float4 matDiff = tex2D(diffuseMap, input.uv);
  float4 matSpec = tex2D(specularMap, input.uv);

  // Normalize interpolated vertex normal
  float3 N = normalize(input.normal);

  // Compute vector to light
  float3 L = omniPos - input.modelPos;

  // Normalize it, and save off distance to use later
  // for attenuation
  float dist = length(L);
  L /= dist;

  // Compute view vector and halfway vector
  float3 V = normalize(viewPos - input.modelPos);
  float3 H = normalize(V + L);

  // Compute attenuated light color.
  float3 lightColor = omniColor * max(1 - dist*invOmniRad,0);

  // Compute diffuse and specular factors
  float diffFactor = max(dot(N, L),0);
  float specFactor = pow(max(dot(N,H),0), specExponent);

  // Compute effective light colors
  float3 diffColor = lightColor*diffFactor + ambientLightColor;
  float3 specColor = lightColor*specFactor;

  // Sum up colors.  Note that HLSL has a very flexible swizzling system
  // which allows us to access a portion of a vector as if were a
  // "member" of the vector
  float4 result = matDiff; // RGB and opacity from diffuse map
  result.rgb *= diffColor; // modulate by diffuse+ambient lighting
  result.rgb += matSpec.rgb*specColor; // add spec, ignoring map alpha

  // Modulate it by the constant and output
  return result*constantColor;
}
```

Listing 10.14
Pixel shader for per-pixel lighting of a single omni plus ambient

Of course, several of the values needed in this calculation could be computed in the vertex shader, and we could use the interpolated results in the pixel shader. This is *usually* a performance win because we assume that most of our triangles fill more than a pixel or two, so that number of pixels to fill is significantly more than the number of vertices to shade. However, a precise analysis can be complicated because the number of vertices and pixels is not the only factor; the number of execution units available for vertex and pixel shading is also important. Furthermore, on some hardware, a pool of generic execution units are shared between vertex and pixel shading. There can also be performance implications for increasing the number of interpolated values. Still, dividing up the labor to do more calculations per vertex is a speedup on most platforms and in most situations. Listings 10.15 and 10.16 show one way we could shift work up to the vertex shader.

```
// Mesh inputs.
struct Input {
  float4 pos    : POSITION;  // position in model space
  float3 normal : NORMAL;    // vertex normal in model space
  float2 uv     : TEXCOORD0; // texture coords for diffuse, spec maps
};

// Vertex shader output
struct Output {
  float4 clipPos    : POSITION;  // clip-space position
  float2 uv         : TEXCOORD0; // texture coords for diffuse, spec maps
  float3 normal     : TEXCOORD1; // vertex normal in model space
  float3 L          : TEXCOORD2; // vector to light
  float3 H          : TEXCOORD3; // halfway vector
  float3 lightColor : TEXCOORD4; // light color + attenuation factor
};

// Model->clip transform matrix.
uniform float4x4 modelToClip;

// Omni light position, in MODEL space,
uniform float3 omniPos;

// Reciprocal of omni light radius. (The light will falloff
// linearly to zero at this radius). Note that it's common to tuck
// this into the w component of the position, to reduce the number of
// constants, since each constant usually takes a full 4D vector slot.
uniform float invOmniRad;

// Unattenuated omni light color
uniform float3 omniColor;

// View position, in MODEL space
uniform float3 viewPos;

// The body of our vertex shader
Output main(Input input) {
  Output output;

  // Transform vertex position to clip space.
  output.clipPos = mul(input.pos, modelToClip);
```

```
// Compute vector to light
float3 L = omniPos − input.pos;

// Normalize it , and save off distance to use later
// for attenuation
float dist = length(L);
output.L = L / dist;

// Compute view and halfway vector.
float3 V = normalize(viewPos − input.pos);
output.H = normalize(V + output.L);

// Compute attenuation factor.  Note that we do NOT clamp
// to zero here, we will do that in the pixel shader.  This
// is important in case the falloff reaches zero in the middle
// of a large polygon.
float attenFactor = 1 − dist*invOmniRad;
output.lightColor = omniColor * attenFactor;

// Pass through other vertex inputs without modification
output.normal = input.normal;
output.uv = input.uv;

return output;
}
```

Listing 10.15
Alternate vertex shader for per-pixel lighting of a single omni plus ambient

Now the pixel shader has less work to do. According to the DirectX 10 FXC compiler, the pixel shader in Listing 10.16 compiles to approximately 25 instruction slots, compared to 33 instruction slots for Listing 10.14.

```
// Interpolated inputs from the vertex shader.
struct Input {
  float2 uv         : TEXCOORD0; // texture coords for diffuse , spec maps
  float3 normal     : TEXCOORD1; // vertex normal in model space
  float3 L          : TEXCOORD2; // vector to light
  float3 H          : TEXCOORD3; // halfway vector
  float3 lightColor : TEXCOORD4; // light color + attenuation factor
};

// A global constant RGB and opacity
uniform float4 constantColor;

// Constant ambient light color
uniform float3 ambientLightColor;

// Material glossiness (phong exponent)
uniform float specExponent;

// Diffuse and specular map samplers.  Note we assume that diffuse
// and spec maps use the same UV coords
sampler2D diffuseMap;
sampler2D specularMap;

// Pixel shader body
float4 main(Input input): COLOR {

  // Fetch the texels to get the material colors
  float4 matDiff = tex2D(diffuseMap , input.uv);
```

```
float4 matSpec = tex2D(specularMap, input.uv);

// Normalize interpolated vectors
float3 N = normalize(input.normal);
float3 L = normalize(input.L);
float3 H = normalize(input.H);

// Compute diffuse and specular factors
float diffFactor = max(dot(N, L),0);
float specFactor = pow(max(dot(N,H),0), specExponent);

// Clamp the light color, (Note that this max is applied
// component-wize)
float3 lightColor = max(input.lightColor,0);

// Compute effective light colors
float3 diffColor = lightColor*diffFactor + ambientLightColor;
float3 specColor = lightColor*specFactor;

// Sum up colors. Note that HLSL has a very flexible swizzling system
// which allows us to access a portion of a vector as if were a
// "member" of the vector
float4 result = matDiff; // RGB and opacity from diffuse map
result.rgb *= diffColor; // modulate by diffuse+ambient lighting
result.rgb += matSpec.rgb*specColor; // add spec, ignoring map alpha

// Modulate it by the constant and output
return result*constantColor;
}
```

Listing 10.16
Alternate pixel shader for per-pixel lighting of a single omni plus ambient

Finally, we present one last variation on this example. Notice that in the previous pixel shader, Listing 10.16, the code does not assume that the lighting is taking place in any particular coordinate space. We have been performing the lighting calculations in model space, but it is also common to do it in camera space. The advantage is that we do not need to resend shader constants for lighting data for each object that is rendered, as we do when those values are specified in modeling space (which will vary for each object). Listing 10.17 is a vertex shader that illustrates this technique.

```
// Mesh inputs.
struct Input {
  float4 pos     : POSITION;  // position in model space
  float3 normal : NORMAL;     // vertex normal in model space
  float2 uv      : TEXCOORD0; // texture coords for diffuse, spec maps
};

// Vertex shader output
struct Output {
  float4 clipPos    : POSITION;  // clip-space position
  float2 uv         : TEXCOORD0; // texture coords for diffuse, spec maps
  float3 normal     : TEXCOORD1; // vertex normal in camera space
  float3 L          : TEXCOORD2; // vector to light in camera space
  float3 H          : TEXCOORD3; // halfway vector in camera space
  float3 lightColor : TEXCOORD4; // light color + attenuation factor
};
```

```
// Model->view transform matrix.  (The "modelview" matrix)
uniform float4x4 modelToView;

// Clip matrix.  (The "projection" matrix).
uniform float4x4 viewToClip;

// Omni light position, in VIEW space, and reciprocal of
// falloff in the w component
uniform float4 omniPosAndInvRad;

// Unattenuated omni light color
uniform float3 omniColor;

// The body of our vertex shader
Output main(Input input) {
  Output output;

  // Transform vertex position to view space.
  float4 vPos = mul(input.pos, modelToView);

  // And into clip space.  Note that the clip matrix
  // often has a simple structure which can be exploited
  // and the number of vector operations can be reduced.
  output.clipPos = mul(vPos, viewToClip);

  // Transform normal to camera space.  We "promote" the normal
  // to float4 by setting w to 0, so it will receive any translation
  output.normal = mul(float4(input.normal,0), modelToView);

  // Compute vector to light
  float3 L = omniPosAndInvRad.xyz - vPos;

  // Normalize it, and save off distance to use later
  // for attenuation
  float dist = length(L);
  output.L = L / dist;

  // Compute view and halfway vector.
  // Note that the view position is the origin,
  // in view space, by definition
  float3 V = normalize(-vPos);
  output.H = normalize(V + output.L);

  // Compute attenuation factor.  Note that we do NOT clamp
  // to zero here, we will do that in the pixel shader.  This
  // is important in case the falloff reaches zero in the middle
  // of a large polygon.
  float attenFactor = 1 - dist*omniPosAndInvRad.w;
  output.lightColor = omniColor * attenFactor;

  // Pass through UV's without modification
  output.uv = input.uv;

  return output;
}
```

Listing 10.17
Vertex shader for per-pixel lighting of a single omni plus ambient, calculated in camera space

World space ("upright space") is an attractive option for lighting calculations in many circumstances because shadow cube maps or lighting probes are usually rendered in this orientation; it also has the advantage that we do not need to resend lighting-related shader constants due to change of the model reference frame for each object.

10.11.3   Gouraud Shading

Even modest modern hardware has plenty of beef for Phong shading; indeed, the previous examples are relatively cheap shaders. However, it's very instructive to consider how to implement Gouraud shading. Even though the results are inferior to Phong shading, and Gouraud shading precludes bump mapping, Gouraud shading can still be useful on the PC to emulate the results of other hardware.

Listing 10.18 is a vertex shader that performs the same lighting calculations as just demonstrated in Section 10.11.2, only they are done at the vertex level. Compare this shader code to Equation (10.15).

```
// Mesh inputs.
struct Input {
  float4 pos    : POSITION;   // position in model space
  float3 normal : NORMAL;     // vertex normal in model space
  float2 uv     : TEXCOORD0;  // texture coords for diffuse, spec maps
};

// Vertex shader output
struct Output {
  float4 clipPos   : POSITION;   // clip-space position
  float2 uv        : TEXCOORD0;  // texture coords for diffuse, spec maps
  float3 diffColor : TEXCOORD1;  // diffuse lighting RGB
  float3 specColor : TEXCOORD2;  // specular lighting RGB
};

// Model->clip transform matrix.
uniform float4x4 modelToClip;

// Omni light position, in MODEL space, and reciprocal of
// falloff in the w component
uniform float4 omniPosAndInvRad;

// Unattenuated omni light color
uniform float3 omniColor;

// Constant ambient light color
uniform float3 ambientLightColor;

// View position, in MODEL space
uniform float3 viewPos;

// Material glossiness (phong exponent)
uniform float specExponent;

// The body of our vertex shader
Output main(Input input) {
  Output output;

  // Transform vertex position to clip space.
  output.clipPos = mul(input.pos, modelToClip);

  // Compute vector to light
```

```
float L = omniPosAndInvRad.xyz − input.pos;

// Normalize it, and save off distance to use later
// for attenuation
float dist = length(L);
L /= dist;

// Compute view and halfway vector
float3 V = normalize(viewPos − input.pos);
float3 H = normalize(V + L);

// Compute attenuated light color.
float3 lightColor = omniColor * max(1 − dist*omniPosAndInvRad.w,0);

// Compute diffuse and specular factors
float diffFactor = max(dot(input.normal, L),0);
float specFactor = pow(max(dot(input.normal,H),0), specExponent);

// Compute effective light colors
output.diffColor = lightColor*diffFactor + ambientLightColor;
output.specColor = lightColor*specFactor;

// Pass through the supplied UV coordinates without modification
output.uv = input.uv;

return output;
}
```

Listing 10.18
Vertex shader for Gouraud shading of a single omni plus ambient

Now the pixel shader (Listing 10.19) simply takes the lighting results and modulates by the material diffuse and specular colors, from the texture maps.

```
// Interpolated inputs from the vertex shader.
struct Input {
    float2 uv        : TEXCOORD0; // texture coords for diffuse, spec maps
    float3 diffColor : TEXCOORD1; // diffuse lighting RGB
    float3 specColor : TEXCOORD2; // specular lighting RGB
};

// A global constant RGB and opacity
uniform float4 constantColor;

// Diffuse and specular map samplers. Note that we assume that diffuse
// and spec maps are mapped the same, and so they use the same UV coords
sampler2D diffuseMap;
sampler2D specularMap;

// Pixel shader body
float4 main(Input input): COLOR {

    // Fetch the texels to get the material colors
    float4 materialDiff = tex2D(diffuseMap, input.uv);
    float4 materialSpec = tex2D(specularMap, input.uv);

    // Sum up colors. Note that HLSL has a very flexible swizzling system
    // which allows us to access a portion of a vector as if were a
    // "member" of the vector
    float4 result = materialDiff; // RGB *and* opacity from diffuse map
```

```
  result.rgb *= input.diffColor; // modulate by diffuse+ambient lighting
  result.rgb +=
    materialSpec.rgb*input.specColor; // add spec, ignore map alpha

  // Modulate it by the constant and output
  return result*constantColor;
}
```

Listing 10.19
Pixel shader for Gouraud shading for any lighting environment

As the caption for Listing 10.19 indicates, this pixel shader does not
depend on the number of lights, or even the lighting model, since all lighting
calculations are done in the vertex shader. Listing 10.20 shows a vertex
shader that could be used with this same pixel shader, but it implements a
different lighting environment: ambient plus three directional lights. This
is a very useful lighting environment in editors and tools, since it's easy to
create one lighting rig that works decently well for practically any object
(although we would usually use it with per-pixel shading).

```
// Mesh inputs.
struct Input {
  float4 pos    : POSITION;  // position in model space
  float3 normal : NORMAL;    // vertex normal in model space
  float2 uv     : TEXCOORD0; // texture coords for diffuse, spec maps
};

// Vertex shader output
struct Output {
  float4 clipPos   : POSITION;  // clip-space position
  float2 uv        : TEXCOORD0; // texture coords for diffuse, spec maps
  float3 diffColor : TEXCOORD1; // diffuse lighting RGB
  float3 specColor : TEXCOORD2; // specular lighting RGB
};

// Model->clip transform matrix.
uniform float4x4 modelToClip;

// Three light directions (in MODEL space).  These point
// in the opposite direction that the light is shining.
uniform float3 lightDir[3];

// Three light RGB colors
uniform float3 lightColor[3];

// Constant ambient light color
uniform float3 ambientLightColor;

// View position, in MODEL space
uniform float3 viewPos;

// Material glossiness (phong exponent)
uniform float specExponent;

// The body of our vertex shader
Output main(Input input) {
  Output output;

  // Transform vertex position to clip space.
  output.clipPos = mul(input.pos, modelToClip);
```

```
// Compute the V vector
float3 V = normalize(viewPos − input.pos);

// Clear accumulators.
output.diffColor = ambientLightColor;
output.specColor = 0;

// Sum up lights.  Note that the compiler is *usually* pretty
// good at unrolling small loops like this, but to ensure
// the fastest code, it's best not to depend on the compiler,
// and unroll the loop yourself
for (int i = 0 ; i < 3 ; ++i) {

    // Compute lambert term and sum diffuse contrib
    float nDotL = dot(input.normal, lightDir[i]);
    output.diffColor += max(nDotL,0) * lightColor[i];

    // Compute halfway vector
    float3 H = normalize(V + lightDir[i]);

    // Sum  specular contrib
    float nDotH = dot(input.normal,H);
    float s = pow(max(nDotH,0), specExponent);
    output.specColor += s*lightColor[i];
}

// Pass through the supplied UV coordinates without modification
output.uv = input.uv;

return output;
}
```

Listing 10.20
Vertex shader for Gouraud shading, using constant ambient plus three directional lights

10.11.4 Bump Mapping

Next, let's look at an example of normal mapping. We will be performing the lighting in tangent space, and we'll stick with the lighting environment of a single omni light plus constant ambient to make the examples easier to compare. In the vertex shader (Listing 10.21), we synthesize the binormal from the normal and tangent. Then, we use the three basis vectors to rotate L and H into tangent space, after first computing them as usual in model space. Notice the use of the three dot products, which is equivalent to multiplication by the transpose of the matrix. We also perform the attenuation calculations in the vertex shader, passing the unclamped attenuated light color, as we have done in previous examples.

```
// Mesh inputs.
struct Input {
    float4 pos         : POSITION;    // position in model space
    float3 normal      : NORMAL;      // vertex normal in model space
    float4 tangentDet  : TANGENT;     // tangent in model space, det in w
    float2 uv          : TEXCOORD0;   // texture coords for diffuse, spec maps
};
// Vertex shader output
struct Output {
    float4 clipPos     : POSITION;    // clip−space position
```

```
    float2 uv          : TEXCOORD0;  // texture coords for all maps
    float3 L           : TEXCOORD1;  // vector to light, in TANGENT space
    float3 H           : TEXCOORD2;  // halfway vector, in TANGENT space
    float3 lightColor  : TEXCOORD3;  // light color & attenuation factor
};

// Model->clip transform matrix.
uniform float4x4 modelToClip;

// Omni light position, in MODEL space, and reciprocal of
// falloff in the w component
uniform float4 omniPosAndInvRad;

// Unattenuated omni light color
uniform float3 omniColor;

// View position, in MODEL space
uniform float3 viewPos;

// The body of our vertex shader
Output main(Input input) {
  Output output;

    // Transform vertex position to clip space.
    output.clipPos = mul(input.pos, modelToClip);

    // Compute vector to light (in model space)
    float3 L_model = omniPosAndInvRad.xyz - input.pos.xyz;

    // Normalize it, and save off distance to use later
    // for attenuation
    float dist = length(L_model);
    float3 L_model_norm = L_model / dist;

    // Compute view and halfway vector
    float3 V_model = normalize(viewPos - input.pos);
    float3 H_model = normalize(V_model + L_model_norm);

    // Reconstruct the third basis vector
    float3 binormal =
      cross(input.normal, input.tangentDet.xyz) * input.tangentDet.w;

    // Rotate lighting-related vectors into tangent space
    output.L.x = dot(L_model, input.tangentDet.xyz);
    output.L.y = dot(L_model, binormal);
    output.L.z = dot(L_model, input.normal);

    output.H.x = dot(H_model, input.tangentDet.xyz);
    output.H.y = dot(H_model, binormal);
    output.H.z = dot(H_model, input.normal);

    // Compute UNCLAMPED color + attenuation factor.
    float attenFactor = 1 - dist*omniPosAndInvRad.w;
    output.lightColor = omniColor * attenFactor;

    // Pass through mapping coords without modification
    output.uv = input.uv;

    return output;
}
```

Listing 10.21
Vertex shader for omni lighting of normal mapped object, with lighting done in tangent space

The pixel shader (Listing 10.22) is quite compact, since most of the prep work has been done in the vertex shader. We unpack the normal and normalize the interpolated L and H vectors. Then we perform the Blinn-Phong lighting equation, just as in the other examples.

```
// Interpolated inputs from the vertex shader.
struct Input {
  float2 uv          : TEXCOORD0; // texture coords for all maps
  float3 L           : TEXCOORD1; // vector to light, in TANGENT space
  float3 H           : TEXCOORD2; // halfway vector, in TANGENT space
  float3 lightColor  : TEXCOORD3; // light color + and attenuation factor
};

// A global constant RGB and opacity
uniform float4 constantColor;

// Constant ambient light color
uniform float3 ambientLightColor;

// Material glossiness (phong exponent)
uniform float specExponent;

// Diffuse, spec, and normal map samplers
sampler2D diffuseMap;
sampler2D specularMap;
sampler2D normalMap;

// Pixel shader body
float4 main(Input input): COLOR {

  // Fetch the texels to get the material colors
  float4 matDiff = tex2D(diffuseMap, input.uv);
  float4 matSpec = tex2D(specularMap, input.uv);

  // Decode the tangent-space normal
  float3 N = tex2D(normalMap, input.uv).rgb * 2 - 1;

  // Normalize interpolated lighting vectors
  float3 L = normalize(input.L);
  float3 H = normalize(input.H);

  // Compute diffuse and specular factors
  float diffFactor = max(dot(N, L),0);
  float specFactor = pow(max(dot(N,H),0), specExponent);

  // Clamp the light color and attenuation
  float3 lightColor = max(input.lightColor,0);

  // Compute effective light colors
  float3 diffColor = lightColor*diffFactor + ambientLightColor;
  float3 specColor = lightColor*specFactor;

  // Sum up colors.
  float4 result = matDiff; // RGB & opacity from the diffuse map
  result.rgb *= diffColor; // modulate by diffuse+ambient lighting
  result.rgb += matSpec.rgb*specColor; // add spec, ignore map alpha

  // Modulate it by the constant and output
  return result*constantColor;
}
```

Listing 10.22
Pixel shader for omni lighting of normal mapped object, with lighting done in tangent space

10.11.5   Skinned Mesh

Now for some examples of skeletal rendering. All of the skinning happens in the vertex shaders, and so we will not need to show any pixel shaders here; the vertex shaders here can be used with the pixel shaders given previously. This is not unusual: skinned and unskinned geometry can usually share the same pixel shader. We give two examples. The first example (Listing 10.23) illustrates per-pixel lighting of our omni + ambient lighting rig. We will do all the lighting in the pixel shader (Listing 10.14), so that we can focus on the skinning, which is what is new.

```
// Mesh inputs.
struct Input {
    float4 pos      : POSITION;      // model space position (binding pose)
    float3 normal   : NORMAL;        // model space vertex normal (ditto)
    byte4  bones    : BLENDINDICES;  // Bone indices. Unused entries are 0
    float4 weight   : BLENDWEIGHT;   // Blend weights. Unused entries are 0
    float2 uv       : TEXCOORD0;     // texture coords for diffuse, spec maps
};

// Vertex shader output.
struct Output {
    float4 clipPos  : POSITION;      // clip-space position (for rasterization)
    float2 uv       : TEXCOORD0;     // texture coords for diffuse, spec maps
    float3 normal   : TEXCOORD1;     // vertex normal in model space
    float3 modelPos : TEXCOORD2;     // position in model space (for lighting)
};

// Model->clip transform matrix.
uniform float4x4 modelToClip;

// Declare an arbitrary max number of bones.
#define MAX_BONES 40

// Array of ''binding pose -> current'' pose matrices for each bone.
// These are 4x3 matrices, which we iterpret as 4x4 matrices with the
// rightmost column assumed to be [0,0,0,1]. Note we are assuming
// that column_major is the default storage —— meaning each column
// is stored in a 4D register. Thus each matrix takes 3 registers.
uniform float4x3 boneMatrix[MAX_BONES];

// The body of our vertex shader
Output main(Input input) {
    Output output;

    // Generate a blended matrix. Notice that we always blend 4 bones,
    // even though most vertices will use fewer bones. Whether its
    // faster to use conditional logic to try to bypass this extra logic,
    // or if it's better to just to all of calculations (which can be
    // easily scheduled by the assembler to hide any instruction
    // latency) will depend on the hardware.
    float4x3 blendedMat =
    boneMatrix[input.bones.x]*input.weight.x
      + boneMatrix[input.bones.y]*input.weight.y
      + boneMatrix[input.bones.z]*input.weight.z
      + boneMatrix[input.bones.w]*input.weight.w;

    // Perform skinning to transform position and normal
    // from their binding pose position into the position
```

```
// for the current pose.  Note the matrix  multiplication
// [1x3] = [1x4] x [4x3]
output.modelPos = mul(input.pos, blendedMat);
output.normal = mul(float4(input.normal,0), blendedMat);
output.normal = normalize(output.normal);

// Transform vertex position to clip space.
output.clipPos = mul(float4(output.modelPos,1), modelToClip);

// Pass through UVs
output.uv = input.uv;

    return output;
}
```

Listing 10.23
Vertex shader for skinned geometry

We have declared the vertices as an array of vertex shader constants, and sending all these matrices to the hardware can be a significant performance bottleneck. On certain platforms there are more efficient ways of doing this, such as indexing into an auxiliary "vertex" stream.

Next, let's show how to use normal mapping on a skinned mesh. The vertex shader in Listing 10.24 could be used with the pixel shader in Listing 10.22.

```
// Mesh inputs.
struct Input {
    float4 pos          : POSITION;      // model space posn (binding pose)
    float3 normal       : NORMAL;        // vertex normal in model space
    float4 tangentDet   : TANGENT;       // model space tangent, det in w
    byte4  bones        : BLENDINDICES;  // Bone indices.  Unused entries 0
    float4 weight       : BLENDWEIGHT;   // Blend weights.  Unused entries 0
    float2 uv           : TEXCOORD0;     // texture coords for diff, spec maps
};

// Vertex shader output
struct Output {
    float4 pos          : POSITION;    // clip−space position
    float2 uv           : TEXCOORD0;   // texture coords for all maps
    float3 L            : TEXCOORD1;   // vector to light, in TANGENT space
    float3 H            : TEXCOORD2;   // halfway vector, in TANGENT space
    float3 lightColor   : TEXCOORD3;   // light color + and attenuation factor
};

// Model−>clip transform matrix.
uniform float4x4 modelToClip;

// Array of ''binding pose −> current'' pose matrices for each bone.
#define MAX_BONES 40
uniform float4x3 boneMatrix[MAX_BONES];

// Omni light position, in MODEL space, and reciprocal of
// falloff in the w component
uniform float4 omniPosAndInvRad;

// Unattenuated omni light color
uniform float3 omniColor;
```

```
// View position, in MODEL space
uniform float3 viewPos;

// The body of our vertex shader
Output main(Input input) {
  Output output;

  // Generate a blended matrix.
  float4x3 blendedMat =
    boneMatrix[input.bones.x]*input.weight.x
    + boneMatrix[input.bones.y]*input.weight.y
    + boneMatrix[input.bones.z]*input.weight.z
    + boneMatrix[input.bones.w]*input.weight.w;

  // Perform skinning to get values in model space,
  // in the current pose
  float3 pos = mul(input.pos, blendedMat);
  float3 normal = normalize(mul(float4(input.normal,0), blendedMat));
  float3 tangent =
    normalize(mul(float4(input.tangentDet.xyz,0), blendedMat));

  // Transform vertex position to clip space.
  output.pos = mul(float4(pos,1), modelToClip);

  // Compute vector to light (in model space)
  float3 L_model = omniPosAndInvRad.xyz - pos;

  // Normalize it, and save off distance to use later
  // for attenuation
  float dist = length(L_model);
  float3 L_model_norm = L_model / dist;

  // Compute view and halfway vector
  float3 V_model = normalize(viewPos - pos);
  float3 H_model = normalize(V_model + L_model_norm);

  // Reconstruct the third basis vector
  float3 binormal = cross(normal, tangent) * input.tangentDet.w;

  // Rotate lighting-related vectors into tangent space
  output.L.x = dot(L_model, tangent);
  output.L.y = dot(L_model, binormal);
  output.L.z = dot(L_model, normal);

  output.H.x = dot(H_model, tangent);
  output.H.y = dot(H_model, binormal);
  output.H.z = dot(H_model, normal);

  // Compute UNCLAMPED color + attenuation factor.
  float attenFactor = 1 - dist*omniPosAndInvRad.w;
  output.lightColor = omniColor * attenFactor;

  // Pass through mapping coords without modification
  output.uv = input.uv;

  return output;
}
```

Listing 10.24
Vertex shader for skinned, normal-mapped geometry

10.12 Further Reading

A student seeking a good background in graphics is encouraged to divide his or her reading across the spectrum, from "ivory tower" theoretical principles at one end, to "here is some source code that runs on a particular platform and probably will be obsolete in 5 years" on the other. We have made an attempt here to select, from the large body of graphics literature, just a few sources that are especially recommended.

Fundamentals of Computer Graphics [61] by Shirley provides a solid introductory survey of the fundamentals. Written by one of the field's founding fathers, it is used as the first-year textbook for graphics courses at many universities, and is our recommendation for those near the beginning of their graphics education.

Glassner's magnum opus *Principles of Digital Image Synthesis* [23] has stood out among the theoretical works for its comprehensive scope and continued relevance since it was first published in 1995. For a reader wishing to learn "how graphics really works," as we described at the start of this chapter, this masterwork is required reading, even though it is inexplicably underappreciated in the video game industry. Best of all, both volumes have recently been made available in electronic form for free (legitimately). You can find them on books.google.com. A consolidated, corrected PDF should be available soon.

Phar and Humphreys' *Physically Based Rendering* [53] is an excellent way to learn the proper theoretical framework of graphics. Shorter and more recent than Glassner's, this text nonetheless provides a broad theoretical foundation of rendering principles. Although this is an excellent book for theoretical purposes, a unique feature of the book is the source code for a working raytracer that is woven throughout, illustrating how the ideas can be implemented.

Real-Time Rendering [1], by Akenine-Möller et al., gives a very broad survey of issues specific to real-time rendering, such as rendering hardware, shader programs, and performance. This classic, in its third edition at the time of this writing, is essential reading for any intermediate or advanced student interested in real-time graphics.

The OpenGL [49] and DirectX [14] API documentations are certainly important sources. Not only is such reference material necessary from a practical standpoint, but a surprising amount of knowledge can be gained just by browsing. Nearly a generation of OpenGL users have grown up on the "red book" [50].

The subtleties related to radiometry and color spaces that we glossed over are explained in more detail by Glassner [23] and also by Phar and Humphreys [53]. Ashdown [3] and Poynton [55] have written papers that are approachable and freely available.

10.13   Exercises

(Answers on page 779.)

1. On the Nintendo Wii, a common frame buffer resolution is 640×480. This same frame buffer resolution is used for 4:3 and 16:9 televisions.

   (a) What is the pixel aspect on a 4:3 television?

   (b) What is the pixel aspect on a 16:9 television?

2. Continuing the previous exercise, assume we are making a split-screen co-operative game, and we assign one player the left 320×480, and the other player the right 320×480. We always want the horizontal field of view to be $60°$. Assume the system settings tell us that the console is connected to a 4:3 television.

   (a) What is the window aspect?

   (b) What should the horizontal zoom value be?

   (c) What should the vertical zoom value be?

   (d) What is the resulting vertical field of view, in degrees?

Figure 10.43
Texture mapped quads for Exercise 4

(e) Assume the near and far clip planes are 1.0 and 256.0. What is the clip matrix, assuming all OpenGL conventions?

(f) What about the DirectX conventions?

3. Repeat parts (a)–(d) from Exercise 2, but assume a 16:9 television.

4. For each set of UV coordinates (a)–(f), match it with the corresponding texture-mapped quad in Figure 10.43. The upper-left vertex is numbered 0, and the vertices are enumerated clockwise around the quad.

(a) $0 : (0.20, -0.30)$ $1 : (1.30, -0.30)$ $2 : (1.30, 1.20)$ $3 : (0.20, 1.20)$

(b) $0 : (5.00, -1.00)$ $1 : (6.00, -1.00)$ $2 : (6.00, 0.00)$ $3 : (5.00, 0.00)$

(c) $0 : (1.00, 0.00)$ $1 : (-0.23, -0.77)$ $2 : (0.00, 1.00)$ $3 : (1.24, 1.77)$

(d) $0 : (2.00, 0.00)$ $1 : (1.00, 1.00)$ $2 : (0.00, 1.00)$ $3 : (1.00, 0.00)$

(e) $0 : (-0.10, 1.10)$ $1 : (-0.10, 0.10)$ $2 : (0.90, 0.10)$ $3 : (0.90, 1.10)$

(f) $0 : (0.00, -1.00)$ $1 : (3.35, 0.06)$ $2 : (1.00, 2.00)$ $3 : (-2.36, 0.94)$

5. For each entry (a)–(j) in the table on the next page, match the Blinn-Phong material diffuse color, specular color, and specular exponent with the corresponding creepy floating head in Figure 10.44. There is a single white omni light in the scene. Diffuse and specular colors are given as (red, green, blue) triples.

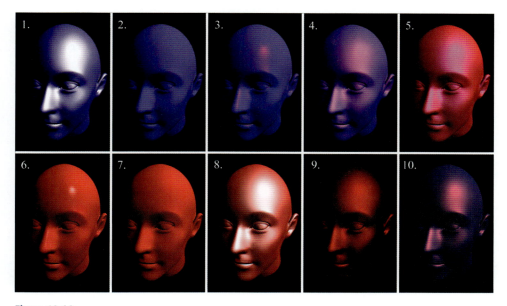

Figure 10.44
Creepy floating heads for Exercise 5

	Diffuse Color	Specular Color	Specular Exponent
(a)	(210,40,50)	(0,0,0)	1
(b)	(65,55,200)	(150,0,0)	16
(c)	(65,55,200)	(230,230,230)	2
(d)	(50,50,100)	(210,40,50)	4
(e)	(65,55,200)	(210,40,50)	2
(f)	(65,55,200)	(0,0,0)	64
(g)	(0,0,0)	(210,40,50)	1
(h)	(210,40,50)	(100,100,100)	64
(i)	(210,40,50)	(230,230,230)	2
(j)	(210,40,50)	(65,55,200)	2

6. How would the following normals be encoded in a 24-bit normal map with the usual conventions?

   (a) $[-1.00, 0.00, 0.00]$ (b) $[0.267, -0.535, 0.805]$

   (c) $[0.00, 0.00, 1.00]$ (d) $[0.00, 0.857, 0.514]$

7. For each row (a)–(d) in the table below, decode the texel from the normal map to obtain the tangent-space surface normal. Determine the binormal from the vertex normal, tangent, and determinant. Then calculate the model-space coordinates of the per-texel surface normal.

	Normal map texel RGB	Vertex normal	Vertex tangent	Determinant (Mirror flag)
(a)	(128,255,128)	$[0.408, -0.408, -0.816]$	$[0.707,0.707,0.000]$	1
(b)	(106,155,250)	$[0.000, 1.000, 0.000]$	$[1.000,0.000,0.000]$	-1
(c)	(128,218,218)	$[1.000, 0.000, 0.000]$	$[0.000,0.447,-0.894]$	1
(d)	(233,58,145)	$[0.154, -0.617, 0.772]$	$[0.986,0.046,-0.161]$	-1

There are too many special effects in all these movies today.

— Steven Spielberg (1946–)

Chapter 11

Mechanics 1: Linear Kinematics and Calculus

Always in motion is the future.

— Yoda in *Star Wars Episode V: The Empire Strikes Back* (1980)

"Ladies and gentlemen, may I direct your attention to the center ring. Witness before you two ordinary textbooks, one labeled *College Physics* and the other *Calculus*. Their combined 2,500+ pages weigh over 25 lbs. Yet in this chapter and the next, your brave stunt-authors will attempt a most death-defying and impossible spectacle of mysticism and subterfuge: to reduce these two massive books into a mere 150 pages!"

Just like any good circus act, this one is prefaced with a lot of build up to set your expectations. The difference here is that the purpose of our preface is to *lower* your expectations.

11.1 Overview and Other Expectation-Reducing Remarks

OK, there's no way we can really cover all of physics and calculus in two chapters. As any politician knows, the secret to effectively communicate complicated subject matter in a short amount of time is to use lies, both the omission and commission kind. Let's talk about each of these kinds of lies in turn, so you will know what's really in store.

11.1.1 What is Left Out?

Just about everything—let's talk about what we are leaving out of physics first. To put the word "physics" on this chapter would be even more of an insult to people who do real physics than this chapter already is. We are concerned only with *mechanics*, and very simple mechanics of rigid bodies at that. Some topics traditionally found in a first-year physics textbook that are *not* discussed in this book include:

- energy and work

- temperature, heat transfer, thermodynamics, entropy

- electricity, magnetism, light

- gases, fluids, pressure

- oscillation and waves.

A note about energy and work is in order, because even in the limited context of mechanics, the fundamental concept of energy plays a central role in traditional presentations. Many problems are easier to solve by using conservation of energy than by considering the forces and applying Newton's laws. (In fact, an alternative to the Newtonian dynamics that we study in this book exists. It is known as *Lagrangian dynamics* and focuses on energy rather than forces. When used properly, both systems produce the same results, but Lagrangian dynamics can solve certain problems more elegantly and is especially adept at handling friction, compared to Newtonian dynamics.) However, at the time of this writing, basic general purpose digital simulations are based on Newtonian dynamics, and energy does not play a direct role. That isn't to say an understanding of energy is useless; indeed disobedience of the conservation of energy law is at the heart of many simulation problems! Thus, energy often arises more as a way to understand the (mis)behavior of a digital simulation, even if it doesn't appear in the simulation code directly.

Now let's talk about the ways in which this book will irritate calculus professors. We think that a basic understanding of calculus is really important to fully grasp many of the concepts from physics. Conversely, physics provides some of the best examples for explaining calculus. Calculus and physics are often taught separately, usually with calculus coming first. It is our opinion that this makes calculus harder to learn, since it robs the student of the most intuitive examples—the physics problems for which calculus was invented to solve! We hope interleaving calculus with physics will make it easier for you to learn calculus.

Our calculus needs are extremely modest in this book, and we have left out even more from calculus than we did from physics. After reading this chapter, you should know:

- The basic idea of what a derivative measures and what it is used for.

- The basic idea of what an integral measures and what it is used for.

- Derivatives and integrals of trivial expressions containing polynomials and trig functions.

Of course, we are aware that a large number of readers may already have this knowledge. Take a moment to put yourself into one of the following categories:

1. I know absolutely nothing about derivatives or integrals.

2. I know the basic idea of derivatives and integrals, but probably couldn't solve any freshman calculus problems with a pencil and paper.

3. I have studied some calculus.

Level 2 knowledge of calculus is sufficient for this book, and our goal is to move everybody who is currently in category 1 into category 2. If you're in category 3, our calculus discussions will be a (hopefully entertaining) review. We have no delusions that we can move anyone who is not already there into category 3.

11.1.2   Some Helpful Lies about Our Universe

The universe is commonly thought to be discrete in both space and time. Not only is matter broken up into discrete chunks called atoms, but there is evidence that the very fabric of space and time is broken up into discrete pieces also. Now, there is a difference of opinion as to whether it's really that way or just appears that way because the only way we can interact with space is to throw particles at it, but it's our opinion that if it looks like a duck, walks like a duck, quacks like a duck, has webbed feet and a beak, then it's a good working hypothesis that it tastes good when put into eggrolls with a little dark sauce.

For a long time, the mere thought that the universe might not be continuous had not even considered the slightest possibility of crossing anybody's mind, until the ancient Greeks got a harebrained and totally unjustified idea that things might be made up of atoms. The fact that this later turned out to be true is regarded by many as being good luck rather then good judgment. Honestly, who would have thought it? After all, everyday objects, such as the desk on which one of the authors is currently resting his wrists as he types this sentence, give every appearance of having smooth, continuous surfaces. But who cares? Thinking of the desk as having a smooth, continuous surface is a harmless but useful delusion that lets the author rest his wrists comfortably without worrying about atomic bond energy and quantum uncertainty theory at all.

Not only is this trick of thinking of the world as continuous a handy psychological rationalization, it's also good mathematics. It turns out that the math of continuous things is a lot less unwieldy than the math of discrete things. That's why the people who were thinking about how the world works in the 15th century were happy to invent a mathematics for a continuous universe; experimentally, it was a good approximation to reality, and theoretically the math worked out nicely. Sir Isaac Newton was thus able to discover a lot of fundamental results about continuous mathematics, which we call "calculus," and its application to the exploration of a continuous universe, which we call "physics."

Now, we're mostly doing this so that we can model a game world inside a computer, which is inherently discrete. There's a certain amount of cognitive dissonance involved with programming a discrete simulation of a continuous model of a discrete universe, but we'll try not to let it bother us. Suffice it to say that we are in complete control of the discrete universe inside our game, and that means that we can choose the kind of physics that applies inside that universe. All we really need is for the physical laws to be sufficiently like the ones we're used to for the player to experience willing suspension of disbelief, and hopefully say, "Wow! Cool!" and want to spend more money. For almost all games that means a cozy Newtonian universe without the nasty details of quantum mechanics or relativity. Unfortunately, that means also that there are a pair of nasty trolls lurking under the bridge, going by the names of chaos and instability, but we will do our best to appease them.

For the moment, we are concerned about the motion of a small object called a "particle." At any given moment, we know its position and velocity.[1] The particle has mass. We do not concern ourselves with the orientation of the particle (for now), and thus we don't think of the particle as spinning. The particle does not have any size, either. We will defer adding those elements until later, when we shift from particles to rigid bodies.

We are studying classical mechanics, also known as Newtonian mechanics, which has several simplifying assumptions that are incorrect in general but true in everyday life in most ways that really matter to us. So we can darn well make sure they are true inside our computer world, if we please. These assumptions are:

- Time is absolute.

- Space is Euclidian.

- Precise measurements are possible.

- The universe exhibits causality and complete predictability.

---

[1] Thanks to Heisenberg, we know that's not possible on the atomic level, but when we said "small" we didn't mean *that* small.

The first two are shattered by relatively, and the second two by quantum mechanics. Thankfully, these two subjects are not necessary for video games, because your authors do not have more than a pedestrian understanding of them.

We will begin our foray into the field of mechanics by learning about *kinematics*, which is the study of the equations that describe the motion of a particle in various simple but commonplace situations. When studying kinematics, we are *not* concerned with the causes of motion—that is the subject of *dynamics*, which will be covered in Chapter 12. For now, "ours is not to question why," ours is just to do the math to get equations that predict the position, velocity, and acceleration of the particle at any given time t, or die. Well, forget about the last part anyway.

Because we are treating our objects as particles and tracking their position only, we will not consider their orientation or rotational effects until Chapter 12. When rotation is ignored, all of the ideas of linear kinematics extend into 3D in a straightforward way, and so for now we will be limiting ourselves to 2D (and 1D). This is convenient, since the authors do not know how to design those little origami-like things that lay flat and then pop up when you open the book, and the publisher wouldn't let us even if we were compulsive enough to learn how to do it. Later we'll see why treating objects as particles is perfectly justifiable.

11.2   Basic Quantities and Units

Mechanics is concerned with the relationship among three fundamental quantities in nature: *length*, *time*, and *mass*. Length is a quantity you are no doubt familiar with; we measure length using units such as centimeters, inches, meters, feet, kilometers, miles, and astronomical units.[2] Time is another quantity we are very comfortable with measuring, in fact most of us probably learned how to read a clock before we learned how to measure distances.[3] The units used to measure time are the familiar second, minute, day, week, fortnight,[4] and so on. The month and the year are often not good units to use for time because different months and years have different durations.

---

[2]An astronomical unit is equal to the average distance between the Earth and the sun, approximately 150 million kilometers or 93 million miles. That's a big number, but I wouldn't say it's astronomical.

[3]In fact, it's even easier now. Only dinosaurs like the authors know how to read an analog clock that has hands.

[4]OK, maybe that one's not so familiar. It is to one of the authors, but that's because he's British.

The quantity *mass* is not quite as intuitive as length and time. The measurement of an object's mass is often thought of as measuring the "amount of stuff" in the object. This is not a bad (or at least, not completely terrible) definition, but its not quite right, either [57]. A more precise definition might be that mass is a measurement of *inertia*, that is, how much resistance an object has to being accelerated. The more massive an object is, the more force is required to start it in motion, stop its motion, or change its motion.

Mass is often confused with *weight*, especially since the units used to measure mass are also used to measure weight: the gram, pound, kilogram, ton, and so forth. The mass of an object is an intrinsic property of an object, whereas the weight is a local phenomenon that depends on the strength of the gravitational pull exerted by a nearby massive object. Your mass will be the same whether you are in Chicago, on the moon, near Jupiter, or light-years away from the nearest heavenly body, but in each case your weight will be very different. In this book and in most video games, our concerns are confined to a relatively small patch on a *flat* Earth, and we approximate gravity by a constant downward pull. It won't be too harmful to confuse mass and weight because gravity for us will be a constant. (But we couldn't resist a few cool exercises about the International Space Station.)

In many situations, we can discuss the relationship between the fundamental quantities without concern for the units of measurement we are using. In such situations, we'll find it useful to denote length, time, and mass by L, T, and M, respectively. One important such case is in defining *derived quantities*. We've said that length, time, and mass are the fundamental quantities—but what about other quantities, such as area, volume, density, speed, frequency, force, pressure, energy, power, or any of the numerous quantities that can be measured in physics? We don't give any of these their own capital letter, since each of these can be defined in terms of the fundamental quantities.

For example, we might express a measurement of area as a number of "square feet." We have created a unit that is in terms of another unit. In physics, we say that a measurement of area has the unit "length squared," or L^2. How about speed? We measure speed using the units such as miles per hour or meters per second. Thus speed is the ratio of a distance per unit time, or L/T.

One last example is frequency. You probably know that *frequency* measures how many times something happens in a given time interval (how "frequently" it happens). For example, a healthy adult has an average heart rate of around 70 beats per minute (BPM). The motor in a car might be rotating at a rate of 5,000 revolutions per minute (RPM). The NTSC television standard is defined as 29.97 frames per second (FPS). Note that in each of these, we are counting how many times something happens within

Quantity	Notation	SI unit	Other units
Length	L	m	cm, km, in, ft, mi, light year, furlong
Time	T	s	min, hr, ms
Mass	M	kg	g, slug, lb (pound-mass)
Velocity	L/T	m/s	ft/s, m/hr, km/hr
Acceleration	L/T^2	m/s^2	ft/s^2, (m/hr)/s, (km/hr)/s
Force	ML/T^2	N (Newton) $= \text{kg} \cdot \text{m/s}^2$	lb (pound-force), poundal
Area	L^2	m^2	mm^2, cm^2, km^2, in^2, ft^2, mi^2, acre, hectare
Volume	L^3	m^3	mm^3, cm^3, L (liter), in^3, ft^3, teaspoon, fl oz (fluid ounce), cup, pint, quart, gallon
Pressure	Force/Area $= (ML/T^2)/L^2$ $= M/(LT^2)$	Pa (Pascal) $= \text{N/m}^2$ $= \text{kg}/(\text{m} \cdot \text{s}^2)$	psi (lbs/in^2), millibar, inch of mercury, atm (atmosphere)
Energy	Force $\times$ Length $= (ML/T^2) \cdot L$ $= ML^2/T^2$	J (Joule) $= \text{N} \cdot \text{m}$ $= \frac{\text{kg} \cdot \text{m}}{\text{s}^2} \cdot \text{m}$ $= \frac{\text{kg} \cdot \text{m}^2}{\text{s}^2}$	kW $\cdot$ hr (kilowatt-hour), foot-pound, erg, calorie, BTU (British thermal unit), ton of TNT
Power	Energy / Time $= (ML^2/T^2)/T$ $= ML^2/T^3$	W (Watt) $= \text{J/s}$ $= \frac{\text{kg} \cdot \text{m}^2}{\text{s}^2} \cdot \text{s}^{-1}$ $= \frac{\text{kg} \cdot \text{m}^2}{\text{s}^3}$	hp (horsepower)
Frequency	$1/T = T^{-1}$	Hz $= 1/\text{s} = \text{s}^{-1}$ $=$ "per second"	KHz $= 1{,}000$ Hz, MHz $= 1{,}000{,}000$ Hz, "per minute", "per annum"

Table 11.1
Selected physical quantities and common units of measurements

a given duration of time. So we can write frequency in generic units as $1/T$ or T^{-1}, which you can read as "per unit time." One of the most important measurements of frequency is the *Hertz*, abbreviated Hz, which means "per second." When you express a frequency in Hz, you are describing the number of events, oscillations, heartbeats, video frames, or whatever *per second*. By definition, 1 Hz $= 1$ s^{-1}.

Table 11.1 summarizes several quantities that are measured in physics, their relation to the fundamental quantities, and some common units used to measure them.

Of course, any real measurement doesn't make sense without attaching specific units to it. One way to make sure that your calculations always make sense is to carry around the units at all times and treat them like algebraic variables. For example, if you are computing a pressure and your answer comes out with the units m/s, you know you have done something wrong; pressure has units of force per unit area, or $ML/(T^2L^2)$. On the other hand, if you are solving a problem and you end up with an answer in pounds per square inch (psi), but you are looking for a value in Pascals, your answer is probably correct, but just needs to be converted to the desired units. This sort of reasoning is known as *dimensional analysis*. Carrying around the units and treating them as algebraic variables quite often highlights mistakes caused by different units of measurement, and also helps make unit conversion a snap.

Because unit conversion is an important skill, let's briefly review it here. The basic concept is that to convert a measurement from one set of units to another, we multiply that measurement by a well-chosen fraction that has a value of 1. Let's take a simple example: how many feet is 14.57 meters? Looking up the conversion factor,[5] we see that $1 \text{ m} \approx 3.28083 \text{ ft}$. This means that $1 \text{ m}/3.28083 \text{ ft} \approx 1$. So let's take our measurement and multiply it by a special value of "1:"

$$14.57 \text{ m} = 14.57 \text{ m} \times 1 \approx 14.57 \text{ m} \times \frac{3.28083 \text{ ft}}{1 \text{ m}} \approx 47.80 \text{ ft.} \qquad (11.1)$$

Our conversion factor tells us that the numerator and denominator of the fraction in Equation (11.1) are equal: 3.28083 feet is *equal to* 1 meter. Because the numerator and denominator are equal, the "value" of this fraction is 1. (In a physical sense, though, certainly numerically the fraction doesn't equal 1.) And we know that multiplying anything by 1 does not change its value. Because we are treating the units as algebraic variables, the m on the left cancels with the m in the bottom of the fraction.

Of course, applying one simple conversion factor isn't too difficult, but consider a more complicated example. Let's convert 188 km/hr to ft/s. This time we need to multiply by "1" several times:

$$188 \frac{\text{km}}{\text{hr}} \times \frac{1 \text{ hr}}{3600 \text{ s}} \times \frac{1000 \text{ m}}{1 \text{ km}} \times \frac{3.28083 \text{ ft}}{1 \text{ m}} \approx 171 \frac{\text{ft}}{\text{s}}.$$

11.3   Average Velocity

We begin our study of kinematics by taking a closer look at the simple concept of speed. How do we measure speed? The most common method is to measure how much time it takes to travel a fixed distance. For example, in a race, we say that the fastest runner is the one who finishes the race in the shortest amount of time.

Consider the fable of the tortoise and the hare. In the story, they decide to have a race, and the hare, after jumping to an early lead, becomes

---

[5]By "looking it up," we mean using the Internet. There isn't room in this book for tables of information that are easily found online. We needed the space for all of our opinions, jokes, and useless footnotes.

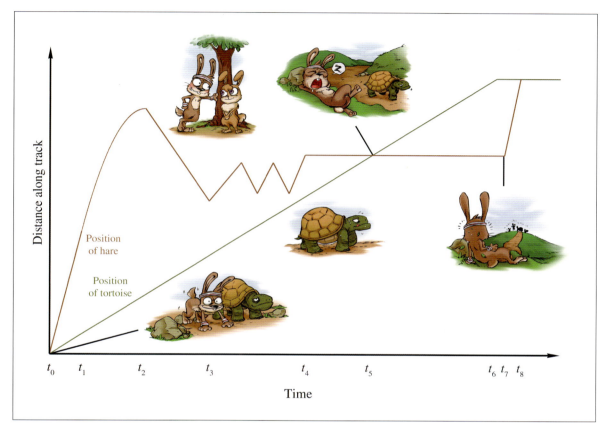

Figure 11.1
Plot of position versus time during the race between the tortoise and the hare

overconfident and distracted. He stops during the race to take a nap, smell the flowers, or some other form of lollygagging. Meanwhile, the tortoise plods along, eventually passing the hare and crossing the finish line first. Now this is a math book and not a self-help book, so please ignore the moral lessons about focus and perseverance that the story contains, and instead consider what it has to teach us about average velocity. Examine Figure 11.1, which shows a plot of the position of each animal over time.

A play-by-play of the race is as follows. The gun goes off at time t_0, and the hare sprints ahead to time t_1. At this point his hubris causes him to slow his pace, until time t_2 when a cute female passes by in the opposite direction. (Her position over time is not depicted in the diagram.) At this point a different tragic male trait causes the hare to turn around and walk with her, and he proceeds to chat her up. At t_3, he realizes that his advances are getting him nowhere, and he begins to pace back and forth along the track dejectedly until time t_4. At that point, he decides to take a nap. Meanwhile, the tortoise has been making slow and steady progress, and at time t_5, he catches up with the sleeping hare. The tortoise plods along and crosses the tape at t_6. Quickly thereafter, the hare, perhaps awakened by the sound of the crowd celebrating the tortoise's victory, wakes up at time t_7 and hurries in a frenzy to the finish. At t_8, the hare crosses the finish line, where he is humiliated by all his peers, and the cute girl bunny, too.

To measure the *average velocity* of either animal during any time interval, we divide the animal's displacement by the duration of the interval. We'll be focusing on the hare, and we'll denote the position of the hare as x, or more explicitly as $x(t)$, to emphasize the fact that the hare's position varies as a function of time. It is a common convention to use the capital Greek letter delta ("Δ") as a prefix to mean "amount of change in." For example, Δx would mean "the change in the hare's position," which is a displacement of the hare. Likewise Δt means "the change in the current time," or simply, "elapsed time between two points." Using this notation, the average velocity of the hare from t_a to t_b is given by the equation

Definition of average velocity

$$\text{average velocity} = \frac{\text{displacement}}{\text{elapsed time}} = \frac{\Delta x}{\Delta t} = \frac{x(t_b) - x(t_a)}{t_b - t_a}.$$

This is the definition of average velocity. No matter what specific units we use, velocity always describes the ratio of a length divided by a time, or to use the notation discussed in Section 11.2, velocity is a quantity with units L/T.

If we draw a straight line through any two points on the graph of the hare's position, then the slope of that line measures the average velocity of the hare over the interval between the two points. For example, consider the average velocity of the hare as he decelerates from time t_1 to t_2, as shown in Figure 11.2. The slope of the line is the ratio $\Delta x / \Delta t$. This slope is also equal to the tangent of the angle marked α, although for now the values Δx and Δt are the ones we will have at our fingertips, so we won't need to do any trig.

Returning to Figure 11.1, notice that the hare's average velocity from t_2 to t_3 is *negative*. This is because velocity is defined as the ratio of *net displacement* over time. Compare this to speed, which is the *total distance* divided by time and cannot be negative. The sign of displacement and velocity are sensitive to the direction of travel, whereas distance and speed are intrinsically nonnegative. We've already spoken about these distinctions way back in Section 2.2. Of course it's obvious that the average velocity is negative between t_2 and t_3, since the hare was going backwards during the entire interval. But average velocity can also be negative on an interval

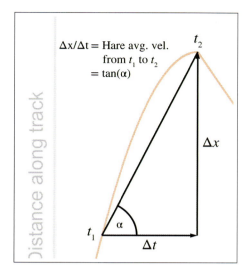

Figure 11.2
Determining average velocity
graphically

even in situations where forward progress is being made for a portion of the interval, such as the larger interval between t_2 and t_4. It's a case of "one step forward, two steps back."

Average velocity can also be zero, as illustrated during the hare's nap from t_4 to t_7. In fact, the average velocity will be zero any time an object starts and ends at the same location, even if it was it motion during the entire interval! ("Two steps forward, two steps back.") Two such intervals are illustrated in Figure 11.3.

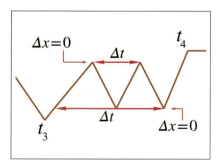

Figure 11.3
Two intervals during which the hare has no net displacement, and thus his average velocity is zero

And, of course, the final lesson of the fable is that the average velocity of the tortoise is *greater* than the average velocity of the hare, at least from t_0 to t_7, when the tortoise crosses the finish line. This is true despite the fact that the hare's average *speed* was higher, since he certainly traveled a larger distance with all the female distractions and pacing back and forth.

One last thing to point out. If we assume the hare learned his lesson and congratulated the tortoise (after all, let's not attribute to the poor animal *all* the negative personality traits!), then at $t = t_8$ they were standing at the same place. This means their net displacements from t_0 to t_8 are the same, and thus they have the *same* average velocity during this interval.

11.4   Instantaneous Velocity and the Derivative

We've seen how physics defines and measures the average velocity of an object over an interval, that is, between two time values that differ by some finite amount Δt. Often, however, it's useful to be able to speak of an object's *instantaneous velocity*, which means the velocity of the object for one value of t, a single moment in time. You can see that this is not a trivial question because the familiar methods for measuring velocity, such as

$$\text{average velocity} = \frac{\text{displacement}}{\text{elapsed time}} = \frac{\Delta x}{\Delta t} = \frac{x(t_b) - x(t_a)}{t_b - t_a},$$

don't work when we are considering only a single instant in time. What are t_a and t_b, when we are looking at only one time value? In a single instant, displacement and elapsed time are both zero; so what is the meaning of the ratio $\Delta x/\Delta t$? This section introduces a fundamental tool of calculus known as the *derivative*. The derivative was invented by Newton to investigate precisely the kinematics questions we are asking in this chapter. However, its applicability extends to virtually every problem where one quantity varies as a function of some other quantity. (In the case of velocity, we are interested in how position varies as a function of time.)

Because of the vast array of problems to which the derivative can be applied, Newton was not the only one to investigate it. Primitive applications of integral calculus to compute volumes and such date back to ancient Egypt. As early as the 5th century, the Greeks were exploring the building blocks of calculus such as infinitesimals and the method of exhaustion. Newton usually shares credit with the German mathematician Gottfried Leibniz[6] (1646–1716) for inventing calculus in the 17th century, although Persian and Indian writings contain examples of calculus concepts being used. Many other thinkers made significant contributions, including Fermat, Pascal, and Descartes.[7] It's somewhat interesting that many of

---

[6] Ian Parberry is conflicted by this. Although he is British and feels that he should consequently support Newton's case, his PhD adviser's adviser's ... adviser back 14 generations ago was Leibniz, and hence he feels he owes him some "familial" loyalty.

[7] Pascal and Descartes are PhD adviser "cousins" of Ian Parberry's back to the 16th generation, but nonetheless he can't help thinking of Monty Python's *Philosopher's Song* whenever he thinks of Descartes.

the earlier applications of calculus were integrals, even though most calculus courses cover the "easier" derivative before the "harder" integral.

We first follow in the steps of Newton and start with the physical example of velocity, which we feel is the best example for obtaining intuition about how the derivative works. Afterwards, we consider several other examples where the derivative can be used, moving from the physical to the more abstract.

11.4.1   Limit Arguments and the Definition of the Derivative

Back to the question at hand: how do we measure instantaneous velocity? First, let's observe one particular situation for which it's easy: if an object moves with constant velocity over an interval, then the velocity is the same at every instant in the interval. That's the very definition of constant velocity. In this case, the average velocity over the interval must be the same as the instantaneous velocity for any point within that interval. In a graph such as Figure 11.1, it's easy to tell when the object is moving at constant velocity because the graph is a straight line. In fact, almost all of Figure 11.1 is made up of straight line segments,[8] so determining instantaneous velocity is as easy as picking any two points on a straight-line interval (the endpoints of the interval seem like a good choice, but any two points will do) and determining the average velocity between those endpoints.

But consider the interval from t_1 to t_2, during which the hare's over-confidence causes him to gradually decelerate. On this interval, the graph of the hare's position is a curve, which means the slope of the line, and thus the velocity of the hare, is changing continuously. In this situation, measuring instantaneous velocity requires a bit more finesse.

For concreteness in this example, let's assign some particular numbers. To keep those numbers round (and also to stick with the racing theme), please allow the whimsical choice to measure time in minutes and distance in furlongs.[9] We will assign $t_1 = 1$ min and $t_2 = 3$ min, so the total duration is 2 minutes. Let's say that during this interval, the hare travels from $x(1) = 4$ fur to $x(3) = 8$ fur.[10] For purposes of illustration, we will set our sights on the answer to the question: what is the hare's instantaneous velocity at $t = 2.5$ min? This is all depicted in Figure 11.4.

---

[8] Mostly because that's the easiest thing for lazy authors to create in *Photoshop*.

[9] The speed chosen for the hare bears some semblance to reality, but for pedagogical reasons and to make Figure 11.1 fit nicely on a page, the speed of the tortoise is totally fudged. Sticklers should remind themselves that this is a story with a talking bunny rabbit and turtle. Oh, and a furlong is 1/8 of a mile.

[10] The abbreviation "fur" means "furlongs" and has nothing to do with the fur on the bunny.

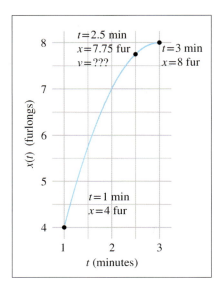

Figure 11.4
What is the hare's velocity at $t = 2.5$ min?

It's not immediately apparent how we might measure or calculate the velocity at the *exact moment* $t = 2.5$, but observe that we can get a good approximation by computing the average velocity of a very small interval near $t = 2.5$. For a small enough interval, the graph is nearly the same as a straight line segment, and the velocity is nearly constant, and so the instantaneous velocity at any given instant within the interval will not be too far off from the average velocity over the whole interval.

In Figure 11.5, we fix the left endpoint of a line segment at $t = 2.5$ and move the right endpoint closer and closer. As you can see, the shorter the interval, the more the graph looks like a straight line, and the better our approximation becomes. Thinking graphically, as the second endpoint moves closer and closer to $t = 2.5$, the slope of the line between the endpoints will converge to the slope of the line that is *tangent* to the curve at this point. A tangent line is the graphical equivalent of instantaneous velocity, since it measures the slope of the curve just at that one point.

Let's carry out this experiment with some real numbers and see if we cannot approximate the instantaneous velocity of the hare. In order to do this, we'll need to be able to know the position of the hare at any given time, so now would be a good time to tell you that the position of the hare is given by the function[11]

$$x(t) = -t^2 + 6t - 1.$$

---

[11] While it may seem nicely contrived that the hare's motion is described by a quadratic equation with whole number coefficients, we'll see later that it isn't as contrived as you might think. Nature apparently likes quadratic equations. But you do have us on the whole number coefficients, which were cherry-picked.

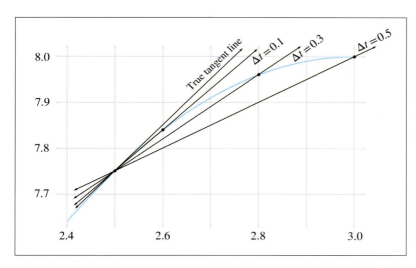

Figure 11.5. Approximating instantaneous velocity as the average velocity of a smaller and smaller interval

Table 11.2 shows tabulated calculations for average velocity over intervals with a right hand endpoint $t + \Delta t$ that moves closer and closer to $t = 2.5$.

The right-most column, which is the average velocity, appears to be converging to a velocity of 1 furlong/minute. But how certain are we that this is the correct value? Although we do not have any calculation that will produce a resulting velocity of exactly 1 furlong/minute, for all practical purposes, we may achieve any degree of accuracy desired by using this approximation technique and choosing Δt sufficiently small. (We are ignoring issues related to the precision of floating point representation of numbers in a computer.)

This is a powerful argument. We have essentially assigned a value to an expression that we cannot evaluate directly. Although it is mathematically illegal to substitute $\Delta t = 0$ into the expression, we can argue that for

t	Δt	$t + \Delta t$	$x(t)$	$x(t + \Delta t)$	$x(t + \Delta t) - x(t)$	$\dfrac{x(t + \Delta t) - x(t)}{\Delta t}$
2.500	0.500	3.000	7.750	8.0000	0.2500	0.5000
2.500	0.100	2.600	7.750	7.8400	0.0900	0.9000
2.500	0.050	2.550	7.750	7.7975	0.0475	0.9500
2.500	0.010	2.510	7.750	7.7599	0.0099	0.9900
2.500	0.005	2.505	7.750	7.7549	0.0049	0.9950
2.500	0.001	2.501	7.750	7.7509	0.0009	0.9990

Table 11.2. Calculating average velocity for intervals of varying durations

smaller and smaller values of Δt, we converge to a particular value. In the parlance of calculus, this value of 1 furlong/minute is a *limiting value*, meaning that as we take smaller and smaller positive values for Δt, the result of our computation approaches 1, but does not cross it (or reach it exactly).

Convergence arguments such as this are defined with rigor in calculus by using a formalized tool known as a *limit*. The mathematical notation for this is

$$v(t) = \lim_{\Delta t \to 0} \frac{x(t + \Delta t) - x(t)}{\Delta t}. \tag{11.2}$$

The notation '$\to$' is usually read as "approaches" or "goes to." So the right side of Equation (11.2) might be read as

"The limit of $\dfrac{x(t + \Delta t) - x(t)}{\Delta t}$ as Δt approaches zero,"

or

"The limit as Δt approaches zero of $\dfrac{x(t + \Delta t) - x(t)}{\Delta t}$."

In general, an expression of the form $\lim_{a \to k}[\text{blah}]$ is interpreted to mean "The value that [blah] converges to, as a gets closer and closer to k."

This is an important idea, as it defines what we mean by instantaneous velocity.

Instantaneous velocity at a given time t may be interpreted as the average velocity of an interval that contains t, in the limit as the duration of the interval approaches zero.

We won't have much need to explore the full power of limits or get bogged down in the finer points; that is the mathematical field of *analysis*, and would take us a bit astray from our current, rather limited,[12] objectives. We are glossing over some important details[13] so that we can focus on one particular case, and that is the use of limits to define the *derivative*.

The derivative measures the *rate of change* of a function. Remember that "function" is just a fancy word for any formula, computation, or procedure that takes an input and produces an output. The derivative quantifies the rate at which the output of the function will change in response to a change to the input. If x denotes the value of a function at a specific time t,

---

[12]No pun intended. Regretfully.
[13]Including such things as continuity, limits taken from the left, from the right, etc.

the derivative of that function at t is the ratio dx/dt. The symbol dx represents the change in the output produced by a very small change in the input, represented by dt. We'll speak more about these "small changes" in more detail in just a moment.

For now, we are in an imaginary racetrack where rabbits and turtles race and moral lessons are taught through metaphor. We have a function with an input of t, the number of minutes elapsed since the start of the race, and an output of x, the distance of the hare along the racetrack. The rule we use to evaluate our function is the expression $x(t) = -t^2 + 6t - 1$. The derivative of this function tells us the rate of change of the hare's position with respect to time and is the *definition* of instantaneous velocity. Just previously, we defined instantaneous velocity as the average velocity taken over smaller and smaller intervals, but this is essentially the same as the definition of the derivative. We just phrased it the first time using terminology specific to position and velocity.

When we calculate a derivative, we won't end up with a single number. Expecting the answer to "What is the velocity of the hare?" to be a single number makes sense only if the velocity is the same everywhere. In such a trivial case we don't need derivatives, we can just use average velocity. The interesting situation occurs when the velocity varies over time. When we calculate the derivative of a position function in such cases, we get a velocity *function*, which allows us to calculate the instantaneous velocity at any point in time.

The previous three paragraphs express the most important concepts in this section, so please allow us to repeat them.

A derivative measures a rate of change. Since velocity is the rate of change of position with respect to time, the derivative of the position function is the velocity function.

The next few sections discuss the mathematics of derivatives in a bit more detail, and we return to kinematics in Section 11.5. This material is aimed at those who have not had[14] first-year calculus. If you already have a calculus background, you can safely skip ahead to Section 11.5 unless you feel in need of a refresher.

Section 11.4.2 lists several examples of derivatives to give you a better understanding of what it means to measure a rate of change, and also to

---

[14] The authors define "had" to mean that you passed it, you understood it, and you remember it.

back up our claim that the derivative has very broad applicability. Section 11.4.3 gives the formal mathematical definition of the derivative[15] and shows how to use this definition to solve problems. We also finally figure out how fast that hare was moving at $t = 2.5$. Section 11.4.4 lists various commonly used alternate notations for derivatives, and finally, Section 11.4.5 lists just enough rules about derivatives to satisfy the very modest differential calculus demands of this book.

11.4.2   Examples of Derivatives

Velocity may be the easiest introduction to the derivative, but it is by no means the only example. Let's look at some more examples to give you an idea of the wide array of problems to which the derivative is applied.

The simplest types of examples are to consider other quantities that vary with time. For example, if $R(t)$ is the reading of a rain meter at a given time t, then the derivative, denoted $R'(t)$, describes how hard it was raining at time t. Perhaps $P(t)$ is the reading of a pressure valve on a tank containing some type of gas. Assuming the pressure reading is proportional to the mass of the gas inside the chamber,[16] the rate of change $P'(t)$ indicates how fast gas is flowing into or out of the chamber at time t.

There are also physical examples for which the independent variable is not time. The prototypical case is a function $y(x)$ that gives the height of some surface above a reference point at the horizontal position x. For example, perhaps x is the distance along our metaphorical racetrack and y measures the height at that point above or below the altitude at the starting point. The derivative $y'(x)$ of this function is the slope of the surface at x, where positive slopes mean the runners are running uphill, and negative values indicate a downhill portion of the race. This example is not really a new example, because we've looked at graphs of functions and considered how the derivative is a measure of the slope of the graph in 2D.

Now let's become a bit more abstract, but still keep a physical dimension as the independent variable. Let's say that for a popular rock-climbing wall, we know a function $S(y)$ that describes, for a given height y, what percentage of rock climbers are able to reach that height or higher. If we assume the climbers start at $y = 0$, then $S(0) = 100\%$. Clearly $S(y)$ is a nonincreasing function that eventually goes all the way down to 0% at some maximum height y_{max} that nobody has ever reached.

Now consider the interpretation of derivative $S'(y)$. Of course, $S'(y) \leq 0$, since $S(y)$ is nonincreasing. A large negative value of $S'(y)$ is an indi-

---

[15]Spoiler alert: we already gave it to you in this section!

[16]This would be true if we are operating in a range where the ideal gas law is a valid approximation, and the temperature remains constant.

cation that the height y is an area where climbers are likely to drop[17] out. Perhaps the wall at that height is a challenging area. $S'(y)$ closer to zero is an indication that fewer climbers drop out at height y. Perhaps there is a plateau that climbers can reach, and there they rest. We might expect $S'(y)$ to decrease just after this plateau, since the climbers are more rested. In fact, $S'(y)$ might also become closer to zero just *before* the plateau, because as climbers begin to get close to this milestone, they push a bit harder and are more reluctant to give up.[18]

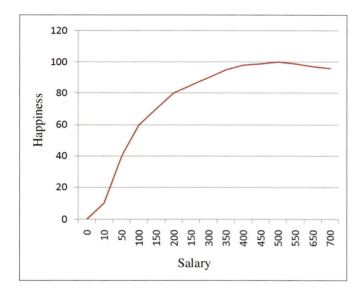

Figure 11.6
Happiness versus salary

One last example. Figure 11.6 shows happiness as a function of salary. In this case, the derivative is essentially the same thing as what economists would call "marginal utility." It's the ratio of additional units of happiness per additional unit of income. According this figure, the marginal utility of income decreases, which of course is the famous law of diminishing returns. According to our research,[19] it even becomes negative after a certain point, where the troubles associated with high income begin to outweigh the psychological benefits. The economist-speak phrase "negative marginal utility" is translated into everyday language as "stop doing that."

---

[17]So to speak.

[18]The alert reader may have noticed that, because there are only a finite number of climbers to sample, $S(y)$ comes from samples data, and therefore has discontinuous "steps" in it, rather than being a smooth line. At those steps, the derivative isn't really defined. For purposes of this illustration, let's assume that we have fit a smooth curve through the experimental data in order to yield a continuous function.

[19]We just made quotes in the air with our fingers as we wrote that word.

11.4.3   Calculating Derivatives from the Definition

Now we're ready for the official[20] definition of the derivative found in most math textbooks, and to see how we can compute derivatives using the definition. A derivative can be understood as the limiting value of $\Delta x / \Delta t$, the ratio of the change in output divided by the change in input, taken as we make Δt infinitesimally small. Let's repeat this description using mathematical notation. It's an equation we gave earlier in the chapter, only this time we put a big box around it, because that's what math books do to equations that are definitions.

The Definition of a Derivative

$$\frac{dx}{dt} = \lim_{\Delta t \to 0} \frac{\Delta x}{\Delta t} = \lim_{\Delta t \to 0} \frac{x(t + \Delta t) - x(t)}{\Delta t}. \tag{11.3}$$

Here the notation for the derivative dx/dt is known as *Leibniz's notation*. The symbols dx and dt are known as *infinitesimals*. Unlike Δx and Δt, which are variables representing finite changes in value, dx and dt are symbols representing "an infinitesimally small change." Why is it so important that we use a very small change? Why can't we just take the ratio $\Delta x / \Delta t$ directly? Because the rate of change is varying *continuously*. Even within a very small interval of $\Delta t = .0001$, it is not constant. This is why a limit argument is used, to make the interval as small as we can possibly make it—infinitesimally small.

In certain circumstances, infinitesimals may be manipulated like algebraic variables (and you can also attach units of measurement to them and carry out dimensional analysis to check your work). The fact that such manipulations are often correct is what gives Leibniz notation its intuitive appeal. However, because they are infinitely small values, they require special handling, similar to the symbol ∞, and so should not be tossed around willy-nilly. For the most part, we interpret the notation $\frac{dx}{dt}$ not as a ratio of two variables, but as a single symbol that means "the derivative of x with respect to t." This is the safest procedure and avoids any chance of the aforementioned willy-nilliness. We have more to say later on Leibniz and other notations, but first, let's finally calculate a derivative and answer the burning question: how fast was the hare traveling at $t = 2.5$?

---

[20]We use the word "official" here because there are other ways to define the derivative that lead to improved methods for approximating derivatives numerically with a computer. Such methods are useful when an analytical solution is too difficult or slow to compute.

Differentiating a simple function by using the definition Equation (11.3) is an important rite of passage, and we are proud to help you cross this threshold. The typical procedure is this:

1. Substitute $x(t)$ and $x(t + \Delta t)$ into the definition. (In our case, $x(t) = -t^2 + 6t - 1$).

2. Perform algebraic manipulations until it is legal to substitute $\Delta t = 0$. (Often this boils down to getting Δt out of the denominator.)

3. Substitute $\Delta t = 0$, which evaluates the expression "at the limit," removing the limit notation.

4. Simplify the result.

Applying this procedure to our case yields

$$
\begin{aligned}
v(t) = \frac{dx}{dt} &= \lim_{\Delta t \to 0} \frac{x(t + \Delta t) - x(t)}{\Delta t} \\
&= \lim_{\Delta t \to 0} \frac{[-(t + \Delta t)^2 + 6(t + \Delta t) - 1] - (-t^2 + 6t - 1)}{\Delta t} \\
&= \lim_{\Delta t \to 0} \frac{(-t^2 - 2t(\Delta t) - (\Delta t)^2 + 6t + 6(\Delta t) - 1) + (t^2 - 6t + 1)}{\Delta t} \\
&= \lim_{\Delta t \to 0} \frac{-2t(\Delta t) - (\Delta t)^2 + 6(\Delta t)}{\Delta t} \\
&= \lim_{\Delta t \to 0} \frac{\Delta t \, (-2t - \Delta t + 6)}{\Delta t} \\
&= \lim_{\Delta t \to 0} -2t - \Delta t + 6. \tag{11.4}
\end{aligned}
$$

Now we are at step 3. Taking the limit in Equation (11.4) is now easy; we simply substitute $\Delta t = 0$. This substitution was not legal earlier because there was a Δt in the denominator:

$$
\begin{aligned}
v(t) = \frac{dx}{dt} &= \lim_{\Delta t \to 0} -2t - \Delta t + 6 \\
&= -2t - (0) + 6 \\
&= -2t + 6. \tag{11.5}
\end{aligned}
$$

Finally! Equation (11.5) is the velocity function we've been looking for. It allows us to plug in any value of t and compute the instantaneous velocity of the hare at that time. Putting in $t = 2.5$, we arrive at the answer to our question:

$$
\begin{aligned}
v(t) &= -2t + 6, \\
v(2.5) &= -2(2.5) + 6 = 1.
\end{aligned}
$$

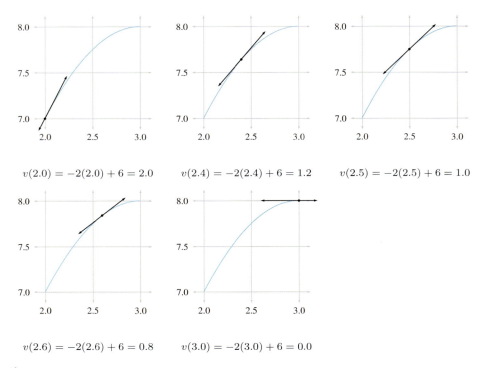

$$v(2.0) = -2(2.0) + 6 = 2.0 \qquad v(2.4) = -2(2.4) + 6 = 1.2 \qquad v(2.5) = -2(2.5) + 6 = 1.0$$

$$v(2.6) = -2(2.6) + 6 = 0.8 \qquad v(3.0) = -2(3.0) + 6 = 0.0$$

Figure 11.7
The hare's velocity and corresponding tangent line at selected times

So the instantaneous velocity of the hare at $t = 2.5$ was precisely 1 furlong per minute, just as our earlier arguments predicted. But now we can say it with confidence.

Figure 11.7 shows this point and several others along the interval we've been studying. For each point, we have calculated the instantaneous velocity at that point according to Equation (11.5) and have drawn the tangent line with the same slope.

It's very instructive to compare the graphs of position and velocity side by side. Figure 11.8 compares the position and velocity of our fabled racers.

There are several interesting observations to be made about Figure 11.8.

- When the position graph is a horizontal line, there is zero velocity, and the velocity graph traces the $v = 0$ horizontal axis (for example, during the hare's nap).

- When the position is increasing, the velocity is positive, and when the position is decreasing (the hare is moving the wrong way) the velocity is negative.

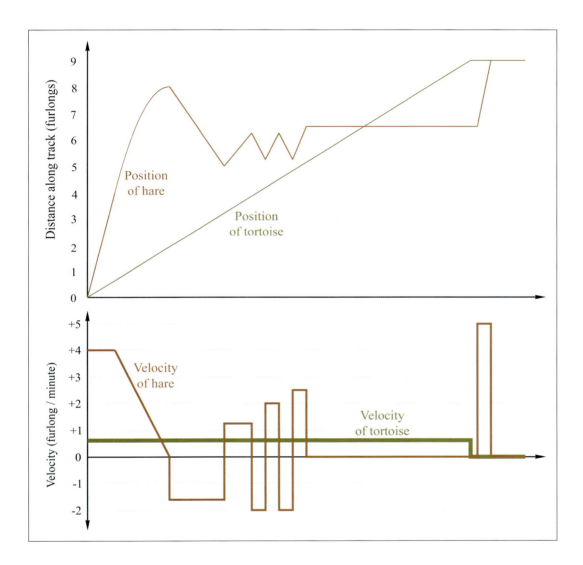

Figure 11.8
Comparing position and velocity

- When the position graph is a straight line, this constant velocity is indicated by a horizontal line in the velocity graph.
- When the position graph is curved, the velocity is changing continuously, and the velocity graph will not be a horizontal line. In this case, the velocity graph happens to be a straight line, but later we'll examine situations where the velocity graph is curved.

- When the position function changes slope at a "corner," the velocity graph exhibits a discontinuity. In fact, the derivative at such points does not exist, and there is no way to define the instantaneous velocity at those points of discontinuity. Fortunately, such situations are nonphysical—in the real world, it is impossible for an object to change its velocity instantaneously. Changes to velocity always occur via an acceleration over a (potentially brief, but finite) amount of time.[21] Later we show that such rapid accelerations over short durations are often approximated by using *impulses*.

- There are sections on the velocity graph that look identical to each other even though the corresponding intervals on the position graph are different from one another. This is because the derivative measures only the rate of change of a variable. The absolute value of the function does not matter. If we add a constant to a function, which produces a vertical shift in the graph of that function, the derivative will not be affected. We have more to say on this when we talk about the relationship between the derivative and integral.

At this point, we should acknowledge a few ways in which our explanation of the derivative differs from most calculus textbooks. Our approach has been to focus on one specific example, that of instantaneous velocity. This has led to some cosmetic differences, such as notation. But there were also many finer points that we are glossing over. For example, we have not bothered defining continuous functions, or given rigorous definitions for when the derivative is defined and when it is not defined. We have discussed the *idea* behind what a limit is, but have not provided a formal definition or considered limits when approached from the left and right, and the criteria for the existence of a well-defined limit. We feel that leading off with the best intuitive example is always the optimum way to teach something, even if it means "lying" to the reader for a short while. If we were writing a calculus textbook, at this point we would back up and correct some of our lies, reviewing the finer points and giving more precise definitions.

However, since this is *not* a calculus textbook, we will only warn you that what we said above is the big picture, but isn't sufficient to handle many edge cases when functions do weird things like go off into infinity or exhibit "jumps" or "gaps." Fortunately, such edge cases just don't happen too often for functions that model physical phenomena, and so these details won't become an issue for us in the context of physics.

---

[21] For all you math sticklers who object to the vertical lines at the discontinuities where the derivative is mathematically undefined, the engineer's justification in this and other similar situations is that the mathematical formula is only a *model* for what is actually a physical situation.

We *do* have room, however, to mention alternate notations for the derivative that you are likely to encounter.

11.4.4   Notations for the Derivative

Several different notations for derivatives are in common use. Let's point out some ways that other texts might look different from what we've said here. First of all, there is a trivial issue of naming. Most calculus textbooks define the derivative in very general terms, where the output variable is named y, the symbol x refers to the input variable rather than the output variable, and the function is simply named f. In other words, the function being differentiated is $y = f(x)$. Furthermore, many will assign the shrinking "step amount" to the variable h rather than using the Δ notation, which has advantages when solving the equations that result when you work out derivatives from the definition.[22] With these variables, they would define the derivative as

$$\frac{dy}{dx} = \lim_{h \to 0} \frac{y(x+h) - y(x)}{h}. \tag{11.6}$$

Definition of a derivative using variables in most calculus textbooks

The differences between Equations (11.3) and (11.6) are clearly cosmetic.

A variation on the Leibniz notation we prefer in this book is to prefix an expression with d/dt to mean "the derivative with respect to t of this thing on the right." For example

$$\frac{d}{dt}(t^2 + 5t)$$

can be read as "the derivative with respect to t of $t^2 + 5t$." This is a very descriptive and intuitive notation. If we call the expression on the right x, and interpret the juxtaposition of symbols as multiplication, we can pull the x back on top of the fraction to get our original notation, as in

$$\frac{d}{dt}(t^2 + 5t) = \frac{d}{dt}x = \frac{dx}{dt}.$$

It's important to interpret these manipulations as notational manipulations rather than having any real mathematical meaning. The notation is attractive because such algebraic manipulations with the infinitesimals often work out. But we reiterate our warning to avoid attaching much mathematical meaning to such operations.

Another common notation is to refer to the derivative of a function $f(x)$ with a prime: $f'(x)$. This is known as *prime notation* or *Lagrange's*

---

[22] Notice our klunky need for parentheses with $(\Delta t)^2$ to avoid the potentially confusing notation Δt^2.

notation. It's used when the independent variable that we are differentiating with respect to is implied or understood by context. Using this notation we would define velocity as the derivative of the position function by $v(t) = x'(t)$.

One last notation, which was invented by Newton and is used mostly when the independent variable is time (such as in the physics equations Newton invented), is *dot notation*. A derivative is indicated by putting a dot over the variable; for example, $v(t) = \dot{x}(t)$.

Here is a summary of the different notations for the derivative you will see, using velocity and position as the example:

$$v(t) = \frac{dx}{dt} = \frac{d}{dt}x(t) = x'(t) = \dot{x}(t).$$

11.4.5   A Few Differentiation Rules and Shortcuts

Now let's return to calculating derivatives. In practice, it's seldom necessary to go back to the definition of the derivative in order to differentiate an expression. Instead, there are simplifying rules that allow you to break down complicated functions into smaller pieces that can then be differentiated. There are also special functions, such as $\ln x$ and $\tan x$, for which the hard work of applying the definition has already been done and written down in those tables that line the insides of the front and back covers of calculus books. To differentiate expressions containing such functions, one simply refers to the table (although we're going to do just a bit of this "hard work" ourselves for sine and cosine).

In this book, our concerns are limited to the derivatives of a very small set of functions, which luckily can be differentiated with just a few simple rules. Unfortunately, we don't have the space here to develop the mathematical derivations behind these rules, so we are simply going to accompany each rule with a brief explanation as to how it is used, and a (mathematically nonrigorous) intuitive argument to help you convince yourself that it works.

Our first rule, known as the *constant rule*, states that the derivative of a constant function is zero. A constant function is a function that always produces the same value. For example, $x(t) = 3$ is a constant function. You can plug in any value of t, and this function outputs the value 3. Since, the derivative measures how fast the output of a function changes in response to changes in the input t, in the case of a constant function, the output *never changes*, and so the derivative is $x'(t) = 0$.

The Constant Rule

$$\frac{d}{dt}\, k = 0, \quad k \text{ is any constant.}$$

The next rule, sometimes known as the *sum rule*, says that differentiation is a *linear* operator. The meaning of "linear" is essentially identical to our definition given in Chapter 5, but let's review it in the context of the derivative. To say that the derivative is a linear operator means two things. First, to take the derivative of a sum, we can just take the derivative of each piece individually, and add the results together. This is intuitive—the rate of change of a sum is the total rate of change of all the parts added together. For example, consider a man who moves about on a train. His position in world space can be described as the sum of the train's position, plus the man's position in the body space of the train.[23] Likewise, his velocity relative to the ground is the sum of the train's velocity relative to the ground, plus his velocity relative to the train.

Derivative of a Sum

$$\frac{d}{dt}\, [f(t) + g(t)] = \frac{d}{dt}\, f(t) + \frac{d}{dt}\, g(t). \tag{11.7}$$

The second property of linearity is that if we multiply a function by some constant, the derivative of that function gets scaled by that same constant. One easy way to see that this must be true is to consider unit conversions. Let's return to our favorite function that yields a hare's displacement as a function of time, measured in furlongs. Taking the derivative of this function with respect to time yields a velocity, in furlongs per minute. If somebody comes along who doesn't like furlongs, we can switch from furlongs to meters, by scaling the original position function by a factor of 201.168. This *must* scale the derivative by the same factor, or else the hare would suddenly change speed just because we switched to another unit.

---

[23] Assume that the train tracks are straight, so that the train's body axes are aligned with the world axes, and no rotation is needed.

Derivative of a Function Times a Constant

$$\frac{d}{dt}\left[kf(t)\right] = k\left[\frac{d}{dt}\,f(t)\right], \quad k \text{ is any constant.} \tag{11.8}$$

If we combine Equations (11.7) and (11.8), we can state the linearity rule in a more general way.

The Sum Rule

$$\frac{d}{dt}\left[af(t) + bg(t)\right] = a\left[\frac{d}{dt}\,f(t)\right] + b\left[\frac{d}{dt}\,g(t)\right].$$

The linear property of the derivative is very important since it allows us to break down many common functions into smaller, easier pieces.

One of the most important and common functions that needs to be differentiated also happens to be the easiest: the polynomial. Using the linear property of the derivative, we can break down, for example, a fourth-degree polynomial with ease:

$$x(t) = c_4 t^4 + c_3 t^3 + c_2 t^2 + c_1 t + c_0,$$
$$\frac{dx}{dt} = \frac{d}{dt}[c_4 t^4 + c_3 t^3 + c_2 t^2 + c_1 t + c_0]$$
$$= c_4\left[\frac{d}{dt}\,t^4\right] + c_3\left[\frac{d}{dt}\,t^3\right] + c_2\left[\frac{d}{dt}\,t^2\right] + c_1\left[\frac{d}{dt}\,t\right] + \left[\frac{d}{dt}\,c_0\right]. \tag{11.9}$$

The last derivative $\frac{d}{dt}c_0$ is zero by the constant rule, since c_0 does not vary. This leaves us with four simple derivatives, each of which can be plugged into the definition of a derivative, Equation (11.3), without too much trouble. Solving each of these four individually is considerably easier than plugging the original polynomial into Equation (11.3). If you do go through this exercise (like every first-year calculus student does), you notice two things. First of all, the algebraic tedium increases as the power of t gets higher. Second, a quite obvious pattern is revealed, known as the *power rule*.

The Power Rule

$$\frac{d}{dt} t^n = nt^{n-1}, \quad n \text{ is an integer.}$$

This rule gives us the answers to the four derivatives needed above:

$$\frac{d}{dt} t^4 = 4t^3, \qquad\qquad \frac{d}{dt} t^3 = 3t^2,$$

$$\frac{d}{dt} t^2 = 2t^1 = 2t, \qquad\qquad \frac{d}{dt} t = 1t^0 = 1.$$

Notice in the last equation we used the identity $t^0 = 1$. However, even without that identity,[24] it should be very clear that $\frac{d}{dt}t$ *must* be unity. Remember that the derivative answers the question, "What is the rate of change of the output, relative to the rate of change of the input?" In the case of $\frac{d}{dt}t$, the "output" and the "input" are both the variable t, and so their rates of change are equal. Thus the ratio that defines the derivative is equal to one.

One last comment before we plug these results into Equation (11.9) to differentiate our polynomial. Using the identity $t^0 = 1$, the power rule is brought into harmony with the constant rule:

$$\frac{d}{dt} k = \frac{d}{dt} (kt^0) \qquad\qquad \text{Using } t^0 = 1,$$

$$= k \left[\frac{d}{dt} t^0 \right] \qquad\qquad \text{Linear property of derivative,}$$

$$= k[0(t^{-1})] \qquad\qquad \text{Power rule for } n = 0,$$

$$= 0.$$

Derivative of a constant, using the power rule

Let's get back to our fourth-degree polynomial. With the sum and power rule at our disposal, we can make quick work of it:

$$x(t) = c_4 t^4 + c_3 t^3 + c_2 t^2 + c_1 t + c_0,$$

$$\frac{dx}{dt} = 4c_4 t^3 + 3c_3 t^2 + 2c_2 t + c_1.$$

Below are several more examples of how the power rule can be used.

---

[24]Be careful, t^0 is *undefined* when $t = 0$.

Notice that the power rule works for negative exponents as well:

$$\frac{d}{dt}\left(3t^5 - 4t\right) = 15t^4 - 4,$$

$$\frac{d}{dt}\left(\frac{t^{100}}{100} + \sqrt{\pi}\right) = t^{99},$$

$$\frac{d}{dt}\left(\frac{1}{t} + \frac{4}{t^3}\right) = \frac{d}{dt}\left(t^{-1} + 4t^{-3}\right) = -t^{-2} - 12t^{-4} = \frac{-1}{t^2} - \frac{12}{t^4}.$$

11.4.6 Derivatives of Some Special Functions with Taylor Series

This section looks at some very special examples of differentiating polynomials. Given any arbitrary function $f(x)$, the *Taylor series* of f is a way to express f as a polynomial. Each successive term in the polynomial is determined by taking a higher order derivative of the function, which is perhaps the main point of Taylor series that you should learn when you take a real calculus class, but right now we're not interested in where Taylor series come from, just that they exist. The Taylor series is a very useful tool in video games because it provides polynomial approximations, which are "easy" to evaluate in a computer, for functions that are otherwise "hard" to evaluate. We don't have the space to discuss much of anything about Taylor series in general, but we would like to look at a few important examples of Taylor series. The Taylor series for the sine and cosine functions are

Taylor series for sin(x) and cos(x)

$$\sin x = x - \frac{x^3}{3!} + \frac{x^5}{5!} - \frac{x^7}{7!} + \frac{x^9}{9!} + \cdots,$$

$$\cos x = 1 - \frac{x^2}{2!} + \frac{x^4}{4!} - \frac{x^6}{6!} + \frac{x^8}{8!} + \cdots. \qquad (11.10)$$

This pattern continues forever; in other words, to compute the exact value of $\sin x$ would require us to evaluate an infinite number of terms. However, notice that the denominators of the terms are growing very rapidly, which means we can approximate $\sin x$ simply by stopping after a certain number of terms, and ignore the rest.

   This is exactly the process by which trigonometric functions are computed inside a computer. First, trig identities are used to get the argument into a restricted range (since the functions are periodic). This is done because when the Taylor series is truncated, its accuracy is highest near a particular value of x, and in the case of the trig functions, this point is usually chosen to be $x = 0$.[25] Then the Taylor series polynomial with, say, four terms is evaluated. This approximation is highly accurate. Stopping at the x^7 term is sufficient to calculate $\sin x$ to about five and a half decimal digits for $-1 < x < +1$.

All this trivia concerning approximations is interesting, but our real reason for bringing up Taylor series is to use them as nontrivial examples of differentiating polynomials with the power rule, and also to learn some interesting facts about the sine, cosine, and exponential functions. Let's use the power rule to differentiate the Taylor series expansion of $\sin(x)$. It's not that complicated—we just have to differentiate each term by itself. We're not even intimidated by the fact that there are an infinite number of terms:

$$\frac{d}{dx}\sin x = \frac{d}{dx}\left(x - \frac{x^3}{3!} + \frac{x^5}{5!} - \frac{x^7}{7!} + \frac{x^9}{9!} + \cdots\right)$$

Differentiating Taylor series for sin(x)

$$= \frac{d}{dx}x - \frac{d}{dx}\frac{x^3}{3!} + \frac{d}{dx}\frac{x^5}{5!} - \frac{d}{dx}\frac{x^7}{7!} + \frac{d}{dx}\frac{x^9}{9!} + \cdots \quad \text{(Sum rule)}$$

$$= 1 - \frac{3x^2}{3!} + \frac{5x^4}{5!} - \frac{7x^6}{7!} + \frac{9x^8}{9!} + \cdots \quad \text{(Power rule)}$$

$$= 1 - \frac{x^2}{2!} + \frac{x^4}{4!} - \frac{x^6}{6!} + \frac{x^8}{8!} + \cdots \quad (11.11)$$

In the above derivation, we first used the sum rule, which says that to differentiate the whole Taylor polynomial, we can differentiate each term individually. Then we applied the power rule to each term, in each case multiplying by the exponent and decrementing it by one. (And also remembering that $\frac{d}{dx}x = 1$ for the first term.) To understand the last step, remember the definition of the factorial operator: $n! = 1 \times 2 \times 3 \times \cdots \times n$. Thus the constant in the numerator of each term cancels out the highest factor in the factorial in the denominator.

Does Equation (11.11) the last look familiar? It should, because it's the same as Equation (11.10), the Taylor series for $\cos x$. In other words, we now know the derivative of $\sin x$, and by a similar process we can also obtain the derivative of $\cos x$. Let's state these facts formally.[26]

---

[25] In this special case, the Taylor series is given the more specific name of *Maclaurin series*.

[26] We emphasize that we have not proven that these are the correct derivatives, because we started with the Taylor series expansion, which is actually *defined* in terms of the derivatives.

Derivatives of Sine and Cosine

$$\frac{d}{dx}\sin x = \cos x, \qquad\qquad \frac{d}{dx}\cos x = -\sin x.$$

The derivatives of the sine and cosine functions will become useful in later sections.

Now let's look at one more important special function that will play an important role later in this book, which will be convenient to be able to differentiate, and which also happens to have a nice, tidy Taylor series. The function we're referring to is the *exponential* function, denoted e^x. The mathematical constant $e \approx 2.718282$ has many well known and interesting properties, and pops up in all sorts of problems from finance to signal processing. Much of e's special status is related to the unique nature of the function e^x. One manifestation of this unique nature is that e^x has such a beautiful Taylor series:

Taylor series of e^x

$$e^x = 1 + x + \frac{x^2}{2!} + \frac{x^3}{3!} + \frac{x^4}{4!} + \frac{x^5}{5!} + \cdots \qquad (11.12)$$

Taking the derivative gives us

$$\frac{d}{dx}e^x = \frac{d}{dx}\left(1 + x + \frac{x^2}{2!} + \frac{x^3}{3!} + \frac{x^4}{4!} + \frac{x^5}{5!} + \cdots\right)$$

$$= 0 + 1 + \frac{x}{1!} + \frac{x^2}{2!} + \frac{x^3}{3!} + \frac{x^4}{4!} + \cdots$$

$$= 1 + x + \frac{x^2}{2!} + \frac{x^3}{3!} + \frac{x^4}{4!} + \cdots$$

But this result is equivalent to the definition of e^x in Equation (11.12); the only difference between them is the cosmetic issue of when to stop listing terms explicitly and end with the "$\cdots$". In other words, the exponential function is its own derivative: $d/dx\ e^x = e^x$. The exponential function is the only function that can boast this unique property. (To be more precise, any *multiple* of the exponential function, including zero, has this quality.)

The Exponential Function Is Its Own Derivative

$$\frac{d}{dx}e^x = e^x.$$

It is this special property about the exponential function that makes it unique and causes it to come up so frequently in applications. Anytime the rate of change of some value is proportionate to the value itself, the exponential function will almost certainly arise somewhere in the math that describes the dynamics of the system.

The example most of us are familiar with is compound interest. Let $P(t)$ be the amount of money in your bank account at time t; assume the amount is accruing interest. The rate of change per time interval—the amount of interest earned—is proportional to the amount of money in your account. The more money you have, the more interest you are earning, and the faster it grows. Thus, the exponential function works its way into finance with the equation $P(t) = P_0 e^{rt}$, which describes the amount of money at any given time t, assuming an initial amount P_0 grows at an interest rate of r, where the interest is compounded continually.

You might have noticed that the Taylor series of e^x is strikingly similar to the series representation of $\sin x$ and $\cos x$. This similarity hints at a deep and surprising relationship between the exponential functions and the trig functions, which we explore in Exercise 11.

We hope this brief encounter with Taylor series, although a bit outside of our main thrust, has sparked your interest in a mathematical tool that is highly practical, in particular for its fundamental importance to all sorts of approximation and numerical calculations in a computer. We also hope it was an interesting non-trivial example of differentiation of a polynomial. It also has given us a chance to discuss the derivatives of the sine, cosine, and exponential functions; these derivatives come up again in later sections.

11.4.7   The Chain Rule

The chain rule is the last rule of differentiation we discuss here. The chain rule tells us how to determine the rate of change of a function when the argument to that function is itself some other function we know how to differentiate.

In the race between the tortoise and hare, we never really thought much about exactly what our function $x(t)$ measured, we just said it was the "position" of the hare. Let's say that the course was actually a winding track with hills and bridges and even a vertical loop, and that the function that we graphed and previously named $x(t)$ actually measures the *linear distance* along this winding path, rather than, say, a horizontal position. To avoid the horizontal connotations associated with the symbol x, let's introduce the variable s, which gives the distance along the track (in furlongs, of course).

Let's say that we have a function $y(s)$ that describes the altitude of the track at a given distance. The derivative dy/ds tells us very basic things about the track at that location. A value of zero means the course is flat at that location, a positive value means the runners are running uphill, and a large positive or negative value indicates a location where the track is very steep.

Now consider the composite function $y(s(t))$. You should be able to convince yourself that this tells us the hare's altitude for any given time t. The derivative dy/dt tells us how fast the hare was moving vertically, at a given time t. This is very different from dy/ds. How might we calculate dy/dt? You might be tempted to say that to make this determination, we simply find out where the hare was on the track at time t, and then the answer is the slope of the track at this location. In math symbols, you are saying that the vertical velocity is $y'(s(t))$. But that isn't right. For example, while the hare was taking a nap ($ds/dt = 0$), it doesn't matter what the slope of the track was; since he wasn't moving along it, his vertical velocity is zero! In fact, at a certain point in the race he turned around and ran on the track in the wrong direction ($ds/dt < 0$), so his vertical velocity dy/dt would be *opposite* of the track slope dy/ds. And obviously if he sprints quickly over a place in the track, his vertical velocity will be higher than if he strolled slowly over that same spot. But likewise, where the track is flat, it doesn't matter how fast he runs across it, his vertical velocity will be zero. So we see that the hare's vertical velocity is the *product* of his speed (measured parametrically along the track) and the slope of the track at that point.

This rule is known as the *chain rule*. It is particularly intuitive when written in Leibniz notation, because the ds infinitesimals appear to "cancel."

The Chain Rule of Differentiation

$$\frac{dy}{dt} = \frac{dy}{ds}\frac{ds}{dt}.$$

Here are a few examples, using functions we now know how to differentiate:

$$\frac{d}{dt}\sin 3x = 3\cos 3x,$$

$$\frac{d}{dt}\sin(x^2) = 2x\cos(x^2),$$

$$\frac{d}{dt}e^{\cos x + 3x} = (-\sin x + 3)e^{\cos x + 3x},$$

$$\frac{d}{dt}e^{\sin 3x + \sin(x^2)} = (3\cos 3x + 2x\cos(x^2))e^{\sin 3x + \sin(x^2)}.$$

Examples of the chain rule

We're going to put calculus from a purely mathematical perspective on the shelf for a while and return our focus to kinematics. (After all, our purpose in discussing calculus was, like Ike Newton, to improve our understanding of mechanics.) However, it won't be long before we will return to calculus with the discussion of the integral and the fundamental theorem of calculus.

11.5   Acceleration

We've made quite a fuss about the distinction between instantaneous velocity and average velocity, and this distinction is important (and the fuss is justified) when the velocity is changing continuously. In such situations, we might be interested to know the *rate* at which the velocity is changing. Luckily we have just learned about the derivative, whose *raison d'être* is to investigate rates of change. When we take the derivative of a velocity function $v(t)$ we get a new function describing how quickly the velocity is increasing or decreasing at that instant. This instantaneous rate of change is an important quantity in physics, and it goes by a familiar name: *acceleration*.

In ordinary conversation, the verb "accelerate" typically means "speed up." However, in physics, the word "acceleration" carries a more general meaning and may refer to *any* change in velocity, not just an increase in speed. In fact, a body can undergo an acceleration even when its speed is constant! How can this be? Velocity is a vector value, which means it has both magnitude and direction. If the direction of the velocity changes, but the magnitude (its speed) remains the same, we say that the body is experiencing an acceleration. Such terminology is not mere nitpicking with words, the acceleration in this case is a very real sensation that would be felt by, say, two people riding in the back seat of a swerving car who find

themselves pressed together to one side. We have more to say about this
particular situation in Section 11.8.

We can learn a lot about acceleration just by asking ourselves what sort
of units we should use to measure it. For velocity, we used the generic units
of L/T, unit length per unit time. Velocity is a rate of change of position
(L) per unit time (T), and so this makes sense. Acceleration is the rate of
change of velocity per unit time, and so it must be expressed in terms of
"unit velocity per unit time." In fact, the units used to measure velocity
are L/T^2. If you are disturbed by the idea of "time squared," think of it
instead as $(L/T)/T$, which makes more explicit the fact that it is a unit of
velocity (L/T) per unit time.

For example, an object in free fall near Earth's surface accelerates at a
rate of about 32 ft/s^2, or 9.8 m/s^2. Let's say that you are dangling a metal
bearing off the side of Willis Tower.[27] You drop the bearing, and it begins
accelerating, adding 9.8 m/s to its downward velocity each second. (We are
ignoring wind resistance.) After, say, 2.4 seconds, its velocity will be

$$2.4 \text{ s} \times 32 \ \frac{\text{ft}}{\text{s}^2} = 76.8 \ \frac{\text{ft}}{\text{s}}.$$

More generally, the velocity at an arbitrary time t of an object under constant acceleration is given by the simple linear formula

$$v(t) = v_0 + at, \tag{11.13}$$

where v_0 is the initial velocity at time $t = 0$, and a is the constant acceleration. We study the motion of objects in free fall in more detail in
Section 11.6, but first, let's look at a graphical representation of acceleration. Figure 11.9 shows plots of a position function and the corresponding
velocity and acceleration functions.

You should study Figure 11.9 until it makes sense to you. In particular,
here are some noteworthy observations:

- Where the acceleration is zero, the velocity is constant and the position is a straight (but possibly sloped) line.

- Where the acceleration is positive, the position graph is curved like
  ∪, and where it is negative, the position graph is curved like ∩.
  The most interesting example occurs on the right side of the graphs.
  Notice that at the time when the acceleration graph crosses $a = 0$,
  the velocity curve reaches its apex, and the position curve switches
  from ∪ to ∩.

---

[27]The building that everybody still calls the Sears Tower.

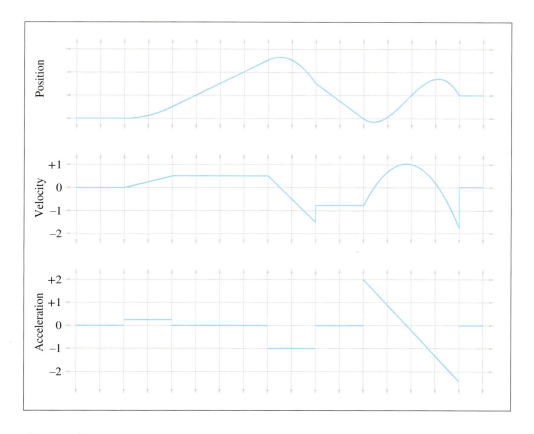

Figure 11.9
Plots of position, velocity, and acceleration over time.

- A discontinuity in the velocity function causes a "kink" in the position graph. Furthermore, it causes the acceleration to become infinite (actually, undefined), which is why, as we said previously, such discontinuities don't happen in the real world. This is why the lines in the velocity graph are connected at those discontinuities, because the graph is of a physical situation being approximated by a mathematical model.

- A discontinuity in the acceleration graph causes a kink in the velocity graph, but notice that the position graph is still smooth. In fact, acceleration *can* change instantaneously, and for this reason we have chosen not to bridge the discontinuities in the acceleration graph.

The accelerations experienced by an object can vary as a function of time, and indeed we can continue this process of differentiation, resulting in yet another function of time, which some people call the "jerk" function. We stick with the position function and its first two derivatives in this book. Furthermore, it's very instructive to consider situations in which the acceleration is constant (or at least has constant magnitude). This is precisely what we're going to do in the next few sections.

Section 11.6 considers objects under constant acceleration, such as objects in free fall and projectiles. This will provide an excellent backdrop to introduce the integral, the complement to the derivative, in Section 11.7. Then Section 11.8 examines objects traveling in a circular path, which experience an acceleration that has a constant magnitude but a direction that changes continually and always points towards the center of the circle.

11.6   Motion under Constant Acceleration

Let's look now at the trajectory an object takes when it accelerates at a constant rate over time. This is a simple case, but a common one, and an important one to fully understand. In fact, the equations of motion we present in this section are some of the most important mechanics equations to know by heart, especially for video game programming.

Before we begin, let's consider an even simpler type of motion—motion with constant velocity. Motion with constant velocity is a special case of motion with constant acceleration—the case where the acceleration is constantly zero. The motion of a particle with constant velocity is an intuitive linear equation, essentially the same as Equation (9.1), the equation of a ray. In one dimension, the position of a particle as a function of time is

$$x(t) = x_0 + vt, \tag{11.14}$$

where x_0 is the position of the particle at time $t = 0$, and v is the constant velocity.

Now let's consider objects moving with constant acceleration. We've already mentioned at least one important example: when they are in free fall, accelerating due to gravity. (We'll ignore wind resistance and all other forces.) Motion in free fall is often called *projectile motion*. We start out in one dimension here to keep things simple. Our goal is a formula $x(t)$ for the position of a particle at a given time.

Take our example of illegal ball-bearing-bombing off of Willis Tower. Let's set a reference frame where x increases in the downward direction, and $x_0 = 0$. In other words, $x(t)$ measures the distance the object has fallen from its drop height at time t. We also assume for now that initial velocity is $v_0 = 0$ ft/s, meaning you merely release the ball bearing and don't throw it.

At this point, we don't even know what form $x(t)$ should take, so we're a bit stuck. The "front door" to this solution seems to be locked for us at the moment, so instead we try to sneak around and enter through the back, using an approach similar to the one we used earlier to define instantaneous velocity. We'll consider ways that we might approximate the answer and then watch what happens as the approximations get better and better.

Let's make our example a bit more specific. Earlier, we computed that after being in free fall for 2.4 seconds, the ball bearing would have a velocity of $v(2.4) = 76.8$ ft/s. However, we didn't say anything about how far it had traveled during that time. Let's try to compute this distance, which is $x(2.4)$. To do this, we chop up the total 2.4 second interval into a number of smaller "slices" of time, and approximate how far the ball bearing travels during each slice. We can approximate the total distance traveled as the sum of the distances traveled during each slice. To approximate how far the ball bearing travels during one single slice, we first compute the velocity of the ball bearing at the start of the slice by using Equation (11.13). Then we approximate the distance traveled during the slice by plugging this velocity as the constant velocity for the slice into Equation (11.14).

6 Slices, $\Delta t = 0.40$

t_0	v_0	Δx
0.00	0.00	0.00
0.40	12.80	5.12
0.80	25.60	10.24
1.20	38.40	15.36
1.60	51.20	20.48
2.00	64.00	25.60
	Total	76.80

12 Slices, $\Delta t = 0.20$

t_0	v_0	Δx
0.00	0.00	0.00
0.20	6.40	1.28
0.40	12.80	2.56
0.60	19.20	3.84
0.80	25.60	5.12
1.00	32.00	6.40
1.20	38.40	7.68
1.40	44.80	8.96
1.60	51.20	10.24
1.80	57.60	11.52
2.00	64.00	12.80
2.20	70.40	14.08
	Total	84.48

24 Slices, $\Delta t = 0.10$

t_0	v_0	Δx
0.00	0.00	0.00
0.10	3.20	0.32
0.20	6.40	0.64
0.30	9.60	0.96
0.40	12.80	1.28
0.50	16.00	1.60
0.60	19.20	1.92
0.70	22.40	2.24
0.80	25.60	2.56
0.90	28.80	2.88
1.00	32.00	3.20
1.10	35.20	3.52
1.20	38.40	3.84
1.30	41.60	4.16
1.40	44.80	4.48
1.50	48.00	4.80
1.60	51.20	5.12
1.70	54.40	5.44
1.80	57.60	5.76
1.90	60.80	6.08
2.00	64.00	6.40
2.10	67.20	6.72
2.20	70.40	7.04
2.30	73.60	7.36
	Total	88.32

Table 11.3. Values for different numbers of slices

Table 11.3 shows tabulated values for 6, 12, and 24 slices. For each slice, t_0 refers to the starting time of the slice, v_0 is the velocity at the start of the slice (computed according to Equation (11.13) as $v_0 = t_0 \times 32$ ft/s^2), Δt is the duration of the slice, and Δx is our approximation for the displacement during the slice (computed according to Equation (11.14) as $\Delta x = v_0 \Delta t$).

Since each slice has a different initial velocity, we are accounting for the fact that the velocity changes over the entire interval. (In fact, the computation of the starting velocity for the slice is not an approximation—it is exact.) However, since we ignore the change in velocity *within* a slice, our answer is only an approximation. Taking more and more slices, we get better and better approximations, although it's difficult to tell to what value these approximations are converging. Let's look at the problem graphically to see if we can gain some insight.

In Figure 11.10, each rectangle represents one time interval in our approximation. Notice that the distance traveled during an interval is the same as the area of the corresponding rectangle:

$$\begin{aligned}
(\text{area of rectangle}) &= (\text{width of rectangle}) \times (\text{height of rectangle}) \\
&= (\text{duration of slice}) \times (\text{velocity used for slice}) \\
&= (\text{displacement during slice}).
\end{aligned}$$

Now we come to the key observation. As we increase the number of slices, the total area of the rectangles becomes closer and closer to the area of the triangle under the velocity curve. In the limit, if we take an infinite number of rectangles, the two areas will be equal. Now, since total displacement of the falling ball bearing is equal to the total area of the rectangles, which is equal to the area under the curve, we are led to an important discovery.

The distance traveled is equal to the area under the velocity curve.

We have come to this conclusion by using a limit argument very similar to the one we made to define instantaneous velocity—we consider how a series of approximations converges in the limit as the approximation error goes to zero.

Notice that we have made no assumptions in this argument about $v(t)$. In the example at hand, it is a simple linear function, and the graph is a straight line; however, you should be able to convince yourself that this

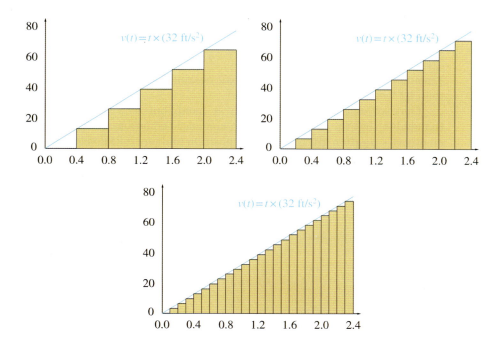

Figure 11.10
Graphical representation of Table 11.3

procedure will work for any arbitrary velocity function.[28] This limit argument is a formalized tool in calculus known as the *Riemann integral*, which we will consider in Section 11.7. That will also be the appropriate time to consider the general case of any $v(t)$. However, since there is so much we can learn from this specific example, let's keep it simple as long as possible.

Remember the question we're trying to answer: how far does an object travel after being dropped at an initial zero velocity and then accelerated due to gravity for 2.4 seconds at a constant rate of 32 ft/s²? How does this new realization about the equivalence of distance traveled and the area under the graph of $v(t)$ help us? In this special case, $v(t)$ is a simple linear function, and the area under the curve from $t = 0$ to $t = 2.4$ is a triangle. That's an easy shape for us to compute an area. The base of this triangle

---

[28]Well, almost. There are certain limitations we must place on $v(t)$. For example, if it blows up and goes to infinity, it's likely, though not certain, that the displacement will be infinite or undefined. In this book, we are focused on physical phenomena and so we sidestep these issues by assuming our functions will be well-behaved.

has length 2.4 s, and the height is $v(2.4) = 76.8$ ft/s, so the area is

$$\frac{\text{base} \times \text{height}}{2} = \frac{2.4 \text{ s} \times 76.8 \text{ ft/s}}{2} = 92.16 \text{ ft}.$$

Thus after a mere 2.4 seconds, the ball bearing had already dropped more than 92 feet!

That solves the specific problem at hand, but let's be more general. Remember that the larger goal was a kinematic equation $x(t)$ that predicts an object's position given any initial position and any initial velocity. First, let's replace the constant 2.4 with an arbitrary time t. Next, let's remove the assumption that the object initially has zero velocity, and instead allow an arbitrary initial velocity v_0. This means the area under the curve $v(t)$ is no longer a triangle—it is a triangle on top of a rectangle, as shown in Figure 11.11.

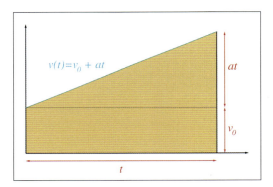

Figure 11.11
Calculating displacement at time t, given initial velocity v_0 and constant acceleration a

The rectangle has base t and height v_0, and its area represents the distance that would be traveled if there were no acceleration. The triangle on top of the rectangle also has base t, and the height is at, the difference in $v(t)$ compared to the initial velocity as a result of the acceleration at the rate a over the duration of t seconds. Summing these two parts together yields the total displacement, which we denote as Δx:

$$\Delta x = (\text{Area of rectangle}) + (\text{Area of triangle})$$

$$= \left(\begin{array}{c} \text{Rectangle} \\ \text{base} \end{array} \right) \left(\begin{array}{c} \text{Rectangle} \\ \text{height} \end{array} \right) + \frac{1}{2} \left(\begin{array}{c} \text{Triangle} \\ \text{base} \end{array} \right) \left(\begin{array}{c} \text{Triangle} \\ \text{height} \end{array} \right)$$

$$= (t)(v_0) + (1/2)(t)(at)$$

$$= v_0 t + (1/2)at^2.$$

We have just derived a very useful equation, so let's highlight it so that people who are skimming will notice it.

Formula for Displacement Given Initial Velocity and Constant Acceleration

$$\Delta x = v_0 t + (1/2)at^2. \tag{11.15}$$

Equation (11.15) is one of only a handful of equations in this book that are worth memorizing. It is very useful for solving practical problems that arise in physics simulations.[29]

It's common that we only need the displacement Δx, and the absolute position $x(t)$ doesn't matter. However, since the function $x(t)$ was our stated goal, we can easily express $x(t)$ in terms of Equation (11.15) by adding the displacement to our initial position, which we denote as x_0:

$$x(t) = x_0 + \Delta x = x_0 + v_0 t + (1/2)at^2.$$

Let's work through some examples to show the types of problems that can be solved by using Equation (11.15) and its variants. One tempting scenario is to let our ball bearing hit the ground. The observation deck on the 103rd floor of Willis Tower is 1,353 ft above the sidewalk. If it is dropped from that height, how long will it take to fall to the bottom? Solving Equation (11.15) for t, we have

$$\Delta x = v_0 t + (1/2)at^2$$

$$0 = (a/2)t^2 + v_0 t - \Delta x$$

Solving for time

$$t = \frac{-v_0 \pm \sqrt{v_0^2 - 4(a/2)(-\Delta x)}}{2(a/2)} \qquad \text{(quadratic formula)}$$

$$t = \frac{-v_0 \pm \sqrt{v_0^2 + 2a\Delta x}}{a}. \tag{11.16}$$

Equation (11.16) is a very useful general equation. Plugging in the numbers specific to this problem, we have

$$t = \frac{-v_0 \pm \sqrt{v_0^2 + 2a\Delta x}}{a}$$

$$= \frac{-(0) \pm \sqrt{(0)^2 + 2(32 \text{ ft/s}^2)(1\,353 \text{ ft})}}{32 \text{ ft/s}^2}$$

---

[29]It comes up frequently in job interviews, too.

$$= \pm \frac{\sqrt{86\,592\ (\mathrm{ft/s})^2}}{32\ \mathrm{ft/s}^2}$$

$$\approx \pm \frac{294.3\ \mathrm{ft/s}}{32\ \mathrm{ft/s}^2}$$

$$\approx \pm 9.197\ \mathrm{s}.$$

The square root in Equation (11.16) introduces the possibility for two solutions. We always use the root that results in a positive value for t.[30] Naturally, a person in the business of dropping ball bearings from great heights is interested in how much damage he can do, so the next logical question is, "How fast is the ball bearing traveling when it hits the sidewalk?" To answer this question, we plug the total travel time into Equation (11.13):

$$v(t) = v_0 + at = 0\ \mathrm{ft/s} + (32\ \mathrm{ft/s}^2)(9.197\ \mathrm{s}) = 294.3\ \mathrm{ft/s}.$$

If we ignore wind resistance, at the moment of impact, the ball bearing is traveling at a speed that covers a distance of roughly a football field in one second! You can see why the things we are doing in our imagination are illegal in real life. Let's keep doing them.

Now let's assume that instead of just dropping the ball bearing, we give it an initial velocity (we toss it up or down). It was our free choice to decide whether up or down is positive in these examples, and we have chosen $+x$ to be the downward direction, so that means the initial velocity will be negative. What must the initial velocity be in order for the ball bearing to stay in the air only a few seconds longer, say a total of 12 seconds? Once again, we'll first manipulate Equation (11.15) to get a general solution; this time we'll be solving for v_0:

Solving for initial velocity

$$\Delta x = v_0 t + (1/2)at^2,$$
$$-v_0 t = -\Delta x + (1/2)at^2,$$
$$v_0 = \Delta x/t - (1/2)at.$$

And now plugging in the numbers for our specific problem, we have

$$v_0 = \Delta x/t - (1/2)at$$
$$= (1\,353\ \mathrm{ft})/(12.0\ \mathrm{s}) - (1/2)(32\ \mathrm{ft/s}^2)(12.0\ \mathrm{s})$$
$$= 112.8\ \mathrm{ft/s} - 192\ \mathrm{ft/s}$$
$$= -79.2\ \mathrm{ft/s}.$$

---

[30] The negative root tells us the other point where the infinite parabola containing the ball bearing's trajectory crosses the sidewalk.

Notice that the result is negative, indicating an upwards velocity. If we give the ball bearing this initial velocity, we might wonder how long it takes for the bearing to come back down to its initial position. Using Equation (11.16) and letting $\Delta x = 0$, we have

$$t = \frac{-v_0 \pm \sqrt{v_0^2 + 2a\Delta x}}{a}$$

$$= \frac{-(-79.2 \text{ ft/s}) \pm \sqrt{(-79.2 \text{ ft/s})^2 + 2(32 \text{ ft/s}^2)(0 \text{ ft})}}{32 \text{ ft/s}^2}$$

$$= \frac{79.2 \text{ ft/s} \pm \sqrt{(-79.2 \text{ ft/s})^2}}{32 \text{ ft/s}^2}$$

$$= \frac{79.2 \text{ ft/s} \pm 79.2 \text{ ft/s}}{32 \text{ ft/s}^2}$$

$$= 0 \text{ or } 4.95 \text{ s}.$$

It's no surprise that $t = 0$ is a solution; we were solving for the time values when the ball bearing was at its initial position.

Examine the graph in Figure 11.12, which plots the position and velocity of an object moving under constant velocity a with an initial velocity v_0, where v_0 and a have opposite signs. Let's make three key observations. Although we use terms such as "height," which are specific to projectile motion, similar statements are true anytime the signs of v_0 and a are opposite.

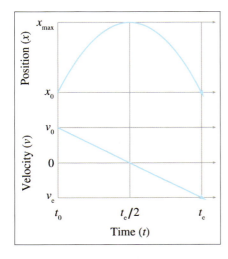

Figure 11.12
Projectile motion

The first observation is that the projectile reaches its maximum height, denoted $x_{\max}$, when the acceleration has consumed all of the velocity and $v(t) = 0$. It's easy to solve for the time when this will occur by using Equation (11.13), $v(t) = v_0 + at$:

Time to reach apex

$$v(t) = 0,$$
$$v_0 + at = 0,$$
$$t = -v_0/a.$$

Right now we are in one dimension and considering only the height. But if we are in more than one dimension, only the velocity parallel to the acceleration must vanish. There could be horizontal velocity, for example. We discuss projectile motion in more than one dimension in just a moment.

The second observation is that the time it takes for the object to travel from its maximum altitude to its initial altitude, denoted t_e in Figure 11.12, is the same as the time taken to reach the maximum. In other words, the projectile reaches its apex at $t_e/2$.

The third and final observation is that the velocity at $t = t_e$, which we have denoted v_e, has the same magnitude as the initial velocity v_0, but the opposite sign.

Before we look at projectile motion in more than one dimension, let's summarize the formulas we have derived in this section. The first two are the only ones worth memorizing; the others can be derived from them.

Summary of Kinematics Equations Dealing with Constant Acceleration

$$v(t) = v_0 + at,$$
$$\Delta x = v_0 t + (1/2)at^2,$$
$$x(t) = x_0 + \Delta x = x_0 + v_0 t + (1/2)at^2,$$
$$v_0 = \Delta x/t - (1/2)at,$$
$$t = \frac{-v_0 \pm \sqrt{v_0^2 + 2a\Delta x}}{a}, \tag{11.17}$$
$$a = 2\frac{\Delta x - v_0 t}{t^2}.$$

Extending the ideas from the previous section into 2D or 3D is mostly just a matter of switching to vector notation; x, v, and a become $\mathbf{p}$, $\mathbf{v}$, and

a, respectively.[31] Of course, the time t remains a scalar:

$$\mathbf{v}(t) = \mathbf{v}_0 + t\mathbf{a},$$

$$\Delta\mathbf{p} = \mathbf{v}_0 t + (t^2/2)\mathbf{a}, \tag{11.18}$$

$$\mathbf{p}(t) = \mathbf{p}_0 + \Delta\mathbf{p} = \mathbf{p}_0 + t\mathbf{v}_0 + (t^2/2)\mathbf{a}, \tag{11.19}$$

$$\mathbf{v}_0 = \Delta\mathbf{p}/t - (1/2)at,$$

$$\mathbf{a} = 2\frac{\Delta\mathbf{p} - t\mathbf{v}_0}{t^2}.$$

Equations for motion under constant acceleration, in vector form

Note that we didn't make a vector version of Equation (11.17); we'll get to that in a moment.

This seemingly trivial change in notation is actually hiding two rather deep facts. First, in the algebraic sense, the vector notation is really just shorthand for sets of parallel scalar equations for x, y, and z. The important point is that the three (Cartesian) coordinates are *completely independent of one another*. For example, we can make calculations regarding y and completely ignore the other dimensions, provided that the hypothesis of constant acceleration is met for the object's motion. If it were not for the independence of the coordinates, we could not make this change in notation. The second fact hidden in this notation is that, when we view the vectors in the equations above as geometric rather than algebraic entities, the particular coordinate system used to describe those vectors is irrelevant. We don't even need to specify one. Of course, this is a basic principle of physics: Mother Nature doesn't know what coordinate system you are using.

We were able to leap from 1D to 3D mostly just by bolding a few letters due to the independence of the coordinates. However, there is a bit more to say about projectile motion in multiple dimensions because there are situations where we need to consider the effects of all the coordinates at the same time. One situation has already been alluded to by the lack of a vector equation corresponding to Equation (11.17). In other words, how could we solve for time t given a displacement $\Delta\mathbf{p}$, acceleration $\mathbf{a}$, and initial velocity $\mathbf{v}_0$? In one dimension, the projectile is "confined" and basically cannot help but hitting the target implied by Δx.[32] But in two or more dimensions, the situation is more complicated. The increase in complexity that attends the increase in dimensions is analogous to computing the intersection of two rays (see Section A.8). In 2D, any two rays must intersect unless they are parallel, whereas in 3D, the possibility exists for *skew* rays, which are not parallel but do not intersect.

---

[31] We use **p**, which is short for "position," rather than **x**, to avoid the assumption that the x-coordinate is special compared to y or z.

[32] With one exception—see Exercise 8.

For example, earlier we computed how long it would take for a ball bearing dropped from a great height to hit the sidewalk below, which is a one-dimensional problem. The corresponding three-dimensional problem would be to try to drop the ball bearing into a bucket which is free to move around on the sidewalk. Let's say that the bucket is off to our left. Our initial velocity had better have some leftward component then, or else the ball bearing won't land in the bucket. Another indication that the multi-dimensional case is more complicated than 1D is that a direct translation of Equation (11.17) into vector form results in nonsensical operations of taking the square root of a vector and dividing one vector by another.

The key to solving this problem is to realize that any horizontal changes (either to the bucket's position or the initial velocity of the ball bearing) do *not* affect how long it takes the ball bearing to reach the sidewalk. This is because the coordinates are independent from one another. The horizontal velocity and acceleration do not interact with the vertical velocity and acceleration. To be specific, let's switch to our standard 3D coordinate system, which has $+y$ pointing up and x and z in the horizontal plane. The time it takes the ball bearing to reach the altitude of the bucket depends *only* on the equations having to do with y; the x- and z-coordinates can be ignored for this purpose.[33] In other words, calculating the time when a projectile will reach a target is still a one-dimensional calculation—we just need to chose which direction to use. We can apply Equation (11.17) to solve for a time of impact t. But this solution is just a proposal. We know that *if* the projectile were to hit the target, it would do so at this time. To make sure we really did hit the target, we must plug this time of alleged impact into Equation (11.19) to see where the projectile will be at that location, and verify that the position of the projectile is within appropriate tolerances.

Let's talk a bit more about exactly what it means to "chose which direction to use," as was stated in the previous paragraph. In cases of simple projectile motion, such as the ball-bearing example, where gravity is the constant acceleration, the direction to choose is obvious: use the direction of gravity. Furthermore, because coordinate systems are chosen such that "up" is one of the cardinal axes, the process of solving a one-dimensional problem in that direction is a trivial matter of plucking out the appropriate Cartesian coordinate and discarding the others. In general, however, the situation can be more complicated. But before we discuss the details of the general case, there are a few more things we can say about this very important and common special situation.

---

[33]This is all assuming *ideal* projectile motion, which ignores wind resistance. Of course we can't crash into an adjacent building, or else the horizontal motion certainly would be relevant.

To study projectile motion where acceleration is solely due to gravity, which is a constant and acts along a cardinal axis, let's establish a 2D coordinate space where $+y$ is up and x is the horizontal axis. Without loss of generality we can rotate our plane such that it contains the initial velocity, and thus the entire trajectory of the particle. We choose $+x$ in the horizontal direction of the initial velocity. We also simplify things

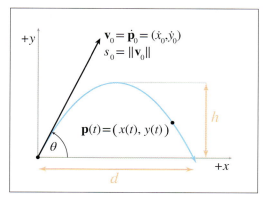

Figure 11.13
Projectile motion

by setting the origin at the object's initial position. This notation (along with a few other items that we'll need in a moment) are illustrated in Figure 11.13.

Reviewing the notation in Figure 11.13, we see that we can express the position of the particle as a function of time either as $\mathbf{p}(t)$, or we can refer to an individual coordinate with $x(t)$ and $y(t)$. Instantaneous velocity (not shown on the diagram), can be denoted in vector form either as $\mathbf{v}(t)$ or using derivative notation as $\dot{\mathbf{p}}(t)$. The scalar velocity components will be denoted using derivative notation as $\dot{x}(t)$ and $\dot{y}(t)$. The initial position and velocity will be denoted by adding a subscript 0 ($\dot{y}_0$ is the initial vertical velocity). We denote the acceleration due to gravity as either g or $\mathbf{g}$.

Let's state the equations for velocity and position using the notation just described:

$$\dot{\mathbf{p}}(t) = \mathbf{v}_0 + t\mathbf{g}, \qquad \dot{x}(t) = \dot{x}_0, \qquad \dot{y}(t) = \dot{y}_0 + gt, \qquad (11.20)$$

$$\mathbf{p}(t) = t\mathbf{v}_0 + (t^2/2)\mathbf{g}, \qquad x(t) = t\dot{x}_0, \qquad y(t) = t\dot{y}_0 + (1/2)gt^2. \qquad (11.21)$$

The distances labeled h and d in Figure 11.13 are often of interest; they are the apex height and horizontal travel distance, respectively. As discussed earlier in a one-dimensional context, the maximum height is reached when all of the initial velocity in the upwards direction has been consumed by gravity, in other words when $\dot{y}(t) = 0$. This occurs at time

$$t_a = -\dot{y}_0/g, \qquad \qquad \text{Time to reach apex}$$

and at this time, the height is equal to

Altitude at apex

$$
\begin{aligned}
h = y(t_a) &= t_a \dot{y}_0 + (1/2)gt_a^2 \\
&= (-\dot{y}_0/g)\dot{y}_0 + (1/2)g(-\dot{y}_0/g)^2 \\
&= (-\dot{y}_0^2/g) + (1/2)(\dot{y}_0^2/g) \\
&= -\dot{y}_0^2/2g.
\end{aligned}
$$

We stated earlier that the time for the object to come back down to its initial height (which we denoted t_e) was twice the time needed to reach its apex; however, at that time we merely appealed to a diagram. This time, let's verify it algebraically:

Time to return to initial altitude.

$$
\begin{aligned}
y(t) &= t\dot{y}_0 + (1/2)gt^2, \\
0 &= t_e\dot{y}_0 + (1/2)gt_e^2, \qquad \text{(initial position is at the origin)} \\
-(1/2)gt_e^2 &= t_e\dot{y}_0, \\
t_e &= -2\dot{y}_0/g. \qquad\qquad \text{(divide by } -(1/2)gt_e)
\end{aligned}
$$

As expected, the flight time t_e is twice the time needed to reach the apex. Now, let's compute d, the horizontal distance traveled:

Horizontal travel distance

$$
d = x(t_e) = t_e\dot{x}_0 = -2\dot{y}_0\dot{x}_0/g.
$$

Of course, t_e and d are based on the assumption that we want to know when the projectile returns to its initial altitude. This is important when launching a projectile from a flat ground plane. If the projectile isn't launched from the ground, or if the ground isn't flat, then we'll need to consider where the parabola intersects the ground plane.

We often wish to specify the initial velocity in terms of an *angle* and *speed*, rather than velocities along each axes. In other words, we wish to use polar coordinates rather than Cartesian. As shown in Figure 11.13, we denote the initial launch speed as s_0 (which is equal to the magnitude of $\mathbf{v}_0$) and the launch angle as θ. Converting the initial velocity from Cartesian to polar coordinates (see Section 7.1.3 if you don't remember how), we get

$$
\dot{x}_0 = s_0 \cos\theta, \qquad\qquad \dot{y}_0 = s_0 \sin\theta.
$$

Plugging this into our kinematics Equations (11.20) and (11.21), we get the equations of motion for a projectile in terms of its launch angle and speed:

$$
\begin{aligned}
\dot{x}(t) &= s_0 \cos\theta, & \dot{y}(t) &= s_0 \sin\theta + gt, \\
x(t) &= ts_0 \cos\theta, & y(t) &= ts_0 \sin\theta + (1/2)gt^2.
\end{aligned}
$$

We can also express t_e, h, and d in terms of s_0 and θ:

$$
\begin{aligned}
t_a &= -\dot{y}_0/g = -(s_0 \sin\theta)/g & &= -s_0(\sin\theta)/g, \\
t_e &= -2\dot{y}_0/g = -2(s_0 \sin\theta)/g & &= -2s_0(\sin\theta)/g, \\
d &= -2\dot{y}_0\dot{x}_0/g = -2(s_0 \sin\theta)(s_0 \cos\theta)/g & &= -2s_0^2(\sin\theta)(\cos\theta)/g, \\
h &= -(1/2)\dot{y}_0^2/g = -(1/2)(s_0 \sin\theta)^2/g & &= -s_0^2(\sin^2\theta)/2g.
\end{aligned}
$$

Important quantities in projectile motion, expressed in terms of launch angle and speed

These equations are highly practical because they directly capture the relationship between the "user-friendly" quantities of launch speed, launch angle, flight time, and flight distance.

At this point, let's pause to make an interesting observation about the relationship between the initial speed s_0 and the horizontal distance traveled d. It's a quadratic relationship, meaning when we increase s_0 by a factor of k, we increase d by a factor of k^2. It might seem more natural for the relationship to be linear, meaning that d would increase by the same factor k. We can understand the quadratic relationship by breaking the initial velocity into its horizontal and vertical components, denoted earlier as $\dot{x}_0$ and $\dot{y}_0$, respectively. It's not difficult to see that increasing $\dot{x}_0$ will increase d by the same factor. Less obvious is that the same is true for $\dot{y}_0$. This is true because the duration that the object is airborne is proportional to $\dot{y}_0$. So if we increase the vertical velocity, we give the object more time to travel. Thus any scale factor we apply to s will affect the distance twice, once as a result of the increased ground velocity due to $\dot{x}_0$, and again as a result of the increased travel time due to $\dot{y}_0$. This produces a quadratic relationship between s and d.

Now let's return to a question we put on hold from earlier: how might we determine the point of impact for any arbitrary vectors $\Delta\mathbf{p}$, $\mathbf{a}$, and $\mathbf{v}_0$? We said before that the key was to "choose a direction" and solve a one-dimensional problem in that direction. If a cardinal direction is chosen, we just throw out the other coordinates. For an arbitrary direction, we project the problem onto a line in that direction. Any component of displacement, velocity, or acceleration perpendicular to that line is discarded during the projection. We learned how to project onto a line and measure displacement in a particular direction by using the dot product in Section 2.11. All that is left is to select a direction.

Assuming the projectile hits the target, we will get the same value for t no matter what direction we choose. But that doesn't mean the choice is irrelevant. For example, in the ball-bearing example, it would be a disaster to chose the $+x$ or $+z$ directions, since there is no acceleration in either of those directions and application of Equation (11.17) would result in a division by zero. This suggests the strategy of simply using $\mathbf{a}$ itself as the direction of projection. To do this, we dot each vector quantity with $\mathbf{a}$, making the substitutions $\Delta x = \Delta \mathbf{p} \cdot \mathbf{a}$, $v = \mathbf{v} \cdot \mathbf{a}$, and $a = \mathbf{a} \cdot \mathbf{a}$. Then these scalar quantities can be plugged into Equation (11.17). Exercise 10 explores this in more detail.

11.7   The Integral

We have just showed that the total displacement of an object in a time interval is equal to the area under the plot of the object's velocity. We used the example of constant acceleration, which has a simple graph, and the area was easy to solve geometrically. We did not pursue in further generality the limit argument that led us to the surprising equivalence, because this special case has such compelling applications. Now we are ready to discuss more general cases. The need to compute a "continuous summation," where the rate of growth is a known function, is a common concept in engineering and science. The calculus tool used to compute these sums is the *integral*.

If you have already studied integral calculus and have a good intuition about what the integral is used for, then you can safely skip ahead to Section 11.8, when our focus returns to the subject of mechanics. However, if you've never had integral calculus or if your intuition about the integral is a bit shaky, keep reading.

There are two important ways of approaching the integral. The first way is essentially to make the notion of "summing up many tiny elements" a bit more precise and introduce some mathematical formalism. The other way is to compare the integral to the derivative. It's important to understand both interpretations. The integral is a bit more difficult to grasp than the derivative, but for reasons that become apparent later, it plays a much greater role in physics simulations and many other areas of video game programming. Understanding what the integral *does* is very important, even if the vast assortment of pen-and-paper techniques to compute integrals analytically is not very useful in our case, being replaced instead by techniques of numerical integration.

Let's turn our informal summation into mathematics notation, in which we compute the area under the curve $f(x)$ in the interval $a \leq x \leq b$. We partition this interval into n slices, each having the width $\Delta x = (b-a)/n$. The ith rectangle will have a left-hand coordinate x_i, a height equal to $f(x_i)$, and an area of $f(x_i)\,\Delta x$. Using summation notation, we add up all these rectangles:

$$\text{Area} \approx \sum_{i=1}^{n} f(x_i)\Delta x.$$

The error in this approximation decreases as we increase the number of slices n, and by now, unless you're the new kid in town, you know that we need to take it to the limit one more time.[34] By taking the limit as n increases without bound and the slices become infinitesimally small, we get our definition of the *definite integral.*

Definite Integral

$$\int_a^b f(x)\ dx = \lim_{n \to \infty} \sum_{i=1}^{n} f(x_i)\Delta x. \qquad (11.22)$$

In this equation,

$$\Delta x = (b-a)/n,$$
$$x_i = a + i\Delta x.$$

Equation (11.22) is read as "The integral from a to b of $f(x)\ dx$." Some people read dx as "with respect to x." The great similarity in notation between the left- and right-hand sides of Equation (11.22) is by design. Just like with the derivative, the finite step size Δx becomes the infinitesimal dx. The sigma symbol $\sum$ used for discrete summations is replaced with the symbol $\int$, which is an elongated S that Leibniz intended to stand for "summation."[35] The a and b are known as the "limits of integration" and define the starting and ending points. The function being integrated is called the *integrand.*

An integral defined as a sum of "vertical slices" like this is known as a *Riemann integral.* It's the most common definition, but not the most general. In fact, our definition is not quite as general as the typical definition of a Riemann integral. The astute reader may notice that Δx is a constant, and could be pulled in front of the summation, making it $\Delta x \sum_{i=1}^{n} f(x_i)$. That works in this case because we are using a *regular partition,* and all the slices are the same width. In general, however, this restriction is not necessary. The traditional definition of the Riemann integral takes the limit as the width of the largest slice goes to zero. Our definition is certainly powerful enough for well-behaved functions we deal with, but more powerful definitions are needed to integrate more esoteric functions. Furthermore,

---

[34] A shout out to all the Eagles fans out there who got that joke. The rest of you will find out in the long run. Just take it easy, get over it, and you'll get a peaceful, easy feeling.

[35] Actually, "summierung" since he spoke German; we're just lucky it works in English, too.

you may wonder why we calculate the area of the rectangle by using the function value at the left-hand side of the rectangle, instead of, say the center point. Surely that would be more accurate. For theoretical purposes of defining the Riemann integral, these choices become identical in the limit and so we are free to make any choice we want. However, when approximating integrals numerically, it is useful to consider such options.

11.7.1   Examples of Integrals

At this point, we have introduced just enough notation and terminology that we can look at some examples of integrals. We would like to do this before going any deeper into the mathematical details. Many applications of the integral in video game programming (and many other engineering disciplines) are more directly thought of not as an area under the curve, but as a "running total." Think of an electric meter. At any given time, the meter is increasing at a rate that is determined by the amount of electricity being used at that instant. The meter is a continuous running total, and we say that it *integrates* the consumption rate. When the air-conditioner kicks in, the consumption rate increases, and the meter counts up faster; at night, when all the lights are out and the windows are open because the weather is nice outside, the consumption is lowest and the meter turns slowly. The consumption rate is a function that varies with time and is the function being integrating. A definite integral of this function between two time values a and b would give us the total amount of energy consumed during that time interval:

Calculating electricity usage

$$\begin{pmatrix} \text{Total energy} \\ \text{used} \end{pmatrix} = \int_{\text{Start time}}^{\text{End time}} \begin{pmatrix} \text{Instantaneous} \\ \text{consumption rate} \end{pmatrix} dt. \qquad (11.23)$$

Although it's not important for our discussion here, we might as well mention what the proper physics terms and units are. Going back to our dimensional analysis from Section 11.2, energy is a derived quantity; it is the product of force and length. Chapter 12 shows that force is itself a derived quantity that has units ML/T^2 and is measured in the SI system using the Newton (N). Thus, energy has abstract units ML^2/T^2, a combination of fundamental units that truly boggles the mind. In the SI system, energy is measured in *Joules* (J), and 1 J = 1 N m = 1 kg m^2/s^2. The proper physics term for "rate of energy transfer per unit time" is *power*, and the SI unit for power is the *watt* (W), which is equal to one joule per second.

Denoting the total energy consumption as E and the instantaneous consumption rate as $P(t)$, we can rewrite Equation (11.23) as

Calculating electricity usage, this time with more dignified notation

$$E = \int_{\text{Start time}}^{\text{End time}} P(t)\ dt. \qquad (11.24)$$

Although the details of how to quantify energy are not core to our discussion, there is one very important observation to make: Equation (11.24) is dimensionally consistent. On the left, the quantity measured is energy, which in the SI system is measured in joules. But on the right, the consumption rate is measured in watts. How can this be? Remember that the integral represents a summation, and the infinitesimal items being summed are the *product* of the integrand (in this case, $P(t)$) and a infinitesimal bit of the domain of integration (in this case, dt). In terms of a Riemann integral, the former determines the height of each slice, and the latter determines its width. Here, dt represents an infinitesimally small step in time, measured in seconds, so the units on the right are $W \times s = (J/s) \times s = J$. Thus, the left- and right-hand sides of Equation (11.24) are measured in joules.

We can extend this example by calculating the electricity bill, rather than just the total usage. Of course, if the price for energy is fixed, then we simply multiply the consumption by the price. But what if the price varied on a moment-by-moment basis? (This shouldn't be too hard to imagine nowadays.) In this case, we would be integrating the *cost* rather than the energy. We determine how to calculate the cost of a single interval of duration dt (a "differential" slice of time) and then sum over all the intervals:

$$\left(\begin{array}{c} \text{Total} \\ \text{cost} \end{array}\right) = \int_{\text{Start time}}^{\text{End time}} \text{RateOfExpendature}(t) \; dt$$

Calculating electricity cost

$$= \int_{\text{Start time}}^{\text{End time}} \text{ConsumptionRate}(t) \, \text{Price}(t) \; dt.$$

Moving on to another example of the integral, imagine a man using a sewing machine with a foot pedal that has variable-speed response. If he depresses the pedal just a bit, the sewing machine advances the fabric slowly, and if he "puts the pedal to the metal," the sewing machine moves at its fastest rate. Now, imagine his daughter sitting under the table watching her father sew. She can only see the pedal, but not the sewing machine or the fabric. The only information available to the girl is the amount of depression of the pedal, and we assume that, based on her knowledge of sewing machines and foot pedals, she can infer a function $f(t)$ that describes the rate that the fabric is moving at time t. The girl watches the pedal for a minute or so, and then her father stops and asks her, "How far have I traveled along the fabric?" Let's say this girl is particularly bright and knows some integral calculus, so she integrates the function $f(t)$, to yield the total amount of fabric that has passed under the needle. As we see later, this sort of question is actually quite close to the types of mechanics problems that are solved with integrals in video games!

One last helpful analogy: think of a derivative as a speedometer that tells you an instantaneous rate of change, and the integral as an odometer describing the continuous summation of this rate of change. Notice that the reading on the speedometer does not depend on that road trip last summer, or even what happened two seconds ago. The speedometer reading is only affected by what is happening *at that instant.* The odometer, on the other hand, is a running tally, and the entire history since the car was first driven off the lot is included in its reading. Our girl under the sewing table must pay attention to the pedal the entire time if she is going to make an accurate estimate of the total amount of fabric consumed at any given time.

Many types of engineering problems solved with integrals are couched in terms of continuous summations such as these: What is the total displacement, when I know the velocity function $v(t)$? What is the total amount of water in the bathtub, given the history of the deflection angle of the faucet? How much fuel is remaining, given the burn rate as a function of time? To set up the integral for problems like these, we can first imagine approximating the value we wish to calculate by using a finite sum $(\sum)$ and a finite step size (Δx). We then use a limit argument to replace the $\sum$ with a $\int$, and the Δx with a dx (review Equation (11.22)). This is the essence of what is meant by a "continuous summation."

Of course, we can use the integral to calculate the area under a curve, as calculus textbooks are so fond of pointing out. As we sweep a line from left to right, the function being integrated determines the rate at which we are accumulating area. Where the function has a large value, our total area is adding up more rapidly, because the "slices" in that area are tall. However, from the viewpoint of a video game programmer, calculus textbooks seem to focus on this particular application of the integral in great disproportion to its application to real world problems.

11.7.2   The Relationship between the Derivative and the Integral

Let's see how we calculate integrals now that the purpose of an integral is (we hope) firmly grounded in your mind. Looking at the definition Equation (11.22), one wonders how in the world you can evaluate this limit. For the derivative, we were able to manipulate the expression being taken to the limit such that we could simply substitute $\Delta t = 0$, but this doesn't seem possible in Equation (11.22). As it turns out, Equation (11.22) is mostly useful as a way to recognize when the problem you have is an integral, and is helpful to properly turn that problem into integral notation. It's also used when we approximate integrals numerically, where instead of taking the slice width down to zero, we just stop at some small but finite Δx. But this definition is not used to solve integrals with pen and paper.

Let's poke around with Equation (11.15), the integral we were able to solve through a simple geometric argument. Since this is a function that describes position as a function of time, we should be able to take its derivative and get a function describing the velocity function $v(t)$, and then take the derivative again to get the acceleration function $a(t)$. Let's make sure this is true:

$$x(t) = x_0 + v_0 t + (1/2)at^2, \hspace{3cm} (11.25)$$
$$\dot{x}(t) = v(t) = v_0 + at, \hspace{2cm} \text{(taking the derivative)}$$
$$\ddot{x}(t) = \dot{v}(t) = a(t) = a. \hspace{1.5cm} \text{(taking the derivative again)}$$

OK, that turned out as expected. No surprises here, but it's comforting to confirm that math and physics do actually work. We knew that the derivative of the position function is the velocity function. The question is: why didn't we use this knowledge earlier? Remember that we knew $v(t)$ and were trying to figure out what $x(t)$ was. We were able to get at the answer through a graphical argument, but it seems like there may be another way. Instead of looking for a function to calculate the area under the curve, we could have instead looked for *a position function whose derivative was the velocity function we already knew.* Such a function is known as an *antiderivative.*

Let's investigate this idea of "integration as an antiderivative" a bit further. To do so, essentially all we need to do is apply the rules of differentiation, including the small subset we learned in Section 11.4.5, in reverse.[36] Assume that we start with the velocity function $v(t) = v_0 + at$, and we are looking for an $x(t)$ whose derivative is $v(t)$. Pretend for the moment that you don't already know the answer. To find $x(t)$, we break up $v(t)$ into its terms (using the sum rule in reverse), then take the antiderivative of each term (using the power rule in reverse). Remember that the power rule of differentiation basically says, "Multiply by the exponent, and then decrease the exponent by one." So the power rule for antidifferentiation is "Increase the exponent by one, and then divide by the new exponent." Applying these two rules to $v_0 + at$ leads us to write

$$v(t) = \dot{x}(t) = v_0 + at,$$
$$x(t) = v_0 t + (1/2)at^2.$$

But compare this result to Equation (11.25); you'll notice that it's missing an x_0 term. What happened? There is a certain amount of "information

---

[36]This statement applies more generally than is usually acknowledged. For example, if you've had some calculus, notice that the technique of integration known as "integration by parts" is actually just the product rule of differentiation in reverse!

loss" that occurs when we take the derivative. If we know how fast we were going, we can always figure out how far we traveled. However, we cannot know where we ended up unless we know where we started. This extra term x_0 is the "starting point" that the derivative throws out, because any constant value has a derivative of zero. For this reason, it's not entirely accurate to refer to "the" antiderivative of $v(t)$, since there is not a unique function whose derivative is $v(t)$, but infinitely many. All the different antiderivatives are really just copies of one another, shifted on the graph vertically according to their particular value of x_0.

We've stated in a general way that there is some relationship between the (definite) integral and the antiderivative. So we know that in a certain sense the operations of integration and differentiation are inverse operations. The theorem of calculus that summarizes these relationships precisely goes by an important-sounding name: *the fundamental theorem of calculus.* The theorem actually consists of two parts. (Sources don't always list them in the same order.)

The first part shows how a definite integral may be computed by using an antiderivative.

Fundamental Theorem of Calculus, Part 1

Let $f(t) = F'(t)$. (In other words, $F(t)$ is any antiderivative of $f(t)$.) Then the definite integral $\int_a^b f(x)\,dx$ can be computed as

$$\int_a^b f(t)\,dx = F(b) - F(a). \tag{11.26}$$

Equation (11.26) can seem a bit mystifying in abstract terms, but when we replace the generic $F(t)$ and $f(t)$ with notation specific to position and velocity, the first part of the fundamental theorem of calculus seems to state the obvious:

$$\int_a^b v(t)\,dt = x(b) - x(a).$$

This says that the cumulative effect of velocity from time a to time b (the net displacement during that interval), is equal to the difference in the position at time b and the position at time a.

Notice how *any* antiderivative will work—it doesn't matter which one. That's because the constant offset x_0 inside of $x(t)$ cancels itself out when we do the subtraction $x(b) - x(a)$. To see this, consider the metaphor of the electric meter. You can think of the raw numeric readout on the meter as an antiderivative of your consumption rate. The readings on the dial at the beginning and end of the month correspond to $F(a)$ and $F(b)$, respectively. Note that the raw numeric value of the reading is mostly irrelevant. It could contain data that was influenced by somebody who lived in the house before you. The *difference* between the two readings, however, is quite relevant. It corresponds to the definite integral, and will determine how much your electric bill is for the month.

Or consider the odometer on a car. Let's say you wanted to measure the length of a particular journey. To do so, at the start of the trip you would reach over to the dedicated trip odometer that every car has had since about 1980 and press the reset button, and then at the end of the trip you just read off the value of the trip odometer. Then you would rejoice in not having to exercise your brain one iota or utilize a single principle from calculus. But what if the trip odometer was broken and all you had was the master odometer? This cannot be easily reset.[37] In this case, armed with the calculus knowledge you gleaned from this book (or maybe just common sense you could have picked up anywhere), you would subtract the odometer reading at the end of the journey from the reading at the start of the journey to obtain the distance of the journey. The actual readings of the odometer are $F(a)$ and $F(b)$, the values of the antiderivative. Just as with the electric meter, the raw values are not useful[38]—only their difference matters.

The first part of the fundamental theorem of calculus is very important because it's how we actually compute integrals, at least with a pen and paper. Remember that we defined the definite integral as a sum of a large number of slices in the limit as the number of slices approached infinity and the slices became infinitesimally thin. This definition is not amenable to algebraic manipulation, like the definition of the derivative was. The first part of the fundamental theorem of calculus says that although we may formulate problems using the definition of the integral, we compute definite integrals by finding an antiderivative of the function being integrated (with pen-and-paper, at least).

The second part of the fundamental theorem of calculus is the flip side of the first part. The first part said that definite integrals can be calculated by using antiderivatives; the second part shows how to define an antiderivative in terms of a definite integral.

---

[37] Nor, as we learned from *Ferris Bueller's Day Off*, can it be easily rolled back.

[38] At least not for this purpose. When the timing chain breaks 5 miles past your warranty expiration, those raw values are *very* important.

Fundamental Theorem of Calculus, Part 2

Let $F(t)$ be defined by

$$F(t) = \int_{t_0}^{t} f(u) \; du. \tag{11.27}$$

Then the derivative of $F(t)$ is given by

$$F'(t) = f(t).$$

It can take some effort to decipher this terse elegance, so let's restate it in English. We start with a given function f. We then form a new function F, whose value is determined by taking the definite integral of f from any arbitrary starting point t_0, and an ending point t. Note that the argument to F is used to define *when to stop* the integration of f. The variable u is a notational dummy variable of integration; it is not seen outside of the integral. The second fundamental theorem of calculus says that if we take the derivative of this new function F, the result is our original function f. In this sense, integration and differentiation are inverse operations.

It can be difficult to grasp the reason why t ends up in what may seem to be an odd location, defining the upper limit of the integration, but that is the essential point. The second theorem is saying that a function defined as an integral such as Equation (11.27) will grow at a rate determined by the integrand. If we adjust the upper limit of integration a tiny bit, the change in the result of the overall sum will be proportional to the value of integrand. Thinking of an integral as calculating an area, the upper limit of integration, t, determines the right-hand boundary. If we push this boundary to the right a bit, the increase in the amount of area will depend on the height of the function at t.

Let's rewrite the theorem using notation particular to displacements and velocities:

$$x(t) = \int_{t_0}^{t} v(u) \; du,$$

$$x'(t) = v(t).$$

Now we see that, to define the displacement $x(t)$ in terms of $v(t)$, there's really only one logical place we could put t. The velocity *before* t is relevant to the displacement that had occurred by time t, and the history *after* t

is not relevant. We use t to define the stopping point of the time range of velocities to integrate.

Where does t_0 come from? It is an arbitrary starting point, reflecting a degree of uncertainty (or freedom) very similar to the unknown (or irrelevant) starting position x_0. We can pick t_0 to be whatever we want our measurements to be relative to. The value of t_0 defines the point where $x(t) = 0$. It's probably more precise to say that $x(t)$ describes our *relative* position. Relative to where? Wherever we were at time t_0.

Now we're ready to clear up the sometimes confusing relationship between the definite and indefinite integral. The adjective "definite" in "definite integral" comes from the fact that we have specified the limits of integration. Because of this, the "answer" to a definite integral can be a single number. When we evaluate a definite integral, such as

$$\int_{t_{\text{start}}}^{t_{\text{end}}} v(t) \ dt,$$

the t gets "integrated out" and does not appear in the result. The meaning of the above is "the continuous summation of the velocity during the time interval t_{start} to t_{end}." It wouldn't make sense for the result to contain t—which t would we be talking about? Thus, if all the other variables in $v(t)$ are known, and the limits t_{start} and t_{end} are known, we can boil down the answer to a simple number. If, however, $v(t)$ contains some other unknown quantities (perhaps some variable density ρ), or the limits of integration themselves are parameters, then the result will be function in terms of those variables. In any case, in a definite integral the t will *not* be part of the result. If you're a programmer, then you can think of the t as a "local variable" to the definite integral.

An indefinite integral, on the other hand, since it is an antiderivative, will have an "answer" that is *function*, not a single number. It is denoted simply by dropping the limits of integration, such as

$$\int v(t) \ dt.$$

Again, we stress that while this may look very similar to the notation used to denote a definite integral, its meaning is actually quite different. The result of evaluating this integral should not be a number, but an antiderivative of $v(t)$; that is, we should get a *function* of t. Furthermore, a proper result will have some arbitrary constant added, known as the *constant of integration*, which reminds us that there is a whole family of functions whose derivative is $v(t)$. Thus, the meaning of the indefinite integral above is "some function that expresses the continuous summation of the velocity as a function of time, from some unknown starting point." We have been denoting this constant offset as x_0, but in a more general setting it is typically written with a capital C. For example,

Constant of integration

$$\int v(t)\ dt = x(t) + x_0 \qquad \text{(displacement and velocity notation)},$$

$$\int f(t)\ dt = F(t) + C \qquad \text{(common abstract notation)}.$$

We do not need to write the limits of integration in an indefinite integral because they are implicit. As we saw in the second part of the fundamental theorem of calculus, the interpretation of an antiderivative in terms of a definite integral is to use the argument of the antiderivative as the upper limit of the range of integration. In other words, an indefinite integral is simply a definite integral with implied limits of integration of the form in Equation (11.27). The degree of freedom in Equation (11.27) connecting the set of possible antiderivatives was captured by the unknown lower limit of integration (t_0). In an indefinite integral we don't write the limits of integration, and instead the uncertainty is contained in the constant of integration $(x_0$ or $C)$. We can summarize this (written using both naming schemes) by

The indefinite integral

$$\int v(t)\ dt = \int_{t_0}^{t} v(u)\ du = x(t) + x_0, \qquad x_0 = -x(t_0),$$

$$\int f(t)\ dt = \int_{t_0}^{t} v(u)\ du = F(t) + C, \qquad C = -F(t_0).$$

11.7.3   Summary of Calculus

We have completed our main presentation of calculus in this book, aside from a few small bits that come up in later sections. Our goal has been to take a reader with absolutely no knowledge of calculus to a point where that reader understands the big picture of what derivatives and integrals are used for. We have whizzed right past the many, many details and techniques that arise in practical situations—these details fill up thousands of pages in calculus textbooks.

Let's summarize the important points that you need to know about calculus to fully utilize the remainder of this book.

- The basic purpose of a derivative is to measure a rate of change.

- The derivative is defined by using a limit argument. We form an approximation of the result, and then watch what happens as we take better and better approximations in the limit as our error approaches zero.

- We presented just a few pen-and-paper rules for differentiation. Differentiation is a linear operator, which allows us to differentiate sums. The power rule tells us how to evaluate expressions of the form $\frac{d}{dt} t^n$. Together, these rules allow us to take derivatives of polynomials. We also presented the derivatives for the sine, cosine, and exponential function. The chain rule tells us how to differentiate a function of the form $f(g(t))$.

- An integral is a "continuous summation," or "running total." These sums are also equivalent to the area under the graph of the function being summed.

- A Riemann integral defines an integral using a limit argument. We take the sum of a large number of small elements, which in general is an approximation to the true sum when the number of elements is finite. The true sum is obtained by considering what happens as we increase the number of elements to infinity, causing the error in our approximation to vanish.

- Riemann integrals are usually not directly solvable in the same way that derivatives are. They are used to recognize when the problem we are solving is an integral, and to help set up the integral properly. It's also how we solve them numerically (we have not yet discussed the details of how to do this).

- The fundamental theorem of calculus says that integration and differentiation are inverse operations. On paper, definite integrals are computed by looking for an antiderivative, not by evaluating the Riemann integral at the limit. A function whose argument defines the upper limit of integration will be an antiderivative of the integrand.

- An indefinite integral is a function that is an antiderivative of the integrand. A definite integral produces a number representing the continuous summation of the integrand over the interval identified by the limits of integration. A definite integral can be calculated by evaluating any antiderivative at the starting and ending points, and taking the difference between these two values (by subtracting the value at the start of the interval from the value at the end of the interval). An indefinite integral is actually a definite integral where the limits of integration are implied.

11.8   Uniform Circular Motion

Enough calculus—let's get back to physics. This section studies the motion of a particle moving in a circle at a constant speed. We study the motion of a particle because many physics calculations can be simplified by representing a rigid body as a point mass at its so-called center of mass. Since a circular path is inherently restricted to a plane, Section 11.8.1 begins our investigation in two dimensions. After establishing the basic relations, Section 11.8.2 shows how to apply these in a world where the plane of orbit is arbitrarily oriented in three dimensions.

11.8.1   Uniform Circular Motion in the Plane

A particle traveling in a circle with constant speed does *not* have constant velocity; if it did, it would travel in a straight line. Since the object's velocity is changing over time, it must be under some sort of acceleration. Let's see if we can determine what that is. Consider an object moving at constant speed s in a circular path of radius r. To make our calculations easier, and without loss of generality, we establish a two-dimensional reference frame that lies in the plane of motion and has its origin at the center of the circle. Remember that the instantaneous velocity $\mathbf{v}(t)$ of a particle is always tangent to its trajectory, so the velocity vector at any given point will always be tangent to the circle at that point. Also, from the definition of speed, we know that $\|\mathbf{v}(t)\| = s$.

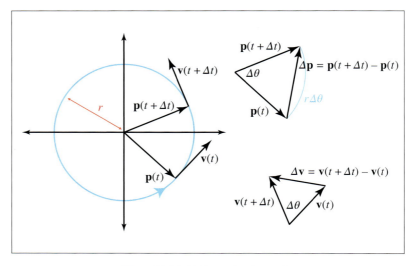

Figure 11.14. Uniform circular motion

On the left side of Figure 11.14 we see a particle moving in uniform circular motion during a finite time step Δt. The figure examines the state of the particle at time t and also at a later time $t + \Delta t$.

Let's consider instantaneous velocity and acceleration, starting with a geometric tack. Examine the triangles on the right-hand side of Figure 11.14. The triangle on the top shows the change in position over some time interval Δt, as a result of the angular change $\Delta \theta$. It is an isosceles triangle in which the legs have length r, the radius of the circle, and the base is $\Delta \mathbf{p}$, which is the net change in position during the interval. The bottom triangle depicts the change in velocity over this same interval, and it is also an isosceles triangle. The legs of the bottom triangle have length s, since we are hypothesizing that the velocity has constant magnitude, and the base is $\Delta \mathbf{v}$. The two triangles are similar, since both triangles are isosceles with the included angle $\Delta \theta$, so we can write

$$\frac{\|\Delta \mathbf{v}\|}{s} = \frac{\|\Delta \mathbf{p}\|}{r}.$$

In general, the length of $\Delta \mathbf{p}$ measures a "shortcut" through the circle, rather than the actual distance traveled around the perimeter of the circle, which is $r\Delta\theta = s\Delta t$. But consider what happens as Δt and $\Delta \theta$ become very small, as shown in Figure 11.15.

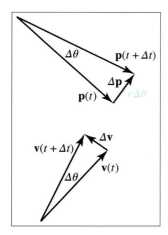

Figure 11.15
A small rotation.

Notice that as $\Delta \theta$ grows smaller and smaller, the length of $\Delta \mathbf{p}$ approaches the true distance, and in the limit, the two distances are equal:

$$\lim_{\Delta t \to 0} \|\Delta \mathbf{p}\| = s\Delta t.$$

Plugging this into our result from similar triangles, we have

$$\frac{\|\Delta \mathbf{v}\|}{s} = \frac{\|\Delta \mathbf{p}\|}{r},$$

$$\lim_{\Delta t \to 0} \frac{\|\Delta \mathbf{v}\|}{s} = \frac{s \Delta t}{r},$$

$$\lim_{\Delta t \to 0} \frac{\|\Delta \mathbf{v}\|}{\Delta t} = \frac{s^2}{r}. \tag{11.28}$$

The left-hand side of Equation (11.28) is a change in velocity over an interval as the length of the interval approaches zero. This is the definition of instantaneous acceleration! Thus the magnitude of the acceleration is s^2/r.

Of course, acceleration is a vector quantity, and all we have determined so far is its (constant) magnitude. What is the direction? To see this, compare the vectors $\mathbf{p}(t)$ and $\Delta \mathbf{v}$ in Figure 11.15. Notice that they point in opposite directions. In fact, in the limit as $\Delta \theta$ goes to zero, they point in *exactly* the opposite direction. That is, the acceleration is always towards the center of the circle, which is why it is called *centripetal* ("center-seeking") acceleration.

Velocity and Acceleration of Uniform Circular Motion

When an object moves with constant speed s in a circular path with radius r, the velocity $\mathbf{v}$ is tangent to the circle. The acceleration at any instant is pointed towards the center of the circle and has magnitude

$$a = s^2/r. \tag{11.29}$$

By combining some elementary geometry with some ideas of calculus, we have obtained the most important facts about uniform circular motion. A slightly different combination of geometry and calculus will yield the actual kinematics equations. To this end, it will be helpful to refer to $\theta(t)$, the angle that the vector $\mathbf{p}$ makes with with $+x$ axis using the traditional mathematical conventions, as shown in Figure 11.16.

Previously, we were concerned with the $\Delta \theta$, the *change* in this angle, but now we consider its value as a function of time. We denote the initial angle as $\theta(0) = \theta_0$. We also define the *angular frequency* as $\omega = s/r$, which

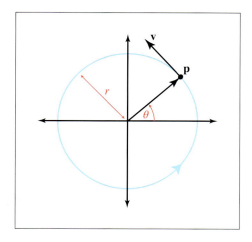

Figure 11.16
The position of the particle can be
identified by the angle θ

is measured in radians per second.[39] Thus, we can express the angle at any
given time as

$$\theta(t) = \theta_0 + \omega t.$$

Angle as a function of time

We've seen the parametric equation for a circle before in Section 9.1, so we
know how to express the kinematics equations for the particle's position in
terms of the radius r and the angle $\theta(t)$, as

$$x(t) = r\cos(\theta(t)) = r\cos(\theta_0 + \omega t),$$
$$y(t) = r\sin(\theta(t)) = r\sin(\theta_0 + \omega t).$$

Position as a function of time

Since the velocity function is the derivative of the position function, we
can differentiate these equations to obtain the velocity equations. Luckily,
we learned the derivatives of the sine and cosine functions in Section 11.4.6
and the chain rule in Section 11.4.7. Differentiating gives us

$$\dot{x}(t) = \frac{d}{dt}\left(r\cos(\theta_0 + \omega t)\right) = -r\omega\sin(\theta_0 + \omega t),$$
$$\dot{y}(t) = \frac{d}{dt}\left(r\sin(\theta_0 + \omega t)\right) = r\omega\cos(\theta_0 + \omega t).$$

Velocity as a function of time

---

[39]Omega (ω) is the traditional letter for angular frequency. To see where the calculation s/r comes from, consider that the circumference of the circle is $2\pi r$, and this
distance is traversed at a speed of s. Therefore the angular frequency is $2\pi r/s$ revolutions per second. But one revolution is equal to 2π radians, so the factor of 2π cancels
out. This is an example of why the use of radians is often so convenient (provided we
are working symbolically and aren't concerned with the numeric values of any angles).

Differentiating once more to get the acceleration, we have

$$\ddot{x}(t) = \frac{d}{dt}\left(-r\omega\sin(\theta_0 + \omega t)\right) = -r\omega^2\cos(\theta_0 + \omega t),$$

$$\ddot{y}(t) = \frac{d}{dt}\left(r\omega\cos(\theta_0 + \omega t)\right) = -r\omega^2\sin(\theta_0 + \omega t).$$

These results agree with our earlier findings. Comparing the acceleration functions to the position, we confirm that they do indeed point in opposite directions. Furthermore, recalling that $\omega = s/r$, we note that, as predicted, acceleration has a length of s^2/r.

Sometimes ω is more immediately accessible than s. In these situations, it's useful to be able to express the magnitude of the centripetal acceleration just in terms of ω and r. Solving $\omega = s/r$ for s gives us $s = r\omega$. Plugging this in to Equation (11.29), we have

$$a = s^2/r = (r\omega)^2/r = r\omega^2. \tag{11.30}$$

Let's work through an interesting example, the results of which will be useful in later sections. All of us are aboard a spinning centrifuge right now: Earth! Earth's rotation creates an apparent centrifugal force, which tends to throw us away from the Earth's center. Luckily, Earth's gravity is strong enough to keep us here. Given that Earth's mean radius is 6,371 km, what is the centripetal acceleration experienced at the equator?

To answer this question, we use Equation (11.30). The radius was given as $r = 6,371$ km, and the rotation rate is $\omega = 2\pi/\text{day}$.

$$a = r\omega^2 = (6\,371 \text{ km})(2\pi/\text{day})^2 = (6.371 \times 10^6 \text{ m})(2\pi/(86\,400 \text{ s}))^2$$
$$\approx (6.371 \times 10^6 \text{ m})(5.2885 \times 10^{-9} \text{ s}^{-2}) \approx 0.03369 \text{ m/s}^2.$$

What about the magnitude of the centripetal acceleration at the poles? Is it the same? Keep this question in mind; we return to it in Section 12.2.1.

11.8.2   Uniform Circular Motion in Three Dimensions

So far, we've essentially been working in two dimensions, operating "in the plane" and not concerning ourselves with how this plane might be oriented in three dimensions. Now let us consider the more general case. We wish to describe the position, velocity, and acceleration of the particle as three-dimensional vectors, where the axis of rotation (which is perpendicular to the plane containing the circular path) is arbitrarily oriented.

Suppose a particle at position $\mathbf{p}$ is moving in a circular path around point $\mathbf{o}$. Since there are many different circular paths that contain both $\mathbf{o}$ and $\mathbf{p}$, we must also specify an axis of rotation perpendicular to the plane.

As we've done in earlier chapters (see Section 5.1.3 and Section 8.4), we describe the direction of the axis by using a unit vector $\hat{\mathbf{n}}$, and, as before, the sign of $\hat{\mathbf{n}}$ tells us which direction is considered positive rotation using the left-hand rule. The scalar ω defines the rate of rotation, in radians per unit time. The question we want to answer is this: What is the velocity $\mathbf{v}$ of the particle at that instant?

Let's review what we already know. First of all, from the relationship between speed and angular frequency observed earlier, we know that the speed $s = \|\mathbf{v}\|$ must be ωr, where r is the radius of the circle, or the distance between $\mathbf{o}$ and $\mathbf{p}$. Second, $\mathbf{v}$ must be perpendicular to $\hat{\mathbf{n}}$, or else the particle will stray from the plane containing the circular path, and $\mathbf{v}$ must also be tangent to this path. Thus, we know both the magnitude and direction of the velocity $\mathbf{v}$, we just need a way to express it algebraically. To do so, let's introduce the vector

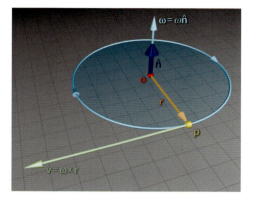

Figure 11.17
Uniform circular motion in three dimensions

$\mathbf{r} = \mathbf{p} - \mathbf{o}$, the radial vector from $\mathbf{o}$ to $\mathbf{p}$. Note that $\mathbf{r}$ lies in the plane of rotation and has a constant length, the radius of the circular path, as shown in Figure 11.17.

Now, $\mathbf{v}$ is perpendicular to both $\mathbf{r}$ (since it's tangent to the path) and $\hat{\mathbf{n}}$ (since it lies in the plane of orbit). You may remember that we have a tool that can compute a vector that is perpendicular to two other given vectors: the cross product. Perhaps $\hat{\mathbf{n}} \times \mathbf{r} = \mathbf{v}$? The direction works out correctly,[40] but let's consider the length. Remember from Section 2.12.2 that the length of the cross product is equal to the product of the magnitudes of the inputs, times the sine of the angle between the two vectors. Well $\hat{\mathbf{n}}$ is a unit vector by assumption, and $\hat{\mathbf{n}}$ and $\mathbf{r}$ are perpendicular, so the sine of the angle between them is unity. Thus the length of the cross product $\hat{\mathbf{n}} \times \mathbf{r}$ is simply $\|\mathbf{r}\|$. The correct speed is $\omega r = \omega\|\mathbf{r}\|$, so we are just missing a factor of ω.

---

[40]Don't just trust Figure 11.17, use your left hand to verify this. Your thumb is the first argument, $\hat{\mathbf{n}}$; index finger is the second argument, $\mathbf{r}$; and middle finger is the result, $\mathbf{v}$. Look at your hand with your thumb pointing at you (it's the axis of rotation, remember), and rotate your hand in the direction of $\mathbf{v}$ (your middle finger), performing the rotation about your thumb. (The particle is at the end of your index finger.) Your hand will rotate clockwise from your perspective, which is the definition of positive rotation according to the left-hand rule.

Calculating linear point velocity from angular velocity

Putting this all together, we have the formula for the velocity of a particle with radial vector $\mathbf{r} = \mathbf{p} - \mathbf{o}$ rotating about the axis $\hat{\mathbf{n}}$ at an angular rate of ω radians per unit time:

$$\mathbf{v} = \omega\hat{\mathbf{n}} \times \mathbf{r}.$$

As we discussed in Section 8.4, angular velocity is often described in exponential map form by a single vector $\boldsymbol{\omega} = \omega\hat{\mathbf{n}}$ (note the boldface $\boldsymbol{\omega}$ to indicate a vector quantity). In this case, the formula is even simpler.

Calculating Linear Point Velocity from Angular Velocity

$$\mathbf{v} = \boldsymbol{\omega} \times \mathbf{r}. \tag{11.31}$$

Now let's consider the opposite problem. Assume we know $\mathbf{p}$ and $\mathbf{v}$, and we wish the measure the angular velocity relative to $\mathbf{o}$. Again, we can use the cross product, but this time, we need a division to get the right length:

Angular velocity of a particle relative to an arbitrary point

$$\boldsymbol{\omega} = \frac{\mathbf{r} \times \mathbf{v}}{\|\mathbf{r}\|^2}. \tag{11.32}$$

To understand the division by $\|\mathbf{r}\|^2$, consider two points on a rigid disk that is rotating around its center. Assume that angular velocity is measured relative to this center. One point has the radial vector $\mathbf{r}$, and another point has a radial vector $k\mathbf{r}$, which is in the same direction from the center, but at a distance scaled by the factor k. These two points (indeed, *all* the points on the disk) should have the same angular velocity. Thus one division by $\|r\|$ is necessary to compensate for the change in $\mathbf{r}$ as we adjust the radius. The extra division is necessary because the outer points have a higher velocity; if we move on the disk by scaling $\mathbf{r}$ by k, the new point will have a velocity that also is scaled by k.

Although thus far we have been assuming that $\mathbf{p}$ is actually rotating about $\mathbf{o}$, it may not be. It might be rotating about some other point, or moving in a straight line. However, we can still calculate the angular velocity of $\mathbf{p}$ relative to $\mathbf{o}$. Essentially, what Equation (11.32) tells us is what the angular velocity would be if $\mathbf{p}$ were indeed orbiting around $\mathbf{o}$, in the plane containing both $\mathbf{r}$ and $\mathbf{v}$. The axis of rotation, which is parallel to $\boldsymbol{\omega}$, is perpendicular to this plane. Actually, there is one slight wrinkle—$\mathbf{r}$ and $\mathbf{v}$ might not be perpendicular, which, of course, they would be if the particle were orbiting around $\mathbf{o}$. The cross product in Equation (11.32) essentially discards any velocity parallel to **r**; only velocity perpendicular to **r** contributes to the results.

If the particle **p** is indeed orbiting **o** at constant speed, then the angular velocity computed by Equation (11.32) will be constant. In general, however, the angular velocity measured relative to any arbitrary point is not constant. For example, consider a particle moving with constant linear velocity. The angular velocity measured relative to a stationary point **o** will grow as the particle approaches **o**, reaches a maximum at the point of closest approach, and then decreases. Furthermore, even if the particle *is* moving in an orbital path, the angular velocity will be a constant only when measured relative to the center of the orbit.

One extremely important example of orbital motion in 3D is a particle attached to a rigid body rotating about an axis. Let's choose **o** to be at the intersection of the axis of rotation and the plane that contains the circular orbit of **p**; This causes **r** to be perpendicular to the axis of rotation. Under these assumptions, the orbital angular velocity computed by Equation (11.32) is the same for every particle, and it's also the same as the *spin angular velocity* of the rigid body. We have more to say about this in the next chapter.

We don't often need to calculate angular velocity of a point relative to some point that isn't the center of the orbit. (However, Equation (11.31) is used frequently to compute a linear point velocity based on its orbital velocity.) So why do we talk about this? Because the computation is similar to the way we measure *torque* (see Section 12.5) for a force applied at an arbitrary direction at an arbitrary location.

11.9   Exercises

(Answers on page 781.)

1. The Pascal is a unit of measurement for pressure, defined as one Newton per square meter. One Pascal is equal to how many psi? (The psi is one pound of force per square inch.)

2. The 1D position of a particle is described piecewise by

$$x(t) = \begin{cases} 2t - t^2 & 0 \le t < 2, \\ 0 & 2 \le t < 4, \\ \sin(\pi t) & 4 \le t < 7, \\ 7 - t & 7 \le t. \end{cases}$$

Plot a graph of the particle's motion.

3. What is the average velocity of the particle from Exercise 2, over the following intervals?

    (a) $t = 0 \ldots 1$?

    (b) $t = 1 \ldots 2$?

    (c) $t = 0 \ldots 2$?

    (d) $t = 5.5 \ldots 6.5$?

    (e) $t = 0 \ldots 9$?

4. Write a similar piece-wise function $v(t)$ that describes the velocity of the the particle from Exercise 2 at time t. In this case, the velocity is not defined at the "junctions" between the pieces, so only worry about what happens in the middle of each piece. (This is unfortunately one of those finer points we had to skip over.)

5. What is the instantaneous velocity of the particle from Exercise 2 at the following times?

    (a) $t = 0.1$

    (b) $t = 1.0$

    (c) $t = 1.9$

    (d) $t = 4.1$

    (e) $t = 5$

    (f) $t = 6.5$

    (g) $t = 8$

    (h) $t = 9$

6. Write a similar piece-wise function $a(t)$ that describes the acceleration of the particle from Exercise 2 at time t. Once again, don't worry about what happens at the junction points.

7. What is the particle's acceleration at the following times?

    (a) $t = 0.1$

    (b) $t = 1.0$

    (c) $t = 1.9$

    (d) $t = 4.1$

    (e) $t = 5$

    (f) $t = 6.5$

    (g) $t = 8$

    (h) $t = 9$

8. What physical situation is signified by a negative discriminant in Equation (11.16), resulting in complex solutions? What if the discriminant is zero and there is only one solution?

9. A projectile is launched with an initial speed of 150 ft/s, with an angle of inclination of $40°$ from the initial position $\mathbf{p}_0 = (0 \text{ ft}, 10 \text{ ft})$.

   (a) What is the initial velocity in vector form?

   (b) At what time will the projectile reach its apex?

   (c) What are the coordinates of the projectile at the apex?

   (d) How long will it take the projectile to come back to an altitude of $y = 10$?

   (e) What will the horizontal displacement be at this time?

10. At the end of our projectile discussion in Section 11.6, we posed the problem of solving for the time of intersection when the acceleration is an arbitrary vector $\mathbf{a}$. Take Equation (11.18) and dot both sides with $\mathbf{a}$, and then solve for t. (Use the quadratic formula, as before.)

11. *Complex exponentials* such as e^{ix} (were i is the imaginary number such that $i^2 = -1$) are very important in differential equations, control systems, and signal processing. Although it seems odd to put a complex number into the exponent, *Euler's formula* gives a meaningful interpretation. To find this interpretation, you could just go to wikipedia.com and look it up (which is why the answer is in the back of the book, anyway). But before you do, expand the Taylor series of e^{ix} and see whether you can figure it out for yourself. (*Then* go online and read about the surprising importance of this expression.)

12. The International Space Station orbits Earth at approximately 340 km above Earth's surface. (The orbit is actually elliptical, but ignore that and assume it moves with uniform circular motion.) Given that the average speed is about 27,740 km/hr, what is the orbital period? What is the centripetal acceleration in m/s^2? Think carefully!

Of course the mathematicians know how to write all those numbers. You can read someday in a mathematics book how to write them all in a high-class and elegant form, but it is first a good idea to know in a rough way what it is that you are trying to write about.

— Richard Feynman (1918–1988) from
The Feynman Lectures on Physics

Chapter 12

Mechanics 2: Linear and Rotational Dynamics

The Force is what gives a Jedi his power.
It's an energy field created by all living things.
It surrounds us and penetrates us.
It binds the galaxy together.

— Obi-Wan Kenobi in
Star Wars Episode IV: A New Hope (1977)

Chapter 11 was about linear kinematics—how to describe the motion of object, without concerning ourselves with the "cause" of the motion, its orientation, or how we might go about simulating that object on a computer. The main goals of this chapter are to address those three topics.

- Section 12.1 identifies and quantifies the "cause" of motion, *force*, and presents three physical laws formalized more than 400 years ago by Isaac Newton in his masterwork the *Principia*.

- Section 12.2 discusses a few particularly important and simple types of forces.

- Section 12.3 introduces *momentum* and presents the important relationship between force and momentum.

- Section 12.4 is about collisions and *impulses*, which are large forces that act for short durations.

- Section 12.5 considers the rotation of objects and the angular analogs of the linear concepts introduced up to this point.

- Section 12.6 turns to matters of implementation, looking at some basic problems that a digital simulation needs to address. It gives an overview of how contemporary real-time rigid body simulations solve them.

12.1   Newton's Three Laws

Sir Isaac Newton established three simple laws that provide a framework, commonly known as *Newtonian mechanics*, for understanding such diverse physical systems as an apple falling from a tree, the motion of the planets, and the physical interactions that happen in a video game. Newtonian mechanics is also called *classical mechanics*, and that name should alert you to the fact that the laws we are about to study are *wrong*, in the sense that they do not agree with the result of experiments conducted at very high speed (which require relativistic mechanics) or at very small scale (which require quantum mechanics).[1] For everyday phenomena (and for the phenomena we need to simulate in a video game), the discrepancy between the results predicted by Newtonian mechanics and the correct results (as correctly predicted by quantum-relativistic mechanics) is generally less than can be detected with the most accurate instruments. The differences in the predictions become significant only at speeds very close to the speed of light and at scales approaching the size of an atom; otherwise, all the theories are in great agreement with each other and with experimental results. It was precisely because Newtonian mechanics has such a long and decorated history of accurate predictions that it was so shocking to find out that the laws were in need of correction. It should be clear that these laws, having been sufficient to describe the motions of the heavenly bodies to a great deal of accuracy, will also be quite sufficient for our purposes here.

12.1.1   Newton's First Two Laws: Force and Mass

Chapter 11 noted that *mass* measures the degree to which an object resists being accelerated. This resistance is called *inertia*, and the physical quantity needed to overcome it and create an acceleration is called *force*. In other words, all of those "causes of motion" that we so scrupulously avoided mentioning in the previous chapter actually go by the collective name *force*.

The idea that objects resist acceleration is summarized by Newton's first law.

Newton's First Law

Every body persists in its state of being at rest or of moving uniformly straight forward, except insofar as it is compelled to change its state by force impressed.

---

[1]Some of these experiments took place in the imaginations of physicists.

This seems like quite a simple statement, even in the anachronistic translation of Newton's original Latin. But consider how audacious it was for Newton to assert this, when it so clearly is at odds with the commonsense observations we all have from our daily lives! A more "commonsense" way to think about force is to assume that force is needed not only to start an object in motion, but also to maintain its motion. (This is the rule under so-called *Aristotelian dynamics*.) After all, once we stop applying the force, eventually the object will stop moving, right? According to Newton, once an object is set in motion, it does *not* require any force to continue this motion. In fact, Newton claims that the force is required to *stop* the object, and absent this stopping force, the object will continue on indefinitely.

Of course, the reason Newton's first law seems counterintuitive is that in our everyday experience, when we set objects in motion, they are always brought to a stop by the ubiquitous force of friction. But we can argue that Newton's law is correct, even though objects always come to a stop through friction, with a simple thought experiment. Imagine we apply a certain amount of force and set an object in motion across a surface. The object will travel a certain distance and eventually come to a stop. Did it stop due the lack of continued pushing force, or due to some force that acted to slow it down? If we perform the same experiment on different surfaces, performing the initial push in the same manner in each case, we find that the object travels farther on a smoother surface, and less distance on a rougher surface. You probably aren't surprised at these "commonsense" results, but notice how they actually contradict the notion that a force is required to keep the object in motion and validate Newton's laws.

Newton clarified the precise relationship among mass, acceleration, and net force in his second law.

Newton's Second Law

The acceleration of a body is proportional to (and in the same direction as) the net external force acting on the body, and inversely proportional to the mass of the body:

$$\mathbf{f} = m\mathbf{a}. \tag{12.1}$$

This simple equation is the most important one in this chapter. You should certainly memorize it. It basically says that whenever a particle with mass m is seen accelerating at a rate $\mathbf{a}$, you can be sure that there is a net force $\mathbf{f}$ acting on the particle. Likewise, whenever there is a net force, the object will accelerate, because net force and acceleration always go together. *There are no exceptions to this rule.* The acceleration of an object is always proportional to the net force acting on it at that moment.

Now this does not mean that when there are *any* forces on an object, it will necessarily accelerate. Nor does it mean that if an object is not accelerating, then there are no forces acting on it. The **f** in **f** = *m***a** is the *net* force. Consider the tremendous forces exerted on the beams at the bottom of a skyscraper. Clearly there is a force that wants to accelerate the beam downward. However, since the beams do not in fact accelerate downward, we know by Newton's second law that this downward force must be exactly opposed by some other force acting in the opposite direction.

What sort of quantity is force? First of all, force has magnitude and direction, and so it is a vector quantity, just like acceleration (although at times it is easier to study force in a one-dimensional setting, just like we did with acceleration). And force must have the same dimensions (1D, 2D, or 3D, depending on the "world" in which we are working) as the acceleration **a** for Equation (12.1) to make sense, because m is a scalar quantity.

Let's use dimensional analysis to determine the physical units that we should use to measure force. Mass is one of our fundamental quantities, denoted M, and from the previous chapter we know that acceleration has units L/T^2. Therefore (dropping the bold to indicate vector quantities), force must have units

Dimensional analysis of force

$$f = ma = (M)(L/T^2).$$

When measuring with the SI units—mass in kilograms, length in meters, and time in seconds—force has the units of "kilogram meter per second squared." This is quite a mouthful, so it goes by a special name, the *Newton*, denoted N:

The Newton is an SI unit of force

$$1\text{ N} = 1\text{ kg}\frac{\text{m}}{\text{s}^2}.$$

If you're having trouble grasping just what a "kilogram meter per second squared" is, just remember that a Newton is the amount of force required to accelerate a mass of 1 kg at a rate of 1 m/s^2.

There's a common misunderstanding that we'd like to cut off as early as possible. Force creates an acceleration on a body, and it acts *over time*. For example, the question, "How much force does it take to get a 100 lb object to go 100 mi/hr?" does not make sense. Force doesn't produce velocity directly, it causes the velocity to change over time. This can be especially confusing when you consider collisions, such as a ball bouncing on the floor or being struck by a bat. Although the velocity appears to have changed instantaneously, what is really happening is that a very large force is acting for a very short (but finite) duration. We study collisions in more detail in

Section 12.3. Typically in digital simulations impulsive forces are handled differently from more persistent forces that act over several simulation steps, so for now, don't think of a force as an impact; instead, think of it as more of a gradual push or pull that could be provided by, for example, a spring, the wind, or gravity.

We said that Equation (12.1) is the traditional way to express the relationship among force, mass, and acceleration. However, written in that way, with force on the left-hand side, you might get the idea that the common situation is for us to know the mass and acceleration, and use Newton's laws to compute the force. In fact, especially in digital simulations, the more common scenario is that we have calculated the forces acting on a body, and we wish to predict the body's response to those forces. In other words, we'll usually use Newton's second law in the form

$$\mathbf{a} = \mathbf{f}/m. \tag{12.2}$$

We usually use this form of Newton's second law

Most physics textbooks teach the important conceptual tool known as a *free-body diagram*. Newton's second law, especially as expressed in Equation (12.2), is at the heart of this exercise. The basic procedure is as follows, starting with a representation of the object.

1. Draw and label all the forces acting on it.

2. Sum up those forces (using vector addition) to compute the net force.

3. Use Newton's second law (Equation (12.2)) to compute the acceleration of the object.

4. Integrate the acceleration to determine the motion of the object. When solving problems analytically, this means solving differential equations. We don't use any differential equations in this book because there are only a few simple cases that we will look at analytically. Numerical methods of integration must be used. Later, we examine Euler integration, which is the most simple method imaginable, but also the one used by most real-time rigid body simulators.

The above procedure is a very important tool that we use several times in Section 12.2; it's also essentially how most digital physics simulations work inside a computer. Of course, the simplicity with which we've described this 4-step process hides many troublesome difficulties. The forces in Equation (12.2) may vary continuously over time; be dependent on time, position, and velocity; exhibit nonlinearities or discontinuities; and in general be difficult to compute exactly or express and integrate in closed form. Section 12.6 deals with physics simulations, but for now the key point that we want to emphasize is that Newton's second law is the fundamental driving equation.

12.1.2   Inertial Reference Frames

If we take the special case where $\mathbf{f} = \mathbf{0}$, then according to Newton's second law, $\mathbf{a} = \mathbf{0}$. This is a restatement of his first law. So we see that if Newton had been just a *bit* more clever, he could have said the same thing in only two laws instead of three. Of course, Newton not only broke through the barrier of "common sense" to create elegant formulas that explain the workings of every physical system in the entire universe, he also simultaneously invented a complete branch of the mathematics needed to fully explore these ideas—calculus. So perhaps he was a clever guy after all. We assume he had a good reason for keeping his first law; we interpret it as a statement about *reference frames*.

The vectors $\mathbf{a}$ and $\mathbf{f}$ are specified in some reference frame, and if we choose a bad reference frame, the equation does not hold. Reference frames in which the basic mechanical laws hold (especially $\mathbf{f} = m\mathbf{a}$) are known as *inertial* reference frames. Coordinate spaces for which this law does not hold unless we invent fictional forces are called *noninertial* frames.

For example, imagine a robot eating a herring sandwich in an elevator. Someone cuts the elevator cables, and the elevator, robot, and sandwich begin to fall. Now, this robot has been programmed with the knowledge that it likes to eat herring sandwiches,[2] but without any general sense of self-preservation, so it does not panic. It looks at the herring sandwich floating in mid-air instead of falling to the elevator floor, as it would reasonably expect. The robot, having also been programmed with an incomplete understanding of Newton's laws, thinks to itself, "My goodness, this is quite unusual! I know gravity must be pulling this sandwich downwards and I know $\mathbf{f} = m\mathbf{a}$, and since the sandwich is not accelerating downwards, the net force acting on it must be zero. Therefore, there must be some *upward* force acting on this sandwich. Quite fascinating! What might be the source of this force? Now, if I calculate..." *CRASH!*

Figure 12.1
A robot in a falling elevator is in a non-inertial frame. He must invent a fictitious upward force to counteract gravity to explain why his herring sandwich doesn't fall.

---

[2]You will, of course, recall the herring-sandwich-loving robot from Section 3.3. This robot is a newer model with improved programming that allows him to pick up the sandwich without a scoop—a major innovation in the herring-sandwich-eating robot business.

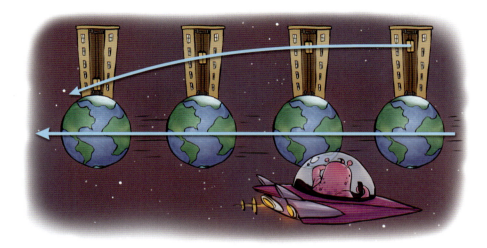

Figure 12.2. An alien watching the elevator fall as Earth moves past will not see anything to contradict Newton's three laws, assuming the time period is short enough so that Earth's rotation and curved path are not significant factors.

A viewer on the ground would not see any need to invent a fictitious force to explain the sandwich's behavior. Using a reference frame with the origin fixed at the bottom of the building, the viewer sees the sandwich as accelerating downward, and has no reason to think anything is amiss.[3] The person driving by in a car also doesn't see any problems. In the reference frame of the car, the sandwich appears to travel in a parabolic motion. But the relation $\mathbf{f} = m\mathbf{a}$ seems to hold, and so the driver observes that Newton's laws are valid in her reference frame. Likewise for the advanced alien civilization watching from their cloaked spaceship as Earth whizzes past (see Figure 12.2). From their[4] perspective, everything seems to be obeying Newton's laws. To the alien, Earth is moving with a constant linear velocity and the elevator's trajectory is parabolic, just as we would predict using the projectile equations developed in Section 11.6. (Actually, we are ignoring some finer points, such as Earth's rotation, the curved path it takes as it orbits the sun, and the do-si-do it does with the moon. These deviations from constant linear velocity are the exceptions that prove the

---

[3] Aside from a falling elevator that is about to crash to the ground.

[4] His/her/its. . . since aliens are just as likely to have three genders as two, the limitations of the English language are not up to the demands of interstellar political correctness. If there are any aliens reading this book, please note that we *tried*, and therefore do not atomize our planet. We are quite fond of it.

rule: they are the reason why a reference frame fixed to Earth's surface is not quite an inertial frame. Experiments such as Foucault's pendulum can detect the difference, even though it is slight.)

In summary, if a reference frame is accelerating or rotating, the motion of objects described using that reference frame will not be consistent with mechanical laws. An inertial reference frame must be stationary or moving at a constant linear velocity.

12.1.3   Newton's Third Law

Newton's third law is often misunderstood in spite of being the one most often quoted. It has a certain zen-like justice to it.[5]

Newton's Third Law

To every action there is always an equal and opposite reaction. Or, the forces of two bodies on each other are always equal and are directed in opposite directions.

This law basically says that there is no such thing as a single unilateral force. If object A pushes or pulls on object B, then object B always pushes or pulls back on object A with a force of the same magnitude but opposite direction. If gravity is pulling me towards Earth, it's also pulling Earth towards me! A force is always part of an *interaction* between two bodies.

In diagrams, we often draw a force as an arrow, since it is a vector. But really, these diagrams would be more accurate if both ends of the arrow had arrowheads. When we leave off the other side of the arrow, it's because it is acting on an object in which we have no interest. When you see a single-sided arrow that represents a force in a diagram, you can always fill in the other half in your mind.

One source of misunderstanding of Newton's third law is the word "reaction." The purpose of this word is to describe the forces as being in opposition to one another. It is not meant to imply a causal link between them; neither force is a "cause" or "effect." The two opposing forces act simultaneously and, so far as the laws of physics are concerned, have equal status.

But aside from this mistaken inference of cause and effect, the third law is just plain counterintuitive. Let's say a guy named Moe pushes a

---

[5] One could easily imagine Yoda saying "To every action, always an equal and opposite reaction, there is."

box forward on the ground. The box weighs twice as much as Moe, and he has placed it on a cart that rolls with very little friction. According to Newton's third law, the box pushes back on Moe. But then why does the box accelerate and Moe doesn't? It doesn't look like there are "equal and opposite actions" happening here.

Conundrums such as these are always resolved by considering *all* the forces acting on both bodies. In the example just discussed, Moe is not floating in midair, or else he *would* have been accelerated backwards just as Newton's third law predicts he would. (Consider what would happen if Moe and the box were on ice.) No, Moe is standing on the ground. Through the force of friction, Moe pushes against the Earth and the Earth pushes back on Moe. In fact, if we assume that Moe makes some forward progress instead of just being stuck there grunting, then the force of the Earth pushing against him must exceed the force of the box pushing back against him, and he accelerates *forward*. This is illustrated in Figure 12.3.

Figure 12.3
The four forces involved in Moe pushing the box

An inquisitive reader might wonder about the previous scenario, "Why doesn't the Earth then accelerate?" The short answer is, "It does!" A medium-length answer is, "It does, in the short run." For the full length answer, we have to wait until Section 12.3, which tells us a little bit about *momentum*.

Of course, these theoretical questions are certainly interesting to ponder, but what practical application is there for Newton's third law? The most important application, for our purposes, is the justification to simplify a rigid body and treat it as a single particle. For example, earlier we considered the forces acting on a large beam in a skyscraper. What if the beam is not a single solid piece, but instead it is really two beams that have been bolted together? Then really what is happening is that forces push down on the top part of the beam, which pushes down on the bottom part of the beam, which pushes down on the Earth. Likewise, the Earth is pushing back up on the bottom part of the beam, which pushes up on the top part of the beam.

But why stop there? Isn't any object actually composed of not just two or three pieces, but trillions of molecules? How can we possibly calculate all these complicated quantum-electrical forces? This is where Newton's third law comes in. We are justified in the treatment of this spliced beam as a single rigid body, and we can ignore all the internal forces, *provided that the body stays rigid*, which means that all pairs of points within the object maintain a fixed distance from each other. In this situation, the parts are not accelerating relative to each other, and this means that the internal forces must be exactly balanced. In other words, all the internal forces cancel each other out and thus make no contribution to the net force, which is why we can ignore them. Of course, to the extent that the pieces *do* accelerate relative to each other, any calculations we make ignoring the internal forces will be inaccurate. If the bending or compression of the object is very slight, then our calculations will not be perfect, but they will be very close; if the object breaks apart, then our calculations will be meaningless.

We can generalize arguments such as this even further to the case where the parts *are* moving relative to each other. Of course, an object with moving internal parts is the opposite of a rigid body; however, we'll see that in many respects we are still able treat these complicated systems as "particles." Section 12.3 discusses this idea and how it allows us to resolve the conundrum of Moe and his box.

12.2   Some Simple Force Laws

Many different types of forces are at work in our universe.[6] In a real-time simulation, we often ignore certain forces, make approximations to them, and even invent fictional[7] forces to achieve a desired effect (such as forcing a trajectory to obey an animator's constraints, or helping the AI or the

---

[6] Actually, this is not true. At the time of this writing, physicists believe that there are four fundamental forces. Almost all forces caused by matter bumping up against other matter are essentially *electrical* because it is electrical forces that keep atoms separated from one another. However, the situations in which matter pushes away from other matter are so diverse that at the macroscopic level it is useful to have many different force laws to describe the behavior.

[7] As opposed to frictional.

player hit the target). Although our guiding principle is always $\mathbf{f} = m\mathbf{a}$, the methods used to define $\mathbf{f}$ can vary greatly.

This section discusses three important forces that exist in the real world and are often used in physics simulations. Gravity, friction, and springs are the subjects of Section 12.2.1, Section 12.2.2, and Section 12.2.3, respectively. Of course, a computer simulation may need to consider many more real-world forces, such as buoyancy, drag, or lift. The goal of this book is to give an overview of the most important topics and not to be exhaustive; however, sources that cover these types of forces are listed in the suggested reading in Section 12.7.

One other extremely important force that appears in physics simulations is the *contact force*, also known as a *normal force*. This is the force that prevents objects from penetrating each other. When a box is resting on a table, the force the table exerts on the box, counteracting the force of gravity and preventing the box from accelerating downwards, is called a *contact force*. Contact forces in a physics engine are inherently tied up with the engine's method for resolving collisions and are usually handled in a way that forms a compromise between the stability of the simulation and physical reality. As such, the details for how contact forces are computed can vary from one physics engine to another; indeed, resolving collisions is a very active area of research.

12.2.1 Gravitational Force

In *Principia*, Newton stated all sorts of laws in addition to the three for which he is the most famous. One such law, which he discovered through analysis of the motions of the planets, is the *law of universal gravitation*, which states that all objects in the universe feel an attractive force to each other. This force is proportionate to the product of their masses and inversely proportionate to the square of the distance between the objects and can be calculated by Equation (12.3).

Law of Universal Gravitation

$$f = G\frac{m_1 m_2}{d^2}. \tag{12.3}$$

In this equation, f is the magnitude of the force, m_1 and m_2 are the masses of the two objects, and d is the distance between their centers of mass. (We'll have more to say about exactly what the center of mass is in Section 12.3.2.) G is a physical constant of the universe, approximately equal to 6.673×10^{-11} N m^2 kg^{-2}.

The law of universal gravitational attraction is very helpful if you want to understand planetary motion or the tides, or just need a cheesy pick-up line.[8] However, most simulations are confined to a fairly small region close to Earth's surface. When we make the typical assumption that one Cartesian axis points "down," we are ignoring the curvature of Earth and also locking in the direction of the force of gravity to a constant. It's also common to ignore the slight decrease in the strength of gravity that occurs at higher altitudes, and assume a constant value for d. Thus, if we let m_1 stand for Earth's mass, then the only variable in Equation (12.3) is m_2, the mass of the object being simulated. In most video games, the force of gravity is computed using Equation (12.4).

Video Game Gravity

$$\mathbf{f} = m\mathbf{g}. \tag{12.4}$$

In Equation (12.4), m is the mass of the object and $\mathbf{g}$ is a constant vector pointing in the downward direction. Notice that the force of gravity is proportional to the mass, but Newton's second law says that the acceleration due to any force is *inversely* proportional to mass. Therefore, $\mathbf{g}$ specifies the acceleration due to gravity for all objects in free fall. (Notice the similarity between Equation (12.4) and Newton's second law, $\mathbf{f} = m\mathbf{a}$.)

Chapter 11 told you what the magnitude of $\mathbf{g}$ is in the real world, but let's see if we can derive it from the universal law of gravitation. Earth's mass is approximately $m_1 = 5.98 \times 10^{24}$ kg, and its mean radius is 6,371 km:

Calculating the force of gravity near Earth's surface from the law of universal gravitation

$$f = G\frac{m_1 m_2}{d^2} = \left(6.673 \times 10^{-11}\frac{\text{N m}^2}{\text{kg}^2}\right)\frac{(5.98 \times 10^{24}\text{ kg})m_2}{(6.371 \times 10^6\text{ m})^2}$$

$$\approx (9.83\text{ N})\frac{m_2}{\text{kg}} \approx \left(9.83\frac{\text{m}}{\text{s}^2}\right)m_2.$$

But wait, this value is larger than the value of 9.81 quoted earlier! The reason for the difference is that, while Earth's gravity provides a centripetal force, its rotation creates an apparent *centrifugal* force, which partly counteracts gravity. We calculated the magnitude of the acceleration required to keep objects from spinning out into space in Section 11.8. At the equator,

---

[8]Hint: make sure and work in the phrase "heavenly body" while you're at it.

Earth's rotation requires that gravity provide a centripetal acceleration of 0.03369 m s^{-2}. Thus a small part of the force of gravity that makes objects feel heavy is being counteracted by the spinning of Earth, which makes them feel lighter.

Subtracting this apparent centrifugal force from the force of gravity gives us $9.83 - 0.03369 \approx 9.796$, but now the value is too small, not the 9.81 we were looking for. The reason is that gravity exhibits variations in magnitude over Earth's surface. The biggest source of this variation is the centripetal acceleration we have just calculated; it varies with latitude. We computed its magnitude assuming r was Earth's radius; the resulting value for gravity of 9.796 is actually the correct strength of gravity at the equator. As the latitude increases and we move towards the poles, the radius r of the circular path (which has constant latitude) decreases. At the poles, the radius shrinks to zero, and objects rotate but do not move in a circular path. Thus there is no apparent centrifugal force at the poles, and the force of gravity is equal to the 9.83 value we computed above. The value 9.81 is known as the "standard value," and is the average force of gravity at sea level at a latitude of about $45°$.

Now that we've discussed at some length the strength of gravity in the real world, let's talk about how this number is often completely irrelevant in video games. In certain genres, such as racing or flight simulators, realism is important. However, in most other video games, the *first law of video game physics* applies. (Hey, Newton made up some laws, so why can't we?)

First Law of Video Game Physics

Reality is overrated.

For example, first-person shooters are notorious for poor jumping mechanics. The most important reason is probably the fundamental fact that you cannot see your feet, yet some first-person games have for some reason added jumping puzzles. But even many third-person shooters that adopt an over-the-shoulder camera also have jumping mechanics that just don't feel right. Why? In most first-person shooters, when you jump, you are given an initial burst of upward velocity, and then your position is simulated just like every other airborne object in the world, using gravity, which causes your motion to be parabolic. Compare this to the jump mechanic in most third-person action games. Most of these games do *not* simulate jumps using a constant acceleration. Instead, your character will spring up almost instantaneously after you hit the button, and reach a maximum height very quickly. In many games, the character will *hover* at that maximum height for a duration, and then slam back down on the ground as quickly as it rose up, perhaps leaving a crater behind. This is clearly not physically accurate, but then again, neither is being able to jump two or three times your own height, steer in midair, or double jump. When it comes to jumping in video games, reality is not just overrated, it's completely ignored. It just doesn't feel right.

If simulating a jump mechanic using gravity makes for a bad jump mechanic, simulating a jump mechanic using a value of 9.8 m/s^2 is even worse. The basic problem is that most players expect a jump to take a certain amount of time but also expect to be capable of jumping to unrealistic heights. When real-world gravity is used to attain these heights, the player is in the air too long, and it feels "floaty." Many arcade racing games also increase gravity to get the car back on the ground more quickly. Whether it be racing games or character games, the player wants to be in full control again as quickly as possible, and waiting for real-world gravity to get them back down usually takes too long. And then there are other racing games that use a value of gravity that is *less* than the real world value, to facilitate unrealistic jumps at realistic vehicle speeds.

There are also reasons to fiddle with gravity for non-player-character objects as well. Sometimes real-world gravity can create an "objects made of styrofoam" feeling for simulated objects in general,[9] so gravity is increased to get an object to tip over and come to rest more quickly. In other situations, an artificially low value of gravity can make a large object seem even more massive (especially when accompanied by the right sound effects), because acceleration on Earth is constant and is one of a few cues humans instinctively use to establish an absolute scale for objects in the distance.[10]

Hopefully, while reading the preceding design discussion you absorbed a general message rather than focusing on our specific opinions. What "feels right" is a subjective matter; furthermore—and this is the key point—it is based more on player expectation than physical reality. In the end, what matters most in a video game is not what's going on in the CPU or even on the screen, but what is going on in the player's mind. And the human mind is highly susceptible to suggestion. When creating video games, always remember that the quest for realism should never be an end unto itself, but rather a successful video game will harness realism only where it serves the ultimate goal, which is entertainment. In fact, realism is quite often *opposed* to this goal. Video game makers (especially programmers!) often

---

[9]This is often caused by excessive damping in the physics system used to help mask instability.

[10]Fletch: I like to call this technique "*Lord of the Rings* gravity." It reminds me of the giant staircases that get destroyed during the escape from the Mines of Moria.

get these priorities confused and end up creating an impressive technical demo that isn't any fun.

12.2.2   Frictional Forces

If we take an object such as a bowl of petunias and slide it along a surface, we know that it will eventually come to a stop. We also know that if we place this bowl on a surface that isn't quite level, it won't necessarily slide downhill unless the angle of inclination exceeds a certain threshold. These two phenomena are slightly different aspects of the force of *friction*. We are accustomed to thinking of friction as an onerous enemy of productivity, the evil cause of wear on machines and more frequent trips to the gas station. But keep in mind that without friction, we wouldn't be able to walk across a room or pick up a child (or a bowl of petunias). Without friction, our cars might have better fuel efficiency, but the transmission wouldn't work and the tires would spin in place instead of propelling the car forward.

Here we consider the two modes of the standard dry friction model, which is sometimes called *Coulomb friction*. Although several thinkers contributed to our understanding of friction, Charles-Augustin de Coulomb (1736–1806) is the guy who got his name to stick. When an object is at rest on top of another object, a certain amount of force is required to get it unstuck and set it in motion. If any less force is applied to the object, the force of friction will push back with a counteracting force up to some maximum amount. This type of friction is known as *static friction*, and it prevents bowls of petunias sitting on slightly inclined tables from sliding off. Once static friction is overcome and the object is moving, friction continues to push against the relative motion of the two surfaces, but the magnitude of this force, known as *kinetic friction*, is less than that of static friction. Kinetic friction is what causes a bowl of petunias to eventually come to a stop after we set it in motion.

Friction is the result of complicated interactions at the microscopic level, and so it is somewhat surprising that its macroscopic behavior can be described by relatively simple equations. Let's consider static friction first. Like any force, static friction is a vector. The direction of static friction is always in the direction that opposes any forces that would otherwise cause objects to move relative to each other. This might seems a bit like cheating ("How does the friction always know the correct direction to push?"), but remember that the force is actually the aggregate result of many electrical forces acting at the microscopic level. The forces are the result of molecular bonds that have formed between the objects as they came in contact, and these bonds need a force to pull them apart.

A good approximation for the maximum magnitude of static friction is computed with Equation (12.5).

Static Friction

$$f_s = \mu_s n. \tag{12.5}$$

The dimensionless constant μ_s is known as the *coefficient of static friction*, and n is the magnitude of the normal force. Let's talk about each of these in more detail.

From our perspective, μ_s is certainly the easier of the two to deal with: just look it up in a table! Table 12.1 shows just such a table. Note that we are jumping ahead a bit and showing coefficients for both static and kinetic friction. Ignore the kinetic friction column for now.

Of course, somebody actually has to fill out these tables! The methods for obtaining these data are interesting and rather elegant but they are not our primary concern here. What is very important for us is that the coefficients of static and kinetic friction depend on the properties of *both* interacting surfaces. In other words, Table 12.1 is indexed not by a single surface type, but by a *pair* of interacting surfaces. So, for example, although using this table we can find the coefficient of static friction for rubber against asphalt, we cannot use this information to say anything about, for example, rubber against ice, or wood against asphalt. The coefficient of static friction for each pair of surfaces has to be measured experimentally because of the complexity of the microscopic interactions.

Also, note that Equation (12.5) tells us the *maximum* strength of the static friction force. The actual force exerted at any instant will meet the

Material 1	Material 2	μ_s (Static)	μ_k (Kinetic)
Aluminum	Steel	0.61	0.47
Copper	Steel	0.53	0.36
Leather	Metal	0.4	0.2
Rubber	Asphalt (dry)	0.9	0.5–0.8
Rubber	Asphalt (wet)		0.25–0.75
Rubber	Concrete (dry)	1.0	0.6–0.85
Rubber	Concrete (wet)	0.30	0.45–0.75
Steel	Steel	0.80	
Steel	Teflon	0.04	
Teflon	Teflon	0.04	
Wood	Concrete	0.62	
Wood	Clean metal	0.2–0.6	
Wood	Ice	0.05	
Wood	Wood	0.25–0.5	
Wood (waxed)	Dry snow		0.04

Table 12.1. Static and kinetic coefficients of friction

magnitude of any forces acting on the objects that tend to induce lateral relative motion, up to this maximum. Once this maximum is exceeded, the static friction ceases to operate, and kinetic friction takes over.

The other factor in Equation (12.5) is the magnitude of the *normal force*, which is the force acting perpendicular to the surfaces that prevent them from penetrating each other. One common situation occurs when an object (such as a bowl of petunias) is resting on top of another object (such as a table). The normal force in this case is simply the force required to counteract gravity. To be more precise, it is the force required to counteract the component of gravity that acts perpendicular to the surfaces and wants to smash them together. If the table is at an incline, then we can separate gravity into a normal component and a lateral component, as shown in Figure 12.4. (Inside a computer, we'd probably describe the orientation of the table with a normal vector, and use the dot product to separate gravity into the relative and normal components, as we described in Section 2.11.2.) Since the bowl and the table do not accelerate relative to each other, we know that the normal force of the table pushing against the bowl must be exactly equal to the normal component of the force of gravity pulling the bowl towards the table.

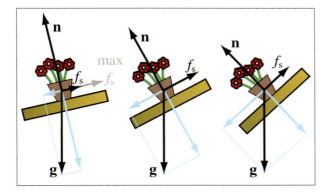

Figure 12.4. Free-body diagrams of a bowl of petunias on a table at various angles of inclination.

Figure 12.4 shows several free-body diagrams of identical bowls of petunias resting on tables at various angles of inclination. Notice that in each figure, the force of gravity acting on the bowl, labeled **g**, is the same. The normal and lateral components of gravity have been broken out in blue. The actual force of static friction is the black vector labeled f_s. On the left, the maximum amount of friction available is labeled "max f_s," in the middle and right-hand images the maximum amount of friction is being applied.

In the first scenario, there is more friction available than necessary to stop the sliding. However, as the angle of inclination increases, the normal component of gravity decreases, which reduces the amount of friction available. Meanwhile, as the perpendicular component of gravity is decreasing, the lateral component is increasing, which makes the bowl want to slide. The force of static friction must counter this lateral component if the bowl is to remain in equilibrium. The center image shows the critical angle at which the lateral force of gravity is exactly equal to the to the maximum amount of friction. On the right, we imagine that we have tilted the table while holding the bowl in place, and then let go of the bowl. The maximum available friction is being applied, but it's less than was available in the center picture due to the decrease in the normal force and isn't enough to overcome the increased lateral component of gravity. Just after this picture was taken, friction switched from static mode into kinetic mode, the bowl slid off the table and shattered, and a cartoon cleaning robot scurried in through a little door in the wall to clean up the mess.

Calculating kinetic friction is essentially identical to static friction. The only difference is that we replace the subscript s with a k.

Kinetic Friction

$$f_k = \mu_k n. \tag{12.6}$$

The direction of the force of kinetic friction is always opposed to the relative motion of the surfaces. (Remember, according to Newton's third law, there are actually *two* forces, one pushing against the bowl, and the other against the table, and they are in opposite directions.) As we said earlier, the coefficient of kinetic friction is usually *less* than the coefficient of static friction. Thus, if we increase the angle of the table ever so slowly so that static friction is *just* overcome, the petunias will begin to accelerate, based on this difference between kinetic and static friction. Coulomb's primary contribution to the theory, sometimes called *Coulomb's law of friction*, was that the force of kinetic friction does *not* depend on the relative velocities of the surfaces, so, unlike static friction, there is no distinction between the effective force and the maximum force.

Notice that the amount of area that the two objects are in contact does *not* appear in Equations (12.5) or (12.6). For example, let's say we repot the petunias in a taller bowl with a smaller footprint but the same weight. We have reduced the surface area where the bowl and table are in contact, but all of the forces depicted in the free-body diagrams in Figure 12.4 remain the same. Doing this would *not* change the angle at which the bowl would begin to slide! Although it might seem like a larger surface area would give the objects more to "grab" with, this is offset by the decrease in pressure, since the same total normal force is now distributed over a smaller contact area. Now, a very tall bowl may begin to *tip over* before it begins to slide. But this is a matter of rotation, the increase in the tendency to rotate being caused by an increase in the lever arm resulting in a greater torque. We cover these issues in Section 12.5.

12.2.3   Spring Forces

There's one more class of force that is important enough to discuss in its own section: the forces exerted by a spring disturbed from its equilibrium position. Why do we discuss this admittedly peculiar class of force? Have springs suddenly become prominent features in video games and their accurate simulation an important gameplay feature? Actually, yes. Even if you don't see very many *literal* springs in a video game, there are likely very many "virtual springs" at work. Springs exhibit a general behavior that is *very* useful for enforcing constraints, preventing objects from penetrating, and the like.

   This section presents the classic equations of motion for damped and undamped oscillation. It covers undamped oscillation first, and then damped. It's often the case in a video game that programmers use a virtual spring (often in the form of a spring-damper system) when really what they are using is a *control system*. There are certain advantages to be had when the physical nature of the problem is dropped and we think of it purely in mathematical terms. (Indeed, many times the problem was never really physical to begin with, and was only recast in physical terms so that the spring-damper apparatus could be applied.)

   Like the friction law, the force law for springs is a surprisingly accurate approximation to the macroscopic behavior that is the result of complicated microscopic interactions. Consider a spring with one end fixed and the other end free to move in one dimension. When the spring is at equilibrium with no external forces on it, it has a natural length, called the *rest length*. If we stretch the spring, then it will pull back to try to regain its rest length. Likewise, if we compress the spring, it will push back. But how do we know the strength of the force in each case? That's what the force law tells us.

   The force law for springs is known as *Hooke's law*, and it basically says that the magnitude of the restorative force is proportional to the difference from the current length and the rest length (provided the force does not exceed a value called the *elastic limit*, which varies with the material used to construct the spring). If we let l be the current length of the spring and

l_{rest} denote the rest length, then the magnitude of the restoring force f_{r} is calculated by Equation (12.7).

Hooke's Law for Spring Forces

$$f_{\mathrm{r}} = k(l_{\mathrm{rest}} - l). \tag{12.7}$$

The constant k is known as the *spring constant* and essentially describes how "stiff" the spring is. The constant is not dimensionless. In order for Equation (12.7) to make sense, we must have

$$[ML/T^2] = k[L],$$

$$[ML/T^2]/[L] = k,$$

$$[M/T^2] = k,$$

or you can just think of k as having units of "unit force per unit length."

The really interesting thing about springs is how they behave over time. To see this, let's restate Hooke's law in a way that focuses on the kinematics of a particle that is being acted on by restorative forces. Specifically, we're interested in functions for the position, velocity, and acceleration of a particle.

Things get easier if we adopt a reference frame where the position $x = 0$ designates the "rest" position, where there are no restorative forces. Furthermore, since we are interested in the acceleration of the particle rather than the forces acting on it, we will introduce a constant $K = k/m$, and since K contains both the spring constant k and the mass of the particle m, it measures the spring's ability to accelerate the specific particle of interest to us. With those notational changes, we can rewrite Equation (12.7) as

Acceleration due to Hooke's law

$$a(t) = -Kx(t). \tag{12.8}$$

You should convince yourself that this is equivalent to Equation (12.7) before continuing.

Equation (12.8) makes a statement about the relationship between the position function and the acceleration function; but what we really want is the function $x(t)$ itself. Equations like this are called *differential equations*; they describe the relationship between some unknown function (in this case, $x(t)$) and one or more of its derivatives (remember that acceleration is the second derivative of position). To "solve" a differential equation is to find the unknown function $x(t)$ that satisfies the equation. We have been able to just barely scratch the surface of basic differential and integral calculus in this book, so we're not going to be able to cover the techniques of solving differential equations. Luckily, you don't need to know differential equations in order to verify that a proposed function $x(t)$ is a solution—that requires only the ability to differentiate the function $x(t)$. As it turns out, this will be sufficient in the few cases in which we bump up against differential equations in this book.

We can make a pretty good guess at the form of $x(t)$ by looking at a graph. We are not engaging in circular logic here; we don't need to know $x(t)$ in order to get a graph, all we need is a spring with some sort of marking device attached to it.[11] Such a graph is shown in Figure 12.5.

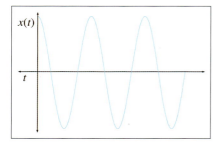

Figure 12.5
The graph of the motion of a spring. Hey, that looks familiar...

This function ought to look familiar to you: it's the graph of the cosine function. Let's see what happens if we just try $x(t) = \cos(t)$ as our position function. Differentiating twice to get the velocity and acceleration functions (remember, we learned about the derivative of the sine and cosine functions in Section 11.4.6), we get

$$x(t) = \cos(t),$$

$$\dot{x}(t) = -\sin(t),$$

$$\ddot{x}(t) = -\cos(t),$$

Close, but not quite right

which is very close, but we're missing the factor of K.

To understand where K should appear in $x(t)$, consider what happens to the graph of $x(t)$ when we change the value of K. In other words, we repeat our physical experiment and vary the stiffness of the spring or the mass of the marking device attached to the end of the spring. The result is that larger values of K (stiffer springs or less massive marking devices) result

---

[11]Professor Walter Lewin does this classroom demonstration in his physics class at MIT. All of the lectures can be downloaded free through MIT OpenCourseWare at http://ocw.mit.edu.

in a graph that is horizontally "compressed": the frequency of oscillation is increased. Likewise, smaller values of K cause the spring to oscillate more slowly, and the graph is expanded. Furthermore, we observe that the frequency is proportional to the *square root* of K—when we increase K by a factor of four, the frequency doubles. This gives us a hint as to where K should appear, since all we are doing is scaling the time axis.

A solution, but is it the only one?

$$x(t) = \cos(\sqrt{K}\, t),$$
$$\dot{x}(t) = -\sqrt{K} \sin(\sqrt{K}\, t),$$
$$\ddot{x}(t) = -K \cos(\sqrt{K}\, t).$$

One verifies that this is a solution to the differential equation by plugging it into Equation (12.8). Remembering that $a(t) = \ddot{x}(t)$, we have

$$a(t) = -Kx(t),$$
$$-K \cos(\sqrt{K}\, t) = -K(\cos(\sqrt{K}\, t)).$$

The quantity $\sqrt{K}$ is the *angular frequency* and comes up often enough that we find it helpful to introduce the notation

Angular frequency

$$\omega = \sqrt{K} = \sqrt{k/m},$$

and we can write the solution as

$$x(t) = \cos(\omega t); \tag{12.9}$$

hence the reason for the name "angular frequency" becomes apparent.

So we have found the kinematics equation for the spring. Or, perhaps we should say that we have found *a* solution to the differential equation. There are some degrees of freedom inherent in the motion of the spring that are not accounted for in Equation (12.9). First, we are not accounting for the maximum displacement, known as the *amplitude* of the oscillations and denoted A. Our equation always has an amplitude of 1. Second, we are assuming that $x(0) = A$, meaning the spring was initially stretched to the maximum displacment A and released with zero initial velocity. However, in general, we could have pulled it to some displacement $x_0 \neq A$ and then given it a shove so it has some initial velocity v_0.

It would appear that we have three more variables which need to be somehow accounted for in our equation if it is going to be completely general. As it turns out, the three variables we have just discussed—the amplitude, initial position, and initial velocity—are interrelated. If we pick any two, the value for the third is locked in. We'll keep A as is, but we'll replace

x_0 and v_0 with the *phase offset* θ_0, which describes where in the cycle the spring is at $t = 0$. Adjustments to the phase offset have the simple effect of shifting the graph horizontally on the time axis. Adding these two variables, we arrive at the general solution, the equations of *simple harmonic oscillation*.

Simple Harmonic Motion

$$x(t) = A\cos(\omega t + \theta_0), \qquad\qquad (12.10)$$
$$\dot{x}(t) = -A\omega\sin(\omega t + \theta_0),$$
$$\ddot{x}(t) = -A\omega^2\cos(\omega t + \theta_0).$$

Now let's make some observations. First, remember that the sine and cosine functions are just shifted versions of each other: $\sin(t + \pi/2) = \cos(t)$. Thus we could have written $x(t)$ using sine instead of cosine, the choice being mostly a matter of preference and an adjustment in the phase by $\pi/2$. The term "sinusoidal" can be used to refer to the shape of the sine and cosine functions, and we use it when either function will do.

Second, consider the frequency of oscillation. The sine and cosine functions have a period of 2π; thus the oscillator will complete one cycle in the time it takes for ωt to increase by 2π. The angular frequency ω is measured in radians per unit time, but we can also measure the frequency F, which is in cycles per unit time, as

$$F = \frac{\omega}{2\pi} = \frac{\sqrt{K}}{2\pi} = \frac{1}{2\pi}\sqrt{\frac{k}{m}}.$$

Frequency of simple harmonic motion

Notice that the frequency of oscillation depends only on ratio of the spring stiffness to the mass. In particular, it *does not depend* on the initial displacement x_0: if we stretch the spring farther before letting it go, the amplitude increases, but the frequency will not change.

In many situations, the frequency is the important number we wish to control. This is especially the case for "virtual springs," which are really control systems in disguise. In these situations, we don't need to bother with spring constants or masses, and we can write the equation of motion directly in terms of frequency, as

$$x(t) = A\cos(2\pi F t + \theta_0).$$

Simple harmonic motion in terms of frequency

So far, we have been studying a physically nonexistent situation in which the restorative force is the only force present, and the spring will oscillate forever. In reality, there are usually at least two more interesting forces. The first of these forces is an external force, sometimes called the *driving force*, that acts as the "input" to the system and causes the motion to begin in the first place. The other force is the friction that any real spring experiences, which eventually causes the motion to cease. The general term used to describe any effect that tends to reduce the amplitude of an oscillatory system is *damping*, and we call oscillation where the amplitude decays over time *damped oscillation*. Damping forces are particularly important for our purposes, so let's discuss them in more detail.

The most common model for the damping force is a simple one that acts proportional to velocity but in the opposite direction, similar to the friction law. (Unlike the friction laws from the previous section, we don't have any of the business concerning the normal force.) The force is simply

Damping force
$$f_\mathrm{d} = -c\dot{x},$$

where f_d indicates the instantaneous magnitude and direction of the damping force, $\dot{x}$ is the instantaneous velocity, and c is a constant that describes the viscosity, roughness, etc.

The damping force has an extremely simple form, but just as with the restorative force, things get interesting when we study the motion over time. Qualitatively, we can make some basic predictions about how damped oscillation of a spring would differ from undamped oscillation of the same spring. The more obvious prediction is that we would expect the amplitude of oscillation to decay over time, meaning the maximum displacement at the crest of each cycle is a bit less than the previous one. Like the force of friction, damping tends to remove energy from the system. The second observation is only slightly less obvious: since damping in general slows the velocity of the mass on the end of the spring, we would expect the frequency of oscillation to be reduced compared to undamped oscillation. Those two intuitive predictions turn out to be correct, although, of course, to be more specific we will need to analyze the math.

Combining the restorative and damping forces, the net force can be written as
$$f_\mathrm{net} = f_\mathrm{r} + f_\mathrm{d} = -kx - c\dot{x}.$$

To derive the equation of motion, we will need accelerations, not forces. Applying Newton's second law and dividing both sides by the mass, we have

$$\ddot{x} = \frac{f_\mathrm{net}}{m} = -\frac{k}{m}x - \frac{c}{m}\dot{x}, \tag{12.11}$$

Next we rewrite this in terms of two new quantities. The first quantity, ω_0, is the *undamped angular frequency* and is not really new. It is identical to the $\omega = \sqrt{k/m}$ introduced earlier; we are adding the zero subscript just to emphasize that it is the frequency that would occur without the damping rather than the actual frequency. (Remember, our prediction is that the actual frequency will be slower in some way.)

The second quantity is called the *damping ratio*, not to be confused with the damping coefficient c. The damping ratio is traditionally denoted by ζ, the Greek letter zeta, which looks weird and takes some practice to write by hand. The damping ratio is related to the damping coefficient, mass, and undamped angular frequency by the formula

$$\zeta = \frac{c}{2\sqrt{mk}} = \frac{c}{2m\omega_0}.$$

Damping ratio

In just a moment, when we explain the qualitative meaning of ζ, the utility in using this arbitrary formula will become apparent.

Substituting the undamped frequency ω_0 and damping ratio ζ into Equation (12.11), we have

$$\ddot{x} = -\omega_0^2 x - 2\zeta\omega_0\dot{x}. \qquad (12.12)$$

Differential equation for damped harmonic oscillation

Readers with training in differential equations should recognize Equation (12.12) as a second-order linear homogenous differential equation with constant coefficients, which is one of the very nicest differential equations we could hope for, meaning we can actually solve it with pencil and paper. Readers without this training shouldn't worry, because it won't be needed to understand the answer, to which we now fast forward, skipping the derivation. There are three distinct cases: underdamping, critical damping, and overdamping.

When $0 \leq \zeta < 1$, we say that the system is *underdamped*. In this case, as we have been predicting, the motion will continue to oscillate indefinitely with an amplitude that decays exponentially over time. The equation that describes this motion is

$$x(t) = (k_1 \cos(\omega_d t) + k_2 \sin(\omega_d t))\, e^{-\zeta\omega_0 t}, \qquad (12.13)$$

Kinematic equation for underdamped system

where ω_d is the actual frequency of the damped oscillation and is related to the undamped frequency ω_0 by

$$\omega_d = \omega_0 \sqrt{1 - \zeta^2}. \qquad (12.14)$$

Damped angular frequency

The constants k_1 and k_2 are determined by the initial position and velocity:

$$k_1 = x(0), \qquad\qquad k_2 = \frac{\zeta\omega_0 x(0) + \dot{x}(0)}{\omega_d}.$$

Using $\zeta = 0$ produces undamped oscillation, and Equation (12.13) is equivalent to Equation (12.10).

Your common sense tells you that as we increase the damping ratio, the frequency of oscillation decreases; consulting Equation (12.14), we see that at $\zeta = 1$ the frequency completely vanishes. At this threshold, known as *critical damping*, the behavior of the system changes qualitatively. The system no longer oscillates, but instead decays exponentially. The kinematic equation in this situation is

Equation of motion at critical damping

$$x(t) = (k_1 + k_2 t)\, e^{-\omega_0 t}, \tag{12.15}$$

where k_1 and k_2 are again determined by the initial conditions:

$$k_1 = x(0), \qquad\qquad k_2 = \omega_0 x(0) + \dot{x}(0).$$

Critical damping is just the right amount such that the system decays as quickly as possible without oscillation. If the damping is decreased, the system is underdamped, as previously described, and will oscillate. If the damping is increased, the system is *overdamped*; it will not oscillate, and the rate of decay will be slower than the rate at critical damping. Figure 12.6 shows how the damping value affects the behavior of a system.

Now that we've reviewed the classic equations that may be found in any physics textbook or on wikipedia.org, let's say a few words about how spring-damper systems are used in video games as *control systems*. In

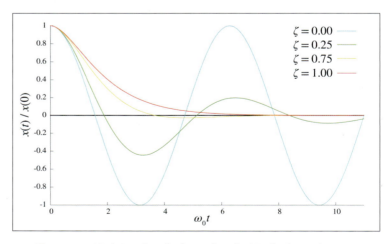

Figure 12.6. Undamped, underdamped, and critically damped systems.

general, a control system[12] takes as input a function of time that represents some target value. For example, our camera code might compute a desired camera position based on the player's position each frame; some AI code might determine an exact targeting angle for an enemy; we may have a desired player character velocity based on the instantaneous amount of control stick deflection; or we might have a desired screen-space position for some highlight effect, based on the currently selected choice in a menu. In any case, the current value of the input signal is known as the *set point* in control system terminology. The set point is essentially the rest position of the spring, and the input signal is like somebody taking the other end of the spring and yanking it around. (So it's similar to a driving force, only usually what we have is a function describing a *position* rather than a force or acceleration.)

The job of any control system is to take this input signal and produce an output signal. To go back to our earlier examples, the output signal might be the actual camera position to use for each frame, or the actual animated targeting angle the enemy will use to aim the weapon, the actual player character velocity, or the actual screen-space position of the highlight. For many control systems, the actual position and set point are not used; rather, only the *error* is needed. Of course, an obvious question is, if we know the "desired" value, why don't we just use that directly? Because it's too jerky. In the same way that the shocks and springs on a car (a classic example of a spring-damper system) don't just pass along the elevation of the road directly to the car, a control system in a video game is often designed to "smooth out the bumps" caused by sudden state changes that might make the camera snap to a new position or the player jerk into motion. The camera or screen-space highlight are nonphysical examples in which the quantity of "mass" is not really appropriate and is dropped. But the differential equations are still the same, and they have the same solution. Stripped of the spring metaphor, we are left with what is known as a *PD controller*. The P stands for proportional, and this is the spring part of the controller, since it acts proportional to the current error. The damper is the D part, which stands for derivative, because the action of the damper at any given instant is proportional to the derivative (the velocity). PD controllers (and their more robust cousin, the *PID controller*, where the I stands for integral and is used to remove steady-state error) are broadly applicable tools; they have been standard engineering tools for decades (centuries?) and are well understood. Nevertheless, they are one of the most frequently reinvented wheels in video game programming.

---

[12]This is not the broadest possible definition of "control system," but it is the most common one.

In practice, the simulation code is a very simple Euler integration of Equation (12.11). As shown in Section 12.6.3, this is a fancy way of describing code that looks like Listing 12.1.

```
struct SpringDamper {
    float value;      // current value
    float setPoint;   // "desired" value
    float velocity;   // current ''velocity'' (derivative of value)
    float c;          // damping coefficient
    float k;          // spring constant

    // Update the current value and velocity, stepping forward
    // in time by the given time step
    void update(float dt) {

        // Compute acceleration
        float error = value - setPoint;
        float accel = -error*k - c*velocity;

        // Euler integration
        velocity += accel*dt;
        value += velocity*dt;
    }
};
```

Listing 12.1
A simple spring-damper control system

Different cars have suspensions that are tuned differently; sports cars are "tighter" and the cars retirees like to drive are smoother. In the same way, we tune our control systems to get the response we like. Notice that the simulation uses the k and c from Equation (12.11). However, most people don't find those to be the most intuitive dials to have for tweaking. Instead, the damping ratio and frequency of oscillation are used for the designer interface, while k and c are computed as derived quantities. To tune the frequency, we might adjust either the damped or undamped version, using either angular frequency or simply Hertz; the units and absolute value are often not important because the value that feels good will be determined experimentally anyway. For many systems in video games, oscillation is undesirable, so it's common to assume a critically damped system and fix $\zeta = 1$, leaving just the "frequency" (we put it in quotes since the system doesn't oscillate) as the only tunable value. A higher frequency is the sports car (more responsive, but jerkier), and a lower frequency is smoother, but can feel "laggy."

Notice that the kinematic equations (12.13) and (12.15) are not needed directly by the simulation, nor do we need to explicitly distinguish between underdamped, critically damped, or overdamped.

Before we leave this discussion, we must mention that the second-order systems we have described here are certainly not the only type of control system, nor even the simplest, but they do behave nicely under a very broad set of circumstances and are easy to implement and tune. Another commonly used control system is a simple first order lag, $\dot{x} = kx$, under which the error decays exponentially. This is similar to a critically damped second-order system, but with a bit jerkier response to a sudden change in the set point. Another important and common technique is to "chase" the set point at a fixed velocity. A *filter* is another broad class of control system, in which the output is computed by taking some linear combination of set points or values on previous frames.

12.3   Momentum

Let's say that Moe's box from Section 12.1.3 has a mass m, and at a certain instant we observe it moving with a velocity $\mathbf{v}$. Coming in late to the story, we cannot tell what magnitude of forces were used to achieve that motion, or how long the forces were applied, or what the history of the box's velocity was. For example, it could have been that the box was accelerated as a result of a constant net force $\mathbf{f}$ being applied over a duration Δt. But we have no way of knowing the values of $\mathbf{f}$ and Δt. Was a large force used for a small duration, or a small force for a longer duration? In fact, we have no reason to assume that the force was constant at all! Moe could have given the box a good shove and set it in motion, and then gave it another shove to speed it up.

While we don't know the exact history of Moe's pushes, we do know what the "total" was, in the sense about to be described. Assume that Moe did make one push with a constant force $\mathbf{f}$ applied for a duration Δt. Then according to Newton's second law, the acceleration was $\mathbf{a} = \mathbf{f}/m$. Assuming the initial velocity was zero, we know that

$$\mathbf{v} = \mathbf{a}\,\Delta t.$$

Substituting $\mathbf{a} = \mathbf{f}/m$ and rearranging, we get

$$\mathbf{v} = (\mathbf{f}/m)\,\Delta t,$$
$$m\mathbf{v} = \mathbf{f}\,\Delta t. \qquad (12.16)$$

Two ways to think of momentum

The left- and right-hand sides of Equation (12.16) illustrate two different ways of thinking about the important concept of *momentum*. Momentum is the correct quantity to track in order to quantify the "total amount of pushing."

Let's do dimensional analysis on Equation (12.16), first just to verify that it makes physical sense—it isn't intuitively obvious that these two products would bear the same physical meaning—and also to see what the units of momentum should be:

$$m\mathbf{v} = \mathbf{f}\,\Delta t,$$
$$M(L/T) = (ML/T^2)\,T,$$
$$ML/T = ML/T.$$

Note that momentum is a vector quantity, having both magnitude and direction.

To understand what momentum is, let's look at the two sides of Equation (12.16). First consider the left side, which interprets momentum as a product of mass and velocity. In fact, somewhere in almost every physics textbook you can find Equation (12.17).

Momentum as Product of Mass and Velocity

$$\mathbf{P} = m\mathbf{v}. \tag{12.17}$$

The variable $\mathbf{P}$ is the traditional variable used to represent momentum. (Despite the capital letter, $\mathbf{P}$ is a vector quantity. We use capital $\mathbf{P}$ to avoid confusion with the notation $\mathbf{p}$, which we sometimes use to refer to the position of a particle.)

Equation (12.17) makes it clear that the momentum of an object is an *instantaneous* property of an object. By saying this, we mean that we can define its value knowing only its instantaneous state, without worrying about how it got into that state. Furthermore, if you think of momentum as the "total amount of pushing" required to stop a moving object, then it certainly is intuitively appealing that it should be the product of mass and velocity. If the object is small and moving slowly (a pencil rolling on a desktop), only a small total force will suffice. If it's fast (a bullet) or heavy (a car that somebody left parked on an incline without the emergency brake set), a larger amount will be needed. If it's fast *and* heavy (an airplane coming in for a landing), then you'd better get out of the way. The equation $\mathbf{P} = m\mathbf{v}$ quantifies the idea of "hard to stop."

Although the memorable equation $\mathbf{P} = m\mathbf{v}$ from the left-hand side of Equation (12.16) is perhaps the more common way of explaining momentum, the right-hand side actually provides the most insight. The relation $\mathbf{P} = \mathbf{f}\,\Delta t$ shows that momentum, as the product of force and time, is what results when force acts over time. This is what was meant by the sloppy phrase "total amount of pushing." We don't mean that the magnitude of the force pushing itself is changing or accumulating, but rather that the continued application of a net force always results in a buildup of momentum (or a reduction of momentum, when the directions of the force and momentum are opposed).

In fact, if we generalize the equation $\mathbf{P} = \mathbf{f}\,\Delta t$, we can discover an even deeper relationship between force and momentum. What if, instead of pushing the box with a constant force, Moe pushed it with a force that varied over time? Then we can express the acceleration at any given time t as

$$\mathbf{a}(t) = \mathbf{f}(t)/m, \tag{12.18}$$

which is just Newton's second law to which we've added the notation "(t)" to be more explicit that $\mathbf{a}$ and $\mathbf{f}$ vary with time. We learned in Chapter 11 that if we integrate acceleration over time, we get the velocity as a function of time:

$$\mathbf{v}(t) = \int \mathbf{a}(t)\,dt. \tag{12.19}$$

Velocity is the time integral of acceleration, remember?

Substituting Equation (12.18) into Equation (12.19), assuming that the mass does not vary over time, we have

$$\mathbf{v}(t) = \int \mathbf{f}(t)/m\,dt,$$

$$m\mathbf{v}(t) = \int \mathbf{f}(t)\,dt.$$

Finally, if we let $\mathbf{P}(t)$ be the momentum of a body as a function of time, then by substituting $\mathbf{P}(t) = m\mathbf{v}(t)$, we arrive at the important relation

$$\mathbf{P}(t) = \int \mathbf{f}(t)\,dt. \tag{12.20}$$

Momentum as force accumulated over time

Since integration is a "summing up" process, Equation (12.20) confirms our interpretation of momentum as the result of continued application of force over time. (Note: in the preceding integrals we omitted the constant of integration, essentially assuming the initial velocity was zero.)

Remember that integration and differentiation are inverse operations. By taking the derivative of both sides with respect to t, we state the flip side of the relationship between momentum and force.

Force as the Derivative of Momentum

$$\frac{d}{dt}\,\mathbf{P}(t) = \mathbf{f}(t). \tag{12.21}$$

The net external force on a system is equal to the rate of change of momentum of the system.

Equation (12.21) is not just an interesting observation about force and momentum, it's a completely valid way to *define* force. In fact, although the modern presentation of Newton's laws is in terms of forces and masses, when Newton himself first expressed the laws, he wrote in terms of momentum. He used the word "motion," but from his writings we understand that he used that word in a very particular sense, and he really was talking about momentum. (The word momentum hadn't been attached to that concept yet. Remember, he was the guy laying down all the ground rules.) Newton's second law was originally expressed in a form that more closely resembles Equation (12.21) than the $\mathbf{f} = m\mathbf{a}$ form you will more commonly see.

12.3.1   Conservation of Momentum

Let's return to our investigation into what happens when Moe pushes against the Earth to get his box moving. Newton's law tells us that the Earth, not having anything else to push back on, receives a net force, and thus an acceleration (and a torque, which we discuss later). Yes, you cause the Earth to accelerate when you push boxes as well as when you take each and every step! Of course, the Earth's mass is so large compared to Moe's force that this acceleration is small. Not only that, but Moe's force pushing the box to the east might be canceled out by Joe's force in North Dakota pushing his box to the west at the same time. An issue even more important than these two facts involves the "accounting laws" of physics: "there is no such thing as free momentum." Moe doesn't need Joe to balance out his force; as it turns out, he can't help but do it all by himself!

Observe that once Moe sets the box in motion, he will need to eventually stop it. According to Newton's first law, the only way to stop a moving box is through a force, and according to the third law, this can happen only if there is some other object involved to receive the opposite force. Perhaps the box bumps into a tree and comes to a stop. (We consider the tree to be part of the Earth. Remember that Newton's third law justifies our treatment of connected objects as a single object, provided that they remain rigidly connected.) To stop the box, the Earth must push against it with a force that is in the opposite direction that Moe pushed to start it moving. However, we know that the "total amount" of pushing must be the same, meaning the Earth must push back with a strong enough force, or for a long enough duration (perhaps Moe's box rolls into a patch of tall grass) to bring the momentum of the box down to zero. So you see that whatever acceleration the Earth received as a result of getting Moe's box in motion must always be exactly canceled by the force required to bring the box to a stop.

But perhaps Moe's box does not come to a stop by pushing directly against the Earth. Let's say it bumps up against Joe's box. Voila! We have stopped Moe's box, and no force has been applied to the Earth. But now, by Newton's third law, *Joe's box* must begin accelerating, and we are back to where we started with a moving box that will continue moving unless it receives a force to bring it to a stop. Eventually, the only way we can stop this chain reaction started by Moe's push against his box is for something, eventually, to push against the Earth.

We can generalize this idea even further. We are justified in treating the entire Earth, and all of its moving parts, as a single particle with all of its mass centered at some location known as the *center of mass*. (We talk more about this special point in Section 12.3.2.) The pushing against the Earth of people like Moe results in transfers of momentum between the objects in the system. Each part will move around within this very complicated system relative to the other parts and relative to the center of mass of the system. However, the total amount of momentum of the entire system is always a constant, unless there are external forces acting on the system. This is known as the *law of conservation of momentum*.

The Law of Conservation of Momentum

The momentum of a system is constant unless external forces act on that system.

The conservation of momentum is precisely what Equation (12.21) is saying. It's certainly an experimentally verified fact, but it also follows naturally as a result of Newton's laws. Section 12.4 discusses how to use this important law to simulate the collision of objects. However, before we get to that, we need to take a closer look at the center of mass.

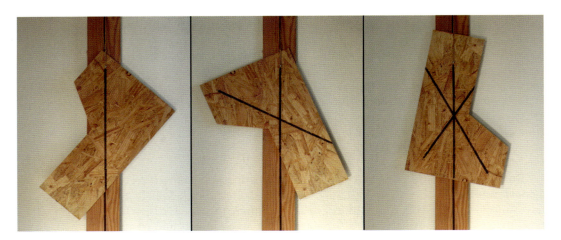

Figure 12.7
Experimentally measuring the center of mass of a odd-shaped piece of particle board

12.3.2   The Center of Mass

Our discussion of momentum has led us to consider the center of mass of
an object. Let's say a few more words concerning this important concept.
For everyday purposes, the center of mass is equivalent to the *center of
gravity*, which is essentially the point around which the object is perfectly
balanced. If we balance an object on the tip of a very thin rod or hang it
from a wire, then the rod or wire will be in a line that contains the center
of mass.

Before we discuss how to compute the center of mass mathematically,
let's see how we can measure it experimentally. Imagine that we have some
object with an odd shape, or of an irregular density. We can determine
its center of gravity by hanging the object from any arbitrary point on the
surface of the object. This defines a vertical line upon which the center
of gravity must lie. By repeating the experiment with a different point on
the object and finding the intersection of those two lines, we can locate the
center of gravity.

The authors performed this experiment on a piece of particle board,
as shown in Figure 12.7. First, the board was cut into a purposefully
asymmetric shape. Next, we chose three arbitrary locations from which
to hang the board, and when the board had finished swinging around, we
drew a heavy line on it, coincident with the string by which the board
was suspended. Sure enough, physics worked, and the third line passed
right through the intersection of the first two lines, at the board's center of
mass.

To compute the center of mass mathematically, we imagine the object being divided up into a very large number of small "mass elements." If there are n such elements, and we denote the mass and position of the ith element as m_i and $\mathbf{r}_i$, respectively, then the center of mass $\mathbf{r}_c$ is simplify the weighted average of the positions of all the mass elements.

Calculating the Center of Mass

$$\mathbf{r}_c = \frac{1}{M} \sum_i^n m_i \mathbf{r_i}. \tag{12.22}$$

In Equation (12.22), M is the total mass of the object

$$M = \sum_i^n m_i.$$

For our purposes, the most important property of the center of mass is that if the object rotates, it will rotate about its center of mass. This assumes, of course, that the object is freely rotating and there isn't a constraint compelling it to rotate about some other point.

As an example, consider a sledge hammer. Clearly, the center of mass of the sledge hammer is close to the heavy end of the sledge, not in the middle of the handle. Assume we throw the hammer across the room. As it tumbles through space, any arbitrary point on the hammer will trace out a complicated spiraling shape. The center of mass, however, moves in a parabola, in perfect agreement with the kinematics equations we learned in Chapter 11.

The authors couldn't resist the opportunity to chuck big objects around, so we verified this hypothesis experimentally, and you can, too.[13] We started with the odd-shaped piece of particle board, whose center of mass had been experimentally located and clearly marked. Next, the fun part:

---

[13]If you do decide to do this, take this advice: (1) Please be safe. Seriously, we assume no liability for people being dumb. (Speaking of dumb, one author had to replace his father's lawn mower wheel, which was found to have a perfectly parabolic trajectory, but alas unable to withstand the landing impact.) (2) We took still pictures at about 4 Hz, but using a nicer camera capable of taking more frames per second, or perhaps extracting frames from video might work better. (3) Please send us your pictures at gamemath.com!

Figure 12.8
The center of mass is the special point that obeys the simple kinematic equations from Chapter 11. Any other point traces out a spiraling path as the object rotates.

we threw it in the air and took a series of pictures of its trajectory with a camera mounted on a tripod. Finally, we merged these frames together into a single image, and used least-squares to fit a parabola through the points marking the center of mass. The result of the experiment is Figure 12.8.

One small note: When fitting the parabola, we did not include the first frame in the data set. As you can see, on the first frame the board is still in the assistant's hand, and thus has not yet begun its (parabolic) free fall trajectory.

Because an object will rotate about its center of mass when allowed to rotate freely, in a physics simulation is it highly advantageous to select the origin of your object to be at its center of mass. Of course, you may have good reasons to put the origin of the object elsewhere. For example, you may have a graphical representation of an object with the origin placed somewhere that made sense to the artist who made that model. In general, if you place the origin somewhere other than the center of mass, then you will likely have to deal with two "positions" of the object: one position within the physics system that describes the world coordinates of the center of mass, and another, perhaps in the rendering system, for the origin of your graphical model. The code that translates between these two conventions is likely to be found either in the interface with the physics engine (for example, the code that updates the position of the objects after the physics simulation has run), or the code that sets up the reference frame for the object during rendering.

The center of mass is fixed for a rigid body such as a sledge hammer; this assumption has been implicit in the whole discussion; otherwise, it wouldn't make sense to advise setting the origin at the center of mass. However, for a general system with moving parts, such as Earth, the center of mass is a dynamic property, not a constant. *The center of mass shifts around within the object* as the parts are reconfigured.

For example, imagine if all of the people in the world decided to visit the North Pole at the same time. Assuming we would all fit and there were enough earmuffs to go around, the Earth's center of mass would shift towards the North Pole. This new center of mass, however, would trace out the exact same trajectory as the old one would have. In other words, while the trajectory of the Earth's geometric center would be slightly "southward" from where it would have been if we all stayed at home, the trajectory traced out by the center of mass is the same in either case.

Or, let's say that instead of visiting the North Pole, we all decided to go to the Galapagos Islands, which is very near the equator. Would Earth's rotation suddenly get all "wobbly" like an out-of-balance ceiling fan? No! Instead, the center of mass would shift towards the Galapagos Islands, and the Earth would rotate about this new center of mass. So although the rotation, when viewed from above, might appear asymmetrical, since the rotation would not be about the center of the spherical Earth (assuming the Earth were perfectly spherical), the rotation would be smooth. An unbalanced ceiling fan is wobbly because it is not free to choose its axis of rotation, and so it must be balanced in order to align the center of mass with the fixed axis of rotation. Earth, however, is not connected to anything, and it is free to rotate about its center of mass, wherever that center of mass may be.

Of course, all the people on Earth put together have less mass than our moon, so our discussion has been misleading. The point that traces an ellipse as we orbit the sun isn't the Earth's center of mass at all! It is the center of mass of the *entire Earth-moon system*. This point isn't really close to Earth's geometric center, although it is beneath the surface, but only because Earth is so much more massive than the moon. As the moon orbits the Earth, the center of mass of the system shifts around within Earth. It is this imaginary point that orbits the sun, not the center of mass of Earth itself.

12.4   Impulsive Forces and Collisions

In video games, things are always ramming into each other, so it seems appropriate for us to spend some time talking about *collisions*. As we've mentioned, in the real world, momentum does not change instantaneously; rather, a large force acts over a very small period of time. However, despite reality (remember the first law of video game physics), it is frequently the case that the interval during which these forces act is below the resolution of our physics time step, and for practical purposes we can consider the change in momentum to have happened instantaneously. The most important and common scenario is when the object is involved in a collision. Since the mass of most objects is constant, an instantaneous change in momentum usually boils down to an instantaneous change in velocity.

Consider two objects traveling towards each other in one dimension, with masses m_1 and m_2 and velocities v_1 and v_2, as illustrated in Figure 12.9.

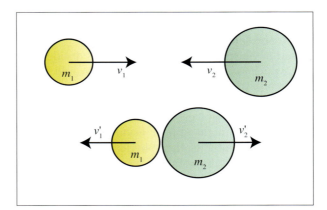

Figure 12.9
A collision

Using the momentum relation $p = mv$, we can calculate the momentum of the two objects (denoted p_1 and p_2) and the system as a whole (denoted as simply p) before and after the collision. We assume that the masses remain constant and put primes on the symbols that refer to the values after the collision:

$$p_1 = m_1 v_1, \qquad p_2 = m_2 v_2, \qquad p = p_1 + p_2 = m_1 v_1 + m_2 v_2,$$
$$p'_1 = m_1 v'_1, \qquad p'_2 = m_2 v'_2, \qquad p' = p'_1 + p'_2 = m_1 v'_1 + m_2 v'_2.$$

The change in momentum of each object, according to the law of conservation of momentum, is actually the result of a force acting over time. However, we think of the collision here as producing an instantaneous change in momentum of the two objects. A force that is treated in this way is known as an *impulsive force*, or more simply, an *impulse*. Since an impulse is an immediate change in momentum, it has the same units as momentum: ML/T. Note that an impulse is a very different beast from a regular force, which has units ML/T^2. It is very common programmer mistake to use impulses and forces incorrectly, so be sure to watch your units.

When two objects collide, many things can happen, even if we assume they remain intact. A likely scenario is for them to bounce off of each other, changing the signs of both v_1 and v_2. Or they may stick together. The former is known as an *elastic* collision and the latter an *inelastic* collision. (Actually, only a "perfect bounce" is considered truly elastic. The terms "perfectly inelastic" and "perfectly elastic" are used to refer to the two extremes, while an intermediate collision is described as simply "inelastic." Section 12.4.2 defines these terms a bit more precisely by using the coefficient of restitution, but to fully understand the distinction requires an understanding of kinetic energy. As we mentioned at the beginning of Chapter 11, energy is certainly an important concept in physics, but it actually doesn't play a central role in the Newton-Euler dynamics used by most real-time simulations, and it isn't discussed much in this book.) The velocity (and momentum) of each object is likely to change, but the law of conservation of momentum says that the total momentum of the *system* of both objects must remain constant. That is, $p = p'$.

In general, we cannot predict the individual velocities v_1 and v_2 using just the law of conservation of momentum, since the conservation of momentum law gives us one equation ($p = p'$) and there are two unknowns. Before we consider what other bit of information we need, let's look at some simpler cases of collisions and conservation of momentum. Assume for the moment that the collision is perfectly inelastic, that is, the objects stick together upon impact. This gives us the other equation we needed to solve the system of equations: $v_1' = v_2'$.

12.4.1   Perfectly Inelastic Collisions

A classic example of an inelastic collision is a gun firing a bullet into a block. Assume that, as illustrated in Figure 12.10, a block of wood weighing 2.00 kg is at rest, hanging from a wire whose mass is neglected. We fire a gun at the block, and the bullet, which has a mass of 10 g, strikes the block with a speed of 350 m/s. The bullet remains stuck in the block. What is the horizontal speed of the block immediately after the impact?

First, we compute the initial momentum of the system, which is all contained in the bullet:

$$p = m_1 v_1 + m_2 v_2 = (2.00 \text{ kg})(0) + (10.0 \text{ g})(350 \text{ m/s}) = 3.50 \text{ kg m/s}.$$

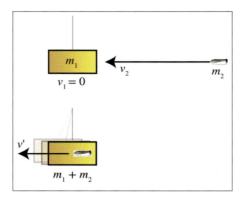

Figure 12.10
A bullet fired into a block suspended by a wire (in other words, a typical episode of *MythBusters*)

Now we look for the resulting common velocity, which we'll denote simply as v', using the law of conservation of momentum, $p = p'$, and the knowledge that this is an inelastic collision, $v' = v'_1 = v'_2$:

$$p' = m_1 v'_1 + m_2 v'_2,$$
$$3.50 \text{ kg m/s} = (2.00 \text{ kg})v' + (10.0 \text{ g})v',$$
$$3.50 \text{ kg m/s} = (2.00 \text{ kg} + 10.0 \text{ g})v',$$
$$(3.50 \text{ kg m/s})/(2.01 \text{ kg}) = v',$$
$$1.74 \text{ m/s} = v'.$$

Let's look at one more example of an inelastic collision, this time in 2D. Consider a driver who runs a red light and crashes into a car crossing the intersection. Let's say that Grant is the safe driver, and at the time of the collision, Grant and his fuel-efficient hybrid have a combined mass of 1,500 kg and are traveling west at 35 km/hr. Kari,[14] who is not paying attention, sees Grant's car too late, and swerves to the left. She and her car have a combined mass of 2,500 kg. At impact, she is traveling at 65 km/hr, heading 25° west of north, as shown in Figure 12.11. Assume that we can treat the collision as inelastic. What is the velocity of the crash just after the collision?[15]

To solve this problem, let's set up a 2D coordinate space where $+x$ is east and $+y$ is north. We compute the total momentum before the crash

---

[14]We choose to make Kari the bad driver not because of gender bias or because she's a redhead, but on the assumption that there are fewer Karis than Grants out there to offend.

[15]Another *MythBusters* moment.

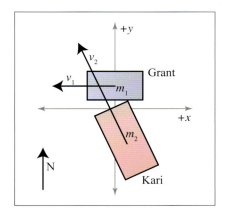

Figure 12.11
A car crash

as

$$\mathbf{P} = m_1\mathbf{v}_1 + m_2\mathbf{v}_2 = (1,500 \text{ kg})(35 \text{ km/hr})\begin{bmatrix} -1 \\ 0 \end{bmatrix}$$

$$+ (2,500 \text{ kg})(65 \text{ km/hr})\begin{bmatrix} \cos 115° \\ \sin 115° \end{bmatrix}$$

$$= (52,500 \text{ kg km/hr})\begin{bmatrix} -1 \\ 0 \end{bmatrix} + (162,500 \text{ kg km/hr})\begin{bmatrix} -0.423 \\ 0.906 \end{bmatrix}$$

$$= \left(\begin{bmatrix} -52,500 \\ 0 \end{bmatrix} + \begin{bmatrix} -68,700 \\ 147,000 \end{bmatrix} \right) \text{ kg km/hr} = \begin{bmatrix} -121,200 \\ 147,000 \end{bmatrix} \text{ kg km/hr}.$$

The resulting velocity of the glob of two cars is simply the momentum we have just computed divided by the total combined mass:

$$\mathbf{v}' = \mathbf{P}'/(m_1 + m_2) = \left(\begin{bmatrix} -121,200 \\ 147,000 \end{bmatrix} \text{ kg km/hr} \right) /(1,500 \text{ kg} + 2,500 \text{ kg})$$

$$= \left(\begin{bmatrix} -121,200 \\ 147,000 \end{bmatrix} \text{ kg km/hr} \right) /(4,000 \text{ kg}) = \begin{bmatrix} -30.3 \\ 36.8 \end{bmatrix} \text{ km/hr}.$$

12.4.2 General Collision Response

Simple inelastic collisions can be solved by using the principle of conservation of momentum, but how do we compute the velocities in the general case? Before we can fully answer that question, we need to consider the context in which it is asked. Dealing with collisions is typically a two-step process. First, we must detect that a collision has occurred, meaning the objects are already penetrating, or that a collision is about to occur in this time step. The second step is to take measures to resolve or prevent the collision. The former task is known as *collision detection*, and the latter as *collision response*. Our purpose at this time is to discuss collisions in theoretical terms (we touch on a few practical issues related to how physics simulations really work in Section 12.6), so for now let us merely attempt a general explanation of how the momentum of rigid bodies changes in response to collision. We assume that we already know that two objects have collided, and we wish to predict their behavior after the collision.

To do this, either in the abstract in a physics problem, or in collision response code in a digital simulation, we typically need to know not just that two objects have collided, but also *where* they collided, and how the two objects were oriented relative to each other at the point of contact. For example, in the car crash between Grant and Kari, we need to know that Grant's car was hit near the left door, and Kari's in her right front fender. Of course, a collision doesn't necessarily happen at a single point. Often an edge of one object may touch a surface of another, or perhaps entire surfaces are in contact. At the time of this writing, most real-time collision detection systems do not return collisions in such a descriptive manner, nor are collision response systems really capable of making use of that extra information. The closest we get is for the detection system to locate several points of contact (or penetration) and then to process that list in some way (for example, by finding their convex hull or looking for an average surface normal). In any case, the best way to do this quickly is very much at the forefront of research, and as such doesn't belong in an introductory book like this. Here we just consider the principles involved in a *single* point of contact. More advanced techniques build upon these principles.

Figure 12.12 shows some example collision results that might be returned from a collision detection system. Note that each collision result (black arrow) has a point of contact (at the tail of the arrow), and a surface normal, usually assumed to be a unit vector. An arbitrary convention is chosen for which way the normal will point; in Figure 12.12 it points away from the "first" object. Furthermore, the two objects may be arbitrarily assigned the roles of "first" and "second" objects, perhaps by the luck of the draw that the objects happened to be in some space-partitioning structure. These assignments can even vary frame-by-frame, so the response calculation must be symmetric. What is not depicted in the diagram is that a penetration depth is often returned if the objects are already penetrating.

And always, we should bear in mind the first law of video game physics. Not all collisions in video games must be between "real" objects, meaning those objects that are represented in the physics system. For example,

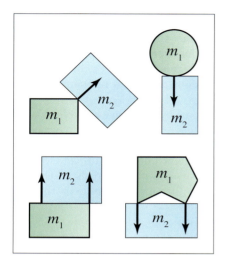

Figure 12.12
Examples of the types of results returned by collision detection.

consider a game with a kick mechanic. It may be helpful to treat a kick as a "collision" with an object, even if the character's foot is not in the physics system or if this collision was determined by simple proximity tests or raycasting that were designed based on gameplay goals and not reality. (The player character may have been too far away or too close for the kick to really have hit the target; but that is irrelevant.) It might be highly useful for the response to this action to be handled in a manner similar to ordinary collisions. For example, a sound and particle effect plays, the object receives a reduction in hit points, and its visual appearance changes, etc. In this case, in order to use the same collision response code for both ordinary collisions and "virtual" collisions such as player kicks (even if it is just to get the cosmetic effects right), you may need to synthesize values ordinarily provided by the collision detection system for your "virtual" collisions, such as the mass and velocity of the foot, and the point of contact and surface normal.

One final caveat is that we are concerned only with linear momentum for now; our objects do not rotate. This means that the explanation we give in this section will be incomplete. However, it will be helpful to cover the general principles here in a context free of the extra complexity that comes with rotation.

Now to the heart of the matter. Assuming we have somehow detected a collision and obtained a position and normal, how do we determine the resulting velocities for the collision response? Here we show the usual method, following an article by Chris Hecker [34]. We start by reviewing our guiding principles.

Our first guiding principle is that, although we know that in reality a very large force acts for a short period of time, the period of time is so short relative to our time step that we will consider the collision response to occur instantaneously. That is, we will not be calculating a force, but rather an *impulse*, which will result in an instantaneous change in momentum of the objects.

The second guiding principle is given to us by Newton's third law: whatever (impulsive) force is applied to one object, the opposite force must be applied to the other object. The conservation of momentum law essentially says the same thing: if we change the momentum of one object, we must make the opposite change in momentum to the other object so that the total momentum of the system after the collision is the same as the total momentum before the collision.

Thus, to resolve a collision between two objects, the game plan is to compute an impulse with the proper magnitude and apply that impulse to *both* objects, but in opposite directions. An impulse is a vector quantity, and so we need to know its magnitude and direction. The direction is given to us already: it is the surface normal provided by the collision detection system. The details of selecting a surface normal is a matter of collision detection, not response, and will not be discussed here. But notice that if the objects move parallel to this normal, they are either making the problem worse (penetrating further) or better (moving apart and resolving the penetration). In contrast, if we assume the penetration distance is relatively small and the surfaces are locally flat and perpendicular to the normal near the point of contact, then any motion *perpendicular* to the surface normal does not cause the penetration distance to change. So the surface normal is really the only direction that matters.

In summary, our task is to determine the proper magnitude of an impulse that will be directed along the surface normal and will resolve (or prevent) the penetration. To merely prevent a penetration that has not yet occurred, we need only remove any relative velocity acting parallel to the surface normal. This portion of the relative velocity is the velocity that, if applied to move the objects forward in time, would result in penetration. Any relative velocity acting perpendicular to the normal is OK and does not need to be counteracted, according to our assumption that the surfaces are locally flat near the point of contact. As illustrated in Figure 12.13, the velocity of m_1 relative to m_2 is computed as $\mathbf{v}_{\mathrm{rel}} = \mathbf{v}_1 - \mathbf{v}_2$, and the length of this projection onto the normal is given by $\mathbf{n} \cdot \mathbf{v}_{\mathrm{rel}}$.

Canceling the relative velocity will prevent penetration, but it's not always the correct response. When objects collide, they don't just come to a stop next to each other—they bounce off each other. So we're missing an ingredient that describes the difference in the collision responses of a dropped beanbag and a dropped SuperBall. A simple and popular collision

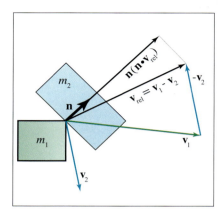

Figure 12.13
Computing the amount of relative velocity acting parallel to the surface normal

law (though not the only one) that can be used to discriminate between these cases is *Newton's collision law*. This law introduces the *coefficient of restitution*, denoted e, which is a fractional number that relates the magnitude of the relative velocity along the surface normal after the collision with the same value measured just before the collision. When $e = 0$, the post-collision velocity along the normal is zero, and we have a perfectly inelastic collision. Using $e = 1$ produces a perfectly elastic collision, where the relative velocity along the normal has the same magnitude (but opposite sign) as before the collision. Dropping a beanbag onto carpet is a close example of an inelastic collision, whereas dropping a SuperBall is a highly elastic collision. Using the formulas from Section 11.6, we can also show that, if an object is dropped and allowed to bounce multiple times, the coefficient of restitution gives the ratio of apex heights at successive bounces.

Let's denote the magnitude of the collision response impulse as k. The first mass, m_1, will receive the (vector) impulse $-k\mathbf{n}$, while m_2 undergoes the opposite change in momentum of $k\mathbf{n}$. The signs are based on our arbitrary choice of the direction of the surface normal. Now that we know what we want, calculating the proper k to cancel the relative velocity is a straightforward algebraic exercise. As before, we denote the post-collision values with primes. An impulse is an instantaneous change of momentum, so the post-collision momentum of the first object is $\mathbf{P}'_1 = \mathbf{P}_1 - k\mathbf{n}$. Dividing by the mass and remembering $\mathbf{P} = m\mathbf{v}$, we express the post-collision velocities as

$$\mathbf{v}'_1 = \mathbf{v}_1 - k\mathbf{n}/m_1, \qquad\qquad \mathbf{v}'_2 = \mathbf{v}_2 + k\mathbf{n}/m_2.$$

Post-collision velocities

The post-collision relative velocity is simply their difference, $\mathbf{v}'_{\text{rel}} = \mathbf{v}'_1 - \mathbf{v}'_2$. We solve for k by expressing the resulting relative velocity along the normal as the desired multiple of the relative velocity along the normal before impact.

$$\mathbf{v}'_{\text{rel}} \cdot \mathbf{n} = -e\,\mathbf{v}_{\text{rel}} \cdot \mathbf{n},$$

$$(\mathbf{v}'_1 - \mathbf{v}'_2) \cdot \mathbf{n} = -e\,\mathbf{v}_{\text{rel}} \cdot \mathbf{n},$$

$$[(\mathbf{v}_1 - k\mathbf{n}/m_1) - (\mathbf{v}_2 + k\mathbf{n}/m_2)] \cdot \mathbf{n} = -e\,\mathbf{v}_{\text{rel}} \cdot \mathbf{n},$$

$$[(\mathbf{v}_1 - \mathbf{v}_2) - (k\mathbf{n}/m_1 + k\mathbf{n}/m_2)] \cdot \mathbf{n} = -e\,\mathbf{v}_{\text{rel}} \cdot \mathbf{n},$$

$$[\mathbf{v}_{\text{rel}} - k(1/m_1 + 1/m_2)\mathbf{n}] \cdot \mathbf{n} = -e\,\mathbf{v}_{\text{rel}} \cdot \mathbf{n},$$

$$\mathbf{v}_{\text{rel}} \cdot \mathbf{n} - k(1/m_1 + 1/m_2)\,\mathbf{n} \cdot \mathbf{n} = -e\,\mathbf{v}_{\text{rel}} \cdot \mathbf{n},$$

$$k(1/m_1 + 1/m_2)\,\mathbf{n} \cdot \mathbf{n} = (e+1)\,\mathbf{v}_{\text{rel}} \cdot \mathbf{n},$$

$$k = \frac{(e+1)\mathbf{v}_{\text{rel}} \cdot \mathbf{n}}{(1/m_1 + 1/m_2)\,\mathbf{n} \cdot \mathbf{n}}. \tag{12.23}$$

Equation (12.23) can be simplified slightly in the common case that $\mathbf{n}$ is known to have unit length. If $\mathbf{n}$ is not a unit vector, then the change to k as a result of the length of $\mathbf{n}$ is balanced by the calculation of the (vector) impulse $k\mathbf{n}$. Thus k is the true magnitude of the impulse only when $\mathbf{n}$ is a unit vector.

Let's work through a few examples of Equation (12.23). First, let's see how the coefficient of restitution can be used to describe the difference between dropping a beanbag and a SuperBall. We'll be dropping these objects onto a concrete floor, which is an enlightening example because it shows how immovable objects can be easily handled in most physics engines by acting as if they have infinite mass. As it turns out, the *inverse mass* is the quantity we usually work with in calculations involving such special objects (as illustrated in Equation (12.23)). Furthermore, the inverse mass (and its analog, the inverse inertia tensor, to be discussed later) are derived quantities that are needed so frequently that they are often precomputed. This means that physics code can often work with immovable[16] objects without treating them as a special case, simply by setting the inverse mass equal to zero. When one of the inverse masses is zero, we could actually deal directly with velocities and bypass Equation (12.23), since k will be proportional to the mass but the velocity change as a result of applying k is inversely proportional to the mass. Later, we solve a general case example where no such simplifications are possible.

---

[16] Actually, an object don't necessarily have to be stationary to be "special" like this. Consider a platform that moves along a spline path created by a level designer, or some other hand-animated object that is not allowed to deviate from its prescribed trajectory. These so-called *kinematically controlled* objects do move around in the world and must be known to the physics engine if other (nonkinematically controlled!) objects need to interact with them, but such objects do not respond to forces, and their position is not updated by the physics engine. Although the mass of these objects is treated as infinite, proper collision response requires knowledge of the (kinematically determined) velocity.

Let's say we have one SuperBall and one beanbag, each weighing 50 grams. To avoid making the solution to our example completely obvious without Equation (12.23), we will arrange for the point of impact to occur on an inclined surface, so that $\mathbf{n}$ is not a trivial cardinal axis. Since we are not bound to use a unit vector, let's say that $\mathbf{n} = [1, 3]$. In our imagination, we throw both objects in exactly the same way, such that the impact velocity in both cases is $\mathbf{v}_2 = [-4, -4]$. This is illustrated in Figure 12.14.

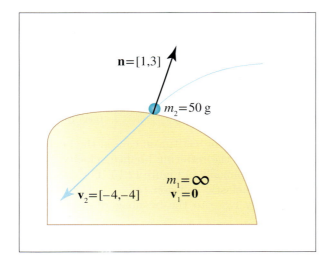

Figure 12.14
Bouncing a beanbag or SuperBall off of an inclined surface

To determine the resulting velocity, we first solve for k, the scale factor for the impulse. To do this, we must choose a coefficient of restitution. For the SuperBall collision, we'll use $e = 0.9$, which is near the advertised value. Solving for k (remember that $\mathbf{v}_{\mathrm{rel}} = \mathbf{v}_1 - \mathbf{v}_2$) we get

$$k = \frac{(e+1)\mathbf{v}_{\mathrm{rel}} \cdot \mathbf{n}}{(1/m_1 + 1/m_2)\,\mathbf{n} \cdot \mathbf{n}}$$

Calculating the impulse multiplier k for the SuperBall

$$= \frac{(0.9+1)\begin{bmatrix} 4\ \mathrm{m/s} \\ 4\ \mathrm{m/s} \end{bmatrix} \cdot \begin{bmatrix} 1 \\ 3 \end{bmatrix}}{(0 + 1/(50\ \mathrm{g}))\begin{bmatrix} 1 \\ 3 \end{bmatrix} \cdot \begin{bmatrix} 1 \\ 3 \end{bmatrix}} = \frac{1.9\,(16\ \mathrm{m/s})}{10/(50\ \mathrm{g})} = 152\ \mathrm{g\ m/s}.$$

To compute the post-collision velocity, we add an impulse of $k\mathbf{n}$ to the momentum of the SuperBall. Since momentum is mass times velocity, the change in velocity is equal to this impulse divided by m_2, the mass of the SuperBall:

Calculating the post-impulse velocity for the SuperBall

$$\mathbf{v}_2' = \mathbf{v}_2 + \frac{k\mathbf{n}}{m_2} = \begin{bmatrix} -4 \text{ m/s} \\ -4 \text{ m/s} \end{bmatrix} + \frac{(152 \text{ g m/s}) \begin{bmatrix} 1 \\ 3 \end{bmatrix}}{50 \text{ g}}$$

$$= \begin{bmatrix} -4 \text{ m/s} \\ -4 \text{ m/s} \end{bmatrix} + \begin{bmatrix} 3.04 \text{ m/s} \\ 9.12 \text{ m/s} \end{bmatrix} = \begin{bmatrix} -0.96 \text{ m/s} \\ 5.12 \text{ m/s} \end{bmatrix}.$$

Before we show this velocity graphically, let's look at the beanbag. We treat the beanbag collision as almost completely inelastic and use $e = 0.01$. Other than the change to e, the procedure is the same as for the SuperBall:

Calculating the impulse multiplier k and post-impulse velocity for the beanbag

$$k = \frac{(e+1)\mathbf{v}_{\text{rel}} \cdot \mathbf{n}}{(1/m_1 + 1/m_2)\,\mathbf{n} \cdot \mathbf{n}} = \frac{1.01\,(16 \text{ m/s})}{10/(50 \text{ g})} = 80.8 \text{ g m/s},$$

$$\mathbf{v}_2' = \mathbf{v}_2 + \frac{k\mathbf{n}}{m_2} = \begin{bmatrix} -4 \text{ m/s} \\ -4 \text{ m/s} \end{bmatrix} + \frac{(80.8 \text{ g m/s}) \begin{bmatrix} 1 \\ 3 \end{bmatrix}}{50 \text{ g}} = \begin{bmatrix} -2.38 \text{ m/s} \\ 0.85 \text{ m/s} \end{bmatrix}.$$

Notice that the reduced coefficient of restitution caused the beanbag to receive a smaller impulse scale, and the resulting bounce velocity was also lower. This difference is shown graphically in Figure 12.15. For comparison, we've also included a perfectly inelastic collision ($e = 1$). The perfectly inelastic collision ($e = 0$) is very close to the beanbag result and is not depicted.

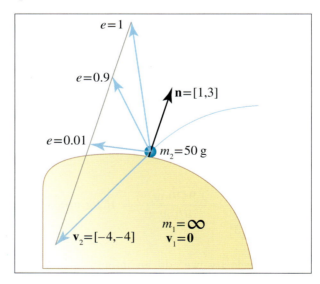

Figure 12.15
Post-impact velocities for different values of e, the coefficient of restitution

It is obvious from the beanbag trajectory in Figure 12.15 that an important aspect of collisions is not captured by the model: friction. The velocity perpendicular to the normal is the same before and after the collision. We would expect an object like a beanbag to scrub off a great deal of horizontal velocity as a result of the collision—in fact, it might come to a stop completely. The correct thing to do is to add in a perpendicular component to the impulse. However, correct handling of friction is, in general, a tricky business, and it remains at the forefront of real-time simulation research. Many current methods for calculating the horizontal impulse (especially for sliding contact) fall under the category of "completely fudged."

In the previous example, we only had one "live" object and the other object was inert. This is a special case, and the equations could be simplified. (As we mentioned earlier, the mass of the projectile cancels itself out and is not needed.) In Exercise 9, we ask you to consider what would happen if Grant and Kari's collision were not perfectly inelastic; the full power of this law is needed in this case. We show how to handle collisions where the objects are free to rotate in Section 12.5.4.

We have been assuming thus far that the objects are in contact, but are not yet penetrating. This might not be the case, depending on the overall strategy used to resolve collisions. One technique is to attempt to reverse the simulation in time back to the point of contact. This can be difficult to do efficiently because there are frequently many, many collisions that happen at different times within a single time step. Furthermore, defining "exactly in contact" is difficult when using floating point math. Another strategy is to simply allow penetration, and apply the impulse to objects that are already penetrating. In this case, the impulse must do more than remove the relative velocity to prevent (further) penetration. The resulting relative velocity must be sufficiently large to separate the objects by the end of the time step, after it has been integrated into the position. In other words, the positions of the objects will be advanced at a rate according to the calculated velocities; after this update, the penetration needs to be resolved. Or at least it needs to be *mostly* resolved. There are some advantages to allowing some small penetration. All of these issues are a bit outside the realm of established principles and fall more under the heading of current research,[17] and they will not be discussed here.

12.4.3 The Dirac Delta

Before we leave linear dynamics and talk about rotational dynamics, let's mention briefly one bit of mathematical notation you might see, especially in the context of impulses. As we've said, many natural phenomena (such as momentum) do not change instantaneously in theory, but for practical purposes we treat them as changing instantaneously. Furthermore, it is

---

[17] As Werner von Braun, who really was a rocket scientist, said, "Research is what I'm doing when I don't know what I'm doing."

often the case where a body of mathematical tools exists to handle continuous functions (or, equivalently, "signals"), and we wish to apply those tools to signals with discontinuities. There is a handy mathematical kludge that can be used to encode a discontinuity in a function such that it can be integrated. It is known as the *Dirac delta*, and is usually denoted with the lowercase letter delta, for example $\delta(a)$.

The symbol $\delta(t)$ is a special sort of function, which is a spike, or impulse. Its value is zero everywhere except at $t = 0$, where it is infinite. But the actual values of the Dirac delta are not to be taken too literally—what is more important is that its integral (the "area" of this infinite spike) is equal to 1. The best way to think about the Dirac delta is as a box centered at 0 with width w and height $1/w$. (Other shapes may be chosen, but the important point is that the shape must have unit area.) The Dirac delta is defined as the limit of such a function, as the width approaches zero, and the height approaches infinity, all the while maintaining the unit area. We like the advice of Bracewell [9], which is to avoid using the term "function" when referring to special, uh, functions such as the Dirac delta, and instead use the word "symbol." This is because whenever we see the Dirac delta symbol, the *entire surrounding expression* should be interpreted as a limit. We are considering the limit as the width approaches zero of some shape with unit area centered at the origin.

Armed with the Dirac delta, we can differentiate functions with discontinuities. For example, let's say the velocity of a baseball being struck by a bat at time $t = 2$ is approximated by the discontinuous function

$$v(t) = \begin{cases} -130 \text{ ft/s} & t < 2, \\ 130 \text{ ft/s} & t \geq 2. \end{cases}$$

We can differentiate this expression by using the Dirac delta as

$$v'(t) = 260\,\delta(t - 2),$$

which can be read as "an impulse of magnitude 260 at 2." Remember that the δ symbol is not an ordinary function, and so we interpret the above as "the limit of a total change in velocity of 260 ft/s that takes place over an interval of duration Δt surrounding the time $t = 2$, as Δt approaches zero."

The Dirac delta comes up in a variety of contexts where tools from continuous math are applied to discontinuous signals. For example, in graphics, the screen-space image is a signal with inherent discontinuities, and we need to sample this signal and reconstruct it. The user's input via the controller is another signal that exhibits discontinuities. The Dirac delta and other related symbols (such as the ramp function and Heaviside's step function) are helpful in discussing and manipulating such signals.

12.5   Rotational Dynamics

We are now ready to extend the ideas we have learned about particles to rigid bodies. Particles have position, but until now we have not concerned ourselves with orientation. Likewise, particles have mass, but until now we have not thought about the size or shape of a particle and how that mass was distributed. The key linear quantities and laws each have rotational analogs, and there is a certain beautiful correspondence between them. This correspondence is certainly pedagogically convenient and will be leveraged in our discussion. As we did for linear dynamics, we first define the basic kinematics quantities and consider those issues related to describing the rotation without worrying about the causes of the rotation. We then examine the rotational analogs to mass, force, and momentum, although we will discuss these topics in a different order.

You might notice that this section is surprisingly brief, both compared to our discussion of linear matters, and also similar presentations of other sources. There are two reasons for this. First, we spent considerable time in the previous chapter building up intuition about derivatives and linear dynamics, and these ideas need not be repeated here—though there will be some important differences concerning integration of angular displacement. Second, there are certain prerequisites that are usually bundled in this discussion in traditional physics books; in this book, it has been more appropriate to place these prerequisites elsewhere. You should make sure you have read and understood these prerequisites before reading the rest of this section. In particular, we use the cross product, which we covered in Section 2.12, and basic methods for describing rotation in three dimensions, which were the subject of Chapter 8.

12.5.1   Rotational Kinematics

Chapter 11 was about linear kinematics: we considered a function $\mathbf{p}(t)$ that described the position of a particle as a function of time. We also considered its first and second derivatives, the velocity and acceleration functions, which we denoted $\mathbf{v}(t) = \dot{\mathbf{p}}(t)$ and $\mathbf{a}(t) = \ddot{\mathbf{p}}(t)$, respectively. The rotational analog of position is, of course, orientation. Several methods can be used to describe the orientation of a body. A considerable number of pages were spent explaining and comparing these methods in Chapter 8, and in this chapter we assume you are familiar with the basics. In a rigid body simulator, it's common to keep on hand redundant copies of the orientation in alternate formats. Typically, both a quaternion and rotation matrix are maintained. We will adopt a similar policy here with our notation. We let $\mathbf{R}(t)$ be the object-to-upright rotation matrix at time t; it is the orientation of the body expressed in matrix form. We also use $\mathbf{q}(t)$ to refer to that same rotation as a quaternion. Although both functions express the same value, they are different "data types."

The analog of linear velocity and acceleration are called *angular velocity* and *angular acceleration*. We denote them as $\boldsymbol{\omega}(t)$ and $\boldsymbol{\alpha}(t)$, respectively, and both of these quantities are 3D vectors, or infinitesimal exponential maps (see Section 8.4) if you will. We were able to define linear velocity as the time derivative of position, but things are a bit more complicated with orientation. In general, $\boldsymbol{\omega}(t)$ is not the derivative of the orientation in any format (even exponential map). We have more to say on this when we look at what the derivatives of angular values really are in Section 12.6.4.

So our first order of business is to understand how to express and measure angular velocity. This is tricky, not just because rotation in 3D is more complicated than position, but also because there are two slightly different types of angular velocity. The first is known as *spin* angular velocity, and the second is *orbital* angular velocity. Spin angular velocity describes the rate of change of orientation of an object and is not affected by translation of the object. Orbital angular velocity is actually not concerned with orientation at all; instead, it measures the rate at which the *position* of an object traces out an orbit around some other point. We already introduced orbital angular velocity in Section 11.8.2, so if you skipped that section, now would be a good time to review it.

To see the relationship between spin angular velocity and orbital angular velocity, let's look at an example. Consider an object that is rotating about its center of mass $\mathbf{c}$, which is fixed in space. To describe this rotation, we must specify two things. First, we describe the direction of the rotation; we choose to do this by naming $\hat{\mathbf{n}}$, a unit vector parallel to the axis of rotation whose sign (in combination with the left hand rule) establishes a direction of positive rotation. Note that $\hat{\mathbf{n}}$ tells us only the direction of the axis; the position comes from our assumption that the axis passes through the center of mass $\mathbf{c}$. The other element necessary to describe the rotation is, of course, the rate of rotation, which we measure in radians per unit time and denote by the scalar ω. Now, we can define the spin angular velocity of the rigid body by the angular velocity vector $\boldsymbol{\omega}$ (note the boldface), which is simply the multiplication of the rotation rate with the axis:

$$\boldsymbol{\omega} = \omega\hat{\mathbf{n}}.$$

These ideas should be familiar to you, if you read Section 8.4, which talked about exponential maps, and Section 11.8.2, which discussed uniform circular velocity of a particle and defined orbital velocity. If so, then you can probably already see the connection between spin angular velocity and orbital angular velocity.

Spin and Orbital Angular Velocity

The spin angular velocity of a rigid body is equal to the orbital angular velocity of every point on the rigid body.

When talking about orbital angular velocity, we must be clear about what point **o** we are measuring the angular velocity relative to. We do *not* measure the orbital angular velocity relative to the center of mass **c**! We measure the orbital angular velocity relative to the point that is actually being orbited, and only those points on the "equator" of the object are actually orbiting around **c**. Given any other arbitrary point, it will orbit a point **o** that lies on the axis of rotation, as shown in Figure 12.16.

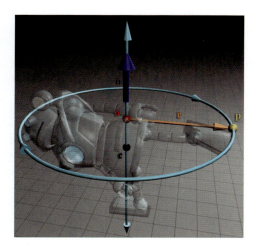

Figure 12.16
The spin angular velocity of the robot is the same as the orbital velocity of every point on the robot, provided that we choose **o** carefully.

Now, the astute reader may have noticed some circularity in the definitions just given. We said that the spin angular velocity of the rigid body is equal to the orbital angular velocity of each and every point, provided that the orbital velocity is measured relative to the closest point on the axis of rotation. But how did we know the axis of rotation in the first place? The question is typically moot because, both in analytical kinematics equations and in a digital simulation in a computer, the angular velocity vector $\boldsymbol{\omega}$ is simply one of the fundamental state variables that we track, so we do not need to infer it from the point velocities. Still, it is worth pointing out that this axis is uniquely determined (up to the reversal of signs). Remember that the axis of rotation is perpendicular to the velocity an orbiting particle. (We must measure the velocity of the particle relative to the center of mass of the rigid object, if the center of mass is moving.) There is only one direction that is simultaneously perpendicular to all the velocities of all the particles, and this direction is the axis of rotation.

We started with "a simple example" of an object rotating about an axis that passes through its center of mass, but, as it turns out, this is the general case, at least if we consider instantaneous velocity. The only simplification we made is to fix the center of mass, but, in general, an object can translate as well as rotate. It is somewhat surprising to realize, when you imagine an object tumbling through space, that it will always rotate about an axis passing through the center of mass (though the axis can be arbitrarily oriented). When an object receives a force that induces rotation (known as a *torque*, to be discussed shortly), the induced rotation will always occur about the center of mass. In fact, to rotate an object about an axis that does *not* pass through the center of mass requires continual application of some sort of constraint force. In the absence of any external torque (say, the constraint force is removed), the object will rotate about an axis passing through its center of mass, and the angular velocity will be constant—the axis of rotation will not change direction, and the rate of rotation will not change. We are getting a bit ahead of ourselves talking about torques, but we wanted to make it clear that this situation of angular velocity is, in fact, the only situation we need to understand.

Of course, if torques are acting on the object, then the axis and rate of rotation will change over time. This leads us to consider *angular acceleration*, which is a vector quantity that we denote $\boldsymbol{\alpha}$. Angular velocity was not simply the derivative of orientation, as one might naïvely expect by analogy with the linear counterparts. However, the analogy *does* work for angular acceleration, which is the vector time derivative of angular velocity:

$$\boldsymbol{\alpha}(t) = \dot{\boldsymbol{\omega}}(t).$$

The analogy to the linear equation $\mathbf{a}(t) = \dot{\mathbf{v}}(t)$ is clear.

12.5.2   2D Rotational Dynamics

Now that we've defined the simple kinematics quantities involved—which was primarily an exercise in notation and reusing the ideas developed elsewhere—let's consider the dynamics of rotation. We first simplify the situation to the case of rotation in the plane (or alternatively, we can think about this as fixing the axis of rotation). In this situation, the angular velocity and acceleration are scalar rather than vector quantities, since there is only one degree of freedom. After we develop some basic ideas in two dimensions in this section, we extend these ideas into three dimensions in Section 12.5.3.

Imagine a point with mass m that is attached to a rigid disk whose mass is neglected. The center of the disk is fixed at a pivot, and this constrains the mass to a circular path. Let $\mathbf{r}$ be the vector from the pivot to the mass. Thus the orbital radius of the mass is $r = \|\mathbf{r}\|$. We draw a line on the disk outward from its center, and we assume that we can affix the mass to the disk at any distance r along this line. Note that the radius of the disk itself (being massless) is not relevant, so try to shut off your intuition that a really big disk would be hard to rotate. The only source of resistance to the rotation—the only source of inertia—is the point mass. We are neglecting the inertia of the disk.

Consider what happens when a force $\mathbf{f}$ is applied directly at m. According to Newton's second law, the mass wants to accelerate in the direction of $\mathbf{f}$. However, any portion of $\mathbf{f}$ parallel to $\mathbf{r}$ will be repelled by a contact force from the disk and will have no effect on the mass. In contrast, the portion of $\mathbf{f}$ perpendicular to $\mathbf{r}$, tangential to the orbit of the mass, will cause the point mass to accelerate. Let F denote the magnitude of $\mathbf{f}$, and $F_\perp$ denote the amount of $\mathbf{f}$ that is perpendicular $\mathbf{r}$. This is depicted in Figure 12.17.

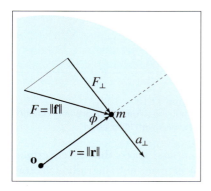

Figure 12.17
A disk is constrained to rotate about $\mathbf{o}$. A mass m is attached to the disk at a radius r, and a force $\mathbf{f}$ is applied on the mass.

With a bit of trig, we can compute $F_\perp$ from ϕ, which is the angle between $\mathbf{f}$ and $\mathbf{r}$:

$$F_\perp/F = \sin(\pi - \phi), \qquad (\pi - \phi \text{ is the interior angle})$$
$$F_\perp/F = \sin \pi \cos \phi - \cos \pi \sin \phi, \qquad (\text{using Equation (1.1)})$$
$$F_\perp/F = (0)\cos \phi - (-1)\sin \phi,$$
$$F_\perp = F \sin \phi.$$

By applying Newton's second law, we can compute the magnitude of the tangential acceleration as
$$a_\perp = F_\perp/m.$$

(Don't confuse tangential acceleration, which causes a change in the rate of rotation, with centripetal acceleration, which is created by contact force with the disk and maintains the orbital path.)

By definition, the linear tangential acceleration $\mathbf{a} = \ddot{\mathbf{p}}$ is simultaneously an angular acceleration $\alpha = \ddot{\theta}$ about the pivot, and they are related by

$$\alpha = a_\perp / r.$$

To understand where this comes from, remember from Section 11.8.1 the relationship between linear speed and angular velocity: $v_\perp = s = r\omega$.

Now imagine that we keep the mass fixed at r, but instead of applying our force directly onto the mass, we push against a peg that sticks up out of the disk at some other location, at a distance $l = \|\mathbf{l}\|$. The vector $\mathbf{l}$ goes from the fulcrum $\mathbf{o}$ to the point where the force is applied and is known as the *lever arm*. This is shown in Figure 12.18.

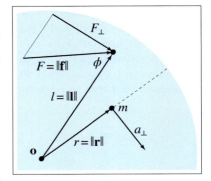

Allow us to clarify a point which is potentially confusing. Earlier, in Figure 12.17, the vectors $\mathbf{r}$ and $\mathbf{l}$ were the same, since we were applying the force directly onto the (one and only) point mass. But in general, mass is distributed around the disk, so there are

Figure 12.18
Applying a force at an arbitrary point on the disk with lever arm $\mathbf{l}$.

many m's and $\mathbf{r}$'s. The portion of the applied force that is effective at rotating the disk is perpendicular to the *lever arm*, not the radius vector $\mathbf{r}$, as you might infer from Figure 12.17.

Let's consider how changes to the lever arm affect the resulting accelerations. First, we note that if we apply the force anywhere on the circle with the same radius as the mass, meaning $l = r$, then the whole apparatus gains angular momentum, just as if we had pushed against the mass directly. In other words, the only thing that matters about $\mathbf{l}$ is the distance from the fulcrum $l = \|\mathbf{l}\|$ and the angle ϕ between the lever arm and the applied force. Rotating both the lever arm and the applied force does not change the resulting angular acceleration.

If the peg is closer to the fulcrum than the mass ($l < r$), then we must push *harder* in order to produce the same acceleration of the mass. If $l > r$, a smaller force will accomplish the job. Thus, increasing the lever arm l has a proportionate effect on the tangential acceleration $a_\perp$ of the mass. This is the basic principle of mechanical advantage of a lever, which

Archimedes discovered. But what effect does changing l have on α, the *angular* acceleration of the apparatus? It, too, is directly proportional: if we push on the disk at a point twice as far out, the angular acceleration is twice as much. This is clear from the relation $\alpha = a_\perp / r$. If this seems too obvious to bother pointing out, then keep reading.

Let's summarize what we've found out. When a force is applied to an object, it has a tendency to rotate that object. This "cause" of angular acceleration is known as *torque*, which we denote by using the Greek letter tau:[18] $\boldsymbol{\tau}$. Although the linear acceleration of a body as a result of an applied force does not depend on where the force is applied, the amount of torque that results from an applied force depends on how effectively the force is applied. The effectiveness of the force to create rotational acceleration—the magnitude of the torque—depends on several factors:

- It is proportional to the magnitude of the applied force $\mathbf{f}$.

- It is proportional to the length of the lever arm $\mathbf{l}$, which is the vector from the fulcrum to the point of application of the force.

- Only the portion of the force perpendicular to the lever arm counts. Equivalently, the torque is proportional to $\sin\phi$, where ϕ is the angle between $\mathbf{f}$ and $\mathbf{l}$.

In two dimensions, we can state this succinctly by

$$\tau = Fl\sin\phi. \tag{12.24}$$

Torque in two dimensions

The dimensions of torque *are not the same as force.* Torque has units of "force times length." The SI unit for torque is the Newton meter. (This is dimensionally equivalent to the joule, but torque and energy are distinct concepts, and the joule is not really the proper unit to use for torque.)

In might not be intuitively obvious that torque increases with the length of the lever arm. For example, your intuition might be inclined to tell you that it would be more difficult to push at an increased radius because you have to push faster just to keep up. If so, your intuition is wrong, but don't feel bad. Our experience with force is often via an everyday type of physical push like we might make with our hands against some object. But this is not necessarily a good example because the push must move faster and faster as the object accelerates in order to maintain contact. But the speed of some source of force has nothing to do with the magnitude of the force itself. So try replacing the physical push with either a gust of wind or a quick "thump" (an impulse). You can conduct this experiment on a door.

---

[18] It rhymes with "cow"

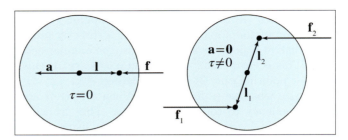

Figure 12.19. On the left, the line of action passes through the center of mass, resulting in linear acceleration but no torque. On the right, a couple produces a torque but no linear acceleration.

If you blow on the door or whack it close to the hinge (shorter lever arm), your action won't cause the door to rotate as easily if you do the same thing nearer to side with the handle (longer lever arm). That's exactly why we put door handles on the side opposite the hinge: to make it easier to open the door!

The relationship between torque and force is an important one to understand. Any force applied to a body can produce both a linear acceleration *and* a torque. Of course, it is the *net* force and torque that determine the acceleration(s) of the body. Two key examples in Figure 12.19 illustrate this point. On the left, a force is acting along a line that passes through the center of mass, resulting in *no torque*. In this case, $\sin\phi = 0$ since the line of action of the force is parallel to the lever arm. The right side of Figure 12.19 shows a different extreme case: two forces with equal magnitudes but opposite directions are acting on opposite lever arms. A pair of forces coordinated like this are known as a *couple*, and they result in a net torque, but *zero* net linear force. When you turn a bolt with a wrench, what you are really doing is supplying two or more contact forces. The direction and lever arms of these contact forces are coordinated in a circular pattern to produce torque, but (nearly) zero net linear force.

In a digital simulation, torques can come from multiple sources. One common source is an applied (linear) force at some lever arm, with collisions being the most common source of torques of all. An impulsive force applied to an object can result in an *angular impulse* (also known as *impulsive torque*). Similar to a linear impulse, an angular impulse is an instantaneous change in angular momentum, and we can think of this as the result of a large torque acting for a small duration. We might also instruct the physics engine to automatically apply torques (perhaps limited to some maximum magnitude) on an object in order to enforce some angular constraint. Angular springs and motors are examples of this. Finally, we might have a reason to add a torque at will to any object without there being any corresponding linear force.

Back to our thought experiment with the mass attached to the rotating disk. What if we fix the lever arm l, and instead vary r, the radial distance between the mass and the pivot? The same principle of the lever is at work, but in reverse. The force (and therefore the tangential acceleration $a_\perp$) experienced by the mass will be *inversely* proportional to r. Said another way, the effective inertia of the mass—its resistance to linear acceleration—is proportional to r. But what about the ability of the apparatus to resist angular acceleration? How does its *moment of inertia* change as we vary r? The moment of inertia is *not* proportional to r, it is proportional to r^2! To see why, consider that if we increase l and r by the same factor, then the tangential acceleration $a_\perp$ experienced by the mass is unchanged. However, at this increased radius, the same tangential acceleration now corresponds to a reduced angular acceleration, due to the relation $\alpha = a_\perp/r$.

In summary, the moment of inertia of an object, which must be measured relative to some particular pivot (in this case, it's the fixed pivot point, but for a rigid body we usually measure it relative to its center of mass), quantifies the degree to which the object will resist angular acceleration about that pivot. The moment of inertia J of a point mass is proportional to its mass and proportional to the *square* of the distance from the mass to the pivot.

Moment of Inertia of a Point Mass in the Plane

$$J = mr^2.$$

Now imagine that the disk in our thought experiment has multiple masses placed on it. Each of these masses contributes to the disk's resistance to rotation, and their contribution is the same, regardless of where any force is applied. To compute the moment of inertia of an arbitrary rigid body, we break up the object into "mass elements" such that for each element we know the mass m_i and radial distance to the center of mass r_i. We then sum up the moments of inertia of each individual mass element:

$$J = \sum_i J_i = \sum_i m_i r_i^2. \qquad (12.25)$$

Moment of inertia of a rigid body

Let's work through a few instructive examples.

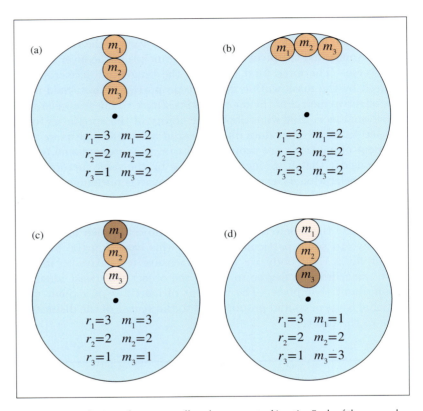

Figure 12.20. Distribution of mass can affect the moment of inertia. Each of the examples above has the same total mass.

Consider the four disks in Figure 12.20. Since each disk has three masses, we start by expanding the sum in Equation (12.25) to $J = m_1 r_1^2 + m_2 r_2^2 + m_3 r_3^2$:

(a) $J = 2 \cdot 3^2 + 2 \cdot 2^2 + 2 \cdot 1^2 = 18 + 8 + 2 = 28$;

(b) $J = 2 \cdot 3^2 + 2 \cdot 3^2 + 2 \cdot 3^2 = 18 + 18 + 18 = 54$;

(c) $J = 3 \cdot 3^2 + 2 \cdot 2^2 + 1 \cdot 1^2 = 27 + 8 + 1 = 36$;

(d) $J = 1 \cdot 3^2 + 2 \cdot 2^2 + 3 \cdot 1^2 = 9 + 8 + 3 = 20$.

Aside from the rote practice this example provides, it highlights a crucial fact: the *distribution* of mass within the object can have a profound effect on the moment of inertia. Notice the widely differing moments of inertia, despite the fact that each of these disks has a total mass of 6. Compare this to linear inertia, where shifting the mass around will *not* make the object any easier or harder to accelerate. The only thing that matters for linear acceleration is the total mass. Furthermore, in three dimensions, although linear mass can still be quantified with a scalar, the moment of inertia cannot be described by a single number, due to its dependence on the distribution of mass.

When we take the limit as the number of mass elements reaches infinity, Equation (12.25) turns into an integral. Luckily, formulas for the moments of inertia for many common shapes, such as spheres, cylinders, rings, and so forth, can be found on the Internet.

Newton's second law, $F = ma$, has a straightforward rotational equivalent.

Rotational Equivalent of Newton's Second Law

$$\tau = J\alpha. \tag{12.26}$$

For rotation in a plane, all of the variables in Equation (12.26) are scalars. In three dimensions, however, τ and α become vector quantities, and J becomes a matrix. Section 12.5.3 discusses this.

We've considered torque, the rotational analog to linear force. Now let's turn our attention to momentum. Remember that linear momentum is the "quantity of motion" contained in an object. Its analog, *angular momentum*, has a similar interpretation. Intuitively, angular momentum describes how hard it will be to stop the rotation of the object. If the angular momentum is large, then the magnitude of the applied torque, or the duration for which it is applied, or both, must be large.

In our discussion of linear momentum, we encountered two ways of thinking about momentum. The first was to interpret momentum in an "instantaneous respect" as the product of mass and velocity by using $P = mv$. The rotational equivalent is shown in Equation (12.27).

Spin Angular Momentum

Angular momentum (L) is the product of the moment of inertia J and angular velocity (ω):

$$L = J\omega. \tag{12.27}$$

Alternatively, we can compute the angular momentum for an individual point mass directly from its linear momentum by

Relationship between linear and angular momentum

$$L = rP \sin \phi, \tag{12.28}$$

where P is the linear momentum. The purpose of the $\sin \phi$ term is the same as in the computation of torque: it isolates the tangential motion. If the trajectory is known to be orbital, this term can be omitted since it will always be unity. Notice that since Equation (12.28) contains a factor r, it really is only for orbital angular momentum. Equation (12.27) can also be used for orbital angular momentum, provided that J and ω are both measured relative to the same pivot, but Equation (12.27) is probably more appropriate for spin angular momentum of a rigid body, where J measures the moment of inertia of the entire body. In any case, the spin angular momentum of a rigid body can be computed by breaking the object into mass elements and taking the sum of the orbital angular momenta of these elements. Here are several ways in which the sum could be accomplished:

Spin angular momentum of a rigid body

$$L = \omega J = \omega \left(\sum_i J_i \right) = \sum_i J_i \omega = \sum_i (r_i{}^2 m_i) \omega$$
$$= \sum_i r_i m_i (r_i \omega) = \sum_i r_i m_i v_i = \sum_i r_i P_i = \sum_i L_i.$$

Here the subscripted variables refer to values for a particular particle. Note that we dropped the $\sin \phi$ term under the assumption that each particle is in an orbital trajectory.

The second interpretation of linear momentum that we discussed is that of momentum as the time integral of force. When we apply force over time, linear momentum is accumulated. A similar relationship exists between angular momentum and torque. As torque is applied over time, it builds up angular momentum; equivalently, torque is equal to the rate of change (derivative) of angular momentum. As with linear momentum, this can be interpreted as a conservation law.

Torque and the Conservation of Angular Momentum

$$L = \int \tau \, dt, \qquad\qquad \tau = \frac{dL}{dt}.$$

In the absence of external torque, angular momentum is conserved.

12.5.3   3D Rotational Dynamics

Now let's extend the basic principles developed in Section 12.5.2 into three dimensions. First, let's review the 3D rotational kinematics quantities. The single angle θ is replaced by a rotation tensor of some kind, with a rotation matrix $\mathbf{R}$ or a quaternion $\mathbf{q}$ being the most common methods of describing orientation in general rigid body simulations. The angular velocity ω and acceleration α become vector quantities and get bolded as $\boldsymbol{\omega}$ and $\boldsymbol{\alpha}$, respectively.

To extend the dynamics principles into three dimensions, we start with torque. Not surprisingly, torque becomes a vector quantity denoted $\boldsymbol{\tau}$, and the direction of this vector indicates the axis about which the torque is tending to induce rotation. (Later we consider what happens if the object is already rotating about a different axis.) The formula for computing the torque for an applied force $\mathbf{f}$ and lever arm $\mathbf{l}$ is actually simpler in 3D than the corresponding 2D formula!

Torque in Three Dimensions

$$\boldsymbol{\tau} = \mathbf{l} \times \mathbf{f}. \tag{12.29}$$

Compare Equation (12.29) to $\tau = Fl \sin \phi$ (Equation (12.24)), and notice that the cross product has the magnitude and $\sin \phi$ terms built in.

Angular momentum likewise becomes a vector $\mathbf{L}$, with a similar formula for its relation to the linear quantity:

$$\mathbf{L} = \mathbf{r} \times \mathbf{P}.$$

Compare this to Equation (12.28).

A reader who is paying attention might note that Equation (12.28) is only one of two equations we gave for angular momentum in the plane—the one we deemed to be more appropriate for orbital angular velocity of a particle—and wonder about the other formula, Equation (12.27), which was more appropriate for spin angular velocity. That formula was $L = J\omega$, and to get its three-dimensional equivalent, we must understand how to extend J, the moment of inertia, into three dimensions. Luckily, the link between the two momentum equations is an excellent way to get this understanding. Let's start by expanding $\mathbf{L} = \mathbf{r} \times \mathbf{P}$, with the goal of ending up with

Orbital angular momentum of a particle in three dimensions with radial vector r

something that looks like $L = J\omega$:

$$\mathbf{L} = \mathbf{r} \times \mathbf{P} = \mathbf{r} \times (m\mathbf{v}) = (m\mathbf{r}) \times \mathbf{v} = (m\mathbf{r}) \times (\omega \times \mathbf{r})$$

$$= m \begin{bmatrix} r_x \\ r_y \\ r_z \end{bmatrix} \times \begin{bmatrix} \omega_y r_z - \omega_z r_y \\ \omega_z r_x - \omega_x r_z \\ \omega_x r_y - \omega_y r_x \end{bmatrix} = m \begin{bmatrix} r_y(\omega_x r_y - \omega_y r_x) - r_z(\omega_z r_x - \omega_x r_z) \\ r_z(\omega_y r_z - \omega_z r_y) - r_x(\omega_x r_y - \omega_y r_x) \\ r_x(\omega_z r_x - \omega_x r_z) - r_y(\omega_y r_z - \omega_z r_y) \end{bmatrix}$$

$$= m \begin{bmatrix} r_y\omega_x r_y - r_y\omega_y r_x - r_z\omega_z r_x + r_z\omega_x r_z \\ r_z\omega_y r_z - r_z\omega_z r_y - r_x\omega_x r_y + r_x\omega_y r_x \\ r_x\omega_z r_x - r_x\omega_x r_z - r_y\omega_y r_z + r_y\omega_z r_y \end{bmatrix}$$

$$= m \begin{bmatrix} (r_y^2 + r_z^2)\omega_x - r_y r_x \omega_y - r_z r_x \omega_z \\ -r_x r_y \omega_x + (r_z^2 + r_x^2)\omega_y - r_z r_y \omega_z \\ -r_x r_z \omega_x - r_y r_z \omega_y + (r_x^2 + r_y^2)\omega_z \end{bmatrix}$$

$$= \left(m \begin{bmatrix} r_y^2 + r_z^2 & -r_y r_x & -r_z r_x \\ -r_x r_y & r_z^2 + r_x^2 & -r_z r_y \\ -r_x r_z & -r_y r_z & r_x^2 + r_y^2 \end{bmatrix} \right) \begin{bmatrix} \omega_x \\ \omega_y \\ \omega_z \end{bmatrix}.$$

The key point is in the last line, which bears a striking resemblance to $L = J\omega$. In fact, the quantity in parenthesis is the three-dimensional moment of inertia.

Inertia Tensor

The inertia tensor of a mass m with radial vector $\mathbf{r}$ is

$$\mathbf{J} = m \begin{bmatrix} r_y^2 + r_z^2 & -r_y r_x & -r_z r_x \\ -r_x r_y & r_z^2 + r_x^2 & -r_z r_y \\ -r_x r_z & -r_y r_z & r_x^2 + r_y^2 \end{bmatrix}. \tag{12.30}$$

Notice that this quantity is a *matrix*, not a vector. In recognition of this, in three dimensions we sometimes refer to the moment of inertia as the *inertia tensor*. This mathematical artifact is a result of the physical fact that an object's resistance to rotational acceleration is anisotropic: it can be easier to rotate an object about one axis compared to another. For example, compare the torque required to spin a piece of rebar around like a helicopter versus rolling it about its lengthwise axis.

Notice that $\mathbf{J}$ is a symmetric matrix. One trivial (but thankful) consequence of this is that we can be sloppy with row and column vectors, while ordinarily we must be very careful in order to avoid transposed results.[19]

---

[19]Remember, our convention in this book is to use row vectors, since the majority of

Before we look at some other properties of $\mathbf{J}$, let's finish extending the basic formulas in three dimensions with perhaps the most important one.

In the plane, the rotational equivalent of Newton's second law, $F = ma$, is $\tau = J\alpha$. Extending this into three dimensions is straightforward.

3D Rotational Analog of Newton's Second Law

$$\tau = \alpha \mathbf{J}.$$

As we've said before, in computer simulations it's often the case that the force and mass are known, and we compute the acceleration using $a = F/m$. A similar situation exists for rotational dynamics, where division by m is replaced with multiplication by $\mathbf{J}^{-1}$:

$$\alpha = \tau \mathbf{J}^{-1}.$$

As it turns out, the inverse inertia tensor is needed more frequently in digital simulations than is $\mathbf{J}$, and it is usually precomputed and stored.

Equation (12.30) tells us how to compute the inertia tensor for a point mass, but what about more complicated shapes? In a manner similar to how we computed the center of mass, we can imagine breaking up a compound object into a large number of mass elements and taking the sum of their individual moments of inertia. Taking the limit as the volume of the largest element approaches zero, this sum becomes a multidimensional integral. Such integrals are typically difficult or impossible to solve analytically, except for abstract primitives such as boxes, disks, cylinders, spheres, cones, and the like. Fortunately, such primitives arise commonly in practice and can make adequate approximations. Even more fortunate for us is the fact that the hard work of solving the integral has already been done for a large variety of primitives. For such primitives, the best method for obtaining the moment of inertia is to look up the formula in a table. (At the time of this writing, such a table can be found on wikipedia.org under "List of moment of inertia tensors.")

More complicated objects are typically approximated by breaking down the object into primitives with known formulas, calculating their moments

---

matrices encountered in 3D interactive applications are transformation matrices, and the left-to-right reading order is a useful advantage. In the derivation above, we used column vectors for pedagogical and aesthetic purposes (the inertia tensor is not a transform matrix), but shame on us for being inconsistent. From here on out, we'll stick with our convention and put the vectors on the left.

of inertia individually, and then summing together the results. There is just one complication. The formulas for these primitives assume the origin of the coordinate space is at some auspicious location, such as at the center of the sphere. But imagine we are computing the moment of inertia of a human body and approximating the head by a sphere. Odds are low that we chose to place the origin in the center of the head. Luckily the *parallel axis theorem* tells us how the moment of inertia changes if we translate a mass.

Assume that $\mathbf{J}_{\text{cm}}$ is the inertia tensor of some object with mass m, measured relative to its center of mass. The inertia tensor $\mathbf{J}'$ of the mass measured relative to some arbitrary pivot that has a displacement $[x, y, z]$ from this center of mass is given by Equation (12.31).

Parallel Axis Theorem

$$\mathbf{J}' = \mathbf{J}_{\text{cm}} + m \begin{bmatrix} y^2 + z^2 & -xy & -xz \\ -xy & x^2 + z^2 & -yz \\ -xz & -yz & x^2 + y^2 \end{bmatrix}. \tag{12.31}$$

12.5.4   Collision Response with Rotations

Now let us complete the collision response calculation we started in Section 12.4.2. At that time we did not consider rotational effects, but now we know better. We continue, following Hecker [35].

Remember the basic strategy:

1. Compute the relative velocities at the point of contact.

2. Project the relative velocities onto the surface normal. This is the velocity that must be counteracted in order to prevent (further) penetration.

3. Compute k, the magnitude of an impulse, such that when we apply the impulse to both objects (in opposite directions) parallel to the surface normal, the post-collision velocities, measured along the surface normal, have the desired magnitude according to some collision law. This discussion will be based upon Newton's collision law and the coefficient of restitution e.

4. Apply the impulse $k\mathbf{n}$ to one object and $-k\mathbf{n}$ to the other.

Dealing with objects that can rotate adds a few complications. First, we earlier referred to "the" velocity of an object. But when an object is rotating, the different points on the object have different linear velocities. The velocity we need is not the velocity of the object's center of mass, but rather the velocity *at the point of contact*. So we must extend our velocity calculation to account for the angular velocity. Likewise, our prediction of the post-impulse velocities at the points of impact must also take into consideration rotational effects. Application of an impulse at the point of contact will create an angular impulse, which will change the rate of rotation. In general, when objects are free to rotate, a smaller impulse will suffice to resolve the collision, since the impulse has two ways by which to reduce the point velocities. The change in angular velocity will cause the points of contact to move away from each other faster than their centers of mass do. Indeed, the centers of mass may still be moving towards each other after the collision.

In Section 12.4.2, we started with the equation

$$\mathbf{v}'_{\text{rel}} \cdot \mathbf{n} = -e\,\mathbf{v}_{\text{rel}} \cdot \mathbf{n}$$

and then expanded the math and solved for k, the magnitude of the impulse. (It's hidden in the left hand side.) The same strategy works here, only we need to derive new expressions for the point velocities before and after the impulse that take into consideration the rotational effects. First, let's check the notation. We use $\mathbf{r}_i$ to denote the position of the point of impact on object i relative to its center of mass, $\boldsymbol{\omega}_i$ for the angular velocity of the object, m_i for the mass, and $\mathbf{J}_i$ for the inertia tensor. For linear velocity, we need to introduce some new notation to distinguish between the linear velocities at the point of contact, which we denote as $\mathbf{u}_i$, and the linear velocities of the center of mass, denoted $\mathbf{v}_i$. As before, primes on the quantities refer to their values after the collision.

With this notation, we compute the point velocity of each object by adding the velocity due to the motion of the center of mass, plus that portion caused by the rotation, according to Equation (11.31):

$$\mathbf{u}_1 = \mathbf{v}_1 + \boldsymbol{\omega}_1 \times \mathbf{r}_1, \qquad\qquad \mathbf{u}_2 = \mathbf{v}_2 + \boldsymbol{\omega}_2 \times \mathbf{r}_2.$$

Computing the point velocities

The velocities after the collision depend on the change in linear velocity of the center of mass, and also the change in angular velocity. These are computed by

$$\mathbf{v}'_1 = \mathbf{v}_1 - k\mathbf{n}/m_1, \qquad\qquad \mathbf{v}'_2 = \mathbf{v}_2 + k\mathbf{n}/m_2,$$
$$\boldsymbol{\omega}'_1 = \boldsymbol{\omega}_1 - (\mathbf{r}_1 \times k\mathbf{n})\mathbf{J}_1^{-1}, \qquad\qquad \boldsymbol{\omega}'_2 = \boldsymbol{\omega}_2 + (\mathbf{r}_2 \times k\mathbf{n})\mathbf{J}_2^{-1},$$

Post-impulse linear and angular velocities of the bodies

where the sign conventions are determined by our arbitrary choice to have the collision normal $\mathbf{n}$ point from the first object to the second. Combining the equations above, we get the post-impulse point velocities:

$$\mathbf{u}_1' = \mathbf{v}_1' + \boldsymbol{\omega}_1' \times \mathbf{r}_1$$

Post-impulse velocities
at the point of contact

$$= \mathbf{v}_1 - k\mathbf{n}/m_1 + (\boldsymbol{\omega}_1 - (\mathbf{r}_1 \times k\mathbf{n})\mathbf{J}_1^{-1}) \times \mathbf{r}_1$$

$$= \mathbf{v}_1 - k\mathbf{n}/m_1 + \boldsymbol{\omega}_1 \times \mathbf{r}_1 - k((\mathbf{r}_1 \times \mathbf{n})\mathbf{J}_1^{-1}) \times \mathbf{r}_1$$

$$= (\mathbf{v}_1 + \boldsymbol{\omega}_1 \times \mathbf{r}_1) - k\mathbf{n}/m_1 - k((\mathbf{r}_1 \times \mathbf{n})\mathbf{J}_1^{-1}) \times \mathbf{r}_1$$

$$= \mathbf{u}_1 - k\mathbf{n}/m_1 - k((\mathbf{r}_1 \times \mathbf{n})\mathbf{J}_1^{-1}) \times \mathbf{r}_1,$$

$$\mathbf{u}_2' = \mathbf{u}_2 + k\mathbf{n}/m_2 + k((\mathbf{r}_2 \times \mathbf{n})\mathbf{J}_2^{-1}) \times \mathbf{r}_2.$$

Defining $\mathbf{u}_{\text{rel}} = \mathbf{u}_1 - \mathbf{u}_2$ as the relative point velocity, we are now ready to grind through the algebra:

$$-e\,\mathbf{u}_{\text{rel}} \cdot \mathbf{n} = \mathbf{u}_{\text{rel}}' \cdot \mathbf{n},$$

$$-e\,\mathbf{u}_{\text{rel}} \cdot \mathbf{n} = (\mathbf{u}_1' - \mathbf{u}_2') \cdot \mathbf{n},$$

$$-e\,\mathbf{u}_{\text{rel}} \cdot \mathbf{n} = [\,(\mathbf{u}_1 - k\mathbf{n}/m_1 - k((\mathbf{r}_1 \times \mathbf{n})\mathbf{J}_1^{-1}) \times \mathbf{r}_1)$$
$$- (\mathbf{u}_2 + k\mathbf{n}/m_2 + k((\mathbf{r}_2 \times \mathbf{n})\mathbf{J}_2^{-1}) \times \mathbf{r}_2)\,] \cdot \mathbf{n},$$

$$-e\,\mathbf{u}_{\text{rel}} \cdot \mathbf{n} = [\,(\mathbf{u}_1 - \mathbf{u}_2) - k\mathbf{n}/m_1 - k\mathbf{n}/m_2$$
$$- k((\mathbf{r}_1 \times \mathbf{n})\mathbf{J}_1^{-1}) \times \mathbf{r}_1) - k((\mathbf{r}_2 \times \mathbf{n})\mathbf{J}_2^{-1}) \times \mathbf{r}_2)\,] \cdot \mathbf{n},$$

$$-e\,\mathbf{u}_{\text{rel}} \cdot \mathbf{n} = \mathbf{u}_{\text{rel}} \cdot \mathbf{n} - k[\,(1/m_1 + 1/m_2)\mathbf{n} + ((\mathbf{r}_1 \times \mathbf{n})\mathbf{J}_1^{-1}) \times \mathbf{r}_1)$$
$$+ ((\mathbf{r}_2 \times \mathbf{n})\mathbf{J}_2^{-1}) \times \mathbf{r}_2)\,] \cdot \mathbf{n},$$

$$-(e+1)\,\mathbf{u}_{\text{rel}} \cdot \mathbf{n} = -k[\,(1/m_1 + 1/m_2)\mathbf{n} + ((\mathbf{r}_1 \times \mathbf{n})\mathbf{J}_1^{-1}) \times \mathbf{r}_1)$$
$$+ ((\mathbf{r}_2 \times \mathbf{n})\mathbf{J}_2^{-1}) \times \mathbf{r}_2)\,] \cdot \mathbf{n}.$$

With one more step, we have the formula we're seeking.

Collision Response with Rotation

The magnitude of the collision impulse can be calculated from the relative point velocity, masses, moments of inertia, surface normal, and coefficient of restitution, by

$$k = \frac{(e+1)\,\mathbf{u}_{\text{rel}} \cdot \mathbf{n}}{[(1/m_1 + 1/m_2)\mathbf{n} + ((\mathbf{r}_1 \times \mathbf{n})\mathbf{J}_1^{-1}) \times \mathbf{r}_1) + ((\mathbf{r}_2 \times \mathbf{n})\mathbf{J}_2^{-1}) \times \mathbf{r}_2)] \cdot \mathbf{n}}.$$

$$(12.32)$$

Equation (12.32) assumes that all vectors are measured in the same co-ordinate space and that the inverse inertia tensors operate on vectors in that same space. However, the inverse inertia tensor is a constant only in *body space*, and the same matrix in world space may not be updated con-tinually as the object rotates. Our convention is to describe the orientation of the object by using $\mathbf{R}$, a rotation matrix that transforms row vectors on the left from body space to upright space. Under these assumptions, in Equation (12.32) we would replace $\mathbf{J}^{-1}$ with $\mathbf{R}^T \mathbf{J}^{-1} \mathbf{R}$.

12.6   Real-Time Rigid Body Simulators

This section presents an overview of real-time rigid body simulators such as PhysX, Havok, Bullet Physics, and the Open Dynamics Engine. Few game programmers will work directly on the physics engine, and certainly far fewer will write one from scratch. Most of us just need to know how to *use* the thing. Luckily, in this regard the physics engines previously listed have much in common. However, a physics engine is like many other programming tools: even if you don't intend to write one, you can use it more effectively if you have a basic understanding of how things work under the hood.

We regret that we won't be able to go into great detail, for a few reasons. First, any attempt to describe "how a physics engine works" is complicated by the fact that currently there is still great diversity and rapid innovation in the field. Second, the math quickly becomes more advanced than is appropriate for this introductory book. Frankly, your authors simply do not have enough expertise to succinctly summarize the state of the art from top to bottom. However, from a user's perspective, there are many similarities between physics engines that can be covered in an introductory manner, so we start with an overview of a typical physics engine interface. There are also a few choice mathematics that are discussed near the end of this chapter.

12.6.1   Physics Engine State Variables

An old computer science adage attributed to Fred Brooks says, "Show me your flow charts and conceal your tables, and I shall continue to be mystified. Show me your tables, and I won't usually need your flowcharts; they'll be obvious." Despite the image of a banner printed on Z-fold paper using *The Print Shop* that is evoked by the arcane terminology, the essence of the message is still true today: to write or understand software, a good place to start is a description of the *data* that are being operated on. From the perspective of the user of a physics engine, there are three main types of data objects within a rigid body simulator: the dynamics bodies, collision geometry data, and constraints. Let's examine each of these in turn.

Dynamics body state. Perhaps the most fundamental type of data object within a real-time rigid-body simulator is the dynamics body. You can think of the dynamics body as the "soul" of a rigid body: it tells it where to go, but it has no outward appearance, so you can see it only indirectly via the effects it has. The collision geometry (to be discussed later) is what gives a rigid body its "earthly form," meaning its shape. The part of a rigid body that we see is usually some sort of graphical model, which doesn't have anything to do with the physics engine and is not discussed here.

At this point, it's probably important to discuss the relationship between a dynamics body and an "object" in the nontechnical (and nonprogramming) sense of the word. A simple rigid object, such as the ubiquitous video game crates Old Man Murray was so fond of reporting sighting, can be simulated by using a single dynamics body. A more complicated object with moving parts, such as a car or human body, is not a rigid body and thus cannot be simulated with a single dynamics body. Instead, the object must be broken down into rigid parts, and then the dynamics bodies corresponding to those parts connected via *joint constraints*, to be discussed later. Of course, the graphical representation for such a compound object like this need not be "rigid," but within the physics simulation, each dynamics body is a rigid body.

Another way that a dynamics body is like a soul is that it's easier to just list its properties rather that attempt a precise definition. So let's enumerate the variables that comprise a dynamics body. We classify these variables by their life cycle, meaning when those variables are initialized and how often their values change. Certain properties are (mostly) constant, some change continuously over the life of the object, and some are working variables, coming into existence (or reset) at the start of a simulation time step and being thrown away at the end of the time step.

The first class of properties are those that are initialized when the body is instantiated by the application, and typically (but not necessarily) remain constant during the simulation.

- The mass and inertia tensor are certainly critical properties of a rigid body, and often these do not change over the course of the simulation, although there is no inherent reason why they cannot be changed by the application over time, for example, to simulate a car burning fuel. As stated earlier, it's frequently the *inverse* mass and tensor that are actually needed by a simulation, so these are often kept on hand as additional derived quantities so that they do not need to be recomputed every time they are needed. Also, the inertia tensor, being a rotation, is a constant only if it operates on vectors expressed relative to the body axes of the rigid body. Relative to the world axes, the inertia tensor varies continuously as the body rotates.

- Sometimes the physics engine can store an offset of the center of mass of the object (which, like the inertia tensor, is constant only in the body space of the object). We noted earlier that for reasons of numeric stability and simplicity, internally a physics engine might prefer to assume the origin of the body space is its center of mass, and this center of mass offset is taken into consideration in the client interface routines.

- Each body is associated with one or more collision geometry objects, the union of which defines the shape of the rigid body, as discussed below.

The second class of variables that define the state of the dynamics body are those that evolve over time, and contain some "history" in them and *must* be carried forward from one frame to the next. If we need to save the complete state of the simulation and resume it later (for example, in a save game), these variables cannot be derived from any other source and must be serialized:

- position,

- orientation,

- linear velocity,

- angular velocity.

This list reflects a choice to make velocities the primary state variables, in which case the momentums are easily derived quantities. An alternate strategy is to use the momentum as the primary state variable and velocity as the derived quantity. The former approach has an advantage in dealing with kinematically controlled objects, whose inertia (and therefore momentum) is considered infinite. The latter can more elegantly deal with the situation in which the mass changes over time, since conservation of momentum is automatically enforced.

Finally, each dynamics body has certain properties that are stored in working variables. These quantities change over time but do not inherently have "history" integrated into them. These sorts of variables are often reset at some point within the simulation step. If we wanted to save the state of the simulation and resume it later, it is usually *not* necessary[20] to include these variables in the state description:

- force and torque accumulators,

- linear and angular impulse accumulators,

- list of current contact constraints.

We use the word "accumulators" here to reflect what often happens in practice. These values are often reset to zero at some point within the frame, and then different sources of external forces are polled, and the net result stored in these variables.

Listing 12.2 is a structure that summarizes the essentials of a dynamics body state and hints at how it might be implemented in C++. These variable names are used in the pseudocode later in this chapter. You should compare this to the corresponding class in a real physics engine implementation to see what additional data that engine chose to keep on hand, or what different choices were made.

```cpp
struct DynamicsBody {

    //
    // Primary quantities
    //

    // Position of center of mass
    Vector3 pos;

    // Orientation in quaternion format
    Quaternion rotQuat;

    // Mass
    float mass;

    // Moment of inertia, expressed in body space.
    Matrix3x3 jBody;

    // Velocities
    Vector3 linVel;
    Vector3 angVel;

    //
    // Derived quantities
    //

    // Orientation, in matrix form
    Matrix3x3 rotMat;

    // Inverse mass and inertia tensor
    float oneOverMass;
    Matrix3x3 invJBody;
```

---

[20]The precise details depend on the method of numerical integration. Some methods of integration utilize historical values in order to approximate higher derivatives.

```
//
// Temporary / working variables
//

// Force accumulators.  Cleared to zero each time step
Vector3 force;
Vector3 torque;
Vector3 linImpulse;
Vector3 angImpulse;


//
// Collision and constraint lists
//
vector<UserConstraint*> userConstraints;
vector<CollisionShape*> collisionShapes;
};
```

Listing 12.2
Dynamics body state variables

Collision Geometry. If the dynamics body is the soul of a rigid object, the
collision geometry is its earthly manifestation. The collision geometry is
used to define the shape of dynamics bodies, and also of other "soulless" or
static objects. Typically, a physics engine will support a number of different
primitives. In order of complexity, these are

- basic abstract primitives such as spheres, boxes, planes, cylinders,
  cones, and the like;

- convex polyhedra;

- arbitrary collision mesh, sometimes called a "triangle soup."

The simpler shapes have advantages in both speed and stability, which
is why it's often best to build up a concave or complicated shape from
multiple primitives. These primitives are allowed to penetrate each other;
only their union matters, since two geometry objects attached to the same
dynamics body will not collide with each other. Indeed, it's important for
the physics engine to provide flexibility in deciding which pairs of geometry
objects collide; for example, on a character, the thigh part might not collide
with the connected torso and shin parts, but it might collide against every
other part in the body. Or, under the auspices of the first law of video
game physics, we might create a barrier that enemies can pass through but
the player character cannot.

A collision geometry object either will be associated with a *single* dynamics body, to which it has a fixed relative position and orientation, or
will not be associated with any dynamics body and is part of the static
"world."

Constraints. The third and final main type of object within a rigid body simulator is a *constraint*. Constraints are used to enforce relationships between pairs of rigid bodies, or between a rigid body and the world. Applications can create two types of *user constraints*: joints and motors. A third type of constraint is the *contact constraint*, which is involved in collision response.

User constraints are the "regular" kind of constraints that are specified by the application in order to maintain some desired relationship. The more common and easily understood type of constraint maintains a spatial relationship between two parts and is known as a *joint*. Some examples of the types of joints that come built-in to most physics engines are the following.

- A *ball-and-socket joint* constrains two objects such that a shared point maintains a fixed position relative to each set of body axes. Alternatively, you can think of one object having a ball at a fixed location in its body space, and the other as having a socket in its body space, and the constraint attempts to force those points to be coincident.

- A *hinge joint* is a ball-and-socket joint with an additional constraint that two axes, one connected to the ball and the other to the socket, must be collinear. Thus the two objects can rotate about the shared axis like a hinge. Additionally, limits may be set on the hinge rotation angle.

- A *slider joint* or *prismatic joint* operates on two axes that are fixed relative to the body space of the two objects, constraining them to be collinear. The objects may only slide back and forth along this axis, or perhaps twist along it. Limits may be applied to the range of the translation.

- A *universal joint* is similar to a ball and socket joint, but allows for limits to be specified on the angles of rotation. The angular limits are Euler angles (think heading and pitch), resulting in a rectangular range of motion. Limits on the twisting ("bank") can also be enforced.

- A *conical joint* is similar to a universal joint, but the rotation limits are conical rather than rectangular.

As an example, in a human skeleton, each "bone" may be simulated as a separate rigid body, with constraints used to attach each bone to its parent. Hinge joints might be used at the knees, with limits set up to prevent the knee from bending backwards. Conical joints or perhaps universal joints might be used for the hips and shoulders.

Whereas a joint constraint is concerned with the position and orientation of the bodies, a *motor* is a type of constraint that attempts to enforce a requested relative velocity between two bodies. For example, by using the proper kind of motor, an application can instruct the physics engine: "Body A should maintain an angular speed ω, relative to the axis $\hat{\mathbf{n}}$, which is fixed in the reference frame of body B."

Most physics engines have a variety of constraints, and even mechanisms for adding your own types of constraints. Furthermore, constraints need not be absolute, but the physics engine can be given instructions to limit the force that may be applied to enforce the constraint. Most physics engines provide a mechanism by which a constraint may be queried for the amount of force that was applied in the attempt to satisfy the constraint. This query can be quite useful, for example, to play a sound if a motor is straining, or perhaps destroy a joint if some threshold is exceeded. Joints and their limits can also be "soft." For example, in the Open Dynamics Engine, values known as the error reduction parameter (ERP) and constraint force mixing (CFM) parameter can be tuned to cause the joint to behave like a spring-damper system.

Although application constraints don't necessarily have any "memory" and can be created and destroyed at will by the application—for example to detach a wheel from a car or an arm from a body—they typically persist across time steps. *Contact constraints*, in contrast, are instantiated by the physics engine and are always destroyed within the same physics time step. (Conceptually, at least. They may survive internally for performance or stability reasons.) They are used to enforce nonpenetration between the collision geometry of two bodies (or of a body and some static geometry). These constraints are the primary output of the collision detection system and the mechanism by which collision response is performed.

Although contact constraints are created by the physics engine during collision detection, that doesn't mean the application cannot be involved in the process. Physics engines provide numerous hooks to customize the creation of these contact points and notify the application when contact constraints are applied to a dynamics body. These hooks are a powerful means for fine-grained customization of how particular pairs of objects interact, and collision notifications are essential for implementing feedback such as sound and particle effects or a reduction in hit points.

This section presented the three main "tables" of data in a physics engine. The next section covers the "flowcharts"—and we hope that you and Fred Brooks don't think they are too "obvious."

12.6.2   HighLevel Overview

The APIs presented by modern physics engines, through which you can manipulate the "tables" from the previous section, show significant similarity. However, their inner workings are more diverse. This is the point at which we must really begin using generalities and waving our hands. This section starts by presenting some high-level pseudocode of how a physics engine fits into the game loop. Afterwards, we briefly discuss a few general strategies for what happens inside of the heart of the physics engine.

We begin with the game loop itself, which is summarized by Listing 12.3.

```
void gameLoop() {
    getReady();
    while (!gameOver) {
        simPrePhysics();
        physicsEngine->update();
        simPostPhysics();
        render();
    }
}
```

Listing 12.3
A very simple game loop

Let's describe the physics-related work that goes on in each of the "functions" in this listing. First, in `getReady()`, along with the usual loading of textures and models, we also create objects within the physics system of the three main types discussed in Section 12.6.1. The world will likely have a large amount of collision geometry. Each simple simulated object might have one dynamics body and one or a few collision shapes. A complex articulated model might require several dynamics bodies, linked together with joint constraints, and collision geometry for each body. For the player character, we might set up a carefully tuned constraint used to pull the character towards a desired position each frame.

Inside the game loop, we have broken the simulation into three steps. First, there is a glob of activities we do before calling the physics engine, which we have lumped together under `simPrePhysics()`. Here we prepare the inputs to physics processing. We might process kinematically controlled objects and notify the physics system of their new positions and velocities. We would read the player input; determine where those controls indicate the player should be; and update the constraints, forces, or torques used by the physics engine to attempt to move the player character into position. In a network game, we might poll the network for objects for which the local host is not the authority, and update a constraint that tells the physics engine, "Try to get the object to go here." Also, the physics objects do not all need to be created once in `getReady()` and live forever. We can certainly add and remove objects from the physics universe at any time.

Next, we invoke the physics engine to do its thing. Although most of this code is in the physics engine itself, it will communicate back with the game code for several purposes, either just for notification, or perhaps to provide customization opportunities to the application. We review a few different approaches that are used by physics engines later in this section.

When the physics engine is complete, some miscellaneous steps need to happen that we have grouped together under the function `simPost Physics()`. Perhaps the most important step is to update our game objects with the new positions and orientations that have been determined by the physics engine. We might do this by looping through all the objects and polling the physics engine for the updated position. Or we might receive notification in the form of a callback. The updated object positions are not the only output of the physics update. We might also be interested in the forces that were required to maintain constraints, or the list of collisions that occurred. Depending on the game design, and how the camera is simulated, we often update the camera after the physics has completed, so that it moves in response to the player movement.

Finally, of course, at some point we need to draw the scene, as indicated by the presence of the function `render()`.

In our pseudocode, `physicsEngine->update()` represents the heart of the physics engine. As we've mentioned, no two physics engines work exactly the same, but there are some common themes. Here we briefly outline a few strategies, summarizing the more in-depth survey of Erleben et al. [19].

Penalty Methods. Penalty methods resolve collisions with a spring-like mechanism. The collision detection provides a list of penetrating collision shapes. For each pair, we locate the dynamics bodies that own these shapes and apply a repulsive force to each, where the magnitude of this force is proportionate to the penetration depth. In other words, the penalty method does not attempt to resolve collisions on the same time step that they are detected; rather, over time the force will cause the objects to separate. Of course, we must tune our "springs" carefully; for stacked objects, the spring force will balance with gravity when the objects are in equilibrium, so in general the penalty method does not attempt to completely eliminate penetration, but rather just to limit it to an acceptable level. Listing 12.4 shows a simplified version of how this could be done.

```
void PhysicsEngine::update() {

    // Gather up external forces acting on the dynamics bodies
    // (i.e. gravity, springs, etc)
    computeForces();

    // Locate penetrating collision geometry, and their owner bodies
```

```
struct Collision {
    DynamicsBody *body1, *body2; // the bodies involved
    Vector3 p; // location of collision
    Vector3 n; // contact normal
    float penetrationDepth;
};
vector<Collision> collisions = collisionDetection();

// Treat each collision as a spring
for (each collision) {

    // Calculate a force based on the penetration depth
    Vector3 f = calculateForce(collisions[i]);

    // Add this force to the two bodies
    collisions[i].body1->addForceAtPoint(collisions[i].p, f);
    collisions[i].body2->addForceAtPoint(collisions[i].p, -f);
}

// Integrate the forces (accelerations) into velocity, and
// velocities into position, to move the simulation forward
integrateForces();
}
```

Listing 12.4
Pseudocode for a physics simulation based on the penalty method

Sequential impulse simulations. These methods were popularized by Mirtich [46]. Both resting and colliding contact are modeled as (possibly very high frequency) collisions. When a collision is detected between two objects A and B, a collision law (such as the simple Newton collision model shown in Section 12.5.4) is used to calculate an impulse that will prevent penetration. However, this might result in a different collision, either elsewhere between A and B, or perhaps between B and C, so the process must be repeated until all of the relative velocities at the contacts are resting or separating. Listing 12.5 illustrates the basic idea.

```
void PhysicsEngine::update() {

    // Gather up external forces acting on the dynamics bodies
    // (i.e. gravity, springs, etc)
    computeForces();

    // Integrate force (acceleration) into velocities, but don't yet
    // update positions
    updateVelocities();

    // Locate colliding bodies
    struct Collision {
        DynamicsBody *body1, *body2; // the bodies involved
        Vector3 p; // location of collision
        Vector3 n; // contact normal
        float penetrationDepth;
    };
    vector<Collision> collisions = collisionDetection();

    // Keep applying impulses until all
```

```
    // relative contact velocity is resolved
    for (;;) {

        // Find a collision where the relative point velocities at
        // the point of contact are such that the bodies are moving
        // towards each other along the normal. (Colliding contact,
        // rather than resting or separating contact)
        Collision *c = nextUnresolvedCollision(collisions);
        if (c == NULL) break;

        // Calculate an impulse using a collision law
        Vector3 impulse = calculateCollisionImpulse(c);

        // Apply the impulse to the two bodies. (This produces an
        // immediate change in the linear and angular velocities)
        c->body1->addImpulseAtPoint(c->p, impulse);
        c->body2->addImpulseAtPoint(c->p, -impulse);

        // Keep looping until all collision velocities are resting
        // or separating
    }

    // Now step the positions forward based on the velocities
    integrateVelocities();
}
```

Listing 12.5
Pseudocode for a sequential impulse physics simulation

The first difficulty is that the order that the contacts are processed (which are often an arbitrary artifact of collision detection) can cause different simulation results. A second is that the simulation can get caught in an infinite loop, so care must be taken to ensure termination.

Velocity-based simulations. These techniques currently represent the state of the art in real-time simulations. Preventing penetration and resolving collisions is viewed as a constraint to be satisfied. As mentioned earlier, these constraints are treated in a standardized way with user constraints such as joints and motors. For each constraint, the simulation examines the ratio of the rate of change of satisfaction of the constraint versus changes in linear and angular velocity. Using this information (a matrix of partial derivatives is known as a *Jacobian* matrix), the simulator solves for the velocities that satisfy the constraints. This is illustrated in Listing 12.6.

```
void PhysicsEngine::update() {

    // Gather up external forces acting on the dynamics bodies
    // (excluding those from the constraints)
    computeForces();

    // Tentatively apply forces, to compute proposed
    // (unconstrained) positions and velocities
    integrateForcesTentatively();

    // Build up list of constraints. These come from two
    // sources: collision detection (contact constraints) and
```

```
// the user constraints.
vector<ConstraintRow> constraintRows;
collisionDetection(constraintRows);
processApplicationConstraints(constraintRows);

// Solve for velocities which satisfy the constraints
solveConstraints(constraintRows);

// Now really update the positions, based on the computed
// velocities that satisfied the constraints
integrateVelocities();
}
```

Listing 12.6
Pseudocode for a velocity-based physics simulation

The resulting matrix problem is not a standard system of linear equations; rather, it is a system of inequalities. For example, if a collision law predicts that two objects should bounce away with a certain velocity, that velocity is interpreted as a *minimum* velocity. If some other constraint (say one object is being pulled away by a spring) causes the objects to bounce away faster than predicted by the collision law, this is not considered a violation of the contact constraint. This type of system can be put into a standard form, known as the *linear complimentary problem* (LCP).

Various implementations of velocity-based methods borrow ideas from penalty and sequential impulse simulations, so it can sometimes be difficult to classify a simulation as strictly one or the other. The incremental matrix solvers that are used to solve the linear complimentary problem make small adjustments to the velocity of the bodies; these adjustments can be physically interpreted as a series of impulses. Thus, there are some similarities between a velocity-based solver and a sequential impulse solver. The distinction is that in a sequential impulse solver, the collision law is applied multiple times. In a velocity-based simulation, the collision law is applied once in order to determine the ideal relative velocity, and then this velocity is treated as a goal or constraint. Some velocity-based solvers, notably the Open Dynamics Engine, allow constraints to be treated as "soft," using techniques similar to penalty-based methods.

12.6.3   Euler Integration

Every physics engine needs to be able to "step forward" in time. We assume that we know the position and velocity values (both linear and angular) at some time t, and we want to determine their values at some future time $t+\Delta t$. The outcome will depend on the net forces experienced by the object, and, of course, the initial velocity at time t. The forces themselves may vary depending on the position or the velocity of the body. For example, a body connected to a fixed point by a spring experiences forces that vary based on the position of the body, and a body moving through a fluid such as air experiences a drag force that has some (potentially nonlinear) relationship with the relative velocity of the body and the fluid. Furthermore, the forces can be the result of *interactions* between bodies (e.g., four bodies connected by springs forming a tetrahedron, with one corner submerged in water). In short, the problem is not trivial.

The mathematical term for this is *numerical integration*. There are two main parts to our discussion. This section ignores rotation and discusses the basic concepts of integration in terms of linear acceleration and velocity. Section 12.6.4 considers how to integrate angular acceleration and velocity.

Recall that integration is the process of determining a function from its derivative. In our case, we are working with three different functions of time: position, velocity, and acceleration. A reader who was not asleep will remember that velocity is the derivative of position and acceleration is the derivative of velocity. We are dealing with *numerical* integration because we are not symbolically solving the differential equations. Instead, the derivative function that we know (the acceleration, determined using Newton's second law from the force) is being sampled only at a discrete number of time values. To appreciate the difficulties, we begin with a naïve approach and see where it fails.

Let h denote our step size in seconds ($1/h$ is the simulation frequency, so, for example, if we were running at 60 Hz, h would be $1/60$). The simplest method of integration is to assume that the derivative is constant during the step. Assume that the current time step is the kth time step. Then the position for the next time step, $k + 1$, is determined by

$$\mathbf{p}_{k+1} = \mathbf{p}_k + h\mathbf{v}_k.$$

Euler integration of velocity into position

This strategy is known as *Euler integration*. Although numerical integration may be an "advanced" subject often taught after calculus, Euler integration is easier for most people to understand than true (analytic) integration. It is common practice to use Euler integration to introduce analytic integration, which is exactly what we did in Figure 11.10 to determine the movement of a rabbit, although we didn't call the technique by its name. The key point that was brought out was that simple Euler integration ignores changes in the derivative during the time step, and this is the source of error. As we saw, the most obvious way to reduce the error in the answer is to decrease the step size h. In some cases, we can decrease it to the limit through symbolic manipulations and arrive at a perfect answer (i.e., we use analytic integration). But sometimes we have a complicated function and all we can do is evaluate ("sample") the function.

Consider three different simulations currently of interest in video games: hair, cloth, and fluid. In each case, we break up the thing being simulated into pieces ("discretize" the problem), and then simulate each piece using a simplified set of force laws. A basic strategy for cloth is to model the cloth as a set of vertices, where each vertex is connected to nearby vertices by "springs" and is affected by drag. The forces experienced by a given vertex in this simulation depend on both the position and velocity of the vertex and its neighbors. Although it is not difficult to calculate the active forces at any given instant, the result cannot be directly integrated because of the dependence on position and velocity. This is what makes it a problem of differential equations rather than simple integration. These differential equations are usually not vulnerable to the frontal assault of analytic solution; we must use numerical integration.

Suppose we express the velocity of a single cloth vertex as a function of time, approximated as a polynomial (by using its Taylor series expansion; see Section 11.4.6). Assume we know the value at time t_k and are interested in only a small interval surrounding t_k. Then we could write our approximation in the form

$$\mathbf{v}(t_k + h) = \mathbf{v}(t_k) + h\mathbf{c}_1 + h^2\mathbf{c}_2 + h^3\mathbf{c}_3 + \cdots . \qquad (12.33)$$

Euler integration is known as a *first-order* method of numerical integration because it matches only up to the first-degree term in this expression, $h\mathbf{c}_1$, which in this case is simply the acceleration $h\mathbf{a}(t_k)$. It is true that h is on the order of $1/60$ and in general the higher-order terms decrease rapidly, but we don't know anything about the magnitude of the $\mathbf{c}$'s, so the more terms we can match, the better accuracy can be obtained.[21] If the acceleration is constant, then Euler integration would calculate the velocity exactly, but if the forces depended on the position or velocity, then the higher order terms in the expansion, which are ignored by Euler integration, would be nonzero.

In order to improve our results, we will need to use a higher-order method of numerical integration that is able to match more of the terms in Equation (12.33). One common and important technique is Runge-Kutta, and the idea is to take one or more trial steps and sample the forces at different locations. Then the samples from the trial steps are combined in a smart way. The field of numerical integration is a mature one with a large amount of literature and many different well-studied techniques. It would be appropriate to discuss several of these methods in an introductory book that focused only on physics simulation, but we regret that it is just beyond what we can squeeze into a few chapters in this broad introductory book. However, we want to stress two important lessons concerning numerical integration.

---

[21]Usually. Press et al. [56] make a point of stressing that higher order does not always guarantee higher accuracy.

The first message has already been delivered: we simply want you to appreciate the shortcomings in Euler integration and be aware that other methods exist that offer different trade-offs among accuracy, stability, and performance. Some integrators are better suited for different purposes. Almost any integrator will work better than Euler integration for the simulations people are currently interested in solving in real-time: cloth, hair, fluid, and soft body simulation, combining dynamic physics and pre-generated animation for articulated (especially humanoid) characters, and others. Anyone interested in these more advanced simulations will find an understanding of integration methods to be a prerequisite. The sources listed in the suggested reading in Section 12.7 are a good start.

The second message is that although the higher-order methods are better choices for the "high class" simulations we have named, for simulation of rigid bodies, the trivial Euler integration is still often used. Why? In a word: constraints. In some simulations, the force is a continuous function of time or position, and has a nice Taylor series expansion that is well met by the higher-order integration methods. Simulations based on spring-like connections or those that deal with collisions in a soft, penalty-method-like fashion behave like this. For such simulations, using a higher-order integrator (which increases the number of "inner loop" trial steps and samples per time step) offers a better reduction in error for a given amount of CPU time, compared to reducing h (which takes more "outer loop" iterations). However, the constraints in rigid-body simulations are often discontinuous and inherently require an LCP-based approach. When these discontinuous functions are approximated by Hooke's law, the spring constant must be very large, and this leads to instability. In fact, the resulting differential equations are referred to as "stiff" equations, which are known to require implicit rather than explicit integration. Unfortunately, velocity-based constraint solvers—as we mentioned, these are currently a popular type of simulation—operate by adjusting the velocities to satisfy the constraints, and essentially they need to be able to "see into the future" to know how changes to the proposed adjusted velocity will affect the integrated position and thus the satisfaction of the constraint. More sophisticated methods of integration make this fortune-telling more complicated. (The implicit integration methods used to solve stiff equations essentially operate by "seeing into the future.") In short, due to the current methods of resolving contact and joint constraints, decreasing the stepsize h offers a more attractive use of CPU time than higher-order integrators, at least according to the votes cast by designs of the more popular real-time rigid-body simulators in use in video games right now. Perhaps put more bluntly: stability is currently valued more than accuracy in real-time simulation.

Having brought up numerous complications and then promptly set them aside, we are now ready to present the basic equations for Euler integration of position, velocity, and acceleration.

Euler Integration of Acceleration and Velocity

$$\mathbf{a}_k = \mathbf{f}_k/m \qquad \text{(Newton's second law)},$$
$$\mathbf{v}_{k+1} = \mathbf{v}_k + h\mathbf{a}_k \qquad \text{(integrate acceleration)},$$
$$\mathbf{p}_{k+1} = \mathbf{p}_k + h\mathbf{v}_{k+1} \qquad \text{(integrate velocity)}.$$

These are key operations, so let's see how they might get implemented in C++. The code in Listing 12.7 is a bit easier to read than the equations, since the order of operations makes the subscripting unnecessary.

```
struct Particle {
    Vector3 pos;    // world position of center of mass
    Vector3 linVel; // velocity
    Vector3 force;  // current forces
    float   mass;   // mass of object

    // Take a simple Euler step forward in time by the time step dt
    void eulerIntegrate(float dt) {
        Vector3 acceleration = force / mass;

        linVel += acceleration * dt;
        pos += linVel * dt;
    }
}
```

Listing 12.7
Simple Euler integration

12.6.4   Integration of Rotation

Now let's say a few words about integration of 3D rotational data. What is the equivalent of $\mathbf{p}_{k+1} = \mathbf{p}_k + h\mathbf{v}_k$ for rotation? First, let's present some well-known results from mechanics concerning the relationship between angular velocity and the derivative of orientation values.

Consider an arbitrary point $\mathbf{r}$ fixed on a body that is rotating about its center of mass at an instantaneous angular velocity $\boldsymbol{\omega}$. We assume the coordinate space used to describe $\mathbf{r}$ has its origin at the body's center of mass, but the axes do not rotate with the body. (In this book, we say that we are expressing the coordinates of the point in "upright space." These coordinates are also sometimes called "center of mass coordinates.")

Section 11.8.2 showed how to compute the velocity of this vector: $\mathbf{v} = \boldsymbol{\omega} \times \mathbf{r}$. This can be written equivalently as

$$\dot{\mathbf{r}} = \boldsymbol{\omega} \times \mathbf{r}. \tag{12.34}$$

Section 11.8 stated that is it a basic feature of uniform circular motion that the velocity varies continuously due to centripetal acceleration. Since the velocity is not constant, as we've just seen, a simple Euler-step is not accurate: $\mathbf{r}(t+h) \neq \mathbf{r}(t) + h\dot{\mathbf{r}}$. However, if we zoom in close enough, a small segment of the circular path starts to look very much like a straight line, and the Euler step isn't that bad. In other words, with a small enough h, or a slow enough angular velocity $\boldsymbol{\omega}$, the approximation might be acceptable: $\mathbf{r}(t + h) \approx \mathbf{r}(t) + h\dot{\mathbf{r}}$.

Everything we've said so far works for any vector $\mathbf{r}$, so let's apply these ideas to the body axes themselves. Remember that the rotation matrix $\mathbf{R}$ describes the orientation of the object and rotates row vectors on the left from body space to upright space. The rows of $\mathbf{R}$ are formed by the body axes expressed in upright space. What we want to do is apply Equation (12.34) to each body axis (i.e., take the cross product of $\boldsymbol{\omega}$ with each row). Luckily, we can write the cross product operation as a matrix multiplication (see Exercise 4.8) as

$$\dot{\mathbf{r}} = \boldsymbol{\omega} \times \mathbf{r} = \mathbf{r} \begin{bmatrix} 0 & \omega_z & -\omega_y \\ -\omega_z & 0 & \omega_x \\ \omega_y & -\omega_x & 0 \end{bmatrix}.$$

Now, the derivative of the rotation matrix can be expressed as the matrix product

$$\dot{\mathbf{R}} = \mathbf{R} \begin{bmatrix} 0 & \omega_z & -\omega_y \\ -\omega_z & 0 & \omega_x \\ \omega_y & -\omega_x & 0 \end{bmatrix}.$$

Derivative of orientation matrix R for an object with angular velocity $\boldsymbol{\omega}$

What does this mean? Just as with a single vector $\mathbf{r}$ and its derivative $\dot{\mathbf{r}}$, each element in $\dot{\mathbf{R}}$ gives the derivative of the corresponding element in $\mathbf{R}$. In reality, any particular element of the matrix function $\mathbf{R}(t)$ will oscillate within the range $[-1, 1]$. But as before, for small values of h (and small angular velocities $\boldsymbol{\omega}$!), a small section of this curved, oscillatory pattern looks like a straight line, and the simple Euler step $\mathbf{R}_{k+1} \approx \mathbf{R}_k + h\dot{\mathbf{R}}_k$ might be acceptable. However, with rotation matrices, a new wrinkle is present: the resulting matrix is unlikely to be orthonormal. In essence, we are taking Euler steps on each component in isolation, ignoring their interdependence. The solution is to re-orthogonalize the matrix (see Section 6.3.3) after each step.

If the orientation of the body is specified by using a quaternion rather than a rotation matrix, the same basic technique can still be used: find the (component-wise) derivative of the orientation, take a simple Euler step on each component independently, and then correct the orientation. With quaternions, the derivative is given by

**Derivative of orientation
quaternion q for an
object with angular
velocity $\boldsymbol{\omega}$**

$$\dot{\mathbf{q}} = \frac{1}{2}\boldsymbol{\omega}\mathbf{q},$$

where the 3D angular velocity vector $\boldsymbol{\omega}$ has been extended into quaternion space with $w = 0$. (Eberly [17, Section 10.6] derives this result; we will not prove it here.) Note that we do not expect $\dot{\mathbf{q}}$ to be a rotation (unit) quaternion. Neither do we expect the result of Euler integration, $\mathbf{q}_{k+1} \approx \mathbf{q}_k + h\dot{\mathbf{q}}_k$ to have unit length, and thus it must be normalized.

The technique of integration of orientation just described is a standard one. When using this technique, there are two sources of error. The first is caused by the Euler integration itself, in which we ignore angular acceleration (and higher derivatives) and proceed as if the angular velocity were constant. This error existed with linear data as well as angular; but when integrating linear quantities it was the only source of error. The second source of error is due to the use of component-wise derivatives, which does not take into consideration the interdependence of the components of the rotation matrix or quaternion. This type of error is unique to angular data, since the components of positional data *are* independent (at least when Cartesian coordinates are used). Fortunately, this source of error can be eliminated.

Assume for the moment that the object whose orientation is described by the rotation matrix $\mathbf{R}$ or quaternion $\mathbf{q}$ is rotating with constant angular velocity $\boldsymbol{\omega}$. In this common case, no accuracy is lost by ignoring angular acceleration, but there is a loss in accuracy, which increases with larger $h\boldsymbol{\omega}$. The solution is straightforward: determine the finite rotation that would occur in this time step, and then apply the appropriate angular displacement. We already have the tools at our disposal. We convert the rotation $h\boldsymbol{\omega}$ from exponential map form into axis-angle form (see Section 8.4). This angular displacement can then be converted to a rotation matrix (see Section 5.1.3) or quaternion (see Section 8.5.2), and concatenated with the current rotation. Essentially, what we have done is to choose a better coordinate system in which to perform the Euler integration.

Since it is slightly more expensive to perform the second method, a valid question is: Is it worth it? In the common situation of constant angular velocity, the second method is exact for any step size h, ignoring errors due to floating-point roundoff. In this case, if the angular velocity is high, or the accuracy required is high (e.g., the hands of a clock), then switching to this alternative method will probably be a win. However, the error introduced by Euler integration can interfere with the error introduced by the component-wise derivative method either constructively or destructively, so there is no guarantee that reducing one source of error will actually improve the accuracy of the final result in all cases. This is why the method of integrating angular displacement is an option in some physics engines.

12.7 Suggested Reading

Our discussion has been necessarily compact, and we cringe at the derision actual physics experts must feel towards our presentation. Clearly, a student with a serious interest in computer simulation will need a more thorough background in physics than was provided here. The basics of mechanics are usually covered in the first semester of a traditional physics course, for which there are numerous textbooks of high quality. In preparing this book, we used Resnick and Halliday's venerable textbook [57], which has a distinct advantage of being incredibly inexpensive, and also a textbook by Knight [38]. No one should ever learn this material without in-class demonstrations; a student engaged in self-study (or stuck with a lame physics teacher) need not miss out, as many demonstrations can be found online. We recommend Professor Walter Lewin's lectures, available from MIT OpenCourseWare at ocw.mit.edu.

Three books are recommended for their discussion of physics simulation tailored to the needs of video games. Bourg's *Physics for Game Developers* [8] is an introductory text, with good coverage of the basics and a unique presentation of numerous force laws applied towards different types of vehicle simulations. *Physics-Based Animation* [19] contains a wealth of information for both rigid-body and continuous simulations, including a survey of several different approaches to multibody simulation; this text is your best bet for filling in the gaps left by the vague hand-waving we've had to do here. Eberly's *Game Physics* [17] considers physics engines for games in a slightly more academic and mathematically oriented manner. It contains good discussions on techniques of numerical integration and a unique section concerning the potential (currently unrealized) advantages of Lagrangian dynamics. For Bourg's book, the calculus covered in this book is sufficient, but exposure to differential equations is recommended before tackling the two more advanced books.

Another excellent way to learn about a real-time physics simulation is to study the code of one. Two well-designed and documented open source physics engines that have been used in commercial video games were influential in the writing of this book and are worth mentioning. Russell Smith's *Open Dynamics Engine* [65], available online at http://ode.org/, is slightly older and is not under active development, but has been influential in the industry and is a useful resource. A newer collaboration by various industry experts called *Bullet Physics* (http://bulletphysics.org/) is actively maintained and has been used in many games and even some Dreamworks movies. The engine and the website are both useful resources.

Collision detection is a large portion of any physics engine, both in terms of lines of code and CPU time consumed. Unfortunately, it's difficult to say "a little" about collision detection, and we haven't had the space to do it justice in this book. Ericson's *Real-Time Collision Detection* [18] is our top recommendation, but van den Bergen's text [70] is also useful. A significant amount of material on collision detection material can be found in Eberly's books on physics engines [17] and geometric tools for games [59].

Many of the mathematical problems that arise in computer simulation fall under the broad category of *scientific computing*. (Older names for this same basic subject area are "applied mathematics" and "numerical analysis.") *Numerical Recipes in C* [56] is a classic work for engineers, with clear explanations and a large toolkit of source code. Several good textbooks exist on the subject; we can recommend *Scientific Computing* by Heath [32]. Strang's textbook [67] has the compelling feature that an entire course of accompanying lectures are available free from MIT OpenCourseWare at ocw.mit.edu.

Chris Hecker has a collection of resources for real-time physics at http://chrishecker.com/Physics_References.

12.8   Exercises

(Answers on page 784.)

1. In a cartoon universe, a sailboat can be propelled by placing a fan in the sailboat and pointing it at the sail. Explain why this doesn't work in the real world, by using Newton's laws.

2. A boy and a girl are playing tug of war. The girl begins to win. Name all of the important forces involved, and describe which force imbalance is causing the girl to begin winning.

3. True or false: Lighter objects fall faster than heavier ones because the force of gravity is a constant near Earth's surface.

4. The International Space Station orbits Earth at approximately 340 km above Earth's surface at a speed of approximately 27,740 km/hr. (The orbit is actually elliptical, but ignore that for now.) What is the acceleration caused by Earth's gravity in this "zero gravity" environment? Also, if the Earth's gravity still has a significant effect, why are astronauts in the space station "weightless"? (Note: see also Exercise 11.12.)

5. A concrete block is placed on a wooden ramp. According to Table 12.1, what is the critical angle of inclination of the ramp at which the block will begin to slide? If we conducted the experiment on the moon, would the critical angle increase, decrease, or stay the same?

6. (a) A weight with mass m is hung from a spring with a stiffness of k, causing the spring to increase in length by a distance x_0. What is the formula that relates m, x_0, and k?

   (b) A 5.00 kg object is suspended from a spring, causing the spring to increase in length by 10.0 cm. What is the spring constant k? (Make sure to include the proper units.)

   (c) A different object is hung from the same spring, this time causing the spring to lengthen by 17.0 cm. What is the mass of this other object?

   (d) Later, in a different environment, a 1.00 kg weight is hung from this same spring, this time causing an increase in length of 8.0 cm. What can you say about this environment that is different from the original environment? What are some possible explanations for these differences?

7. A horizontal spring with stiffness 1.00×10^2 N/m is fixed at one end, and a 5.00 kg mass is connected on the other end, such that the mass slides back and forth on a frictionless surface. The spring is extended from its rest position by a distance of 14.7 cm.

   (a) What is the frequency of the oscillation?

   (b) What is the amplitude of the oscillation?

   (c) What is the speed of the mass as it crosses the rest position?

8. A man weighing 75.0 kg is standing at one end of a train car. The car weighs 1.00×10^3 kg, is 20.0 m long, and is from the future, where they have invented a special type of wheel that rolls over the tracks with zero friction. Use a coordinate space where the forward direction is $+x$. The man walks from the back end of the car to the front end at a rate of 1.25 m/s.

   (a) What are velocities of the man and the car, relative to Earth, during the man's walk?

   (b) When the man reaches the end of the car, how far have the man and the car moved, in world coordinates?

   (c) What if, rather than walking at a constant velocity, the man builds up as much speed as he can and then comes to an abrupt stop at the end of the car. What would change about the motion of the car? What about the final positions?

   The experiment is repeated (the man walks with a constant velocity), only this time the car and man have an initial velocity of $+5.00$ m/s.

   (d) What are velocities of the man and the car, relative to Earth, during the man's walk?

   (e) At the moment the man reaches the end of the car, how far have the man and the car been displaced relative to Earth?

9. Consider the car crash between Grant and Kari in Section 12.4.1. Calculate the magnitude of the collision impulse, only this time instead of assuming a perfectly inelastic collision, use a coefficient of restitution of $e = 0.1$. (Assume a contact normal of $20°$ west of south.) What are the resulting velocities of the two cars?

10. How does the monkey in Figure 12.21 stay balanced on the tightrope? How does the bend in the bar help?

Figure 12.21
How does this monkey stay balanced?

11. Two cylinders have the same shape and mass. One is hollow, and the other is solid with uniform density. Which do you expect would be harder to roll?

12. The mass distribution of a truck is approximated by using three boxes for the body and four cylinders for the wheels, as shown in Figure 12.22.

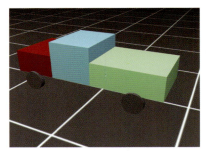

Figure 12.22
Using primitives to approximate the distribution of mass in a truck

Description	Mass (kg)	Center of mass (cm)	Dimensions (cm, $x \times y \times z$)
Body front	1000	$(0, 100, 225)$	$200 \times 80 \times 150$
Body middle	600	$(0, 125, 75)$	$200 \times 130 \times 150$
Body rear	400	$(0, 100, -120)$	$200 \times 80 \times 240$
Front left wheel	50	$(-100, 35, 230)$	$20 \times 70 \times 70$
Front right wheel	50	$(100, 35, 230)$	$20 \times 70 \times 70$
Rear left wheel	50	$(-100, 35, -150)$	$20 \times 70 \times 70$
Rear right wheel	50	$(100, 35, -150)$	$20 \times 70 \times 70$

Table 12.2. Primitives used to approximate the mass of a truck

The position, mass, and dimensions of these primitives are given by Table 12.2.

(a) What are the coordinates of the center of mass of the truck?

(b) Calculate the inertia tensor for each primitive, relative to its center of mass. Assume all the primitives have uniform density. (You will need to find the appropriate formulas online.) Hint: convert the measurements to meters first.

(c) Use the parallel axis theorem to compute the inertia tensor of the truck relative to its center of mass.

We love force and we care very little how it is exhibited.

— Ralph Waldo Emerson (1808–1882)

Chapter 13

Curves in 3D

I didn't discover curves; I only uncovered them.

— Mae West (1892–1980)

This chapter talks about how to represent curves mathematically in 3D. Recreating a curve from its mathematical definition is relatively easy; the tricky part is obtaining a curve with desired properties, or alternatively, making a tool that designers can use to draw such curves. Our goal in this chapter is to provide a graceful and intuitive introduction to the mathematics of curves. In comparison with most of the other books on the subject, our aim is to hit the most important points, without stopping every other paragraph to prove that what we are saying is true. (We will, however, stop periodically to discuss correct pronunciation, which is probably appropriate considering that most of the people who developed the math we'll be using in this chapter were French.) Curves and splines are very useful for all sorts of reasons. There are obvious applications such as moving objects around on curved trajectories. But then the coordinates of our curve need not have a spatial interpretation; essentially, any time we wish to fit a function for a color, intensity, or other property to given data points, we have a potential application for curves and splines.

The chapter is divided roughly into two parts. The first part is about simple, "short" curves that can be described by one equation.

- Section 13.1 introduces the specific type of curve we focus on almost exclusively: the parametric polynomial curve. (It pays special attention to cubic polynomials.)

- Section 13.2 describes polynomial interpolation, whereby a curve is threaded through specified control points.

- Section 13.3 discusses Hermite form, which describes a curve in terms of its endpoints and the derivatives at those endpoints.

- Section 13.4 shows how the Bézier form specifies the curve endpoints, plus interior control points that influence the shape of the curve but are not interpolated.

- Section 13.5 shows how to subdivide a curve into smaller pieces.

The second half of the chapter covers *splines*, which are longer curves created by joining together multiple curves in succession.

- Section 13.6 introduces some basic notation, terminology, and concepts.

- Section 13.7 discusses how to join together Hermite or Bézier curves into a spline.

- Section 13.8 considers continuity (smoothness) conditions for splines.

- Section 13.9 ends the discussion on splines by considering various methods for automatically determining the tangents of a spline at the control points.

13.1   Parametric Polynomial Curves

We focus here almost exclusively on one particular type of curve, the *parametric polynomial* curve. It's important to understand what the two adjectives *parametric* and *polynomial* mean, so Section 13.1.1 and Section 13.1.2. discuss them in detail. Section 13.1.3 reviews some useful alternate notation. Section 13.1.4 examines the straight line, which is a particularly instructive example of a parametric polynomial curve. Section 13.1.5 considers the relationship between the endpoints of the curve and polynomial coefficients. Section 13.1.6 discusses derivatives, such as velocity and acceleration, and shows how they are related to tangent vectors and local curvature.

13.1.1   Parametric Curves

The word *parametric* in the phrase "parametric polynomial curve" means (not altogether surprisingly) that the curve can be described by a function of an independent parameter, which is often assigned the symbol t. This curve function is of the form $\mathbf{p}(t)$, taking a scalar input (the parameter t) and returning the point on the curve corresponding to that parameter value as a vector output. The function $\mathbf{p}(t)$ traces out the shape of the curve as

t varies. For example, consider the classic parametric description of a unit circle,

$$x(t) = \cos(2\pi t),$$
$$y(t) = \sin(2\pi t).$$

(13.1)

Parametric description of a circle

We briefly introduced parametric representation of geometric primitives in Section 9.1. Let's take a moment to review some of the alternative forms from that section so we can understand ways of describing a curve that are *not* parametric. An *implicit* representation is a relation that is true for all points in the shape being described; for example, the unit circle can be described implicitly as the set of points satisfying $x^2 + y^2 = 1$. Another alternative to parametric form is the *functional* form, in which one coordinate is expressed as a function of the other coordinate or coordinates; for example, the top half of a unit circle can be described in functional form as $y = \sqrt{1 - x^2}$.

The curve $\mathbf{p}(t)$ could be infinite, particularly if we place no limits on the range of t. Often it's useful to select a finite segment by restricting t to a particular bounded domain, most commonly the domain $[0, 1]$. It's natural to designate the "forward" direction as the direction of increasing t, so the curve "starts" at $t = 0$, "ends" at $t = 1$, and consists of all of the points between.

Sometimes we think of the position function $\mathbf{p}(t)$ as a single function that yields a vector result; other times it will be helpful to extract the function for a specific coordinate. For example, the scalar function $x(t)$ specifies the x-coordinate of $\mathbf{p}(t)$, so in two dimensions $\mathbf{p}(t) = (x(t), y(t))$. Notice that each coordinate is specified by a function that depends only on the parameter value so that each coordinate is independent of the others. We work in the plane for the majority of this chapter because almost every important aspect of parametric curves can be demonstrated in 2D and, in general, extension into three dimensions is straightforward.

13.1.2   Polynomial Curves

Now that we know what the adjective *parametric* means, let's turn our attention to the second important word, *polynomial*. A polynomial parametric curve is a parametric curve function $\mathbf{p}(t)$ that can be written as a polynomial in t:

$$\mathbf{p}(t) = \mathbf{c}_0 + \mathbf{c}_1 t + \mathbf{c}_2 t^2 + \cdots + \mathbf{c}_{n-1} t^{n-1} + \mathbf{c}_n t^n.$$

Polynomial parametric form of arbitrary degree n

The number n is called the *degree* of the polynomial. Higher degree polynomials are more flexible in the sense that they can describe curves with more "wiggles." However, sometimes extra "wiggles" come in that we don't want;[1] more on this in Section 13.6.

We've already seen an example of a curve function that is parametric but not polynomial—the parametric circle given by Equation (13.1). The expressions for $x(t)$ and $y(t)$ are not polynomials because they use trig functions. A complete circle can't be described in parametric polynomial form, although a circular arc can be described by a *rational curve*. A rational curve is essentially the result of dividing one curve by another, sort of like the projective geometry of homogeneous coordinates (see Section 6.4.1). The curve in the denominator is a 1D curve. Rational curves are not as common in video games as simple polynomial curves and are not discussed in this book.

Of most interest to us are the parametric polynomial curves of degree 3, known as *cubic* curves. Cubic curves are those that can be expressed in the form shown in Equation (13.2).

Cubic Curve in Monomial Form

$$\mathbf{p}(t) = \mathbf{c}_0 + \mathbf{c}_1 t + \mathbf{c}_2 t^2 + \mathbf{c}_3 t^3. \tag{13.2}$$

This method of describing curves is often called the *monomial* form or the *power* form, to emphasize the fact that the curve is specified by listing the coefficients of the powers of t. Sections 13.2–13.4 discuss other methods of describing a curve with more direct geometric data, such as a list of control points that the curve is to pass through or nearby. These other forms are still polynomial curves in the sense that they can be converted to monomial form.

Once we have the coefficients, it's easy to reconstruct the curve by evaluating the function $\mathbf{p}(t)$ for different values of t. For example, let's say we wish to move a platform along a path in a video game. Our platform actor would have a state variable to remember its parametric position t along the path, and at each simulation time step, we would update t and set the position of the platform to $\mathbf{p}(t)$.

Suppose we need to render a curve. One simple way to do this is to approximate it with, say, 10 line segments, sampling the curve at $t = 0, \frac{1}{10}, \frac{2}{10}, \ldots, \frac{9}{10}, 1$ and drawing line segments between consecutive sample points. We can reduce the error in the approximation to any desired

---

[1] This is not intended as a comment on a certain Australian children's musical group, but may be misinterpreted as such.

threshold simply by using more sample points. We can do much better than this naïve approach by adaptively subdividing the curve, using more segments in the "curvier" parts and fewer in the "straighter" parts.

But where do the coefficients $\mathbf{c}_0, \mathbf{c}_1, \mathbf{c}_2, \mathbf{c}_3$ come from? How can we set them to design a particular curve? In general, the monomial form is particularly ill-suited to this task, so we use other forms and convert to monomial form when appropriate. (In many cases, we don't need the monomial form at all!) Before we discuss these other forms, however, we need to introduce some more notation and concepts about curves.

13.1.3   Matrix Notation

We can rewrite the monomial form (Equation (13.2)) in several different ways. It's useful to be able to refer to a coefficient for a particular coordinate. For example, in 2D let's use the notation $\mathbf{c}_i = \begin{bmatrix} c_{1,i} & c_{2,i} \end{bmatrix}$ so we have one polynomial per coordinate:

$$x(t) = c_{1,0} + c_{1,1}t + c_{1,2}t^2 + c_{1,3}t^3,$$
$$y(t) = c_{2,0} + c_{2,1}t + c_{2,2}t^2 + c_{2,3}t^3.$$

2D cubic curve in expanded monomial form

Some books are fond of writing this more compactly by using matrix notation. Let's put the coefficients into a matrix $\mathbf{C}$ and create a column vector $\mathbf{t}$ from the powers of t, such that $\mathbf{t}_i = t^{i-1}$:

$$\mathbf{C} = \begin{bmatrix} c_{1,0} & c_{1,1} & c_{1,2} & c_{1,3} \\ c_{2,0} & c_{2,1} & c_{2,2} & c_{2,3} \end{bmatrix}, \qquad \mathbf{t} = \begin{bmatrix} t^0 \\ t^1 \\ t^2 \\ t^3 \end{bmatrix} = \begin{bmatrix} 1 \\ t \\ t^2 \\ t^3 \end{bmatrix}.$$

Now we can express our curve function $\mathbf{p}(t)$ as a single matrix product:

$$\mathbf{p}(t) = \mathbf{Ct} = \begin{bmatrix} c_{1,0} & c_{1,1} & c_{1,2} & c_{1,3} \\ c_{2,0} & c_{2,1} & c_{2,2} & c_{2,3} \end{bmatrix} \begin{bmatrix} 1 \\ t \\ t^2 \\ t^3 \end{bmatrix}.$$

2D cubic curve in monomial form, expressed as a matrix product

Don't try to apply any geometric interpretations just yet. The vector $\mathbf{t}$ is not to be interpreted as a point in space, and the matrix $\mathbf{C}$ is not a transformation matrix. Although we're about to learn how to extract geometric meaning from $\mathbf{C}$, the techniques are very different from those learned in previous chapters. For now, let's just be happy to use matrix notation purely for sake of compactness.

The matrix $\mathbf{C}$ must be as "tall" as the number of dimensions the data have; for example, three if we have 3D data. However, we don't need to refer to specific x, y, or z coordinates much in this chapter because most of the ideas work the same in 3D or 2D (or 1D!). We can just leave each coefficient $\mathbf{c}_i$ in vector form and assume that it is a vector of the appropriate dimension, so that each $\mathbf{c}_i$ corresponds to a single column of $\mathbf{C}$:

**Coefficients as column
vectors**

$$\mathbf{C} = \begin{bmatrix} | & | & | & | \\ \mathbf{c}_0 & \mathbf{c}_1 & \mathbf{c}_2 & \mathbf{c}_3 \\ | & | & | & | \end{bmatrix}, \quad \mathbf{p}(t) = \mathbf{C}\mathbf{t} = \begin{bmatrix} | & | & | & | \\ \mathbf{c}_0 & \mathbf{c}_1 & \mathbf{c}_2 & \mathbf{c}_3 \\ | & | & | & | \end{bmatrix} \begin{bmatrix} 1 \\ t \\ t^2 \\ t^3 \end{bmatrix}.$$

When dealing with a higher degree polynomial, the matrix $\mathbf{C}$ is wider and the power vector $\mathbf{t}$ is taller, since we have more coefficients and more powers of t. This not only makes sense, it's the law: for the product $\mathbf{C}\mathbf{t}$ to be legal according to linear algebra rules, the number of columns in $\mathbf{C}$ must match the number of rows in $\mathbf{t}$.

13.1.4   Two Trivial Types of Curves

Although you're reading this section because you want to learn how to draw a curve, allow a brief digression to mention two trivial types of "curves": a straight line segment and a point.

We showed how to represent a line segment parametrically in Section 9.2 when we discussed rays. Consider a ray from the point $\mathbf{p}_0$ to the point $\mathbf{p}_1$. If we let $\mathbf{d}$ be the delta vector $\mathbf{p}_1 - \mathbf{p}_0$, then the ray is expressed parametrically as

Parametric line segment

$$\mathbf{p}(t) = \mathbf{p}_0 + \mathbf{d}t. \tag{13.3}$$

Observe that this is a polynomial of the type we've been considering, where $\mathbf{c}_0 = \mathbf{p}_0$, $\mathbf{c}_1 = \mathbf{d}$, and the other coefficients are zero. In other words, this *linear curve* is a polynomial curve of degree 1.

As boring as lines are, there's an even less interesting shape that can be represented in parametric polynomial form: the point. Lowering the degree of the polynomial from 1 to 0 results in a so-called *constant curve*. In this case, the function $\mathbf{p}(t) = \mathbf{c}_0$ always returns the same value, resulting in a "curve" that is a single stationary point.

13.1.5   Endpoints in Monomial Form

Clearly, one of the most basic properties of a curve that we want to control are the locations of its start and end, $\mathbf{p}(0)$ and $\mathbf{p}(1)$, respectively. Let's see what $\mathbf{p}(t)$ looks like at the endpoints. We'll use the cubic case as our example. At $t = 0$, we have

$$\mathbf{p}(0) = \mathbf{c}_0 + \mathbf{c}_1(0) + \mathbf{c}_2(0)^2 + \mathbf{c}_3(0)^3 = \mathbf{c}_0.$$

$\mathbf{c}_0$ specifies the start point

In other words, $\mathbf{c}_0$ specifies the start point of the curve. Now let's see what happens at the end of the curve at $t = 1$:

$$\mathbf{p}(1) = \mathbf{c}_0 + \mathbf{c}_1(1) + \mathbf{c}_2(1)^2 + \mathbf{c}_3(1)^3 = \mathbf{c}_0 + \mathbf{c}_1 + \mathbf{c}_2 + \mathbf{c}_3.$$

The endpoint is the sum of the coefficients

So the endpoint of the curve is given by the sum of the coefficients.

13.1.6 Velocities and Tangents

We can think of curves as being either static or dynamic. In the static sense, a curve defines a shape. We operate in this mode of thinking when we use a curve to describe the cross section of an airplane wing or a portion of the letter "S" in the Times Roman font. In the dynamic sense, a curve can be a trajectory or path of an object over time, with the parameter t as "time" and the position function $\mathbf{p}(t)$ describing the position of a particle at time t as it moves along the path.

If we consider only the static shape of the curve, then the timing of the curve doesn't matter and our task is a bit easier. For example, when defining a shape, it doesn't matter which endpoint is considered the "start" and which is the "end"; but if we are using the curve to define a path traversed over time, then it matters very much where the path starts and where it ends.

Using the dynamic mental framework and thinking about curves as paths and not just shapes, some natural questions to ask are, "In what direction is the particle moving at a given point in time?" "How fast is it moving?" These questions can be answered if we create another function $\mathbf{v}(t)$ that describes the *instantaneous velocity* of the particle at time t.

The phrase "instantaneous velocity" implies that the velocity changes over time. So the next logical step is to ask, "How fast is the velocity changing?" Thus it is also helpful to define an *instantaneous acceleration* function $\mathbf{a}(t)$ that describes the rate at which the velocity of the particle is changing at time t.

If you've had at least a semester of calculus, or if you read Chapter 11, you should recognize that the velocity function $\mathbf{v}(t)$ is the first derivative of the position function $\mathbf{p}(t)$ because velocity measures the rate of change in position over time. Likewise, the acceleration function $\mathbf{a}(t)$ is the derivative of the velocity function $\mathbf{v}(t)$ because acceleration measures the rate of change of velocity over time.

We're considering curves where $\mathbf{p}(t)$ is a polynomial of t here, so the derivatives are trivially obtained. The position, velocity, and acceleration functions for polynomials of arbitrary degree n are

$$\mathbf{p}(t) = \mathbf{c}_0 + \mathbf{c}_1 t + \mathbf{c}_2 t^2 + \cdots + \mathbf{c}_{n-1} t^{n-1} + \mathbf{c}_n t^n,$$
$$\mathbf{v}(t) = \dot{\mathbf{p}}(t) = \mathbf{c}_1 + 2\mathbf{c}_2 t + \cdots + (n-1)\mathbf{c}_{n-1} t^{n-2} + n\mathbf{c}_n t^{n-1},$$
$$\mathbf{a}(t) = \dot{\mathbf{v}}(t) = \ddot{\mathbf{p}}(t) = 2\mathbf{c}_2 + \cdots + (n-1)(n-2)\mathbf{c}_{n-1} t^{n-3} + n(n-1)\mathbf{c}_n t^{n-2}.$$

The derivatives of cubic curves are especially notable and appear several times in this chapter.

Velocity and Acceleration of Cubic Monomial Curve

$$\mathbf{p}(t) = \mathbf{c}_0 + \mathbf{c}_1 t + \mathbf{c}_2 t^2 + \mathbf{c}_3 t^3, \tag{13.4}$$
$$\mathbf{v}(t) = \dot{\mathbf{p}}(t) = \mathbf{c}_1 + 2\mathbf{c}_2 t + 3\mathbf{c}_3 t^2, \tag{13.5}$$
$$\mathbf{a}(t) = \dot{\mathbf{v}}(t) = \ddot{\mathbf{p}}(t) = 2\mathbf{c}_2 + 6\mathbf{c}_3 t. \tag{13.6}$$

Now let's examine velocity and acceleration in the special case of a parametric ray. Applying the velocity and acceleration functions of Equations (13.5) and (13.6) to the original parameterization of a ray from Equation (13.3) yields

$$\mathbf{p}(t) = \mathbf{p}_0 + \mathbf{d}t,$$
$$\mathbf{v}(t) = \mathbf{c}_1 + 2\mathbf{c}_2 t + 3\mathbf{c}_3 t^2 = \mathbf{d},$$
$$\mathbf{a}(t) = 2\mathbf{c}_2 + 6\mathbf{c}_3 t = \mathbf{0}.$$

As we'd expect, the velocity is constant; there is no acceleration.

Sometimes two curves define the same shape but different paths (see Figure 13.1). We've already mentioned one example of this: if we traverse the path backwards it still traces out the same shape. A more general way to generate alternate paths that trace out the same shape is to *reparameterize* the curve. For example, let's reparameterize our line segment $\mathbf{p}(t) = \mathbf{p}_0 + \mathbf{d}t$. We'll make a new function $s(t) = t^2$ and see what $\mathbf{p}(s(t))$ looks like:

$$\mathbf{p}(s(t)) = \mathbf{p}(t^2) = \mathbf{p}_0 + \mathbf{d}t^2.$$

Notice that both curves in Figure 13.1 define the same static shape, but different paths. On the left, the particle moves with constant velocity, but on the right it starts out slowly and accelerates to the finish.

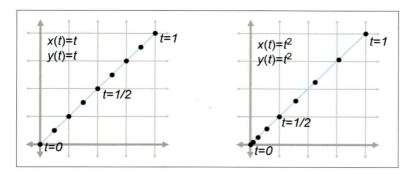

Figure 13.1. Two curves that define the same "shape," but not the same "path"

If we are using a curve as a shape and not a path, then this reparameterization doesn't have a visible effect. But that doesn't mean that the derivatives of the curve are irrelevant in the context of shape design. Imagine that we are creating a font using a curve to define a segment of the letter S. In this instance, we might not care about the velocity at any point, but we would care very much about the *tangent* of the line at any given point. The tangent at a point is the direction the curve is moving at that point, the line that just touches the curve. The tangent is basically the normalized velocity of the curve. Let's formally define the tangent of a curve to be the unit vector pointing in the same direction as the velocity:

$$\mathbf{t}(t) = \hat{\mathbf{v}}(t) = \frac{\mathbf{v}(t)}{\|\mathbf{v}(t)\|}.$$

The tangent vector

Higher derivatives also have geometric meaning. The second derivative is related to *curvature*, which is sometimes denoted κ, the lowercase Greek letter kappa. We can define a measure of curvature by considering a circle of a given radius. A circle with radius r has curvature equal to $\kappa = 1/r$ everywhere on the circle. A straight portion of a curve has zero curvature, which can be interpreted as the curvature of a circle with infinite radius. The curvature is computed by the formula

$$\kappa(t) = \frac{\|\mathbf{v}(t) \times \mathbf{a}(t)\|}{\|\mathbf{v}(t)\|^3}.$$

Curvature

13.2   Polynomial Interpolation

You are probably already familiar with *linear interpolation*. Given two "endpoint" values, create a function that transitions at a constant rate (spatially, in a straight line) from one to the other. We say that the function *interpolates* the two control points, meaning that it passes through the control points and can be used to compute intermediate values.

Polynomial interpolation is similar. Given a series of control points, our goal is to construct a polynomial that interpolates them. The degree of the polynomial depends on the number of control points. A polynomial of degree n can be made to interpolate $n + 1$ control points. For example, linear interpolation is simply first-degree polynomial interpolation. We're primarily interested in cubic (third-degree) curves in this chapter, so we are creating polynomials that interpolate four control points.

In the context of curve design, to say that a curve *interpolates* control points is to place specific emphasis on the fact that the curve *passes through* the control points. This is to be contrasted with a curve that merely *approximates* the control points, meaning it doesn't pass through the points but is attracted to them in some way. We use the word "knot" to refer to control points that are interpolated, invoking the metaphor of a rope with knots in it. It would seem at first glance that the availability of an interpolation scheme would make any approximation scheme obsolete, but we'll see that approximation techniques do have their advantages.

Polynomial interpolation is a classic problem with several well-studied solutions. Since this is a book on 3D math we cast the discussion primarily in geometric terms, but be aware that most of the literature on polynomial interpolation adopts a more general view, because the task of fitting a function to a set of data points has broad applicability.

To facilitate the discussion we use a particular example curve, shown in Figure 13.2. It's somewhat like an S turned on its side. We've marked the four control points on the curve that we are attempting to interpolate. We've chosen to place the y coordinates on the interval $[2, 3]$ for reasons that will be useful later.

Notice that we must specify not only the *position* of each control point (the x and y coordinates), but the *time* when we want the curve to reach that control point (the t value). We use the notation that the independent value (the "time values") of the control points are named $t_1, t_2, \ldots, t_n$ and the dependent variables (the spatial coordinate values at those times) are $y_1, y_2, \ldots, y_n$. The symbol P stands for the polynomial function that we seek: $y_i = P(t_i)$.

The array of time values $t_1 \ldots t_n$ is known in other contexts as the *knot vector* or *knot sequence*. The word "vector" indicates that the sequence of t values is an array of numbers, not that these numbers form a vector in the geometric sense of the word. If the ts are spaced evenly like they are in our example, then we have a *uniform* knot vector; otherwise, we say that the knot vector is *nonuniform*. (Because it might be confusing, let us clarify

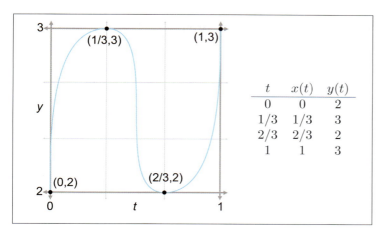

The table shown in the figure:

t	$x(t)$	$y(t)$
0	0	2
1/3	1/3	3
2/3	2/3	2
1	1	3

Figure 13.2. An example curve and four control points. Can we draw this shape?

that the knot vector is the sequence of t values, not the sequence of control points.)

What about the x-coordinate? Because the x and y coordinates are independent of one another, a general 2D curve-fitting application involves two separate one-dimensional problems. Aside from the fact that the two problems use the same knot vector, the coordinates are otherwise unrelated. Even though Figure 13.2 may look like a 2D curve, it is more properly interpreted as a graph of one coordinate (the y-coordinate) as a function of time. We chose as the example curve an S turned on its side, rather than an S in its regular orientation, since the latter is not the graph of a function (technically it's called a *relation* because it associates more than one value of y with each value of x).

With that said, there are two ways of interpreting Figure 13.2. We can interpret it either as a 1D function of $y(t)$, or as a 2D curve, where one of the coordinates has a trivial form $x = t$. This is a common source of confusion when looking at diagrams of curves in this book and elsewhere. Make sure you pay special attention to the horizontal axis to make sure you know whether it is a graph of one coordinate over time or a plot of the 2D curve that includes the behavior of both spatial coordinates. The traditional literature on polynomial interpolation is mostly in abstract terms of any function of the form $y = f(x)$. In this context, x would be the independent variable rather than a dependent value as it is for us. The notation we have chosen avoids the symbol x and its associated baggage.

Now we are ready to answer a question some readers might be thinking: "I don't care what time the curve reaches the points, I just want a smooth shape that goes through the points." Unfortunately, this doesn't unambiguously define a curve—we need to provide some other criteria to nail down the shape, and one way to do this is to associate time values with each control point. In typical applications of polynomial interpolation, we *want* to be able to specify the values of the dependent variable, because we are trying to fit a function to some known data points. There are some reasonable ways we can synthesize this information if we don't have it—for example, by making the difference between adjacent t values proportional to the Euclidian distance between the corresponding control points. However, the general fact that polynomial interpolation needs us to provide the t values when we often don't have a good way to decide what they should be is a harbinger of later discoveries.

Now that we've set the ground rules, let's try to create this curve. We first take a geometric approach in Section 13.2.1. Then, in Section 13.2.2, we look at the problem from a slightly more abstract mathematical perspective.

13.2.1 Aitken's Algorithm

Our first approach to polynomial interpolation is a recursive technique due to Alexander Aitken (1895–1967). Like many recursive algorithms, it works on the principle of *divide and conquer*. To solve a difficult problem, we first divide it into two (or more) easier problems, solve the easier problems independently, and then combine the results to get the solution to the harder problem. In this case, the "hard" problem is to create a curve that interpolates n control points. We split this curve into two "easier" curves: (1) one that interpolates only the first $n-1$ points, disregarding the last point; and (2) another that interpolates the last $n-1$ points without worrying about the first point. Then, we blend these two curves together.

Let's take the important cubic (third-degree) case as an example. A cubic curve has four control points $y_1 \ldots y_4$ that we wish to interpolate at the corresponding times $t_1 \ldots t_4$. Applying the "divide-and-conquer" approach, we split this up into two smaller problems: one curve to interpolate $y_1 \ldots y_3$, and another curve to interpolate $y_2 \ldots y_4$. Since each of these curves has three control points, they are quadratic (second-degree) curves. Of course, quadratic curve-fitting is still a "hard" problem for us, and so each curve must be further subdivided.

Consider the first quadratic curve, between y_1, y_2, and y_3. We further divide this curve into two parts, the first part between y_1 and y_2 and the other part between y_2 and y_3. These two curves have only two control points each; they are straight line segments. Finally, a problem that is truly "easy"!

Since we have lots of curves at this point, we should invent some notation for them. We let $y_i^1(t)$ denote the linear curve between y_i and y_{i+1}, the notation $y_i^2(t)$ denote the quadratic curve between y_i and y_{i+2}, and so on. In other words, the superscript indicates the recursion level in the divide-and-conquer algorithm (and also the degree of the polynomial), and the subscript indexes along the length of the curve.

Take a look at the first quadratic curve $y_1^2(t)$ that interpolates y_1, y_2, and y_3. It is formed by blending together the two lines containing the first two linear segments. An example of such blending is shown in Figure 13.3. (This figure doesn't use the data from our S example; it's a less symmetric case that better illustrates the blending process.) Notice that each curve segment is an interval from an infinite curve that is defined for any value of t.

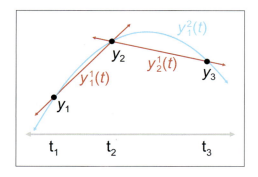

Figure 13.3
Creating a quadratic curve as a blend of two linear segments according to Aitken's algorithm

Now let's look at the math behind this. It's all linear interpolation. The easiest are the linear segments, which are defined by linear interpolation between the adjacent control points:

$$y_1^1(t) = \frac{(t_2 - t)y_1 + (t - t_1)y_2}{t_2 - t_1}, \qquad y_2^1(t) = \frac{(t_3 - t)y_2 + (t - t_2)y_3}{t_3 - t_2}.$$

Linear interpolation between two control points

The quadratic curve is only slightly more complicated. We just linearly interpolate between the line segments:

$$y_1^2(t) = \frac{(t_3 - t)\left[y_1^1(t)\right] + (t - t_1)\left[y_2^1(t)\right]}{t_3 - t_1}.$$

Linear interpolation of lines yields a quadratic curve

Hopefully you can see the pattern—each curve is the result of linearly interpolating two curves of lesser degree. Aitken's algorithm can be summarized succinctly as a recurrence relation.

Aitken's Algorithm

$$y_i^0(t) = y_i,$$

$$y_i^j(t) = \frac{(t_{i+j} - t)\left[y_i^{j-1}(t)\right] + (t - t_i)\left[y_{i+1}^{j-1}(t)\right]}{t_{i+j} - t_i}.$$

Aitken's algorithm works because, at each level both curves being blended already touch the middle control points. The two outermost control points are touched by only one curve or the other, but for those values of t, the blend weights reach their extreme values and all the weight is given to the curve that touches the control point.

Now that we have the basic idea, let's apply it to our sideways S. Figure 13.4 shows Aitken's algorithm at work with our four data points. On the left, the three linear segments are blended to form two quadratic segments. On the right, the two quadratic curves are blending, yielding the final result that we've been seeking: a cubic spline that interpolates all four control points.

So we've successfully interpolated the four control points, and accomplished the goal set out at the start of this section, right? Well, not exactly.

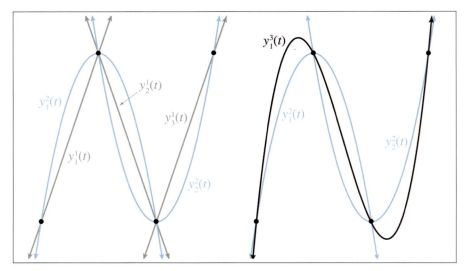

Figure 13.4
Two levels of Aitken's algorithm

Although our curve does pass through the control points, it isn't really the curve we wanted. If we compare the curve on the right side of Figure 13.4 with the curve we set out to create at the start of this section in Figure 13.2, we see that the curve produced by Aitken's algorithm overshoots the y value of the two middle control points. We have discovered an inconvenient truth.[2]

Polynomial interpolation doesn't really give us the type of control we want for curve design in geometric settings.

But don't despair! We've learned several important ideas that will be helpful when we discuss Bézier curves in Section 13.4 and splines in Section 13.6. In fact, we're going to beg your patience to allow us to extend the discussion on polynomial interpolation just a bit further. It's sort of like watching the movie *Titanic*; even though you know that the journey will end tragically, you still might find something useful along the way. We promise that the other techniques in this chapter will have practical as well as educational value.

By the way, you might have noticed that we didn't actually compute the polynomial P that produces the curve. Working through this math is straightforward, but a bit tedious and not all that enlightening. The important point is that Aitken's algorithm is a recursive process of blending curves together and works by repeated linear interpolation. Besides, why bother with the details when we have computers to solve algebra problems for us?[3] However, you needn't feel short-changed by lazy authors. If you really want to know what the polynomial is (or just want to feel like you're getting your money's worth), keep reading. We'll discover it in the next section by using a different method that's less tedious mathematically.

13.2.2   Lagrange Basis Polynomials

Section 13.2.1 applied geometric intuition to the problem of polynomial interpolation and came up with Aitken's algorithm. Now we approach the subject from a more abstract mathematical point of view.

One mathematical approach to the interpolation problem comes from linear algebra.[4] Each control point gives us one equation, and each coeffi-

---

[2]Aitken's Al Gore rhythm, if you will.

[3]Don't try this excuse with your professor, but it's been known to work in job interviews.

[4]We're talking about *real* linear algebra, not the geometry-focused subset of it we study in this book.

cient gives us one unknown. This system of equations can be put into an $n \times n$ matrix,[5] which can be solved by standard techniques such as Gaussian elimination or LU decomposition. Such techniques are outside the scope of this book, but you can learn about them in practically any good book on linear algebra or numerical methods.

Solving a matrix is a relatively time-consuming computational process, requiring $O(n^3)$ time for an $n \times n$ matrix in the worst case. Luckily there are more efficient approaches. As we did with Aitken's algorithm, we solve a large complicated problem by dividing it into a series of smaller, simpler problems, and then combining those results. Aitken's algorithm is a recursive procedure, but here we will make one "simple" problem per control point.

Let's ignore the y's for now and think only about the t's. What if we could create a polynomial for each knot t_i such that the polynomial evaluates to unity at that knot, but for all the other knots it evaluates to zero? If we denote the ith polynomial as ℓ_i, then this idea can be expressed in mathspeak: $\ell_i(t_i) = 1$, and $\ell_i(t_j) = 0$ for all $j \neq i$. For example, let's assume $n = 4$. Then our polynomials would have the following values at the knots:

$$\begin{array}{llll}
\ell_1(t_1) = 1, & \ell_1(t_1) = 0, & \ell_3(t_1) = 0, & \ell_4(t_1) = 0, \\
\ell_1(t_2) = 0, & \ell_2(t_2) = 1, & \ell_3(t_2) = 0, & \ell_4(t_2) = 0, \\
\ell_1(t_3) = 0, & \ell_2(t_3) = 0, & \ell_3(t_3) = 1, & \ell_4(t_3) = 0, \\
\ell_1(t_4) = 0, & \ell_2(t_4) = 0, & \ell_3(t_4) = 0, & \ell_4(t_4) = 1.
\end{array}$$

If we were able to create polynomials with the above properties, we would be able to use them as *basis polynomials*. We would scale each basis polynomial ℓ_i by the corresponding coordinate value y_i, and add all the scaled polynomials together:

Interpolating polynomial in Lagrange basis form

$$P(t) = \sum_{i=1}^{n} y_i \ell_i(t) = y_1 \ell_1(t) + y_2 \ell_2(t) + \cdots + y_{n-1} \ell_{n-1}(t) + y_n \ell_n(t). \quad (13.7)$$

You might want to take a moment to convince yourself that this polynomial actually interpolates the control points, meaning $P(t_i) = y_i$.

Notice that the basis polynomials depend only on the knot vector (the t's) and not on the coordinate values (the y's). Because of this, a set of basis polynomials can be used to quickly construct multiple curves with the same knot vector. This is precisely the situation we find ourselves in when

---

[5]This type of matrix, in which each row or column is a geometric series of the powers of some term, is known as a *Vandermonde* matrix, after the French mathematician Alexandre-Théophile Vandermonde (1735-1796).

dealing with a 3D curve, which is really three one-dimensional curves that share the same knot sequence.

Of course, all of this would work only if we knew the basis polynomials, and finding ℓ_i is itself a problem of polynomial interpolation. However, the "data points" we wish ℓ_i to interpolate are all either 0 or 1, so ℓ_i can be expressed in a simple form. Such basis polynomials are called *Lagrange basis polynomials*.[6] A Lagrange[7] basis polynomial ℓ_i for knot vector $t_1 \ldots t_n$ looks like Equation (13.8):

Lagrange Basis Polynomial

$$\ell_i(t) = \prod_{\substack{1 \le j \le n, \\ j \ne i}} \frac{t - t_j}{t_i - t_j} = \frac{t - t_0}{t_i - t_0} \cdots \frac{t - t_{i-1}}{t_i - t_{i-1}} \frac{t - t_{i+1}}{t_i - i_{i+1}} \cdots \frac{t - t_n}{t_i - t_n}. \quad (13.8)$$

This trick works because at the knot t_i, all the terms in the product equal 1, causing the entire expression to evaluate to 1, and at any other knot, one of the terms in the product is 0, which causes the entire expression to evaluate to 0.

Let's apply this to our example S curve. Recall that it used the uniform knot vector $(0, \frac{1}{3}, \frac{2}{3}, 1)$. Here, we work through the first basis polynomial and just present the results for the others:

$$\ell_1(t) = \left(\frac{t - t_2}{t_1 - t_2} \right) \left(\frac{t - t_3}{t_1 - t_3} \right) \left(\frac{t - t_4}{t_1 - t_4} \right) = \left(\frac{t - 1/3}{0 - 1/3} \right) \left(\frac{t - 2/3}{0 - 2/3} \right) \left(\frac{t - 1}{0 - 1} \right)$$

$$= \left(\frac{3t - 1}{-1} \right) \left(\frac{3t - 2}{-2} \right) \left(\frac{t - 1}{-1} \right) = \frac{(3t - 1)(3t - 2)(t - 1)}{-2}$$

$$= -(9/2)t^3 + 9t^2 - (11/2)t + 1,$$

[6] Although they are named for Joseph Louis Lagrange (1736–1813), Lagrange basis polynomials were discovered in 1779 by Edward Waring (1736–1798). It may be interesting to some readers that Lagrange is Ian Parberry's PhD adviser's PhD adviser's,..., PhD adviser back 10 iterations.

[7] It's important to pronounce the name of this French mathematician "*luh-GRAWNGE*". Otherwise, people might think you are talking about the small Texas town of La Grange (pronounced "*luh-GRAYNGE*"). To the authors' knowledge, La Grange, Texas is not the namesake of any basis polynomials, although ZZ Top did name a song after the town in honor of a nearby brothel.

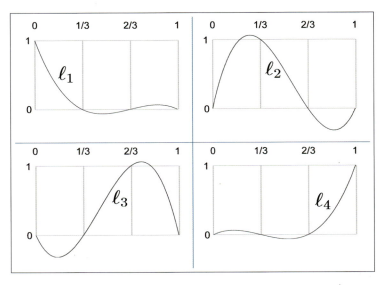

Figure 13.5. Cubic Lagrange basis polynomials for uniform knot vector

$$\ell_2(t) = (27/2)t^3 - (45/2)t^2 + 9t,$$
$$\ell_3(t) = -(27/2)t^3 + 18t^2 - (9/2)t,$$
$$\ell_4(t) = (9/2)t^3 - (9/2)t^2 + t.$$

Figure 13.5 shows what these basis polynomials look like.

Now that we have the Lagrange basis polynomials for the knot vector, let's plug in the y values from our example S curve (Figure 13.2) into Equation (13.7) to get the complete interpolating polynomial:

$$
\begin{aligned}
P(t) &= y_1\ell_1(t) + y_2\ell_2(t) + y_3\ell_3(t) + y_4\ell_4(t) \\
&= 2[-(9/2)t^3 + 9t^2 - (11/2)t + 1] + 3[(27/2)t^3 - (45/2)t^2 + 9t] \\
&\quad + 2[-(27/2)t^3 + 18t^2 - (9/2)t] + 3[(9/2)t^3 - (9/2)t^2 + t] \\
&= -9t^3 + 18t^2 - 11t + 2 + (81/2)t^3 - (135/2)t^2 + 27t \\
&\quad - 27t^3 + 36t^2 - 9t + (27/2)t^3 - (27/2)t^2 + 3t \\
&= 18t^3 - 27t^2 + 10t + 2.
\end{aligned}
$$

Let's show these results graphically. First, we scale each basis polynomial by the corresponding coordinate value, as shown in Figure 13.6.

Finally, adding the scaled basis vectors together yields the interpolating polynomial P, the blue curve at the top of Figure 13.7.

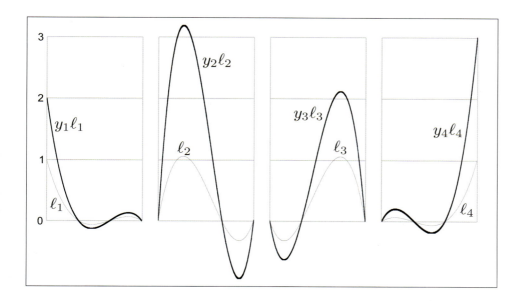

Figure 13.6. Scaling each Lagrange basis polynomial by the corresponding coordinate value

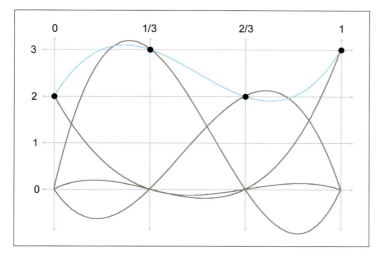

Figure 13.7. The interpolating curve is the sum of the scaled basis polynomials

We use the word *basis* in *basis polynomial* to emphasize the fact that we can use these polynomials as building blocks to reconstruct absolutely any polynomial whatsoever, given the values of the polynomial at the knots. It's the same basic concept as a basis vector (see Section 3.3.3): any arbitrary vector can be described as a linear combination of the basis vectors. In our case, the space being spanned by the basis is not a geometric 3D space, but the vector space of all possible polynomials of a certain degree, and the scale values for each curve are the known values of the polynomial at the knots.

But there's an alternate way to understand the multiplication and summing that's going on. Instead of thinking about the polynomials as the building blocks and the control points as the scale factors, we can view each point on the curve as a result of taking a weighted average of the control points, where the basis polynomials provide the blending weights. So the control points are the building blocks and the basis polynomials provide the scale factors, although we prefer to be more specific and call these scale factors *barycentric coordinates*. We introduced barycentric coordinates in the context of triangles in Section 9.6.3, but the term refers to a general technique of describing some value as a weighted average of data points.

We can think of basis polynomials as functions yielding barycentric coordinates (blending weights).

Notice that some values are negative or greater than 1 on certain intervals, which explains why direct polynomial interpolation overshoots the control points. When all barycentric coordinates are inside the $[0, 1]$ range, the resulting point is guaranteed to lie inside the *convex hull* of the control points. (The convex hull is the smallest polygon that contains all the control points. It "shrink wraps" the control points, sort of like if you were to stretch a rubber band around the control points and then release it.) But when we have any one coordinate outside this interval, the resulting point could extend outside the convex hull. For purposes of geometric curve design, the convex hull guarantee is a very nice one to have. Section 13.4 shows that Bézier curves *do* provide this guarantee through the *Bernstein basis*.

13.2.3   Polynomial Interpolation Summary

We've approached polynomial interpolation from two perspectives. Aitken's algorithm is a geometric approach based on repeated linear interpolation, and with it we can compute a point on the curve for a given t without knowing the polynomial for the curve. Lagrange interpolation works by creating basis functions that depend only on the knot vector. We can view the use of the basis polynomials in two ways. Either we can think about scaling each basis polynomial by the corresponding coordinate value and then adding them all together, or we can think about the polynomials as functions that compute barycentric coordinates that are used as blending weights in a simple weighted average of the coordinate points.

Both methods yield the same curve when given the same data. Furthermore, this polynomial is unique—no other polynomial of the same degree interpolates the data points. An informal argument for why this is true goes like this: A polynomial of degree n has $n + 1$ degrees of freedom, corresponding to the $n + 1$ coefficients in monomial form. Therefore, the degree n polynomial that interpolates $n + 1$ control points must be unique. (Farin [20, Section 6.2] gives a more rigorous argument.)

For purposes of curve design, polynomial interpolation is not ideal, primarily because of our inability to control the overshoot. The overshoot is guaranteed by the fact that the underlying Lagrange basis polynomials are not restricted to the unit interval $[0, 1]$, and the curve escapes the convex hull of the control points.

Direct polynomial interpolation finds limited application in video games, but our study has introduced the themes of repeated linear interpolation and basis polynomials. We've also seen a bit of the beautiful duality between the two techniques.

13.3   Hermite Curves

Polynomial interpolation tries to control the interior of the curve by threading the curve through specified knots. This doesn't work as well as we would like, because of the tendency to oscillate and overshoot, so let's try a different approach. We're still going to want to specify the endpoint positions, of course. But instead of specifying the interior positions to interpolate, let's control the shape of the curve through the tangents at the endpoints. A curve thus specified is said to be a *Hermite curve* or a curve in *Hermite form*, named in honor of Charles Hermite[8] (1822–1901).

The Hermite form specifies a curve by listing its starting and ending positions and derivatives. A cubic curve has only four coefficients, which allows for the specification of just the first derivatives, the velocities at the

---

[8]He's another French guy, and his mother probably pronounced his name "air-MEET." But many English speakers, even some we know with PhDs, pronounce it "HUR-mite," so you can probably do the same.

endpoints. So describing a cubic curve in Hermite form boils down to the following four pieces of information:

- The start position at $t = 0$,

- The first derivative (initial velocity) at $t = 0$,

- The end position at $t = 1$,

- The first derivative (final velocity) at $t = 1$.

Let's call the desired start and end positions $\mathbf{p}_0$ and $\mathbf{p}_1$ and the start and end velocities $\mathbf{v}_0$ and $\mathbf{v}_1$. Figure 13.8 shows some examples of cubic Hermite curves. Please note that the velocity vectors $\mathbf{v}_0$ and $\mathbf{v}_1$ have been drawn at one-third their actual length. One reason for doing this is to save space, and another will make sense later once we learn about Bézier curves in Section 13.4.

Determining the monomial coefficients from the Hermite values is a relatively straightforward algebraic process of combining equations previously discussed in this chapter. The four Hermite values can be translated into

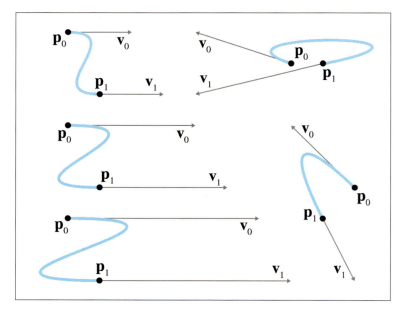

Figure 13.8. Some cubic Hermite curves

the following system of equations:

$$
\begin{aligned}
\mathbf{p}(0) = \mathbf{p}_0 \quad &\Longrightarrow & \mathbf{c}_0 &= \mathbf{p}_0, & (13.9) \\
\mathbf{v}(0) = \mathbf{v}_0 \quad &\Longrightarrow & \mathbf{c}_1 &= \mathbf{v}_0, & (13.10) \\
\mathbf{v}(1) = \mathbf{v}_1 \quad &\Longrightarrow & \mathbf{c}_1 + 2\mathbf{c}_2 + 3\mathbf{c}_3 &= \mathbf{v}_1, & (13.11) \\
\mathbf{p}(1) = \mathbf{p}_1 \quad &\Longrightarrow & \mathbf{c}_0 + \mathbf{c}_1 + \mathbf{c}_2 + \mathbf{c}_3 &= \mathbf{p}_1. & (13.12)
\end{aligned}
$$

System of equations for Hermite conditions

Equations (13.9) and (13.12), which specify the endpoints, just repeat what we said in Section 13.1.5. Equations (13.10) and (13.11), which specify velocities, follow directly from the velocity equations for a cubic polynomial (Equation (13.5) on page 652). The order in which these equations are listed is a convention used in other literature on curves, and the utility of this convention will become apparent later in this chapter.

Solving this system of equations results in a method to compute the monomial coefficients from the Hermite positions and derivatives:

$$
\begin{aligned}
\mathbf{c}_0 &= \mathbf{p}_0, & (13.13) \\
\mathbf{c}_1 &= \mathbf{v}_0, & (13.14) \\
\mathbf{c}_2 &= -3\mathbf{p}_0 - 2\mathbf{v}_0 - \mathbf{v}_1 + 3\mathbf{p}_1, & (13.15) \\
\mathbf{c}_3 &= 2\mathbf{p}_0 + \mathbf{v}_0 + \mathbf{v}_1 - 2\mathbf{p}_1. & (13.16)
\end{aligned}
$$

Converting Hermite form to monomial form

We can also write these equations in the compact matrix notation introduced in Section 13.1.2. Remember that when we put the coefficients as columns in a matrix $\mathbf{C}$, and the powers of t into the column vector $\mathbf{t}$, we can express a polynomial curve as the matrix product $\mathbf{Ct}$,

$$
\mathbf{p}(t) = \mathbf{Ct} = \begin{bmatrix} | & | & | & | \\ \mathbf{c}_0 & \mathbf{c}_1 & \mathbf{c}_2 & \mathbf{c}_3 \\ | & | & | & | \end{bmatrix} \begin{bmatrix} 1 \\ t \\ t^2 \\ t^3 \end{bmatrix},
$$

We can write monomial form using matrix notation, remember?

where $\mathbf{p}(t)$ and each of the coefficient vectors $\mathbf{c}_i$ are column vectors whose height matches the number of geometric dimensions (1D, 2D, or 3D). The height of $\mathbf{t}$ matches the number of $\mathbf{c}$'s, which depends on the degree of the curve.

The coefficient matrix $\mathbf{C}$ may be expressed as a matrix product by putting the Hermite positions and velocities as columns in a matrix $\mathbf{P}$ and multiplying by a conversion matrix $\mathbf{H}$:

$$
\mathbf{p}(t) = \mathbf{Ct} = \mathbf{PHt} = \begin{bmatrix} | & | & | & | \\ \mathbf{p}_0 & \mathbf{v}_0 & \mathbf{v}_1 & \mathbf{p}_1 \\ | & | & | & | \end{bmatrix} \begin{bmatrix} 1 & 0 & -3 & 2 \\ 0 & 1 & -2 & 1 \\ 0 & 0 & -1 & 1 \\ 0 & 0 & 3 & -2 \end{bmatrix} \begin{bmatrix} 1 \\ t \\ t^2 \\ t^3 \end{bmatrix}.
$$

Cubic Hermite curve using matrix notation

We can interpret the product $\mathbf{PHt}$ in two ways. If we group it like $\mathbf{P(Ht)}$, then the matrix product $\mathbf{Ht}$ can be interpreted as Hermite basis functions; we'll have more to say about this basis shortly. Or, we can think about $\mathbf{C} = \mathbf{PH}$, in which case, multiplication by $\mathbf{H}$ can be considered a conversion from the Hermite basis to the monomial basis, essentially a restatement of Equations (13.13)–(13.16).

We emphasize that the adjectives "monomial," "Hermite," and "Bézier" refer to different ways of describing the same set of polynomial curves; they are not different sets of curves. We convert a curve from Hermite form to monomial form by using Equations (13.13)–(13.16), and from monomial form to Hermite form with Equations (13.9)–(13.12).

Let's take a closer look at the Hermite basis and hopefully gain some geometric intuition as to why it works. Remember that we can interpret basis functions as functions of t yielding barycentric coordinates. For cubic Hermite curves, four values are being blended: the two positions and the two velocity vectors.[9] Thus, we have four basis functions that are the elements of the column result of the matrix product $\mathbf{Ht}$. Expanding the product, we have

$$\mathbf{p}(t) = \mathbf{P(Ht)}$$

$$= \begin{bmatrix} | & | & | & | \\ \mathbf{p}_0 & \mathbf{v}_0 & \mathbf{v}_1 & \mathbf{p}_1 \\ | & | & | & | \end{bmatrix} \left(\begin{bmatrix} 1 & 0 & -3 & 2 \\ 0 & 1 & -2 & 1 \\ 0 & 0 & -1 & 1 \\ 0 & 0 & 3 & -2 \end{bmatrix} \begin{bmatrix} 1 \\ t \\ t^2 \\ t^3 \end{bmatrix} \right)$$

$$= \begin{bmatrix} | & | & | & | \\ \mathbf{p}_0 & \mathbf{v}_0 & \mathbf{v}_1 & \mathbf{p}_1 \\ | & | & | & | \end{bmatrix} \begin{bmatrix} 1 - 3t^2 + 2t^3 \\ t - 2t^2 + t^3 \\ -t^2 + t^3 \\ 3t^2 - 2t^3 \end{bmatrix}.$$

Next, we name these basis functions (the rows of $\mathbf{Ht}$) as $H_0(t) \ldots H_3(t)$ (you may see these same functions indexed with different subscripts in other sources):

The cubic Hermite basis functions

$$H_0(t) = 1 - 3t^2 + 2t^3,$$

$$H_1(t) = t - 2t^2 + t^3,$$

$$H_2(t) = -t^2 + t^3,$$

$$H_3(t) = 3t^2 - 2t^3.$$

---

[9]If you're one of those purists who objects to the idea of "blending" points with vectors (see Section 2.4), don't worry. It's possible to interpret the equations such that the offensive comingling does not occur.

Now, expanding the matrix multiplication makes it explicit that these functions serve as blending weights:

$$\mathbf{p}(t) = \begin{bmatrix} | & | & | & | \\ \mathbf{p}_0 & \mathbf{v}_0 & \mathbf{v}_1 & \mathbf{p}_1 \\ | & | & | & | \end{bmatrix} \begin{bmatrix} H_0(t) \\ H_1(t) \\ H_2(t) \\ H_3(t) \end{bmatrix}$$

$$= H_0(t)\,\mathbf{p}_0 + H_1(t)\,\mathbf{v}_0 + H_2(t)\,\mathbf{v}_1 + H_3(t)\,\mathbf{p}_1.$$

Interpreting the Hermite basis functions as blending weights

Figure 13.9 shows a graph of the Hermite basis functions.

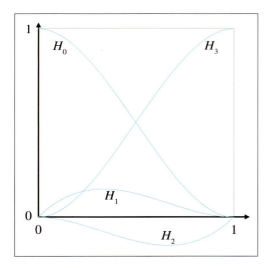

Figure 13.9
The Hermite basis functions

Now let's make a few observations. First, notice that $H_0(t) + H_3(t) = 1$, so those who object to the idea of adding "points" together can breath a sigh of relief, as we can interpret the situation as a proper barycentric combination of the points.

The curve $H_3(t)$ deserves special attention. It is also is known as the *smoothstep* function and is truly a gem that every game programmer should know. This function is found in many places, including the Renderman shading language and HLSL. To remove the rigid, robotic feeling from any linear interpolation (especially camera transitions), simply compute the normalized interpolation fraction t as usual (in the range $0 \leq t \leq 1$), and then replace t with $3t^2 - 2t^3$. *Voila!* Everything will suddenly feel more polished. The reason for this is that the smoothstep function eliminates the sudden jump in velocity at the endpoints: $H_3'(0) = H_3'(1) = 0$.

Smoothstep is Your Friend

The Hermite basis function $H_3(t)$ is also known as the smoothstep function. Almost any transition based on linear interpolation, especially a camera transition, feels better when replaced with the smoothstep function.

One final word about Hermite curves. Like the other forms for polynomial curves, it's possible to design a scheme for Hermite curves of higher degree, although the cubic polynomial is the most commonly used in computer graphics and animation. With the cubic spline, we specified the position (the "0th" derivative) and velocities (first derivatives) at the end points. A quintic (fifth-degree) Hermite curve happens when we also specify the accelerations (second derivatives).

13.4   Bézier Curves

This chapter has so far discussed a number of ideas about curves that were enlightening, but it has yet to describe a fully practical way to design a curve. All of that will change in this section.[10] Bézier curves were invented by Pierre Bézier (1910–1999), a French[11] engineer, while he was working for the automaker Renault. Bézier curves have many desirable properties that make them well suited for curve design. Importantly, Bézier curves *approximate* rather than interpolate: although they do pass through the first and last control points, they only pass near the interior points. For this reason, the Bézier control points are called "control points" rather than "knots." Some example cubic Bézier curves are shown in Figure 13.10.

Recall from Section 13.2 that the problem of polynomial interpolation had two solutions that produced the same result. Aitken's algorithm was a recursive construction technique that appealed to our geometric sensibilities, and a more abstract approach yielded the Lagrange basis polynomials. Bézier curves exhibit a similar duality. The counterpart of Aitken's algorithm for Bézier curves is the *de Casteljau* algorithm, a recursive geometric technique for constructing Bézier curves through repeated linear interpolation; this is the subject of Section 13.4.1. The analog to the Lagrange basis is the Bernstein basis, which is discussed in Section 13.4.2. After consider-

---

[10]Well, just some of that is going to change—we hope your reading will still be enlightening. You know what we mean.

[11]See, we told you a lot of these guys were French! By the way, it's pronounced "BEZ-ee-ay."

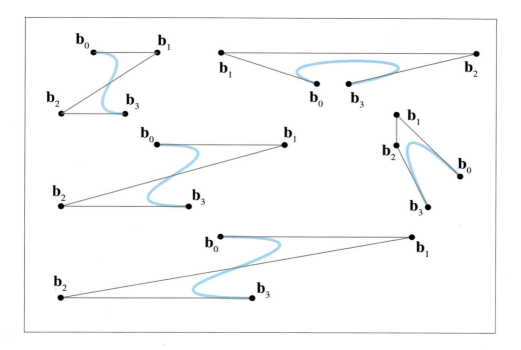

Figure 13.10
Some cubic Bézier curves

ing both sides of this coin, Section 13.4.3 investigates the derivatives[12] of Bézier curves and reveals the relationship to Hermite form.

   We've seen some beautiful cooperation between math and geometry in this book, but the convergence is particularly elegant for Bézier curves. It seems as if almost every important property of Bézier curves was independently discovered multiple times by researchers in different fields. Rogers' book [58] includes an interesting look at this story.

13.4.1   The de Casteljau Algorithm

The *de Casteljau algorithm* defines a method for constructing Bézier curves through repeated linear interpolation. It was created in 1959 by physicist and mathematician Paul de Casteljau (1910–1999).[13] We show how the algorithm works for the important cubic case as our example. First, a bit of notation is necessary. A cubic curve is defined by four control points, $b_0 \ldots b_3$. Notice that Bézier control points traditionally are indexed start-

---

[12] "Rate of exchange," if you will pardon the pun.
[13] Yep, he's French, too, and that means you'd better pronounce his name correctly: "duh CAS-tul-jho." He worked for Renault's rival, Citroen.

ing at zero (which will appeal to the C programmers amongst us). Also, as with Aitken's algorithm, we add a superscript to indicate the level of recursion. The original control points are assigned level 0, thus $\mathbf{b}_i^0 = \mathbf{b}_i$.

With that out of the way, let's consider a specific parameter value t from 0 to 1. The de Casteljau algorithm geometrically constructs the corresponding point on the curve $\mathbf{p}(t)$ as follows. Between each pair of consecutive control points, we interpolate according to the fraction t to obtain a new point. So, starting with the original four control points $\mathbf{b}_0^0 \dots \mathbf{b}_3^0$, we derive three new points $\mathbf{b}_0^1$, $\mathbf{b}_1^1$, and $\mathbf{b}_2^1$. Another round of interpolation between each pair of these three points gives us two points $\mathbf{b}_0^2$ and $\mathbf{b}_1^2$, and a final interpolation yields the point $\mathbf{b}_0^3 = \mathbf{p}(t)$ we're looking for. Figure 13.11 shows the de Casteljau algorithm applied to the same curve at $t = .25$, $t = .50$, and $t = .75$.

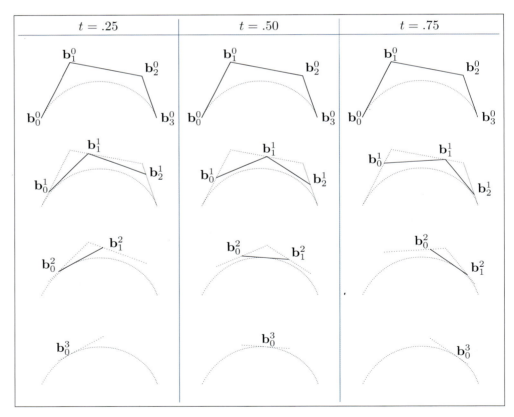

Figure 13.11
The de Casteljau algorithm applied to a cubic curve

It's helpful to write out all the **b**s in a triangular fashion, as shown in Figure 13.12. Each intermediate point is the result of linearly interpolating between two points on the row above.

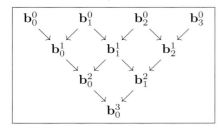

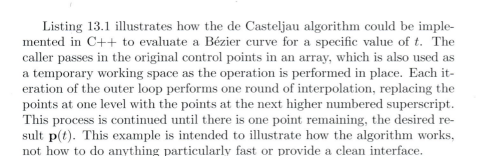

Figure 13.12
Hierarchical relationships in the de Casteljau algorithm for a cubic curve

If we combine these recursive relationships with the basic linear interpolation formula, we obtain the de Casteljau recurrence relation.

De Casteljau Recurrence Relation

$$\mathbf{b}_i^0(t) = \mathbf{b}_i,$$
$$\mathbf{b}_i^n(t) = (1 - t)[\mathbf{b}_i^{n-1}(t)] + t[\mathbf{b}_{i+1}^{n-1}(t)].$$

Listing 13.1 illustrates how the de Casteljau algorithm could be implemented in C++ to evaluate a Bézier curve for a specific value of t. The caller passes in the original control points in an array, which is also used as a temporary working space as the operation is performed in place. Each iteration of the outer loop performs one round of interpolation, replacing the points at one level with the points at the next higher numbered superscript. This process is continued until there is one point remaining, the desired result $\mathbf{p}(t)$. This example is intended to illustrate how the algorithm works, not how to do anything particularly fast or provide a clean interface.

```
Vector3 deCasteljau(
    int n,              // order of the curve, the number of points
    Vector3 points[],   // array of points.  Overwritten, as
                        // the algorithm works in place
    float t             // parameter value we wish to evaluate
) {

    // Perform the conversion in place
    while (n > 1) {
        --n;

        // Perform the next round of interpolation, reducing the
```

```
        // degree of the curve by one.
        for (int i = 0 ; i < n ; ++i) {
            points[i] = points[i]*(1.0f-t) + points[i+1]*t;
        }
    }

    // Result is now in the first slot.
    return points[0];
}
```

Listing 13.1
Evaluating a point on a Bézier curve using de Casteljau's algorithm

This gives us a method for locating a point at any given t through repeated interpolation, but it doesn't directly give us a closed form expression to calculate the point in terms of the control points. We emphasize that such a closed form expression is often not needed, but let's derive it in monomial form anyway. We're looking for a polynomial grouped by powers of t. We'll work our way up from the linear and quadratic cases to the cubic. Section 13.4.2 presents a general pattern leading us to the expression for arbitrary degree curves.

The linear case comes straight from the recurrence relation without any real work:

$$\mathbf{b}_i^0(t) = \mathbf{b}_i,$$
$$\mathbf{b}_i^1(t) = (1-t)[\mathbf{b}_i^0(t)] + t[\mathbf{b}_{i+1}^0(t)]$$
$$= (1-t)\mathbf{b}_i + t\mathbf{b}_{i+1}$$
$$= \mathbf{b}_i + t(\mathbf{b}_{i+1} - \mathbf{b}_i).$$

Applying one more level gives us a quadratic polynomial:

$$\mathbf{b}_i^2(t) = (1-t)[\mathbf{b}_i^1(t)] + t[\mathbf{b}_{i+1}^1(t)]$$
$$= (1-t)[\mathbf{b}_i + t(\mathbf{b}_{i+1} - \mathbf{b}_i)] + t[\mathbf{b}_{i+1} + t(\mathbf{b}_{i+2} - \mathbf{b}_{i+1})]$$
$$= [\mathbf{b}_i + t(\mathbf{b}_{i+1} - \mathbf{b}_i)] - t[\mathbf{b}_i + t(\mathbf{b}_{i+1} - \mathbf{b}_i)]$$
$$\quad + t[\mathbf{b}_{i+1} + t(\mathbf{b}_{i+2} - \mathbf{b}_{i+1})]$$
$$= \mathbf{b}_i + t(\mathbf{b}_{i+1} - \mathbf{b}_i) - t\mathbf{b}_i - t^2(\mathbf{b}_{i+1} - \mathbf{b}_i)$$
$$\quad + t\mathbf{b}_{i+1} + t^2(\mathbf{b}_{i+2} - \mathbf{b}_{i+1})$$
$$= \mathbf{b}_i + t(2\mathbf{b}_{i+1} - 2\mathbf{b}_i) + t^2(\mathbf{b}_i - 2\mathbf{b}_{i+1} + \mathbf{b}_{i+2}).$$

In other words, quadratic Bézier curves, which have three control points, can be expressed in monomial form as

$$\mathbf{p}(t) = \mathbf{b}_0^2(t) = \mathbf{b}_0 + t(2\mathbf{b}_1 - 2\mathbf{b}_0) + t^2(\mathbf{b}_0 - 2\mathbf{b}_1 + \mathbf{b}_2). \tag{13.17}$$

Quadratic Bézier curve in monomial form

Before we do the last round of interpolation to get the cubic curve, let's take a closer look at the quadratic expression in Equation (13.17). This conversion from Bézier form to monomial basis can be written with fewer letters by using the matrix form introduced earlier in this chapter. After putting the control points $\mathbf{b}_0$, $\mathbf{b}_1$, $\mathbf{b}_2$ as columns into a matrix $\mathbf{B}$, we can write

$$\mathbf{p}(t) = \mathbf{Ct} = \mathbf{BMt} = \begin{bmatrix} | & | & | \\ \mathbf{b}_0 & \mathbf{b}_1 & \mathbf{b}_2 \\ | & | & | \end{bmatrix} \begin{bmatrix} 1 & -2 & 1 \\ 0 & 2 & -2 \\ 0 & 0 & 1 \end{bmatrix} \begin{bmatrix} 1 \\ t \\ t^2 \end{bmatrix}. \quad (13.18)$$

Quadratic Bézier curve using matrix notation

As we saw in Section 13.3 with Hermite curves, the two different ways to group the product $\mathbf{BMt}$ lead to two different interpretations. If we perform the multiplication $\mathbf{BM}$ first, we get the matrix of monomial coefficients $\mathbf{C}$, meaning $\mathbf{M}$ is a conversion matrix from Bézier form to monomial form. Direct evaluation of the monomial form is faster than implementing the de Casteljau algorithm, and so this form might be preferable in situations where we need to evaluate the same curve for many different values of t, for example, when moving an object over time along a path described by a Bézier curve. (However, one must be aware of issues related to precision. For example, we can ensure that performing de Casteljau using $t = 1.0$ produces a result that matches the last control point *exactly*. However, substituting $t = 1.0$ into the polynomial in monomial form, the coefficients might not sum exactly to this value due to floating point representation.)

The other way to group the product $\mathbf{BMt}$ is to perform the right-hand multiplication first: $\mathbf{B(Mt)}$. When we plug in a specific value of t, the product $\mathbf{Mt}$ yields a column vector of barycentric coordinates. If we perform this multiplication leaving t as a variable, we get a column vector of polynomials that can be interpreted as a basis. The basis polynomials for Bézier curves are the Bernstein basis, discussed in Section 13.4.2.

Back to repeated interpolation. One last round gives us the cubic polynomial:

$$\begin{aligned} \mathbf{b}_i^3(t) &= (1-t)[\mathbf{b}_i^2(t)] + t[\mathbf{b}_{i+1}^2(t)] \\ &= (1-t)[\mathbf{b}_i + t(2\mathbf{b}_{i+1} - 2\mathbf{b}_i) + t^2(\mathbf{b}_i - 2\mathbf{b}_{i+1} + \mathbf{b}_{i+2})] \\ &\quad + t[\mathbf{b}_{i+1} + t(2\mathbf{b}_{i+2} - 2\mathbf{b}_{i+1}) + t^2(\mathbf{b}_{i+1} - 2\mathbf{b}_{i+2} + \mathbf{b}_{i+3})] \\ &= [\mathbf{b}_i + t(2\mathbf{b}_{i+1} - 2\mathbf{b}_i) + t^2(\mathbf{b}_i - 2\mathbf{b}_{i+1} + \mathbf{b}_{i+2})] \\ &\quad - t[\mathbf{b}_i + t(2\mathbf{b}_{i+1} - 2\mathbf{b}_i) + t^2(\mathbf{b}_i - 2\mathbf{b}_{i+1} + \mathbf{b}_{i+2})] \\ &\quad + t[\mathbf{b}_{i+1} + t(2\mathbf{b}_{i+2} - 2\mathbf{b}_{i+1}) + t^2(\mathbf{b}_{i+1} - 2\mathbf{b}_{i+2} + \mathbf{b}_{i+3})] \\ &= \mathbf{b}_i + t(2\mathbf{b}_{i+1} - 2\mathbf{b}_i) + t^2(\mathbf{b}_i - 2\mathbf{b}_{i+1} + \mathbf{b}_{i+2}) \\ &\quad - t\mathbf{b}_i - t^2(2\mathbf{b}_{i+1} - 2\mathbf{b}_i) - t^3(\mathbf{b}_i - 2\mathbf{b}_{i+1} + \mathbf{b}_{i+2}) \\ &\quad + t\mathbf{b}_{i+1} + t^2(2\mathbf{b}_{i+2} - 2\mathbf{b}_{i+1}) + t^3(\mathbf{b}_{i+1} - 2\mathbf{b}_{i+2} + \mathbf{b}_{i+3}) \end{aligned}$$

One last iteration of de Casteljau iteration yields the cubic polynomial.

Whew, expanding it all out like this is pretty exhausting!

$$= \mathbf{b}_i + t(3\mathbf{b}_{i+1} - 3\mathbf{b}_i) + t^2(3\mathbf{b}_i - 6\mathbf{b}_{i+1} + 3\mathbf{b}_{i+2})$$
$$+ t^3(-\mathbf{b}_i + 3\mathbf{b}_{i+1} - 3\mathbf{b}_{i+2} + \mathbf{b}_{i+3}).$$

Writing the last line again, but this time assuming the cubic level is the final level of recursion, we have

Cubic Bézier curve in monomial form

$$\mathbf{p}(t) = \mathbf{b}_0^3(t) = \mathbf{b}_0 + t(3\mathbf{b}_1 - 3\mathbf{b}_0) + t^2(3\mathbf{b}_0 - 6\mathbf{b}_1 + 3\mathbf{b}_2)$$
$$+ t^3(-\mathbf{b}_0 + 3\mathbf{b}_1 - 3\mathbf{b}_2 + \mathbf{b}_3). \tag{13.19}$$

Just to make sure you didn't miss it, Equation (13.19) tells us how to convert a cubic Bézier curve to monomial form. Since this is important, let's write it a bit more deliberately as

Cubic monomial coefficients from Bézier control points

$$\mathbf{c}_0 = \mathbf{b}_0,$$
$$\mathbf{c}_1 = -3\mathbf{b}_0 + 3\mathbf{b}_1,$$
$$\mathbf{c}_2 = 3\mathbf{b}_0 - 6\mathbf{b}_1 + 3\mathbf{b}_2,$$
$$\mathbf{c}_3 = -\mathbf{b}_0 + 3\mathbf{b}_1 - 3\mathbf{b}_2 + \mathbf{b}_3.$$

We can now put this conversion into a matrix like we did with the quadratic case in Equation (13.18). The cubic equation for a specific point on the curve $\mathbf{p}(t)$ is written in matrix notation as

Cubic Bézier curve using matrix notation

$$\mathbf{p}(t) = \mathbf{Ct} = \mathbf{BMt} = \begin{bmatrix} | & | & | & | \\ \mathbf{b}_0 & \mathbf{b}_1 & \mathbf{b}_2 & \mathbf{b}_3 \\ | & | & | & | \end{bmatrix} \begin{bmatrix} 1 & -3 & 3 & -1 \\ 0 & 3 & -6 & 3 \\ 0 & 0 & 3 & -3 \\ 0 & 0 & 0 & 1 \end{bmatrix} \begin{bmatrix} 1 \\ t \\ t^2 \\ t^3 \end{bmatrix}.$$

We can also invert this process, meaning we can convert any polynomial curve from monomial form to Bézier form. Given any polynomial curve, the Bézier control points that describe the curve are uniquely determined:

Computing Bézier control points from monomial coefficients

$$\mathbf{b}_0 = \mathbf{c}_0, \tag{13.20}$$
$$\mathbf{b}_1 = \mathbf{c}_0 + (1/3)\mathbf{c}_1, \tag{13.21}$$
$$\mathbf{b}_2 = \mathbf{c}_0 + (2/3)\mathbf{c}_1 + (1/3)\mathbf{c}_2, \tag{13.22}$$
$$\mathbf{b}_3 = \mathbf{c}_0 + \mathbf{c}_1 + \mathbf{c}_2 + \mathbf{c}_3. \tag{13.23}$$

And, of course, we can write this in matrix form:

Converting from monomial to Bézier form, in matrix notation

$$\begin{bmatrix} | & | & | & | \\ \mathbf{b}_0 & \mathbf{b}_1 & \mathbf{b}_2 & \mathbf{b}_3 \\ | & | & | & | \end{bmatrix} = \begin{bmatrix} | & | & | & | \\ \mathbf{c}_0 & \mathbf{c}_1 & \mathbf{c}_2 & \mathbf{c}_3 \\ | & | & | & | \end{bmatrix} \begin{bmatrix} 1 & 1 & 1 & 1 \\ 0 & 1/3 & 2/3 & 1 \\ 0 & 0 & 1/3 & 1 \\ 0 & 0 & 0 & 1 \end{bmatrix}.$$

13.4.2 The Bernstein Basis

Section 13.4.1 ended with a bit of algebra to calculate the polynomial for a curve from the Bézier control points. This polynomial was expressed in monomial form, meaning the coefficients were for the powers of t. We can also write the polynomial in Bézier form by collecting the terms on the control points rather than the powers of t. When written this way, each control point has a coefficient that represents the barycentric weight as a function of t that the control point contributes to the curve.

Let's repeat the algebra exercise from Section 13.4.1, only this time we'll be writing things in a slightly different way that will lead us to some observations. As we did before, we start with the linear case (remember, $\mathbf{b}_i^0 = \mathbf{b}_i$):

$$\begin{aligned} \mathbf{b}_i^1(t) &= (1-t)[\mathbf{b}_i^0(t)] + t[\mathbf{b}_{i+1}^0(t)] \\ &= (1-t)\mathbf{b}_i + t\mathbf{b}_{i+1}. \end{aligned}$$

Next comes the quadratic:

$$\begin{aligned} \mathbf{b}_i^2(t) &= (1-t)\mathbf{b}_i^1(t) + t\mathbf{b}_{i+1}^1(t) \\ &= (1-t)[(1-t)\mathbf{b}_i + t\mathbf{b}_{i+1}] + t[(1-t)\mathbf{b}_{i+1} + t\mathbf{b}_{i+2}] \\ &= (1-t)^2\mathbf{b}_i + t(1-t)\mathbf{b}_{i+1} + t(1-t)\mathbf{b}_{i+1} + t^2\mathbf{b}_{i+2} \\ &= (1-t)^2\mathbf{b}_i + 2t(1-t)\mathbf{b}_{i+1} + t^2\mathbf{b}_{i+2}. \end{aligned}$$

And finally, we have the cubic case:

$$\begin{aligned} \mathbf{b}_i^3(t) &= (1-t)[\mathbf{b}_i^2(t)] + t[\mathbf{b}_{i+1}^2(t)] \\ &= (1-t)[(1-t)^2\mathbf{b}_i + 2t(1-t)\mathbf{b}_{i+1} + t^2\mathbf{b}_{i+2}] \\ &\quad + t[(1-t)^2\mathbf{b}_{i+1} + 2t(1-t)\mathbf{b}_{i+2} + t^2\mathbf{b}_{i+3}] \\ &= (1-t)^3\mathbf{b}_i + 2t(1-t)^2\mathbf{b}_{i+1} + t^2(1-t)\mathbf{b}_{i+2} \\ &\quad + t(1-t)^2\mathbf{b}_{i+1} + 2t^2(1-t)\mathbf{b}_{i+2} + t^3\mathbf{b}_{i+3} \\ &= (1-t)^3\mathbf{b}_i + 3t(1-t)^2\mathbf{b}_{i+1} + 3t^2(1-t)\mathbf{b}_{i+2} + t^3\mathbf{b}_{i+3}. \end{aligned}$$

You might see a pattern emerging, but just to make it even more clear, let's show the curves up to degree 5 (we'll skip over the algebra; it's similar to what we did above):

$$\mathbf{b}_0^1(t) = (1-t)\mathbf{b}_0 + t\mathbf{b}_1, \tag{13.24}$$

$$\mathbf{b}_0^2(t) = (1-t)^2\mathbf{b}_0 + 2t(1-t)\mathbf{b}_1 + t^2\mathbf{b}_2, \tag{13.25}$$

$$\mathbf{b}_0^3(t) = (1-t)^3\mathbf{b}_0 + 3t(1-t)^2\mathbf{b}_1 + 3t^2(1-t)\mathbf{b}_2 + t^3\mathbf{b}_3, \tag{13.26}$$

Bézier curves of degree 1–5

$$\mathbf{b}_0^4(t) = (1 - t)^4 \mathbf{b}_0 + 4t(1 - t)^3 \mathbf{b}_1 + 6t^2(1 - t)^2 \mathbf{b}_2$$
$$+ 4t^3(t - 1)\mathbf{b}_3 + t^4 \mathbf{b}_4, \tag{13.27}$$

$$\mathbf{b}_0^5(t) = (1 - t)^5 \mathbf{b}_0 + 5t(1 - t)^4 \mathbf{b}_1 + 10t^2(1 - t)^3 \mathbf{b}_2$$
$$+ 10t^3(1 - t)^2 \mathbf{b}_3 + 5t^4(1 - t)\mathbf{b}_4 + t^5 \mathbf{b}_5. \tag{13.28}$$

Now the pattern is more clear. Each term has a constant coefficient, a power of $(1 - t)$, and a power of t. The powers of t are numbered in increasing order, so $\mathbf{b}_i$ has a coefficient t^i. The powers of $(1 - t)$ follow the opposite pattern and are numbered in decreasing order.

The pattern for the constant coefficients is a bit more complicated. Please permit a brief, but hopefully interesting, detour into combinatorics. Let's write out the first eight levels in a triangular form to make the pattern a bit easier to see:

Pascal's triangle

```
0                           1
1                       1       1
2                    1      2      1
3                 1     3      3      1
4              1     4     6      4     1
5           1     5    10     10     5     1
6        1     6    15     20     15    6     1
7     1     7    21     35     35    21    7    1
```

With the exception of the 1s on the outer edge of the triangle, all other numbers are the sum of the two numbers above it. You are looking at a very famous number pattern that has been studied for centuries, known as the *binomial coefficients* because the nth row gives the coefficients when expanding the binomial $(a + b)^n$. The compulsion to organize these numbers in a triangular manner like this has struck many people, including the mathematician and physicist Blaise Pascal (1623–1662).[14] This triangular arrangement of the binomial coefficients is known as *Pascal's triangle*.[15]

Binomial coefficients have a special notation. We can refer to the kth number on row n in Pascal's triangle (where the indexing starts at 0 for both n and k) using binomial coefficient notation as

Binomial coefficient notation

$$\binom{n}{k}.$$

---

[14] Yes, he was French, too. He appears in Ian Parberry's PhD adviser tree somewhat off to the left back 16 generations.

[15] In addition to his triangle, Pascal has an SI unit of pressure, a law, a programming language, and a wager named after him, although the latter two are no longer in serious use.

For example, $\binom{6}{2} = 15$. We read $\binom{n}{k}$ as "n choose k," because the value of $\binom{n}{k}$ also happens to be the number of subsets of k objects that can be chosen from a set of n objects, disregarding the order.

Now let's look at the general formula for computing binomial coefficients. (We emphasize that this formula is primarily for entertainment purposes, since our use of binomial coefficients in this chapter on curves will be restricted to the first few lines of Pascal's triangle.) Remember from Section 11.4.6 the *factorial* operator, denoted $n!$, which is the product of all the whole numbers up to and including n:

$$n! = \prod_{i=1}^{n} i = 1 \times 2 \times 3 \times \cdots \times n.$$

Factorial operator

Using factorials, and defining $0! \equiv 1$, we compute a binomial coefficient as

$$\binom{n}{k} = \frac{n!}{k!(n-k)!}.$$

Binomial coefficient

Binomial coefficients arise frequently in applications dealing with combinations and permutations, such as probability and analysis of algorithms. Because of their importance, and the amazingly large number of patterns that can be found in them, they have been the subject of quite a large amount of study. A very thorough discussion of binomial coefficients, especially regarding their use in computer algorithms, is presented by Knuth [39].

Back to curves. We've analyzed the pattern of the barycentric weights. Now let's rewrite a Bézier curve, replacing each control point weight with a function $B_i^n(t)$, and using the cubic curve formula (Equation (13.26)) as our example:

$$\begin{aligned} \mathbf{b}_0^3(t) &= (1-t)^3 \mathbf{b}_0 + 3t(1-t)^2 \mathbf{b}_1 + 3t^2(1-t)\mathbf{b}_2 + t^3 \mathbf{b}_3 \\ &= [B_0^3(t)]\mathbf{b}_0 + [B_1^3(t)]\mathbf{b}_1 + [B_2^3(t)]\mathbf{b}_2 + [B_3^3(t)]\mathbf{b}_3. \end{aligned}$$

More generally, we can write a Bézier curve of degree n (having $n+1$ control points) as

$$\mathbf{b}_0^n(t) = \sum_{i=0}^{n} [B_i^n(t)]\mathbf{b}_i.$$

Bézier curve of arbitrary degree

The function $B_i^n(t)$ is a *Bernstein polynomial*, named after Sergei Bernstein (1880–1968).[16] We've already figured out the pattern of these polynomials, but here's the precise formula:

Bernstein polynomial

$$B_i^n(t) = \binom{n}{i} t^i (1-t)^{n-i} , \qquad\qquad 0 \leq i \leq n.$$

Figure 13.13 shows the graphs for the Bernstein polynomials up to the quartic case.

The properties of the Bernstein polynomials tell us a lot about how Bézier curves behave. Let's discuss a few properties in particular.

Sum to one. The Bernstein polynomials sum to unity for all values of t, which is nice because if they didn't, then they wouldn't define proper barycentric coordinates. This fact is not immediately obvious, neither from visual inspection of Figure 13.13 nor from a cursory examination of the equations, but it can be proven. If you relish the idea of working through such a proof for the quadratic case, check out Exercise 4.

Convex hull property. The range of the Bernstein polynomials is $0 \ldots 1$ for the entire length of the curve, $0 \leq t \leq 1$. Combined with the previous property, this means that Bézier curves obey the *convex hull* property: the curve is bounded to stay within the convex hull of the control points. Compare this with the Lagrange basis polynomials, which do not stay within the $[0, 1]$ interval, causing polynomial interpolation to *not* obey the convex hull property. One manifestation of this is the undesirable "overshooting" witnessed in Figure 13.4.

Endpoints interpolated. The first and last polynomials attain unity when we need them to. Because $B_0^n(0) = 1$ and $B_n^n(1) = 1$, the curve touches the endpoints. Notice that $t = 0$ and $t = 1$ are the *only* places where any of the basis polynomials reach 1, which is why the other control points are only approximated and not interpolated.

Global support. All the polynomials are nonzero on the open interval $(0, 1)$—that is, the entire curve excluding the endpoints. The region where the blending weight for a control point is nonzero is called the *support* of the control point. Wherever the control point has support, it exerts some influence on the curve.

Bézier control points have *global support* because the Bernstein polynomials are nonzero everywhere other than the endpoints. The practical result is that when any one control point is moved, the entire curve is affected. This is *not* a desirable property for curve design. Once we have a

---

[16]Russian, not French.

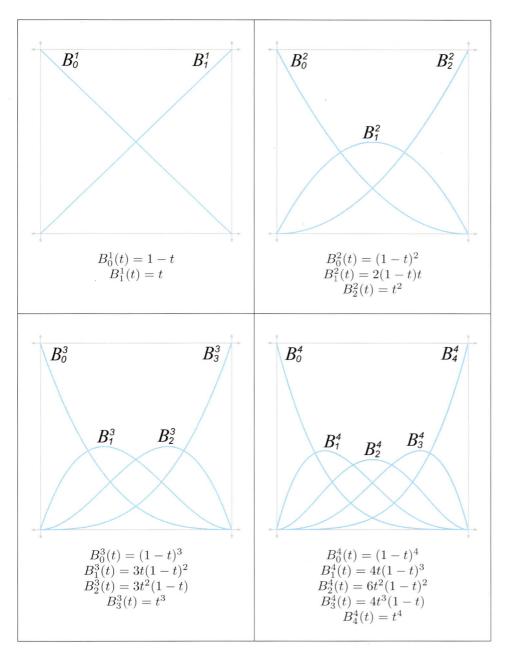

Figure 13.13
Bernstein polynomials of degrees 1–4

section of the curve that looks how we want, we would prefer that editing of some other distant control point not disturb the section that was shaped the way we liked it. This envious situation, known as *local support*, occurs when we move a particular control point and only the part of the curve near that control point is affected, for some definition of "near."

Local support means that the basis function is nonzero only in some interval, and outside this interval it is zero. Unfortunately, such a basis function cannot be described as a polynomial, and thus no polynomial curve can achieve local control. However, local support is possible by piecing together small curves that fit together just right to form a *spline*, as Section 13.6 discusses.

One local maximum. Although each control point exercises influence over the entire curve, each exerts the *most* influence at *one* particular point along the curve. Each Bernstein polynomial $B_i^n(t)$, which serves as the blend weight for the control point $\mathbf{b}_i$, has one maximum at the auspicious time $t = i/n$. Furthermore, at that time, $\mathbf{b}_i$ exerts more weight than any other control point.

Thus, although every point on the interior of the curve is influenced to some degree by all the control points (because Bézier control points have global support), the nearest control point has the *most* influence.

13.4.3   Bézier Derivatives and Their Relationship to the Hermite Form

Let's take a look at the derivatives of a Bézier curve. Since we like to use the cubic curve as our example, we're talking about the velocity and acceleration of the curve. Remember that the velocity is related to the tangent (direction) of the curve, and the acceleration is related to its curvature.

Section 13.1.6 showed how to get the velocity function of a curve from the monomial coefficients:

Position and velocity of a cubic curve

$$\mathbf{p}(t) = \mathbf{c}_0 + \mathbf{c}_1 t + \mathbf{c}_2 t^2 + \mathbf{c}_3 t^3,$$
$$\mathbf{v}(t) = \dot{\mathbf{p}}(t) = \mathbf{c}_1 + 2\mathbf{c}_2 t + 3\mathbf{c}_3 t^2. \tag{13.29}$$

And Section 13.4.1 showed how to extract the monomial coefficients from a cubic Bézier curve:

$$\mathbf{c}_0 = \mathbf{b}_0,$$
$$\mathbf{c}_1 = 3\mathbf{b}_1 - 3\mathbf{b}_0,$$
$$\mathbf{c}_2 = 3\mathbf{b}_0 - 6\mathbf{b}_1 + 3\mathbf{b}_2,$$
$$\mathbf{c}_3 = -\mathbf{b}_0 + 3\mathbf{b}_1 - 3\mathbf{b}_2 + \mathbf{b}_3.$$

Plugging these coefficients into the velocity formula (Equation (13.29)), we obtain a formula for the instantaneous velocity of a curve in terms of the Bézier control points:

$$\mathbf{v}(t) = \mathbf{c}_1 + 2\mathbf{c}_2 t + 3\mathbf{c}_3 t^2$$
$$= (3\mathbf{b}_1 - 3\mathbf{b}_0) + 2(3\mathbf{b}_0 - 6\mathbf{b}_1 + 3\mathbf{b}_2)t + 3(-\mathbf{b}_0 + 3\mathbf{b}_1 - 3\mathbf{b}_2 + \mathbf{b}_3)t^2.$$

First derivative (velocity) of a cubic Bézier curve

Now consider the velocity at the endpoints $t = 0$ and $t = 1$:

$$\mathbf{v}(0) = (3\mathbf{b}_1 - 3\mathbf{b}_0) + 2(3\mathbf{b}_0 - 6\mathbf{b}_1 + 3\mathbf{b}_2)(0)$$
$$+ 3(-\mathbf{b}_0 + 3\mathbf{b}_1 - 3\mathbf{b}_2 + \mathbf{b}_3)(0)^2$$
$$= 3(\mathbf{b}_1 - \mathbf{b}_0), \tag{13.30}$$
$$\mathbf{v}(1) = (3\mathbf{b}_1 - 3\mathbf{b}_0) + 2(3\mathbf{b}_0 - 6\mathbf{b}_1 + 3\mathbf{b}_2)(1)$$
$$+ 3(-\mathbf{b}_0 + 3\mathbf{b}_1 - 3\mathbf{b}_2 + \mathbf{b}_3)(1)^2$$
$$= 3\mathbf{b}_1 - 3\mathbf{b}_0 + 6\mathbf{b}_0 - 12\mathbf{b}_1 + 6\mathbf{b}_2 - 3\mathbf{b}_0 + 9\mathbf{b}_1 - 9\mathbf{b}_2 + 3\mathbf{b}_3$$
$$= 3(\mathbf{b}_3 - \mathbf{b}_2). \tag{13.31}$$

Velocity at the endpoints of a cubic Bézier curve

This is interesting. Observe that $\mathbf{b}_1 - \mathbf{b}_0$ gives us the vector from the first control point to the second control point, and $\mathbf{b}_3 - \mathbf{b}_2$ is the vector from the third control point to the last control point. So the tangent at the start of the curve at $t = 0$ is "aimed towards" the first control point, and the tangent at the end of the curve at $t = 1$ is "aimed towards" the third control point. (Actually, the tangent at $t = 1$ points directly *away* from the third control point, if we think about moving along the curve in the direction of increasing t). This is a key point.

The first edge of the Bézier control polygon completely determines the tangent at the start of the curve, and the last edge determines the tangent at the end of the curve.

Another way to illustrate the role of the middle control points in a cubic Bézier curve is to examine the relationship between the Bézier and Hermite forms. Remember that the cubic Hermite form contains the initial position $\mathbf{p}_0$ and velocity $\mathbf{p}_1$ and the final position $\mathbf{p}_1$ and velocity $\mathbf{v}_1$. Now that we know the relationship between the Bézier control points and the curve velocity, it's easy to convert from Bézier to Hermite form:

Converting cubic curve
from Bézier form to
Hermite form

$$\mathbf{p}_0 = \mathbf{b}_0, \tag{13.32}$$

$$\mathbf{v}_0 = 3(\mathbf{b}_1 - \mathbf{b}_0), \tag{13.33}$$

$$\mathbf{v}_1 = 3(\mathbf{b}_3 - \mathbf{b}_2), \tag{13.34}$$

$$\mathbf{p}_1 = \mathbf{b}_3. \tag{13.35}$$

Or, we can convert from Hermite to Bézier:

Converting cubic curve
from Hermite form to
Bézier form

$$\mathbf{b}_0 = \mathbf{p}_0,$$

$$\mathbf{b}_1 = \mathbf{p}_0 + (1/3)\mathbf{v}_0,$$

$$\mathbf{b}_2 = \mathbf{p}_1 - (1/3)\mathbf{v}_1,$$

$$\mathbf{b}_3 = \mathbf{p}_1.$$

Thus, Hermite and Bézier forms are very closely related, and it is very easy to convert between them. Their relationship is depicted graphically in Figure 13.14.

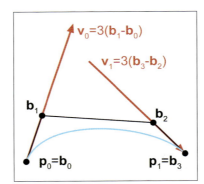

Figure 13.14
Relationship between Bézier and Hermite forms

We've said that the first derivative at either endpoint is completely determined by the nearest two Bézier control points. We can actually make a more general statement. The nth derivative at either endpoint is completely determined by the nearest $n + 1$ control points. The "0th derivative" (the position of the curve) is completely determined by the interpolated control point. The first derivative has been discussed. The second derivative (acceleration) at the end of the curve is determined by the closest three control points. In fact, let's see exactly what the acceleration is in terms of the Bézier control points for a cubic curve. Converting the acceleration function (Equation (13.6)) from monomial to Bézier form, we get

$$\mathbf{a}(t) = 2\mathbf{c}_2 + 6\mathbf{c}_3 t$$
$$= 2(3\mathbf{b}_0 - 6\mathbf{b}_1 + 3\mathbf{b}_2) + 6(-\mathbf{b}_0 + 3\mathbf{b}_1 - 3\mathbf{b}_2 + \mathbf{b}_3)t$$
$$= (6\mathbf{b}_0 - 12\mathbf{b}_1 + 6\mathbf{b}_2) + (-6\mathbf{b}_0 + 18\mathbf{b}_1 - 18\mathbf{b}_2 + 6\mathbf{b}_3)t.$$

Acceleration of a cubic Bézier curve

At the endpoints, the acceleration is given by

$$\mathbf{a}(0) = (6\mathbf{b}_0 - 12\mathbf{b}_1 + 6\mathbf{b}_2) + (-6\mathbf{b}_0 + 18\mathbf{b}_1 - 18\mathbf{b}_2 + 6\mathbf{b}_3)0$$
$$= 6\mathbf{b}_0 - 12\mathbf{b}_1 + 6\mathbf{b}_2,$$
$$\mathbf{a}(1) = (6\mathbf{b}_0 - 12\mathbf{b}_1 + 6\mathbf{b}_2) + (-6\mathbf{b}_0 + 18\mathbf{b}_1 - 18\mathbf{b}_2 + 6\mathbf{b}_3)1$$
$$= 6\mathbf{b}_1 - 12\mathbf{b}_2 + 6\mathbf{b}_3.$$

Acceleration of a cubic Bézier curve at the endpoints

As expected, the acceleration at the start is completely determined by the first three control points, and the acceleration at the end is determined by the last three control points.

Let's define $\mathbf{d}_i = \mathbf{b}_{i+1} - \mathbf{b}_i$ as shorthand for the delta between consecutive control points, the vector of the ith edge of the Bézier control polygon. With this notation, the acceleration formulas bear a striking resemblance to the velocity formulas:

$$\mathbf{a}(0) = 6\mathbf{b}_0 - 12\mathbf{b}_1 + 6\mathbf{b}_2 = 6\mathbf{b}_0 - 6\mathbf{b}_1 - 6\mathbf{b}_1 + 6\mathbf{b}_2$$
$$= 6\left((\mathbf{b}_2 - \mathbf{b}_1) - (\mathbf{b}_1 - \mathbf{b}_0)\right)$$
$$= 6(\mathbf{d}_1 - \mathbf{d}_0), \tag{13.36}$$
$$\mathbf{a}(1) = 6\mathbf{b}_1 - 12\mathbf{b}_2 + 6\mathbf{b}_3 = 6\mathbf{b}_1 - 6\mathbf{b}_2 - 6\mathbf{b}_2 + 6\mathbf{b}_3$$
$$= 6\left((\mathbf{b}_3 - \mathbf{b}_2) - (\mathbf{b}_2 - \mathbf{b}_1)\right)$$
$$= 6(\mathbf{d}_2 - \mathbf{d}_1). \tag{13.37}$$

Acceleration of a cubic Bézier curve at the endpoints, in terms of the delta between consecutive control points

The above discussion applies to Bézier curves of any degree. In general, the pattern is this: if we move control point $\mathbf{b}_i$, we affect the ith derivative and higher at the start of the curve, but not lower-numbered derivatives. (Similar statements apply at the end of the curve, regarding control point $\mathbf{b}_i$ and the derivative $n-i$ and higher.) Of course, for a cubic spline that's just about the end of the story, since we cannot move *any* control point without potentially changing the third derivative at every point on the spline, since the third derivative is constant for a cubic, and all higher derivatives are zero. We come back to these ideas in Section 13.8.1 when we talk about the continuity conditions of two or more Bézier curve segments joined in a spline.

13.5   Subdivision

Beginning with Section 13.6, this chapter addresses the topic of joining together curves into a *spline*, which we can make as long and as complex as we want. Before we do that, this section considers the opposite problem: how to take a curve and chop it up into smaller pieces.

Why would we ever want to do this? There are a couple of reasons.

- *Curve refinement.* In the process of designing a curve interactively, we may find that we *almost* have the shape we want, but one curve can't quite give us the flexibility that we need. So we cut the curve into two pieces (forming a spline), which gives us greater flexibility.

- *Approximation techniques.* Another reason to subdivide a curve is that a piece of a curve is generally simpler than the whole curve, where "simpler" means "more like a straight line." So we can cut it into a sufficiently large number of pieces, and then do something with those pieces as if they were straight line segments, such as render them or raytrace them. In this way, we can approximate the result we would get if we were able to render or raytrace the curve analytically.

  Strictly speaking, we don't need subdivision to do piecewise linear approximation—we already discussed one simple technique that evaluates the curve at fixed-size intervals and draws lines between those sample points. But subdivision allows us to choose the number of line segments *adaptively* by using fewer line segments on the straighter parts of the curve and more line segments on the curvier parts.

So that's the "why" of curve subdivision. Before we learn the "how," let's be a bit more precise about the "what." Consider a parametric polynomial curve P defined by the function $\mathbf{p}(t)$, adopting the usual conventions that the curve starts at $t = 0$ and ends at $t = 1$. Now consider a segment Q that starts at an arbitrary time $t = a$ and ends at $t = b$. This is illustrated in Figure 13.15.

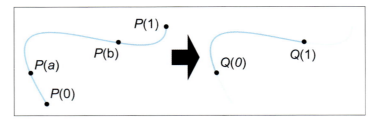

Figure 13.15. Extracting a segment of a curve by using subdivision

The goal of subdivision is a mathematical description for Q in some form (monomial, Hermite, or Bézier). But don't we already have that? After all, we assume that have a mathematical description of P in some form, and so it's perfectly valid to define Q by saying, "Take the curve defined

by P, but instead of starting at 0 and ending at 1, start at a and end at b."
That's not really what we want. We want Q to be a fully independent and
"regular" curve that makes no reference to P, not subordinate or qualified
in some way. For example, if we are using Bézier form, then we want new
Bézier control points that define Q.

The following sections present two different methods for subdividing
curves. Section 13.5.1 presents a straightforward algebraic approach in
monomial form. Section 13.5.2 considers Bézier curve subdivision, which
is geometrically based and lends itself towards rather elegant and efficient
implementations.

Hermite form doesn't lend itself naturally to subdivision. If we wish
to subdivide a Hermite form, we first convert the curve to another form
(probably Bézier) and do the subdivision in that form.

13.5.1   Subdividing Curves in Monomial Form

Extracting a segment from a curve in monomial form is a straightforward
algebraic task. Remember that monomial form is just an explicit polynomial on t. Although we are typically interested only in the part where
$0 \leq t \leq 1$, the polynomial is defined for *all* values of t and so it actually
defines an *infinite* curve. The smaller segment Q that we wish to extract
is just a different subsection of the same infinite curve.

With this in mind, we realize that the problem of subdivision can easily
be viewed as a simple problem of reparameterization. Rather than trying
to muck directly with the monomial coefficients, we perform some algebra
on the parameter value. Let's introduce a *local parameter* s that varies
from 0 to 1 as $\mathbf{q}(s)$ traces out the curve Q. Given this, we can define the
curve $\mathbf{q}(s)$ in terms of $\mathbf{p}(t)$ as

$$t = F(s), \qquad\qquad \mathbf{q}(s) = \mathbf{p}(t) = \mathbf{p}(F(s)),$$

where the function $F(s)$ is our reparameterization function that returns
the global parameter t corresponding to the local parameter s. It's not too
hard to see what form F should be, since we wish to satisfy the endpoint
conditions $F(0) = a$ and $F(1) = b$. Adopting a straightforward linear
relation between t and s yields

$$t = F(s) = a + s(b - a).$$

You might want to verify that this does behave correctly at the endpoints.

Of course, all we have really accomplished is to define Q in terms of
P, which is precisely what we said was *not* sufficient at the start of this
section. The difference is that if we continue working through the math, and
substitute for $\mathbf{p}(t)$ and eliminate t, we can get a direct equation for $\mathbf{q}(s)$,

which *is* a "regular" and independent curve satisfying the goals outlined at the start of this section.

However, the ensuing algebra gruntwork produces a messy result without revealing any insight. The main thing we wish to communicate here is that subdivision of a curve in monomial form is a simple matter of reparameterization, which can be accomplished algebraically. Furthermore, because we can convert between monomial forms and other forms, we now have a surefire method for subdividing any polynomial curve in any format.

But we need not be satisfied with this "brute force" approach; as it turns out, in Bézier form, we can do better.

13.5.2 Subdividing Curves in Bézier Form

Subdivision of a Bézier curve can be done geometrically through a variant of the de Casteljau algorithm. The full algorithm of extracting any subsection for arbitrary endpoint parameters a and b is not immediately grasped, so we follow Farin [20, Section 7.2ff] and start off with a simple case.

We begin by restricting ourselves to extracting only the "left side" of a curve. In other words, we fix $a = 0$. Clearly, the first Bézier control point on the smaller curve (at $s = 0$) is the same as the first control point on the larger curve (at $t = 0$). Equally clear is that the endpoint at $t = b$ is obtainable by the basic de Casteljau algorithm from Section 13.4.1. An example situation with $b = 0.75$ is illustrated in Figure 13.16.

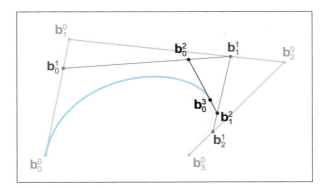

Figure 13.16
Locating the interior endpoint using the de Casteljau algorithm

We have the endpoints—now for those tricky interior points. Surprisingly, if you look closely at Figure 13.16, you'll notice that we already constructed them! As it turns out, each round of de Casteljau interpolation produces one of our Bézier control points. Figure 13.17 makes this clearer, showing the selected Bézier points and the control polygon.

Why does this work? Recall the relationship between the Bézier form and the Hermite form from Section 13.4.3. The first interior control point

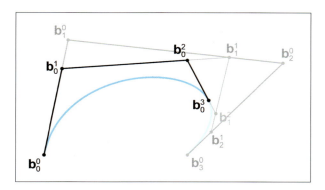

Figure 13.17
The de Casteljau
algorithm gives us *all*
the Bézier control
points of the
extracted curve
segment

$\mathbf{b}_1$ completely determines the first derivative (the velocity) at $t = 0$. Now, the subcurve that we are extracting is part of the same infinite curve, and thus its position and derivatives match everywhere, in a geometric sense. However, the derivative is a rate of change *relative to the rate of change of the parameter*. By subdividing, we have made the parameter t move "faster," since it goes from 0 to 1 over a smaller spatial interval. Thus, the derivative of the subcurve is in the same direction, but it is shorter according to the fraction of the curve that we are extracting, in our case the value b.

Let's summarize our findings. To extract the left half of a curve, $0 \le t \le b$, we perform de Casteljau subdivision as if we were trying to locate the endpoint at $t = b$. The first control point from each round of interpolation gives us another control point for our subdivided curve. Extracting the right half of a curve is analogous, so we won't go into detail here.

There is one important special case of Bézier subdivision that we can do armed only with what we know so far: subdividing a curve "in half" at $t = 1/2$. This computation makes possible rather elegant recursive algorithms for adaptive subdivision. Let's use our standard notation $\mathbf{b}_i$ for the original Bézier control points. For the two halves, we pick two letters at random and call the control points for the left and right halves of the curve $\mathbf{q}_i$ and $\mathbf{r}_i$, respectively. The seven control points are given by

$$\mathbf{q}_0 = \mathbf{b}_0,$$

$$\mathbf{q}_1 = \mathbf{b}_0/2 + \mathbf{b}_1/2,$$

$$\mathbf{q}_2 = \mathbf{b}_0/4 + \mathbf{b}_1/2 + \mathbf{b}_2/4,$$

$$\mathbf{q}_3 = \mathbf{r}_0 = \mathbf{b}_0/8 + 3\mathbf{b}_1/8 + 3\mathbf{b}_2/8 + \mathbf{b}_3/8,$$

$$\mathbf{r}_1 = \mathbf{b}_1/4 + \mathbf{b}_2/2 + \mathbf{b}_3/4,$$

**Subdividing a Bézier
curve at $t = 1/2$**

$$\mathbf{r}_2 = \mathbf{b}_2/2 + \mathbf{b}_3/2,$$

$$\mathbf{r}_3 = \mathbf{b}_3.$$

The general case is obtained through *blossoming*, which is a general term referring to a number of techniques involving repeated de Casteljau steps taken with different interpolation fractions. To determine each control point, we take three de Casteljau steps (for a cubic curve, at least). For each control point $\mathbf{b}_i$ we take i of those steps using $t = b$, and the rest using $t = a$. As it turns out, it doesn't matter which of the interpolation steps use a and which use b, but the number of steps using a or b is important. Let's consider each point on the cubic curve to make this clear. To compute $\mathbf{b}_0$, at each round we use $t = a$ as the interpolation fraction. For $\mathbf{b}_1$, we use $t = a$ for two of the rounds, and $t = b$ for one round. To calculate $\mathbf{b}_2$, we use $t = a$ for the interpolation fraction in only one round, and $t = b$ for the other two. And of course, for the last control point $\mathbf{b}_3$, we use $t = b$ for all three rounds, exactly as we described at the start of this section.

13.6   Splines

So far we have been focusing on cubic curves, and for good reason; they are the most commonly used type of curves in 3D. Such curves inherently have four degrees of freedom, whether we are using Bézier curves with four control points, monomial curves with four coefficients, or Hermite curves with two ending points plus two derivatives. Because there are only four degrees of freedom, the set of curves that can be represented by using only the techniques of cubic curves is sharply limited.

Additional freedom is obtained by joining smaller curves together in a *spline*, which is the subject of the remainder of this chapter. Before we discuss splines, let's pause for a moment to discuss one potential alternative: using a higher degree polynomial. Obviously any degree n curve can be converted to a degree $n + 1$ curve; such a conversion is known as *degree elevation*. In monomial form, of course, this is trivial, we just add a new leading coefficient of zero.

In Bézier form, degree elevation adds a new control point and, as you might have guessed, the positions of the new control points can be constructed geometrically by using linear interpolation. Given a curve of degree n, which has $n + 1$ control points denoted $\mathbf{b}_i$, degree elevation produces a degree $n+1$ curve with $n+2$ control points, denoted $\mathbf{b}'_j$. To determine these new control points, we linearly interpolate using an interpolation fraction proportional to the index of the control point:

Degree elevation in Bézier form

$$\mathbf{b}'_j = \frac{j}{n+1}\,\mathbf{b}_{j-1} + \left(1 - \frac{j}{n+1}\right)\mathbf{b}_j, \qquad 0 \le j \le n+1. \qquad (13.38)$$

(Note that the computation of $\mathbf{b}'_j$ will "blend" the nonexistent point $\mathbf{b}'_{-1}$ with a weight of zero.)

For Hermite curves, we usually are interested only in odd values for n, so that we have the same number of derivatives at each endpoint.

A higher degree polynomial has the ability to describe a curve with more "wiggles," but, unfortunately, in general it suffers from several short-comings:

- The curve has *global support*. Each control point exerts some nonzero weight on every point along the curve, with the exception of the endpoints.

- The curve has extraneous "wiggles" that sometimes show up in places we don't want, oscillating back and forth between the control points. This is known as the *Runge*[17] *phenomenon*.

- Somewhat related to the extra wiggles is the fact that higher degree curves are very sensitive. Due to the curve's global support, a change to any one of the control points will result in a change over the entire curve; due to the high sensitivity, this response can be very large.

- Having ruled out polynomial interpolation as a viable curve design tool, we cannot directly specify a point that we want the curve to interpolate, other than the endpoints.

The basic problem is that we are asking too much from a single polynomial. Splines do not have these shortcomings.

Here's what's in store. First, to facilitate the discussion, we must expand our notation and introduce a level of indirection between the local and global parameterization, which we do in Sections 13.6.1 and 13.6.2. Then, in Section 13.7, we talk about Hermite and Bézier splines, which are used in many software packages, such as *Adobe Photoshop* and *Autodesk 3DS Max*. From there, our focus naturally gravitates towards deciding what to do at the "seams." The first hurdle is to define the criteria that must be met so that the curve is smooth at these junction points. Such continuity conditions are the subject of Section 13.8. Once we understand these issues, we will have finally reached our goal set at the start of this chapter, a spline system that provides an intuitive means to define a curved shape.

Having developed a flexible design tool where the user can specify the position and tangent at each control point, Section 13.9 then investigates methods by which the designer need specify only the positions of the control points, and the tangents are computed automatically based on a set of intuitive user controls.

---

[17] Pronounced "RUN-guh."

13.6.1   Rules of the Game

Our spline is composed of n segments, denoted $\mathbf{q}_0$, $\mathbf{q}_1$, ..., $\mathbf{q}_{n-1}$. The ith segment $\mathbf{q}_i$ is a function that accepts a *local parameter*, named s_i, which is normalized to vary from 0 to 1 over the length of the segment. In other words, for each segment there is a curve function $\mathbf{q}_i(s_i)$ exactly like the ones we studied in the first part of this chapter; the only differences are the cosmetic renaming of the function from $\mathbf{p}$ to $\mathbf{q}_i$ and the argument from t to s_i.

We use two different notations to refer to the entire spline. One way is to just drop the subscripts from the notation above, so the function $\mathbf{q}(s)$ refers to the entire spline, and the parameter s (without subscript) is a *global parameter*. As s varies from 0 to n, the function $\mathbf{q}(s)$ traces out the entire spline.

The composite function $\mathbf{q}(s)$ is very simple. Basically we take the integer portion of s to get the index i, describing which segment we are on, and then the fractional portion is used as s_i and plugged into the segment $\mathbf{q}_i$. So the first segment $\mathbf{q}_0(s_0)$ defines the spline on the interval between $\mathbf{q}(0)$ and $\mathbf{q}(1)$, the second segment defines the spline from $\mathbf{q}(1)$ to $\mathbf{q}(2)$, and so on. More formally,

A composite curve with a simple global parameterization

$$i = \lfloor s \rfloor, \qquad \text{(select segment by using the floor function)}$$
$$s_i = s - i, \qquad \text{(calculate local parameter)}$$
$$\mathbf{q}(s) = \mathbf{q}_i(s_i). \qquad \text{(evaluate segment)}$$

Note that, given a particular value for s, we can unambiguously identify the point $\mathbf{q}(s)$ along the spline. However, a particular value of s_i is meaningful only within the context of segment i; this is emphasized by the subscript.

If we are not concerned with the timing of our curve, then this notation may be all we need. However, when defining an animation path, we usually need a level of indirection. We introduce the notation $\mathbf{p}(t)$ to refer to the final curve, a function that returns our position at a given "time" t. It's just a different parameterization of the same curve; $\mathbf{p}(t)$ and $\mathbf{q}(s)$ trace out the same shape, but the s and t values for a particular point along the path are usually not be the same. We can parameterize the curve so that some sections are traversed quickly and others more slowly. The range of s is fixed by the number of knots, but we are free to assign the range of t, the total duration of the curve, to anything we wish.

In general, we can define $\mathbf{p}(t)$ in terms of $\mathbf{q}(s)$ by creating a function that maps a time value t to a parameter value s. When we want to be explicit that s is a function of t, we use the notation $s(t)$, and this function is called the *time-to-parameter* function. If you're a computer programmer, you can think of $\mathbf{p}(t)$ as the public interface, and $\mathbf{q}(s)$ as an internal implementation detail. We are engaging in a fundamental practice of computer science: breaking down complexity by introducing a level of indirection.

With the above notation established, the basic game plan for evaluating $\mathbf{p}(t)$ is as follows:

1. Map the time value t into a value of s by evaluating the time-to-parameter function $s(t)$.

2. Extract the integer portion of s as i, and the fractional portion as s_i.

3. Evaluate the curve segment $\mathbf{q}_i(s_i)$.

Of course, if we don't care about the timing of the spline (perhaps we only care about its shape), then we have no need of the first step, and we can just use the trivial mapping of $s(t) = t$. Unfortunately, due to space constraints, this is precisely what we're going to do in this book. We don't discuss the subtleties of dealing with the timing.

With the assumption for now that $s = t$, the first step is trivial. The second step is also easy, and we devoted the first part of this chapter to the third step. So we really already know how to *evaluate* a spline; let's look at how we might *create* one.

13.6.2   Knots

Think about the juncture between two segments. For the curve to be continuous, clearly the ending point of one segment must be coincident with the starting point of the next segment. (Section 13.8 addresses additional desirable criteria.) These shared control points that are interpolated by the spline are called the *knots* of the spline. The knot at index i is denoted $\mathbf{k}_i$, and since there is one more knot than the number of segments, the knots are numbered $\mathbf{k}_0 \ldots \mathbf{k}_n$.

We assume that the segments are connected at the knots. In other words, $\mathbf{q}(s)$ passes through the knots at integer values of s. With this assumption, there's no need for separate notation (or separate storage space in a computer program) for the beginning point and ending point of each segment. Instead, each interior knot $\mathbf{k}_i$ serves a dual role as the starting point of segment $\mathbf{q}_i$ and the ending point of segment $\mathbf{q}_{i-1}$. Thus, we establish the following relations:

$$\mathbf{q}(i) = \mathbf{k}_i, \qquad \mathbf{q}_i(0) = \mathbf{k}_i, \qquad \mathbf{q}_i(1) = \mathbf{k}_{i+1}.$$

Note that $\mathbf{k}_i$ specifies a single point, whereas the notation $\mathbf{q}_i$ refers to an entire segment, which is a function of a local parameter s_i that yields a point. All of this notation is depicted in Figure 13.18.

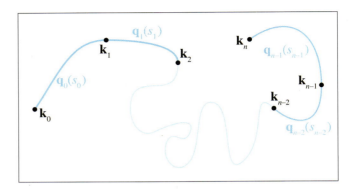

Figure 13.18. A spline with n segments has $n+1$ knots, named $\mathbf{k}_0 \ldots \mathbf{k}_n$.

In animation contexts, the knots are sometimes called *keys*. This is a reference to the old-school animation methods where a master animator would create the *key frames*, or frames where the characters reached important poses. The in-between frames would be filled in by a less experienced (and less expensive) apprentice. In computer animation, a key can be any position, orientation, or other piece of data whose value at a particular time is specified by a human animator (or any other source). The role of the apprentice to "fill in the missing frames" is played by the animation program, using interpolation methods such as the ones being discussed in this chapter. As we've noted before, most of the early research on splines was aimed at defining static shapes, not animated trajectories, and so the term "knot" is more prevalent.

13.7   Hermite and Bézier Splines

A spline is made by patching together curve segments so that they fit together smoothly. What sorts of curve segments? For reasons that will soon become apparent, it is most convenient for us to use the Hermite representation for the individual segments. When we say convenient "for us," we mean the people writing the code for an animation system or carrying out the mathematical discussion in the following sections. When it comes to depicting or manipulating splines graphically, the Bézier form is typically preferred. Of course, the Hermite and Bézier forms are closely related, and it is easy to convert between the two forms. If you don't remember this relationship, we review it in just a moment.

Remember that a Hermite curve segment is defined by its starting and ending positions and velocities. When we were focused on a single segment, we denoted the positions by $\mathbf{p}_0$ and $\mathbf{p}_1$, and the velocities by $\mathbf{v}_0$ and $\mathbf{v}_1$. In the context of a spline, we use a notation organized around a knot rather than a segment. For positions, we don't use the $\mathbf{p}$s because, as we've said earlier, the knot $\mathbf{k}_i$, which is the starting position of the segment $\mathbf{q}_i(0)$, also serves as the ending position of the previous segment at $\mathbf{q}_{i-1}(1)$. For velocities, the notation $\mathbf{v}_i^{\text{out}}$ refers to the outgoing velocity at knot i and defines the starting velocity for the segment $\mathbf{q}_i$. Likewise, the incoming velocity from the left side of $\mathbf{k}_i$ is denoted $\mathbf{v}_i^{\text{in}}$ and defines the ending velocity of the previous segment $\mathbf{q}_{i-1}$. We also refer to these velocity vectors as *tangents*.

Figure 13.19 shows a spline with five Hermite segments. All of the knots, segments, and tangents are labeled according to the notation just described.

Be warned that the tangents in Figure 13.19—and all the figures of Hermite curves in this chapter—are drawn at one-third scale. Officially we'd like to tell you that this was done so that the diagrams would be smaller and this book would consume less of the Earth's natural resources. A more accurate reason is that we draw the tangents at one-third length so the tangents will be the same as the edges of the Bézier control polygon. Matching the Bézier control polygon has some educational benefits, but, more importantly, it facilitates laziness on the part of the authors: the tools we used to create the curves in the diagrams are based on Bézier splines.

The splines in the diagrams in this book were created in Adobe *Photoshop* by making a path and then "stroking" the path. The arrows for the tangent vectors were drawn by putting one end at a knot and the other end

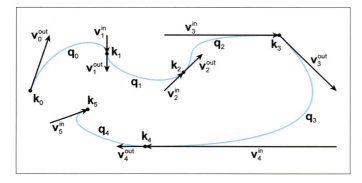

Figure 13.19. Our notation for splines with segments in Hermite form

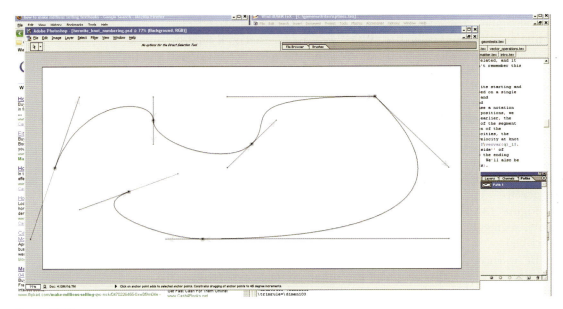

Figure 13.20
Creating Figure 13.19 with Adobe *Photoshop*.

at the "handle" used to control the shape of the curve, which is essentially the same as the Bézier control point. (*Photoshop* calls the knots the "anchor points" and refers to the interior Bézier control points that are not interpolated as "control points.")

For example, Figure 13.20 is a screen capture taken while one author was hard at work creating Figure 13.19. (The opacity of the actual figure has been decreased to make it easier to see the *Photoshop* controls.)

While we're on the subject of Bézier curves, let's take this opportunity to introduce the notation we use for Bézier splines. When we were dealing with only a single Bézier segment, we referred to the ith control point on that segment as $\mathbf{b}_i$. Here we use the notation $\mathbf{f}_i$ to refer to the control point "in front" of the ith knot, and $\mathbf{a}_i$ for the control point "after" it.[18] This notation is illustrated in Figure 13.21.

The important relationship between Hermite and Bézier forms was introduced in Section 13.4.3. Let's restate it here in the newly-introduced

---

[18]Note that by using knot-centric notation and assigning different letters to the control points (based on handy mnemonic memory aids!), we are locking in the degree of the segments to cubic. In other sources you'll find notation such as $\mathbf{b}_i^j$ to refer to the ith point on segment j, or just refer to all the points on the polygon as $\mathbf{b}_i$, where the knots are $\mathbf{b}_0$, $\mathbf{b}_3$, $\mathbf{b}_7$. This notation has the advantage of being more general, but to read it requires more mental effort—something we definitely want to minimize.

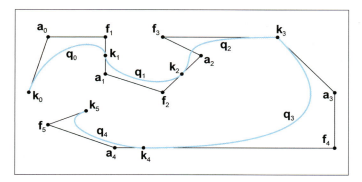

Figure 13.21. A spline with its Bézier control polygon, and the notation we use for Bézier splines

spline notation:

$$\mathbf{v}_i^{\text{in}} = 3(\mathbf{k}_i - \mathbf{f}_i), \qquad\qquad \mathbf{f}_i = \mathbf{k}_i - \mathbf{v}_i^{\text{in}}/3,$$
$$\mathbf{v}_i^{\text{out}} = 3(\mathbf{a}_i - \mathbf{k}_i), \qquad\qquad \mathbf{a}_i = \mathbf{k}_i + \mathbf{v}_i^{\text{out}}/3.$$

Converting between Bézier and Hermite forms

13.8   Continuity

For a few sections now we've been promising to tell you how you can piece together segments into a spline such that they fit together smoothly. All this lead-up may have given the impression that it's a mysterious secret. But if you take a closer look at Figure 13.19, you'll see that the criterion is relatively obvious: if the incoming and outgoing velocity vectors are equal at a knot, as they are at $\mathbf{k}_1$ and also $\mathbf{k}_2$, then the curve will be smooth. Notice that at $\mathbf{k}_3$, the tangents are not equal, and the curve has a kink in it. Pretty obvious, eh? Actually, as it turns out, there's quite a bit more to say on this subject.

Consider the curve near $\mathbf{k}_4$ in Figure 13.19. Notice that the curve is "smooth," yet the incoming velocity vector $\mathbf{v}_4^{\text{in}}$ is much longer than $\mathbf{v}_4^{\text{out}}$. Now, you might be thinking, "That curve isn't smooth there! If you were traveling along the curve, you would slam on the brakes just as you crossed the key." But take the tangent vectors out of the diagram and just look at the shape of the curve. It's a smooth shape, right? We're back to a recurring theme: animation paths are more "demanding" than static shapes. (Notice that in the objection you just raised, you used animation-oriented terminology when you said "key" instead of "knot." You're really catching on fast!)

Speaking of smooth animations, we just said that the curve is smooth at $\mathbf{k}_1$ and $\mathbf{k}_2$. But is it? We can see that the *shape* is smooth, but we've just pointed out how there is a difference between a smooth shape and a smooth animation. In general, we cannot tell if the animation is smooth without knowing more about the time-to-parameter function $s(t)$. If the shape is not smooth, the animation will not be smooth (with one exception to be discussed momentarily). But even if the shape is smooth, discontinuities in $s(t)$ can result in discontinuities in the animation. When $s(t) = t$, no discontinuities are introduced by this trivial mapping, so if the tangents are equal, the motion will be smooth.

Finally, consider a knot for which the incoming and outgoing velocities are both zero. In this case, even though the tangents are continuous, most people would agree that the shape is not smooth at this knot. What about the motion? Is the motion smooth when we come to a complete stop and then accelerate away in a potentially different direction? That will depend on your needs.

It looks like the answer to the question "Is it smooth?" is a bit fuzzy. This is a mathematics book, and it's really bad form to be putting quotation marks around vague words such as "smooth." We really need some more precise terminology. In the context of curves, the most important smoothness criteria are *parametric continuity* and the closely related *geometric continuity*. Let's look at each of these in turn, starting with parametric continuity, which is easier to define mathematically.

13.8.1   Parametric Continuity

A curve is said to have C^n continuity if its first n derivatives are continuous. A C^0 curve is one in which the position (the "0th derivative") is continuous. C^0 continuity means that we can draw a shape on a piece of paper in one stroke without lifting our pencil, or we can move along an animation path without "teleporting."[19] A C^1 curve has a continuous first derivative, which means the velocity doesn't jump instantaneously. This doesn't mean the velocity cannot change *rapidly*, but it never jumps from a velocity at one instant to a different velocity at the next instant without passing through velocities in between. For example, the curve in Figure 13.19 forms one connected line, so it is C^0 continuous everywhere. It is C^1 continuous everywhere except at $\mathbf{k}_3$ and $\mathbf{k}_4$, where the velocity jumps suddenly.

Higher numbers for n just mean the curve's higher-order derivatives are continuous. A curve is C^2 if its second derivative (acceleration) is continuous. Continuity conditions beyond C^1 are not that important for our purposes in this book. The lack of C^1 continuity (a sudden change in

---

[19]Oops, there are the quotation marks that we just said were bad form in a math book!

velocity) corresponds to an infinite acceleration, and this can create many problems. If the path is used to control a physical object, such as a robot or cutting tool, then we are asking for the motors driving the object to do something that is physically impossible. Even if the animation is taking place entirely inside of the virtual world of a computer, when such paths are observed by humans, they are usually perceived as "jerky." Thus it's usually desirable to avoid (or at least control) velocity discontinuities. In contrast, a sudden change in acceleration does not create such a jarring sensation and for most purposes is perfectly acceptable.

Any individual polynomial curve segment by itself has C^∞ continuity, since we can take the derivative of a polynomial as many times as we want and we always get a real-valued, continuous function. (Eventually, the derivatives become the constant zero function.) This is why the question of continuity didn't arise earlier in the chapter—the only places we have to worry about continuity are at the knots.

One last comment regarding higher derivatives. When we say that a curve is C^n continuous, this implies continuity for all lower derivatives as well. For example, if the acceleration is continuous, then the velocity and position must also be continuous. A discontinuity in a function means that the function's derivative is undefined where the discontinuity occurs.

Now that we've discussed parametric continuity informally, let's define the criteria mathematically for Hermite and Bézier curves. To do so, we make use of some observations concerning the derivatives of Bézier curves from Section 13.4.3; our findings from that section are summarized here.

- The nth derivative at an endpoint of a Bézier curve segment is completely determined by the endpoint and the nearest n control points.

- The velocity at an endpoint is proportional to the vector between the endpoint and the adjacent control point (Equations (13.30) and (13.31)).

- The acceleration at an endpoint is proportional to the difference of the delta vectors along the nearest two segments of the control polygon (Equations (13.36) and (13.37)).

Let's start with C^0, which is a no-brainer due to our choice of notation. In our scheme, the ending point of one segment is the same as the starting point of the next segment by definition. Moving on to C^1 continuity, we've said that it occurs when the tangents are equal at a key. This translates directly to Hermite form as

$$\mathbf{v}_i^{\text{in}} = \mathbf{v}_i^{\text{out}},$$

C^1 continuity condition for Hermite splines

and with just a little effort we can also express it in Bézier form as

$$\mathbf{k}_i - \mathbf{f}_i = \mathbf{a}_i - \mathbf{k}_i.$$

With a quick application of algebra, we see that geometrically this means that the knot is at the midpoint of the line between $\mathbf{f}_i$ and $\mathbf{a}_i$:

$$\mathbf{k}_i - \mathbf{f}_i = \mathbf{a}_i - \mathbf{k}_i,$$
$$2\mathbf{k}_i = \mathbf{f}_i + \mathbf{a}_i,$$
$$\mathbf{k}_i = (\mathbf{f}_i + \mathbf{a}_i)/2.$$

Most curve design tools will automatically enforce this rule for you. For example, when you move a control point in *Photoshop*, it automatically moves the opposing control point like a seesaw, and if you pull the control point away from the anchor point (the knot), the opposing control point will mirror your movements to maintain the C^1 continuity relationship. (If you want to force a corner in the curve, you can hold a modifier key to tell *Photoshop* not to do this).

Now let's look at C^2 continuity. It's is easier to visualize in Bézier form than Hermite. We just need to apply what we learned in Section 13.4.3 to make the ending acceleration of one segment (the left side of the equations below) match the starting acceleration of the next segment (on the right side):

$$6\mathbf{a}_{i-1} - 12\mathbf{f}_i + 6\mathbf{k}_i = 6\mathbf{k}_i - 12\mathbf{a}_i + 6\mathbf{f}_{i+1},$$
$$\mathbf{a}_{i-1} - 2\mathbf{f}_i + \mathbf{k}_i = \mathbf{k}_i - 2\mathbf{a}_i + \mathbf{f}_{i+1}$$
$$2\mathbf{f}_i - \mathbf{a}_{i-1}, = 2\mathbf{a}_i - \mathbf{f}_{i+1},$$
$$\mathbf{f}_i + (\mathbf{f}_i - \mathbf{a}_{i-1}) = \mathbf{a}_i + (\mathbf{a}_i - \mathbf{f}_{i+1}).$$

The geometric interpretation of this is as follows: Take the two Bézier control polygon segments that are not direct neighbors of the knot, but one segment away, and "double" them. If they meet at a common point, the curve is C^2 continuous. To visualize this, compare the two Bézier curves in Figure 13.22. Both have C^1 continuity, since the knot $\mathbf{k}_i$ is on the midpoint of the line between $\mathbf{f}_i$ and $\mathbf{a}_i$ for both curves. However, the top curve is C^2 continuous because the extensions of the neighboring control polygon lines meet at the common point; the curve on the bottom is not C^2 continuous.

13.8.2   Geometric Continuity

Geometric continuity is a broader criterion of continuity. Different authors use different definitions for geometric continuity, but a very general one is that a curve has G^n continuity if there exists *some* way to parameterize the curve such that the curve has C^n continuity. Let's look at an example.

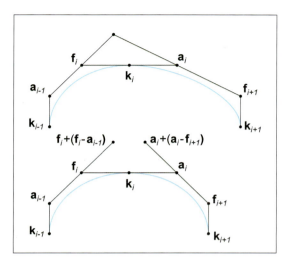

Figure 13.22
Continuity conditions for cubic Bézier splines.

In Figure 13.19 the curve is not C^1 continuous at $\mathbf{k}_4$ because the tangents are not equal. However, the curve *is* G^1 continuous at this location. The hint, of course, is that the tangents are parallel at the knot. If the tangents at a knot are not parallel, then there's no way to move along the curve in a smooth way. However, if the tangents are parallel, then the discontinuity is purely a change in speed, not a change in direction. We could remove this discontinuity by carefully introducing an offsetting discontinuity in the time-to-parameter function $s(t)$ that exactly "undoes" the jump in speed.

Higher-order geometric continuity extends this idea, although it is a bit more difficult to visualize. We say that a curve is G^2 continuous if its curvature changes continuously.

13.8.3 How Smooth Can a Curve Be?

We end our discussion on continuity by asking an important question: what's the highest level of continuity we can expect from a polynomial spline? We said earlier that any particular curve segment has C^∞ continuity, because we can differentiate it as many times as we want and the result is always a continuous function. Can we achieve this same level of smoothness with a spline?

Consider two adjacent cubic Bézier segments. Let's fix the first segment and consider what happens to the second segment as we demand higher and higher levels of continuity at the knot. When we demand C^0 continuity, we lock in the first Bézier control point. Clearly, the first endpoint must match the last endpoint of the first segment for the spline to be C^0 continuous.

What about C^1 continuity? Remember that the velocity at an endpoint is completely determined by the endpoint and the adjacent control point. This means if we want to match the velocity, we are locking in the position of the second control point as well.

Continuing this pattern, we see that for a Bézier segment to match n derivatives requires us to "lock in" $n+1$ control points. For a cubic curve, if we ask for C^4 continuity or higher, we can get it, but only by making every segment be a piece of the same infinite polynomial. We have gained continuity, but we have lost the flexibility that was the very reason we used splines in the first place!

The bottom line is that, practically speaking, a polynomial curve of degree n (a Bézier curve with $n+1$ control points) can really achieve only C^{n-1} continuity. For example, a piecewise linear (degree 1) polynomial can only achieve C^0 continuity. We can make a curve which is connected, but with straight lines, we cannot match the tangents. A quadratic (degree 2) polynomial can match tangents (C^1), but not accelerations. A cubic curve, the type of curve we have been focusing on in this book, can achieve C^2 continuity by reducing the number of degrees of freedom per segment to one. Continuity beyond C^2 can be achieved only by eliminating all degrees of freedom (other than the curve timing), and setting each segment to be a section of the same polynomial.

13.9   Automatic Tangent Control

At the start of this chapter, we began our investigation into curves with the plan of defining a curve just by listing points that we wanted the curve to pass through. We tried basic polynomial interpolation in Section 13.2, but found that it didn't give us what we wanted. We then developed the Bézier forms, which require the user to specify two endpoints, which are interpolated, and two (in the case of a cubic Bézier) interior control points, which are not interpolated but instead define the derivatives at the endpoints. So far in this chapter, we've learned how to piece together those Bézier segments in a smooth spline.

This section investigates various methods whereby a spline can be determined by just the knots, without the need for the user to specify any additional criteria. This is useful to generate a curve that looks "natural" and passes through some points, or any other time we wish to smoothly interpolate some data points.

For the moment, let's ignore the first and last knots and focus our attention on the interior knots. The problem at hand is to compute an appropriate $\mathbf{v}_i^{\text{in}}$ and $\mathbf{v}_i^{\text{out}}$ using only the positions of the knots. Notice that we are posing the problem in Hermite form, which turns out to be the easiest form to use for this problem. The situation is depicted in Figure 13.23,

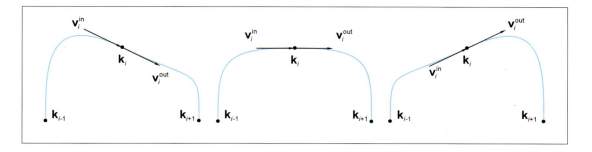

Figure 13.23
Three different choices of tangents for the middle knot, leading to three different interpolating splines

which shows three control points and three different choices we could use for the tangents.

The following sections discuss a family of techniques that can be used to pick tangents that result in "good" interpolating splines. First, Section 13.9.1 discuss the *Catmull-Rom* spline, which is a simple and straightforward technique. Then Section 13.9.2 considers *TCB* splines, a generalization of the Catmull-Rom form and a hybrid that exposes additional "sliders" to the user to adjust the shape of the curve in a (hopefully) more intuitive manner without resorting to direct geometric specification of the tangents. Finally, Section 13.9.3 lists a few options for dealing with the endpoints.

When reading the following sections, keep in mind that all of these splines are still Hermite splines. We are just introducing various techniques for autocalculating the tangents. Once the tangents have been determined, the spline is no different than any other Hermite spline.

13.9.1 Catmull-Rom Splines

Looking at Figure 13.23, it seems obvious which of the three choices of tangents is the most natural: the one in the middle. Why is this? The vector from the previous knot $\mathbf{k}_{i-1}$ to the next knot $\mathbf{k}_{i+1}$ is a horizontal line, and therefore it makes sense that our tangents should be horizontal. So it looks like one heuristic we could use to pick good tangents would be to make the tangents at a knot be parallel to the line between the previous and next knot. (Note that our example is slightly contrived in that the middle knot happens to be halfway between its neighbors, which is a special case. However, the fact that the neighbors lie on a horizontal line is *not* a special case, since we can always rotate our perspective to view the points in this configuration.)

But how long should the tangents be? Perhaps we should again use the vector between the previous and next knots as our guide. It seems as though the farther apart our neighbors are, the larger the curve, and so making our tangents be a constant multiple of this vector would be a good idea. In other words, we would set $\mathbf{v}_i^{in} = \mathbf{v}_i^{out} = a(\mathbf{k}_{i+1} - \mathbf{k}_{i-1})$. But what should we use for the value of a?

One way would be just to experiment and find a nice round number that seems to give results that are aesthetically pleasing. The constant $a = 1/2$ is a nice round number and works moderately well, so let's go with that. Figure 13.24 shows a spline loop generated by this technique.

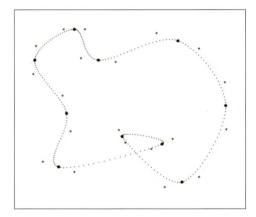

Figure 13.24
A Catmull-Rom spline

Although $a = 1/2$ gives "medium" results, there's definitely an argument to be made that it is a matter of preference. Sometimes we want a "tighter" curve, which would correspond to smaller values of a, and sometimes we want a "looser" curve. This is a good idea, but let's put it on ice for a moment to say two more quick things about the method we've stumbled upon.

First, let's give a formal definition and name to this technique. A spline with the tangents derived according to the relation

Tangent computation for the Catmull-Rom spline and its Bézier control polygon

$$\mathbf{v}_i^{in} = \mathbf{v}_i^{out} = \frac{\mathbf{k}_{i+1} - \mathbf{k}_{i-1}}{2} \tag{13.39}$$

is known as a *Catmull-Rom* spline. The name comes from the two people who invented it, one of whom is Edwin Catmull (1945–). He later went on to become the president of Walt Disney Animation Studios and Pixar Animation Studios.

The other thing we'd like to discuss is an alternative way to derive Equation (13.39). Just a bit of algebraic manipulation yields

$$
\mathbf{v}_i^{\text{in}} = \mathbf{v}_i^{\text{out}} = \frac{\mathbf{k}_{i+1} - \mathbf{k}_{i-1}}{2}
$$

$$
= \frac{\mathbf{k}_{i+1} - \mathbf{k}_i + \mathbf{k}_i - \mathbf{k}_{i-1}}{2}
$$

$$
= \frac{(\mathbf{k}_{i+1} - \mathbf{k}_i) + (\mathbf{k}_i - \mathbf{k}_{i-1})}{2}.
$$

Catmull-Rom spline as average of adjacent delta vectors

The geometric interpretation of the last line states that to compute a tangent at a knot, we take the two neighboring difference vectors of the control polygon and average them.

13.9.2   TCB Splines

Section 13.9.1 showed that the tangent at a knot can be computed by multiplying the vectors of the adjacent edges of the control polygon by an appropriate constant, which we called a, and adding the result. By varying a, we had an intuitive "dial" we could turn to adjust the shape of the curve. We can generalize this idea further by having not just one scaling factor, but two. In other words, we can take an arbitrary linear combination of the adjacent edge vectors. Taking the straightforward approach of assigning one "dial" for each of the two scale factors doesn't quite work out as an intuitive system. Instead, a standard technique is to provide three intuitive dials, known as *tension*, *continuity*, and *bias*, and derive the two scale factors from these dials. A spline with the tangents thus derived is known as *Kochanek-Bartels* spline, often called a *TCB* spline for obvious reasons.[20]

Kochanek and Bartels [40] designed the equations so that if we turn all three dials to zero, we get the standard Catmull-Rom curve. The typical useful range for all of the parameters is $[-1, +1]$, although there's no problem in going outside this range. Thus, you can think of each setting as a way to start with a Catmull-Rom curve and tweak it in a particular direction. First, let's show how each of these settings could be implemented by itself, and then let's present the full formulas that combine all three settings together.

The *tension* setting is related to the a value we discovered in the previous section. We the symbol t to refer to tension, and luckily there won't be any situations where this will be confused with the other meaning of t, the time parameter. Like all the TCB settings, a value of $t = 0$ corresponds to the regular Catmull-Rom curve. As we increase the tension, the curve "tightens"—essentially the same effect we got by decreasing the value of a in

---

[20]The most important for us is that TCB is easier to pronounce than koh-CHAN-ick.

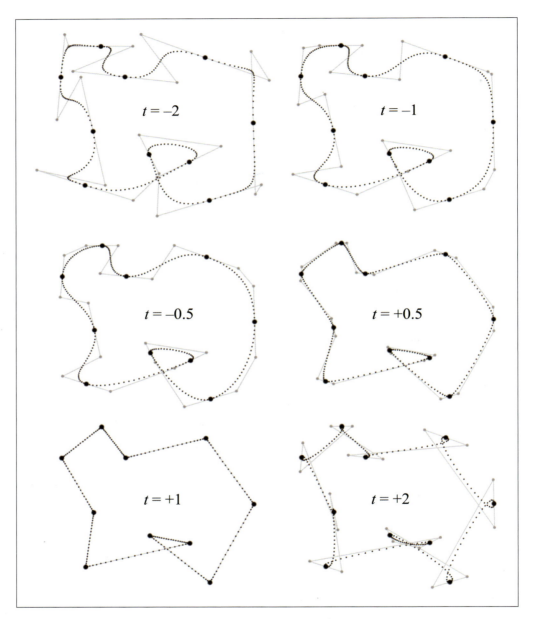

Figure 13.25
A TCB spline with different values for continuity.

the previous section. Figure 13.25 shows the effect of the tension parameter. In each curve, the continuity and bias values are zero. Compare this with the standard Catmull-Rom curve in Figure 13.24, corresponding to $t = 0$.

Note that $t = 1$ results in $\mathbf{v}_i^{in} = \mathbf{v}_i^{out} = \mathbf{0}$, causing the velocity to come to a stop at the knot, creating a cusp in the shape. If we increase t further, the velocities point in the "wrong direction," which creates a loop at the knots. At the other extreme, the value $t = -1$ results in a curve that is "twice as loose" as a Catmull-Rom curve. There's nothing special about this particular value; you can easily make the curve even looser by making t more negative.

We incorporate tension into the Catmul-Rom tangent formula as follows:

$$\mathbf{v}_i^{in} = \mathbf{v}_i^{out} = \frac{(1-t)(\mathbf{k}_{i+1} - \mathbf{k}_{i-1})}{2}$$
$$= \frac{(1-t)}{2}(\mathbf{k}_i - \mathbf{k}_{i-1}) + \frac{(1-t)}{2}(\mathbf{k}_{i+1} - \mathbf{k}_i).$$

Catmull-Rom formula extended to allow tension adjustments

Next let's turn to the *continuity* setting, which can be used to break the smoothness of the curve and force a corner at the knot. The value of zero will result in equal tangent (no matter what values for tension and bias are used), thus ensuring C^2 parametric continuity, as discussed in Section 13.8.1. As we decrease the continuity value, each tangent begins to turn towards its adjacent knot. At $c = -1$, each tangent will point directly to the neighboring knot, causing the "spline" to be composed of linear segments. Figure 13.26 illustrates the effect that different continuity values have on the spline.

One important observation to note is that setting $c = -1$ appears to have an effect on the shape of the curve similar to that of $t = 1$; both result in segments that are shaped like straight line segments. However, they are very different when viewed from an animation perspective. A spline with 100% tension *comes to a stop* at each key, and reaches a maximum value in the middle of the segment. (This is the Hermite smoothstep velocity profile, observable in the nonuniform spacing of the dots in each segment.) Notice that Bézier control points for the $t = 1$ spline in Figure 13.25 are not visible as they are coincident with the knots. Compare this to the $c = -1$ spline in Figure 13.26, where the Bézier control points are spaced equally along each linear segment. We observed earlier that this produces a curve with constant velocity, as evidenced by the equal spacing of the smaller black dots used to draw the curve.

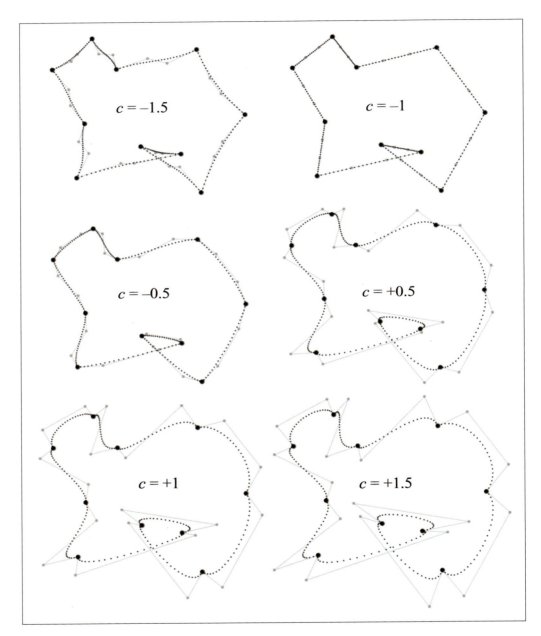

Figure 13.26
A TCB spline with different values for continuity.

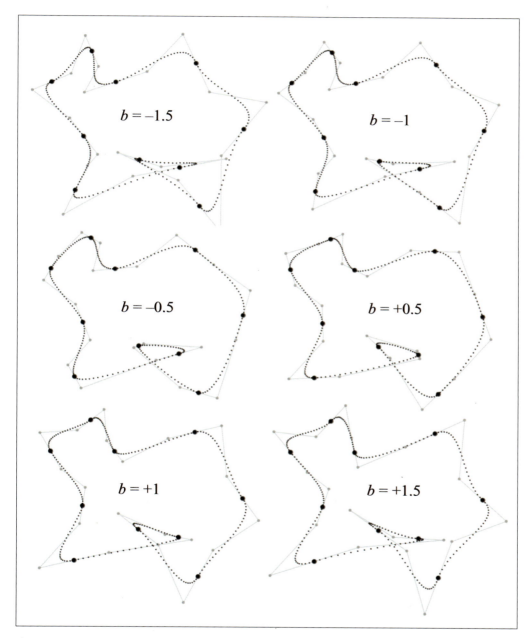

Figure 13.27
A TCB spline with different values for bias.

The math behind TCB continuity is written as

**Catmull-Rom formula
extended to allow
continuity adjustments**

$$\mathbf{v}_i^{\text{in}} = \frac{(1-c)}{2}(\mathbf{k}_i - \mathbf{k}_{i-1}) + \frac{(1+c)}{2}(\mathbf{k}_{i+1} - \mathbf{k}_i),$$

$$\mathbf{v}_i^{\text{out}} = \frac{(1+c)}{2}(\mathbf{k}_i - \mathbf{k}_{i-1}) + \frac{(1-c)}{2}(\mathbf{k}_{i+1} - \mathbf{k}_i).$$

Finally, the *bias* argument can be used to turn the tangents towards one or the other adjacent knots, rather than being parallel to the line between the adjacent knots, as the Catmull-Rom curve does. Consider a sequence of three knots. A negative bias causes the curve to "anticipate" the third knot, turning the curve in the direction of the third knot a bit before the middle knot is reached. In contrast, a positive bias value causes the curve to wait to make the turn towards the third knot, causing some "overshoot" through the middle knot. Figure 13.27 shows our example spline with several different bias values.

The bias value works by scaling the relative weights that the two control polygon edges have on the resultant tangent:

**Catmull-Rom formula
extended to allow bias
adjustments**

$$\mathbf{v}_i^{\text{in}} = \mathbf{v}_i^{\text{out}} = \frac{(1+b)}{2}(\mathbf{k}_i - \mathbf{k}_{i-1}) + \frac{(1-b)}{2}(\mathbf{k}_{i+1} - \mathbf{k}_i).$$

The equations presented thus far have isolated each setting to make it easier to understand the math behind each one. Now let's put all three settings together:

Computing Tangents for TCB Splines

$$\mathbf{v}_i^{\text{in}} = \frac{(1-t)(1+b)(1-c)}{2}(\mathbf{k}_i - \mathbf{k}_{i-1}) + \frac{(1-t)(1-b)(1+c)}{2}(\mathbf{k}_{i+1} - \mathbf{k}_i),$$

$$\mathbf{v}_i^{\text{out}} = \frac{(1-t)(1+b)(1+c)}{2}(\mathbf{k}_i - \mathbf{k}_{i-1}) + \frac{(1-t)(1-b)(1-c)}{2}(\mathbf{k}_{i+1} - \mathbf{k}_i).$$

One last note. The examples in this section used the same values at each knot in the spline, but that need not be the case. The TCB values are often adjusted on a per-knot basis.

13.9.3   Endpoint Conditions

The Catmull-Rom methods rely on the previous and next knots to compute the tangent at a given knot. What should we do at an endpoint when there is no "previous" or "next" knot? Several solutions to this problem have been proposed.

One obvious answer would be to just throw our hands in the air and set the tangent to zero at an endpoint. While this seems like surrendering before the first shot is fired, it actually can be a good choice if the spline is to be used for animation, since it's often natural to want the object being animated to start and end "at rest."

Another idea is to create extra knots $\mathbf{k}_{-1}$ and $\mathbf{k}_{n+1}$, which are used for tangent computations but are not interpolated. Where should we place these so-called *phantom points*? One idea is to duplicate the neighboring endpoint, which produces zero tangents and is equivalent to the "surrender" spline of the previous paragraph. Another idea is simply to ask the user to place the phantom point. When this method is used, the spline is known as a *Cardinal spline*.

One final method is to fit the first (or last) three knots to a quadratic, and use the endpoint tangent of this curve. The curve fitting is an example of polynomial interpolation and can thus be done by using the techniques from earlier in this chapter, such as Aitken's algorithm.

13.10   Exercises

(Answers on page 792.)

1. Compute the Lagrange basis polynomials for the knot sequence $t_1 = 0$, $t_2 = 1$, $t_3 = 2$.

2. The motion of a projectile (see Section 11.6) can be described by the quadratic function

$$\mathbf{p}(t) = \mathbf{p}_0 + t\mathbf{v}_0 + t^2(\mathbf{a}/2),$$

where $\mathbf{p}_0$ is the initial position, $\mathbf{v}_0$ is the initial velocity, and $\mathbf{a}$ is the constant acceleration (typically due to gravity).

Imagine you want to animate the path of a projectile—say, a herring sandwich. Assume you are working in our standard 3D coordinate space (see Section 1.3.4) and the object is launched from the origin, reaches a maximum at $t = 1$ when its position is $\mathbf{p}(1) = (0, h, d/2)$, and finally lands at $t = 2$ at the position $\mathbf{p}(2) = (0, 0, d)$. Derive an expression for $\mathbf{p}(t)$ in monomial form, in terms of the variables h and d.

3. Consider the Bézier curve in the figure below.

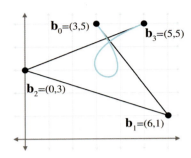

(a) Use de Casteljau to determine the position on the curve at $t = 0.40$.

(b) Convert the curve to Hermite form.

(c) Convert the curve to monomial form.

(d) Check your work on part (a) by substituting $t = 0.40$ into the polynomial computed in part (c).

(e) What is the velocity polynomial function $\mathbf{v}(t)$?

(f) What is the velocity at $t = 0.40$, $t = 0.00$, and $t = 1.00$?

4. Prove that the quadratic Bernstein basis polynomials sum to 1 for any value of t.

5. Where should we put the Bézier control points to get a "constant curve" where $\mathbf{p}(t)$ always returns the same point?

6. Where should we put the Bézier control points to get a linear "curve," which is a straight line segment with constant velocity?

7. Where should we put the Bézier control points to get a straight line shape, but this time the velocity of the curve follows the smoothstep pattern: it starts at zero, accelerates to a maximum velocity at the middle, and then decelerates to end with zero velocity?

8. Describe the motion of a particle that moves along the Bézier curve where $\mathbf{b}_0 = \mathbf{b}_2$ and $\mathbf{b}_1 = \mathbf{b}_3$.

9. Consider the projectile herring sandwich from Exercise 2. Assume you need to animate this sandwich, and the only tools available to you are cubic Bézier curves. Where should you put the four Bézier control points to get physically realistic motion, which is quadratic? Don't worry about the total duration the sandwich is airborne; consider only the shape of the trajectory.

10. To plot the shape of the parabola in Figure 12.8 (page 588), the authors tabulated a list of x, y image-space coordinates of the center of the mass of the board, and then did a least-squares fit to arrive at the equation for the parabola $y = -0.364x^2 + 1.145x + 2.110$. The pen tool in Adobe *Illustrator*,

which was used to draw the parabola, is based on cubic Bézier curves. The starting and ending x-coordinates for our curve were -0.9683 and 4.2253, respectively. What were the (x, y) coordinates for all four control points?

11. Returning to the curve in Exercise 3:

    (a) Compute the Bézier control points for the segment of the curve from 0.2 to 0.5.

    (b) Split this curve into two halves at $t = 1/2$. What are the Bézier control points of the curve on each side?

    (c) Perform degree elevation on this curve to the quartic case. What are the five control points?

My curves are not crazy.

— Henri Matisse (1869–1954)

Chapter 14

Afterword

If you are interested in stories with happy endings
you would be better off reading some other book.

— Lemony Snicket,
A Series of Unfortunate Events: The Bad Beginning

14.1   What Next?

You have reached the end of the book. Where to go from here? Well, if you've stayed with us up to this point, then you probably understand enough to get started with some real code, and you're probably itching to put all this new knowledge to work, right? We've found that the best way to learn is by doing. So don't just sit there, start making something!

As our final act, we leave you with these last few exercises. Some of them might be applicable only for those readers interested in making video games, but we wish you good luck wherever this knowledge takes you. We've learned a lot, worked very hard, and had a lot of fun writing this book, and we hope the same can be said of your experience reading it.

14.2   Exercises

(Answers on page 799.)

1. Download a game engine and make a mod for it.

2. Learn about what makes video games fun. Take three of your favorite games and make a detailed analysis of their mechanics. What makes them fun?

3. Complete a large and challenging project that implements an advanced technique or an experimental gameplay feature.

4. Pick a particular aspect of video game programming that you find interesting and delve deeply into that area.

5. Get a job working for a company that makes the kinds of games you will be proud of. (Hint: This is *greatly* facilitated by taking advantage of your answers to Exercises 1–4. Also, see the index entry for "job interview.")

6. Learn how to get along with other people, work on a team, and use version control and task tracking software. Nobody succeeds alone.

7. Make some great video games. Always use technology as a means to an end, and never lose sight of the end product.

8. Never stop learning.

I'm very well acquainted, too, with matters mathematical,
I understand equations, both the simple and quadratical,
About binomial theorem I'm teeming with a lot o' news,
With many cheerful facts about the square of the hypotenuse.
I'm very good at integral and differential calculus;
I know the scientific names of beings animalculous:
In short, in matters vegetable, animal, and mineral,
I am the very model of a modern Major-General.

— Major-General Stanley from *The Pirates of Penzance*

Appendix A

Geometric Tests

Chapter 9 discussed a number of calculations that can be performed on a single primitive. Here, we present a number of useful calculations that operate on more than one primitive. This appendix is a collection of various geometric calculations that are sometimes useful. It is also instructive to browse these tests, because many illustrate general principles.

A more comprehensive list of fast intersection methods can be found at http://www.realtimerendering.com/intersections.html.

A.1 Closest Point on 2D Implicit Line

Consider an infinite line L in 2D defined implicitly by all points $\mathbf{p}$ such that

$$\mathbf{p} \cdot \hat{\mathbf{n}} = d,$$

where $\hat{\mathbf{n}}$ is a unit vector. Our goal is to find, for any point $\mathbf{q}$, the point $\mathbf{q}'$ that is the closest point on L to $\mathbf{q}$. This is the result of projecting $\mathbf{q}$ onto L. Let us draw a second line M through $\mathbf{q}$, parallel to L, as shown in Figure A.1.

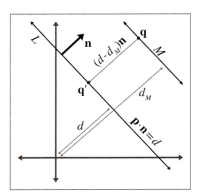

Figure A.1
Finding the closest point on a 2D implicit line

Let $\hat{\mathbf{n}}_M$ and d_M be the normal and d value, respectively, of the line equation for M; since L and M are parallel, they have the same normal: $\hat{\mathbf{n}}_M = \hat{\mathbf{n}}$. Since $\mathbf{q}$ is on M, d_M can be computed as $\mathbf{q} \cdot \hat{\mathbf{n}}$.

Now the signed distance from M to L measured parallel to $\hat{\mathbf{n}}$ is simply

$$d - d_M = d - \mathbf{q} \cdot \hat{\mathbf{n}}.$$

This distance is obviously the same as the distance from $\mathbf{q}$ to $\mathbf{q}'$. (If we need only the distance and not the value of $\mathbf{q}'$, then we can stop here.) To compute the value of $\mathbf{q}'$, we can simply take $\mathbf{q}$ and displace it by a multiple of $\hat{\mathbf{n}}$:

Computing the closest point on a 2D implicit line

$$\mathbf{q}' = \mathbf{q} + (d - \mathbf{q} \cdot \hat{\mathbf{n}})\hat{\mathbf{n}}. \tag{A.1}$$

A.2   Closest Point on a Parametric Ray

Consider the parametric ray R in 2D or 3D defined by

$$\mathbf{p}(t) = \mathbf{p}_{\text{org}} + t\hat{\mathbf{d}},$$

where $\hat{\mathbf{d}}$ is a unit vector, and the parameter t varies from 0 to l, where l is the length of R. For a given point $\mathbf{q}$, we wish to find the point $\mathbf{q}'$ on R that is closest to $\mathbf{q}$.

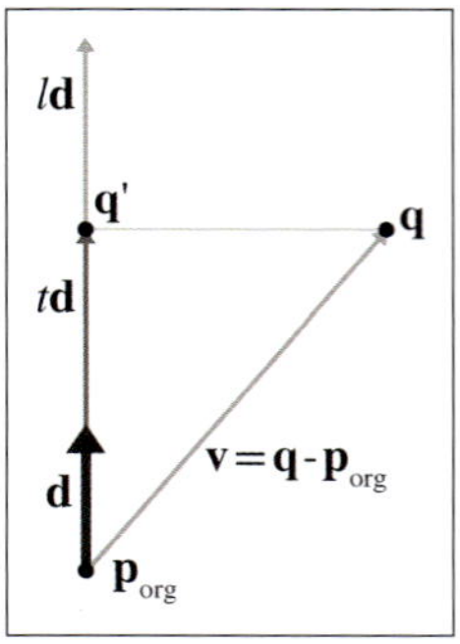

This is just a simple matter of projecting one vector onto another, which was presented in Section 2.11.2. Let $\mathbf{v}$ be the vector from $\mathbf{p}_{\text{org}}$ to $\mathbf{q}$. We wish to compute the result of projecting $\mathbf{v}$ onto $\hat{\mathbf{d}}$—in other words, the portion of $\mathbf{q}$ parallel to $\hat{\mathbf{d}}$. This is illustrated in Figure A.2.

The value of the dot product $\mathbf{v} \cdot \hat{\mathbf{d}}$ is the value t such that $\mathbf{p}(t) = \mathbf{q}'$:

Figure A.2
Finding the closest point on a ray

Computing the closest point on a parametric ray

$$t = \hat{\mathbf{d}} \cdot \mathbf{v} = \hat{\mathbf{d}} \cdot (\mathbf{q} - \mathbf{p}_{\text{org}}),$$

$$\mathbf{q}' = \mathbf{p}(t) = \mathbf{p}_{\text{org}} + t\hat{\mathbf{d}} = \mathbf{p}_{\text{org}} + (\hat{\mathbf{d}} \cdot (\mathbf{q} - \mathbf{p}_{\text{org}}))\hat{\mathbf{d}}. \tag{A.2}$$

Actually, Equation (A.2), for $\mathbf{p}(t)$ computes the point closest to $\mathbf{q}$ on the *infinite line* containing R. If $t < 0$ or $t > l$, then $\mathbf{p}(t)$ is not within the ray R, in which case, the closest point to $\mathbf{q}$ on R will be the ray origin (if $t < 0$) or endpoint (if $t > l$).

If the ray is defined where t varies from 0 to 1 and $\mathbf{d}$ is not necessarily a unit vector, then we must divide by the magnitude of $\mathbf{d}$ to compute the

t value:

$$t = \frac{\mathbf{d} \cdot (\mathbf{q} - \mathbf{p}_{\mathrm{org}})}{\|\mathbf{d}\|}.$$

A.3 Closest Point on a Plane

Consider a plane P defined in the standard implicit manner as all points $\mathbf{p}$ that satisfy

$$\mathbf{p} \cdot \hat{\mathbf{n}} = d,$$

where $\hat{\mathbf{n}}$ is a unit vector. Given a point $\mathbf{q}$, we wish to find the point $\mathbf{q}'$, which is the result of projecting $\mathbf{q}$ onto P. Point $\mathbf{q}'$ is the closest point to $\mathbf{q}$ on P.

We showed how to compute the distance from a point to a plane in Section 9.5.4. To compute $\mathbf{q}'$, we simply displace $\mathbf{q}$ by this distance, parallel to $\hat{\mathbf{n}}$.

$$\mathbf{q}' = \mathbf{q} + (d - \mathbf{q} \cdot \hat{\mathbf{n}})\hat{\mathbf{n}}$$

Computing the closest point on a plane

Notice that this is the same as Equation (A.1), which computes the closest point to an implicit line in 2D.

A.4 Closest Point on a Circle or Sphere

Imagine a 2D point $\mathbf{q}$ and a circle with center $\mathbf{c}$ and radius r. (The following discussion also applies to a sphere in 3D.) We wish to find $\mathbf{q}'$, which is the closest point on the circle to $\mathbf{q}$.

Let $\mathbf{d}$ be the vector from $\mathbf{q}$ to $\mathbf{c}$. This vector intersects the circle at $\mathbf{q}'$. Let $\mathbf{b}$ be the vector from $\mathbf{q}$ to $\mathbf{q}'$, as shown in Figure A.3.

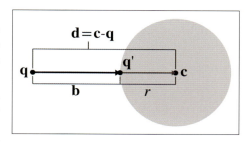

Figure A.3
Finding the closest point on a circle

Now, clearly, $\|\mathbf{b}\| = \|\mathbf{d}\| - r$. Therefore,

$$\mathbf{b} = \frac{\|\mathbf{d}\| - r}{\|\mathbf{d}\|}\mathbf{d}.$$

Computing the closest
point on a circle or
sphere

Adding this displacement to $\mathbf{q}$ to project onto the circle, we get

$$\mathbf{q}' = \mathbf{q} + \mathbf{b}$$
$$= \mathbf{q} + \frac{\|\mathbf{d}\| - r}{\|\mathbf{d}\|}\mathbf{d}.$$

If $\|\mathbf{d}\| < r$, then $\mathbf{q}$ is inside the circle. What should we with this situation? Should $\mathbf{q}' = \mathbf{q}$, or should we project $\mathbf{q}$ outwards onto the surface of the circle? Particular circumstances might call for either behavior. If we decide we wish to project the points onto the surface of the circle, then we'll be forced to make an arbitrary decision on what to do in the degenerate case where $\mathbf{q} = \mathbf{c}$.

A.5   Closest Point in an AABB

Let B be an axially aligned bounding box (AABB) defined by the extreme points $\mathbf{p}_{min}$ and $\mathbf{p}_{max}$. For any point $\mathbf{q}$ we can easily compute $\mathbf{q}'$, the closest point in B to $\mathbf{q}$. This is done by "pushing" $\mathbf{q}$ into B along each axis in turn, as illustrated in Listing A.1. Notice that if the point is already inside the box, this code returns the original point.

```
if (x < minX) {
    x = minX;
} else if (x > maxX) {
    x = maxX;
}

if (y < minY) {
    y = minY;
} else if (y > maxY) {
    y = maxY;
}

if (z < minZ) {
    z = minZ;
} else if (z > maxZ) {
    z = maxZ;
}
```

Listing A.1
Computing the closest point in an AABB to a point

A.6   Intersection Tests

The remaining sections of this chapter present an assortment of *intersection tests*. These tests are designed to determine whether two geometric primitives intersect, and (in some cases) to locate the intersection. We will consider different two types of intersection tests: static and dynamic.

- A *static* test checks two *stationary* primitives and detects whether the two primitives intersect. It is a Boolean test—that is, it usually returns only true (there is an intersection) or false (there is no intersection). If the test returns more details about the intersection, this extra information usually has the purpose of describing where the intersection occurred.

- A *dynamic* test checks two *moving* primitives and detects if and when two primitives intersect. Usually, the movement is expressed parametrically, and therefore the result of such a test is not only a Boolean true/false result but also a time value (the value of the parameter t) that indicates when the primitives intersect. For the tests that we consider here, the movement value is a simple linear displacement—a vector offset by which the primitive moves as t varies from 0 to 1.

  Although each object may have its own displacement over the time interval under consideration, it will be often easier to view the problem from the point of view of one of the primitives, considering that primitive to be "still" while the other primitive does all of the "moving." We can easily do this by combining the two displacement vectors to get a single relative displacement vector that describes how the two primitives move in relation to each other. Thus, the dynamic tests will usually involve one stationary primitive and one moving primitive.

  Notice that many important tests involving rays are actually dynamic tests, since a ray can be viewed as a moving point.

A.7 Intersection of Two Implicit Lines in 2D

Finding the intersection of two lines defined implicitly in 2D is a straightforward matter of solving a system of linear equations. We have two equations (the two implicit equations of the lines) and two unknowns (the x- and y-coordinates of the point of intersection). Our two equations are

$$a_1 x + b_1 y = d_1,$$
$$a_2 x + b_2 y = d_2.$$

Solving this system of equations yields

$$x = \frac{b_2 d_1 - b_1 d_2}{a_1 b_2 - a_2 b_1},$$
$$y = \frac{a_1 d_2 - a_2 d_1}{a_1 b_2 - a_2 b_1}.$$

(A.3)

Computing the intersection of two lines in 2D

Just as for any system of linear equations, there are three solution possibilities (as illustrated in Figure A.4):

- There is *one solution*. In this case, the denominators in (A.3) will be nonzero.

- There are *no solutions*. This indicates that the lines are parallel and not intersecting. The denominators are zero.

- There are an *infinite number of solutions*. This is the case when the two lines are coincident. All the numerators and denominators are zero in this case.

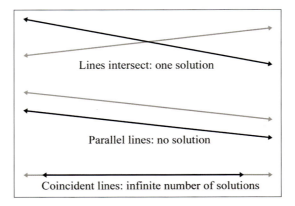

Lines intersect: one solution

Parallel lines: no solution

Coincident lines: infinite number of solutions

Figure A.4
Intersection of two lines in 2D—the three cases

A.8   Intersection of Two Rays in 3D

Given two rays in 3D defined parametrically by

$$\mathbf{r}_1(t_1) = \mathbf{p}_1 + t_1\mathbf{d}_1,$$
$$\mathbf{r}_2(t_2) = \mathbf{p}_2 + t_2\mathbf{d}_2,$$

we can solve for their point of intersection. For a moment, let us not restrict the range of t_1 and t_2; therefore, we consider the infinite lines that contain the rays. The delta vectors $\mathbf{d}_1$ and $\mathbf{d}_2$ do not necessarily have to be unit vectors. If the rays lie in a plane, then we have the same three cases possible as in the previous section:

- The rays intersect at exactly one point.

- The rays are parallel, and there is no intersection.

- The rays are coincident, and there are an infinite number of solutions

However, in 3D we have a fourth case, where the rays are skew and do not share a common plane. An example of skew lines is illustrated in Figure A.5.

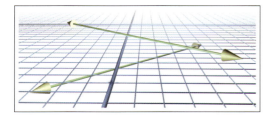

Figure A.5
Skew lines in 3D do not share a common plane or intersect.

We can solve for t_1 and t_2. At the point of intersection,

$$\mathbf{r}_1(t_1) = \mathbf{r}_2(t_2),$$
$$\mathbf{p}_1 + t_1\mathbf{d}_1 = \mathbf{p}_2 + t_2\mathbf{d}_2,$$
$$t_1\mathbf{d}_1 = \mathbf{p}_2 + t_2\mathbf{d}_2 - \mathbf{p}_1,$$
$$(t_1\mathbf{d}_1) \times \mathbf{d}_2 = (\mathbf{p}_2 + t_2\mathbf{d}_2 - \mathbf{p}_1) \times \mathbf{d}_2,$$
$$t_1(\mathbf{d}_1 \times \mathbf{d}_2) = (t_2\mathbf{d}_2) \times \mathbf{d}_2 + (\mathbf{p}_2 - \mathbf{p}_1) \times \mathbf{d}_2,$$
$$t_1(\mathbf{d}_1 \times \mathbf{d}_2) = t_2(\mathbf{d}_2 \times \mathbf{d}_2) + (\mathbf{p}_2 - \mathbf{p}_1) \times \mathbf{d}_2,$$
$$t_1(\mathbf{d}_1 \times \mathbf{d}_2) = t_2\mathbf{0} + (\mathbf{p}_2 - \mathbf{p}_1) \times \mathbf{d}_2,$$
$$t_1(\mathbf{d}_1 \times \mathbf{d}_2) = (\mathbf{p}_2 - \mathbf{p}_1) \times \mathbf{d}_2,$$
$$t_1(\mathbf{d}_1 \times \mathbf{d}_2) \cdot (\mathbf{d}_1 \times \mathbf{d}_2) = ((\mathbf{p}_2 - \mathbf{p}_1) \times \mathbf{d}_2) \cdot (\mathbf{d}_1 \times \mathbf{d}_2),$$
$$t_1 = \frac{((\mathbf{p}_2 - \mathbf{p}_1) \times \mathbf{d}_2) \cdot (\mathbf{d}_1 \times \mathbf{d}_2)}{\|\mathbf{d}_1 \times \mathbf{d}_2\|^2}.$$

We obtain t_2 in a similar fashion:

$$t_2 = \frac{((\mathbf{p}_2 - \mathbf{p}_1) \times \mathbf{d}_1) \cdot (\mathbf{d}_1 \times \mathbf{d}_2)}{\|\mathbf{d}_1 \times \mathbf{d}_2\|^2}.$$

If the lines are parallel (or coincident), then the cross product of $\mathbf{d}_1$ and $\mathbf{d}_2$ is the zero vector, and therefore the denominator of both equations is zero. If the lines are skew, then $\mathbf{r}_1(t_1)$ and $\mathbf{r}_2(t_2)$ are the points of closest approach. To distinguish between skew and intersecting lines, we examine the distance between $\mathbf{r}_1(t_1)$ and $\mathbf{r}_2(t_2)$. Of course, in practice, an exact intersection rarely occurs due to floating point imprecision, and therefore a tolerance must be used.

This discussion has assumed that the range of the parameters t_1 and t_2 were not restricted. If the rays have finite length (or extend in only one direction), then, of course, the appropriate boundary tests would be applied after computing t_1 and t_2.

A.9   Intersection of a Ray and Plane

A ray intersects a plane in 3D at a point. Let the ray be defined parametrically by

$$\mathbf{p}(t) = \mathbf{p}_0 + t\mathbf{d}.$$

The plane will be defined in the standard implicit manner, by all points $\mathbf{p}$ such that

$$\mathbf{p} \cdot \mathbf{n} = d.$$

Although we often restrict the plane normal $\mathbf{n}$ and the ray direction vector $\mathbf{d}$ to be unit vectors, in this case neither restriction is necessary.

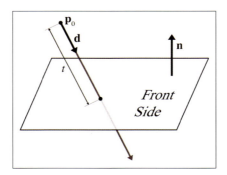

Figure A.6
Intersection of a ray and plane in 3D

Let us solve for t at the point of intersection, assuming an infinite ray for the moment:

Parametric intersection of a ray and a plane

$$(\mathbf{p}_0 + t\mathbf{d}) \cdot \mathbf{n} = d,$$
$$\mathbf{p}_0 \cdot \mathbf{n} + t\mathbf{d} \cdot \mathbf{n} = d,$$
$$t\mathbf{d} \cdot \mathbf{n} = d - \mathbf{p}_0 \cdot \mathbf{n},$$
$$t = \frac{d - \mathbf{p}_0 \cdot \mathbf{n}}{\mathbf{d} \cdot \mathbf{n}}. \tag{A.4}$$

If the ray is parallel to the plane, then the denominator $\mathbf{d} \cdot \mathbf{n}$ is zero and there is no intersection. If the value for t is out of range, then the ray does not intersect the plane. We may also wish to intersect only with the front of the plane. In this case, we say there is an intersection only if the ray points in a direction opposite to the normal of the plane (i.e., $\mathbf{d} \cdot \mathbf{n} < 0$).

A.10   Intersection of an AABB and Plane

Consider a 3D axially aligned bounding box defined by extreme points $\mathbf{p}_{min}$ and $\mathbf{p}_{max}$, and a plane defined in the standard implicit manner by all points $\mathbf{p}$ that satisfy

$$\mathbf{p} \cdot \mathbf{n} = d,$$

where $\mathbf{n}$ is not necessarily a unit vector. The plane must be expressed in the same coordinate space as the AABB.

One obvious implementation strategy for a static test would be to classify each corner point as being on either the front or back side of the plane. We do this by taking the dot products of the corner points with $\mathbf{n}$ and comparing these dot products with d. If all of the dot products are greater than d, then the box is completely on the front side of the plane. If all of the dot products are less than d, then the box is completely on the back side of the plane.

As it turns out, we don't have to check all eight corner points. We'll use a trick similar to the one used in Section 9.4.4 to transform an AABB. For example, if $n_x > 0$, then the corner with the minimal dot product has $x = x_{min}$ and the corner with the maximal dot product has $x = x_{max}$. If $n_x < 0$, then the opposite is true. Similar statements apply to n_y and n_z. We compute the minimum and maximum dot product values. If the minimum dot product value is greater than d, or if the maximum dot product is less than d, then there is no intersection. Otherwise, two corners were found that are on opposite sides of the plane, and therefore an intersection is detected. This strategy is implemented in Listing A.2.

```
// Perform static AABB-plane intersection test.  Returns:
//
// <0    Box is completely on the BACK side of the plane
// >0    Box is completely on the FRONT side of the plane
// 0     Box intersects the plane
int AABB3::classifyPlane(const Vector3 &n, float d) const {

    // Inspect the normal and compute the minimum and maximum
    // D values.
    float minD, maxD;

    if (n.x > 0.0f) {
        minD = n.x*min.x; maxD = n.x*max.x;
    } else {
        minD = n.x*max.x; maxD = n.x*min.x;
    }

    if (n.y > 0.0f) {
        minD += n.y*min.y; maxD += n.y*max.y;
    } else {
        minD += n.y*max.y; maxD += n.y*min.y;
    }

    if (n.z > 0.0f) {
        minD += n.z*min.z; maxD += n.z*max.z;
```

```
    } else {
        minD += n.z*max.z; maxD += n.z*min.z;
    }

    // Check if completely on the front side of the plane
    if (minD >= d) {
        return +1;
    }

    // Check if completely on the back side of the plane
    if (maxD <= d) {
        return −1;
    }

    // We straddle the plane
    return 0;
}
```

Listing A.2
Detecting static intersection of AABB and plane

A dynamic test is just one step further. Let's consider the plane to be stationary (recall from Section A.6 that it is simpler to view the test from the frame of reference of one of the moving objects). The displacement of the box will be defined by a unit vector $\mathbf{d}$ and a length l. As before, we first locate the corner points with the minimum and maximum dot products and check for an intersection at $t = 0$. If the box is not initially intersecting the plane, then it must first strike the plane at the corner point closest to the plane. This will be one of the two corner points identified in the first step. If we are interested only in colliding with the "front" of the plane, then we can always use the corner with the minimum dot product value. Once we have determined which corner will strike the plane, we use the ray-plane intersection test in Section A.9.

A.11   Intersection of Three Planes

In 3D, three planes intersect at a point, as shown in Figure A.7.

Let the three planes be defined implicitly as

$$\mathbf{p} \cdot \mathbf{n}_1 = d_1, \qquad \mathbf{p} \cdot \mathbf{n}_2 = d_2, \qquad \mathbf{p} \cdot \mathbf{n}_3 = d_3.$$

Although we usually use unit vectors for the plane normals, in this case it is not necessary that $\mathbf{n}_i$ be of unit length. These three plane equations give us a system of linear equations with three equations and three unknowns (the x-, y-, and z-coordinates of the point of intersection). Solving this system of equations yields the following result, from Goldman [24]:

Three planes intersect at a point

$$\mathbf{p} = \frac{d_1(\mathbf{n}_2 \times \mathbf{n}_3) + d_2(\mathbf{n}_3 \times \mathbf{n}_1) + d_3(\mathbf{n}_1 \times \mathbf{n}_2)}{(\mathbf{n}_1 \times \mathbf{n}_2) \cdot \mathbf{n}_3}.$$

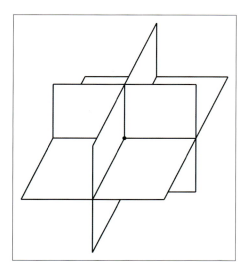

Figure A.7
Three planes intersect at a point in 3D

If any pair of planes is parallel, then the point of intersection either does not exist or is not unique. In either case, the triple product in the denominator is zero.

A.12   Intersection of Ray and a Circle or Sphere

This section discusses how to compute the intersection of a ray and a circle in 2D. This also works for computing the intersection of a ray and a sphere in 3D, since we can operate in the plane that contains the ray and the center of the circle and turn the 3D problem into a 2D one. (If the ray lies on a line that passes through the center of the sphere, the plane is not uniquely defined. This not a problem, however, because any of the infinitely many planes that pass through the ray and the center of the sphere can be used.)

We will use a construction inspired by Hultquist [36]; see Figure A.8. The sphere is defined by its center $\mathbf{c}$ and radius r, and the ray is defined by

$$\mathbf{p}(t) = \mathbf{p}_0 + t\hat{\mathbf{d}}.$$

In this case, we use a unit vector $\hat{\mathbf{d}}$ and vary t from 0 to l, where l is the length of the ray.

We are solving for the value of t at the point of intersection. Clearly, $t = a - f$. We can compute a as follows. Let $\mathbf{e}$ be the vector from $\mathbf{p}_0$ to $\mathbf{c}$:

$$\mathbf{e} = \mathbf{c} - \mathbf{p}_0.$$

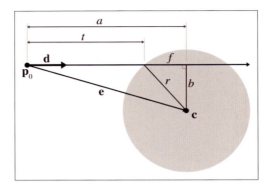

Now we project $\mathbf{e}$ onto $\hat{\mathbf{d}}$ (see Section 2.11.2). The length of this vector is a, and can be computed by

$$a = \mathbf{e} \cdot \hat{\mathbf{d}}.$$

Now all that remains is to compute f. First, by the Pythagorean theorem, we clearly see that

$$f^2 + b^2 = r^2.$$

We can solve for b^2 by using the Pythagorean theorem on the larger triangle:

$$a^2 + b^2 = e^2,$$
$$b^2 = e^2 - a^2,$$

where e is the distance from the origin of the ray to the center, that is, the length of the vector $\mathbf{e}$. Thus, e^2 can be computed by

$$e^2 = \mathbf{e} \cdot \mathbf{e}.$$

Substituting and solving for f, we get

$$f^2 + b^2 = r^2,$$
$$f^2 + (e^2 - a^2) = r^2,$$
$$f^2 = r^2 - e^2 + a^2,$$
$$f = \sqrt{r^2 - e^2 + a^2}.$$

Finally, solving for t, we have

$$t = a - f$$
$$= a - \sqrt{r^2 - e^2 + a^2}.$$

Parametric intersection of a ray and a circle or sphere

If the argument to the square root $(r^2 - e^2 + a^2)$ is negative, then the ray does not intersect the sphere.

The origin of the ray could be inside the sphere. This is indicated by $e^2 < r^2$. Appropriate behavior in this case would vary, depending on the purpose of the test.

A.13   Intersection of Two Circles or Spheres

Detecting the static intersection of two spheres is relatively easy. (The discussion in this section also applies to circles—in fact, we use circles in the diagrams.) Consider two spheres defined by centers c_1 and c_2 and radii r_1 and r_2, as shown in Figure A.9. Let d be the distance between their centers. Clearly, the spheres intersect if $d < r_1 + r_2$. In practice, we can avoid the square root involved in the calculation of d by checking that $d^2 < (r_1 + r_2)^2$.

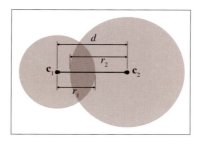

Figure A.9
Intersection of two spheres

Detecting the intersection of two moving spheres is slightly more difficult. Assume, for the moment, that we have two separate displacement vectors d_1 and d_2, one for each sphere, which describe how the spheres will move during the period of time under consideration. This is shown in Figure A.10.

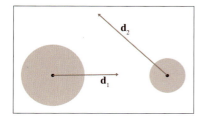

Figure A.10
Two moving spheres

We can simplify the problem by viewing it from the point of view of the first sphere, considering that sphere to be "stationary" while the other sphere is the "moving" sphere. This gives us a single displacement vector $\mathbf{d}$, computed as the difference of the two movement vectors $\mathbf{d}_2 - \mathbf{d}_1$. This is illustrated in Figure A.11.

Let the stationary sphere be defined by its center $\mathbf{c}_s$ and radius r_s. The radius of the moving sphere is r_m. The center of the moving sphere is $\mathbf{c}_m$ at $t = 0$. Rather than varying t from 0 to 1 as described previously, we normalize $\hat{\mathbf{d}}$ and vary t from 0 to l, where l is the length of the total relative displacement. So the position of the center of the moving sphere at time t

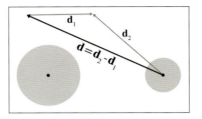

Figure A.11
Combining displacement vectors so that one
sphere is considered stationary

is given by $\mathbf{c}_m + t\hat{\mathbf{d}}$. Our goal is to find t, the time at which the moving
sphere touches the stationary sphere. The geometry involved is illustrated
in Figure A.12.

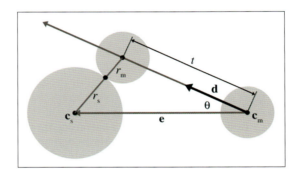

Figure A.12
Dynamic intersection of
circles or spheres

To solve for t, we begin by calculating an intermediate vector $\mathbf{e}$ as the
vector from $\mathbf{c}_m$ to $\mathbf{c}_s$, and set r equal to the sum of the radii:

$$\mathbf{e} = \mathbf{c}_s - \mathbf{c}_m,$$
$$r = r_m + r_s.$$

According to the law of cosines (see Section 1.4.5), we have

$$r^2 = t^2 + \|\mathbf{e}\|^2 - 2t\|\mathbf{e}\| \cos \theta.$$

By applying the geometric interpretation of the dot product (see Section 2.11.2) and simplifying, we get

$$r^2 = t^2 + \|\mathbf{e}\|^2 - 2t\|\mathbf{e}\| \cos \theta,$$
$$r^2 = t^2 + \mathbf{e} \cdot \mathbf{e} - 2t(\mathbf{e} \cdot \hat{\mathbf{d}}),$$
$$0 = t^2 - 2(\mathbf{e} \cdot \hat{\mathbf{d}})t + \mathbf{e} \cdot \mathbf{e} - r^2.$$

Finally, by applying the quadratic formula, we have

$$0 = t^2 - 2(\mathbf{e} \cdot \hat{\mathbf{d}})t + \mathbf{e} \cdot \mathbf{e} - r^2,$$

$$t = \frac{2(\mathbf{e} \cdot \hat{\mathbf{d}}) \pm \sqrt{\left(-2(\mathbf{e} \cdot \hat{\mathbf{d}})\right)^2 - 4(\mathbf{e} \cdot \mathbf{e} - r^2)}}{2},$$

$$t = \frac{2(\mathbf{e} \cdot \hat{\mathbf{d}}) \pm \sqrt{4(\mathbf{e} \cdot \hat{\mathbf{d}})^2 - 4(\mathbf{e} \cdot \mathbf{e} - r^2)}}{2},$$

$$t = \mathbf{e} \cdot \hat{\mathbf{d}} \pm \sqrt{(\mathbf{e} \cdot \hat{\mathbf{d}})^2 + r^2 - \mathbf{e} \cdot \mathbf{e}}.$$

Which root do we pick? The lower number (the negative root) produces the t value when the spheres *begin intersecting*. The greater number (the positive root) is the point where the spheres *cease to intersect*. We are interested in the first intersection:

$$t = \mathbf{e} \cdot \hat{\mathbf{d}} - \sqrt{(\mathbf{e} \cdot \hat{\mathbf{d}})^2 + r^2 - \mathbf{e} \cdot \mathbf{e}}.$$

If $\|\mathbf{e}\| < r$, then the spheres are intersecting at $t = 0$. If $t < 0$ or $t > l$, then the intersection does not occur within the period of time being considered. If the value inside the square root is negative, then there is no intersection.

A.14   Intersection of a Sphere and AABB

To detect the static intersection of a sphere and an AABB, we first find the point on the box that is closest to the center of the sphere by using the techniques from Section A.5. We compute the distance from this point to the center of the sphere and compare this distance with the radius. (Actually, in practice we compare the distance squared against the radius squared to avoid the square root in the distance computation.) If the distance is smaller than the radius, then the sphere intersects the AABB.

Arvo [2] discusses this technique, which he uses for intersecting spheres with "solid" boxes. He also discusses some tricks for intersecting spheres with "hollow" boxes.

Unfortunately, the dynamic test is more complicated than the static one. For details, see Lengyel [42].

A.15   Intersection of a Sphere and a Plane

Detecting the static intersection of a sphere and a plane is relatively easy—we simply compute the distance from the center of the sphere to the plane by using Equation (9.14). If this distance is less than the radius of the sphere, then the sphere intersects the plane. We can actually make a more robust test, which classifies the sphere as being completely on the front, completely on the back, or straddling the sphere. A code snippet is given in Listing A.3.

```
// Given a sphere and plane, determine which side of the plane
// the sphere is on.
//
// Return values:
//
// < 0   Sphere is completely on the back
// > 0   Sphere is completely on the front
// 0     Sphere straddles plane

int classifySpherePlane(
    const Vector3 &planeNormal,    // must be normalized
    float         planeD,          // p * planeNormal = planeD
    const Vector3 &sphereCenter,   // center of sphere
    float         sphereRadius     // radius of sphere
) {

    // Compute distance from center of sphere to the plane
    float d = planeNormal * sphereCenter - planeD;

    // Completely on the front side?
    if (d >= sphereRadius) {
        return +1;
    }

    // Completely on the back side?
    if (d <= -sphereRadius) {
        return -1;
    }

    // Sphere intersects the plane
    return 0;
}
```

Listing A.3
Determining which side of a plane a sphere is on

The dynamic situation is only slightly more complicated. We consider the plane to be stationary, assigning all relative displacement to the sphere.

We define the plane in the usual manner by a normalized surface normal $\hat{\mathbf{n}}$ and distance value d such that all points $\mathbf{p}$ in the plane satisfy the equation $\mathbf{p} \cdot \hat{\mathbf{n}} = d$. The sphere is defined by its radius r and the initial position of the center, $\mathbf{c}$. The displacement of the sphere is given by a unit vector $\hat{\mathbf{d}}$ specifying the direction, and a distance l. As t varies from 0 to l, the motion of the center of the sphere is given by the line equation $\mathbf{c} + t\hat{\mathbf{d}}$. This situation is shown, viewing the plane edge-on, in Figure A.13.

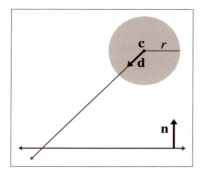

Figure A.13
A sphere moving towards a plane

The problem is greatly simplified by realizing that no matter where on the surface of the plane the intersection occurs, the point of contact on the surface of the sphere is always the same. That point of contact $\mathbf{p}_0$ is given by $\mathbf{c} - r\hat{\mathbf{n}}$, as shown in Figure A.14.

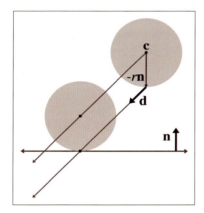

Figure A.14
Point of contact between a sphere and a plane

Now that we know which point on the sphere first contacts the plane, we can use a simple ray-plane intersection test from Section A.9. We start with our solution to the ray-plane intersection test from (A.4) and then substitute $\mathbf{c} - r\hat{\mathbf{n}}$ for $\mathbf{p}_0$:

$$t = \frac{d - \mathbf{p}_0 \cdot \hat{\mathbf{n}}}{\hat{\mathbf{d}} \cdot \hat{\mathbf{n}}},$$

$$= \frac{d - (\mathbf{c} - r\hat{\mathbf{n}}) \cdot \hat{\mathbf{n}}}{\hat{\mathbf{d}} \cdot \hat{\mathbf{n}}},$$

$$= \frac{d - \mathbf{c} \cdot \hat{\mathbf{n}} + r}{\hat{\mathbf{d}} \cdot \hat{\mathbf{n}}}.$$

Dynamic intersection of a sphere and plane

A.16   Intersection of a Ray and a Triangle

The ray-triangle intersection test is very important in graphics and computational geometry. In the absence of a special raytrace test against a given complex object, we can always represent (or at least approximate) the surface of an object as a triangle mesh and then raytrace against this triangle mesh representation.

Here we use a simple strategy from Badouel [4]. The first step is to compute the point where the ray intersects the plane containing the triangle. Section A.9 showed how to compute the intersection of a plane and a ray. Then we test to see whether that point is inside the triangle by computing the barycentric coordinates of the point, as discussed in Section 9.6.4.

To speed up this test, we use a few tricks:

- Detect and return a negative result (no collision) as soon as possible. This is known as "early out."

- Defer expensive mathematical operations, such as division, as long as possible. This is done for two reasons. First, if the result of the expensive calculation is not needed (for example, if we took an early out), then the time we spent performing the operation was wasted. Second, it gives the compiler plenty of room to take advantage of the operator pipeline in modern processors. If an operation such as division has a long latency, then the compiler may be able to look ahead and generate code that begins the division operation early. It then generates code that performs other tests (possibly taking an early out) while the division operation is under way. Then, at execution time, if and when the result of the division is actually needed, the result will be available or at least partially completed.

- Only detect collisions where the ray approaches the triangle from the *front* side. This allows us to take a very early out on approximately half of the triangles. Intersecting with both sides is slightly slower.

Listing A.4 implements these techniques. Although it is commented in the listing, we have chosen to perform some floating point comparisons "backwards" since this behaves better in the presence of invalid floating point input data and NaNs (Not a Number).

```
float rayTriangleIntersect(
    const Vector3 &rayOrg,      // origin of the ray
    const Vector3 &rayDelta,    // ray length and direction
    const Vector3 &p0,          // triangle vertices
    const Vector3 &p1,          // .
    const Vector3 &p2,          // .
    float minT                  // closest intersection found so far.
                                // (Start with 1.0)
```

```
) {

    // We'll return this huge number of no intersection is detected
    const float kNoIntersection = FLT_MAX;

    // Compute clockwise edge vectors.
    Vector3 e1 = p1 - p0;
    Vector3 e2 = p2 - p1;

    // Compute surface normal.  (Unnormalized)
    Vector3 n = crossProduct(e1, e2);

    // Compute gradient, which tells us how steep of an angle
    // we are approaching the *front* side of the triangle
    float dot = n * rayDelta;

    // Check for a ray that is parallel to the triangle, or not
    // pointing towards the front face of the triangle.
    //
    // Note that this also will reject degenerate triangles and
    // rays as well.  We code this in a very particular
    // way so that NANs will bail here.  (I.e. this does NOT
    // behave the same as ''dot >= 0.0f'' when NANs are involved)
    if (!(dot < 0.0f)) {
        return kNoIntersection;
    }

    // Compute d value for the plane equation.  We will
    // use the plane equation with d on the right side:
    // Ax + By + Cz = d
    float d = n * p0;

    // Compute parametric point of intersection with the plane
    // containing the triangle, checking at the earliest
    // possible stages for trivial rejection
    float t = d - n * rayOrg;

    // Is ray origin on the backside of the polygon?  Again,
    // we phrase the check so that NANs will bail
    if (!(t <= 0.0f)) {
        return kNoIntersection;
    }

    // Closer intersection already found?  (Or does
    // ray not reach the plane?)
    //
    // since dot < 0:
    //
    //          t/dot > minT
    //
    // is the same as
    //
    //          t < dot*minT
    //
    // (And then we invert it for NAN checking...)
    if (!(t >= dot*minT)) {
        return kNoIntersection;
    }

    // OK, ray intersects the plane.  Compute actual parametric
    // point of intersection
    t /= dot;
    assert(t >= 0.0f);
    assert(t <= minT);
```

```
// Compute 3D point of intersection
Vector3 p = rayOrg + rayDelta*t;

// Find dominant axis to select which plane
// to project onto, and compute u's and v's
float u0, u1, u2;
float v0, v1, v2;
if (fabs(n.x) > fabs(n.y)) {
    if (fabs(n.x) > fabs(n.z)) {
        u0 = p.y - p0.y;
        u1 = p1.y - p0.y;
        u2 = p2.y - p0.y;

        v0 = p.z - p0.z;
        v1 = p1.z - p0.z;
        v2 = p2.z - p0.z;
    } else {
        u0 = p.x - p0.x;
        u1 = p1.x - p0.x;
        u2 = p2.x - p0.x;

        v0 = p.y - p0.y;
        v1 = p1.y - p0.y;
        v2 = p2.y - p0.y;
    }
} else {
    if (fabs(n.y) > fabs(n.z)) {
        u0 = p.x - p0.x;
        u1 = p1.x - p0.x;
        u2 = p2.x - p0.x;

        v0 = p.z - p0.z;
        v1 = p1.z - p0.z;
        v2 = p2.z - p0.z;
    } else {
        u0 = p.x - p0.x;
        u1 = p1.x - p0.x;
        u2 = p2.x - p0.x;

        v0 = p.y - p0.y;
        v1 = p1.y - p0.y;
        v2 = p2.y - p0.y;
    }
}

// Compute denominator, check for invalid
float temp = u1 * v2 - v1 * u2;
if (!(temp != 0.0f)) {
    return kNoIntersection;
}
temp = 1.0f / temp;

// Compute barycentric coords, checking for out-of-range
// at each step
float alpha = (u0 * v2 - v0 * u2) * temp;
if (!(alpha >= 0.0f)) {
    return kNoIntersection;
}

float beta = (u1 * v0 - v1 * u0) * temp;
if (!(beta >= 0.0f)) {
    return kNoIntersection;
}
```

```
    float gamma = 1.0f - alpha - beta;
    if (!(gamma >= 0.0f)) {
        return kNoIntersection;
    }

    // Return parametric point of intersection
    return t;
}
```

Listing A.4
Ray-triangle intersection test

There is one more significant strategy, not illustrated in Listing A.4, for optimizing expensive calculations: precompute their results. If values such as the polygon normal can be computed ahead of time, then different strategies may be used.

Because of the fundamental importance of this test, programmers are always looking for ways to make it faster. The technique we have given here is a standard one that is easy to understand and produces the barycentric coordinates, often a useful byproduct, as a side effect. It is not the fastest. See Tomas Akenine-Möller's collection of intersection tests on the web page for *Real-Time Rendering* at http://www.realtimerendering.com/intersections.html.

A.17 Intersection of Two AABBs

Detecting the static intersection of two AABBs is an extremely important operation. Luckily, it's rather trivial.[1] We simply check for overlapping extents on each dimension independently. If there is no overlap on a particular dimension, then the two AABBs do not intersect. This technique is used in Listing A.5.

```
bool aabbsOverlap(const AABB3 &a, const AABB3 &b) {

    // Check for a separating axis.
    if (a.min.x >= b.max.x) return false;
    if (a.max.x <= b.min.x) return false;
    if (a.min.y >= b.max.y) return false;
    if (a.max.y <= b.min.y) return false;
    if (a.min.z >= b.max.z) return false;
    if (a.max.z <= b.min.z) return false;

    // Overlap on all three axes, so their
    // intersection must be non-empty
    return true;
}
```

Listing A.5
AABB–AABB overlap test

---

[1]This is one of Fletcher's favorite interview questions. It's surprising how many programmers do not know how to perform this very simple operation. Don't be one of those applicants!

This strategy is actually an instance of a more general strategy known as the *separating axis* test. If two convex polyhedra do not overlap, then there exists a separating axis upon which, if we project the two polyhedra, their projections will not overlap. (In 3D, it's easier to visualize a plane perpendicular to the separating axis that can be placed between the two polyhedra.) The key to the separating axis method is that only a finite number of axes need to be tested: the normals of the faces and certain cross products; for details, see Ericson [18]. If the projections of the polyhedra onto those axes overlap in all cases, then it is safe to assume that no separating axis can be found. In the case of two AABBs, only the three cardinal axes need to be tested. Furthermore, these "projections" simply extract the appropriate coordinate.

The dynamic intersection of AABBs is only slightly more complicated. Consider a stationary AABB defined by extreme points $\mathbf{s}_{\min}$ and $\mathbf{s}_{\max}$, and a moving AABB, which has extreme points $\mathbf{m}_{\min}$ and $\mathbf{m}_{\max}$ in the initial position at $t = 0$. The moving AABB displaces by an amount given by the vector $\mathbf{d}$, as t varies from 0 to 1.

Our task is to compute t, the parametric point in time where the moving box first collides with the stationary box. (We assume that the boxes are not initially intersecting.) To do this, we will attempt to determine the first point in time when the boxes have overlap on all dimensions simultaneously. Since this applies in 2D or 3D, we illustrate the problem here in 2D; extending the technique into 3D is straightforward. We analyze each coordinate separately, solving two (or three, in 3D) separate one-dimensional problems, and then combining these results to give the answer.

The problem is now one-dimensional. We need to know the interval of time when the two boxes overlap on a particular dimension. Imagine projecting the problem onto the x-axis (for example), as shown in Figure A.15.

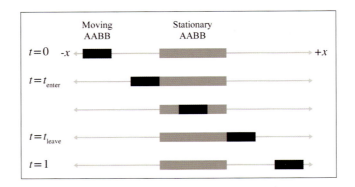

Figure A.15
Projecting the dynamic AABB intersection problem onto one axis

As we advance in time, the line segment representing the moving box will slide along the number line. In the illustration in Figure A.15, at $t = 0$

the moving box is completely to the left of the stationary box, and at $t = 1$, the moving box is completely to the right of the stationary box. There is a point t_{enter} where the boxes first begin to overlap, and a point t_{leave} where the boxes cease to overlap. For the dimension we are considering, let $m_{\min}(t)$ and $m_{\max}(t)$ be the minimum and maximum values, respectively, of the moving box at time t, given by

$$m_{\min}(t) = m_{\min}(0) + td,$$
$$m_{\max}(t) = m_{\max}(0) + td,$$

where $m_{\min}(0)$ and $m_{\max}(0)$ are the initial extents of the moving box and d is the component of the displacement vector $\mathbf{d}$ for this axis. Let $s_{\min}$ and $s_{\max}$ have similar definitions for the stationary box. (Of course, these values are independent of t since the box is stationary.) The time t_{enter} is the t value for which $m_{\max}(t) = s_{\min}$. Solving, we get

$$m_{\max}(t_{\text{enter}}) = s_{\min},$$
$$m_{\max}(0) + t_{\text{enter}}d = s_{\min},$$
$$t_{\text{enter}}d = s_{\min} - m_{\max}(0),$$
$$t_{\text{enter}} = \frac{s_{\min} - m_{\max}(0)}{d}.$$

Likewise, we can solve for t_{leave}:

$$m_{\min}(t_{\text{leave}}) = s_{\max},$$
$$m_{\min}(0) + t_{\text{leave}}d = s_{\max},$$
$$t_{\text{leave}}d = s_{\max} - m_{\min}(0),$$
$$t_{\text{leave}} = \frac{s_{\max} - m_{\min}(0)}{d}.$$

Three important points are noted here:

- If the denominator d is zero, then boxes either always overlap or never overlap.

- If the moving box begins on the right side of the stationary box and moves left, then $t_{\text{enter}} > t_{\text{leave}}$. We handle this scenario by swapping their values to ensure that $t_{\text{enter}} < t_{\text{leave}}$.

- The values for t_{enter} and t_{leave} may be outside the range $[0,1]$. To accommodate t values outside this range, we can think of the moving box as moving along an infinite trajectory parallel to d. If $t_{\text{enter}} > 1$ or $t_{\text{leave}} < 0$, then there is no overlap in the period of time under consideration.

We now have a way to find the interval of time, bounded by t_{enter} and t_{leave}, when the two boxes overlap on a single dimension. The intersection of these intervals on all dimensions gives us the interval of time where the boxes intersect with each other. This is illustrated in Figure A.16 for two time intervals in 2D.

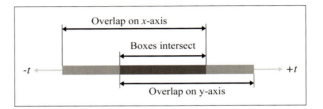

Figure A.16
Intersecting two intervals of time

Don't confuse Figure A.16 with Figure A.15. In Figure A.16, the axis is the time axis; in Figure A.15 the axis is the x-axis.

If the interval is empty, then the boxes never collide. If the interval lies completely outside the range $0 \leq t \leq 1$, then there is no collision over the period of time of interest. Actually, the interval during which the boxes overlap is more information than we wanted, since we are interested only in the point in time when the boxes begin intersecting, not when they cease to intersect. Still, we need to maintain this interval, mainly to determine whether it is empty.

Unfortunately, in practice, bounding boxes for objects are rarely axially aligned in the same coordinate space. However, because this test is relatively fast, it is useful as a preliminary trivial rejection test, to be followed by a more specific (and usually more expensive) test.

A.18   Intersection of a Ray and an AABB

Computing the intersection of a ray with an AABB is an important calculation because the result of this test is commonly used for trivial rejection on more complicated objects. (For example, if we wish to raytrace against multiple triangle meshes, we can first raytrace against the AABBs of the meshes to trivially reject entire meshes at once, rather than having to check each triangle.)

Woo [72] describes a method that first determines which side of the box would be intersected and then performs a ray-plane intersection test against the plane containing that side. If the point of intersection with the plane is within the box, then there is an intersection; otherwise, there is no intersection. This is implemented in Listing A.6.

```
// Return parametric point of intersection 0...1, or a really huge
// number if no intersection is found
float AABB3::rayIntersect(
    const Vector3 &rayOrg,        // orgin of the ray
    const Vector3 &rayDelta,      // length and direction of the ray
    Vector3 *returnNormal         // optionally, the normal is returned
) const {

    // We'll return this huge number if no intersection
    const float kNoIntersection = FLT_MAX;

    // Check for point inside box, trivial reject, and
    // determine parametric distance to each front face
    bool inside = true;

    float xt, xn;
    if (rayOrg.x < min.x) {
        xt = min.x - rayOrg.x;
        if (xt > rayDelta.x) return kNoIntersection;
        xt /= rayDelta.x;
        inside = false;
        xn = -1.0f;
    } else if (rayOrg.x > max.x) {
        xt = max.x - rayOrg.x;
        if (xt < rayDelta.x) return kNoIntersection;
        xt /= rayDelta.x;
        inside = false;
        xn = 1.0f;
    } else {
        xt = -1.0f;
    }

    float yt, yn;
    if (rayOrg.y < min.y) {
        yt = min.y - rayOrg.y;
        if (yt > rayDelta.y) return kNoIntersection;
        yt /= rayDelta.y;
        inside = false;
        yn = -1.0f;
    } else if (rayOrg.y > max.y) {
        yt = max.y - rayOrg.y;
        if (yt < rayDelta.y) return kNoIntersection;
        yt /= rayDelta.y;
        inside = false;
        yn = 1.0f;
    } else {
        yt = -1.0f;
    }

    float zt, zn;
    if (rayOrg.z < min.z) {
        zt = min.z - rayOrg.z;
        if (zt > rayDelta.z) return kNoIntersection;
        zt /= rayDelta.z;
```

```
                inside = false;
                zn = −1.0f;
        } else if (rayOrg.z > max.z) {
                zt = max.z − rayOrg.z;
                if (zt < rayDelta.z) return kNoIntersection;
                zt /= rayDelta.z;
                inside = false;
                zn = 1.0f;
        } else {
                zt = −1.0f;
        }

        // Ray origin inside box?
        if (inside) {
                if (returnNormal != NULL) {
                        *returnNormal = −rayDelta;
                        returnNormal−>normalize();
                }
                return 0.0f;
        }

        // Select farthest plane − this is
        // the plane of intersection.
        int which = 0;
        float t = xt;
        if (yt > t) {
                which = 1;
                t = yt;
        }
        if (zt > t) {
                which = 2;
                t = zt;
        }
        switch (which) {

                case 0: // intersect with yz plane
                {
                        float y = rayOrg.y + rayDelta.y*t;
                        if (y < min.y || y > max.y) return kNoIntersection;
                        float z = rayOrg.z + rayDelta.z*t;
                        if (z < min.z || z > max.z) return kNoIntersection;

                        if (returnNormal != NULL) {
                                returnNormal−>x = xn;
                                returnNormal−>y = 0.0f;
                                returnNormal−>z = 0.0f;
                        }

                } break;

                case 1: // intersect with xz plane
                {
                        float x = rayOrg.x + rayDelta.x*t;
                        if (x < min.x || x > max.x) return kNoIntersection;
                        float z = rayOrg.z + rayDelta.z*t;
                        if (z < min.z || z > max.z) return kNoIntersection;

                        if (returnNormal != NULL) {
                                returnNormal−>x = 0.0f;
                                returnNormal−>y = yn;
                                returnNormal−>z = 0.0f;
                        }

                } break;
```

```
        case 2: // intersect with xy plane
        {
            float x = rayOrg.x + rayDelta.x*t;
            if (x < min.x || x > max.x) return kNoIntersection;
            float y = rayOrg.y + rayDelta.y*t;
            if (y < min.y || y > max.y) return kNoIntersection;

            if (returnNormal != NULL) {
                returnNormal->x = 0.0f;
                returnNormal->y = 0.0f;
                returnNormal->z = zn;
            }

        } break;
    }

    // Return parametric point of intersection
    return t;

}
```

Listing A.6
Ray-box intersection

Appendix B

Answers to the Exercises

I believe that every human has a finite number of heart-beats.
I don't intend to waste any of mine running around doing exercises.

— Buzz Aldrin (1930–)

B.1 Chapter 1

(Page 27.)

1. $\mathbf{a} = (-2.5, 3)$ $\mathbf{b} = (1, 2)$ $\mathbf{c} = (2.5, 2)$
   $\mathbf{d} = (-1, 1)$ $\mathbf{e} = (0, 0)$ $\mathbf{f} = (2, -0.5)$
   $\mathbf{g} = (-0.5, -1.5)$ $\mathbf{h} = (0, -2)$ $\mathbf{i} = (-3, -2)$

2. $\mathbf{a} = (1, 2, 4)$ $\mathbf{b} = (-3, -3, -5)$ $\mathbf{c} = (-3, 6, 2.5)$
   $\mathbf{d} = (3, 0, -1)$ $\mathbf{e} = (0, 0, 0)$ $\mathbf{f} = (0, 0, 3)$
   $\mathbf{g} = (-3.5, 4, 0)$ $\mathbf{h} = (5, -5, -1.5)$ $\mathbf{i} = (4, 1, 5)$

3. See the table below.

Left-handed						Right-handed					
East	**Up**	**North**	**East**	**Up**	**North**	**East**	**Up**	**North**	**East**	**Up**	**North**
$+x$	$+y$	$+z$	$-x$	$-y$	$+z$	$-x$	$-y$	$-z$	$+x$	$+y$	$-z$
$+x$	$-y$	$-z$	$-x$	$+y$	$-z$	$-x$	$+y$	$+z$	$+x$	$-y$	$+z$
$+x$	$+z$	$-y$	$-x$	$-z$	$-y$	$-x$	$-z$	$+y$	$+x$	$+z$	$+y$
$+x$	$-z$	$+y$	$-x$	$+z$	$+y$	$-x$	$+z$	$-y$	$+x$	$-z$	$-y$
$+y$	$+z$	$+x$	$-y$	$-z$	$+x$	$-y$	$-z$	$-x$	$+y$	$+z$	$-x$
$+y$	$-z$	$-x$	$-y$	$+z$	$-x$	$-y$	$+z$	$+x$	$+y$	$-z$	$+x$
$+y$	$+x$	$-z$	$-y$	$-x$	$-z$	$-y$	$-x$	$+z$	$+y$	$+x$	$+z$
$+y$	$-x$	$+z$	$-y$	$+x$	$+z$	$-y$	$+x$	$-z$	$+y$	$-x$	$-z$
$+z$	$+x$	$+y$	$-z$	$-x$	$+y$	$-z$	$-x$	$-y$	$+z$	$+x$	$-y$
$+z$	$-x$	$-y$	$-z$	$+x$	$-y$	$-z$	$+x$	$+y$	$+z$	$-x$	$+y$
$+z$	$+y$	$-x$	$-z$	$-y$	$-x$	$-z$	$-y$	$+x$	$+z$	$+y$	$+x$
$+z$	$-y$	$+x$	$-z$	$+y$	$+x$	$-z$	$+y$	$-x$	$+z$	$-y$	$-x$

4. (a) Right-handed.   (b) Swap y and z.   (c) Swap y and z.

5. (a) Right-handed.

   (b) $x_{\text{us}} \leftarrow y_{\text{aero}}, \quad y_{\text{us}} \leftarrow -z_{\text{aero}}, \quad z_{\text{us}} \leftarrow x_{\text{aero}}$

   (c) $x_{\text{aero}} \leftarrow z_{\text{us}}, \quad y_{\text{aero}} \leftarrow x_{\text{us}}, \quad z_{\text{aero}} \leftarrow -y_{\text{us}}$

6. (a) CW   (b) CCW   (c) CCW   (d) CW

7. (a) 15   (b) 30   (c) 3840   (d) 2016840   (e) 5050

8. (a) $\pi/6$ (b) $-\pi/4$ (c) $\pi/3$ (d) $\pi/2$ (e) $-\pi$
   (f) $5\pi/4$ (g) $-3\pi/2$ (h) 2.923   (i) 9.198   (j) -6π

9. (a) $-30°$ (b) $120°$ (c) $270°$ (d) $-240°$ (e) $360°$
   (f) $1°$ (g) $10°$ (h) $-900°$ (i) $1800°$ (j) $36°$

10. The scarecrow should have said:

    The sum of the squares of the legs of a right triangle is equal to the square of the remaining side.

    since the Pythagorean theorem is $c^2 = a^2 + b^2$, where a and b are the legs of the right triangle and c is the hypotenuse.

11. (a) $(\sin(\alpha)/\csc(\alpha)) + (\cos(\alpha)/\sec(\alpha)) = \sin^2(\alpha) + \cos^2(\alpha) = 1$

    (b) $(\sec^2(\theta) - 1)/\sec^2(\theta) = 1 - (1/\sec^2(\theta)) = 1 - \cos^2(\theta) = \sin^2(\theta)$

    (c) $1 + \cot^2(t) = 1 + (\cos^2(t)/\sin^2(t)) = (\sin^2(t)/\sin^2(t)) + (\cos^2(t)/\sin^2(t)) = (\sin^2(t) + \cos^2(t))/\sin^2(t) = 1/\sin^2(t) = \csc^2(t)$

    (d) $\cos(\phi)(\tan(\phi) + \cot(\phi)) = \sin(\phi) + (\cos^2(\phi)/\sin(\phi)) = (\sin^2(\phi) + \cos^2(\phi))/\sin(\phi) = 1/\sin(\phi) = \csc(\phi)$

B.2   Chapter 2

(Page 71.)

1. (a) **a** is a 2D row vector. **b** is a 3D column vector. **c** is a 4D column vector.

   (b) $b_y + c_w + a_x + b_z = 0 + 6 + (-3) + 5 = 8$

2. (a) *"How much do you weigh?"* Your weight is a scalar quantity. But the force of gravity, which pulls you downwards, is a vector, and so if you said that weight was a vector for that reason, you are also correct. ("My weight is 150 lbs of force in the *downward* direction.")

   (b) *"Do you have any idea how fast you were going?"* The officer is probably referring to the *speed* of your vehicle, which is a scalar quantity.

   (c) *"It's two blocks north of here."* Vector quantity.

   (d) *"We're cruising from Los Angeles to New York at 600 mph, at an altitude of 33,000 ft."* The speed "600 mph" is a scalar quantity. Since New York is east of Los Angeles, you could reasonably infer an eastward direction, so "600 mph eastward" is a velocity, which is a vector quantity. Likewise, "33,000 ft" is a scalar quantity, although if you're a stickler, you might say that a direction of "up" is implied, in which case "33,000 ft up" is a vector quantity.

3. $\mathbf{a} = [0, 2]$ $\mathbf{b} = [0, -2]$ $\mathbf{c} = [0.5, 2]$

   $\mathbf{d} = [0.5, 2]$ $\mathbf{e} = [0.5, -3]$ $\mathbf{f} = [-2, 0]$

   $\mathbf{g} = [-2, 1]$ $\mathbf{h} = [2.5, 2]$ $\mathbf{i} = [6, 1]$

4. (a) *The size of a vector in a diagram doesn't matter; we just need to draw it in the right place.* **False.** This is reversed; for vectors, size matters (meaning the length of the vector), position doesn't.

   (b) *The displacement expressed by a vector can be visualized as a sequence of axially aligned displacements.* **True.**

   (c) *These axially aligned displacements from the previous question must occur in order.* **False.** We can apply them in any order and get the same end result.

   (d) *The vector $[x, y]$ gives the displacement from the point (x, y) to the origin.* **False.** This is reversed; the vector $[x, y]$ gives the displacement from the origin to the point (x, y).

5. (a) $-\begin{bmatrix} 3 & 7 \end{bmatrix} = \begin{bmatrix} -3 & -7 \end{bmatrix}$

   (b) $\left\| \begin{bmatrix} -12 & 5 \end{bmatrix} \right\| = \sqrt{(-12)^2 + 5^2} = \sqrt{169} = 13$

   (c) $\left\| \begin{bmatrix} 8 & -3 & 1/2 \end{bmatrix} \right\| = \sqrt{8^2 + (-3)^2 + (1/2)^2} = \sqrt{64 + 9 + (1/4)}$

   $$= \sqrt{293/4} \approx 8.56$$

   (d) $3\begin{bmatrix} 4 & -7 & 0 \end{bmatrix} = \begin{bmatrix} (3)(4) & (3)(-7) & (3)(0) \end{bmatrix} = \begin{bmatrix} 12 & -21 & 0 \end{bmatrix}$

   (e) $\begin{bmatrix} 4 & 5 \end{bmatrix}/2 = \begin{bmatrix} 2 & 5/2 \end{bmatrix}$

6. (a) $\begin{bmatrix} 12 & 5 \end{bmatrix}_{\text{norm}} = \dfrac{\begin{bmatrix} 12 & 5 \end{bmatrix}}{\left\| \begin{bmatrix} 12 & 5 \end{bmatrix} \right\|} = \dfrac{\begin{bmatrix} 12 & 5 \end{bmatrix}}{13} = \begin{bmatrix} \dfrac{12}{13} & \dfrac{5}{13} \end{bmatrix}$

   $$\approx \begin{bmatrix} 0.923 & 0.385 \end{bmatrix}$$

   (b) $\begin{bmatrix} 0 & 743.632 \end{bmatrix}_{\text{norm}} = \dfrac{\begin{bmatrix} 0 & 743.632 \end{bmatrix}}{\left\| \begin{bmatrix} 0 & 743.632 \end{bmatrix} \right\|} = \dfrac{\begin{bmatrix} 0 & 743.632 \end{bmatrix}}{\sqrt{0^2 + 743.632^2}}$

   $$= \dfrac{\begin{bmatrix} 0 & 743.632 \end{bmatrix}}{743.632} = \begin{bmatrix} 0 & 1 \end{bmatrix}$$

   (c) $\begin{bmatrix} 8 & -3 & 1/2 \end{bmatrix}_{\text{norm}} = \dfrac{\begin{bmatrix} 8 & -3 & 1/2 \end{bmatrix}}{\left\| \begin{bmatrix} 8 & -3 & 1/2 \end{bmatrix} \right\|} \approx \dfrac{\begin{bmatrix} 8 & -3 & 1/2 \end{bmatrix}}{8.56}$

   $$\approx \begin{bmatrix} 0.935 & -0.350 & 0.058 \end{bmatrix}$$

(d) $\begin{bmatrix} -12 & 3 & -4 \end{bmatrix}_{\text{norm}} = \dfrac{\begin{bmatrix} -12 & 3 & -4 \end{bmatrix}}{\left\| \begin{bmatrix} -12 & 3 & -4 \end{bmatrix} \right\|} = \dfrac{\begin{bmatrix} -12 & 3 & -4 \end{bmatrix}}{\sqrt{(-12)^2 + 3^2 + (-4)^2}}$

$= \dfrac{\begin{bmatrix} -12 & 3 & -4 \end{bmatrix}}{13} = \begin{bmatrix} \dfrac{-12}{13} & \dfrac{3}{13} & \dfrac{-4}{13} \end{bmatrix}$

(e) $\begin{bmatrix} 1 & 1 & 1 & 1 \end{bmatrix}_{\text{norm}} = \dfrac{\begin{bmatrix} 1 & 1 & 1 & 1 \end{bmatrix}}{\left\| \begin{bmatrix} 1 & 1 & 1 & 1 \end{bmatrix} \right\|} = \dfrac{\begin{bmatrix} 1 & 1 & 1 & 1 \end{bmatrix}}{\sqrt{1^2 + 1^2 + 1^2 + 1^2}}$

$= \dfrac{\begin{bmatrix} 1 & 1 & 1 & 1 \end{bmatrix}}{2} = \begin{bmatrix} 0.5 & 0.5 & 0.5 & 0.5 \end{bmatrix}$

7. (a) $\begin{bmatrix} 7 & -2 & -3 \end{bmatrix} + \begin{bmatrix} 6 & 6 & -4 \end{bmatrix} = \begin{bmatrix} 7+6 & -2+6 & -3+(-4) \end{bmatrix} = \begin{bmatrix} 13 & 4 & -7 \end{bmatrix}$

(b) $\begin{bmatrix} 2 & 9 & -1 \end{bmatrix} + \begin{bmatrix} -2 & -9 & 1 \end{bmatrix} = \begin{bmatrix} 2+(-2) & 9+(-9) & -1+1 \end{bmatrix} = \begin{bmatrix} 0 & 0 & 0 \end{bmatrix}$

(c) $\begin{bmatrix} 3 \\ 10 \\ 7 \end{bmatrix} - \begin{bmatrix} 8 \\ -7 \\ 4 \end{bmatrix} = \begin{bmatrix} 3-8 \\ 10-(-7) \\ 7-4 \end{bmatrix} = \begin{bmatrix} -5 \\ 17 \\ 3 \end{bmatrix}$

(d) $\begin{bmatrix} 4 \\ 5 \\ -11 \end{bmatrix} - \begin{bmatrix} -4 \\ -5 \\ 11 \end{bmatrix} = \begin{bmatrix} 4-(-4) \\ 5-(-5) \\ -11-11 \end{bmatrix} = \begin{bmatrix} 8 \\ 10 \\ -22 \end{bmatrix}$

(e) $3\begin{bmatrix} a \\ b \\ c \end{bmatrix} - 4\begin{bmatrix} 2 \\ 10 \\ -6 \end{bmatrix} = \begin{bmatrix} 3a \\ 3b \\ 3c \end{bmatrix} - \begin{bmatrix} 8 \\ 40 \\ -24 \end{bmatrix} = \begin{bmatrix} 3a-8 \\ 3b-40 \\ 3c+24 \end{bmatrix}$

8. (a) $\text{distance}\left(\begin{bmatrix} 10 \\ 6 \end{bmatrix}, \begin{bmatrix} -14 \\ 30 \end{bmatrix} \right) = \sqrt{(10-(-14))^2 + (6-30)^2}$

$= \sqrt{24^2 + (-24)^2} = \sqrt{576 + 576}$

$= \sqrt{1152} \approx 33.94$

(b) $\text{distance}\left(\begin{bmatrix} 0 \\ 0 \end{bmatrix}, \begin{bmatrix} -12 \\ 5 \end{bmatrix} \right) = \sqrt{(0-(-12))^2 + (0-5)^2}$

$= \sqrt{12^2 + (-5)^2} = \sqrt{144 + 25}$

$= \sqrt{169} = 13$

(c) $\text{distance}\left(\begin{bmatrix} 3 \\ 10 \\ 7 \end{bmatrix}, \begin{bmatrix} 8 \\ -7 \\ 4 \end{bmatrix} \right) = \sqrt{(3-8)^2 + (10-(-7))^2 + (7-4)^2}$

$= \sqrt{(-5)^2 + 17^2 + 3^2} = \sqrt{25 + 289 + 9}$

$= \sqrt{323} \approx 17.97$

(d) distance $\left(\begin{bmatrix} -2 \\ -4 \\ 9 \end{bmatrix}, \begin{bmatrix} 6 \\ -7 \\ 9.5 \end{bmatrix} \right) = \sqrt{(6 - (-2))^2 + (-7 - (-4))^2 + (9.5 - 9)^2}$

$$= \sqrt{8^2 + (-3)^2 + (0.5)^2} = \sqrt{64 + 9 + 0.25}$$
$$= \sqrt{73.25} \approx 8.56$$

(e) distance $\left(\begin{bmatrix} 4 \\ -4 \\ -4 \\ 4 \end{bmatrix}, \begin{bmatrix} -6 \\ 6 \\ 6 \\ -6 \end{bmatrix} \right) = \sqrt{(-6 - 4)^2 + (6 - (-4))^2 + (6 - (-4))^2 + (-6 - 4)^2}$

$$= \sqrt{(-10)^2 + (10)^2 + (10)^2 + (-10)^2}$$
$$= \sqrt{100 + 100 + 100 + 100}$$
$$= \sqrt{400} = 20$$

9. (a) $\begin{bmatrix} 2 \\ 6 \end{bmatrix} \cdot \begin{bmatrix} -3 \\ 8 \end{bmatrix} = (2)(-3) + (6)(8) = -6 + 48 = 42$

(b) $-7 \begin{bmatrix} 1 & 2 \end{bmatrix} \cdot \begin{bmatrix} 11 & -4 \end{bmatrix} = \begin{bmatrix} -7 & -14 \end{bmatrix} \cdot \begin{bmatrix} 11 & -4 \end{bmatrix}$

$$= (-7)(11) + (-14)(-4)$$
$$= -21$$

(c) $10 + \begin{bmatrix} -5 \\ 1 \\ 3 \end{bmatrix} \cdot \begin{bmatrix} 4 \\ -13 \\ 9 \end{bmatrix} = 10 + ((-5)(4) + (1)(-13) + (3)(9))$

$$= 10 + (-20 + (-13) + 27)$$
$$= 10 + (-6) = 4$$

(d) $3 \begin{bmatrix} -2 \\ 0 \\ 4 \end{bmatrix} \cdot \left(\begin{bmatrix} 8 \\ -2 \\ 3/2 \end{bmatrix} + \begin{bmatrix} 0 \\ 9 \\ 7 \end{bmatrix} \right) = \begin{bmatrix} -6 \\ 0 \\ 12 \end{bmatrix} \cdot \begin{bmatrix} 8 \\ 7 \\ 17/2 \end{bmatrix}$

$$= (-6)(8) + (0)(7) + (12)(17/2) = 54$$

10. $\mathbf{v}_{\parallel} = \hat{\mathbf{n}} \frac{\mathbf{v} \cdot \hat{\mathbf{n}}}{\|\hat{\mathbf{n}}\|^2} = \hat{\mathbf{n}} \frac{\mathbf{v} \cdot \hat{\mathbf{n}}}{1} = \hat{\mathbf{n}} (\mathbf{v} \cdot \hat{\mathbf{n}})$

$$= \begin{bmatrix} \sqrt{2}/2 \\ \sqrt{2}/2 \\ 0 \end{bmatrix} \left(\begin{bmatrix} 4 \\ 3 \\ -1 \end{bmatrix} \cdot \begin{bmatrix} \sqrt{2}/2 \\ \sqrt{2}/2 \\ 0 \end{bmatrix} \right) = \begin{bmatrix} \sqrt{2}/2 \\ \sqrt{2}/2 \\ 0 \end{bmatrix} \left(2\sqrt{2} + \frac{3\sqrt{2}}{2} + 0 \right)$$

$$= \begin{bmatrix} \sqrt{2}/2 \\ \sqrt{2}/2 \\ 0 \end{bmatrix} \frac{7\sqrt{2}}{2} = \begin{bmatrix} 7/2 \\ 7/2 \\ 0 \end{bmatrix}$$

$$\mathbf{v}_\perp = \mathbf{v} - \mathbf{v}_\|$$

$$= \begin{bmatrix} 4 \\ 3 \\ -1 \end{bmatrix} - \begin{bmatrix} 7/2 \\ 7/2 \\ 0 \end{bmatrix} = \begin{bmatrix} 4 - 7/2 \\ 3 - 7/2 \\ -1 - 0 \end{bmatrix} = \begin{bmatrix} 1/2 \\ -1/2 \\ -1 \end{bmatrix}$$

11. Define a triangle using the vectors $\mathbf{a}$, $\mathbf{b}$, and $\mathbf{a} - \mathbf{b}$, and let θ be the angle between $\mathbf{a}$ and $\mathbf{b}$. Then the squared length of the edge $\mathbf{a} - \mathbf{b}$ is:

$$\begin{aligned} \|\mathbf{a} - \mathbf{b}\|^2 &= (\mathbf{a} - \mathbf{b}) \cdot (\mathbf{a} - \mathbf{b}) \\ &= \mathbf{a} \cdot \mathbf{a} - 2\mathbf{a} \cdot \mathbf{b} + \mathbf{b} \cdot \mathbf{b} \\ &= \mathbf{a} \cdot \mathbf{a} + \mathbf{b} \cdot \mathbf{b} - 2\mathbf{a} \cdot \mathbf{b} \\ &= \|\mathbf{a}\|^2 + \|\mathbf{b}\|^2 - 2\|\mathbf{a}\|\|\mathbf{b}\| \cos \theta \end{aligned}$$

which is the law of cosines.

12. First, let's obtain some information about the vector components.

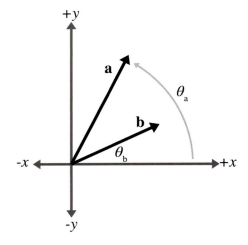

From the figure, we have

$$a_x = \|\mathbf{a}\| \cos \theta_a, \qquad\qquad a_y = \|\mathbf{a}\| \sin \theta_a,$$
$$b_x = \|\mathbf{b}\| \cos \theta_b, \qquad\qquad b_y = \|\mathbf{b}\| \sin \theta_b.$$

Now we can proceed with the algebraic definition of the dot product and the cosine difference identity:

$$\begin{aligned} \mathbf{a} \cdot \mathbf{b} &= a_x b_x + a_y b_y \\ &= \|\mathbf{a}\| \cos \theta_a \|\mathbf{b}\| \cos \theta_b + \|\mathbf{a}\| \sin \theta_a \|\mathbf{b}\| \sin \theta_b \\ &= \|\mathbf{a}\|\|\mathbf{b}\| (\cos \theta_a \cos \theta_b + \sin \theta_a \sin \theta_b) \\ &= \|\mathbf{a}\|\|\mathbf{b}\| \cos (\theta_b - \theta_a) \\ &= \|\mathbf{a}\|\|\mathbf{b}\| \cos \theta. \end{aligned}$$

13. (a) $\begin{bmatrix} 0 \\ -1 \\ 0 \end{bmatrix} \times \begin{bmatrix} 0 \\ 0 \\ 1 \end{bmatrix} = \begin{bmatrix} (-1)(1) - (0)(0) \\ (0)(0) - (0)(1) \\ (0)(0) - (-1)(0) \end{bmatrix} = \begin{bmatrix} -1 - 0 \\ 0 - 0 \\ 0 - 0 \end{bmatrix} = \begin{bmatrix} -1 \\ 0 \\ 0 \end{bmatrix}$

$\begin{bmatrix} 0 \\ 0 \\ 1 \end{bmatrix} \times \begin{bmatrix} 0 \\ -1 \\ 0 \end{bmatrix} = \begin{bmatrix} (0)(0) - (1)(-1) \\ (1)(0) - (0)(0) \\ (0)(-1) - (0)(0) \end{bmatrix} = \begin{bmatrix} 0 - (-1) \\ 0 - 0 \\ 0 - 0 \end{bmatrix} = \begin{bmatrix} 1 \\ 0 \\ 0 \end{bmatrix}$

(b) $\begin{bmatrix} -2 \\ 4 \\ 1 \end{bmatrix} \times \begin{bmatrix} 1 \\ -2 \\ -1 \end{bmatrix} = \begin{bmatrix} (4)(-1) - (1)(-2) \\ (1)(1) - (-2)(-1) \\ (-2)(-2) - (4)(1) \end{bmatrix} = \begin{bmatrix} -4 - (-2) \\ 1 - 2 \\ 4 - 4 \end{bmatrix} = \begin{bmatrix} -2 \\ -1 \\ 0 \end{bmatrix}$

$\begin{bmatrix} 1 \\ -2 \\ -1 \end{bmatrix} \times \begin{bmatrix} -2 \\ 4 \\ 1 \end{bmatrix} = \begin{bmatrix} (-2)(1) - (-1)(4) \\ (-1)(-2) - (1)(1) \\ (1)(4) - (-2)(-2) \end{bmatrix} = \begin{bmatrix} -2 - (-4) \\ 2 - 1 \\ 4 - 4 \end{bmatrix} = \begin{bmatrix} 2 \\ 1 \\ 0 \end{bmatrix}$

(c) $\begin{bmatrix} 3 \\ 10 \\ 7 \end{bmatrix} \times \begin{bmatrix} 8 \\ -7 \\ 4 \end{bmatrix} = \begin{bmatrix} (10)(4) - (7)(-7) \\ (7)(8) - (3)(4) \\ (3)(-7) - (10)(8) \end{bmatrix} = \begin{bmatrix} 40 - (-49) \\ 56 - 12 \\ -21 - 80 \end{bmatrix} = \begin{bmatrix} 89 \\ 44 \\ -101 \end{bmatrix}$

$\begin{bmatrix} 8 \\ -7 \\ 4 \end{bmatrix} \times \begin{bmatrix} 3 \\ 10 \\ 7 \end{bmatrix} = \begin{bmatrix} (-7)(7) - (4)(10) \\ (4)(3) - (8)(7) \\ (8)(10) - (-7)(3) \end{bmatrix} = \begin{bmatrix} -49 - 40 \\ 12 - 56 \\ 80 - (-21) \end{bmatrix} = \begin{bmatrix} -89 \\ -44 \\ 101 \end{bmatrix}$

14. Let $\mathbf{a} = \begin{bmatrix} a_x \\ a_y \\ a_z \end{bmatrix}$ and $\mathbf{b} = \begin{bmatrix} b_x \\ b_y \\ b_z \end{bmatrix}$. Then $\mathbf{a} \cdot \mathbf{b} = \|\mathbf{a}\|\|\mathbf{b}\| \cos\theta$ and $\mathbf{a} \times \mathbf{b} =$

$\begin{bmatrix} a_y b_z - a_z b_y \\ a_z b_x - a_x b_z \\ a_x b_y - a_y b_x \end{bmatrix}$. From $\|\mathbf{a} \times \mathbf{b}\|$, we have:

$$\|\mathbf{a} \times \mathbf{b}\| = \sqrt{(a_y b_z - a_z b_y)^2 + (a_z b_x - a_x b_z)^2 + (a_x b_y - a_y b_x)^2}$$

$$= \sqrt{a_y^2 b_z^2 - 2a_y a_z b_y b_z + a_z^2 b_y^2 + a_z^2 b_x^2 - 2a_x a_z b_x b_z + a_x^2 b_z^2 + a_x^2 b_y^2 - 2a_x a_y b_x b_y + a_y^2 b_x^2}.$$

If we now consider $\|\mathbf{a}\|\|\mathbf{b}\| \sin\theta$, we find that:

$$\|\mathbf{a}\|\|\mathbf{b}\| \sin\theta = \|\mathbf{a}\|\|\mathbf{b}\| \sqrt{1 - \cos^2\theta}$$

$$= \sqrt{a_x^2 + a_y^2 + a_z^2} \sqrt{b_x^2 + b_y^2 + b_z^2} \sqrt{1 - \left(\frac{a_x b_x + a_y b_y + a_z b_z}{\sqrt{a_x^2 + a_y^2 + a_z^2} \sqrt{b_x^2 + b_y^2 + b_z^2}} \right)^2}$$

$$= \sqrt{\left(a_x^2 + a_y^2 + a_z^2\right)\left(b_x^2 + b_y^2 + b_z^2\right)\left(1 - \frac{(a_x b_x + a_y b_y + a_z b_z)^2}{\left(a_x^2 + a_y^2 + a_z^2\right)\left(b_x^2 + b_y^2 + b_z^2\right)}\right)}$$

$$= \sqrt{\left(a_x^2 + a_y^2 + a_z^2\right)\left(b_x^2 + b_y^2 + b_z^2\right) - (a_x b_x + a_y b_y + a_z b_z)^2}$$

$$= \sqrt{a_y^2 b_z^2 - 2a_y a_z b_y b_z + a_z^2 b_y^2 + a_z^2 b_x^2 - 2a_x a_z b_x b_z + a_x^2 b_z^2 + a_x^2 b_y^2 - 2a_x a_y b_x b_y + a_y^2 b_x^2}.$$

Starting from both ends, we have met in the middle, proving that $\|\mathbf{a} \times \mathbf{b}\| = \|\mathbf{a}\|\|\mathbf{b}\|\sin\theta$.

15. (a) (1) $\left\|\begin{bmatrix} 3 & 4 \end{bmatrix}\right\|_1 = |3| + |4| = 7$

$\left\|\begin{bmatrix} 3 & 4 \end{bmatrix}\right\|_2 = \sqrt{|3|^2 + |4|^2} = 5$

$\left\|\begin{bmatrix} 3 & 4 \end{bmatrix}\right\|_3 = \sqrt[3]{|3|^3 + |4|^3} = \sqrt[3]{91} \approx 4.498$

$\left\|\begin{bmatrix} 3 & 4 \end{bmatrix}\right\|_\infty = \max\left(|3|, |4|\right) = 4$

(2) $\left\|\begin{bmatrix} 5 & -12 \end{bmatrix}\right\|_1 = |5| + |-12| = 17$

$\left\|\begin{bmatrix} 5 & -12 \end{bmatrix}\right\|_2 = \sqrt{|5|^2 + |-12|^2} = 13$

$\left\|\begin{bmatrix} 5 & -12 \end{bmatrix}\right\|_3 = \sqrt[3]{|5|^3 + |-12|^3} = \sqrt[3]{1853} \approx 12.283$

$\left\|\begin{bmatrix} 5 & -12 \end{bmatrix}\right\|_\infty = \max\left(|5|, |-12|\right) = 12$

(3) $\left\|\begin{bmatrix} -2 & 10 & -7 \end{bmatrix}\right\|_1 = |-2| + |10| + |-7| = 19$

$\left\|\begin{bmatrix} -2 & 10 & -7 \end{bmatrix}\right\|_2 = \sqrt{|-2|^2 + |10|^2 + |-7|^2} = \sqrt{153} \approx 12.369$

$\left\|\begin{bmatrix} -2 & 10 & -7 \end{bmatrix}\right\|_3 = \sqrt[3]{|-2|^3 + |10|^3 + |-7|^3} = \sqrt[3]{1351} \approx 11.055$

$\left\|\begin{bmatrix} -2 & 10 & -7 \end{bmatrix}\right\|_\infty = \max\left(|-2|, |10|, |-7|\right) = 10$

(4) $\left\|\begin{bmatrix} 6 & 1 & -9 \end{bmatrix}\right\|_1 = |6| + |1| + |-9| = 16$

$\left\|\begin{bmatrix} 6 & 1 & -9 \end{bmatrix}\right\|_2 = \sqrt{|6|^2 + |1|^2 + |-9|^2} = \sqrt{118} \approx 10.863$

$\left\|\begin{bmatrix} 6 & 1 & -9 \end{bmatrix}\right\|_3 = \sqrt[3]{|6|^3 + |1|^3 + |-9|^3} = \sqrt[3]{946} \approx 9.817$

$\left\|\begin{bmatrix} 6 & 1 & -9 \end{bmatrix}\right\|_\infty = \max\left(|6|, |1|, |-9|\right) = 9$

(5) $\left\|\begin{bmatrix} -2 & -2 & -2 & -2 \end{bmatrix}\right\|_1 = |-2| + |-2| + |-2| + |-2| = 8$

$\left\|\begin{bmatrix} -2 & -2 & -2 & -2 \end{bmatrix}\right\|_2 = \sqrt{|-2|^2 + |-2|^2 + |-2|^2 + |-2|^2} = 4$

$\left\|\begin{bmatrix} -2 & -2 & -2 & -2 \end{bmatrix}\right\|_3 = \sqrt[3]{|-2|^3 + |-2|^3 + |-2|^3 + |-2|^3} = \sqrt[3]{32} \approx 3.175$

$\left\|\begin{bmatrix} -2 & -2 & -2 & -2 \end{bmatrix}\right\|_\infty = \max\left(|-2|, |-2|, |-2|, |-2|\right) = 2$

(b) (1) The unit circle for the L^1 norm is a square with sides of length $\sqrt{2}$ rotated by $45°$.

(2) The unit circle for the L^2 norm is the well-known unit circle we all know and love.

(3) The unit circle for the infinity norm is a square with sides of length 2.

Note that all three unit circles include the vectors $[1,0]$, $[0,1]$, $[-1,0]$, $[0,-1]$.

16. The man buys a box or has a piece of luggage that is 2 feet long, 2 feet wide, and 2 feet tall. If the object is very thin, such as a sword, then he can put the object diagonally in the box or luggage. The longest such object he could carry on is $\sqrt{2^2 + 2^2 + 2^2} \approx 3.46$ feet.

17. Let $\mathbf{s} = \mathbf{a} + \mathbf{b} + \mathbf{c} + \mathbf{d} + \mathbf{e} + \mathbf{f}$. From inspection of Figure 2.11 we see that

$$\mathbf{s} = \begin{bmatrix} -5 \\ 3 \end{bmatrix}.$$

We confirm this numerically using the above equation and the values of the other vectors, also obtained from inspection of Figure 2.11:

$$
\begin{aligned}
\mathbf{s} &= \mathbf{a} + \mathbf{b} + \mathbf{c} + \mathbf{d} + \mathbf{e} + \mathbf{f} \\
&= \begin{bmatrix} -1 \\ 3 \end{bmatrix} + \begin{bmatrix} 1 \\ 3 \end{bmatrix} + \begin{bmatrix} 3 \\ -2 \end{bmatrix} + \begin{bmatrix} -1 \\ -2 \end{bmatrix} + \begin{bmatrix} -6 \\ 4 \end{bmatrix} + \begin{bmatrix} -1 \\ -3 \end{bmatrix} \\
&= \begin{bmatrix} (-1) + 1 + 3 + (-1) + (-6) + (-1) \\ 3 + 3 + (-2) + (-2) + 4 + (-3) \end{bmatrix} \\
&= \begin{bmatrix} -5 \\ 3 \end{bmatrix}
\end{aligned}
$$

18. Left-handed.

19. (a) Let $\mathbf{c} = \begin{bmatrix} c_x \\ c_y \end{bmatrix}$ and $\mathbf{r} = \begin{bmatrix} r_x \\ r_y \end{bmatrix}$. Then

$$\mathbf{p}_{\text{UpperLeft}} = \begin{bmatrix} c_x - r_x \\ c_y + r_y \end{bmatrix}, \qquad \mathbf{p}_{\text{UpperRight}} = \begin{bmatrix} c_x + r_x \\ c_y + r_y \end{bmatrix},$$

$$\mathbf{p}_{\text{LowerLeft}} = \begin{bmatrix} c_x - r_x \\ c_y - r_y \end{bmatrix}, \qquad \mathbf{p}_{\text{LowerRight}} = \begin{bmatrix} c_x + r_x \\ c_y - r_y \end{bmatrix}.$$

(b) Let $\mathbf{c} = \begin{bmatrix} c_x \\ c_y \\ c_z \end{bmatrix}$ and $\mathbf{r} = \begin{bmatrix} r_x \\ r_y \\ r_z \end{bmatrix}$. Then

$$\mathbf{p}_{\text{FrontUpperLeft}} = \begin{bmatrix} c_x - r_x \\ c_y + r_y \\ c_z + r_z \end{bmatrix}, \qquad \mathbf{p}_{\text{FrontUpperRight}} = \begin{bmatrix} c_x + r_x \\ c_y + r_y \\ c_z + r_z \end{bmatrix},$$

$$\mathbf{p}_{\text{FrontLowerLeft}} = \begin{bmatrix} c_x - r_x \\ c_y - r_y \\ c_z + r_z \end{bmatrix}, \qquad \mathbf{p}_{\text{FrontLowerRight}} = \begin{bmatrix} c_x + r_x \\ c_y - r_y \\ c_z + r_z \end{bmatrix},$$

$$\mathbf{p}_{\mathrm{BackUpperLeft}} = \begin{bmatrix} c_x - r_x \\ c_y + r_y \\ c_z - r_z \end{bmatrix}, \qquad \mathbf{p}_{\mathrm{BackUpperRight}} = \begin{bmatrix} c_x + r_x \\ c_y + r_y \\ c_z - r_z \end{bmatrix},$$

$$\mathbf{p}_{\mathrm{BackLowerLeft}} = \begin{bmatrix} c_x - r_x \\ c_y - r_y \\ c_z - r_z \end{bmatrix}, \qquad \mathbf{p}_{\mathrm{BackLowerRight}} = \begin{bmatrix} c_x + r_x \\ c_y - r_y \\ c_z - r_z \end{bmatrix}.$$

20. (a) Use the sign of the dot product between $\mathbf{v}$ and $\mathbf{x} - \mathbf{p}$ to determine whether the point $\mathbf{x}$ is in front of or behind the NPC. This follows from the geometric interpretation of the dot product,

$$\mathbf{v} \cdot (\mathbf{x} - \mathbf{p}) = \|\mathbf{v}\| \|\mathbf{x} - \mathbf{p}\| \cos \theta,$$

where θ is the angle between $\mathbf{v}$ and $\mathbf{x} - \mathbf{p}$.

Both $\|\mathbf{v}\|$ and $\|\mathbf{x} - \mathbf{p}\|$ are always positive, leaving the sign of the dot product entirely up to the value of $\cos \theta$. If $\cos \theta > 0$ then θ is less than $90°$ and $\mathbf{x}$ is *in front of* the NPC. Similarly, if $\cos \theta < 0$ then θ is greater than $90°$ and $\mathbf{x}$ is *behind* the NPC.

The special case of $\mathbf{v} \cdot (\mathbf{x} - \mathbf{p}) = 0$ means that $\mathbf{x}$ lies either directly to the left or right of the NPC. If this case does not need to be handled explicitly, it can arbitrarily be assigned to mean either in front of or behind.

(b) (1) $\mathbf{x}$ is in front of the NPC.

$$\begin{bmatrix} 5 \\ -2 \end{bmatrix} \cdot \left(\begin{bmatrix} 0 \\ 0 \end{bmatrix} - \begin{bmatrix} -3 \\ 4 \end{bmatrix} \right) = \begin{bmatrix} 5 \\ -2 \end{bmatrix} \cdot \begin{bmatrix} 3 \\ -4 \end{bmatrix} = (5)(3) + (-2)(-4) = 23$$

(2) $\mathbf{x}$ is in front of the NPC.

$$\begin{bmatrix} 5 \\ -2 \end{bmatrix} \cdot \left(\begin{bmatrix} 1 \\ 6 \end{bmatrix} - \begin{bmatrix} -3 \\ 4 \end{bmatrix} \right) = \begin{bmatrix} 5 \\ -2 \end{bmatrix} \cdot \begin{bmatrix} 4 \\ 2 \end{bmatrix} = (5)(4) + (-2)(2) = 16$$

(3) $\mathbf{x}$ is behind the NPC.

$$\begin{bmatrix} 5 \\ -2 \end{bmatrix} \cdot \left(\begin{bmatrix} -6 \\ 0 \end{bmatrix} - \begin{bmatrix} -3 \\ 4 \end{bmatrix} \right) = \begin{bmatrix} 5 \\ -2 \end{bmatrix} \cdot \begin{bmatrix} -3 \\ -4 \end{bmatrix} = (5)(-3) + (-2)(-4)$$
$$= -7$$

(4) $\mathbf{x}$ is behind the NPC.

$$\begin{bmatrix} 5 \\ -2 \end{bmatrix} \cdot \left(\begin{bmatrix} -4 \\ 7 \end{bmatrix} - \begin{bmatrix} -3 \\ 4 \end{bmatrix} \right) = \begin{bmatrix} 5 \\ -2 \end{bmatrix} \cdot \begin{bmatrix} -1 \\ 3 \end{bmatrix} = (5)(-1) + (-2)(3) = -11$$

(5) $\mathbf{x}$ is in front of the NPC.

$$\begin{bmatrix} 5 \\ -2 \end{bmatrix} \cdot \left(\begin{bmatrix} 5 \\ 5 \end{bmatrix} - \begin{bmatrix} -3 \\ 4 \end{bmatrix} \right) = \begin{bmatrix} 5 \\ -2 \end{bmatrix} \cdot \begin{bmatrix} 8 \\ 1 \end{bmatrix} = (5)(8) + (-2)(1) = 38$$

(6) $\mathbf{x}$ is in front of the NPC.

$$\begin{bmatrix} 5 \\ -2 \end{bmatrix} \cdot \left(\begin{bmatrix} -3 \\ 0 \end{bmatrix} - \begin{bmatrix} -3 \\ 4 \end{bmatrix} \right) = \begin{bmatrix} 5 \\ -2 \end{bmatrix} \cdot \begin{bmatrix} 0 \\ -4 \end{bmatrix} = (5)(0) + (-2)(-4) = 8$$

(7) $\mathbf{x}$ can be either in front of or behind the NPC, depending on how we've decided to handle this special case.

$$\begin{bmatrix} 5 \\ -2 \end{bmatrix} \cdot \left(\begin{bmatrix} -6 \\ -3.5 \end{bmatrix} - \begin{bmatrix} -3 \\ 4 \end{bmatrix} \right) = \begin{bmatrix} 5 \\ -2 \end{bmatrix} \cdot \begin{bmatrix} -3 \\ -7.5 \end{bmatrix} = (5)(-3) + (-2)(-7.5) = 0$$

21. (a) To determine whether the point $\mathbf{x}$ is visible to the NPC, compare $\cos\theta$ to $\cos(\phi/2)$. If $\cos\theta \geq \cos(\phi/2)$, then $\mathbf{x}$ is visible to the NPC.

The value of $\cos(\phi/2)$ can be obtained from the FOV angle. To get $\cos\theta$ use the dot product

$$\cos\theta = \frac{\mathbf{v} \cdot (\mathbf{x} - \mathbf{p})}{\|\mathbf{v}\| \|\mathbf{x} - \mathbf{p}\|}.$$

(b) The NPC's FOV is $90°$, so the value we are interested in is $\cos(45°) \approx 0.707$.

(1) $\mathbf{x}$ is visible to the NPC.

$$\cos\theta = \frac{\begin{bmatrix} 5 \\ -2 \end{bmatrix} \cdot \left(\begin{bmatrix} 0 \\ 0 \end{bmatrix} - \begin{bmatrix} -3 \\ 4 \end{bmatrix} \right)}{\left\| \begin{bmatrix} 5 \\ -2 \end{bmatrix} \right\| \left\| \begin{bmatrix} 0 \\ 0 \end{bmatrix} - \begin{bmatrix} -3 \\ 4 \end{bmatrix} \right\|} = \frac{23}{(\sqrt{29})(\sqrt{(25)})} \approx 0.854 \geq 0.707$$

(2) $\mathbf{x}$ is not visible to the NPC.

$$\cos\theta = \frac{\begin{bmatrix} 5 \\ -2 \end{bmatrix} \cdot \left(\begin{bmatrix} 1 \\ 6 \end{bmatrix} - \begin{bmatrix} -3 \\ 4 \end{bmatrix} \right)}{\left\| \begin{bmatrix} 5 \\ -2 \end{bmatrix} \right\| \left\| \begin{bmatrix} 1 \\ 6 \end{bmatrix} - \begin{bmatrix} -3 \\ 4 \end{bmatrix} \right\|} = \frac{16}{(\sqrt{29})(\sqrt{20})} \approx 0.664 < 0.707$$

(3) $\mathbf{x}$ is not visible to the NPC.

$$\cos\theta = \frac{\begin{bmatrix} 5 \\ -2 \end{bmatrix} \cdot \left(\begin{bmatrix} -6 \\ 0 \end{bmatrix} - \begin{bmatrix} -3 \\ 4 \end{bmatrix} \right)}{\left\| \begin{bmatrix} 5 \\ -2 \end{bmatrix} \right\| \left\| \begin{bmatrix} -6 \\ 0 \end{bmatrix} - \begin{bmatrix} -3 \\ 4 \end{bmatrix} \right\|} = \frac{-7}{(\sqrt{29})(\sqrt{25})}$$

$$\approx -0.260 < 0.707$$

(4) $\mathbf{x}$ is not visible to the NPC.

$$\cos\theta = \frac{\begin{bmatrix} 5 \\ -2 \end{bmatrix} \cdot \left(\begin{bmatrix} -4 \\ 7 \end{bmatrix} - \begin{bmatrix} -3 \\ 4 \end{bmatrix} \right)}{\left\| \begin{bmatrix} 5 \\ -2 \end{bmatrix} \right\| \left\| \begin{bmatrix} -4 \\ 7 \end{bmatrix} - \begin{bmatrix} -3 \\ 4 \end{bmatrix} \right\|} = \frac{-11}{(\sqrt{29})(\sqrt{10})} \approx -0.646 < 0.707$$

(5) $\mathbf{x}$ is visible to the NPC.

$$\cos\theta = \frac{\begin{bmatrix} 5 \\ -2 \end{bmatrix} \cdot \left(\begin{bmatrix} 5 \\ 5 \end{bmatrix} - \begin{bmatrix} -3 \\ 4 \end{bmatrix} \right)}{\left\| \begin{bmatrix} 5 \\ -2 \end{bmatrix} \right\| \left\| \begin{bmatrix} 5 \\ 5 \end{bmatrix} - \begin{bmatrix} -3 \\ 4 \end{bmatrix} \right\|} = \frac{38}{(\sqrt{29})(\sqrt{65})} \approx 0.875 \geq 0.707$$

(6) $\mathbf{x}$ is not visible to the NPC.

$$\cos\theta = \frac{\begin{bmatrix}5\\-2\end{bmatrix}\cdot\left(\begin{bmatrix}-3\\0\end{bmatrix}-\begin{bmatrix}-3\\4\end{bmatrix}\right)}{\left\|\begin{bmatrix}5\\-2\end{bmatrix}\right\|\left\|\begin{bmatrix}-3\\0\end{bmatrix}-\begin{bmatrix}-3\\4\end{bmatrix}\right\|} = \frac{8}{(\sqrt{29})(\sqrt{16})} \approx 0.371 < 0.707$$

(7) $\mathbf{x}$ is not visible to the NPC.

$$\cos\theta = \frac{\begin{bmatrix}5\\-2\end{bmatrix}\cdot\left(\begin{bmatrix}-6\\-3.5\end{bmatrix}-\begin{bmatrix}-3\\4\end{bmatrix}\right)}{\left\|\begin{bmatrix}5\\-2\end{bmatrix}\right\|\left\|\begin{bmatrix}-6\\-3.5\end{bmatrix}-\begin{bmatrix}-3\\4\end{bmatrix}\right\|} = \frac{0}{(\sqrt{29})(\sqrt{65.25})} = 0 < 0.707$$

(c) The NPC can see a distance of only 7 units, so only those points that are both within the FOV and within this distance will be visible.

(1) $\mathbf{x}$ is visible to the NPC.

$$\left\|\begin{bmatrix}0\\0\end{bmatrix}-\begin{bmatrix}-3\\4\end{bmatrix}\right\| = \left\|\begin{bmatrix}3\\-4\end{bmatrix}\right\| = \sqrt{25} = 5 < 7$$

(2) $\mathbf{x}$ is not visible to the NPC; it is outside the FOV.

(3) $\mathbf{x}$ is not visible to the NPC; it is outside the FOV.

(4) $\mathbf{x}$ is not visible to the NPC; it is outside the FOV.

(5) $\mathbf{x}$ is not visible to the NPC.

$$\left\|\begin{bmatrix}5\\5\end{bmatrix}-\begin{bmatrix}-3\\4\end{bmatrix}\right\| = \left\|\begin{bmatrix}8\\1\end{bmatrix}\right\| = \sqrt{65} \approx 8.062 > 7$$

(6) $\mathbf{x}$ is not visible to the NPC; it is outside the FOV.

(7) $\mathbf{x}$ is not visible to the NPC; it is outside the FOV.

22. (a) Let $\mathbf{v}_{ab} = \mathbf{b} - \mathbf{a}$ and $\mathbf{v}_{bc} = \mathbf{c} - \mathbf{b}$. Since the three points lie in the xz-plane, the two vectors also lie in the xz-plane and we have

$$\mathbf{v}_{ab} = \begin{bmatrix}x_{ab}\\0\\z_{ab}\end{bmatrix}, \qquad\qquad \mathbf{v}_{bc} = \begin{bmatrix}x_{bc}\\0\\z_{bc}\end{bmatrix}.$$

Taking the cross product of the vectors in the order that the points are traversed gives.

$$\mathbf{v}_{ab} \times \mathbf{v}_{bc} = \begin{bmatrix}0\\x_{bc}z_{ab} - x_{ab}z_{bc}\\0\end{bmatrix}$$

The sign of $x_{bc}z_{ab} - x_{ab}z_{bc}$ can then be used to determine the NPC's turning direction. Because we are working in a left-handed coordinate system, if the value is negative, the NPC is turning counterclockwise; if it's positive he's turning clockwise. The special case of 0 signifies that the NPC is either walking forward in a straight line or walks forward and then back along the same line.

(b) (1) $\mathbf{v}_{ab} = [-3, 0, 2]$. $\mathbf{v}_{bc} = [-3, 0, -4]$. $x_{bc}z_{ab} - x_{ab}z_{bc} = (-3)(2) - (-3)(-4) = -18 < 0$. Thus, the NPC is turning counterclockwise.

(2) $\mathbf{v}_{ab} = [7, 0, 5]$. $\mathbf{v}_{bc} = [-1, 0, 3]$. $x_{bc}z_{ab} - x_{ab}z_{bc} = (-1)(5) - (7)(3) = -26 < 0$. Thus, the NPC is turning counterclockwise.

(3) $\mathbf{v}_{ab} = [6, 0, -5]$. $\mathbf{v}_{bc} = [-12, 0, -5]$. $x_{bc}z_{ab} - x_{ab}z_{bc} = (-12)(-5) - (6)(-5) = 90 > 0$. Thus, the NPC is turning clockwise.

(4) $\mathbf{v}_{ab} = [3, 0, 1]$. $\mathbf{v}_{bc} = [3, 0, 2]$. $x_{bc}z_{ab} - x_{ab}z_{bc} = (3)(1) - (3)(2) = -3 < 0$. Thus, the NPC is turning counterclockwise.

23. $\mathbf{p}' = \mathbf{p} + (k-1)(\mathbf{p} \cdot \mathbf{n})\mathbf{n}$

$$= \begin{bmatrix} 1 \\ 0 \\ 0 \end{bmatrix} + (k-1)\left(\begin{bmatrix} 1 \\ 0 \\ 0 \end{bmatrix} \cdot \begin{bmatrix} n_x \\ n_y \\ n_z \end{bmatrix} \right) \begin{bmatrix} n_x \\ n_y \\ n_z \end{bmatrix}$$

$$= \begin{bmatrix} 1 \\ 0 \\ 0 \end{bmatrix} + (k-1)(n_x) \begin{bmatrix} n_x \\ n_y \\ n_z \end{bmatrix}$$

$$= \begin{bmatrix} 1 \\ 0 \\ 0 \end{bmatrix} + \begin{bmatrix} (k-1)n_x{}^2 \\ (k-1)n_x n_y \\ (k-1)n_x n_z \end{bmatrix}$$

$$= \begin{bmatrix} 1 + (k-1)n_x{}^2 \\ (k-1)n_x n_y \\ (k-1)n_x n_z \end{bmatrix}$$

24. $\mathbf{p}' = \cos\theta(\mathbf{p} - (\mathbf{p} \cdot \mathbf{n})\mathbf{n}) + \sin\theta(\mathbf{n} \times \mathbf{p}) + (\mathbf{p} \cdot \mathbf{n})\mathbf{n}$

$$= \cos\theta\left(\begin{bmatrix} 1 \\ 0 \\ 0 \end{bmatrix} - \left(\begin{bmatrix} 1 \\ 0 \\ 0 \end{bmatrix} \cdot \begin{bmatrix} n_x \\ n_y \\ n_z \end{bmatrix} \right) \begin{bmatrix} n_x \\ n_y \\ n_z \end{bmatrix} \right) + \sin\theta\left(\begin{bmatrix} n_x \\ n_y \\ n_z \end{bmatrix} \times \begin{bmatrix} 1 \\ 0 \\ 0 \end{bmatrix} \right) + \left(\begin{bmatrix} 1 \\ 0 \\ 0 \end{bmatrix} \cdot \begin{bmatrix} n_x \\ n_y \\ n_z \end{bmatrix} \right) \begin{bmatrix} n_x \\ n_y \\ n_z \end{bmatrix}$$

$$= \cos\theta\left(\begin{bmatrix} 1 \\ 0 \\ 0 \end{bmatrix} - n_x \begin{bmatrix} n_x \\ n_y \\ n_z \end{bmatrix} \right) + \sin\theta \begin{bmatrix} 0 \\ n_z \\ -n_y \end{bmatrix} + n_x \begin{bmatrix} n_x \\ n_y \\ n_z \end{bmatrix}$$

$$= \cos\theta \begin{bmatrix} 1 - n_x{}^2 \\ -n_x n_y \\ -n_x n_z \end{bmatrix} + \sin\theta \begin{bmatrix} 0 \\ n_z \\ -n_y \end{bmatrix} + \begin{bmatrix} n_x{}^2 \\ n_x n_y \\ n_x n_z \end{bmatrix}$$

$$= \begin{bmatrix} \cos\theta - n_x{}^2\cos\theta \\ -n_x n_y\cos\theta \\ -n_x n_z\cos\theta \end{bmatrix} + \begin{bmatrix} 0 \\ n_z\sin\theta \\ -n_y\sin\theta \end{bmatrix} + \begin{bmatrix} n_x{}^2 \\ n_x n_y \\ n_x n_z \end{bmatrix}$$

$$= \begin{bmatrix} \cos\theta - n_x{}^2\cos\theta + n_x{}^2 \\ -n_x n_y\cos\theta + n_z\sin\theta + n_x n_y \\ -n_x n_z\cos\theta - n_y\sin\theta + n_x n_z \end{bmatrix}$$

$$= \begin{bmatrix} n_x{}^2(1-\cos\theta) + \cos\theta \\ n_x n_y(1-\cos\theta) + n_z\sin\theta \\ n_x n_z(1-\cos\theta) - n_y\sin\theta \end{bmatrix}$$

B.3   Chapter 3

(Page 109.)

1. (a) Object space.

    (b) We could compare my world-space x-coordinate with the book's world-space x-coordinate. Or, we just examine the sign of the upright-space x-coordinate.

    (c) World space.

    (d) Object space. Or you might say that we could take a dot product with our facing direction vector—which is equivalent to extracting the object-space z-coordinate.

2. First translate the point by $[-12, 0, 6]$ relative to the axes, and then rotate clockwise around the y-axis $42°$.

3. (a) Linearly dependent. The middle basis vector is the zero vector, which cannot belong to a linearly independent set because it can be expressed as a product of any other basis vector and 0.

    (b) Linearly independent.

    (c) Linearly dependent. For 3D vectors, the largest linearly independent set we could hope for is three vectors, but this set has four.

    (d) Linearly dependent. The last vector is a multiple of the first.

    (e) Linearly dependent. The last vector is the sum of the first two.

    (f) Linearly independent.

4. (a) Orthogonal.

    (b) Not orthogonal. All of the pairs of vectors have nonzero dot products.

    (c) Orthogonal.

    (d) Orthogonal.

    (e) Not orthogonal. The first pair of vectors is perpendicular, but $[7, -1, 5] \cdot [-2, 0, 1] = -9$, and $[5, 5, -6] \cdot [-2, 0, 1] = -16$.

5. (a) No. The second and third basis vectors clearly do not have unit length.

    (b) No. None of the basis vectors have unit length.

    (c) Yes, they are orthonormal.

    (d) No. The first and second basis vectors are not perpendicular.

    (e) Yes, they are orthonormal.

    (f) Yes, they are orthonormal.

    (g) No. The second and third basis vectors do not have unit length.

6. (a) Upright: $(-0.866, 2.000, 0.500)$; World: $(0.134, 12.000, 3.500)$

    (b) Upright: $(0.866, 2.000, -0.500)$; World: $(1.866, 12.000, 2.500)$

    (c) Upright: $(0, 0, 0)$; World: $(1, 10, 3)$

    (d) Upright: $(1.116, 5.000, -0.067)$; World: $(2.116, 15.000, 2.933)$

(e) Upright: $(5.000, 5.000, 8.660)$; World: $(6.000, 15.000, 11.660)$

(f) Upright: $(0.000, 0.000, 0.000)$; Object: $(0.000, 0.000, 0.000)$

(g) Upright: $(-1.000, -10.000, -3.000)$; Object: $(0.634, -10.000, -3.098)$

(h) Upright: $(1.732, 0.000, -1.000)$; Object: $(2.000, 0.000, 0.000)$

(i) Upright: $(1.000, 1.000, 1.000)$; Object: $(0.366, 1.000, 1.366)$

(j) Upright: $(0.000, 10.000, 0.000)$; Object: $(0.000, 10.000, 0.000)$

B.4   Chapter 4

(Page 132.)

1. See the table below.

Matrix	Rows × Columns	Square	Diagonal
A	4×3	No	No
B	3×3	Yes	Yes
C	2×2	Yes	No
D	5×2	No	No
E	1×3	No	No
F	4×1	No	No
G	1×4	No	No
H	3×1	No	No

2. $\mathbf{A}^{\mathrm{T}} = \begin{bmatrix} 13 & 4 & -8 \\ 12 & 0 & 6 \\ -3 & -1 & 5 \\ 10 & -2 & 5 \end{bmatrix}^{\mathrm{T}} = \begin{bmatrix} 13 & 12 & -3 & 10 \\ 4 & 0 & -1 & -2 \\ -8 & 6 & 5 & 5 \end{bmatrix}$

$\mathbf{B}^{\mathrm{T}} = \begin{bmatrix} k_x & 0 & 0 \\ 0 & k_y & 0 \\ 0 & 0 & k_z \end{bmatrix}^{\mathrm{T}} = \begin{bmatrix} k_x & 0 & 0 \\ 0 & k_y & 0 \\ 0 & 0 & k_z \end{bmatrix}$

$\mathbf{C}^{\mathrm{T}} = \begin{bmatrix} 15 & 8 \\ -7 & 3 \end{bmatrix}^{\mathrm{T}} = \begin{bmatrix} 15 & -7 \\ 8 & 3 \end{bmatrix}$ $\mathbf{D}^{\mathrm{T}} = \begin{bmatrix} a & g \\ b & h \\ c & i \\ d & j \\ f & k \end{bmatrix}^{\mathrm{T}} = \begin{bmatrix} a & b & c & d & f \\ g & h & i & j & k \end{bmatrix}$

$\mathbf{E}^{\mathrm{T}} = \begin{bmatrix} 0 & 1 & 3 \end{bmatrix}^{\mathrm{T}} = \begin{bmatrix} 0 \\ 1 \\ 3 \end{bmatrix}$ $\mathbf{F}^{\mathrm{T}} = \begin{bmatrix} x \\ y \\ z \\ w \end{bmatrix}^{\mathrm{T}} = \begin{bmatrix} x & y & z & w \end{bmatrix}$

$\mathbf{G}^{\mathrm{T}} = \begin{bmatrix} 10 & 20 & 30 & 1 \end{bmatrix}^{\mathrm{T}} = \begin{bmatrix} 10 \\ 20 \\ 30 \\ 1 \end{bmatrix}$ $\mathbf{H}^{\mathrm{T}} = \begin{bmatrix} \alpha \\ \beta \\ \gamma \end{bmatrix}^{\mathrm{T}} = \begin{bmatrix} \alpha & \beta & \gamma \end{bmatrix}$

3. $\begin{aligned}\mathbf{AB} &= (4{\times}3)(3{\times}3) &=& 4{\times}3 \\ \mathbf{BB} &= (3{\times}3)(3{\times}3) &=& 3{\times}3 \\ \mathbf{CC} &= (2{\times}2)(2{\times}2) &=& 2{\times}2 \\ \mathbf{EB} &= (1{\times}3)(3{\times}3) &=& 1{\times}3 \\ \mathbf{FE} &= (4{\times}1)(1{\times}3) &=& 4{\times}3 \\ \mathbf{GA} &= (1{\times}4)(4{\times}3) &=& 1{\times}3 \\ \mathbf{HE} &= (3{\times}1)(1{\times}3) &=& 3{\times}3 \end{aligned}$ $\begin{aligned}\mathbf{AH} &= (4{\times}3)(3{\times}1) &=& 4{\times}1 \\ \mathbf{BH} &= (3{\times}3)(3{\times}1) &=& 3{\times}1 \\ \mathbf{DC} &= (5{\times}2)(2{\times}2) &=& 5{\times}2 \\ \mathbf{EH} &= (1{\times}3)(3{\times}1) &=& 1{\times}1 \\ \mathbf{FG} &= (4{\times}1)(1{\times}4) &=& 4{\times}4 \\ \mathbf{GF} &= (1{\times}4)(4{\times}1) &=& 1{\times}1 \\ \mathbf{HG} &= (3{\times}1)(1{\times}4) &=& 3{\times}4 \end{aligned}$

4. (a) $\begin{bmatrix} 1 & -2 \\ 5 & 0 \end{bmatrix} \begin{bmatrix} -3 & 7 \\ 4 & 1/3 \end{bmatrix} = \begin{bmatrix} (1)(-3)+(-2)(4) & (1)(7)+(-2)(1/3) \\ (5)(-3)+(0)(4) & (5)(7)+(0)(1/3) \end{bmatrix}$

$$= \begin{bmatrix} -3+(-8) & 7+(-2/3) \\ -15+0 & 35+0 \end{bmatrix} = \begin{bmatrix} -11 & 19/3 \\ -15 & 35 \end{bmatrix}$$

(b) Not possible; cannot multiply a 2×2 matrix by a 1×2 vector on the right.

(c) $\begin{bmatrix} 3 & -1 & 4 \end{bmatrix} \begin{bmatrix} -2 & 0 & 3 \\ 5 & 7 & -6 \\ 1 & -4 & 2 \end{bmatrix}$

$= \begin{bmatrix} (3)(-2)+(-1)(5)+(4)(1) & (3)(0)+(-1)(7)+(4)(-4) & (3)(3)+(-1)(-6)+(4)(2) \end{bmatrix}$

$= \begin{bmatrix} -6+(-5)+4 & 0+(-7)+(-16) & 9+6+8 \end{bmatrix} = \begin{bmatrix} -7 & -23 & 23 \end{bmatrix}$

(d) $\begin{bmatrix} x & y & z & w \end{bmatrix} \begin{bmatrix} 1 & 0 & 0 & 0 \\ 0 & 1 & 0 & 0 \\ 0 & 0 & 1 & 0 \\ 0 & 0 & 0 & 1 \end{bmatrix} = \begin{bmatrix} x & y & z & w \end{bmatrix}$

(e) Not possible; cannot multiply a 1×4 vector by a 2×1 vector.

(f) $\begin{bmatrix} 1 & 0 \\ 0 & 1 \end{bmatrix} \begin{bmatrix} m_{11} & m_{12} \\ m_{21} & m_{22} \end{bmatrix} = \begin{bmatrix} m_{11} & m_{12} \\ m_{21} & m_{22} \end{bmatrix}$

(g) $\begin{bmatrix} 3 & 3 \end{bmatrix} \begin{bmatrix} 6 & -7 \\ -4 & 5 \end{bmatrix} = \begin{bmatrix} (3)(6) + (3)(-4) & (3)(-7) + (3)(5) \end{bmatrix}$

$$= \begin{bmatrix} 18 + (-12) & -21 + 15 \end{bmatrix} = \begin{bmatrix} 6 & -6 \end{bmatrix}$$

(h) Not possible; cannot multiply a 3×3 matrix by a 2×3 matrix on the right.

5. (a) $\begin{bmatrix} 5 & -1 & 2 \end{bmatrix} \begin{bmatrix} 1 & 0 & 0 \\ 0 & 1 & 0 \\ 0 & 0 & 1 \end{bmatrix}$

$= \begin{bmatrix} (5)(1)+(-1)(0)+(2)(0) & (5)(0)+(-1)(1)+(2)(0) & (5)(0)+(-1)(0)+(2)(1) \end{bmatrix}$

$= \begin{bmatrix} 5 & -1 & 2 \end{bmatrix}$

$\begin{bmatrix} 1 & 0 & 0 \\ 0 & 1 & 0 \\ 0 & 0 & 1 \end{bmatrix} \begin{bmatrix} 5 \\ -1 \\ 2 \end{bmatrix} = \begin{bmatrix} (1)(5) + (0)(-1) + (0)(2) \\ (0)(5) + (1)(-1) + (0)(2) \\ (0)(5) + (0)(-1) + (1)(2) \end{bmatrix} = \begin{bmatrix} 5 \\ -1 \\ 2 \end{bmatrix}$

(b) $\begin{bmatrix} 5 & -1 & 2 \end{bmatrix} \begin{bmatrix} 2 & 5 & -3 \\ 1 & 7 & 1 \\ -2 & -1 & 4 \end{bmatrix}$

$= \begin{bmatrix} (5)(2)+(-1)(1)+(2)(-2) & (5)(5)+(-1)(7)+(2)(-1) & (5)(-3)+(-1)(1)+(2)(4) \end{bmatrix}$

$= \begin{bmatrix} 10+(-1)+(-4) & 25+(-7)+(-2) & -15+(-1)+8 \end{bmatrix} = \begin{bmatrix} 5 & 16 & -8 \end{bmatrix}$

$\begin{bmatrix} 2 & 5 & -3 \\ 1 & 7 & 1 \\ -2 & -1 & 4 \end{bmatrix} \begin{bmatrix} 5 \\ -1 \\ 2 \end{bmatrix} = \begin{bmatrix} (2)(5)+(5)(-1)+(-3)(2) \\ (1)(5)+(7)(-1)+(1)(2) \\ (-2)(5)+(-1)(-1)+(4)(2) \end{bmatrix} = \begin{bmatrix} 10+(-5)+(-6) \\ 5+(-7)+2 \\ -10+1+8 \end{bmatrix} = \begin{bmatrix} -1 \\ 0 \\ -1 \end{bmatrix}$

(c) $\begin{bmatrix} 5 & -1 & 2 \end{bmatrix} \begin{bmatrix} 1 & 7 & 2 \\ 7 & 0 & -3 \\ 2 & -3 & -1 \end{bmatrix}$

$= \begin{bmatrix} (5)(1)+(-1)(7)+(2)(2) & (5)(7)+(-1)(0)+(2)(-3) & (5)(2)+(-1)(-3)+(2)(-1) \end{bmatrix}$

$= \begin{bmatrix} 5+(-7)+4 & 35+0+(-6) & 10+3+(-2) \end{bmatrix} = \begin{bmatrix} 2 & 29 & 11 \end{bmatrix}$

$\begin{bmatrix} 1 & 7 & 2 \\ 7 & 0 & -3 \\ 2 & -3 & -1 \end{bmatrix} \begin{bmatrix} 5 \\ -1 \\ 2 \end{bmatrix} = \begin{bmatrix} (1)(5)+(7)(-1)+(2)(2) \\ (7)(5)+(0)(-1)+(-3)(2) \\ (2)(5)+(-3)(-1)+(-1)(2) \end{bmatrix} = \begin{bmatrix} 5+(-7)+4 \\ 35+0+(-6) \\ 10+3+(-2) \end{bmatrix} = \begin{bmatrix} 2 \\ 29 \\ 11 \end{bmatrix}$

(d) $\begin{bmatrix} 5 & -1 & 2 \end{bmatrix} \begin{bmatrix} 0 & -4 & 3 \\ 4 & 0 & -1 \\ -3 & 1 & 0 \end{bmatrix}$

$= \begin{bmatrix} (5)(0)+(-1)(4)+(2)(-3) & (5)(-4)+(-1)(0)+(2)(1) & (5)(3)+(-1)(-1)+(2)(0) \end{bmatrix}$

$= \begin{bmatrix} 0+(-4)+(-6) & (-20)+0+2 & 15+1+0 \end{bmatrix} = \begin{bmatrix} -10 & -18 & 16 \end{bmatrix}$

$\begin{bmatrix} 0 & -4 & 3 \\ 4 & 0 & -1 \\ -3 & 1 & 0 \end{bmatrix} \begin{bmatrix} 5 \\ -1 \\ 2 \end{bmatrix} = \begin{bmatrix} (0)(5)+(-4)(-1)+(3)(2) \\ (4)(5)+(0)(-1)+(-1)(2) \\ (-3)(5)+(1)(-1)+(0)(2) \end{bmatrix} = \begin{bmatrix} 0+4+6 \\ 20+0+(-2) \\ -15+(-1)+0 \end{bmatrix} = \begin{bmatrix} 10 \\ 18 \\ -16 \end{bmatrix}$

6. (a) $\left(\left(\mathbf{A}^{\mathrm{T}} \right)^{\mathrm{T}} \right)^{\mathrm{T}} = \mathbf{A}^{\mathrm{T}}$

(b) $\left(\mathbf{B} \mathbf{A}^{\mathrm{T}} \right)^{\mathrm{T}} \left(\mathbf{C} \mathbf{D}^{\mathrm{T}} \right) = \left(\left(\mathbf{A}^{\mathrm{T}} \right)^{\mathrm{T}} \left(\mathbf{B} \right)^{\mathrm{T}} \right) \left(\mathbf{C} \mathbf{D}^{\mathrm{T}} \right) = \left(\mathbf{A} \mathbf{B}^{\mathrm{T}} \right) \left(\mathbf{C} \mathbf{D}^{\mathrm{T}} \right) = \mathbf{A} \mathbf{B}^{\mathrm{T}} \mathbf{C} \mathbf{D}^{\mathrm{T}}$

(c) $\left(\left(\mathbf{D}^{\mathrm{T}} \mathbf{C}^{\mathrm{T}} \right) \left(\mathbf{A} \mathbf{B} \right)^{\mathrm{T}} \right)^{\mathrm{T}} = \left(\left(\left(\mathbf{A} \mathbf{B} \right)^{\mathrm{T}} \right)^{\mathrm{T}} \left(\mathbf{D}^{\mathrm{T}} \mathbf{C}^{\mathrm{T}} \right)^{\mathrm{T}} \right) = \left(\mathbf{A} \mathbf{B} \right) \left(\left(\mathbf{C}^{\mathrm{T}} \right)^{\mathrm{T}} \left(\mathbf{D}^{\mathrm{T}} \right)^{\mathrm{T}} \right)$

$= \left(\mathbf{A} \mathbf{B} \right) \left(\mathbf{C} \mathbf{D} \right) = \mathbf{A} \mathbf{B} \mathbf{C} \mathbf{D}$

(d) $\left(\left(\mathbf{A} \mathbf{B} \right)^{\mathrm{T}} \left(\mathbf{C} \mathbf{D} \mathbf{E} \right)^{\mathrm{T}} \right)^{\mathrm{T}} = \left(\left(\left(\mathbf{C} \mathbf{D} \mathbf{E} \right)^{\mathrm{T}} \right)^{\mathrm{T}} \left(\left(\mathbf{A} \mathbf{B} \right)^{\mathrm{T}} \right)^{\mathrm{T}} \right) = \left(\mathbf{C} \mathbf{D} \mathbf{E} \right) \left(\mathbf{A} \mathbf{B} \right)$

$= \mathbf{C} \mathbf{D} \mathbf{E} \mathbf{A} \mathbf{B}$

7. For each of the matrices $\mathbf{M}$, interpret the rows of $\mathbf{M}$ as basis vectors after transformation.

(a) The basis vectors $[1, 0]$ and $[0, 1]$ are transformed to $[0, -1]$ and $[1, 0]$, respectively. Thus, $\mathbf{M}$ performs a $90°$ clockwise rotation.

(b) The basis vectors $[1, 0]$ and $[0, 1]$ are transformed to $[\frac{\sqrt{2}}{2}, \frac{\sqrt{2}}{2}]$ and $[-\frac{\sqrt{2}}{2}, \frac{\sqrt{2}}{2}]$, respectively. Thus, $\mathbf{M}$ performs a $45°$ counterclockwise rotation.

(c) The basis vectors $[1, 0]$ and $[0, 1]$ are transformed to $[2, 0]$ and $[0, 2]$, respectively. Thus, $\mathbf{M}$ performs a *uniform* scale, scaling both the x and y dimensions by 2.

(d) The basis vectors $[1, 0]$ and $[0, 1]$ are transformed to $[4, 0]$ and $[0, 7]$, respectively. Thus, $\mathbf{M}$ performs a *nonuniform* scale, scaling the x dimension by 4 and the y dimension by 7.

(e) The basis vectors $[1, 0]$ and $[0, 1]$ are transformed to $[-1, 0]$ and $[0, 1]$, respectively. Thus, $\mathbf{M}$ performs a reflection across the y axis, negating x values and leaving y values untouched.

(f) The basis vectors $[1, 0]$ and $[0, 1]$ are transformed to $[0, -2]$ and $[2, 0]$, respectively. Thus, $\mathbf{M}$ is performing a combination of transformations: it is rotating clockwise by $90°$ and scaling both dimensions uniformly by 2. This can be confirmed by multiplying the appropriate matrices from parts (a) and (c), which perform these transformations individually:

$$\begin{bmatrix} 0 & -1 \\ 1 & 0 \end{bmatrix} \begin{bmatrix} 2 & 0 \\ 0 & 2 \end{bmatrix} = \begin{bmatrix} 0 & -2 \\ 2 & 0 \end{bmatrix}.$$

8. $\mathbf{M} = \begin{bmatrix} 0 & -b_z & b_y \\ b_z & 0 & -b_x \\ -b_y & b_x & 0 \end{bmatrix}$ This matrix is skew symmetric, as desired, since $\mathbf{M}^{\mathrm{T}} = -\mathbf{M}$.

9. (a) 3   (b) 1   (c) 4   (d) 2

10. The result vector element w_i is the product of the ith row of $\mathbf{M}$ multiplied by the column vector $\mathbf{v}$. To have $w_i = v_i - v_{i-1}$, the ith row of $\mathbf{M}$ needs to capture the ith element of $\mathbf{v}$, as well as the negative of the $(i-1)$th element, but exclude all others. This means that

$$m_{ij} = \begin{cases} 1 & \text{if } j = i, \\ -1 & \text{if } j = i - 1, \\ 0 & \text{otherwise.} \end{cases}$$

Thus,

$$\mathbf{M} = \begin{bmatrix} 1 & 0 & 0 & 0 & 0 & 0 & 0 & 0 & 0 & 0 \\ -1 & 1 & 0 & 0 & 0 & 0 & 0 & 0 & 0 & 0 \\ 0 & -1 & 1 & 0 & 0 & 0 & 0 & 0 & 0 & 0 \\ 0 & 0 & -1 & 1 & 0 & 0 & 0 & 0 & 0 & 0 \\ 0 & 0 & 0 & -1 & 1 & 0 & 0 & 0 & 0 & 0 \\ 0 & 0 & 0 & 0 & -1 & 1 & 0 & 0 & 0 & 0 \\ 0 & 0 & 0 & 0 & 0 & -1 & 1 & 0 & 0 & 0 \\ 0 & 0 & 0 & 0 & 0 & 0 & -1 & 1 & 0 & 0 \\ 0 & 0 & 0 & 0 & 0 & 0 & 0 & -1 & 1 & 0 \\ 0 & 0 & 0 & 0 & 0 & 0 & 0 & 0 & -1 & 1 \end{bmatrix}.$$

11. The result vector element w_i is the product of the ith row of $\mathbf{N}$ multiplied by the column vector $\mathbf{v}$. To have $w_i = \sum_{j=1}^{i} v_j$, the ith row of $\mathbf{N}$ needs to capture all elements of $\mathbf{v}$ up to and including the ith element, but exclude all others. This means that

$$n_{ij} = \begin{cases} 1 & \text{if } j \leq i, \\ 0 & \text{otherwise.} \end{cases}$$

Thus,

$$\mathbf{N} = \begin{bmatrix} 1 & 0 & 0 & 0 & 0 & 0 & 0 & 0 & 0 & 0 \\ 1 & 1 & 0 & 0 & 0 & 0 & 0 & 0 & 0 & 0 \\ 1 & 1 & 1 & 0 & 0 & 0 & 0 & 0 & 0 & 0 \\ 1 & 1 & 1 & 1 & 0 & 0 & 0 & 0 & 0 & 0 \\ 1 & 1 & 1 & 1 & 1 & 0 & 0 & 0 & 0 & 0 \\ 1 & 1 & 1 & 1 & 1 & 1 & 0 & 0 & 0 & 0 \\ 1 & 1 & 1 & 1 & 1 & 1 & 1 & 0 & 0 & 0 \\ 1 & 1 & 1 & 1 & 1 & 1 & 1 & 1 & 0 & 0 \\ 1 & 1 & 1 & 1 & 1 & 1 & 1 & 1 & 1 & 0 \\ 1 & 1 & 1 & 1 & 1 & 1 & 1 & 1 & 1 & 1 \end{bmatrix}.$$

12. (a) Note that the structure of $\mathbf{M}$ causes the ith row of $\mathbf{MN}$ to be equivalent to the difference between the ith and $(i-1)$th rows of $\mathbf{N}$.

(b) Note that the structure of $\mathbf{N}$ causes the ith row of $\mathbf{NM}$ to be equivalent to the sum of the first i rows of $\mathbf{M}$.

(c) $\mathbf{MN} = \mathbf{NM} = \mathbf{I}_{10 \times 10} = \begin{bmatrix} 1 & 0 & 0 & 0 & 0 & 0 & 0 & 0 & 0 & 0 \\ 0 & 1 & 0 & 0 & 0 & 0 & 0 & 0 & 0 & 0 \\ 0 & 0 & 1 & 0 & 0 & 0 & 0 & 0 & 0 & 0 \\ 0 & 0 & 0 & 1 & 0 & 0 & 0 & 0 & 0 & 0 \\ 0 & 0 & 0 & 0 & 1 & 0 & 0 & 0 & 0 & 0 \\ 0 & 0 & 0 & 0 & 0 & 1 & 0 & 0 & 0 & 0 \\ 0 & 0 & 0 & 0 & 0 & 0 & 1 & 0 & 0 & 0 \\ 0 & 0 & 0 & 0 & 0 & 0 & 0 & 1 & 0 & 0 \\ 0 & 0 & 0 & 0 & 0 & 0 & 0 & 0 & 1 & 0 \\ 0 & 0 & 0 & 0 & 0 & 0 & 0 & 0 & 0 & 1 \end{bmatrix}.$

B.5   Chapter 5

(Page 159.)

1. Yes, *any* matrix expresses a linear transformation. Furthermore, because all linear transformations are also affine transformations, the transform is also an affine transformation. (There just isn't any translation in the affine transform, or equivalently, the translation portion is zero.)

2. $\begin{bmatrix} 1 & 0 & 0 \\ 0 & \cos(-22°) & \sin(-22°) \\ 0 & -\sin(-22°) & \cos(-22°) \end{bmatrix} = \begin{bmatrix} 1.000 & 0.000 & 0.000 \\ 0.000 & 0.927 & -0.375 \\ 0.000 & 0.375 & 0.927 \end{bmatrix}$

3. $\begin{bmatrix} \cos 30° & 0 & -\sin 30° \\ 0 & 1 & 0 \\ \sin 30° & 0 & \cos 30° \end{bmatrix} = \begin{bmatrix} 0.866 & 0.000 & -0.500 \\ 0.000 & 1.000 & 0.000 \\ 0.500 & 0.000 & 0.866 \end{bmatrix}$

4. $\begin{bmatrix} 0.968 & -0.212 & -0.131 \\ 0.203 & 0.976 & -0.084 \\ 0.146 & 0.054 & 0.988 \end{bmatrix}$

5. $\begin{bmatrix} 2 & 0 & 0 \\ 0 & 2 & 0 \\ 0 & 0 & 2 \end{bmatrix}$

6. $\begin{bmatrix} 1.285 & -0.571 & 0.857 \\ -0.571 & 2.145 & -1.716 \\ 0.857 & -1.716 & 3.573 \end{bmatrix}$

7. $\begin{bmatrix} 0.929 & 0.143 & -0.214 \\ 0.143 & 0.714 & 0.429 \\ -0.214 & 0.429 & 0.356 \end{bmatrix}$

8. $\begin{bmatrix} 0.857 & .286 & -0.428 \\ 0.286 & .428 & 0.858 \\ -0.428 & .858 & -0.286 \end{bmatrix}$

9. (a)

$$\mathbf{M}_{\text{obj}\rightarrow\text{wld}} = \mathbf{R}_y(30°)\mathbf{R}_x(-22°) = \begin{bmatrix} 0.866 & 0.000 & -0.500 \\ 0.000 & 1.000 & 0.000 \\ 0.500 & 0.000 & 0.866 \end{bmatrix} \begin{bmatrix} 1.000 & 0.000 & 0.000 \\ 0.000 & 0.927 & -0.375 \\ 0.000 & 0.375 & 0.927 \end{bmatrix}$$

$$= \begin{bmatrix} 0.866 & -0.187 & -0.464 \\ 0.000 & 0.927 & -0.375 \\ 0.500 & 0.324 & 0.803 \end{bmatrix}$$

(b) Here, we need to take the opposite rotations, in the opposite order.

$$\mathbf{M}_{\text{wld}\rightarrow\text{obj}} = \mathbf{R}_x(22°)\mathbf{R}_y(-30°) = \begin{bmatrix} 1.000 & 0.000 & 0.000 \\ 0.000 & 0.927 & 0.375 \\ 0.000 & -0.375 & 0.927 \end{bmatrix} \begin{bmatrix} 0.866 & 0.000 & 0.500 \\ 0.000 & 1.000 & 0.000 \\ -0.500 & 0.000 & 0.866 \end{bmatrix}$$

$$= \begin{bmatrix} 0.866 & 0.000 & 0.500 \\ -0.187 & 0.927 & 0.324 \\ -0.464 & -0.375 & 0.803 \end{bmatrix}$$

Or, you might have already known that the result would be the transpose of the answer from the previous problem. If so, good for you.

(c) Convert the z-axis from object space to upright space:

$$\begin{bmatrix} 0 & 0 & 1 \end{bmatrix} \begin{bmatrix} 0.866 & -0.187 & -0.464 \\ 0.000 & 0.927 & -0.375 \\ 0.500 & 0.324 & 0.803 \end{bmatrix} = \begin{bmatrix} 0.500 & 0.324 & 0.803 \end{bmatrix}.$$

Of course, this is just the same thing as extracting the last row of the matrix.

B.6   Chapter 6

(Page 189.)

1. $\begin{vmatrix} 3 & -2 \\ 1 & 4 \end{vmatrix} = 3 \cdot 4 - (-2) \cdot 1 = 14$

2. The determinant is

$$\begin{vmatrix} 3 & -2 & 0 \\ 1 & 4 & 0 \\ 0 & 0 & 2 \end{vmatrix} = 3(4 \cdot 2 - 0 \cdot 0) + (-2)(0 \cdot 0 - 1 \cdot 2) + 0(1 \cdot 0 - 4 \cdot 0) = 28.$$

We compute the cofactors

$$C^{\{11\}} = + \begin{vmatrix} 4 & 0 \\ 0 & 2 \end{vmatrix} = 8, \qquad C^{\{12\}} = - \begin{vmatrix} 1 & 0 \\ 0 & 2 \end{vmatrix} = -2, \qquad C^{\{13\}} = + \begin{vmatrix} 1 & 4 \\ 0 & 0 \end{vmatrix} = 0,$$

$$C^{\{21\}} = - \begin{vmatrix} -2 & 0 \\ 0 & 2 \end{vmatrix} = 4, \qquad C^{\{22\}} = + \begin{vmatrix} 3 & 0 \\ 0 & 2 \end{vmatrix} = 6, \qquad C^{\{23\}} = - \begin{vmatrix} 3 & -2 \\ 0 & 0 \end{vmatrix} = 0,$$

$$C^{\{31\}} = + \begin{vmatrix} -2 & 0 \\ 4 & 0 \end{vmatrix} = 0, \qquad C^{\{32\}} = - \begin{vmatrix} 3 & 0 \\ 1 & 0 \end{vmatrix} = 0, \qquad C^{\{33\}} = + \begin{vmatrix} 3 & -2 \\ 1 & 4 \end{vmatrix} = 14,$$

and put them into the classical adjoint:

$$\mathrm{adj} \begin{bmatrix} 3 & -2 & 0 \\ 1 & 4 & 0 \\ 0 & 0 & 2 \end{bmatrix} = \begin{bmatrix} C^{\{11\}} & C^{\{21\}} & C^{\{31\}} \\ C^{\{12\}} & C^{\{22\}} & C^{\{32\}} \\ C^{\{13\}} & C^{\{23\}} & C^{\{33\}} \end{bmatrix} = \begin{bmatrix} 8 & 4 & 0 \\ -2 & 6 & 0 \\ 0 & 0 & 14 \end{bmatrix}.$$

Dividing by the determinant gives us the inverse:

$$\begin{bmatrix} 3 & -2 & 0 \\ 1 & 4 & 0 \\ 0 & 0 & 2 \end{bmatrix}^{-1} = \frac{1}{28} \begin{bmatrix} 8 & 4 & 0 \\ -2 & 6 & 0 \\ 0 & 0 & 14 \end{bmatrix} = \begin{bmatrix} 2/7 & 1/7 & 0 \\ -1/14 & 3/14 & 0 \\ 0 & 0 & 1/2 \end{bmatrix}.$$

3. The matrix is orthogonal within the appropriate tolerance.

4. Because the matrix is orthogonal, its inverse is simply its transpose:

$$\begin{bmatrix} -0.1495 & -0.1986 & -0.9685 \\ -0.8256 & 0.5640 & 0.0117 \\ -0.5439 & -0.8015 & 0.2484 \end{bmatrix}^{-1} = \begin{bmatrix} -0.1495 & -0.1986 & -0.9685 \\ -0.8256 & 0.5640 & 0.0117 \\ -0.5439 & -0.8015 & 0.2484 \end{bmatrix}^{\mathrm{T}}$$

$$= \begin{bmatrix} -0.1495 & -0.8256 & -0.5439 \\ -0.1986 & 0.5640 & -0.8015 \\ -0.9685 & 0.0117 & 0.2484 \end{bmatrix}.$$

5. This matrix is a standard affine transform matrix with a right-most column of $[0, 0, 0, 1]^\mathrm{T}$, as discussed in Section 6.4.3. Thus, it can be decomposed into a linear portion and a translation portion:

$$\mathbf{M} = \begin{bmatrix} -0.1495 & -0.1986 & -0.9685 & 0 \\ -0.8256 & 0.5640 & 0.0117 & 0 \\ -0.5439 & -0.8015 & 0.2484 & 0 \\ 1.7928 & -5.3116 & 8.0151 & 1 \end{bmatrix}$$

$$= \begin{bmatrix} -0.1495 & -0.1986 & -0.9685 & 0 \\ -0.8256 & 0.5640 & 0.0117 & 0 \\ -0.5439 & -0.8015 & 0.2484 & 0 \\ 0 & 0 & 0 & 1 \end{bmatrix} \begin{bmatrix} 1 & 0 & 0 & 0 \\ 0 & 1 & 0 & 0 \\ 0 & 0 & 1 & 0 \\ 1.7928 & -5.3116 & 8.0151 & 1 \end{bmatrix}.$$

Now taking the inverse is easy, especially when we realize that the linear portion is the same matrix as the previous exercise. The only real work is to multiply the translation row by the inverse of the linear portion:

$$\mathbf{M}^{-1} = \left(\begin{bmatrix} -0.1495 & -0.1986 & -0.9685 & 0 \\ -0.8256 & 0.5640 & 0.0117 & 0 \\ -0.5439 & -0.8015 & 0.2484 & 0 \\ 0 & 0 & 0 & 1 \end{bmatrix} \begin{bmatrix} 1 & 0 & 0 & 0 \\ 0 & 1 & 0 & 0 \\ 0 & 0 & 1 & 0 \\ 1.7928 & -5.3116 & 8.0151 & 1 \end{bmatrix} \right)^{-1}$$

$$= \begin{bmatrix} 1 & 0 & 0 & 0 \\ 0 & 1 & 0 & 0 \\ 0 & 0 & 1 & 0 \\ 1.7928 & -5.3116 & 8.0151 & 1 \end{bmatrix}^{-1} \begin{bmatrix} -0.1495 & -0.1986 & -0.9685 & 0 \\ -0.8256 & 0.5640 & 0.0117 & 0 \\ -0.5439 & -0.8015 & 0.2484 & 0 \\ 0 & 0 & 0 & 1 \end{bmatrix}^{-1}$$

$$= \begin{bmatrix} 1 & 0 & 0 & 0 \\ 0 & 1 & 0 & 0 \\ 0 & 0 & 1 & 0 \\ -1.7928 & 5.3116 & -8.0151 & 1 \end{bmatrix} \begin{bmatrix} -0.1495 & -0.8256 & -0.5439 & 0 \\ -0.1986 & 0.5640 & -0.8015 & 0 \\ -0.9685 & 0.0117 & 0.2484 & 0 \\ 0 & 0 & 0 & 1 \end{bmatrix}$$

$$= \begin{bmatrix} -0.1495 & -0.8256 & -0.5439 & 0 \\ -0.1986 & 0.5640 & -0.8015 & 0 \\ -0.9685 & 0.0117 & 0.2484 & 0 \\ 6.976 & 4.382 & -5.273 & 1 \end{bmatrix}.$$

6. $\mathbf{T}([4, 2, 3]) = \begin{bmatrix} 1 & 0 & 0 & 0 \\ 0 & 1 & 0 & 0 \\ 0 & 0 & 1 & 0 \\ 4 & 2 & 3 & 1 \end{bmatrix}$

7. First, calculate the rotation matrix:

$$\mathbf{R}_x(20^\circ) = \begin{bmatrix} 1 & 0 & 0 & 0 \\ 0 & \cos(20^\circ) & \sin(20^\circ) & 0 \\ 0 & -\sin(20^\circ) & \cos(20^\circ) & 0 \\ 0 & 0 & 0 & 1 \end{bmatrix} = \begin{bmatrix} 1.000 & 0.000 & 0.000 & 0.000 \\ 0.000 & 0.940 & 0.342 & 0.000 \\ 0.000 & -0.342 & 0.940 & 0.000 \\ 0.000 & 0.000 & 0.000 & 1.000 \end{bmatrix}.$$

Now concatenate this with the translation matrix from the previous exercise. We know this will simply copy the rotation portion into the upper 3×3, and the translation into the bottom row.

$$\mathbf{R}_x(20°)\,\mathbf{T}([4,2,3]) = \begin{bmatrix} 1.000 & 0.000 & 0.000 & 0.000 \\ 0.000 & 0.940 & 0.342 & 0.000 \\ 0.000 & -0.342 & 0.940 & 0.000 \\ 0.000 & 0.000 & 0.000 & 1.000 \end{bmatrix} \begin{bmatrix} 1.000 & 0.000 & 0.000 & 0.000 \\ 0.000 & 1.000 & 0.000 & 0.000 \\ 0.000 & 0.000 & 1.000 & 0.000 \\ 4.000 & 2.000 & 3.000 & 1.000 \end{bmatrix}$$

$$= \begin{bmatrix} 1.000 & 0.000 & 0.000 & 0.000 \\ 0.000 & 0.940 & 0.342 & 0.000 \\ 0.000 & -0.342 & 0.940 & 0.000 \\ 4.000 & 2.000 & 3.000 & 1.000 \end{bmatrix}$$

8. This time we concatenate the matrices in the opposite order, and the translation portion gets rotated.

$$\mathbf{T}([4,2,3])\,\mathbf{R}_x(20°) = \begin{bmatrix} 1.000 & 0.000 & 0.000 & 0.000 \\ 0.000 & 1.000 & 0.000 & 0.000 \\ 0.000 & 0.000 & 1.000 & 0.000 \\ 4.000 & 2.000 & 3.000 & 1.000 \end{bmatrix} \begin{bmatrix} 1.000 & 0.000 & 0.000 & 0.000 \\ 0.000 & 0.940 & 0.342 & 0.000 \\ 0.000 & -0.342 & 0.940 & 0.000 \\ 0.000 & 0.000 & 0.000 & 1.000 \end{bmatrix}$$

$$= \begin{bmatrix} 1.000 & 0.000 & 0.000 & 0.000 \\ 0.000 & 0.940 & 0.342 & 0.000 \\ 0.000 & -0.342 & 0.940 & 0.000 \\ 4.000 & 0.853 & 3.503 & 1.000 \end{bmatrix}$$

9. $\begin{bmatrix} 1 & 0 & 0 & 1/5 \\ 0 & 1 & 0 & 0 \\ 0 & 0 & 1 & 0 \\ 0 & 0 & 0 & 0 \end{bmatrix}$

10. $\begin{bmatrix} 105 & -243 & 89 & 1 \end{bmatrix} \begin{bmatrix} 1 & 0 & 0 & 1/5 \\ 0 & 1 & 0 & 0 \\ 0 & 0 & 1 & 0 \\ 0 & 0 & 0 & 0 \end{bmatrix} = \begin{bmatrix} 105 & -243 & 89 & \frac{105}{5} \end{bmatrix} \Rightarrow \begin{bmatrix} 5 & \frac{-81}{7} & \frac{89}{21} \end{bmatrix}$

B.7   Chapter 7

(Page 214.)

In some places in this section, we use the notation $(x,y)_c$ to indicate Cartesian coordinates, and $(r, \theta)_p$ to indicate polar coordinates. If plain (a, b) coordinates are used, then the context will make it clear whether the coordinates are Cartesian or polar.

1.

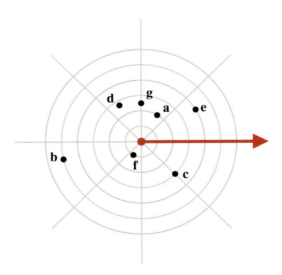

2. (a) $(4, 207°) \equiv (4, 207° - 360°) \equiv (4, -153°)$

(b) $(-5, -720°) \equiv (-5, 0°) \equiv (5, 180°)$

(c) $(0, 45.2°) \equiv (0, 0°)$

(d) $(12.6, 11\pi/4 \text{ rad}) \equiv (12.6, 11\pi/4 \text{ rad} - 2\pi \text{ rad}) \equiv (12.6, 3\pi/4 \text{ rad})$

3. (a) $(1, 45°)_p \equiv (1 \cos 45°, 1 \sin 45°)_c \approx (1 \cdot 0.707, 1 \cdot 0.707)_c = (0.707, 0.707)_c$

(b) $(3, 0°)_p \equiv (3 \cos 0°, 3 \sin 0°)_c = (3 \cdot 1, 3 \cdot 0)_c = (3, 0)_c$

(c) $(4, 90°)_p \equiv (4 \cos 90°, 4 \sin 90°)_c = (4 \cdot 0, 4 \cdot 1)_c = (0, 4)_c$

(d) $(10, -30°)_p \equiv (10 \cos(-30°), 10 \sin(-30°))_c \approx (10 \cdot 0.866, 10 \cdot (-0.500))_c = (8.66, -5.00)_c$

(e) $(5.5, \pi \text{ rad})_p \equiv (5.5 \cos(\pi \text{ rad}), 5.5 \sin(\pi \text{ rad}))_c = (5.5 \cdot (-1), 5.5 \cdot (0))_c = (-5.5, 0)_c$

4. (a) $(4, 207°)_p \equiv (4 \cos 207°, 4 \sin 207°)_c \approx (4 \cdot -.891, 4 \cdot -.454)_c \approx (-3.56, -1.82)_c$

(b) $(-5, -720°)_p \equiv (-5 \cos(-720°), -5 \sin(-720°))_c = (-5 \cdot 1, -5 \cdot 0)_c = (-5, 0)_c$

(c) $(0, 45.2°)_p \equiv (0, 0)_c$. No credit will be awarded if you actually bothered to calculate the sine and cosine of $45.2°$.

(d) $(12.6, 11\pi/4 \text{ rad})_p \equiv (12.6 \cos(11\pi/4 \text{ rad}), 12.6 \sin(11\pi/4 \text{ rad}))_c \approx (12.6 \cdot -.707, 12.6 \cdot .707)_c \approx (-8.91, 8.91)_c$

Notice that it really isn't any different or more difficult to convert noncanonical polar coordinates to Cartesian.

5. (a) $(10, 20)_c$:

$$r = \sqrt{10^2 + 20^2} = \sqrt{100 + 400} = \sqrt{500} \approx 22.36$$
$$\theta = \text{atan2}(20, 10) = \arctan(20/10) \approx 63.43°$$
$$(10, 20)_c \cong (22.36, 63.43°)_p$$

(b) $(-12, -5)_c$:

$$r = \sqrt{(-12)^2 + (-5)^2} = \sqrt{144 + 25} = \sqrt{169} = 13$$
$$\theta = \text{atan2}(-5, -12) = \arctan(5/12) - 180°$$
$$\approx 22.62° - 180° \approx -157.38°$$
$$(-12, -5)_c \cong (13, -157.38°)_p$$

(c) $(0, 4.5)_c$:

$$r = \sqrt{0^2 + 4.5^2} = 4.5$$
$$\theta = \text{atan2}(0, 4.5) = 90°$$
$$(4.5, 0)_c \equiv (4.5, 90°)_p$$

(d) $(-3, 4)_c$:

$$r = \sqrt{(-3)^2 + 4^2} = \sqrt{9 + 16} = \sqrt{25} = 5$$
$$\theta = \text{atan2}(4, -3) = \arctan(4/3) + 180°$$
$$\approx -53.13° + 180° \approx 126.87°$$
$$(-3, 4)_c \equiv (5, 126.87°)_p$$

(e) $(0, 0)_c \equiv (0, 0)_p$

(f) $(-5280, 0)_c$

$$r = \sqrt{(-5280)^2 + 0^2} = 5280$$
$$\theta = \text{atan2}(0, -5280) = 180°$$
$$(-5280, 0)_c \equiv (5280, 180°)_p$$

6. (a) $x = r\cos(\theta) = 4\cos(120°) = 4(-1/2) = -2$
   $y = r\sin(\theta) = 4\sin(120°) = 4(\sqrt{3}/2) = 2\sqrt{3}$
   so $(x, y, z) = (-2, 2\sqrt{3}, 5)$

   (b) $x = r\cos(\theta) = 2\cos(45°) = 2(\sqrt{2}/2) = \sqrt{2}$
   $y = r\sin(\theta) = 2\sin(45°) = 2(\sqrt{2}/2) = \sqrt{2}$
   so $(x, y, z) = (\sqrt{2}, \sqrt{2}, -1)$

(c) $x = r\cos(\theta) = 6\cos(-\pi/6) = 6\cos(\pi/6) = 6(\sqrt{3}/2) = 3\sqrt{3}$
$y = r\sin(\theta) = 6\sin(-\pi/6) = -6\sin(\pi/6) = -6(-1/2) = -3$
so $(x, y, z) = (3\sqrt{3}, -3, -3)$

(d) $x = r\cos(\theta) = 3\cos(3\pi) = 3\cos(\pi) = 3(-1) = -3$
$y = r\sin(\theta) = 3\sin(3\pi) = 3\sin(\pi) = 3(0) = 0$
so $(x, y, z) = (-3, 0, 1)$

7. (a) $r = \sqrt{1^2 + 1^2} = \sqrt{2}$
$\theta = \arctan(1/1) = 45°$
so $(r, \theta, z) = (\sqrt{2}, 45°, 1)$

(b) $r = \sqrt{0^2 + (-5)^2} = 5$
$\theta = -90°$, since $x = 0$ and $y < 0$
so $(r, \theta, z) = (5, -90°, 2)$

(c) $r = \sqrt{(-3)^2 + 4^2} = 5$
$\theta = \arctan(4/(-3)) = 126.87°$
so $(r, \theta, z) = (5, 126.87°, -7)$

(d) $r = \sqrt{0^2 + 0^2} = 0$
$\theta = 0$, since $x = 0$ and $y = 0$
so $(r, \theta, z) = (0, 0, -3)$

8. (a) $x = r\sin(\phi)\cos(\theta) = 4\sin(3\pi/4)\cos(\pi/3) = 4(\sqrt{2}/2)(1/2) = \sqrt{2}$
$y = r\sin(\phi)\sin(\theta) = 4\sin(3\pi/4)\sin(\pi/3) = 4(\sqrt{2}/2)(\sqrt{3}/2) = \sqrt{6}$
$z = r\cos(\phi) = 4\cos(3\pi/4) = 4(-\sqrt{2}/2) = -2\sqrt{2}$
so $(x, y, z) = (\sqrt{2}, \sqrt{6}, -2\sqrt{2})$

(b) $x = r\sin(\phi)\cos(\theta) = 5\sin(\pi/3)\cos(-5\pi/6) = 5(\sqrt{3}/2)(-\sqrt{3}/2) = -15/4$
$y = r\sin(\phi)\sin(\theta) = 5\sin(\pi/3)\sin(-5\pi/6) = 5(\sqrt{3}/2)(-1/2) = -5\sqrt{3}/4$
$z = r\cos(\phi) = 5\cos(\pi/3) = 5(1/2) = 5/2$
so $(x, y, z) = (-15/4, -5\sqrt{3}/4, 5/2)$

(c) $x = r\sin(\phi)\cos(\theta) = 2\sin(\pi)\cos(-\pi/6) = 2(0)(\sqrt{3}/2) = 0$
$y = r\sin(\phi)\sin(\theta) = 2\sin(\pi)\sin(-\pi/6) = 2(0)(-1/2) = 0$
$z = r\cos(\phi) = 2\cos(\pi) = 2(-1) = -2$
so $(x, y, z) = (0, 0, -2)$

(d) $x = r\sin(\phi)\cos(\theta) = 8\sin(\pi/6)\cos(9\pi/4) = 8(1/2)(\sqrt{2}/2) = 2\sqrt{2}$
$y = r\sin(\phi)\sin(\theta) = 8\sin(\pi/6)\sin(9\pi/4) = 8(1/2)(\sqrt{2}/2) = 2\sqrt{2}$
$z = r\cos(\phi) = 8\cos(\pi/6) = 8(\sqrt{3}/2) = 4\sqrt{3}$
so $(x, y, z) = (2\sqrt{2}, 2\sqrt{2}, 4\sqrt{3})$

9. (a1) $(4, \pi/3, 3\pi/4) \implies (4, 4\pi/3, \pi/4) \implies (4, -2\pi/3, \pi/4)$

(a2) $x = r\cos p \sin h = 4\cos(\pi/4)\sin(-2\pi/3) = 4(\sqrt{2}/2)(-\sqrt{3}/2) = -\sqrt{6}$
$y = -r\sin p = -4\sin(\pi/4) = -4(\sqrt{2}/2) = -2\sqrt{2}$
$z = r\cos p\cos h = 4\cos(\pi/4)\cos(-2\pi/3) = 4(\sqrt{2}/2)(-1/2) = -\sqrt{2}$
so $(x, y, z) = (-\sqrt{6}, -2\sqrt{2}, -\sqrt{2})$

(b1) $(5, -5\pi/6, \pi/3)$ is already in the canonical set.

(b2) $x = r\cos p\sin h = 5\cos(\pi/3)\sin(-5\pi/6) = 5(1/2)(-1/2) = -5/4$

$\quad y = -r\sin p = -5\sin(\pi/3) = -5(\sqrt{3}/2) = -(5\sqrt{3})/2$

$\quad z = r\cos p\cos h = 5\cos(\pi/3)\cos(-5\pi/6) = 5(1/2)(-\sqrt{3}/2) = -(5\sqrt{3})/4$

$\quad$ so $(x, y, z) = (-5/4, -(5\sqrt{3})/2, -(5\sqrt{3})/4)$

(c1) $(2, -\pi/6, \pi) \Longrightarrow (2, 5\pi/6, 0)$

(c2) $x = r\cos p\sin h = 2\cos(0)\sin(5\pi/6) = (2)(1)(1/2) = 1$

$\quad y = -r\sin p = -2\sin(0) = (-2)(0) = 0$

$\quad z = r\cos p\cos h = 2\cos(0)\cos(5\pi/6) = (2)(1)(-\sqrt{3}/2) = -\sqrt{3}$

$\quad$ so $(x, y, z) = (1, 0, -\sqrt{3})$

(d1) $(8, 9\pi/4, \pi/6) \Longrightarrow (8, \pi/4, \pi/6)$

(d2) $x = r\cos p\sin h = 8\cos(\pi/6)\sin(\pi/4) = 8(\sqrt{3}/2)(\sqrt{2}/2) = 2\sqrt{6}$

$\quad y = -r\sin p = -8\sin(\pi/6) = -8(1/2) = -4$

$\quad z = r\cos p\cos h = 8\cos(\pi/6)\cos(\pi/4) = 8(\sqrt{3}/2)(\sqrt{2}/2) = 2\sqrt{6}$

$\quad$ so $(x, y, z) = (2\sqrt{6}, -4, 2\sqrt{6})$

10.   (a) $r = \sqrt{x^2 + y^2 + z^2} = \sqrt{(\sqrt{2})^2 + (2\sqrt{3})^2 + (-\sqrt{2})^2} = \sqrt{2 + 12 + 2} = \sqrt{16} = 4$

$\quad\quad h = \arctan(x/z) = \arctan(-\sqrt{2}/\sqrt{2}) = \arctan(-1) = 135°$, given the location of (x, z)

$\quad\quad p = \arcsin(-y/r) = \arcsin(-(2\sqrt{3})/4) = \arcsin(-\sqrt{3}/2) = -60°$

$\quad\quad$ so $(r, h, p) = (4, 135°, -60°)$

$\quad$ (b) $r = \sqrt{x^2 + y^2 + z^2} = \sqrt{(2\sqrt{3})^2 + 6^2 + (-4)^2} = \sqrt{12 + 36 + 16} = \sqrt{64} = 8$

$\quad\quad h = \arctan(x/z) = \arctan(-(2\sqrt{3})/4) = \arctan(-\sqrt{3}/2) = 139.11°$, given the location of (x, z)

$\quad\quad p = \arcsin(-y/r) = \arcsin(-6/8) = \arcsin(-3/4) = -48.59°$

$\quad\quad$ so $(r, h, p) = (8, 139.11°, -48.59°)$

$\quad$ (c) $r = \sqrt{x^2 + y^2 + z^2} = \sqrt{(-1)^2 + (-1)^2 + (-1)^2} = \sqrt{1 + 1 + 1} = \sqrt{3}$

$\quad\quad h = \arctan(x/z) = \arctan((-1)/(-1)) = \arctan(1) = -135°$, given the location of (x, z)

$\quad\quad p = \arcsin(-y/r) = \arcsin(1/\sqrt{3}) = 35.26°$

$\quad\quad$ so $(r, h, p) = (\sqrt{3}, -135°, 35.26°)$

$\quad$ (d) $r = \sqrt{x^2 + y^2 + z^2} = \sqrt{2^2 + (-2\sqrt{3})^2 + 4^2} = \sqrt{4 + 12 + 16} = \sqrt{32} = 4\sqrt{2}$

$\quad\quad h = \arctan(x/z) = \arctan(2/4) = \arctan(1/2) = 26.57°$, given the location of (x, z)

$\quad\quad p = \arcsin(-y/r) = \arcsin((2\sqrt{3})/(4\sqrt{2})) = \arcsin(\sqrt{3}/(2\sqrt{2})) = 37.76°$

$\quad\quad$ so $(r, h, p) = (4\sqrt{2}, 26.57°, 37.76°)$

$\quad$ (e) $r = \sqrt{x^2 + y^2 + z^2} = \sqrt{(-\sqrt{3})^2 + (-\sqrt{3})^2 + (2\sqrt{2})^2} = \sqrt{3 + 3 + 8} = \sqrt{14}$

$\quad\quad h = \arctan(x/z) = \arctan(-\sqrt{3}/(2\sqrt{2})) = -31.48°$, given the location of (x, z)

$\quad\quad p = \arcsin(-y/r) = \arcsin(\sqrt{3}/\sqrt{14}) = 27.58°$

$\quad\quad$ so $(r, h, p) = (\sqrt{14}, -31.48°, 27.58°)$

(f) $r = \sqrt{x^2 + y^2 + z^2} = \sqrt{3^2 + 4^2 + 12^2} = \sqrt{9 + 16 + 144} = 13$

$h = \arctan(x/z) = \arctan(3/12) = \arctan(1/4) = 14.04°$

$p = \arcsin(-y/r) = \arcsin(-4/13) = -17.92°$

so $(r, h, p) = (13, 14.04°, -17.92°)$

11. (a) A sphere with radius r_0.

(b) A vertical plane, obtained by rotating the plane $x = 0$ clockwise about the y axis by the angle h_0.

(c) A "right circular conical surface" (two vertical circular cones meeting tip-to-tip at the origin). The interior angle of the cone is $2p_0$.

12. She was at the north pole, so the bear was white.[1]

B.8   Chapter 8

(Page 291.)

1. (a) 5   (b) 3   (c) 6   (d) 1   (e) 2   (f) 4

2. (a) 3. Yes, they are canonical Euler angles.

(b) 4. Yes, they are canonical Euler angles.

(c) 5. No, this orientation is in Gimbal lock, and in the canonical set, bank should be zero.

(d) 1. Yes, they are canonical Euler angles.

(e) 2. Yes, they are canonical Euler angles.

(f) 3. No, the pitch angle is outside the legal range.

(g) 5. Yes, they are canonical Euler angles.

(h) 2. No, the pitch angle is outside the legal range.

(i) 6. Yes, they are canonical Euler angles.

3. (a) $\begin{bmatrix} \cos(30°/2) \\ \begin{pmatrix} 1 \cdot \sin(30°/2) \\ 0 \cdot \sin(30°/2) \\ 0 \cdot \sin(30°/2) \end{pmatrix} \end{bmatrix} = \begin{bmatrix} 0.966 \\ \begin{pmatrix} .259 \\ 0.000 \\ 0.000 \end{pmatrix} \end{bmatrix}$

(b) *All* rotation quaternions have a magnitude of 1!

---

[1]It was *polar* bear. Get it?! Polar!

(c) $\begin{bmatrix} 0.966 & (-.259 & 0.000 & 0.000) \end{bmatrix}$

(d) This corresponds to a *pitch* of $+30°$.

4. (a) 2   (b) 5   (c) 1   (d) 3   (e) 2   (f) 1   (g) 4   (h) 6   (i) 3

5. (a) 5   (b) 2   (c) 6   (d) 1   (e) 3   (f) 5   (g) 4   (h) 2   (i) 3

6. $(w_1 + x_1 i + y_1 j + z_1 k)(w_2 + x_2 i + y_2 j + z_2 k)$
$$= w_1 w_2 + w_1 x_2 i + w_1 y_2 j + w_1 z_2 k$$
$$+ x_1 w_2 i + x_1 x_2 i^2 + x_1 y_2 ij + x_1 z_2 ik$$
$$+ y_1 w_2 j + y_1 x_2 ji + y_1 y_2 j^2 + y_1 z_2 jk$$
$$+ z_1 w_2 k + z_1 x_2 ki + z_1 y_2 kj + z_1 z_2 k^2$$
$$= w_1 w_2 + w_1 x_2 i + w_1 y_2 j + w_1 z_2 k$$
$$+ x_1 w_2 i + x_1 x_2 (-1) + x_1 y_2 k + x_1 z_2 (-j)$$
$$+ y_1 w_2 j + y_1 x_2 (-k) + y_1 y_2 (-1) + y_1 z_2 i$$
$$+ z_1 w_2 k + z_1 x_2 j + z_1 y_2 (-i) + z_1 z_2 (-1)$$
$$= w_1 w_2 - x_1 x_2 - y_1 y_2 - z_1 z_2$$
$$+ (w_1 x_2 + x_1 w_2 + y_1 z_2 - z_1 y_2)i$$
$$+ (w_1 y_2 + y_1 w_2 + z_1 x_2 - x_1 z_2)j$$
$$+ (w_1 z_2 + z_1 w_2 + x_1 y_2 - y_1 x_2)k$$

7. First, we extract the half-angle and axis of rotation:

$$\alpha = \theta/2 = \arccos w = \arccos 0.965 \approx 15.0°,$$
$$\hat{\mathbf{n}} = \text{normalize}\left(\begin{bmatrix} 0.149 & -0.149 & 0.149 \end{bmatrix}\right) \approx \begin{bmatrix} 0.577 & -0.577 & 0.577 \end{bmatrix}.$$

Now we form a new quaternion using the new half-angle, $\alpha' = 2\alpha \approx 30.0°$:

$$\begin{bmatrix} \cos\alpha' \\ \begin{pmatrix} n_x \sin\alpha' \\ n_y \sin\alpha' \\ n_z \sin\alpha' \end{pmatrix} \end{bmatrix} = \begin{bmatrix} 0.867 \\ \begin{pmatrix} 0.577 \cdot 0.500 \\ -0.577 \cdot 0.500 \\ 0.577 \cdot 0.500 \end{pmatrix} \end{bmatrix} = \begin{bmatrix} 0.867 \\ \begin{pmatrix} 0.288 \\ -0.288 \\ 0.288 \end{pmatrix} \end{bmatrix}.$$

8.   (a) $\mathbf{a \cdot b} = \begin{bmatrix} 0.233 \\ \begin{pmatrix} 0.060 \\ -0.257 \\ -0.935 \end{pmatrix} \end{bmatrix} \cdot \begin{bmatrix} -0.752 \\ \begin{pmatrix} 0.286 \\ 0.374 \\ 0.459 \end{pmatrix} \end{bmatrix} = (0.233)(-0.752) + (0.060)(0.286) + (-0.257)(0.374)$
$$+ (-0.935)(0.459) = -0.683$$

(b) $\mathbf{ab} = \begin{bmatrix} 0.333 \\ \begin{pmatrix} 0.253 \\ -0.015 \\ 0.906 \end{pmatrix} \end{bmatrix}$

(c) $\mathbf{d} = \mathbf{ba}^{-1} = \begin{bmatrix} -0.752 \\ \begin{pmatrix} 0.286 \\ 0.374 \\ 0.459 \end{pmatrix} \end{bmatrix} \begin{bmatrix} 0.233 \\ \begin{pmatrix} 0.060 \\ -0.257 \\ -0.935 \end{pmatrix} \end{bmatrix}^* = \begin{bmatrix} -0.683 \\ \begin{pmatrix} 0.343 \\ -0.401 \\ -0.500 \end{pmatrix} \end{bmatrix}$

9. $\|\mathbf{q}_1\mathbf{q}_2\| = \left\| \begin{bmatrix} w_1 & (x_1 & y_1 & z_1) \end{bmatrix} \begin{bmatrix} w_2 & (x_2 & y_2 & z_2) \end{bmatrix} \right\|$

$$= \left\| \begin{bmatrix} w_1w_2 - x_1x_2 - y_1y_2 - z_1z_2 \\ \begin{pmatrix} w_1x_2 + x_1w_2 + y_1z_2 - z_1y_2 \\ w_1y_2 + y_1w_2 + z_1x_2 - x_1z_2 \\ w_1z_2 + z_1w_2 + x_1y_2 - y_1x_2 \end{pmatrix} \end{bmatrix} \right\|$$

$$= \sqrt{ \begin{aligned} &(w_1w_2 - x_1x_2 - y_1y_2 - z_1z_2)^2 \\ &+ (w_1x_2 + x_1w_2 + y_1z_2 - z_1y_2)^2 \\ &+ (w_1y_2 + y_1w_2 + z_1x_2 - x_1z_2)^2 \\ &+ (w_1z_2 + z_1w_2 + x_1y_2 - y_1x_2)^2 \end{aligned} }$$

After expanding these products and then canceling terms (a step that we have omitted because it is very messy), we then factor:

$$\|\mathbf{q}_1\mathbf{q}_2\| = \sqrt{ \begin{aligned} &w_1{}^2w_2{}^2 + x_1{}^2x_2{}^2 + y_1{}^2y_2{}^2 + z_1{}^2z_2{}^2 \\ &+ w_1{}^2x_2{}^2 + x_1{}^2w_2{}^2 + y_1{}^2z_2{}^2 + z_1{}^2y_2{}^2 \\ &+ w_1{}^2y_2{}^2 + y_1{}^2w_2{}^2 + z_1{}^2x_2{}^2 + x_1{}^2z_2{}^2 \\ &+ w_1{}^2z_2{}^2 + z_1{}^2w_2{}^2 + x_1{}^2y_2{}^2 + y_1{}^2x_2{}^2 \end{aligned} }$$

$$= \sqrt{ \begin{aligned} &w_1{}^2(w_2{}^2 + x_2{}^2 + y_2{}^2 + z_2{}^2) \\ &+ x_1{}^2(w_2{}^2 + x_2{}^2 + y_2{}^2 + z_2{}^2) \\ &+ y_1{}^2(w_2{}^2 + x_2{}^2 + y_2{}^2 + z_2{}^2) \\ &+ z_1{}^2(w_2{}^2 + x_2{}^2 + y_2{}^2 + z_2{}^2) \end{aligned} }$$

$$= \sqrt{(w_1{}^2 + x_1{}^2 + y_1{}^2 + z_1{}^2)(w_2{}^2 + x_2{}^2 + y_2{}^2 + z_2{}^2)}$$

$$= \sqrt{\|\mathbf{q}_1\|^2 \|\mathbf{q}_2\|^2}$$

$$= \|\mathbf{q}_1\| \|\mathbf{q}_2\|$$

B.9   Chapter 9

(Page 339.)

1. First, we convert the ray to implicit form by using Equation (9.5):

$$a = d_y = 5,$$
$$b = -d_x = 7,$$
$$d = x_{\text{org}}d_y - y_{\text{org}}d_x = 5 \cdot 5 - 3 \cdot (-7) = 46.$$

Then, we convert this to slope-intercept form according to Equation (9.6):

$$m = -a/b = -5/7,$$
$$y_0 = d/b = 46/7.$$

So the equation of the line is $y = -(5/7)x + 46/7$.

2. $4x + 7y = 42$

$\qquad 7y = -4x + 42$

$\qquad\quad y = -(4/7)x + 6$

The slope is $-4/7$ and the y-intercept is 6.

3. (a) $\mathbf{p}_{\min} = (-5, -7, -5)$, $\mathbf{p}_{\max} = (7, 11, 8)$

(b) $(x_{\min}, y_{\min}, z_{\min}) = (-5, -7, -5)$ $\qquad$ $(x_{\min}, y_{\min}, z_{\max}) = (-5, -7, 8)$

$\quad\;\;(x_{\min}, y_{\max}, z_{\min}) = (-5, 11, -5)$ $\qquad$ $(x_{\min}, y_{\max}, z_{\max}) = (-5, 11, 8)$

$\quad\;\;(x_{\max}, y_{\min}, z_{\min}) = (7, -7, -5)$ $\qquad$ $(x_{\max}, y_{\min}, z_{\max}) = (7, -7, 8)$

$\quad\;\;(x_{\max}, y_{\max}, z_{\min}) = (7, 11, -5)$ $\qquad$ $(x_{\max}, y_{\max}, z_{\max}) = (7, 11, 8)$

(c) $\mathbf{c} = (\mathbf{p}_{\min} + \mathbf{p}_{\max})/2 = (1, 2, 1.5)$

$\quad\;\;\mathbf{s} = (\mathbf{p}_{\max} - \mathbf{p}_{\min}) = (12, 18, 13)$

(d) $\mathbf{v}_1' = (-2.828, 12.728, -5.000)$ $\qquad$ $\mathbf{v}_2' = (-0.707, 3.5355, 8.000)$

$\quad\;\;\mathbf{v}_3' = (-4.243, 0.000, 1.000)$ $\qquad$ $\mathbf{v}_4' = (1.414, -8.485, 0.000)$

$\quad\;\;\mathbf{v}_5' = (2.121, 6.364, 4.000)$

(e) $\mathbf{p}_{\min} = (-4.243, -8.485, -5)$, $\mathbf{p}_{\max} = (2.121, 12.728, 8)$

(f) First, we determine which products to take by using the technique from Listing 9.4:

$$x_{\min}' = m_{11} \cdot x_{\min} \qquad\qquad x_{\max}' = m_{11} \cdot x_{\max} \qquad\qquad (m_{11} > 0)$$
$$+ m_{21} \cdot y_{\max} \qquad\qquad\quad + m_{21} \cdot y_{\min} \qquad\qquad\quad (m_{21} < 0)$$
$$+ 0 \qquad\qquad\qquad\qquad + 0 \qquad\qquad\qquad\quad (m_{31} = 0)$$
$$y_{\min}' = m_{12} \cdot x_{\min} \qquad\qquad y_{\max}' = m_{12} \cdot x_{\max} \qquad\qquad (m_{12} > 0)$$
$$+ m_{22} \cdot y_{\min} \qquad\qquad\quad + m_{22} \cdot y_{\max} \qquad\qquad\quad (m_{22} > 0)$$
$$+ 0 \qquad\qquad\qquad\qquad + 0 \qquad\qquad\qquad\quad (m_{32} = 0)$$
$$z_{\min}' = 0 \qquad\qquad\qquad\quad z_{\max}' = 0 \qquad\qquad\qquad\;\; (m_{13} = 0)$$
$$+ 0 \qquad\qquad\qquad\qquad + 0 \qquad\qquad\qquad\quad (m_{23} = 0)$$
$$+ z_{\min} \qquad\qquad\qquad\quad + z_{\max} \qquad\qquad\qquad (m_{33} = 1)$$

Summing the appropriate products, we have

$$x_{\min}' = m_{11} \cdot x_{\min} + m_{21} \cdot y_{\max} + 0 = 0.707 \cdot -5 + (-0.707) \cdot 11 + 0 = -11.312,$$
$$y_{\min}' = m_{12} \cdot x_{\min} + m_{22} \cdot y_{\min} + 0 = 0.707 \cdot -5 + 0.707 \cdot -7 + 0 = -8.484,$$
$$z_{\min}' = z_{\min} = -5,$$
$$x_{\max}' = m_{11} \cdot x_{\max} + m_{21} \cdot y_{\min} + 0 = 0.707 \cdot 7 + (-0.707) \cdot -7 + 0 = 9.898,$$
$$y_{\max}' = m_{12} \cdot x_{\max} + m_{22} \cdot y_{\max} + 0 = 0.707 \cdot 7 + 0.707 \cdot 11 + 0 = 12.726,$$
$$z_{\max}' = z_{\max} = 8.$$

Notice how much larger this box is than the one of the transformed points!

4. (a) First, let's find the normal by using Equation (9.12):

$$\mathbf{e}_3 = \mathbf{p}_2 - \mathbf{p}_1 = \begin{bmatrix} 3 \\ -1 \\ 17 \end{bmatrix} - \begin{bmatrix} 6 \\ 10 \\ -2 \end{bmatrix} = \begin{bmatrix} -3 \\ -11 \\ 19 \end{bmatrix},$$

$$\mathbf{e}_1 = \mathbf{p}_3 - \mathbf{p}_2 = \begin{bmatrix} -9 \\ 8 \\ 0 \end{bmatrix} - \begin{bmatrix} 3 \\ -1 \\ 17 \end{bmatrix} = \begin{bmatrix} -12 \\ 9 \\ -17 \end{bmatrix},$$

$$\mathbf{e}_3 \times \mathbf{e}_1 = \begin{bmatrix} (-11)(-17) - (19)(9) \\ (19)(-12) - (-3)(-17) \\ (-3)(9) - (-11)(-12) \end{bmatrix} = \begin{bmatrix} 187 - 171 \\ -228 - 51 \\ -27 - 132 \end{bmatrix} = \begin{bmatrix} 16 \\ -279 \\ -159 \end{bmatrix}.$$

Let's normalize it:

$$\| \mathbf{e}_3 \times \mathbf{e}_1 \| = \sqrt{16^2 + (-279)^2 + (-159)^2} = \sqrt{103378} \approx 321.5,$$

$$\hat{\mathbf{n}} = \frac{\mathbf{e}_3 \times \mathbf{e}_1}{\| \mathbf{e}_3 \times \mathbf{e}_1 \|} \approx \frac{\begin{bmatrix} 16 & -279 & -159 \end{bmatrix}}{321.5}$$

$$\approx \begin{bmatrix} .04976 & -.8677 & -.4945 \end{bmatrix}.$$

Just for kicks, we'll verify that we get the same result with Equation (9.13) from Section 9.5.3:

$$\begin{aligned} n_x &= (z_1 + z_2)(y_1 - y_2) + (z_2 + z_3)(y_2 - y_3) + (z_3 + z_1)(y_3 - y_1) \\ &= ((-2) + 17)(10 - (-1)) + (17 + 0)((-1) - 8) + (0 + (-2))(8 - 10) \\ &= 16, \\ n_y &= (x_1 + x_2)(z_1 - z_2) + (x_2 + x_3)(z_2 - z_3) + (x_3 + x_1)(z_3 - z_1) \\ &= (6 + 3)((-2) - 17) + (3 + (-9))(17 - 0) + ((-9) + 6)(0 - (-2)) \\ &= -279, \\ n_z &= (y_1 + y_2)(x_1 - x_2) + (y_2 + y_3)(x_2 - x_3) + (y_3 + y_1)(x_3 - x_1) \\ &= (10 + (-1))(6 - 3) + ((-1) + 8)(3 - (-9)) + (8 + 10)((-9) - 6) \\ &= -159, \end{aligned}$$

$$\hat{\mathbf{n}} = \frac{\begin{bmatrix} 16 & -279 & 159 \end{bmatrix}}{\sqrt{16^2 + (-279)^2 + 159^2}} \approx \frac{\begin{bmatrix} 16 & -279 & 159 \end{bmatrix}}{321.5} \approx \begin{bmatrix} .04976 & -.8677 & -0.4945 \end{bmatrix}.$$

Now that we have $\hat{\mathbf{n}}$, we can compute d. We'll arbitrarily use $\mathbf{p}_1$:

$$d = \mathbf{n} \cdot \mathbf{p}_1 \approx \begin{bmatrix} .04976 & -.8677 & -.4945 \end{bmatrix} \cdot \begin{bmatrix} 6 & 10 & -2 \end{bmatrix}$$

$$\approx (.04976)(6) + (-.8677)(10) + (-.4945)(-2) \approx -7.389.$$

The plane equation for this triangle is

$$.04976x - .8677y - .4945z = -7.389.$$

(b) To answer both questions, we compute the signed distance by Equation (9.14) from Section 9.5.4:

$$a = \mathbf{q} \cdot \hat{\mathbf{n}} - d$$
$$\approx \begin{bmatrix} 3 & 4 & 5 \end{bmatrix} \cdot \begin{bmatrix} .04976 & -.8677 & -.4945 \end{bmatrix} - (-7.389)$$
$$\approx (.04976)(3) + (-.8677)(4) + (-.4945)(5) + 7.389$$
$$\approx 1.595$$

Since this value is positive, we conclude that the point is on the *front* side of the plane.

(c) Let's first solve this problem by using the 2D projection method. The dominant axis of the normal is y, and so we'll discard the y coordinates of the vertices and project onto the xz plane. Applying notation from Listing 9.6 (but using 1-based subscripts):

$$
\begin{array}{llll}
u_1 = z_1 - z_3 & u_2 = z_2 - z_3 & u_3 = p_z - z_1 & u_4 = p_z - z_3 \\
\quad = -2 - 0 & \quad = 17 - 0 & \quad = 17.11 - (-2) & \quad = 17.11 - 0 \\
\quad = -2 & \quad = 17 & \quad = 19.11 & \quad = 17.11 \\
v_1 = x_1 - x_3 & v_2 = x_2 - x_3 & v_3 = p_x - x_1 & v_4 = p_x - x_3 \\
\quad = 6 - (-9) & \quad = 3 - (-9) & \quad = 13.60 - 6 & \quad = 13.60 - (-9) \\
\quad = 15 & \quad = 12 & \quad = 7.60 & \quad = 22.60
\end{array}
$$

$$
\begin{array}{llll}
\text{denom} & = v_1 u_2 - v_2 u_1 & = (15)(17) - (12)(-2) & = 279 \\
(b_1)(\text{denom}) & = v_4 u_2 - v_2 u_4 & = (22.60)(17) - (12)(17.11) & = 178.9 \\
b_1 & = 178.9/279 & & = 0.641 \\
(b_2)(\text{denom}) & = v_1 u_3 - v_3 u_1 & = (15)(19.11) - (7.60)(-2) & = 301.85 \\
b_2 & = 301.85/279 & & = 1.082 \\
b_3 & = 1 - b_1 - b_2 & = 1 - 0.641 - 1.082 & = -0.723
\end{array}
$$

(d) $\mathbf{c}_{\text{Grav}} = \dfrac{\mathbf{v}_1 + \mathbf{v}_2 + \mathbf{v}_3}{3} = \dfrac{\begin{bmatrix} 6 & 10 & -2 \end{bmatrix} + \begin{bmatrix} 3 & -1 & 17 \end{bmatrix} + \begin{bmatrix} -9 & 8 & 0 \end{bmatrix}}{3}$

$$= \dfrac{\begin{bmatrix} (6+3-9) & (10-1+8) & (-2+17+0) \end{bmatrix}}{3} = \dfrac{\begin{bmatrix} 0 & 17 & 15 \end{bmatrix}}{3}$$

$$= \begin{bmatrix} 0 & 17/3 & 5 \end{bmatrix} \approx \begin{bmatrix} 0 & 5.66 & 5 \end{bmatrix}$$

(e) First, we calculate the side lengths.

$$l_1 = \left\| \begin{bmatrix} -9 & 8 & 0 \end{bmatrix} - \begin{bmatrix} 3 & -1 & 17 \end{bmatrix} \right\| = \left\| \begin{bmatrix} -12 & 9 & -17 \end{bmatrix} \right\| \approx 22.67$$
$$l_2 = \left\| \begin{bmatrix} 6 & 10 & -2 \end{bmatrix} - \begin{bmatrix} -9 & 8 & 0 \end{bmatrix} \right\| = \left\| \begin{bmatrix} 15 & 2 & -2 \end{bmatrix} \right\| \approx 15.26$$
$$l_3 = \left\| \begin{bmatrix} 3 & -1 & 17 \end{bmatrix} - \begin{bmatrix} 6 & 10 & -2 \end{bmatrix} \right\| = \left\| \begin{bmatrix} -3 & -11 & 19 \end{bmatrix} \right\| \approx 22.16$$

$$\mathbf{c}_{\text{In}} = \frac{l_1\mathbf{v}_1 + l_2\mathbf{v}_2 + l_3\mathbf{v}_3}{p}$$

$$= \frac{22.67\begin{bmatrix} 6 & 10 & -2 \end{bmatrix} + 15.26\begin{bmatrix} 3 & -1 & 17 \end{bmatrix} + 22.16\begin{bmatrix} -9 & 8 & 0 \end{bmatrix}}{22.67 + 16.22 + 22.16}$$

$$= \frac{\begin{bmatrix} 136.02 & 226.70 & -45.34 \end{bmatrix} + \begin{bmatrix} 45.78 & -15.26 & 259.42 \end{bmatrix} + \begin{bmatrix} -199.44 & 177.28 & 0 \end{bmatrix}}{60.09}$$

$$= \frac{\begin{bmatrix} -17.64 & 388.72 & 214.08 \end{bmatrix}}{60.09} = \begin{bmatrix} -0.294 & 6.47 & 3.56 \end{bmatrix}$$

(f) $\mathbf{e}_1 = \begin{bmatrix} -9 \\ 8 \\ 0 \end{bmatrix} - \begin{bmatrix} 3 \\ -1 \\ 17 \end{bmatrix} = \begin{bmatrix} -12 \\ 9 \\ -17 \end{bmatrix}$

$\mathbf{e}_2 = \begin{bmatrix} 6 \\ 10 \\ -2 \end{bmatrix} - \begin{bmatrix} -9 \\ 8 \\ 0 \end{bmatrix} = \begin{bmatrix} 15 \\ 2 \\ -2 \end{bmatrix}$

$\mathbf{e}_3 = \begin{bmatrix} 3 \\ -1 \\ 17 \end{bmatrix} - \begin{bmatrix} 6 \\ 10 \\ -2 \end{bmatrix} = \begin{bmatrix} -3 \\ -11 \\ 19 \end{bmatrix}$

$d_1 = -\mathbf{e}_2 \cdot \mathbf{e}_3 = - \begin{bmatrix} 15 \\ 2 \\ -2 \end{bmatrix} \cdot \begin{bmatrix} -3 \\ -11 \\ 19 \end{bmatrix} = -((15 \cdot -3) + (2 \cdot -11) + (-2 \cdot 19)) = 105$

$d_2 = -\mathbf{e}_3 \cdot \mathbf{e}_1 = - \begin{bmatrix} -3 \\ -11 \\ 19 \end{bmatrix} \cdot \begin{bmatrix} -12 \\ 9 \\ -17 \end{bmatrix} = -((-3 \cdot -12) + (-11 \cdot 9) + (19 \cdot -17)) = 386$

$d_3 = -\mathbf{e}_1 \cdot \mathbf{e}_2 = - \begin{bmatrix} -12 \\ 9 \\ -17 \end{bmatrix} \cdot \begin{bmatrix} 15 \\ 2 \\ -2 \end{bmatrix} = -((-12 \cdot 15) + (9 \cdot 2) + (-17 \cdot -2)) = 128$

$c_1 = d_2 d_3 = 386 \cdot 128 = 49408$

$c_2 = d_3 d_1 = 128 \cdot 105 = 13440$

$c_3 = d_1 d_2 = 105 \cdot 386 = 40530$

$c = c_1 + c_2 + c_3 = 49408 + 13440 + 40530 = 103378$

$$\mathbf{c}_{\text{Circ}} = \frac{(c_2 + c_3)\mathbf{v}_1 + (c_3 + c_1)\mathbf{v}_2 + (c_1 + c_2)\mathbf{v}_3}{2c}$$

$$= \frac{(13440 + 40530)\begin{bmatrix} 6 \\ 10 \\ -2 \end{bmatrix} + (40530 + 49408)\begin{bmatrix} 3 \\ -1 \\ 17 \end{bmatrix} + (49408 + 13440)\begin{bmatrix} -9 \\ 8 \\ 0 \end{bmatrix}}{2(103378)}$$

$$= \frac{53970}{206756} \begin{bmatrix} 6 \\ 10 \\ -2 \end{bmatrix} + \frac{89938}{206756} \begin{bmatrix} 3 \\ -1 \\ 17 \end{bmatrix} + \frac{62848}{206756} \begin{bmatrix} -9 \\ 8 \\ 0 \end{bmatrix}$$

$$= 0.261 \begin{bmatrix} 6 \\ 10 \\ -2 \end{bmatrix} + 0.435 \begin{bmatrix} 3 \\ -1 \\ 17 \end{bmatrix} + 0.304 \begin{bmatrix} -9 \\ 8 \\ 0 \end{bmatrix}$$

$$= \begin{bmatrix} 1.566 \\ 2.610 \\ -0.522 \end{bmatrix} + \begin{bmatrix} 1.305 \\ -0.435 \\ 7.395 \end{bmatrix} + \begin{bmatrix} -2.736 \\ 2.432 \\ 0 \end{bmatrix} = \begin{bmatrix} 0.135 \\ 4.607 \\ 6.873 \end{bmatrix}$$

5. Using Equation (9.13):

$$\begin{aligned} n_x =&(12.70 + (-9.22))(13.90 - 12.77) + (-9.22 + 12.67)(12.77 - 2.34) \\ &+ (12.67 + (-7.09))(2.34 - 10.64) + (-7.09 + 18.68)(10.64 - 3.16) \\ &+ (18.68 + 12.70)(3.16 - 13.90) = -256.73 \\ n_y =&(-29.74 + 11.53)(12.70 - (-9.22)) + (11.53 + 9.16)(-9.22 - 12.67) \\ &+ (9.16 + 14.62)(12.67 - (-7.09)) + (14.62 + (-3.31))(-7.09 - 18.68) \\ &+ (-3.31 + (-29.74))(18.68 - 12.70) = -871.27 \\ n_z =&(13.90 + 12.77)(-29.74 - 11.53) + (12.77 + 2.34)(11.53 - 9.16) \\ &+ (2.34 + 10.64)(9.16 - 14.62) + (10.64 + 3.16)(14.62 - (-3.31)) \\ &+ (3.16 + 13.90)(-3.31 - (-29.74)) = -437.40 \end{aligned}$$

Normalizing this result, we have

$$\hat{\mathbf{n}} = [-0.255, -0.864, -0.434].$$

Now the best-fit d value is computed by

$$\begin{aligned} d &= \hat{\mathbf{n}} \cdot (\mathbf{p}_1 + \mathbf{p}_2 + \mathbf{p}_3 + \mathbf{p}_4 + \mathbf{p}_5)/5 \\ &= [-0.255, -0.864, -0.434] \cdot [2.26, 42.81, 27.74]/5 = -9.92. \end{aligned}$$

6. The seven-sided polygon is fanned into five triangles. One possible way to fan the polygon, based on the simple strategy given in Section 9.7.3, is

$$\{\mathbf{v}_1, \mathbf{v}_2, \mathbf{v}_3\}, \{\mathbf{v}_1, \mathbf{v}_3, \mathbf{v}_4\}, \{\mathbf{v}_1, \mathbf{v}_4, \mathbf{v}_5\}, \{\mathbf{v}_1, \mathbf{v}_5, \mathbf{v}_6\}, \{\mathbf{v}_1, \mathbf{v}_6, \mathbf{v}_7\}.$$

B.10   Chapter 10

(Page 476.)

1. This is a straightforward application of Equation (10.2).

   (a) $\dfrac{\text{pixPhys}_x}{\text{pixPhys}_y} = \dfrac{\text{devPhys}_x}{\text{devPhys}_y} \cdot \dfrac{\text{devRes}_y}{\text{devRes}_x} = \dfrac{4}{3} \cdot \dfrac{480}{640} = 1$

(b) $\dfrac{\text{pixPhys}_x}{\text{pixPhys}_y} = \dfrac{\text{devPhys}_x}{\text{devPhys}_y} \cdot \dfrac{\text{devRes}_y}{\text{devRes}_x} = \dfrac{16}{9} \cdot \dfrac{480}{640} = \dfrac{4}{3}$ (width greater than height)

2. (a) $\dfrac{\text{winPhys}_x}{\text{winPhys}_y} = \dfrac{\text{winRes}_x}{\text{winRes}_y} \cdot \dfrac{\text{pixPhys}_x}{\text{pixPhys}_y} = \dfrac{320}{480} \cdot 1 = \dfrac{2}{3}$ (width less than height)

(b) Using the left side of Equation (10.3), we have

$$\text{zoom}_x = \frac{1}{\tan(\text{fov}_x/2)} = \frac{1}{\tan(60^\circ/2)} \approx 1.732.$$

(c) Using Equation (10.4),

$$\frac{\text{zoom}_y}{\text{zoom}_x} = \frac{\text{winPhys}_x}{\text{winPhys}_y},$$

$$\frac{\text{zoom}_y}{1.732} = \frac{2}{3},$$

$$\text{zoom}_y = 1.155.$$

(d) Using the right side of Equation (10.3), we have

$$\text{fov}_y = 2 \arctan(1/\text{zoom}_y) = 2 \arctan(1/1.155) = 81.77^\circ.$$

(e) The correct formula is given by Equation (10.3.4).

$$\begin{bmatrix} \text{zoom}_x & 0 & 0 & 0 \\ 0 & \text{zoom}_y & 0 & 0 \\ 0 & 0 & -\frac{f+n}{f-n} & \frac{-2nf}{f-n} \\ 0 & 0 & -1 & 0 \end{bmatrix} = \begin{bmatrix} 1.732 & 0 & 0 & 0 \\ 0 & 1.155 & 0 & 0 \\ 0 & 0 & -\frac{256.0+1.0}{256.0-1.0} & \frac{-2(1.0)(256.0)}{256.0-1.0} \\ 0 & 0 & -1 & 0 \end{bmatrix}$$

$$= \begin{bmatrix} 1.732 & 0 & 0 & 0 \\ 0 & 1.155 & 0 & 0 \\ 0 & 0 & -1.00784 & -2.00784 \\ 0 & 0 & -1 & 0 \end{bmatrix}$$

(f) This time we use Equation (10.7).

$$\begin{bmatrix} \text{zoom}_x & 0 & 0 & 0 \\ 0 & \text{zoom}_y & 0 & 0 \\ 0 & 0 & \frac{f}{f-n} & 1 \\ 0 & 0 & \frac{-nf}{f-n} & 0 \end{bmatrix} = \begin{bmatrix} 1.732 & 0 & 0 & 0 \\ 0 & 1.155 & 0 & 0 \\ 0 & 0 & \frac{256}{256-1} & 1 \\ 0 & 0 & \frac{-(1)(256)}{256-1} & 0 \end{bmatrix} = \begin{bmatrix} 1.732 & 0 & 0 & 0 \\ 0 & 1.155 & 0 & 0 \\ 0 & 0 & 1.00392 & 1 \\ 0 & 0 & -1.00392 & 0 \end{bmatrix}$$

3. (a) $\dfrac{\text{winPhys}_x}{\text{winPhys}_y} = \dfrac{\text{winRes}_x}{\text{winRes}_y} \cdot \dfrac{\text{pixPhys}_x}{\text{pixPhys}_y} = \dfrac{320}{480} \cdot \dfrac{4}{3} = \dfrac{8}{9}$

(b) Same as before, 1.732.

(c) $\dfrac{\text{zoom}_y}{\text{zoom}_x} = \dfrac{\text{winPhys}_x}{\text{winPhys}_y}$

$\dfrac{\text{zoom}_y}{1.732} = \dfrac{8}{9}$,

$\text{zoom}_y = 1.540$

(d) $\text{fov}_y = 2 \arctan(1/\text{zoom}_y) = 2 \arctan(1/1.540) = 66.00^\circ$

4. (a) 2   (b) 1   (c) 4   (d) 6   (e) 3   (f) 5

5. (a) 7   (b) 3   (c) 1   (d) 10   (e) 4   (f) 2   (g) 9   (h) 6   (i) 8   (j) 5

6. Here we encode each component by multiplying by 127, adding 128, and then rounding to an integer. If any answer is off by 1 pixel, that's probably OK. (It's best to make sure -1 gets encoded as zero.)

   (a)   R=0, G=128, B=128   (b)   R=162, G=60, B=230
   (c)   R=128, G=128, B=255   (d)   R=128, G=237, B=193

7.

	Tangent-space normal	Binormal	Model-space normal
(a)	$[0.000, 1.000, 0.000]$	$[0.577, -0.577, 0.577]$	$[0.577, -0.577, 0.577]$
(b)	$[-0.172, 0.211, 0.953]$	$[0.000, 0.000, 1.000]$	$[-0.172, 0.953, 0.211]$
(c)	$[0.000, 0.703, 0.703]$	$[0.000, 0.894, 0.447]$	$[0.703, 0.628, 0.314]$
(d)	$[0.820, -0.547, 0.133]$	$[-0.064, -0.786, -0.615]$	$[0.864, 0.386, 0.307]$

B.11   Chapter 11

(Page 549.)

1. $1 \frac{\text{lb}}{\text{in}^2} \approx 1 \frac{\text{lb}}{\text{in}^2} \times \frac{4.448 \text{ N}}{1 \text{ lb}} \times \left(\frac{1 \text{ in}}{0.0254 \text{ m}} \right)^2 \approx 6.89 \times 10^3 \frac{\text{N}}{\text{m}^2}$

2.

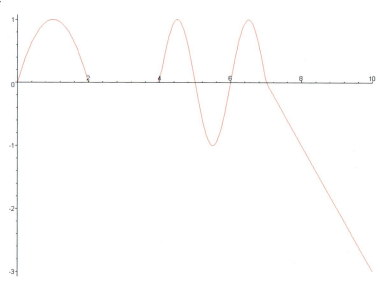

3. (a) $\dfrac{x(1) - x(0)}{1 - 0} = \dfrac{1 - 0}{1} = 1$

(b) $\dfrac{x(2) - x(1)}{2 - 1} = \dfrac{0 - 1}{1} = -1$

(c) $\dfrac{x(2) - x(0)}{2 - 0} = \dfrac{0 - 0}{2} = 0$

(d) $\dfrac{x(6.5) - x(5.5)}{6.5 - 5.5} = \dfrac{1 - (-1)}{1} = 2$

(e) $\dfrac{x(9) - x(0)}{9 - 0} = \dfrac{(-2) - 0}{9} = -\dfrac{2}{9}$

4. $v(t) = \begin{cases} 2 - 2t & 0 < t < 2 \\ 0 & 2 < t < 4 \\ \pi \cos(\pi t) & 4 < t < 7 \\ -1 & 7 < t \end{cases}$

5. (a) $v(0.1) = 2 - 2(0.1) = 1.8$

(b) $v(1.0) = 2 - 2(1.0) = 0.0$

(c) $v(1.9) = 2 - 2(1.9) = -1.8$

(d) $v(4.1) = \pi \cos(4.1\pi) = 2.988$

(e) $v(5) = \pi \cos(5\pi) = -\pi$

(f) $v(6.5) = \pi \cos(6.5\pi) = 0$

(g) $v(8) = -1$

(h) $v(9) = -1$

6. $a(t) = \begin{cases} -2 & 0 < t < 2 \\ 0 & 2 < t < 4 \\ -\pi^2 \sin(\pi t) & 4 < t < 7 \\ 0 & 7 < t \end{cases}$

7. (a) $a(0.1) = -2$

(b) $a(1.0) = -2$

(c) $a(1.9) = -2$

(d) $a(4.1) = -\pi^2 \sin(4.1\pi) = -3.050$

(e) $a(5) = -\pi^2 \sin(5\pi) = 0$

(f) $a(6.5) = -\pi \sin(6.5\pi) = -9.870$

(g) $a(8) = 0$

(h) $a(9) = 0$

8. A negative discriminant indicates that the apex of the movement (the farthest displacement in the direction of the initial velocity) is not large enough to reach the desired displacement Δx. Thus, there is no value of t for which the sought-after displacement will be reached. If the discriminant is zero, then there is exactly one solution to Equation (11.16), and the displacement is equal to the maximum displacement at the apex.

Note that if the acceleration and the displacement have the same sign, then the discriminant can never be negative, and there will always be two solutions except in the trivial case where all the values are zero.

9. (a) $\mathbf{v}_0 = 150[\cos 40°, \sin 40°]$ ft/s $\approx [114.9, 96.4]$ ft/s

(b) $t = -(\mathbf{v}_0)_y/a_y = -(96.4 \text{ ft/s})/(-32.0 \text{ ft/s}^2) = 3.01$ s

(c) $\mathbf{p}(t) = \begin{bmatrix} 0 \text{ ft} \\ 10 \text{ ft} \end{bmatrix} + \begin{bmatrix} 114.9 \text{ ft/s} \\ 96.4 \text{ ft/s} \end{bmatrix} t + \frac{1}{2} \begin{bmatrix} 0 \text{ ft/s}^2 \\ -32.0 \text{ ft/s}^2 \end{bmatrix} t^2$

$\mathbf{p}(3.01 \text{ s}) = \begin{bmatrix} 0 \text{ ft} \\ 10 \text{ ft} \end{bmatrix} + \begin{bmatrix} 114.9 \text{ ft/s} \\ 96.4 \text{ ft/s} \end{bmatrix} (3.01 \text{ s}) + \frac{1}{2} \begin{bmatrix} 0 \text{ ft/s}^2 \\ -32.0 \text{ ft/s}^2 \end{bmatrix} (3.01 \text{ s})^2$

$= \begin{bmatrix} 0 \text{ ft} \\ 10 \text{ ft} \end{bmatrix} + \begin{bmatrix} 345.8 \text{ ft} \\ 290.2 \text{ ft} \end{bmatrix} + \begin{bmatrix} 0 \text{ ft} \\ -145.0 \text{ ft} \end{bmatrix} = \begin{bmatrix} 345.8 \text{ ft} \\ 155.2 \text{ ft} \end{bmatrix}$

(d) It's twice the time to reach the apex, $2(3.01 \text{ s}) = 6.02$ s.

(e) $x(t) = (0 \text{ ft}) + (114.9 \text{ ft/s})t + (1/2)(0 \text{ ft/s}^2)t^2$
$x(6.02 \text{ s}) = (114.9 \text{ ft/s})(6.02 \text{ s}) = 691.7$ ft

10. $\Delta\mathbf{p} = \mathbf{v}_0 t + (1/2)\mathbf{a}t^2$

$\Delta\mathbf{p} \cdot \mathbf{a} = (\mathbf{v}_0 t + (1/2)\mathbf{a}t^2) \cdot \mathbf{a}$

$\Delta\mathbf{p} \cdot \mathbf{a} = (\mathbf{v}_0 \cdot \mathbf{a})t + (1/2)(\mathbf{a} \cdot \mathbf{a})t^2$

$0 = (\mathbf{a} \cdot \mathbf{a}/2)t^2 + (\mathbf{v}_0 \cdot \mathbf{a})t - \Delta\mathbf{p} \cdot \mathbf{a}$

$t = \dfrac{-(\mathbf{v}_0 \cdot \mathbf{a}) \pm \sqrt{(\mathbf{v}_0 \cdot \mathbf{a})^2 - 4(\mathbf{a} \cdot \mathbf{a}/2)(\Delta\mathbf{p} \cdot \mathbf{a})}}{2(\mathbf{a} \cdot \mathbf{a}/2)}$

$t = \dfrac{-(\mathbf{v}_0 \cdot \mathbf{a}) \pm \sqrt{(\mathbf{v}_0 \cdot \mathbf{a})^2 - 2(\mathbf{a} \cdot \mathbf{a})(\Delta\mathbf{p} \cdot \mathbf{a})}}{\mathbf{a} \cdot \mathbf{a}}$

11. Expanding the Taylor series for e^{ix}:

$$e^{ix} = 1 + ix + \frac{(ix)^2}{2!} + \frac{(ix)^3}{3!} + \frac{(ix)^4}{4!} + \frac{(ix)^5}{5!} + \frac{(ix)^6}{6!} + \frac{(ix)^7}{7!} + \frac{(ix)^8}{8!} + \cdots$$

Substituting the powers of i ($i^2 = -1$, $i^3 = -i$, $i^4 = 1$, etc.):

$$e^{ix} = 1 + ix - \frac{x^2}{2!} - \frac{ix^3}{3!} + \frac{x^4}{4!} + \frac{ix^5}{5!} - \frac{x^6}{6!} - \frac{ix^7}{7!} + \frac{x^8}{8!} + \cdots$$

Now we separate the real and imaginary terms:

$$e^{ix} = \left(1 - \frac{x^2}{2!} + \frac{x^4}{4!} - \frac{x^6}{6!} + \frac{x^8}{8!} - \cdots\right) + i\left(x - \frac{x^3}{3!} + \frac{x^5}{5!} - \frac{x^7}{7!} + \cdots\right)$$

The sums can be recognized as the Taylor series expansion for cosine and sine; therefore,

$$e^{ix} = \cos x + i \sin x.$$

This equation is known as *Euler's formula*. Substituting $x = \pi$ and moving everything to the left hand side gives us *Euler's identity*, a beautiful equation that links together five important mathematical constants:

$$e^{i\pi} + 1 = 0.$$

12. This one has a few tricks. First, we need to compute the actual radius of the orbit, taking into account Earth's (average) radius of 6,371 km, as

$$r = 6,371 \text{ km} + 340 \text{ km} = 6,711 \text{ km}.$$

Now the length of the circular orbit is just the circumference of a circle with this radius, which can be computed using elementary geometry:

$$C = 2\pi r = 2\pi(6,711 \text{ km}) = 4.217 \times 10^4 \text{ km}.$$

Finally, we divide this distance by the average speed to get the orbital period:

$$P = C/s = (4.217 \times 10^4 \text{ km})/(27,740 \text{ km/hr}) = 1.520 \text{ hr} = 91.21 \text{ min}.$$

The centripetal acceleration can be computed by Equation (11.29):

$$a = \frac{s^2}{r} = \frac{\left(27,740 \ \frac{\text{km}}{\text{hr}} \times \frac{1 \text{ hr}}{3,600 \text{ s}}\right)^2}{6,711 \text{ km}} = \frac{\left(7.706 \ \frac{\text{km}}{\text{s}}\right)^2}{6,711 \text{ km}} = 0.008849 \ \frac{\text{km}}{\text{s}^2} = 8.849 \ \frac{\text{m}}{\text{s}^2}.$$

B.12 Chapter 12

(Page 640.)

1. We must consider all the forces acting on the fan, the air, and the boat. As the fan rotates, a force exists between the fan and the air, which wants to push the air forward and the fan backwards. Since the fan does not accelerate backwards, we know that there must be some force opposing it, and this force comes from the force of friction provided by the boat. But then this means the boat is receiving a backwards force, and this backwards force counteracts any force eventually received by the wind hitting the sail.

2. First, we identify four bodies: the girl, the boy, the rope, and Earth. Next, we identify the active tension and friction forces:

$T_{g,r}$	Girl pulls on rope	$T_{r,g}$	Rope pulls on girl
$F_{g,e}$	Girl pushes on Earth	$F_{e,g}$	Earth pushes on girl
$T_{b,r}$	Boy pulls on rope	$T_{r,b}$	Rope pulls on boy
$F_{b,e}$	Boy pushes on Earth	$F_{e,b}$	Earth pushes on boy

By Newton's third law, we assume that each force on the left is equal in magnitude but opposite in direction to the corresponding force on the right. Next, since we assume that the stretching of the rope is negligible, the tension at one end must be equal to the tension at the other end, so all the T forces have equal magnitude. Since the children are both accelerating, there must be a net force on each of them causing their displacement. Earth is pushing the girl harder than the rope is pulling her, so she accelerates backwards. For the boy, the opposite is true, and the directions of the forces cause him to move forwards. So the reason that the children, as a system, accelerate relative to Earth is because the girl's pushing force is larger than the boy's pushing force, resulting in a net force on the children in the direction of the girl.

3. False. The acceleration due to gravity is constant, but the force due to gravity increases proportionately with mass.

4. This is a straightforward application of Newton's law of universal gravitation with the distance equal to the radius of Earth plus the orbit altitude.

$$d = 6{,}371 \text{ km} + 340 \text{ km} = 6{,}711 \text{ km}.$$

Plugging this value and the mass of Earth into Equation (12.3), we have

$$f = G\frac{m_1 m_2}{d^2} = \left(6.673 \times 10^{-11} \frac{\text{N m}^2}{\text{kg}^2}\right) \frac{(5.98 \times 10^{24} \text{ kg})m_2}{(6.711 \times 10^6 \text{ m})^2}$$
$$= (8.86 \text{ N})\frac{m_2}{\text{kg}} = \left(8.86 \frac{\text{m}}{\text{s}^2}\right) m_2$$

We observe a few things about this result. First, it most definitely is not zero; in fact, it is only about 10% less than the acceleration due to gravity at Earth's surface. Although the term "zero gravity" is often used to describe the environment of objects orbiting in space, we see that this term is a bit of a misnomer, since gravity is quite alive and well, even at 340 km above Earth's surface. In fact, it is gravity that supplies the necessary centripetal acceleration to maintain the orbit.

Second, we compare this answer to our results from Exercise 11.12, and we see that the numbers are the same. (Well, almost exactly the same. The discrepancy of 0.1% is a result of some slight simplifications to the problem and rounding.) This match leads us to answer the second part of the problem. The apparent weightlessness exists because the space station and all the objects in it are in free fall. Apparent weightlessness occurs in any free-fall situation, no matter what the force of gravity and even if the object isn't orbiting (for example, in a falling elevator or amusement park ride or in NASA's "vomit comet" aircraft).

The difference between a falling elevator and an object orbiting Earth is that the free fall in the space station continues *indefinitely*. The orbit speed and altitude are selected such that the acceleration due to gravity is exactly the same as the centripetal acceleration, and unlike a falling elevator, the space station never gets any closer to the ground. The space station keeps "falling over the horizon" and never hits bottom.

5. At the critical angle, the force of static friction f_s exactly balances the lateral component of the force of gravity, $g_\parallel$. The maximum friction is equal to the magnitude n of the normal force times the coefficient of static friction μ_s. The normal force is equal to $g_\perp$, the

component of gravity perpendicular to the surface, and from Table 12.1, the coefficient of static friction between concrete and wood is 0.62. Thus, we have

$$f_s = g_{\parallel},$$
$$(\mu_s n) = g_{\parallel},$$
$$(0.62 g_{\perp}) = g_{\parallel}.$$

The normal and lateral components of gravity can be expressed in terms of the angle of inclination θ as

$$g_{\perp} = \|\mathbf{g}\| \cos\theta,$$
$$g_{\parallel} = \|\mathbf{g}\| \sin\theta,$$

where $\|\mathbf{g}\|$ is the total magnitude of the force of gravity on the block. Plugging in these values and solving this for θ, we have

$$0.62 g_{\perp} = g_{\parallel},$$
$$0.62 \|\mathbf{g}\| \cos\theta = \|\mathbf{g}\| \sin\theta,$$
$$0.62 = \frac{\|\mathbf{g}\| \sin\theta}{\|\mathbf{g}\| \cos\theta},$$
$$0.62 = \frac{\sin\theta}{\cos\theta},$$
$$0.62 = \tan\theta,$$
$$\arctan 0.62 = \theta,$$
$$32° \approx \theta.$$

Notice that neither the weight of the block nor the acceleration due to gravity was relevant in this experiment. Thus, if you were to conduct this experiment on the moon, you would get the same critical angle.

6. (a) Hooke's law tells us $f = kx_0$. The force in this case is gravity, which is proportional to the mass and given by $f = mg$. Thus, the relation is

$$mg = kx_0.$$

   (b) Substituting into the equation obtained in part (a), we have

$$mg = kx_0,$$
$$(5.00\text{ kg})(9.8\text{ m/s}^2) = k(10.0\text{ cm}),$$
$$\frac{49\text{ N}}{0.100\text{ m}} = k,$$
$$4.9 \times 10^2\text{ N/m} = k.$$

   (c) $mg = kx_0$
$$m(9.8\text{ m/s}^2) = (4.9 \times 10^2\text{ N/m})(17.0\text{ cm})$$
$$m = \frac{(4.9 \times 10^2\text{ N/m})(0.170\text{ m})}{9.8\text{ m/s}^2} = 8.5\text{ kg}$$

(d) Assuming the spring and environment had not changed, we would expect a 1 kg mass to cause an extension of 2 cm. Since the actual change in length was 8 cm, there are only two explanations.[2] Either the spring constant has been reduced or the apparent force of gravity has increased (or both). Maybe the spring was worn out? The increase in gravity could be caused by conducting the experiment on a larger planet or in a noninertial reference frame that is accelerating upwards.

7. (a) $F = \dfrac{1}{2\pi}\sqrt{\dfrac{k}{m}} = \dfrac{1}{2\pi}\sqrt{\dfrac{1.00 \times 10^2 \text{ (kg m/s}^2)/m}{5.00 \text{ kg}}} = \dfrac{\sqrt{20.0 \text{ s}^{-2}}}{2\pi} = 0.712 \text{ Hz}$

(b) The amplitude is simply the initial displacement, 14.7 cm.

(c) We know that the motion of the mass must be of the form $A\cos(\omega t + \theta_0)$. We already determined the amplitude $A = 14.7$ cm, and we know the angular frequency $\omega = 2\pi F = 4.47$ Hz.

When the mass crosses the rest position, $\cos(\omega t + \theta_0) = 0$. Therefore, at this time, $\sin(\omega t + \theta_0) = \pm 1$, and the velocity is

$$v(t) = -A\omega \sin(\omega t + \theta_0)$$
$$= \pm(14.7 \text{ cm})(4.47 \text{ s}^{-1}) = \pm 65.7 \text{ cm/s}$$

Since speed is always positive, we can discard the "$\pm$".

8. Since there are no external forces, the center of mass of the man + car system does not move, and the total momentum of this system must remain zero throughout. We'll let v_m and v_c refer to the velocity of the man and car, respectively, relative to Earth.

(a) The relative velocity of the man and car is expressed by

$$v_m - v_c = 1.25 \text{ m/s},$$
$$v_m = v_c + 1.25 \text{ m/s},$$

and we also know that the combined momentum of the system must remain at zero,

$$P_m + P_c = m_m v_m + m_c v_c = 0.$$

Plugging the first equation into the second, we have

$$(75.0 \text{ kg})(v_c + 1.25 \text{ m/s}) + (1.00 \times 10^3 \text{ kg})v_c = 0,$$
$$(75.0 \text{ kg})v_c + 93.8 \text{ kg m/s} + (1.00 \times 10^3 \text{ kg})v_c = 0,$$
$$(1.08 \times 10^3 \text{ kg})v_c = -93.8 \text{ kg m/s},$$
$$v_c = -0.0869 \text{ m/s}.$$

So we obtain the inertial velocity of the man as

$$v_m = v_c + 1.25 \text{ m/s} = -0.0869 \text{ m/s} + 1.25 \text{ m/s} = 1.16 \text{ m/s}.$$

[2]No credit is given for suggesting "physics stopped working." However, if you answered "we are inside of a video game," give yourself 20 points extra credit.

(b) First, we compute the duration of the journey by considering the man's motion in the coordinate space fixed to the car, as

$$\Delta t = \Delta x / v = (20.0 \text{ m}) / (1.25 \text{ m/s}) = 16.0 \text{ s}.$$

We then multiply the velocities of the car and man by this duration to obtain their displacements.

$$\Delta x_m = v_m \Delta t = (1.16 \text{ m/s})(16.0 \text{ s}) = 18.6 \text{ m}$$
$$\Delta x_c = v_c \Delta t = (-0.0869 \text{ m/s})(16.0 \text{ s}) = -1.39 \text{ m}$$

An alternate approach is to recognize that the center of mass of the system does not move, since there are no external forces, and treat the man and the car as point masses. Since the man walked the length of the car,

$$\Delta x_m = \Delta x_c + 20.0 \text{ m}.$$

Now the movement of the man must be offset by the movement of the car, such that the center of gravity does not shift.

$$\Delta x_m m_m + \Delta x_c m_c = 0$$

One again, the system of equations is solved by plugging the first equation into the second.

$$(\Delta x_c + 20.0 \text{ m}) m_m + \Delta x_c m_c = 0$$
$$(\Delta x_c + 20.0 \text{ m})(75.0 \text{ kg}) + \Delta x_c (1.00 \times 10^3 \text{ kg}) = 0$$
$$\Delta x_c (75.0 \text{ kg}) + (1.50 \times 10^3 \text{ kg m}) + \Delta x_c (1.00 \times 10^3 \text{ kg}) = 0$$
$$\Delta x_c (1.08 \times 10^3 \text{ kg}) = -1.50 \times 10^3 \text{ kg m}$$
$$\Delta x_c = -1.39 \text{ m}$$

(c) The car's velocity would also increase in proportion to the man's. At all times, the total momentum and total displacement of the center of mass would be zero. The ending configuration of the car and the man would be the same as before.

(d) Here all we must do is add $+5.00$ m/s to our earlier results.

$$v_c = -0.0869 \text{ m/s} + 5.00 \text{ m/s} = 4.91 \text{ m/s}$$
$$v_m = 1.16 \text{ m/s} + 5.00 \text{ m/s} = 6.16 \text{ m/s}$$

(e) $\Delta x_m = v_m \Delta t = (6.16 \text{ m/s})(16.0 \text{ s}) = 98.6 \text{ m}$
$\Delta x_c = v_c \Delta t = (4.91 \text{ m/s})(16.0 \text{ s}) = 78.6 \text{ m}$

9. First, we must compute the contact normal $\mathbf{n}$ as

$$\mathbf{n} = \begin{bmatrix} \cos -110° \\ \sin -110° \end{bmatrix} \approx \begin{bmatrix} -0.342 \\ -0.940 \end{bmatrix},$$

and the relative velocity $\mathbf{v}_{\text{rel}}$ as

$$\mathbf{v}_{\text{rel}} = \mathbf{v}_1 - \mathbf{v}_2 = (35 \text{ km/hr}) \begin{bmatrix} -1 \\ 0 \end{bmatrix} - (65 \text{ km/hr}) \begin{bmatrix} \cos 115° \\ \sin 115° \end{bmatrix} \approx \begin{bmatrix} -7.505 \\ -58.890 \end{bmatrix} \text{ km/hr}.$$

Now we plug these values, along with masses and coefficient of restitution, into Equation (12.23), to determine k, the magnitude of the collision impulse. (Note that we also assume $\mathbf{n}$ is a unit vector, so $\mathbf{n} \cdot \mathbf{n} = 1$.)

$$k = \frac{(e+1)\mathbf{v}_{\text{rel}} \cdot \mathbf{n}}{(1/m_1 + 1/m_2)\,\mathbf{n} \cdot \mathbf{n}} = \frac{(1+0.1)\left(\begin{bmatrix} -7.505 \\ -58.890 \end{bmatrix} \text{ km/hr}\right) \cdot \begin{bmatrix} -0.342 \\ -0.940 \end{bmatrix}}{(1/1,500 \text{ kg} + 1/2,500 \text{ kg})}$$

$$= \frac{63.716 \text{ km/hr}}{0.00107 \text{ kg}^{-1}} = 59,533 \text{ kg km/hr}$$

The vector impulse received by Kari (who is m_2) is

$$k\mathbf{n} = (59,533 \text{ kg km/hr}) \begin{bmatrix} -0.342 \\ -0.940 \end{bmatrix} = \begin{bmatrix} -20,360 \\ -55,961 \end{bmatrix} \text{ kg km/hr}.$$

Finally, computing the velocities after the impact, we have

$$\mathbf{v}_1' = \frac{\mathbf{P}_1'}{m_1} = \frac{\mathbf{P}_1 - k\mathbf{n}}{m_1} = \frac{\left(\begin{bmatrix} -52,500 \\ 0 \end{bmatrix} - \begin{bmatrix} -20,360 \\ -55,961 \end{bmatrix}\right) \text{ kg km/hr}}{1,500 \text{ kg}} = \begin{bmatrix} -21.43 \\ 37.31 \end{bmatrix} \text{ km/hr}$$

$$\mathbf{v}_2' = \frac{\mathbf{P}_2'}{m_2} = \frac{\mathbf{P}_2 + k\mathbf{n}}{m_2} = \frac{\left(\begin{bmatrix} -68,700 \\ 147,000 \end{bmatrix} + \begin{bmatrix} -20,360 \\ -55,961 \end{bmatrix}\right) \text{ kg km/hr}}{2,500 \text{ kg}} = \begin{bmatrix} -35.62 \\ 36.42 \end{bmatrix} \text{ km/hr}$$

10. Because the force of gravity is always directly downwards, the bend in the balance bar causes the lever arm of each side to change depending on the angle of the monkey. For example, if the monkey begins to lean to the left, this rotates the weight on the right end of the pole upwards. In this situation, the force of gravity acts more perpendicular to the bar on the right (the lever arm), and the torque is increased. At the same time, the mass on the opposite end rotates downwards, causing the bar to become more parallel with the force of gravity, thus decreasing the torque. In other words, the restorative torque is always greater than the torque that would tend to tip him over, and when he is upright, they are in equilibrium.

11. The hollow cylinder will be harder to roll, because the moment of inertia will be larger. Imagine that the cylinders are made of a compressible substance. Now imagine taking an individual mass element from the center of the solid center and pushing it outwards. As the radius increases, the moment of inertia of this element will increase. This is essentially the difference between the two cylinders, the hollow one has a denser outer ring, with more of its mass pushed outwards.

12. (a) First we determine the total mass, which is 2200 kg. Then we take a weighted average of the mass centers according to Equation (12.22).

$$\mathbf{r}_c = \frac{1}{M} \sum_i^n m_i \mathbf{r_i} = \frac{1}{2200} \left(\begin{array}{l} 1000 \cdot (0, 100, 225) + 600 \cdot (0, 125, 75) \\ + 400 \cdot (0, 100, -120) + 50 \cdot (-100, 35, 230) \\ + 50 \cdot (100, 35, 230) + 50 \cdot (-100, 35, -150) \\ + 50 \cdot (100, 35, -150) \end{array} \right)$$

$$= (0, 101, 105)$$

(b) At the time of this writing, all the formulas are available on the Wikipedia article *List of moment of inertia tensors.*

Body front:

$$J = \frac{1000 \text{ kg}}{12} \begin{bmatrix} (0.80 \text{ m})^2 + (1.50 \text{ m})^2 & 0 & 0 \\ 0 & (2.00 \text{ m})^2 + (1.50 \text{ m})^2 & 0 \\ 0 & 0 & (2.00 \text{ m})^2 + (0.80 \text{ m})^2 \end{bmatrix}$$

$$= \begin{bmatrix} 241 & 0 & 0 \\ 0 & 521 & 0 \\ 0 & 0 & 387 \end{bmatrix} (\text{kg m}^2)$$

Body middle:

$$J = \frac{600 \text{ kg}}{12} \begin{bmatrix} (1.30 \text{ m})^2 + (1.50 \text{ m})^2 & 0 & 0 \\ 0 & (2.00 \text{ m})^2 + (1.50 \text{ m})^2 & 0 \\ 0 & 0 & (2.00 \text{ m})^2 + (1.30 \text{ m})^2 \end{bmatrix}$$

$$= \begin{bmatrix} 197 & 0 & 0 \\ 0 & 313 & 0 \\ 0 & 0 & 285 \end{bmatrix} (\text{kg m}^2)$$

Body rear:

$$J = \frac{400 \text{ kg}}{12} \begin{bmatrix} (0.80 \text{ m})^2 + (2.40 \text{ m})^2 & 0 & 0 \\ 0 & (2.00 \text{ m})^2 + (2.40 \text{ m})^2 & 0 \\ 0 & 0 & (2.00 \text{ m})^2 + (0.80 \text{ m})^2 \end{bmatrix}$$

$$= \begin{bmatrix} 213 & 0 & 0 \\ 0 & 325 & 0 \\ 0 & 0 & 155 \end{bmatrix} (\text{kg m}^2)$$

Each wheel:

$$J = (50 \text{ kg}) \begin{bmatrix} \frac{1}{2}(0.35 \text{ m})^2 & 0 & 0 \\ 0 & \frac{1}{12}(3(0.35 \text{ m})^2 + (0.20 \text{ m})^2) & 0 \\ 0 & 0 & \frac{1}{12}(3(0.35 \text{ m})^2 + (0.20 \text{ m})^2) \end{bmatrix}$$

$$= \begin{bmatrix} 3.06 & 0 & 0 \\ 0 & 1.70 & 0 \\ 0 & 0 & 1.70 \end{bmatrix} (\text{kg m}^2)$$

(c) We apply the parallel axis theorem (Equation (12.31)) to each part. We must first compute the position of each part relative to the center of mass of the truck, which we denote as $\mathbf{r}'$.

Body front:

$$\mathbf{r}' = (0, 1.00, 2.25) - (0, 1.01, 1.05) = (0, -0.01, 1.20)$$

$$J' = \begin{bmatrix} 241 & 0 & 0 \\ 0 & 521 & 0 \\ 0 & 0 & 387 \end{bmatrix} + 1000 \begin{bmatrix} (-0.01)^2 + 1.20^2 & -0 \cdot (-0.01) & -0 \cdot 1.20 \\ -0 \cdot (-0.01) & 0^2 + 1.20^2 & -(-0.01) \cdot 1.20 \\ -0 \cdot 1.20 & -(-0.01) \cdot 1.20 & 0^2 + (-0.01)^2 \end{bmatrix} \ (\text{kg m}^2)$$

$$= \begin{bmatrix} 1680 & 0 & 0 \\ 0 & 1960 & 12.0 \\ 0 & 12.0 & 387 \end{bmatrix} \ (\text{kg m}^2)$$

Body middle:

$$\mathbf{r}' = (0, 125, 75) - (0, 1.01, 1.05) = (0, 0.24, -0.30)$$

$$J' = \begin{bmatrix} 197 & 0 & 0 \\ 0 & 313 & 0 \\ 0 & 0 & 285 \end{bmatrix} + 600 \begin{bmatrix} 0.24^2 + (-0.30)^2 & -0 \cdot 0.24 & -0 \cdot (-0.30) \\ -0 \cdot 0.24 & 0^2 + (-0.30)^2 & -0.24 \cdot (-0.30) \\ -0 \cdot (-0.30) & -0.24 \cdot (-0.30) & 0^2 + 0.24^2 \end{bmatrix} \ (\text{kg m}^2)$$

$$= \begin{bmatrix} 286 & 0 & 0 \\ 0 & 367 & 43.2 \\ 0 & 43.2 & 320 \end{bmatrix} \ (10^2 \ \text{kg m}^2)$$

Body rear:

$$\mathbf{r}' = (0, 1.00, -1.20) - (0, 1.01, 1.05) = (0, -0.01, -2.25)$$

$$J' = \begin{bmatrix} 213 & 0 & 0 \\ 0 & 325 & 0 \\ 0 & 0 & 155 \end{bmatrix} + 400 \begin{bmatrix} (-0.01)^2 + (-2.25)^2 & -0 \cdot (-0.01) & -0 \cdot (-2.25) \\ -0 \cdot (-0.01) & 0^2 + (-2.25)^2 & -(-0.01) \cdot (-2.25) \\ -0 \cdot (-2.25) & -(-0.01) \cdot (-2.25) & 0^2 + 0.01^2 \end{bmatrix} \ (\text{kg m}^2)$$

$$= \begin{bmatrix} 22.4 & 0 & 0 \\ 0 & 23.5 & -0.09 \\ 0 & -0.09 & 1.55 \end{bmatrix} \ (10^2 \ \text{kg m}^2)$$

Front left wheel:

$$\mathbf{r}' = (-1.00, 0.35, 2.30) - (0, 1.01, 1.05) = (-1.00, -0.66, 1.25)$$

$$J' = \begin{bmatrix} 3.06 & 0 & 0 \\ 0 & 1.70 & 0 \\ 0 & 0 & 1.70 \end{bmatrix} \ (\text{kg m}^2)$$

$$+ 50 \begin{bmatrix} (-0.66)^2 + 1.25^2 & -(-1.00) \cdot (-0.66) & -(-1.00) \cdot 1.25 \\ -(-1.00) \cdot (-0.66) & (-1.00)^2 + 1.25^2 & -(-0.66) \cdot 1.25 \\ -(-1.00) \cdot 1.25 & -(-0.66) \cdot 1.25 & (-1.00)^2 + (-0.66)^2 \end{bmatrix} \ (\text{kg m}^2)$$

$$= \begin{bmatrix} 103 & -33.0 & 62.5 \\ -33.0 & 130 & 41.3 \\ 62.5 & 41.3 & 73.5 \end{bmatrix} \ (\text{kg m}^2)$$

Front right wheel:

$$\mathbf{r}' = (1.00, 0.35, 2.30) - (0, 1.01, 1.05) = (1.00, -0.66, 1.25)$$

$$J' = \begin{bmatrix} 3.06 & 0 & 0 \\ 0 & 1.70 & 0 \\ 0 & 0 & 1.70 \end{bmatrix} \ (\text{kg m}^2)$$

$$+ 50 \begin{bmatrix} (-0.66)^2 + 1.25^2 & -(1.00) \cdot (-0.66) & -(1.00) \cdot 1.25 \\ -(1.00) \cdot (-0.66) & (1.00)^2 + 1.25^2 & -(-0.66) \cdot 1.25 \\ -(1.00) \cdot 1.25 & -(-0.66) \cdot 1.25 & (1.00)^2 + (-0.66)^2 \end{bmatrix} \ (\text{kg m}^2)$$

$$= \begin{bmatrix} 103 & 33.0 & -62.5 \\ 33.0 & 130 & 41.3 \\ -62.5 & 41.3 & 73.5 \end{bmatrix} \ (\text{kg m}^2)$$

Rear left wheel:

$$\mathbf{r}' = (-1.00, 0.35, -1.50) - (0, 1.01, 1.05) = (-1.00, -0.66, -2.55)$$

$$J' = \begin{bmatrix} 3.06 & 0 & 0 \\ 0 & 1.70 & 0 \\ 0 & 0 & 1.70 \end{bmatrix} \text{ (kg m}^2)$$

$$+ 50 \begin{bmatrix} (-0.66)^2 + (-2.55)^2 & -(-1.00) \cdot (-0.66) & -(-1.00) \cdot (-2.55) \\ -(-1.00) \cdot (-0.66) & (-1.00)^2 + (-2.55)^2 & -(-0.66) \cdot (-2.55) \\ -(-1.00) \cdot (-2.55) & -(-0.66) \cdot (-2.55) & (-1.00)^2 + (-0.66)^2 \end{bmatrix} \text{ (kg m}^2)$$

$$= \begin{bmatrix} 350 & -33.0 & -128 \\ -33.0 & 377 & -84.2 \\ -128 & -84.2 & 73.5 \end{bmatrix} \text{ (kg m}^2)$$

Rear right wheel:

$$\mathbf{r}' = (1.00, 0.35, -1.50) - (0, 1.01, 1.05) = (1.00, -0.66, -2.55)$$

$$J' = \begin{bmatrix} 3.06 & 0 & 0 \\ 0 & 1.70 & 0 \\ 0 & 0 & 1.70 \end{bmatrix} \text{ (kg m}^2)$$

$$+ 50 \begin{bmatrix} (-0.66)^2 + (-2.55)^2 & -(1.00) \cdot (-0.66) & -(1.00) \cdot (-2.55) \\ -(1.00) \cdot (-0.66) & (1.00)^2 + (-2.55)^2 & -(-0.66) \cdot (-2.55) \\ -(1.00) \cdot (-2.55) & -(-0.66) \cdot (-2.55) & (1.00)^2 + (-0.66)^2 \end{bmatrix} \text{ (kg m}^2)$$

$$= \begin{bmatrix} 350 & 33.0 & 128 \\ 33.0 & 377 & -84.2 \\ 128 & -84.2 & 73.5 \end{bmatrix} \text{ (kg m}^2)$$

Total:

$$\begin{bmatrix} 5110 & 0 & 0 \\ 0 & 5690 & -40 \\ 0 & -40 & 1150 \end{bmatrix} \text{ (kg m}^2)$$

B.13   Chapter 13

(Page 711.)

1. $\ell_1(t) = \left(\dfrac{t - t_2}{t_1 - t_2} \right) \left(\dfrac{t - t_3}{t_1 - t_3} \right) = \left(\dfrac{t - 1}{0 - 1} \right) \left(\dfrac{t - 2}{0 - 2} \right) = (t - 1)(t - 2)/2$

   $= (t^2 - 3t + 2)/2$

   $\ell_2(t) = \left(\dfrac{t - t_1}{t_2 - t_1} \right) \left(\dfrac{t - t_3}{t_2 - t_3} \right) = \left(\dfrac{t - 0}{1 - 0} \right) \left(\dfrac{t - 2}{1 - 2} \right) = -t(t - 2)$

   $= -t^2 + 2t$

   $\ell_3(t) = \left(\dfrac{t - t_1}{t_3 - t_1} \right) \left(\dfrac{t - t_2}{t_3 - t_2} \right) = \left(\dfrac{t - 0}{2 - 0} \right) \left(\dfrac{t - 1}{2 - 1} \right) = t(t - 1)/2$

   $= (t^2 - t)/2$

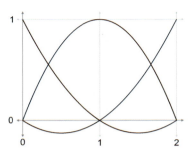

2. We can solve this exercise several ways, but since this is a chapter on curves, we *wanted* you to use the polynomial interpolation techniques from Section 13.2 to fit a parabola through the "control points" given in the problem. Those control points just *happen* to share the knot sequence from the previous exercise, and we were hoping you would take advantage of that work. The math starts by multiplying each Lagrange basis polynomial by the corresponding control point.

$$\mathbf{p}(t) = \mathbf{p}_1 \ell_1(t) + \mathbf{p}_2 \ell_2(t) + \mathbf{p}_3 \ell_3(t)$$

$$= \begin{bmatrix} 0 \\ 0 \end{bmatrix} (t^2 - 3t + 2)/2 + \begin{bmatrix} d/2 \\ h \end{bmatrix} (-t^2 + 2t) + \begin{bmatrix} d \\ 0 \end{bmatrix} (t^2 - t)/2$$

$$= \begin{bmatrix} d/2 \\ h \end{bmatrix} (-t^2 + 2t) + \begin{bmatrix} d \\ 0 \end{bmatrix} (t^2 - t)/2$$

$$= \begin{bmatrix} -d/2 \\ -h \end{bmatrix} t^2 + \begin{bmatrix} d \\ 2h \end{bmatrix} t + \begin{bmatrix} d/2 \\ 0 \end{bmatrix} t^2 + \begin{bmatrix} -d/2 \\ 0 \end{bmatrix} t$$

$$= \begin{bmatrix} 0 \\ -h \end{bmatrix} t^2 + \begin{bmatrix} d/2 \\ 2h \end{bmatrix} t$$

3. (a) Starting with $t = 0.4$. The first round of interpolation:

$$\mathbf{b}_0^1 = 0.60\,\mathbf{b}_0^0 + 0.40\,\mathbf{b}_1^0 = 0.60 \cdot (3,5) + 0.40 \cdot (6,1) = (4.20, 3.40)$$
$$\mathbf{b}_1^1 = 0.60\,\mathbf{b}_1^0 + 0.40\,\mathbf{b}_2^0 = 0.60 \cdot (6,1) + 0.40 \cdot (0,3) = (3.60, 1.80)$$
$$\mathbf{b}_2^1 = 0.60\,\mathbf{b}_2^0 + 0.40\,\mathbf{b}_3^0 = 0.60 \cdot (0,3) + 0.40 \cdot (5,5) = (2.00, 3.80)$$

Round two:

$$\mathbf{b}_0^2 = 0.60\,\mathbf{b}_0^1 + 0.40\,\mathbf{b}_1^1 = 0.60 \cdot (4.20, 3.40) + 0.40 \cdot (3.60, 1.80) = (3.96, 2.76)$$
$$\mathbf{b}_1^2 = 0.60\,\mathbf{b}_1^1 + 0.40\,\mathbf{b}_2^1 = 0.60 \cdot (3.60, 1.80) + 0.40 \cdot (2.00, 3.80) = (2.96, 2.60)$$

And the final round:

$$\mathbf{b}_0^3 = 0.60\,\mathbf{b}_0^2 + 0.40\,\mathbf{b}_1^2 = 0.60 \cdot (3.96, 2.76) + 0.40 \cdot (2.96, 2.60) = (3.56, 2.70)$$

(b) Applying Equations (13.32)–(13.35):

$$\mathbf{p}_0 = \mathbf{b}_0 = (3,5)$$
$$\mathbf{v}_0 = 3(\mathbf{b}_1 - \mathbf{b}_0) = 3[(6,1) - (3,5)] = [9, -12]$$
$$\mathbf{v}_1 = 3(\mathbf{b}_3 - \mathbf{b}_2) = 3[(5,5) - (0,3)] = [15, 6]$$
$$\mathbf{p}_1 = \mathbf{b}_3 = (5,5)$$

(c) Using Equation (13.19):

$$\mathbf{p}(t) = \mathbf{b}_0 + t(3\mathbf{b}_1 - 3\mathbf{b}_0) + t^2(3\mathbf{b}_0 - 6\mathbf{b}_1 + 3\mathbf{b}_2) + t^3(-\mathbf{b}_0 + 3\mathbf{b}_1 - 3\mathbf{b}_2 + \mathbf{b}_3)$$

$$= \begin{bmatrix} 3 \\ 5 \end{bmatrix} + t\left(3\begin{bmatrix} 6 \\ 1 \end{bmatrix} - 3\begin{bmatrix} 3 \\ 5 \end{bmatrix}\right) + t^2\left(3\begin{bmatrix} 3 \\ 5 \end{bmatrix} - 6\begin{bmatrix} 6 \\ 1 \end{bmatrix} + 3\begin{bmatrix} 0 \\ 3 \end{bmatrix}\right)$$

$$+ t^3\left(-\begin{bmatrix} 3 \\ 5 \end{bmatrix} + 3\begin{bmatrix} 6 \\ 1 \end{bmatrix} - 3\begin{bmatrix} 0 \\ 3 \end{bmatrix} + \begin{bmatrix} 5 \\ 5 \end{bmatrix}\right)$$

$$= \begin{bmatrix} 3 \\ 5 \end{bmatrix} + t\begin{bmatrix} 9 \\ -12 \end{bmatrix} + t^2\begin{bmatrix} -27 \\ 18 \end{bmatrix} + t^3\begin{bmatrix} 20 \\ -6 \end{bmatrix}$$

(d) $$\mathbf{p}(t) = \begin{bmatrix} 3 \\ 5 \end{bmatrix} + 0.40\begin{bmatrix} 9 \\ -12 \end{bmatrix} + 0.40^2\begin{bmatrix} -27 \\ 18 \end{bmatrix} + 0.40^3\begin{bmatrix} 20 \\ -6 \end{bmatrix}$$

$$= \begin{bmatrix} 3.00 \\ 5.00 \end{bmatrix} + \begin{bmatrix} 3.60 \\ -4.80 \end{bmatrix} + \begin{bmatrix} -4.32 \\ 2.88 \end{bmatrix} + \begin{bmatrix} 1.28 \\ -0.38 \end{bmatrix} = \begin{bmatrix} 3.56 \\ 2.70 \end{bmatrix}$$

(e) Using Equation (13.5):

$$\mathbf{v}(t) = \mathbf{c}_1 + 2\mathbf{c}_2 t + 3\mathbf{c}_3 t^2 = \begin{bmatrix} 9 \\ -12 \end{bmatrix} + 2t\begin{bmatrix} -27 \\ 18 \end{bmatrix} + 3t^2\begin{bmatrix} 20 \\ -6 \end{bmatrix}$$

$$= \begin{bmatrix} 9 \\ -12 \end{bmatrix} + t\begin{bmatrix} -54 \\ 36 \end{bmatrix} + t^2\begin{bmatrix} 60 \\ -18 \end{bmatrix}$$

(f) $$\mathbf{v}(0.40) = \begin{bmatrix} 9 \\ -12 \end{bmatrix} + 0.40\begin{bmatrix} -54 \\ 36 \end{bmatrix} + 0.40^2\begin{bmatrix} 60 \\ -18 \end{bmatrix} = \begin{bmatrix} 9 \\ -12 \end{bmatrix} + \begin{bmatrix} -21.6 \\ 14.4 \end{bmatrix} + \begin{bmatrix} 9.60 \\ -2.88 \end{bmatrix} = \begin{bmatrix} -3.00 \\ -0.48 \end{bmatrix}$$

$$\mathbf{v}(0.00) = \begin{bmatrix} 9 \\ -12 \end{bmatrix} + 0.00\begin{bmatrix} -54 \\ 36 \end{bmatrix} + 0.00^2\begin{bmatrix} 60 \\ -18 \end{bmatrix} = \begin{bmatrix} 9.00 \\ -12.00 \end{bmatrix}$$

$$\mathbf{v}(1.00) = \begin{bmatrix} 9 \\ -12 \end{bmatrix} + 1.00\begin{bmatrix} -54 \\ 36 \end{bmatrix} + 1.00^2\begin{bmatrix} 60 \\ -18 \end{bmatrix} = \begin{bmatrix} 15.00 \\ 6.00 \end{bmatrix}$$

4. $1 = B_0^2(t) + B_1^2(t) + B_2^2(t)$

$= (1-t)^2 + 2(1-t)t + t^2$

$= (1 - 2t + t^2) + (2t - 2t^2) + t^2$

$= 1$

5. All four control points should be in the same place.

6. It's obvious that $\mathbf{b}_0$ is the starting point of the line, and $\mathbf{b}_3$ is the ending point, but what to do with the interior points $\mathbf{b}_1$ and $\mathbf{b}_2$ is not quite as obvious. One way to solve it is to write out the equation of the ray in monomial form.

$$\mathbf{p}(t) = \mathbf{p}_0 + (\mathbf{p}_1 - \mathbf{p}_0)t$$

Now, if we extract the monomial coefficients

$$\mathbf{c}_0 = \mathbf{p}_0,$$
$$\mathbf{c}_1 = \mathbf{p}_1 - \mathbf{p}_0,$$
$$\mathbf{c}_2 = \mathbf{0},$$
$$\mathbf{c}_3 = \mathbf{0},$$

we can use Equations (13.20)–(13.23) to convert to Bézier form:

$$
\begin{aligned}
\mathbf{b}_0 &= \mathbf{c}_0 & &= \mathbf{p}_0 \\
\mathbf{b}_1 &= \mathbf{c}_0 + (1/3)\mathbf{c}_1 & &= \mathbf{p}_0 + (1/3)(\mathbf{p}_1 - \mathbf{p}_0) \\
\mathbf{b}_2 &= \mathbf{c}_0 + (2/3)\mathbf{c}_1 + (1/3)\mathbf{c}_2 & &= \mathbf{p}_0 + (2/3)(\mathbf{p}_1 - \mathbf{p}_0) \\
\mathbf{b}_3 &= \mathbf{c}_0 + \mathbf{c}_1 + \mathbf{c}_2 + \mathbf{c}_3 = \mathbf{p}_0 + (\mathbf{p}_1 - \mathbf{p}_0) & &= \mathbf{p}_1
\end{aligned}
$$

Note that $\mathbf{b}_0$ and $\mathbf{b}_3$ are mapped to the endpoints, as expected. To achieve a constant velocity, we divide the ray into thirds and place the two intermediate points at the division between these thirds.

This makes sense when you think about what a constant velocity curve looks like in Hermite form. The difference vector $\mathbf{p}_1 - \mathbf{p}_0$ must be traversed over the unit time interval, so the desired velocity vectors $\mathbf{v}_0$ and $\mathbf{v}_1$ are both equal to this difference vector. Recalling the relationship between the Bézier control points and the Hermite vectors (Equations (13.32)–(13.35)) leads us to the same conclusion obtained above.

It also makes sense when you think about the Bernstein basis. Remember that each basis function $B_i^n(t)$ has one local maximum at $t = i/n$ where the corresponding control point $\mathbf{b}_i$ exerts the most influence over the curve.

7. We hope you were able to get this one just by thinking about it. We know that the starting and ending velocities of the curve are zero, and so in Hermite form the vectors $\mathbf{v}_0$ and $\mathbf{v}_1$ are zero. Since the interior Bézier control points are offset from the endpoints by one-third of the velocity, that means the second control point must be the same as the first, and the third control point should be the same as the last:

$$\mathbf{b}_1 = \mathbf{b}_0 + (1/3)\mathbf{v}_0 = \mathbf{b}_0 + (1/3)\mathbf{0} = \mathbf{b}_0,$$
$$\mathbf{b}_2 = \mathbf{b}_3 - (1/3)\mathbf{v}_1 = \mathbf{b}_3 - (1/3)\mathbf{0} = \mathbf{b}_3.$$

8. To solve this one, let's convert to Hermite form and examine the starting and ending velocities:

$$\mathbf{v}_0 = 3(\mathbf{b}_1 - \mathbf{b}_0) = 3(\mathbf{b}_3 - \mathbf{b}_0),$$
$$\mathbf{v}_1 = 3(\mathbf{b}_3 - \mathbf{b}_2) = 3(\mathbf{b}_3 - \mathbf{b}_0).$$

We know from Exercise 6 that if the interior points are distributed equally, dividing the interval into thirds, then the resulting curve has a constant velocity. But now, the interior control points are farther away from their neighboring endpoint than one-third of the total interval; the distance is equal to the total interval length. So the starting and ending velocities are three times as fast. This means we are starting off "too fast" and will have to slow down in the middle somewhere, and then accelerate back up to the high velocity at the end. How slow do we have to go? Let's plot the curve to see.

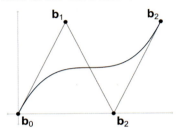

Here the control points are spaced evenly on the x-axis, and so it moves at constant horizontal velocity. The y-coordinates are distributed according to the description for this exercise, and as you can see, the initial vertical velocity is large, it slows down to a minimum at $t = 1/2$, and then accelerates to a high final velocity. Judging from the diagram, the slope is horizontal at the midpoint $t = 1/2$, which means that the vertical velocity is zero.

9. We can solve this algebraically. First, we need to convert our answer from Exercise 2 into a curve in monomial form with a normalized parameter. Remember that the curve is traced out as t varies from 0 to 2, but all of our curves have been using an argument that varies from 0 to 1. So we'll set $s = t/2$ and come up with a new curve in terms of s:

$$\mathbf{p}(t) = \begin{bmatrix} 0 \\ -h \end{bmatrix} t^2 + \begin{bmatrix} d/2 \\ 2h \end{bmatrix} t,$$

$$\mathbf{p}(s) = \begin{bmatrix} 0 \\ -h \end{bmatrix} (2s)^2 + \begin{bmatrix} d/2 \\ 2h \end{bmatrix} (2s),$$

$$= \begin{bmatrix} 0 \\ -4h \end{bmatrix} s^2 + \begin{bmatrix} d \\ 4h \end{bmatrix} s.$$

Writing out the monomial coefficients, we get

$$\mathbf{c}_0 = \mathbf{0}, \qquad \mathbf{c}_1 = \begin{bmatrix} d \\ 4h \end{bmatrix}, \qquad \mathbf{c}_2 = \begin{bmatrix} 0 \\ -4h \end{bmatrix}, \qquad \mathbf{c}_3 = \mathbf{0}.$$

Now we can convert to Bézier form by using Equations (13.20)–(13.23):

$$\mathbf{b}_0 = \mathbf{c}_0 \qquad\qquad\qquad\qquad\qquad\qquad\qquad\qquad = \mathbf{0},$$

$$\mathbf{b}_1 = \mathbf{c}_0 + (1/3)\mathbf{c}_1 = (1/3)\begin{bmatrix} d \\ 4h \end{bmatrix} \qquad\qquad\qquad = \begin{bmatrix} d/3 \\ 4h/3 \end{bmatrix},$$

$$\mathbf{b}_2 = \mathbf{c}_0 + (2/3)\mathbf{c}_1 + (1/3)\mathbf{c}_2$$

$$= (2/3)\begin{bmatrix} d \\ 4h \end{bmatrix} + (1/3)\begin{bmatrix} 0 \\ -4h \end{bmatrix} = \begin{bmatrix} 2d/3 \\ 8h/3 \end{bmatrix} + \begin{bmatrix} 0 \\ -4h/3 \end{bmatrix} \qquad = \begin{bmatrix} 2d/3 \\ 4h/3 \end{bmatrix},$$

$$\mathbf{b}_3 = \mathbf{c}_0 + \mathbf{c}_1 + \mathbf{c}_2 + \mathbf{c}_3 = \begin{bmatrix} d \\ 4h \end{bmatrix} + \begin{bmatrix} 0 \\ -4h \end{bmatrix} \qquad\qquad\qquad = \begin{bmatrix} d \\ 0 \end{bmatrix}.$$

10. Our first task is to convert the curve from functional the form $y = f(x)$ into parametric form $(x(t), y(t))$. We will assume that x is a linear function of t, so we solve $x(t) = mt + b$ given $x(0) = -0.9683$ and $x(1) = 4.2253$. This gives us $x(t) = 5.1936t - 0.9683$. The x-coordinates of the Bézier control points are easily determined using the results of Exercise 6. We chose x to be a linear function of t because we assumed it had constant horizontal velocity, so the x-coordinates of the control points are spaced evenly at $t = 1/3$ and $t = 2/3$. Summarizing all four control points, we have

$$x_0 = x(0) = -0.968,$$
$$x_1 = x(1/3) = 5.1936(1/3) - 0.9683 = 0.763,$$
$$x_2 = x(2/3) = 5.1936(2/3) - 0.9683 = 2.494,$$
$$x_3 = x(1) = 4.225.$$

The x-coordinates were trivial in this problem; all the real work was in the y-coordinates. Plugging $x(t)$ into our functional form, we have

$$y = -0.364x^2 + 1.145x + 2.110$$
$$= -0.364(5.1936t - 0.9683)^2 + 1.145(5.1936t - 0.9683) + 2.110$$
$$= -9.818t^2 + 9.608t + 0.660,$$

which is a perfectly valid 1D cubic curve in monomial form. We convert this to Bézier form by using Equations (13.20)–(13.23):

$$y_0 = c_0 = 0.660,$$
$$y_1 = c_0 + (1/3)c_1 = 0.660 + (1/3)(9.608) = 3.863,$$
$$y_2 = c_0 + (2/3)c_1 + (1/3)c_2 = 0.660 + (2/3)(9.608) + (1/3)(-9.818) = 3.793,$$
$$y_3 = c_0 + c_1 + c_2 + c_3 = 0.660 + 9.608 + (-9.818) + 0 = 0.450.$$

Notice that the two middle y-coordinates are nearly equal, since the starting and ending points we have chosen make our parabola slightly asymmetric.

Putting all this together, our four Bézier control points are

$$\mathbf{b}_0 = (-0.968, 0.660), \quad \mathbf{b}_1 = (0.763, 3.863), \quad \mathbf{b}_2 = (2.494, 3.793), \quad \mathbf{b}_3 = (4.225, 0.450).$$

11. (a) For the first control point, we do regular de Casteljau using 0.2 for each round.

$$\mathbf{b}_0^1 = 0.80\,\mathbf{b}_0^0 + 0.20\,\mathbf{b}_1^0 = 0.80 \cdot (3,5) + 0.20 \cdot (6,1) = (3.60, 4.20)$$
$$\mathbf{b}_1^1 = 0.80\,\mathbf{b}_1^0 + 0.20\,\mathbf{b}_2^0 = 0.80 \cdot (6,1) + 0.20 \cdot (0,3) = (4.80, 1.40)$$
$$\mathbf{b}_2^1 = 0.80\,\mathbf{b}_2^0 + 0.20\,\mathbf{b}_3^0 = 0.80 \cdot (0,3) + 0.20 \cdot (5,5) = (1.00, 3.40)$$
$$\mathbf{b}_0^2 = 0.80\,\mathbf{b}_0^1 + 0.20\,\mathbf{b}_1^1 = 0.80 \cdot (3.60, 4.20) + 0.20 \cdot (4.80, 1.40) = (3.84, 3.64)$$
$$\mathbf{b}_1^2 = 0.80\,\mathbf{b}_1^1 + 0.20\,\mathbf{b}_2^1 = 0.80 \cdot (4.80, 1.40) + 0.20 \cdot (1.00, 3.40) = (4.04, 1.80)$$
$$\mathbf{b}_0' = 0.80\,\mathbf{b}_0^2 + 0.20\,\mathbf{b}_1^2 = 0.80 \cdot (3.84, 3.64) + 0.20 \cdot (4.04, 1.80) = (3.88, 3.27)$$

For the second control point, we do the last round using the fraction 0.5.

$$\mathbf{b}_0^2 = (3.84, 3.64)$$
$$\mathbf{b}_1^2 = (4.04, 1.80)$$
$$\mathbf{b}_1' = 0.50\,\mathbf{b}_0^2 + 0.50\,\mathbf{b}_1^2 = 0.50 \cdot (3.84, 3.64) + 0.50 \cdot (4.04, 1.80) = (3.94, 2.72)$$

For the third control point, we do the last two rounds using the fraction 0.5.

$$\mathbf{b}_0^1 = (3.60, 4.20)$$
$$\mathbf{b}_1^1 = (4.80, 1.40)$$
$$\mathbf{b}_2^1 = (1.00, 3.40)$$
$$\mathbf{b}_0^2 = 0.50\,\mathbf{b}_0^1 + 0.50\,\mathbf{b}_1^1 = 0.50 \cdot (3.60, 4.20) + 0.50 \cdot (4.80, 1.40) = (4.20, 2.80)$$
$$\mathbf{b}_1^2 = 0.50\,\mathbf{b}_1^1 + 0.50\,\mathbf{b}_2^1 = 0.50 \cdot (4.80, 1.40) + 0.50 \cdot (1.00, 3.40) = (2.90, 2.40)$$
$$\mathbf{b}_2' = 0.50\,\mathbf{b}_0^2 + 0.50\,\mathbf{b}_1^2 = 0.50 \cdot (4.20, 2.80) + 0.50 \cdot (2.90, 2.40) = (3.55, 2.60)$$

For the final control point, we do all the rounds using 0.5 as the fraction.

$$\mathbf{b}_0^1 = 0.50\,\mathbf{b}_0^0 + 0.50\,\mathbf{b}_1^0 = 0.50 \cdot (3, 5) + 0.50 \cdot (6, 1) = (4.50, 3.00)$$
$$\mathbf{b}_1^1 = 0.50\,\mathbf{b}_1^0 + 0.50\,\mathbf{b}_2^0 = 0.50 \cdot (6, 1) + 0.50 \cdot (0, 3) = (3.00, 2.00)$$
$$\mathbf{b}_2^1 = 0.50\,\mathbf{b}_2^0 + 0.50\,\mathbf{b}_3^0 = 0.50 \cdot (0, 3) + 0.50 \cdot (5, 5) = (2.50, 4.00)$$
$$\mathbf{b}_0^2 = 0.50\,\mathbf{b}_0^1 + 0.50\,\mathbf{b}_1^1 = 0.50 \cdot (4.50, 3.00) + 0.50 \cdot (3.00, 2.00) = (3.75, 2.50)$$
$$\mathbf{b}_1^2 = 0.50\,\mathbf{b}_1^1 + 0.50\,\mathbf{b}_2^1 = 0.50 \cdot (3.00, 2.00) + 0.50 \cdot (2.50, 4.00) = (2.75, 3.00)$$
$$\mathbf{b}_3' = 0.50\,\mathbf{b}_0^2 + 0.50\,\mathbf{b}_1^2 = 0.50 \cdot (3.75, 2.50) + 0.50 \cdot (2.75, 3.00) = (3.25, 2.75)$$

(b) $\mathbf{q}_0 = \mathbf{b}_0 = (3.0, 5.0)$

$\mathbf{q}_1 = \mathbf{b}_0/2 + \mathbf{b}_1/2 = (3, 5)/2 + (6, 1)/2 = (4.5, 3.0)$

$\mathbf{q}_2 = \mathbf{b}_0/4 + \mathbf{b}_1/2 + \mathbf{b}_2/4 = (3, 5)/4 + (6, 1)/2 + (0, 3)/4 = (3.75, 2.5)$

$\mathbf{q}_3 = \mathbf{r}_0 = \mathbf{b}_0/8 + 3\mathbf{b}_1/8 + 3\mathbf{b}_2/8 + \mathbf{b}_3/8$
$\quad = (3, 5)/8 + 3 \cdot (6, 1)/8 + 3 \cdot (0, 3)/8 + (5, 5)/8 = (3.25, 2.75)$

$\mathbf{r}_1 = \mathbf{b}_1/4 + \mathbf{b}_2/2 + \mathbf{b}_3/4 = (6, 1)/4 + (0, 3)/2 + (5, 5)/4 = (2.75, 3.0)$

$\mathbf{r}_2 = \mathbf{b}_2/2 + \mathbf{b}_3/2 = (0, 3)/2 + (5, 5)/2 = (2.5, 4.0)$

$\mathbf{r}_3 = \mathbf{b}_3 = (5.0, 5.0)$

(c) Using Equation (13.38) with $n = 3$:

$$\mathbf{b}_0' = \frac{0}{4}\mathbf{b}_{-1} + \left(1 - \frac{0}{4}\right)\mathbf{b}_0 = 0\begin{bmatrix} ? \\ ? \end{bmatrix} + 1\begin{bmatrix} 3 \\ 5 \end{bmatrix} = \begin{bmatrix} 3.00 \\ 5.00 \end{bmatrix}$$

$$\mathbf{b}_1' = \frac{1}{4}\mathbf{b}_0 + \left(1 - \frac{1}{4}\right)\mathbf{b}_1 = \frac{1}{4}\begin{bmatrix} 3 \\ 5 \end{bmatrix} + \frac{3}{4}\begin{bmatrix} 6 \\ 1 \end{bmatrix} = \begin{bmatrix} 5.25 \\ 2.00 \end{bmatrix}$$

$$\mathbf{b}_2' = \frac{2}{4}\,\mathbf{b}_1 + \left(1 - \frac{2}{4}\right)\mathbf{b}_2 = \frac{1}{2}\begin{bmatrix} 6 \\ 1 \end{bmatrix} + \frac{1}{2}\begin{bmatrix} 0 \\ 3 \end{bmatrix} = \begin{bmatrix} 3.00 \\ 2.00 \end{bmatrix}$$

$$\mathbf{b}_3' = \frac{3}{4}\,\mathbf{b}_2 + \left(1 - \frac{3}{4}\right)\mathbf{b}_3 = \frac{3}{4}\begin{bmatrix} 0 \\ 3 \end{bmatrix} + \frac{1}{4}\begin{bmatrix} 5 \\ 5 \end{bmatrix} = \begin{bmatrix} 1.25 \\ 3.50 \end{bmatrix}$$

$$\mathbf{b}_4' = \frac{4}{4}\,\mathbf{b}_3 + \left(1 - \frac{4}{4}\right)\mathbf{b}_4 = 1\begin{bmatrix} 5 \\ 5 \end{bmatrix} + 0\begin{bmatrix} ? \\ ? \end{bmatrix} = \begin{bmatrix} 5.00 \\ 5.00 \end{bmatrix}$$

B.14   Chapter 14

(Page 715.)

If only it were that easy!

*For every complex problem
there is an answer that is clear, simple, and
wrong.*

— H. L. Mencken

Bibliography

[1] Tomas Akenine-Möller, Eric Haines, and Natty Hoffman. *Real-Time Rendering*, Third edition. Natick, MA: A K Peters, Ltd., 2008. http://www.realtimerendering.com/.

[2] James Arvo. "A Simple Method for Box-Sphere Intersection Testing." In *Graphics Gems*, edited by Andrew S. Glassner. San Diego: Academic Press Professional, 1990.

[3] Ian Ashdown. "Photometry and Radiometry: A Tour Guide for Computer Graphics Enthusiasts." Adapted from *Radiosity: A Programmer's Perspective*, Ian Ashdown, Wiley, 1994. http://www.helios32.com/.

[4] Didier Badouel. "An Efficient Ray-Polygon Intersection." In *Graphics Gems*, edited by Andrew S. Glassner. San Diego: Academic Press Professional, 1990.

[5] Ronen Barzel. "Lighting Controls for Computer Cinematography." *J. Graph. Tools* 2 (1997), 1–20.

[6] James F. Blinn. "Models of Light Reflection for Computer Synthesized Pictures." *SIGGRAPH Comput. Graph.* 11:2 (1977), 192–198.

[7] James F. Blinn. "A Generalization of Algebraic Surface Drawing." *ACM Trans. Graph.* 1 (1982), 235–256.

[8] David M. Bourg. *Physics for Game Developers*. Sebastapol, CA: O'Reilly Media, 2002.

[9] Ronald N. Bracewell. *The Fourier Transform and Its Applications*, Second edition. New York: McGraw-Hill, 1978.

[10] G. H. Bryan. *Stability In Aviation: An Introduction to Dynamical Stability as Applied to the Motions of Aeroplanes*. London: Macmillan and Co., 1911.

[11] Erik B. Dam, Martin Koch, and Martin Lillholm. "Quaternions, Interpolation and Animation." Technical Report DIKU-TR-98/5, Department of Computer Science, University of Copenhagen, 1998. http://www.diku.dk/students/myth/quat.html.

[12] M. de Berg, M. van Kreveld, M. Overmars, and O. Schwarzkopf. *Computational Geometry—Algorithms and Applications*. Springer-Verlag, 1997.

[13] James Diebel. "Representing Attitude: Euler Angles, Unit Quaternions, and Rotation Vectors." http://citeseerx.ist.psu.edu/viewdoc/summary?doi=10. 1.1.110.5134, 2006.

[14] "DirectX Developer Center." http://msdn.microsoft.com/en-us/directx/ default.aspx.

[15] Tevian Dray and Corinne A. Manogue. "The Geometry of the Dot and Cross Products." *Journal of Online Mathematics and Its Applications* 6. http://mathdl.maa.org/images/upload_library/4/vol6/Dray2/Dray.pdf.

[16] Fletcher Dunn and Ian Parberry. *3D Math Primer for Graphics and Game Development*, First edition. Plano, TX: Wordware Publishing, 2002.

[17] David H. Eberly. *Game Physics*. San Francisco: Morgan Kaufmann Publishers, 2004.

[18] Christer Ericson. *Real-Time Collision Detection*. San Francisco: Morgan Kaufmann Publishers, 2005.

[19] Kenny Erleben, Jon Sporring, Knud Henricksen, and Henrik Dohlmann. *Physics-Based Animation*. Boston: Charles River Media, 2005.

[20] Gerald Farin. *Curves and Surfaces for Computer Aided Geometric Design: A Practical Guide*, Second edition. Boston: Academic Press, 1990.

[21] Frederick Fisher and Andrew Woo. "$R \cdot E$ versus $N \cdot H$ Specular Highlights." In *Graphics Gems IV*, edited by Paul S. Heckbert. San Diego: Academic Press Professional, 1994.

[22] Andrew S. Glassner. "Maintaining Winged-Edge Models." In *Graphics Gems II*, edited by James Arvo. San Diego: Academic Press Professional, 1991.

[23] Andrew S Glassner. *Principles of Digital Image Synthesis*. San Francisco: Morgan Kaufmann Publishers, 1995. http://glassner.com/andrew/writing/ books/podis.htm.

[24] Ronald Goldman. "Intersection of Three Planes." In *Graphics Gems*, edited by Andrew S. Glassner. San Diego: Academic Press Professional, 1990.

[25] Ronald Goldman. "Triangles." In *Graphics Gems*, edited by Andrew S. Glassner. San Diego: Academic Press Professional, 1990.

[26] H. Gouraud. "Continuous Shading of Curved Surfaces." *IEEE Transactions on Computers* 20 (1971), 623–629.

[27] F. Sebastin Grassia. "Practical Parameterization of Rotations Using the Exponential Map." *J. Graph. Tools* 3 (1998), 29–48. http://jgt.akpeters. com/papers/Grassia98/.

[28] Ned Greene. "Environment Mapping and Other Applications of World Projections." *IEEE Comput. Graph. Appl.* 6 (1986), 21–29.

[29] Roy Hall. *Illumination and Color in Computer Generated Imagery*. New York: Springer-Verlag New York, 1989.

[30] Andrew J. Hanson. *Visualizing Quaternions (The Morgan Kaufmann Series in Interactive 3D Technology)*. San Francisco: Morgan Kaufmann, 2006.

[31] John C. Hart, George K. Francis, and Louis H. Kauffman. "Visualizing Quaternion Rotation." *ACM Trans. Graph.* 13:3 (1994), 256–276.

[32] Michael T. Heath. *Scientific Computing: An Introductory Survey*, Second edition. New York: McGraw-Hill, 2002. http://www.cse.illinois.edu/heath/scicomp/.

[33] Paul S. Heckbert. "What Are the Coordinates of a Pixel?" In *Graphics Gems*, edited by Andrew S. Glassner, pp. 246–248. San Diego: Academic Press Professional, 1990. http://www.graphicsgems.org/.

[34] Chris Hecker. "Physics, Part 3: Collision Response." *Game Developer Magazine*, pp. 11–18. http://chrishecker.com/Rigid_Body_Dynamics#Physics_Articles.

[35] Chris Hecker. "Physics, Part 4: The Third Dimension." *Game Developer Magazine*, pp. 15–26. http://chrishecker.com/Rigid_Body_Dynamics#Physics_Articles.

[36] Jeff Hultquist. "Intersection of a Ray with a Sphere." In *Graphics Gems*, edited by Andrew S. Glassner. San Diego: Academic Press Professional, 1990.

[37] James T. Kajiya. "The Rendering Equation." In *SIGGRAPH '86: Proceedings of the 13th Annual Conference on Computer Graphics and Interactive Techniques*, pp. 143–150. New York: ACM, 1986.

[38] Randall D. Knight. *Physics for Scientists and Engineers*. Reading, MA: Addison-Wesley, 2004. http://wps.aw.com/aw_knight_physics_1/.

[39] Donald E. Knuth. *The Art of Computer Programming, Volume 1: Fundamental Algorithms*, Third edition. Reading, MA: Addison-Wesley Longman, 1997.

[40] Doris H. U. Kochanek and Richard H. Bartels. "Interpolating Splines with Local Tension, Continuity, and Bias Control." *SIGGRAPH Comput. Graph.* 18:3 (1984), 33–41.

[41] Jack B. Kuipers. *Quaternions and Rotation Sequences: A Primer with Applications to Orbits, Aerospace, and Virtual Reality*. Princeton, NJ: Princeton University Press, 1999.

[42] Eric Lengyel. *Mathematics for 3D Game Programming and Computer Graphics*, Second edition. Boston: Charles River Media, 2004. http://www.terathon.com/books/mathgames2.html.

[43] William E. Lorensen and Harvey E. Cline. "Marching Cubes: A High Resolution 3D Surface Construction Algorithm." In *Proceedings of the 14th Annual Conference on Computer Graphics and Interactive Techniques, SIGGRAPH '87*, pp. 163–169. New York: ACM, 1987.

[44] T. M. MacRobert. *Spherical Harmonics*, Second edition. New York: Dover Publications, 1948.

[45] John McDonald. "Teaching Quaternions Is Not Complex." *Computer Graphics Forum* 29:8 (2010), 2447–2455.

[46] Brian Vincent Mirtich. "Impulse-based Dynamic Simulation of Rigid Body Systems." Ph.D. thesis, University of California at Berkeley, 1996.

[47] Jason Mitchell, Gary McTaggart, and Chris Green. "Shading in Valve's Source Engine." In *ACM SIGGRAPH 2006 Courses, SIGGRAPH '06*, pp. 129–142. New York: ACM, 2006. http://www.valvesoftware.com/publications.html.

[48] Addy Ngan, Frédo Durand, and Wojciech Matusik. "Experimental Validation of Analytical BRDF Models." In *ACM SIGGRAPH 2004 Sketches, SIGGRAPH '04*, pp. 90–. New York: ACM, 2004.

[49] "OpenGL Software Development Kit." http://www.opengl.org/sdk/docs/man/.

[50] OpenGL Architecture Review Board, Shreiner, Dave, Woo, Mason, Neider, Jackie, and Davis, Tom. *OpenGL(R) Programming Guide: The Official Guide to Learning OpenGL(R), Version 2.1*. Reading, MA: Addison-Wesley Professional, 2007. http://www.opengl.org/documentation/red_book/.

[51] Oliver M. O'Reilly. *Intermediate Dynamics for Engineers: A Unified Treatment of Newton–Euler and Lagrangian Mechanics*. Cambridge, UK: Cambridge University Press, 2008.

[52] Joseph O'Rourke. *Computational Geometry in C*, Second edition. Cambridge, UK: Cambridge University Press, 1994.

[53] Matt Pharr and Greg Humphreys. *Physically Based Rendering: From Theory to Implementation*. San Francisco: Morgan Kaufmann Publishers, 2004. http://www.pbrt.org/.

[54] Bui Tuong Phong. "Illumination for Computer Generated Pictures." *Commun. ACM* 18:6 (1975), 311–317.

[55] Charles Poynton. "Frequently Asked Questions about Color." http://www.poynton.com/ColorFAQ.html.

[56] William H. Press, Saul A. Teukolsky, William T. Vetterling, and Brian P. Flannery. *Numerical Recipes in C*, Second edition. Cambridge, UK: Cambridge University Press, 1992. http://www.nr.com/.

[57] Robert Resnick and David Halliday. *Physics*, Third edition. New York: John Wiley and Sons, 1977.

[58] David F. Rogers. *An Introduction to NURBS: With Historical Perspective*. New York: Academic Press, 2001.

[59] Philip J. Schneider and David H. Eberly. *Geometric Tools for Computer Graphics*. San Francisco: Morgan Kaufmann Publishers, 2003.

[60] Peter Schorn and Frederick Fisher. "Testing the Convexity of a Polygon." In *Graphics Gems IV*, edited by Paul S. Heckbert. San Diego: Academic Press Professional, 1994.

[61] Peter Shirley. *Fundamentals of Computer Graphics*. Natick, MA: A K Peters, Ltd., 2002. http://www.cs.utah.edu/~shirley/books/.

[62] Ken Shoemake. "Quaternions and 4×4 Matrices." In *Graphics Gems II*, edited by James Arvo. San Diego: Academic Press Professional, 1991.

[63] Ken Shoemake. "Euler Angle Conversion." In *Graphics Gems IV*, edited by Paul S. Heckbert. San Diego: Academic Press Professional, 1994.

[64] Peter-Pike Sloan. "Stupid Spherical Harmonics (SH) Tricks." Technical report, Microsoft Cooporation, 2008. http://www.ppsloan.org/publications.

[65] Russell Smith. "Open Dynamics Engine User Guide." http://www.ode.org/ode-latest-userguide.html.

[66] Alvy Ray Smith. "A Pixel Is Not a Little Square, a Pixel Is Not a Little Square, a Pixel Is Not a Little Square! (And a Voxel is Not a Little Cube." Technical report, Technical Memo 6, Microsoft Research, 1995. http://alvyray.com/memos/6_pixel.pdf.

[67] Gilbert Strang. *Computational Science and Engineering*. Cambridge, UK: Wellesley-Cambridge, 2007.

[68] Gilbert Strang. *Introduction to Linear Algebra*, Fourth edition. Cambridge, UK: Wellesley-Cambridge, 2009.

[69] Paul S. Strauss. "A Realistic Lighting Model for Computer Animators." *IEEE Comput. Graph. Appl.* 10:6 (1990), 56–64.

[70] Gino van den Bergen. *Collision Detection in Interactive 3D Environments*. San Francisco: Morgan Kaufmann Publishers, 2004.

[71] David R. Warn. "Lighting Controls for Synthetic Images." In *Proceedings of the 10th Annual Conference on Computer Graphics and Interactive Techniques, SIGGRAPH '83*, pp. 13–21. New York: ACM, 1983.

[72] Andrew Woo. "Fast Ray-Box Intersection." In *Graphics Gems*, edited by Andrew S. Glassner. San Diego: Academic Press Professional, 1990.

*Always read stuff that will make you look good
if you die in the middle of it.*

— P. J. O'Rourke (1947–)

Index

What is it ye would see?
If aught of woe or wonder, cease your search.

— Horatio in *Hamlet*, Act V, scene II